Petroleum Geology of Southeast Asia

Geological Society Special Publications
Series Editor A. J. FLEET

GEOLOGICAL SOCIETY SPECIAL PUBLICATION NO. 126

Petroleum Geology of Southeast Asia

EDITED BY

A. J. FRASER
(BP Exploration, Houston, Texas, USA)

S. J. MATTHEWS
(BP Exploration, Houston, Texas, USA)

AND

R. W. MURPHY
(Consulting Petroleum Geologist, Walton-on-Thames, UK)

1997

Published by

The Geological Society

London

THE GEOLOGICAL SOCIETY

The Society was founded in 1807 as The Geological Society of London and is the oldest geological society in the world. It received its Royal Charter in 1825 for the purpose of 'investigating the mineral structure of the Earth'. The Society is Britain's national society for geology with a membership of around 8000. It has countrywide coverage and approximately 1000 members reside overseas. The Society is responsible for all aspects of the geological sciences including professional matters. The Society has its own publishing house, which produces the Society's international journals, books and maps, and which acts as the European distributor for publications of the American Association of Petroleum Geologists, SEPM and the Geological Society of America.

Fellowship is open to those holding a recognized honours degree in geology or cognate subject and who have at least two years' relevant postgraduate experience, or who have not less than six years' relevant experience in geology or a cognate subject. A Fellow who has not less than five years' relevant postgraduate experience in the practice of geology may apply for validation and, subject to approval, may be able to use the designatory letters C. Geol. (Chartered Geologist).

Further information about the Society is available from the Membership Manager, The Geological Society, Burlington House, Piccadilly, London W1V 0JU, UK. The Society is a Registered Charity, No. 210161.

Published by The Geological Society from:
The Geological Society Publishing House
Unit 7, Brassmill Enterprise Centre
Brassmill Lane
Bath BA1 3JN
UK
(*Orders*: Tel. 01225 445046
Fax 01225 442836)

First published 1997

British Library Cataloguing in Publication Data
A catalogue record for this book is available from the British Library.

ISBN 1-897799-91-8

Typeset by Aarontype Ltd, Unit 47, Easton Business Centre, Felix Road, Easton, Bristol, UK

Printed by The Alden Press,
Osney Mead, Oxford, UK

Distributors

USA
AAPG Bookstore
PO Box 979
Tulsa
OK 74101-0979
USA
(*Orders*: Tel. (918) 584-2555
Fax (918) 560-2652)

Australia
Australian Mineral Foundation
63 Conyngham Street
Glenside
South Australia 5065
Australia
(*Orders*: Tel. (08) 379-0444
Fax (08) 379-4634)

India
Affiliated East-West Press PVT Ltd
G-1/16 Ansari Road
New Delhi 110 002
India
(*Orders:* Tel. (11) 327-9113
Fax (11) 326-0538)

Japan
Kanda Book Trading Co.
Tanikawa Building
3-2 Kanda Surugadai
Chiyoda-Ku
Tokyo 101
Japan
(*Orders*: Tel. (03) 3255-3497
Fax (03) 3255-3495)

Contents

Petroleum geology of SE Asia: an introduction

This volume reflects the current broad interest in exploration and production in the energy-hungry SE Asia region. Analysts currently predict that energy production in SE Asia will increase by 85% over the next 15 years and, by the end of this period, consumption may be double that of Europe. Although fairly modest in a global context, production is still sufficient to underpin economic growth rates of up to 8–12% in regions such as Indonesia, the Philippines, Vietnam, Malaysia and Brunei. In contrast, those economies that do not currently benefit from substantial oil and gas production are more stagnant, e.g. Laos, Cambodia and Burma.

SE Asia, including Indonesia, Thailand, Malaysia, Brunei and the Philippines, contains ten oil and gas provinces that individually contain more than three billion barrels of oil equivalent. All of these provinces are Tertiary basins with Tertiary-aged petroleum systems. This represents a considerable resource that has been actively exploited in the past and continues to expand with further exploration. Two main reservoir systems contribute to the majority of petroleum reserves in the region. These are deltaic clastics, e.g. Baram, Rajang and Mahakam deltas, and carbonates, e.g. Luconia, Terumbu and Nam Con Son platforms. They were charged by two key sources types. Lacustrine source rocks which are associated with 'oily' basins and paralic source beds which are typical of gas-prone basins. Key trap types are inversion anticlines and three-way-dip and fault-controlled traps in low net/gross clastic reservoirs, often with variable fluid types in stacked reservoirs with over 1000 ft plus pay zones. Carbonate traps are characterized by broad regional drape features that are seldom full to spill.

The volume covers all aspects of the petroleum geology of SE Asia from economic background, through plate tectonic models, petroleum charging and reservoir systems to detailed field and reservoir studies. In the first paper **Sladen** provides the economic context for all the other papers. He summarises current energy trends in SE Asia and emphasizes the global importance of the regions coal output. Current SE Asia GDP growth is quite phenomenal and fuel oil, gas and coal will be required to underpin this growth. **Sladen** notes a steady growth in oil and gas production particularly in China, Malaysia, Vietnam and Papua New Guinea. In terms of total oil production, SE Asia is a minor world player, but gas may become increasingly important for both domestic and export markets particularly in the emerging gas production areas of Vietnam, PNG and the Philippines. The next three papers complement this economic overview with regional reviews of plate tectonics, petroleum charging and lacustrine systems. **Hall** reconstructs plate movements in SE Asia over the last 50 Ma based on new palaeomagnetic data. He discusses rotational models for Borneo and the resulting implications for the evolution of the South China Sea regions and the progressive motion of the Australian continent. **Todd *et al*.** develop the idea that there are two dominant petroleum source-rock types in SE Asia: paralic and lacustrine. Their presence in any particular basin is reflected by the distribution of oil and gas in the basin. Lacustrine source rocks are associated with oil-prone basins such as Malay, Bohai and Central Sumatra whereas paralic source rocks are typical of gas-rich basins such as Sabah, Luconia, Nam Con Son and West Natuna. **Sladen**, in his second paper, considers the lacustrine source rocks in the context of SE Asian Mesozoic and Cenozoic lacustrine basins and reviews the development of reservoirs in the basins and the charge issues which link exploration for specific sources and reservoirs in these basins.

The 20 papers that lie at the heart of the volume focus on specific areas of SE Asia and the issues and problems of their petroleum geology. The Bongkot gas field in the Gulf of Thailand is the subject of the first of these case studies. **Leo** describes the reservoir characteristics of the deltaic/fluvial sandstones of Oligocene to mid-Miocene age encountered in the field. He concludes that seismic facies analysis was an essential exploration tool in this area, highlighting the Bongkot accumulation as a complex of largely independent structural and statigraphic closures.

Vietnam is addressed in the next three papers. First **Matthews *et al*.** review the Nam Con Son Basin, offshore South Vietnam, highlighting reservoir issues. **Worden** addresses the controls on porosity reduction with depth in the area of the southern Vietnam continental shelf. He concludes that porosity was mainly controlled by lithic content and in particular compaction of ductile grains and is able to demonstrate a strong relationship between permeability and percentage of ductile grains. The paper by **Mayall *et al*.** complements this contribution focusing on the large carbonate plays of the Nam Con Son Basin.

Indonesia is the focus of the next four papers. **Wight** addresses the petroleum geology and production history of the Sumatra-Java area

From Fraser, A. J., Matthews, S. J. & Murphy, R. W. (eds), 1997, *Petroleum Geology of Southeast Asia*, Geological Society Special Publication No. 126, pp. 1–2.

from which the one billionth barrel of oil has recently been produced and highlights southeast Sumatra. **Schiefelbein & Cameron** argue the case for grouping the oils from the area into a number of families, while **Cole & Crittenden** discuss the development of Tertiary syn-rift lacustrine source rocks. **Beach *et al.*** present an empirical study of the controls on effective fault seal in SE Asia based on the analysis of a database comprising some 190 fields from the region. The study suggests, amongst other conclusions, that fault strike was an apparent control, particularly in some Sumatran basins which have experienced some strike-slip movement.

The next five papers present case studies from around the SE Asia archipelago. **Prosser & Carter** describe an onshore Brunei analogue for Miocene subsurface reservoirs in Brunei. Detailed mini-permeameter data is correlated with different facies types and input to a reservoir water floor simulator. This highlights channel and upper shoreface sands as intervals of high permeability and sites of preferential imbibition of water. **Mat-zin & Swarbrick** describe the petroleum geology of the Sarawak Basin, Malaysia. He suggests the main structural control on hydrocarbon accumulations was the presence of a series of wrench controlled 'flower structures'. Geochemical analysis indicates that the oils were largely land-plant derived, in common with many other oils in the region. **Bosence & Wilson** present a depositional model for the many carbonate platform reservoirs in SE Asia based on fieldwork in Sulawesi. **Boult** discusses the exploration history and petroleum potential of the East Papuan Basin. The main producing areas at Hides and Iagifu were delineated by field mapping over 30 years ago. There are now 13 major discoveries from a total of 26 drilled structures. **Blanche & Blanche** summarize the petroleum potential of the Spratly Archipelago, an area of some 150 isolated reefs, banks and atolls in the South China Sea, which is currently the subject of border disputes involving six countries. A diversity of play types is envisaged being an extension of known plays from offshore southern Vietnam and northern Borneo. Charge is highlighted as the critical technical risk, long distance migration being required. **Blanche & Blanche** estimate a reserves potential of 1.5 billion barrels of oil equivalent and suggest that future exploration of this region could only be progressed with the creation of joint development areas.

The final seven papers deal with structural aspects of SE Asian petroleum geology. **Longley** discusses an alternative view of the tectono-stratigraphic evolution SE Asia. **Roques *et al.*** describe the structural development of the offshore Da Nang area, Vietnam and the links to onshore fault trends identified in Palaeozoic basement. **McCarthy** describes the effects of oblique oceanic subduction at the western Sundaland margin. He discusses the various estimates for the amount of the offset along the Sumatran fault system. These range from a minimum of 20–30 km suggested by offsets on Quaternary river systems to a maximum of 460 km from the extent of opening in the Andaman Sea. **Howells** describes the origins and petroleum potential of the Ombilin Basin, onshore Sumatra. **Chambers & Daley** and **Moss *et al.***, in two complementary papers, discuss the structural development and hydrocarbon habitat of the onshore Kutei Basin in East Kalimantan. The basin covers an area of 70 000 km^2, containing 14–15 km of Tertiary sediments. Both authors note an important Pliocene–Recent uplift event associated with trap formation (Samarinda Anticlinorium). Finally **Milsom** reports on the implications of gravity anomalies on the Sabah margin of the SE South China Sea and suggests a model for the location of the Baram Delta related to a variation in onshore and offshore structural trends.

The volume arises from a two-day conference organized by the Petroleum Group of the Geological Society on 26 and 27 September 1995.

As with any conference there are always a number of people working behind the scenes without whose assistance the event would never take place. Firstly we would like to thank the following companies whose financial assistance helped make the conference possible: The BP and Statoil Alliance, Mitsubishi Oil Group, Mobil North Sea Limited, Petroconsultants (UK) Ltd.

We would also like to thank the conference organising team of Amanda Harrison and Heidi Gould at the Geological Society in London and to June Daniel and Sophie Duffy at BP Exploration in Uxbridge for their untiring efforts in chasing down abstracts and papers.

Finally we would like to thank our session chairmen; Dick Murphy (Consultant), Dick Pegrum (Statoil) and Chris Sladen (BP Exploration), the referees of the papers contained in these proceedings and lastly the authors and presenters who put a terrific effort into both the presentation of their papers and completion of the manuscripts. Thanks are also due to Bruce Blanche for his input to the Introductory chapter of this volume.

A. J. Fraser
S. J. Matthews

Energy trends in SE Asia

CHRIS SLADEN

BP Exploration Operating Company Limited,
Villa A15, An Phu, Thu Duc, Ho Chi Minh City, Vietnam

Abstract: SE Asia is undergoing rapid economic growth, achieving annual increases in gross domestic product of 6–15% in a number of countries. In many instances, the growth has been underpinned by rapid establishment and development of indigenous petroleum resources. Accompanying this growth, is a steady and significant increase in energy consumption. To meet the increasing energy requirement, oil, gas and coal production are steadily increasing. Oil consumption considerably exceeds regional production; the supply gap is widening and the shortfall is principally made up by oil imports from Middle East countries. Replacement of produced oil reserves is increasingly difficult in SE Asia. In the future, gas is expected to become a much more significant energy source. In addition to large gas reserves already found, but currently undeveloped, there are a wide variety of exploration opportunities for further gas. Technological breakthroughs expected in the next 5–30 years will also increase the attractiveness of gas as the preferred energy source in SE Asia during the next 50 years. These include advances in the use of gas which is heavily contaminated with CO_2, improvements which enable more commercial developments of deepwater gas, and a continued improvement in coal-bed methane technology.

The object of this short review is first to examine recent trends in energy production and consumption in SE Asia[1]. This includes the major energy sources, principally oil, gas and coal, and also hydroelectric, nuclear and solar energy[2]. After reviewing recent trends, there is an attempt to identify future trends in these energy sources together with issues that will need to be faced by the energy industry as it moves towards the twenty-first century.

SE Asia is characterized by a large population (>40% of the world) in countries that are undergoing rapid economic change. Rapidly expanding economies in countries such as China, Thailand, Vietnam and Philippines demonstrate yearly growth rates of 6–15% gross domestic product (Fig. 1). Population growth in many countries reaches 2% per year, further creating a strong energy demand. Not surprisingly, in 1994, regional energy consumption increased by over 5%[3].

[1] For the purpose of this review, SE Asia is considered to embrace all the countries eastward from Pakistan and India to the Pacific Ocean, from Japan, China and Korea in the north, to Australia and New Zealand in the south. Although beyond many people's definition of SE Asia, it embraces an area with key energy trade links, energy financing relationships and energy construction that consequently denotes a region of more distinctive energy characteristics.

[2] A number of other energy sources exist including wind, waves, wood, oil shale and batteries. Whilst these are sometimes important very locally, they have not been included in this short review.

[3] Energy statistics throughout this paper are based on those published by BP (1995).

In most instances, expanding economies have been underpinned by rapid development of an indigenous petroleum industry, which has been a key pillar to continued growth. There appears to be little basis to predict a long-term decline in energy demand. Although some countries may experience a period of economic consolidation and flattening in energy demand as their economies mature (for example, South Korea, Taiwan, and later on Malaysia and Thailand), this can be expected to be offset by increases in energy

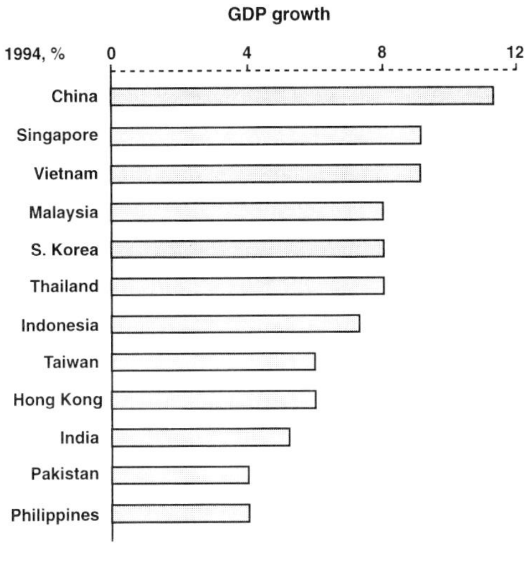

Fig 1. Growth in SE Asia viewed as increases in gross domestic product (GDP) in different countries during 1994 (based on data from Deutsche Bank).

From Fraser, A. J., Matthews, S. J. & Murphy, R. W. (eds), 1997, *Petroleum Geology of Southeast Asia*, Geological Society Special Publication No. 126, pp. 3–10.

demand elsewhere (for example economic growth in China, India, Vietnam and perhaps Myanmar).

Trends in oil

Oil production and consumption

SE Asia contributes just over 10% of annual world oil production, and during the last five years the region has shown a small, but steady growth in production of 1–2% per year (Table 1). Oil production in SE Asia is currently dominated by China (*c.* 42%), with other major producing countries being Indonesia (*c.* 22%), followed by Malaysia, India and Australia (each *c.* 9%).

All of the major producing countries are struggling to deliver significant increases in oil production and in most cases production has been flat during the last five years (Table 1). The challenge has become one of maintaining current production levels and, in the case of India, there are now indications of a downward trend in oil production. A number of new producers have helped maintain the overall growth in production. Although not significant in world terms, both Vietnam and Papua New Guinea have made relevant contributions to local markets.

Oil consumption in SE Asia has grown steadily at around 6% in recent years. Annual consumption of crude oil in SE Asia in 1994 was *c.* 5.8 billion barrels and second only to coal as the major energy source (Fig. 2). Annual oil production was considerably less than consumption at only *c.* 2.5 billion barrels. The supply gap is steadily widening and is currently *c.* 10 million barrels per day. Imports of oil from outside of SE Asia have doubled over the last 10 years. SE Asia has now overtaken the USA in terms of oil imports. Soon it will overtake Europe to become the worlds largest oil importing region. Growth in demand for transport fuels is particularly strong. The shortfall between consumed and produced oil has consistently been balanced by oil imported from the Middle East.

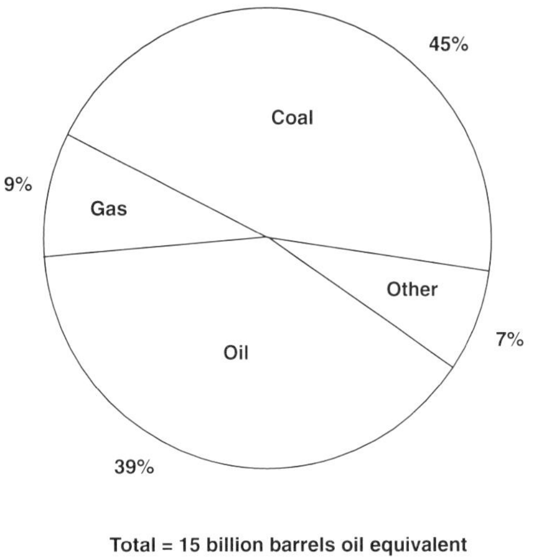

Fig. 2. Primary energy consumption in SE Asia during 1994 (source BP 1995).

Oil reserves and future possibilities

Many currently producing oil-prone basins in SE Asia are struggling to replace produced oil reserves. They are now fairly mature in the extent of exploration that has been carried out. Very few large oilfields have been found in recent years and few giant fields are probably remaining to be found. The significant growth in oil reserves that occurred during the last 40 years is unlikely to be continued at the same rate.

Table 1. *Annual oil production in SE Asia (source BP 1995)*

Country	Annual oil production (million tonnes)					% 1994 world production
	1990	1991	1992	1993	1994	
Australia	28.4	26.9	26.5	24.9	26.9	0.8%
Brunei	7.4	7.9	9.0	8.3	8.5	0.3%
China	138.3	141.0	142.0	144.0	144.9	4.5%
India	34.8	33.0	30.3	29.1	33.6	1.1%
Indonesia	71.9	78.3	74.1	74.3	74.2	2.3%
Malaysia	29.6	30.8	31.3	30.6	30.9	1.0%
Papua New Guinea	—	—	2.5	5.9	5.7	0.2%
Vietnam	2.7	4.0	5.5	6.3	6.9	0.2%
Other SE Asia	8.8	9.3	9.9	9.7	9.3	0.3%
Total	321.9	331.0	331.0	333.0	340.7	10.6%

The current reserves to production ratio (which assumes that no new reserves are found) indicates only 18 years of remaining supply. This is, of course, overly alarmist because some new reserves will undoubtedly be found and recovery from existing accumulations will most likely improve. Nevertheless, it is difficult to envisage any large, sustainable increases in oil production. Meanwhile, future consumption is estimated to increase at between 2% and 4% per year for the next 15 years.

The total reserves base for oil is relatively small compared to the Middle East which currently contains more than 12 times the proven oil reserves of those in SE Asia (Table 2). It is likely that SE Asia will be dependant on the Middle East for crude oil imports for the foreseeable future. Other sources of supply may become more important in the future, for example Alaska, or possibly East Siberia and Sakhalin Island. These sources would introduce more diversity of supply and lower dependence on the Middle East.

Few frontier areas that are thought to be oil-prone remain to be explored in SE Asia. Success in frontier areas will be key to creating any significant growth in oil reserves and production. For the moment, there is some uncertainty in the scale of their contribution to future supply. Possibilities include basins in NW China, Vietnam, E Indonesia and NW Australia, and also perhaps parts of the East China Sea, and Mongolia. The opportunities are often remote, geologically complex, or in deepwater. It seems reasonable to expect very few new oilfields of over 500 million barrels and only a handful of 1000 million barrels or more. Most oilfields are expected be in the 1–100 million barrels range, strongly skewed towards the smaller fields of less than 50 million barrels.

Not only are newly discovered oilfields expected to be mostly small, but they are also expected to be geologically more complex. A combination of complex structural and stratigraphic trap configurations, together with complex reservoir characteristics will make it more difficult to realise their value. To cope effectively with a combination of these kinds of reservoirs and small fields, SE Asia will need to be at the forefront as users of the latest technology if it is to maintain or improve its oil production. Technological improvements will be needed in subsurface imaging techniques for what are expected to often be tectonically complex, highly faulted structures and thin, heterogeneous clastic reservoirs. Reservoir modelling and production technology will need to advance to deliver solutions to producing from these reservoirs.

SE Asia is unusual in having most oil production from non-marine sandstones, principally deposited as lacustrine and deltaic facies, and most oil generated from non-marine lacustrine and deltaic source-rocks. Problems inherent in these hydrocarbon systems are the presence of waxy crude oil in complex reservoirs, characterized by frequently low oil flowrates in complex traps. This creates complicated production issues and often difficult petroleum economics. In contrast, the giant fields of the Middle East are relatively simple traps which contain reservoirs capable of very high oil flowrates. However, there are some advantages of crude oils produced in SE Asia. First, is naturally the close proximity to a very large market, secondly the greater security of an indigenous source of supply, and thirdly the low sulphur content of SE Asia crudes. Overall, these factors enable the oil producers of SE Asia to realise a small price premium against Middle East crudes.

Trends in gas

Gas production and consumption

Steady and significant growth in gas production (*c.* 5–7% per year) has been occurring in SE Asia over the last five years, albeit starting from a fairly low base (Table 3). At present, SE Asia

Table 2. *Proven oil, gas and coal reserves in different regions of the world in 1994 (source BP 1995)*

	North America	South America	Middle East	Europe	Africa	Former Soviet Union	SE Asia
Proven oil reserves (Thousand million barrels)	88	78	660	17	62	59	45
Proven gas reserves (Trillion cubic metres)	9	5	45	5	10	57	10
Proven coal reserves (Thousand million tonnes)	250	10	—	95	61	315	312

Table 3. *Annual gas production in SE Asia (source BP 1995)*

Country	Annual gas production (million tonnes oil equivalent)					% 1994 world production
	1990	1991	1992	1993	1994	
Australia	18.6	19.5	21.1	22.0	25.3	1.4%
Brunei	7.6	7.8	8.1	8.2	8.4	0.4%
China	12.8	13.4	13.6	14.6	14.9	0.8%
India	11.1	12.6	14.3	14.6	15.6	0.8%
Indonesia	40.8	46.4	48.9	50.6	55.8	3.0%
Japan	1.8	1.9	1.9	2.0	2.0	0.1%
Malaysia	16.0	18.3	20.5	22.4	23.4	1.3%
Pakistan	9.9	10.3	10.9	12.1	12.5	0.7%
Other SE Asia	16.6	17.8	19.1	20.7	21.8	1.2%
Total	135.2	148.0	158.4	167.2	179.7	9.7%

annually contributes almost 10% of world gas production. Within the region, gas production is currently dominated by Indonesia *c.* 31%, with Malaysia and Australia *c.* 13% each. China and India are also important gas producers with almost 10% each. All of the major gas producing countries have delivered a steady increase in production over the last five years. In 1994, Indonesia raised production by 10%, Malaysia by 4% and Australia by 15%. In addition there is increasing gas production in many other countries, notably Bangladesh, Pakistan, Thailand and New Zealand. Few countries currently have decreasing gas production; only Afghanistan, where there are production difficulties due to civil unrest, and Taiwan, which has a limited reserves base, have had notable decreases in gas production.

Compared to other regions, SE Asia represents the fastest-growing gas consumer in the world, with consumption having increased seven-fold over the last twenty years. With most gas being consumed within the region, gas consumption and production have remained roughly in balance. Japan continues to be the major gas consumer within the region, importing large volumes of liquefied petroleum gas (LPG) and liquefied natural gas (LNG). Demand grew strongly in 1994 in South Korea (33%), Taiwan (30%) and Indonesia (13%). In contrast to oil, production and consumption of most gas is very much dictated by the need for a guaranteed nearby market to which the gas can be easily transported. Growth in gas use is thus closely tied to growth in local economies, not necessarily the economy of an entire country.

In the strongly developing economies which are now a characteristic of many countries in SE Asia, gas is an ideal alternative energy source to oil for coping with the increasing energy demand. Thailand, Vietnam and Indonesia are each good examples of countries utilizing their gas resources to underpin growing economies. This has included schemes for using gas to generate power and fertilizer, and the use of LPG and LNG.

LPG and LNG have the advantage of being suitable for bulk storage and can be transported for long distances. LNG in particular, has risen rapidly as a key energy source in the more developed countries of SE Asia, particularly in the last 20 years. In 1994, LNG trade within the region grew by 11%. LNG schemes in SE Asia have utilized gas production from Australia, Indonesia and Malaysia and transported the gas to large markets in Japan, South Korea and Taiwan.

Gas reserves and future possibilities

The gas reserves base in SE Asia is currently rather small when compared to other parts of the world such as the Former Soviet Union and Middle East (Table 2). However, considerable new gas reserves are being found in SE Asia and extensive infrastructure is being built, both onshore and offshore. The current reserves to production ratio indicates 50 years of remaining supply, and many new exploration opportunities exist.

Proximity to markets will continue to be a key issue in gas remaining a competitive energy source due to the costs of long distance transport. In the past, many gas discoveries were too small and distant from a market to fund the initial capital intensive infrastructure. Many nearby markets were simply not sufficiently developed to justify gas developments. However, the situation has changed significantly in the last 15 years. As markets have grown and the infrastructure has become more developed, so many smaller gasfields and groupings of gasfields have been developed. Groups of small fields and satellite discoveries can now be aggregated to form viable developments.

Possibilities also exist for long-distance gas pipeline networks linking a number of countries throughout the region and covering many thousands of kilometres. These networks offer a host of new opportunities and development possibilities for gas. As more infrastructure is built, this can be expected to reduce transportation costs, make many more small gasfields commercially attractive, spur more exploration, and consequently more gas can be expected to be found and developed.

Gas is currently only a very limited part (*c.* 9%) of total energy consumption in SE Asia (Fig. 2). However, growth in production and consumption is strong and expected to continue to rise even more strongly as economies develop in the next Century. As a result, gas will become an increasingly significant part of the total energy budget in SE Asia in coming years.

In the future, we should expect gas to be put to a wider number of applications. The principal use is expected to be for power generation, but in particular circumstances gas may be important as a fertiliser feedstock, for methanol production, chemicals manufacturing, or for other industrial uses. Installation of domestic gas distribution networks has numerous logistical problems reaching individual households in many countries in SE Asia. In addition, the large areas of warm climate do not require gas for central heating. Piping of gas to individual households is therefore considered unlikely. A tendency for domestic consumption of electricity as opposed to gas is therefore anticipated. However, LPG is expected to become increasingly attractive, particularly for domestic cooking and light industries where it is gradually replacing wood, oil and coal.

New markets for LNG can also be expected to develop, particularly China and Thailand. There are many potential new suppliers of gas for LNG such as the Philippines, Vietnam and Papua New Guinea. Malaysia currently has vast reserves of undeveloped gas and these too could be developed for additional LNG. The LNG produced in SE Asia is usually very competitive because the principal alternative source of LNG supply is the Middle East which has much higher transportation costs to the markets of Japan, South Korea and Taiwan. LNG is also considered attractive to many economies in the region as it provides a security and diversity of energy supply, and at a very competitive price to oil.

A host of exploration opportunities exist for adding significant new gas reserves. In addition to adding further reserves in proven gas provinces, there are numerous frontier areas, or areas with as yet very limited exploration. Opportunities include the Andaman Sea, Bangladesh, East China Sea, Cambodia, Laos, Tibet, northwest China, eastern Indonesia, the Gulf of Tonkin between China and Vietnam, and various parts of the South China Sea and East Vietnam Sea.

A number of opportunities for finding new gas reserves also exist by drilling deeper in search of gas in the presently producing oil basins. For example, many of the onshore basins that are producing oil in north and east China, in Indonesia, India and Malaysia have gas potential in deeper sequences but these sequences have often not yet been penetrated by the drill bit. Although the non-marine oil-prone source-rocks responsible for generating oil in these basins tend not to yield much gas during maturation, there are opportunities for oil becoming cracked to gas in the deeply buried sequences, together with additional opportunities for gas generation from deeply buried coals that may be present. Improvements in the drilling and formation evaluation of deep, often high-pressure and high-temperature wells will assist exploration for these kinds of deep gas.

Improvements and breakthroughs in gas processing technology can also be expected to add new gas reserves. The ability to utilize hydrocarbon gas with up to 85% CO_2 pollutants, whilst suitably disposing of unwanted CO_2, would allow vast gas reserves to be developed in SE Asia. For example, giant gas fields of this type have been found but remain undeveloped in offshore areas of Myanmar, Indonesia, the Gulf of Thailand and central Vietnam. The technology breakthrough that would allow development of these gas reserves can be expected in 5–25 years. This would then spur further exploration for CO_2 polluted hydrocarbon gas.

Gas produced from coal-bed methane projects is also expected to offer many opportunities in the future. SE Asia with its vast coal resources (see section below) has numerous possibilities for this kind of gas project. Combined with advances in coal-bed methane technology and an improved understanding of the subsurface behaviour of this kind of gas, a large number of coal-bed methane projects can be anticipated. At the moment, however, there are few projects in SE Asia in contrast to more widespread coal-bed methane producers such as the USA. Notably, China has recently begun a number of coal-bed methane projects and more are expected in the near future.

Improvements in deepwater gas production technology are also expected to create new reserves potential for gas. There are a number of deepwater areas in SE Asia that would benefit. These include parts of the South China Sea, East Vietnam Sea and northern Indian Ocean. Much further into the future, improved understanding of gas hydrate deposits that are developed in many deep oceans, combined with a viable technology, may allow us to access this vast potential gas resource. However, such developments are probably many decades away given the abundance of more conventional gas resources.

Trends in coal

Coal production and consumption

Coal represents the principal energy source in SE Asia and the region currently provides about 42% of world coal production (Fig. 2 and Table 4). Coal is frequently the cornerstone of much heavy industry development, and also has widespread domestic use. A number of countries have extensive coal deposits which, together with oil and gas, have helped both satisfy growing energy demand and support the growing economies. Whilst production has overall grown *c.* 3% each year for the last five years, production growth each year is quite variable with a low of *c.* 0.5% and a high of *c.* 5% (Table 4). Production growth is usually tied closely to the annual economic performance of each country.

Most coal production is by simple opencast, drift and adit mining. In SE Asia, deep shaft mining has seldom been used because of abundant resources either at, or very close to the surface. China dominates coal production in SE Asia with *c.* 65%, followed by Australia and India (each *c.* 13%), and then Indonesia (*c.* 2%). All of these countries have been increasing coal production, most notably Indonesia which tripled coal production between 1990 and 1994 (Table 4). Declining coal production is most significant in Japan and South Korea, where the economies are shifting towards using oil and LNG for energy.

Coals formed widely in SE Asia in basins with Jurassic, early Cretaceous and early Tertiary sedimentary sequences. These coals occur both in the entirely non-marine alluvial and lacustrine basins, and also the basins which had large deltas entering the oceans. There is a complete spectrum of coal rank with lower grade brown coals and lignite not surprisingly more typical of the early Tertiary coals, whereas bituminous or anthracitic coals are more typical of the Cretaceous and Jurassic coals, or rare remnants of older Triassic, Permian and Carboniferous coals.

Coal reserves and future possibilities

SE Asia undoubtedly has enormous proven reserves of coal which are roughly equal in size to those of the Former Soviet Union (Table 2). Together with the Former Soviet Union, coal reserves in SE Asia dominate the world's proven reserves. Vast coal resources occur in China, India, Indonesia, Australia and Mongolia. Many other countries have significant coal resources most notably Vietnam, Thailand, Japan, New Zealand, Pakistan and South Korea. The current reserves to production ratio indicates 170 years of remaining supply in SE Asia. Numerous additional deep coal reserves possibilities exist but are either currently unexplored for, or frequently unappraised, due to the enormous reserves of shallow, easily extractable, coal.

Table 4. *Annual coal production in SE Asia (source BP 1995)*

Country	Annual coal production (million tonnes oil equivalent)					% 1994 world production
	1990	1991	1992	1993	1994	
Australia	106.6	110.5	117.0	117.7	118.3	5.5%
China	530.1	520.3	543.7	558.4	592.0	27.4%
India	103.3	110.9	117.3	121.5	122.9	5.7%
Indonesia	6.5	8.7	14.2	17.6	18.8	0.9%
Japan	5.5	5.3	5.1	4.8	4.6	0.2%
New Zealand	1.6	1.8	1.8	2.0	2.2	0.1%
Pakistan	1.3	1.5	1.4	1.5	1.5	0.1%
South Korea	9.1	8.0	6.4	5.0	3.9	0.2%
Other SE Asia	42.0	42.6	42.7	42.8	43.8	2.0%
Total	806.0	809.8	849.6	871.3	908.0	42.1%

A continuing heavy reliance on coal can be expected in the future. The vast reserves offer a number of possible future exploitation opportunities. In addition to advances in coal extraction technology, we may expect technology improvements in the efficiency of burning and using coal. Improvements in coal-conversion technology may also introduce future opportunities if problems develop in regional supply of oil, and oil prices rise significantly.

Trends in other energy sources

To ensure diversity and security of supply, coupled with cost competitiveness, a variety of other energy sources are expected to continue to play a role in most countries in SE Asia. In many cases, they dilute the need for energy imports, particularly oil. However, oil is likely to remain in strong demand for producing transport fuels unless there are radical changes in the mode of public and private travel.

Hydroelectric energy

A number of significant hydroelectric projects have been completed in recent years, and others are planned. The region now contains *c.* 20% of current world hydroelectric capacity. China, Japan, India, Vietnam, New Zealand and Taiwan have each developed significant contributions of power from hydroelectric energy. Construction of further giant projects is now underway in Vietnam, China and Malaysia. There are however problems in SE Asia, particularly due to the distinct 'wet' and 'dry' seasons that occur in many countries. These cause considerable fluctuations in power production capability.

The creation of large dams for hydroelectric power has raised various issues of environmental impact. There is increasing concern and awareness in SE Asia that many hydroelectric projects cause significant modifications to the landscape, altering river systems, agricultural patterns, rural lifestyles and cultures. Most schemes also require extremely large capital investments that result in complex funding issues and a large financial burden. Even so, there is yet to be clear evidence of a slow down in new hydroelectric energy projects. Environmental concerns coupled with a natural limit on the number of opportunities available are expected to cause a decline in projects in future years.

Nuclear energy

Nuclear energy projects are continuing to grow in SE Asia, with various projects recently completed or underway, particularly in China. Many other countries are now considering nuclear energy projects to supplement and diversify energy supply and reduce oil imports, for example Vietnam. Nuclear energy is viewed by many countries as part of their long-term energy solution, both contributing to and diversifying energy supply, thereby reducing reliance on oil and gas. Japan, for example, plans further increases in nuclear energy production during the next 15 years, with more than 40% of energy produced at nuclear plants by 2010. However, continuing environmental concern over nuclear energy is expected to limit increases in the proportion of energy contributed from nuclear sources.

Geothermal energy

Geothermal energy projects are locally significant in SE Asia, and currently concentrated in areas of high geothermal gradients such as Tibet and west Yunnan in China, Java and Sumatra in Indonesia, Taiwan and New Zealand. At the moment, geothermal energy projects are viable only onshore and are fairly small-scale compared to the major energy sources. Offshore geothermal projects are unlikely to become commercial without a fundamental change in the technology and costs. However, there are many possibilities to introduce more projects into other onshore areas of high temperature such as parts of the Philippines and Indonesia. In addition, deep reservoirs that are found to contain very hot water during deep oil and gas exploration may offer possibilities. Opportunities of this type may evolve in areas such as onshore Indonesia and eastern China.

Solar energy

Solar energy is ideal in areas of SE Asia that have no energy distribution network, particularly the more remote parts of China, Mongolia, Vietnam, Australia and India. Production of solar power units is increasing and the application of solar energy to a wide variety of medical, communications, and lighting uses is expected to continue to expand. At the moment, however, overall use is small compared to oil, gas, coal, nuclear and hydroelectric energy.

Support during the preparation and presentation of this paper was provided by BP Exploration. However, the views expressed in this paper represent those of the author not necessarily his employer.

Reference

BP 1995 *BP statistical review of world energy*. *BP Group Media & Publications*.

Cenozoic plate tectonic reconstructions of SE Asia

ROBERT HALL

SE Asia Research Group, Department of Geology, Royal Holloway University of London, Egham, Surrey TW20 0EX, UK (e-mail: robert.hall@gl.rhbnc.ac.uk)

Abstract: A new plate tectonic model for the Tertiary is proposed, based on the integration of new palaeomagnetic data from east Indonesia recording Philippine Sea plate motion, recent revisions of the history of the South China Sea, and previously available geological and palaeomagnetic data from SE Asia. Early Neogene counter-clockwise rotation of Borneo is interpreted to have closed a proto-South China Sea suggesting a strike-slip boundary in NW Borneo before the Neogene. This rotation suggests that the West Philippine Sea, Celebes Sea and Makassar Strait formed a single basin which opened in the late Paleogene, and widened eastwards. At *c.* 25 Ma a major collision, that of Australia with a Philippine Sea plate arc, trapped Indian Ocean lithosphere which later became the Molucca Sea plate. The collision caused clockwise rotation of the Philippine Sea plate, initiated the Sorong Fault system, and then eliminated the Molucca Sea by subduction on its east and west sides. The effects of collision propagated westwards through the region resulting in the initiation of new plate boundaries marked by regional unconformities. The arrival of the Sulawesi ophiolite, which collided with west Sulawesi in the late Oligocene, was the earliest event in collision between Sulawesi and the Bird's Head microcontinent. Continental crust was thrust beneath Sulawesi in the early Miocene, and the Tukang Besi and Sula platforms were sliced in turn from the microcontinent, transferred to the west-moving Philippine Sea/Molucca Sea plate for a few million years, and finally accreted to Sulawesi. Reconstructing the Molucca Sea and Bird's Head microcontinent suggests that most of the Banda Sea has a late Neogene extensional origin. Collision between the Philippine arc and the Eurasian continental margin in Taiwan at *c.* 5 Ma is the key to present regional tectonics.

Plate tectonic reconstructions of SE Asia may help in understanding the development of sedimentary basins, and the distribution of petroleum resources. Reconstructions of the region should also assist in identifying important controls on its tectonic development (e.g. the role of indentor tectonics), critical events (e.g. collisions or extensional events), and the timing and consequences of plate movements and reorganizations (e.g. their expression as unconformities and their regional correlation). Local, particularly two-dimensional, reconstructions often move problems outside the area of immediate interest, and neglect or fail to recognize both inadequacies and important features of models. Reconstructing a large region requires consideration of the implications of local reconstructions, and may indicate new solutions which can reconcile different data or show that commonly accepted views of regional development cannot be sustained.

During the Cenozoic the SE Asia region (Katili 1975; Hamilton 1979) was bounded to the north and west by a Eurasian plate, and to the south by the Indian–Australian plate. The motions of these plates are known, and their positions provide limits to the zone within which the SE Asian collage of microplates and subplate fragments can be moved in reconstructions. However, the past position of the eastern boundary to the region has been much less certain. This boundary is, and has been, the western edge of the Philippine Sea plate but the motion of this plate has been difficult to link to the global plate circuit because it is surrounded by subduction zones.

New palaeomagnetic data from east Indonesia (Ali & Hall 1995; Hall *et al.* 1995*a*, *b*) permit reconstructions of the Philippine Sea plate and its motion since the early Tertiary. The east Indonesian islands of the North Moluccas include a relatively complete upper Mesozoic and Tertiary sequence of rocks which allow palaeomagnetic sampling over a wide stratigraphic range. The palaeomagnetic data from older sequences can be interpreted with confidence because there are results from younger rocks. The rocks are distributed over a large area so it is also possible to distinguish the effects of rotations caused by local deformation at the plate edge from those recording movement of the entire plate. These new palaeomagnetic data provide a clear indication of rotation of the whole Philippine Sea plate. They can also be

From Fraser, A. J., Matthews, S. J. & Murphy, R. W. (eds), 1997, *Petroleum Geology of Southeast Asia*, Geological Society Special Publication No. 126, pp. 11–23.

used to determine poles of rotation for the entire plate, which can account for the new and earlier palaeomagnetic data, and other evidence of motion of the plate, for example, results of magnetic anomaly skewness studies. The calculated poles of rotation can thus be used to reconstruct the position and shape of the plate during the Cenozoic (Hall *et al.* 1995*c*). These reconstructions are linked to recent interpretations of the South China Sea area (Briais *et al.* 1993) to provide boundaries for the regional model of this paper.

During the development of the model an attempt was made to satisfy both geological and palaeomagnetic data. For reconstruction purposes, geological data are essential, but often can be interpreted in different ways, and it is usually impossible to determine amounts of movement between different areas from geological arguments. On the other hand, much of the evidence for movement, and certainly for rotation about vertical axes, can only be acquired from palaeomagnetism and therefore *interpretation* of both geological *and* palaeomagnetic data is essential. Van der Voo (1993) discusses the use and problems of palaeomagnetic results, and provides a summary of data from SE Asia. Distinction between local and regional rotations is often far from clear, particularly because good palaeomagnetic data are sparsely scattered in time and space. In several areas, principally Sundaland–Borneo and the Philippines it is necessary to choose between alternative interpretations of palaeomagnetic data. In such cases decisions were based on regional geological arguments, as discussed by Hall (1996), and account for differences between the models of Rangin *et al.* (1990), Daly *et al.* (1991), and Lee & Lawver (1994). Critical decisions, for example accepting the large counter-clockwise rotation of Borneo in this model, can be tested if reconstructions can be made that are consistent with geological data for the whole region. Thus, the reconstructions here should be seen as a way of distinguishing locally and regionally important data sets, identifying targets for future work, as well as providing a possible model of SE Asia with a new interpretation of its Cenozoic development.

An account of the reconstructions, with a more detailed account of the sources of information, is given by Hall (1996). An animation of the reconstructions, which can be run on either a Windows-based PC or a Mac with adequate hard disc space, is available free from the World Wide Web site at http://glsun2.gl.rhbnc.ac.uk/seasia/welcome.html.

Methods

Philippine Sea Plate motion

Cooperative field investigations by the University of London and the Indonesian Geological Research and Development Centre carried out between 1984 and 1992 (e.g. Hall & Nichols 1990) provided the stratigraphic basis for a systematic palaeomagnetic programme in the north Molucca islands between Halmahera and Waigeo, and the Sula platform. The palaeomagnetic work was the basis for determination of the Philippine Sea Plate motion history (Hall *et al.* 1995*c*), which in turn provided the basis for tectonic reconstructions of SE Asia.

ATLAS model

The reconstructions were made using the ATLAS computer program (Cambridge Paleomap Services 1993). In the ATLAS model the motions of the major plates are defined relative to Africa and its movement is defined relative to magnetic north. There has been little Cenozoic motion of Eurasia and it remains in a similar position in all the reconstructions, although there are small movements of Eurasia due to the plate circuit used in the ATLAS model, particularly for the last 5 Ma. Therefore there are minor differences compared to reconstructions which keep Eurasia fixed in its present position (Lee & Lawver 1994; Rangin *et al.* 1990).

Fragments and tests

Reconstructions of SE Asia (Fig. 1) for the region between 20° S–30° N, and 90° E–160° E, are presented at 5 Ma intervals for 50–5 Ma (Figs 2 to 6). Approximately 60 fragments were used, and they retain their current size in order that they remain recognizable. However, in earlier reconstructions it is likely that fragments had different sizes and shapes or may not have existed, for example in areas of volcanism and extension in the Philippines, Sunda, and Banda arcs. In some cases, this has been incorporated by omitting the fragment before a certain time.

Most minor fragments can be linked to a major plate (Australia, Eurasia, Philippine Sea) and were moved as if attached, or partially coupled to, a plate with known motion. The model attempts to use the simplest possible motion histories. More complex movements were obtained by transferring fragments from one major plate to another. During the reconstruction process possible solutions were tested by asking (1) are palaeomagnetic data satisfied (2) are geological data satisfied, and (3) is there overlap of fragments during movements? Below, the main features of the reconstructions made for different time slices are summarised, with interpretations of the region's development.

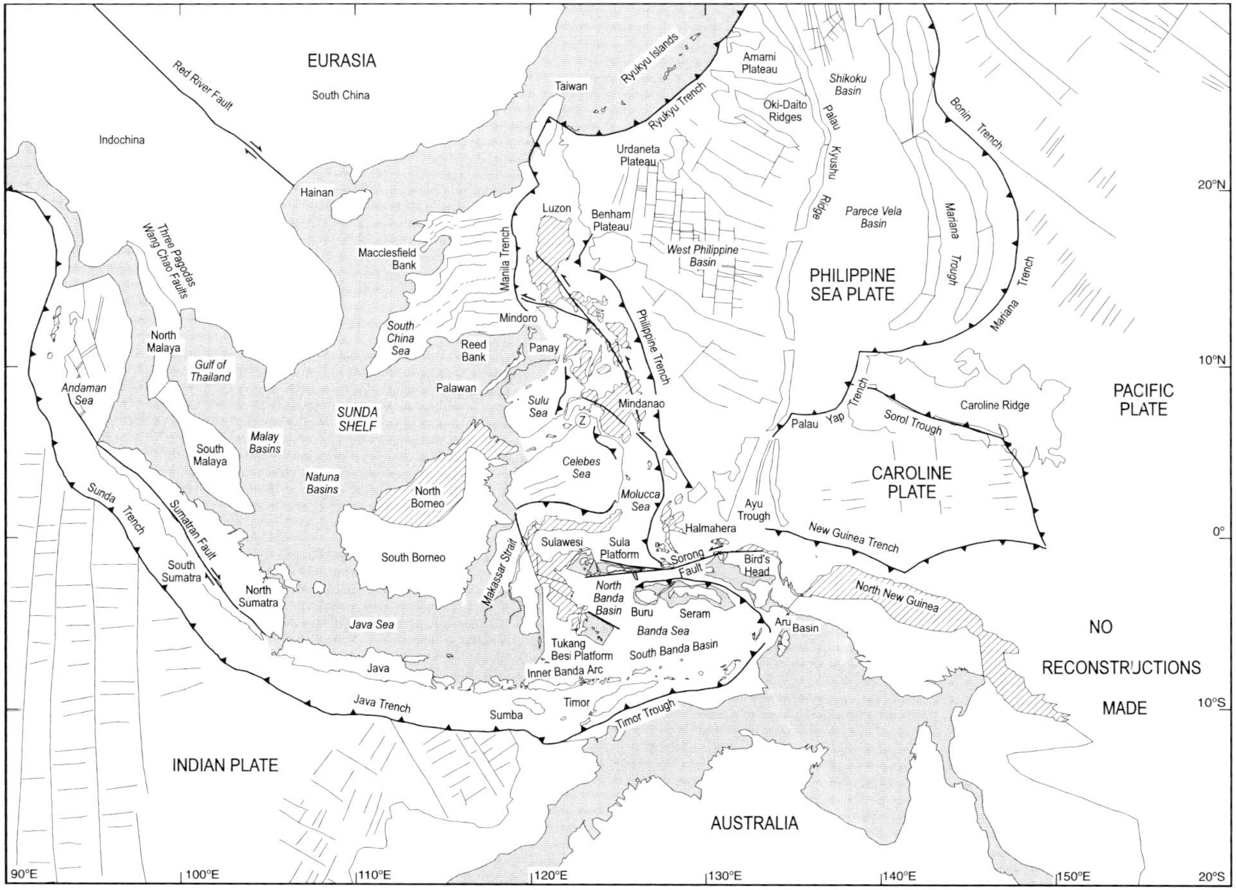

Fig. 1. Simplified present-day tectonic configuration of SE Asia. Shallow marine parts of Eurasian, Sunda and Australian continental shelves are shaded in grey. Hatched areas represent mainly ophiolitic, arc and other accreted material added to Eurasian and Australian margins during the Cenozoic. Principal marine magnetic anomalies are shown schematically for the Indian Ocean, Pacific Ocean, South China Sea, Philippine Sea and Caroline Sea. Circled Z identifies Zamboanga peninsula of Mindanao. Double lines represent active spreading centres. Complexities in the Bismarck–Solomon Sea regions are not shown.

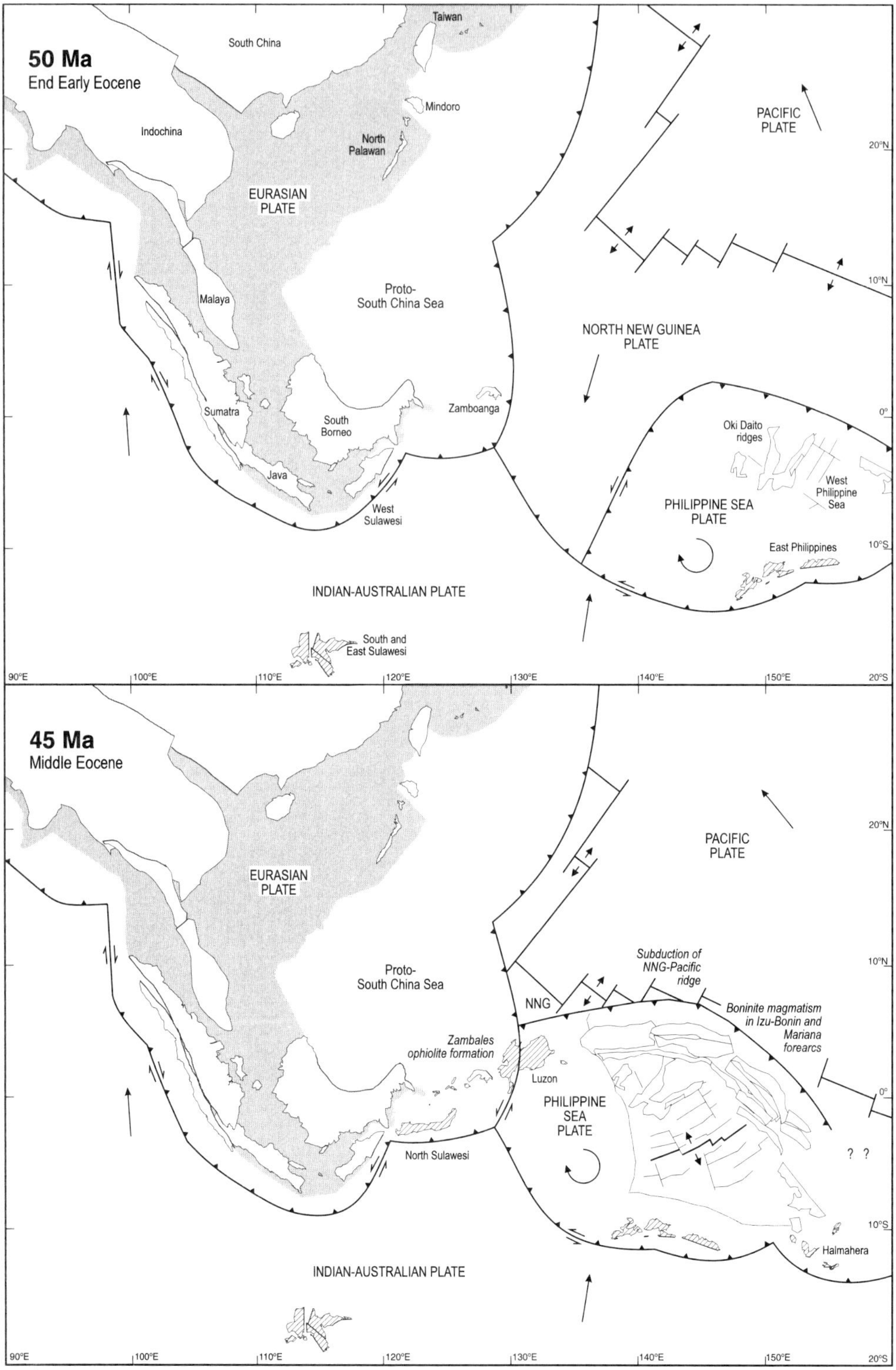

Fig. 2. Reconstructions of the region at 50 and 45 Ma when the Philippine Sea plate was rotating rapidly. Black lines with short paired arrows represent active spreading centres. Long arrows indicate motion directions of major plates. Circular arrows represent rotations.

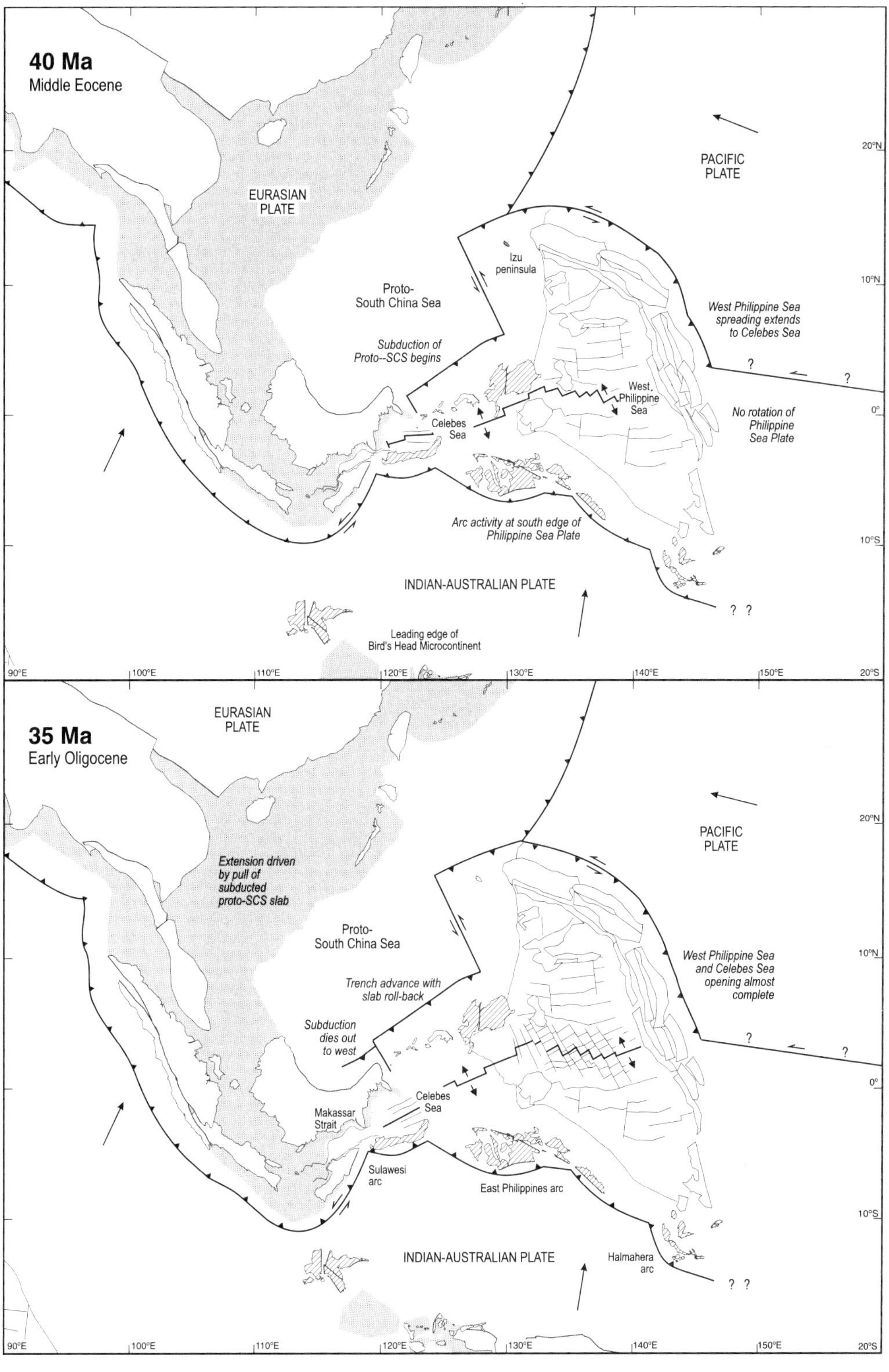

Fig. 3. Reconstructions of the region at 40 and 35 Ma. The Philippine Sea plate had ceased rotation and the West Philippine–Celebes Sea–Makassar Strait basin was opening, accommodated by subduction of the proto-South China Sea.

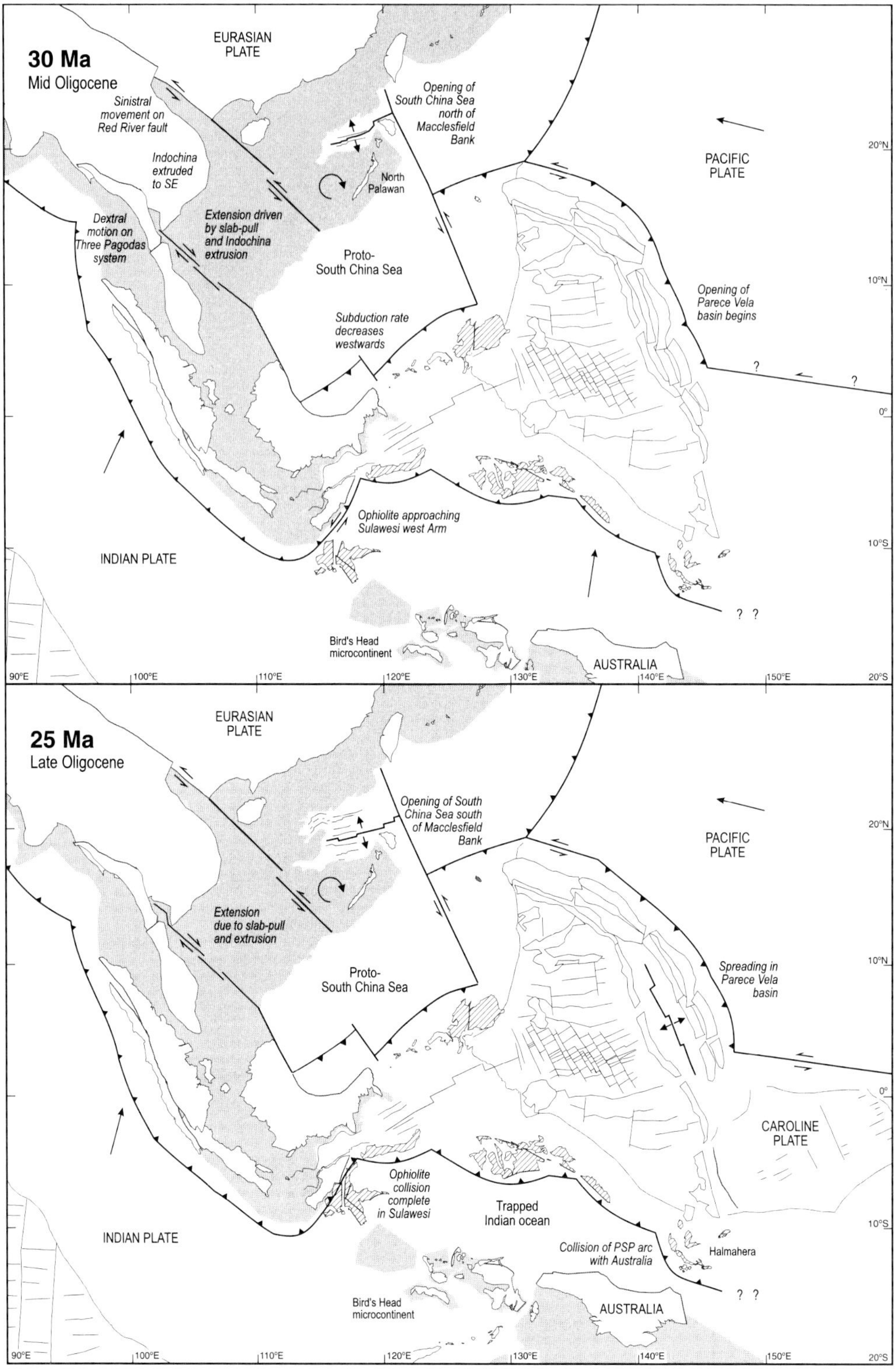

Fig. 4. Reconstructions of the region at 30 and 25 Ma. The South China Sea was opening. Northward movement of Australia trapped Indian oceanic crust in the future Molucca Sea after collisions in the Sulawesi and Halmahera regions.

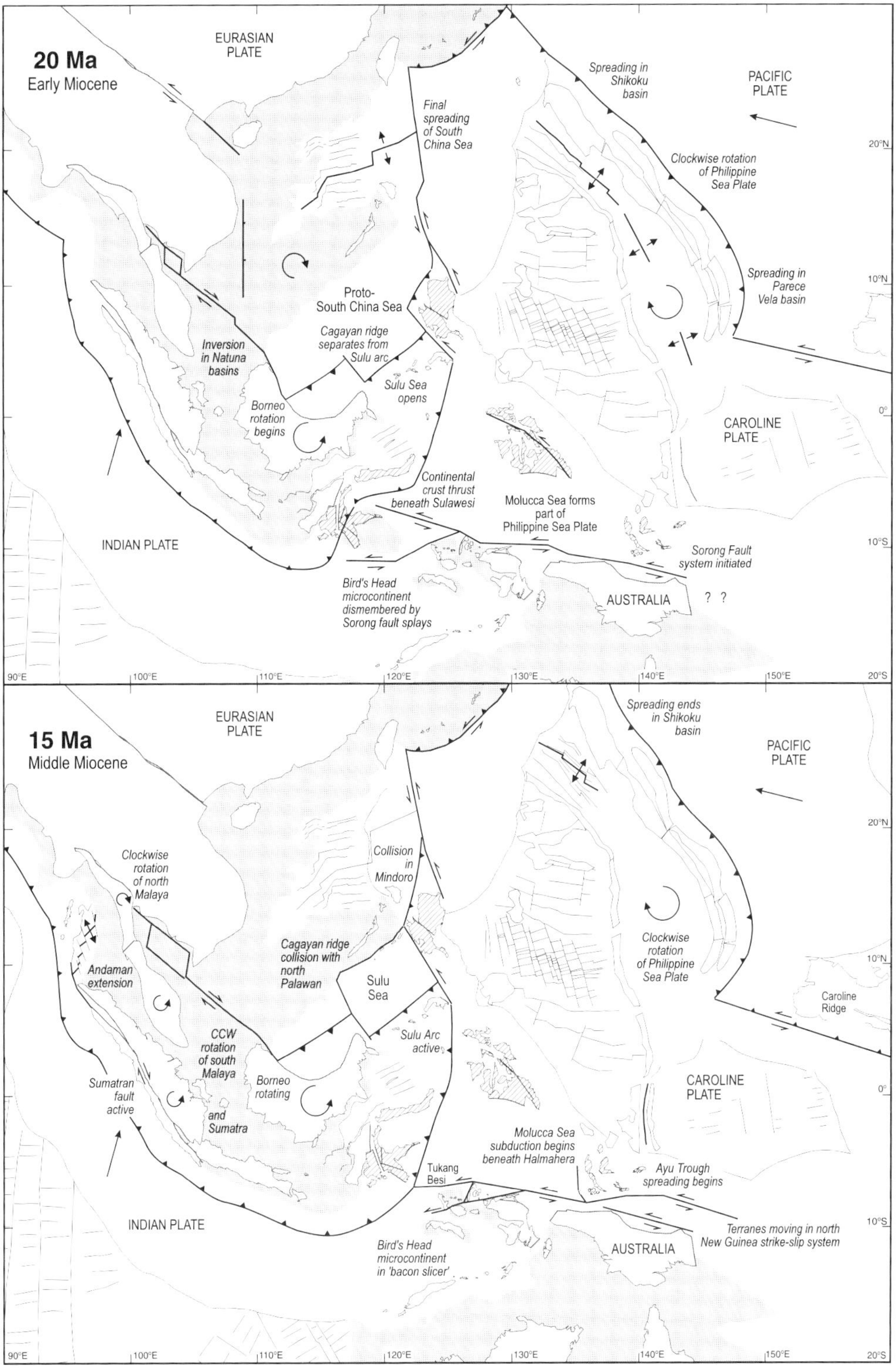

Fig. 5. Reconstructions of the region at 20 and 15 Ma. Borneo was rotating. Renewed rotation of the Philippine Sea plate was accommodated by strike-slip motion on the Sorong fault, with splays moving blocks of the Bird's Head microcontinent west towards Sulawesi.

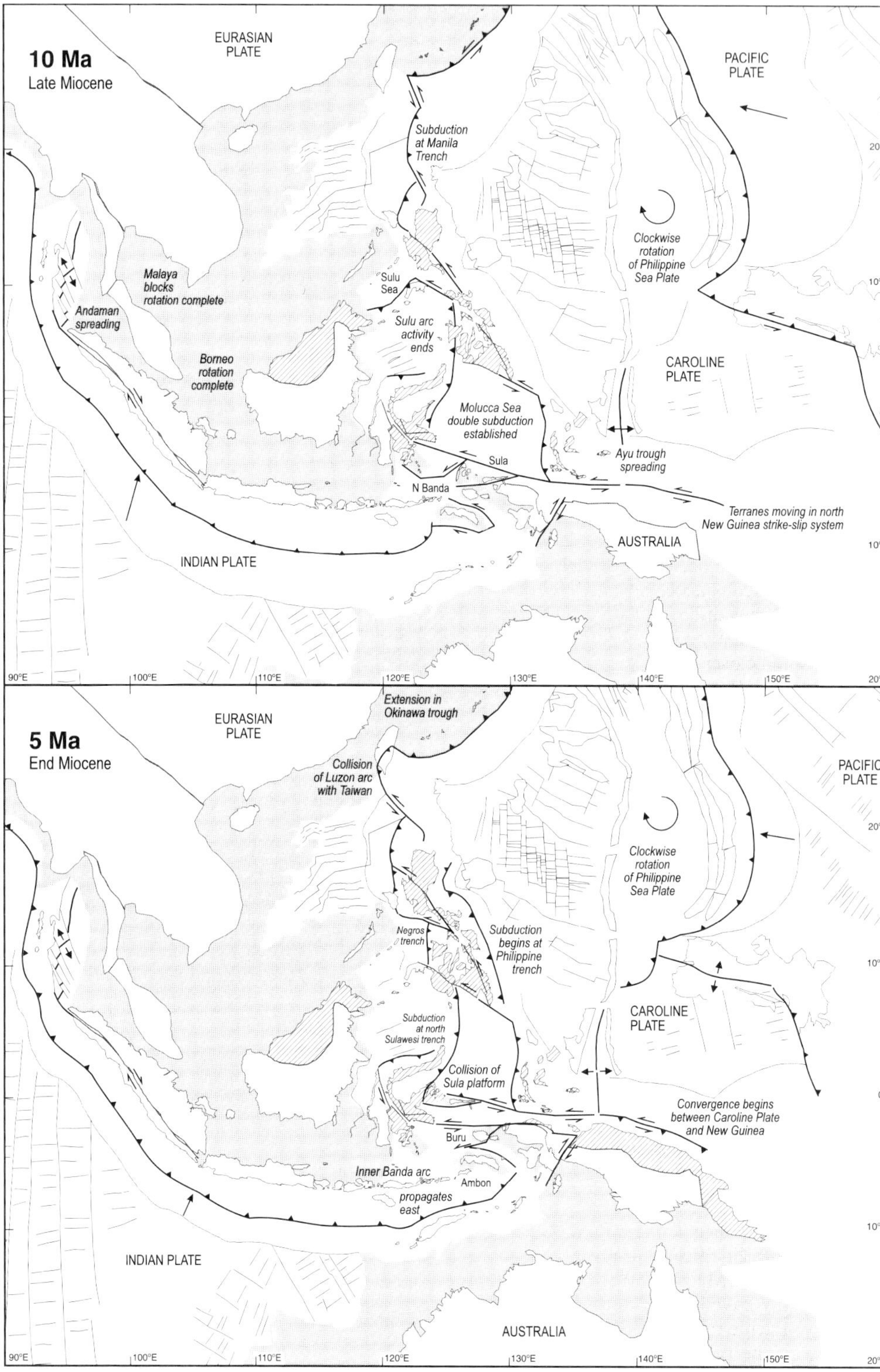

Fig. 6. Reconstructions of the region at 10 and 5 Ma. Borneo rotation was completed. The Molucca Sea was gradually eliminated by subduction on both sides. The Inner Banda arc propagated east, with extension in the North and South Banda Sea regions.

Reconstructions

50 Ma

Eurasia formed a stable continental area, with the Eurasian continental margin oriented broadly NE–SW. Taiwan, north Palawan and the continental shelf currently NW of Borneo were situated on the outer part of a stable passive margin, established during Cretaceous times. Sundaland was separated from Eurasia by a wide proto-South China Sea (SCS) probably floored by Mesozoic ocean crust. The southern edge of this ocean was a passive continental margin NW of a narrow continental promontory extending from Borneo to Zamboanga. North and south Malaya were much closer to Indochina and Sundaland was oriented NW–SE. The southern Sundaland margin had a NW–SE orientation along most of its length, implying a strike-slip and partly active margin resulting from oblique subduction of Indian ocean lithosphere. Because rotation of Borneo is accepted in this model the reconstruction thus differs from those of Rangin *et al.* (1990) and Daly *et al.* (1991) who infer a margin oriented closer to E–W.

Connecting plate boundaries to the Pacific is very speculative. The orientation of the Java trench is interpreted to have changed at its eastern end to NE–SW near west Sulawesi where there was a predominantly strike-slip margin, before linking to subduction zones further east. Alternatively, the Java trench could have continued broadly eastwards to a trench on the south side of the Philippine Sea plate (PSP) formed by the oldest parts of the east Philippines, the West Philippine basin, and Halmahera, which includes arc rocks dating back at least to the Cretaceous. North of the PSP there was a south-dipping subduction zone at the southern edge of a Northern New Guinea (NNG) plate. Spreading in the West Philippine Sea, situated in backarc setting, could have been driven by Indian ocean subduction, NNG plate subduction or both.

45 Ma

There was no significant change in the configuration of Eurasia and Sundaland. However, between 50 and 40 Ma rapid clockwise rotation of the PSP may have been linked to subduction of the NNG-Pacific ridge at the its NE edge. The subduction of the ridge, orientated sub-parallel to the trench, caused massive extension in the Izu–Bonin–Mariana forearc associated with boninite magmatism. Ophiolitic rocks of Zambales complex, Luzon are speculated to be the western end of this boninite arc. On the south side of the PSP Cretaceous arc rocks of Halmahera–east Philippines formed the basement of an arc situated above a north-dipping subduction zone. At 44 Ma a new spreading centre developed in the West Philippine Sea. The Sulawesi ophiolite was located on the Indian plate, although its position south of the bend in the Java trench suggests the possibility of a complex transform setting at this time.

40 Ma

Subduction of the NNG–Pacific ridge and consequent elimination of the NNG plate caused a major change in the length of the subducted slab in the west Pacific, resulting in a change in Pacific plate motion. This required changes in other plate boundaries and the 40 Ma reconstruction is therefore quite different from the older, and more confidently linked to the younger, reconstructions.

A subduction zone at the north edge of the Indian plate is inferred to have extended east from the Sunda–Java trench. Rotation of the PSP finished but the earlier rapid rotation had initiated a transform fault which now separated a formerly continuous arc between Luzon and the Izu–Bonin arc. The arc on the south side of the PSP was stationary. Spreading was well established and produced well-defined magnetic anomalies in the West Philippine Sea. This was situated in a backarc setting, and the width of the basin may be due to subduction beneath the PSP at both its south and NE edges, causing geochemical and tectonic complexity since both Pacific and Indian plates were being subducted.

The new palaeomagnetic data from east Indonesia give a greater confidence in reconstructing the position of the PSP, and suggest that the West Philippine Sea–Celebes Sea formed a single basin, narrowing west, as the spreading rate decreased. At its extreme western end, extension of the Makassar Strait (Situmorang 1982) probably never developed beyond the attenuation of continental crust as indicated by marine seismic studies (Dürbaum & Hinz 1982). The south-dipping subduction zone north of Luzon–Zamboanga, resulting from subduction of the proto-SCS, thus became less important westwards and there was no significant subduction beneath Borneo.

35 Ma

Indian ocean subduction continued at the Sunda–Java trenches, and also at the PSP arc extending from north Sulawesi, south of the east Philippines, to Halmahera. The West Philippine–Celebes Sea basin continued opening until 34 Ma. The southern PSP arc remained stationary, so the whole of the northward motion of the north side of this basin was absorbed by subduction north of Luzon and at the NW edge of the PSP. This subduction system died out westwards so there was no closure of the west end of the proto-SCS. However, further east there was significant subduction and the pull forces of the subducting slab could therefore account for stretching of the Eurasian margin north of Palawan, in the region which was subsequently to become the South China Sea. The major differences between this and previous reconstructions (Rangin *et al.* 1990; Daly *et al.* 1991; Lee & Lawver 1994) are therefore the position of Borneo, the width of the proto-SCS and the link between the Celebes and Philippine Sea basins

30 Ma

Subduction on the south side of the proto-SCS continued, but spreading in the West Philippine–Celebes Sea basin had ceased. However, the transform link at the east end of the subduction zone now linked to the newly established spreading centre of the SCS north of Macclesfield Bank. Movement of blocks driven by indentation of Eurasia by India (Tapponnier *et al.* 1982) contributed further to the closure of the proto-SCS, as Indochina was extruded SE on the sinistral Red River fault and the dextral Three Pagodas and Wang Chao faults (simplified as a single fault at the north end of north Malaya). Rifting of the Palau–Kyushu ridge began, leading to opening of the Parece Vela basin due to Pacific subduction beneath the eastern edge of the PSP.

25 Ma

In the SCS a ridge jump led to ocean floor spreading between Macclesfield Bank and Reed Bank, terminating spreading in the earlier-formed northern basin. Continued movement of Indochina on the Red River fault was also absorbed by extension within the Sunda Shelf. Dextral movement continued on the Three Pagodas and Wang Chao faults, probably partly absorbed in the Gulf of Thailand, Malay and Natuna basins.

PSP motion changed, linked to two important collisions. In the late Oligocene ophiolites, probably situated at the leading edge of the Bird's Head microcontinent, were emplaced on the west Arm of Sulawesi. The area between the ophiolite and the Bird's Head is speculated to have been occupied by continental crust which in the early Miocene was thrust beneath Sulawesi (Coffield *et al.* 1993). The arc on the south side of the PSP collided with the Australian margin in New Guinea. These collisions trapped Indian ocean crust between Sulawesi and Halmahera which then became part of the PSP. Subduction ceased and between 25 Ma and 20 Ma (Fig. 5) the plate boundary became a strike-slip zone, the Sorong Fault system, which subsequently moved terranes of the PSP arc along the New Guinea margin.

Within the PSP, Parece Vela basin opening had propagated both north and south, forming the remnant arc of the Palau–Kyushu ridge. The Caroline plate is shown for the first time on the 25 Ma reconstruction, since although magnetic anomalies indicate Oligocene opening its position of formation and tectonic setting is uncertain. During the Neogene there has been little or no subduction at the Caroline–PSP boundary. Subduction of the Pacific plate beneath the PSP has been accommodated by sinistral motion at the Caroline–Pacific boundary with intervals of transtension and transpression in the region of the Caroline Ridge.

20 Ma

The clockwise rotation of the PSP necessitated changes in plate boundaries throughout SE Asia. These include the re-orientation of spreading in the South China Sea, and the development of new subduction zones at the eastern edge of Eurasia. Borneo started its counter-clockwise rotation, producing the Deep Regional Unconformity (Tan & Lamy 1990) of the north Borneo margin, with counter-clockwise motion of west Sulawesi, and smaller counter-clockwise rotations of most of the adjacent Sundaland blocks. An exception was north Malaya which began to rotate clockwise, thus remaining linked to both Indochina and south Malaya. Because the Borneo rotation pole was close to the NW corner of Borneo there was no major deformation of the Sunda shelf but there was inversion in basins such as West Natuna (Ginger *et al.* 1993). Rotation of Borneo was accommodated on the south side of the proto-SCS by southward subduction linked to the strike-slip boundary in west Borneo, but in the west, close to the Borneo rotation pole, the amount of subduction was small. Further east, the increased rate of subduction caused the Sulu Sea to open as a back-arc basin at 20 Ma (Holloway 1982; Hinz *et al.* 1991; Silver & Rangin 1991) south of the Cagayan ridge, and between 20 and 15 Ma the Cagayan ridge moved northwards across the proto-SCS.

Collision in Sulawesi had almost eliminated the leading edge of the Bird's Head microcontinent. New subduction had begun at the west edge of the Philippine Sea plate below the north Sulawesi–Sangihe arc which extended north to south Luzon. The Philippine islands were carried passively with the PSP towards this subduction zone, as were the Halmahera islands, during a period of widespread carbonate deposition. North of Luzon, sinistral strike-slip movement linked the subducting SW margin of the PSP to subduction at the Ryukyu trench.

15 Ma

The reconstructions exaggerate the width of the western part of the proto-SCS because Borneo north of the Lupar Line is assigned to a single fragment, thus failing to show Neogene northward progradation of the Borneo margin by the addition of sediment. Most of this sediment is interpreted to have been derived from the north across the Sunda shelf, perhaps in part fed along the strike-slip margin from further northwest in Sundaland. As Borneo rotated, the remaining proto-SCS was eliminated.

The model predicts important changes in volcanic and tectonic history beginning at about 20 Ma for both Java and Sumatra. By 15 Ma North Sumatra had rotated counter-clockwise with south Malaya, and as the rotation proceeded the orientation of the Sumatran margin became less oblique to the Indian plate motion vector. This resulted in partitioning of convergence into an orthogonal subduction component and a parallel strike-slip component, leading to formation of the dextral Sumatran strike-slip system, and extension in the Andaman region. This model therefore predicts the development of a strike-slip system in Sumatra during the early-middle Miocene and this is consistent with the history of basins near to the present Sumatran fault, such as the Ombilin Basin. Sedimentation began in the Palaeogene but there is no clear evidence for strike-slip control until the Neogene (C. G. Howells & A. J. McCarthy pers. comm. 1996).

Collision of Luzon and the Cagayan ridge with the Eurasian continental margin in Mindoro and north Palawan resulted in a jump of subduction to the south side of the Sulu Sea. Southward subduction beneath the Sulu arc continued until 10 Ma. The remainder of the Philippines continued to move with the PSP, possibly with intra-plate strike-slip motion and subduction resulting in local volcanic activity.

At the south end of the PSP, splays of the Sorong fault developed sequentially. The Tukang Besi platform separated from the Bird's Head microcontinent and was carried west on the PSP to collide with Sulawesi (Davidson 1991). Locking of splays of the Sorong fault caused subduction to initiate at the eastern margin of the Molucca Sea, followed by volcanic activity in Halmahera. This subduction zone later propagated north to link to a strike-slip zone connecting south Mindanao to the Sangihe trench. Thus the Molucca Sea became a separate plate and the double subduction system developed. At the east edge of the PSP spreading terminated in the Shikoku basin, formed as the PSP rotated accommodated by roll-back of the Izu-Bonin–Mariana trench.

10 Ma

Rotation of Borneo was complete. This, with collision in the central Philippines, the earlier collision in Mindoro, and continued northward movement of Australia, resulted in reorganisation of plate boundaries and intra-plate deformation in the Philippines. At the west end of Sundaland, partitioning of convergence in Sumatra into orthogonal subduction and strike-slip motion had effectively established a south Sumatra forearc sliver plate. Resultant extension led to development of ocean crust in the Andaman Sea (Curray *et al.* 1979).

At the east end of the Java trench the Eurasia–PSP–Australia triple junction decomposed into a zone of microplates, partly as a consequence of development of splays of the Sorong fault. The volcanic inner Banda arc propagated east into the area between Timor and the Bird's Head microcontinent which included trapped Indian ocean crust of probable Mesozoic age. This crust was now subducted and Seram began to move east as the arc propagated into the north Banda Sea, requiring subduction and strike-slip motion at the edges of this microplate. Further north, in the Sorong fault zone, Tukang Besi had been accreted to Sulawesi, locking one strand of the fault and initiating a new splay south of the Sula platform. The Sula platform moved with the Molucca Sea Plate, which was itself partly coupled to the PSP due to the low rate of convergence at the Halmahera trench. The Bird's Head moved north along a strike-slip fault at the Aru basin edge. North of the Bird's Head, and further east in New Guinea, transpressional movements were marked by deformation of arc and ophiolite slivers separated by sedimentary basins. Opening of the Ayu trough separated the Caroline plate and PSP, although the rate of separation at this spreading centre was very low.

Subduction was now established in the whole Halmahera arc, and extended north to a sinistral strike-slip system through west Mindanao. This was one of several strike-slip systems active within the Philippines, probably linked to subduction zones at the western edge of the archipelago, such as the Manila trench. Links between the Sorong system and Sulawesi, with tightening of the knot at the triple junction and cessation of subduction at the Sulu trench, initiated southward subduction of the Celebes Sea beneath north Sulawesi.

5 Ma

The north end of the Philippine arc north of Luzon came into collision with the Eurasian margin in Taiwan. The PSP pole of rotation moved to a position north of the plate; clockwise rotation continued but the change in motion caused re-orientation of existing, and development of new, plate boundaries. Subduction continued at the Manila, Sangihe and Halmahera trenches, and new subduction began at the Negros and Philippine trenches.

In the Banda Sea the eastward movement of Seram relative to the Bird's Head changed to more convergent motion leading to significant subduction at the Seram trough. The Banda volcanic arc propagated east and its eastern end was at the longitude of Ambon, producing the famous ambonites. Since 5 Ma the southern Banda Sea has extended to its present dimensions (compare the 5 Ma reconstruction to Fig. 1), probably by forearc and intra-arc extension, with continental crust now found in the Banda Sea ridges.

The Molucca Sea continued to close by subduction on both sides. At present the Sangihe forearc has overridden the northern end of the Halmahera arc, and is beginning to over-thrust west Halmahera. At the west end of the Sorong system the Sula platform collided with the east arm of Sulawesi, causing rotation of the east and north arms to their present position, and increasing the rate of subduction at the north Sulawesi trench. The northern New Guinea arc terranes have been omitted from the reconstructions before 5 Ma because there are insufficient data to reconstruct them adequately. However, the model suggests their Neogene history was one of movement in a strike-slip zone, and only since 5 Ma has there been significant convergence between the Australian margin and Caroline plate.

Conclusions

A rigid plate model can explain in gross terms the evolution of the Indonesian and wider SE Asian region. Maintaining fragment dimensions fails to portray arc extension, addition of new crust in arc regions, much strike-slip faulting, accretionary growth of continental margins, and neglects the importance of regions of thinned continental crust. Adding new fragments to the model would improve modelling of local movements but dilute the model's simplicity in which driving forces are essentially Australia and PSP motion, and major regional events are linked to arc-continent collisions.

The model provides a coherent basic framework for understanding the evolution of the region. What is clear is that plate boundaries have at certain times changed character rapidly in response to the driving forces of the major plates of the region. Borneo counter-clockwise rotation is still unproven, but is feasible, and the model shows that there are alternative solutions to reconstructing the Philippines and answering other problems of the east Eurasian margin. The model currently has its greatest problems with timing and amounts of rotation of the west Sundaland blocks. Deformation recorded in the Sunda shelf basins suggests earlier rotations than those of the model, and the amount of extension predicted, particularly in the Gulf of Thailand, appears too great. It may be that these problems could be resolved through more complex solutions requiring non-rigid plate tectonics and/or local rotations of rigid blocks. The model shows the need for a major attack on the history of the region though integrated palaeomagnetic and geological studies, to differentiate regional and local movements, to determine their timing, and to correlate regional events.

I am particularly grateful to Simon Baker who has provided me with invaluable support, not only with the work involved in digitising the fragment outlines used in the reconstructions, but with discussion and ideas at many stages. I also thank the numerous people with whom I have discussed the geology of SE Asia over many years who have contributed criticism, ideas and comments, including D. A. Agustiyanto, J. R. Ali, C. D. Anderson, M. G. Audley-Charles, P. D. Ballantyne, S. J. Baker, A. J. Barber, J. C. Briden, T. R. Charlton, E. Forde, R. Garrard, N. Haile, A. S. Hakim, S. Hidayat, T. W. C. Hilde, M.

Fuller, Kusnama, J. A. F. Malaihollo, R. McCabe, J. S. Milsom, A. H. G. Mitchell, S. J. Moss, R. Murphy, G. J. Nichols, C. D. Parkinson, M. Pubellier, R. Quebral, C. Rangin, H. Rickard, S. J. Roberts, N. Sikumbang, T. O. Simandjuntak, M. J. de Smet, B. Taylor, and S. L. Tobing. Financial support has been provided at various stages by NERC award GR3/7149, grants from the Royal Society, the University of London SE Asia Research Group currently supported by Arco, Canadian Occidental, Exxon, Lasmo, Mobil, and Union Texas, and the University of London Central Research Fund. Work in Indonesia was facilitated by GRDC, Bandung and Directors including H. M. S. Hartono, M. Untung, R. Sukamto and I. Bahar.

References

ALI, J. R. & HALL, R. 1995. Evolution of the boundary between the Philippine Sea Plate and Australia: palaeomagnetic evidence from eastern Indonesia. *Tectonophysics*, **251**, 251–275.

BRIAIS, A., PATRIAT, P. & TAPPONNIER, P. 1993. Updated interpretation of magnetic anomalies and seafloor spreading stages in the South China Sea: implications for the Tertiary tectonics of Southeast Asia. *Journal of Geophysical Research*, **98**, 6299–6328.

CAMBRIDGE PALEOMAP SERVICES. 1993. *ATLAS version 3.3*. Cambridge Paleomap Services, PO Box 246, Cambridge, UK.

COFFIELD, D. Q., BERGMAN, S. C., GARRARD, R. A., GURITNO, N., ROBINSON, N. M. & TALBOT, J. 1993. Tectonic and stratigraphic evolution of the Kalosi PSC area and associated development of a Tertiary petroleum system, south Sulawesi, Indonesia. *IPA Proceedings 22nd Annual Convention*, 678–706.

CURRAY, J. R., MOORE, D. G., LAWVER, L. A., EMMEL, F. J., RAITT, R. W., HENRY, M. & KIECKHEFFER, R. 1979. Tectonics of the Andaman Sea and Burma. *American Association of Petroleum Geologists Memoir*, **29**, 189–198.

DALY, M. C., COOPER, M. A., WILSON, I., SMITH, D. G. & HOOPER, B. G. D. 1991. Cenozoic plate tectonics and basin evolution in Indonesia. *Marine and Petroleum Geology*, **8**, 2–21.

DAVIDSON, J. W. 1991. The geology and prospectivity of Buton island, S.E. Sulawesi, Indonesia. *IPA Proceedings 20th Annual Convention*, 209–233.

DÜRBAUM, H.-J. & HINZ, K. 1982. SEATAR-related geophysical studies by BGR in the southwest Pacific. *Transactions 3rd Circum-Pacific Energy and Mineral Resources Conference*, 129–133.

GINGER, D. C., ARDJAKUSUMAH, W. O., HEDLEY, R. J. & POTHECARY, J. 1993. Inversion history of the West Natuna basin: examples from the Cumi-Cumi PSC. *IPA Proceedings 22nd Annual Convention*, 635–658.

HALL, R. 1996. Reconstructing Cenozoic SE Asia. *In*: HALL, R. & BLUNDELL, D. J. (eds) *Tectonic Evolution of SE Asia*. Geological Society, London, Special Publications, **106**, 153–184.

—— & NICHOLS, G. J. 1990. Terrane amalgamation at the boundary of the Philippine Sea Plate. *Tectonophysics*, **181**, 207–222.

——, ALI, J. R. & ANDERSON, C. D. 1995*a*. Cenozoic motion of the Philippine Sea plate: palaeomagnetic evidence from eastern Indonesia. *Tectonics*, **14**, 1117–1132.

——, ——, —— & BAKER, S. J. 1995*b*. Origin and motion history of the Philippine Sea Plate. *Tectonophysics*, **251**, 229–250.

——, FULLER, M., ALI, J. R. & ANDERSON, C. D. 1995*c*. The Philippine Sea Plate: Magnetism and Reconstructions. *In*: TAYLOR, B. & NATLAND, J. H. (eds) *Active Margins and Marginal Basins: A Synthesis of Western Pacific Drilling Results*. American Geophysical Union Monographs, **88**, 371–404.

HAMILTON, W. 1979. *Tectonics of the Indonesian region*. US Geological Survey Professional Paper, **1078**.

HINZ, K., BLOCK, M., KUDRASS, H. R. & MEYER, H. 1991. Structural elements of the Sulu Sea. *Geologische Jahrbuch*, **A127**, 883–506.

HOLLOWAY, N. H. 1982. The stratigraphic and tectonic evolution of Reed Bank, North Palawan and Mindoro to the Asian mainland and its significance in the evolution of the South China Sea. *American Association of Petroleum Geologists Bulletin*, **66**, 1357–1383.

KATILI, J. A. 1975. Volcanism and plate tectonics in the Indonesian island arcs. *Tectonophysics*, **26**, 165–188.

LEE, T. Y. & LAWVER, L. A. 1994. Cenozoic plate tectonic reconstruction of the South China Sea region. *Tectonophysics*, **235**, 149–180.

RANGIN, C., JOLIVET, L. & PUBELLIER, M. 1990. A simple model for the tectonic evolution of southeast Asia and Indonesia region for the past 43 m.y. *Bulletin de la Société géologique de France*, **8 VI**, 889–905.

SILVER, E. A. & RANGIN, C. 1991. Leg 124 tectonic synthesis. *In*: SILVER, E. A., RANGIN, C., VON BREYMANN, M. T. *et al*., *Proceedings of the ODP, Scientific Results*, **124**, 3–9.

SITUMORANG, B. 1982. Formation, evolution, and hydrocarbon prospect of the Makassar basin, Indonesia. *Transactions 3rd Circum-Pacific Energy and Mineral Resources Conference*, 227–231.

TAN, D. N. K. & LAMY, J. M. 1990. Tectonic evolution of the NW Sabah continental margin since the Late Eocene. *Bulletin of the Geological Society of Malaysia*, **27**, 241–260.

TAPPONNIER, P., PELTZER, G., LEDAIN, A., ARMIJO, R. & COBBOLD, P. 1982. Propagating extrusion tectonics in Asia: new insights from simple experiments with plasticine. *Geology*, **10**, 611–616.

VAN DER VOO, R. 1993. *Paleomagnetism of the Atlantic, Tethys and Iapetus Oceans*. Cambridge University Press.

Characterizing petroleum charge systems in the tertiary of SE Asia

S. P. TODD[1], M. E. DUNN[2] & A. J. G. BARWISE[2]

[1] *BP Exploration Operating Company Ltd, Farburn Industrial Estate, Dyce, Aberdeen, AB2 0PD, UK*

[2] *BP Exploration Operating Company Ltd, Uxbridge One, 1 Harefield Road, Uxbridge, Middlesex, UB8 1PD, UK*

Abstract: Most SE Asian Tertiary-aged petroleum has been derived from paralic (lower delta plain to prodelta, higher land plant dominated) source rocks, although the larger proportion of *oil* is from lacustrine (freshwater/brackish algal) sources. Because many SE Asian petroleum provinces have been largely explored without penetrating source rock, the source is inferred from the oil's geochemistry. Prolific lacustrine sources develop mainly in *rift* lakes in the Palaeogene syn-rift megasequence common to many SE Asian basins, but *floodplain* lake sources are also important. Paralic source rocks include both coals and particularly coaly mudrocks developed within the Miocene post-rift megasequence with an oil-prone zone. Oil-prone source rocks are preferentially developed in the paralic realm between the lower coastal plain and lower estuary/delta front facies, perhaps involving a mangrove system. The stratigraphic position and kinetic expulsion behaviour of different source types, migration, seal timing and pressure–temperature–depth relationships converge to make younger plays, including carbonate buildups, more gas prone. The volumetric significance of biogenic gas is still poorly understood. Vertical migration is common in overpressured SE Asian basins; lateral migration is thus commonly restricted to distances of 20 km or less from the kitchen.

SE Asia, including Indonesia, Thailand, Malaysia, Brunei and the Philippines, contains 10 oil and gas provinces that individually contain more than 3×10^9 barrels oil equivalent (boe) (Fig. 1). All of these are Tertiary basins with Tertiary-aged petroleum systems. This represents a considerable resource that has been exploited in the past and continues to grow with further exploration. This paper reviews the nature of petroleum systems in SE Asia. The systems are first placed in their tectonostratigraphic context. The nature of the petroleum fluids and the depositional environment of their source rocks are then described. The possible significant potential of coaly mudrocks rather than coals as oil sources is also considered. Finally the maturation of these source rocks together with migration phenomena are considered in the context of oil and gas distribution in various play types in the region.

SE Asian basin evolution and play systems

The numerous Tertiary basins of SE Asia should be set in the context of the convergence of three tectonic plate systems towards the SE portion of Eurasia which can be considered for convenience as fixed (Daly *et al.* 1991; Fig. 2). The Tertiary convergence of India and Australia, together with continued consumption of Pacific oceanic lithosphere towards the west, caused punctuated pulses of extension and compression on the Eurasian margin, often with a large component of strike slip.

The Palaeogene was dominated by extension and many basins along the Sumatran margin, and the Nam Con Son, West Natuna and Malay basins also originated during this time. Pronounced extension occurred along the China margin of the South China Sea leading to seafloor spreading in the middle Oligocene. Typically, these early rifts or pull apart basins are filled with mainly continental sediments. Because of rapid fault-associated subsidence, tropical climates, and diversion of major sediment-laden river systems away from the rifts by rift-shoulder uplift, prolonged lacustrine deposition was common. A few rifts, particularly those closest to the new oceanic spreading centres, e.g. Palawan (Philippines) and Da Nang (northern offshore Vietnam), contain sediment fills with significant marine components (Fig. 3).

With continued, pulsed extension and increasing regional subsidence, the fault-block topography was depressed and covered in many of the basins. A common SE Asian stratigraphic signature is of a gradual retrogradational megasequence from continental sediments, to paralic, to marine clastics and carbonates to deep marine mudstones (Fig. 3). Major coastal plain, interdistributary bay and tidal swamp systems developed during this period.

From Fraser, A. J., Matthews, S. J. & Murphy, R. W. (eds), 1997, *Petroleum Geology of Southeast Asia*, Geological Society Special Publication No. 126, pp. 25–47.

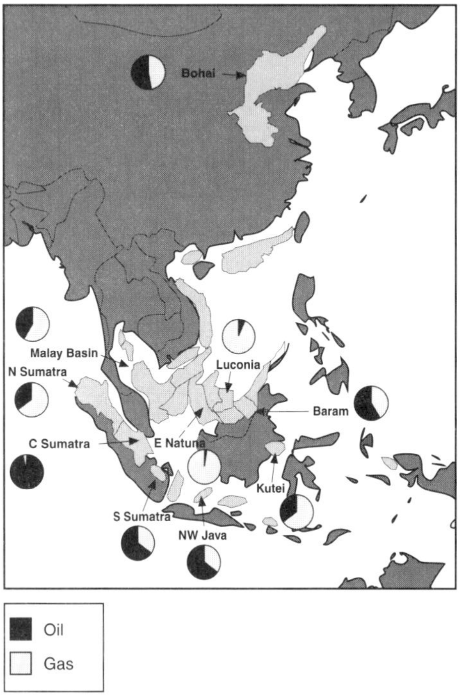

Fig. 1. Principal ($<3 \times 10^9$ boe recoverable) SE Asian petroleum provinces.

Late Mid- to Late Miocene times saw the gentle to pronounced compression of many SE Asian Basins. The compression coincides with the collision of the Australian plate with Eurasia and may represent the 'locking up' of the extensional/strike-slip systems. The result of the compression was inversion of depocentres and generation of many compressional anticlines. Along the southern margin of the East Vietnam Sea convergence between China and Borneo/Palawan culminated during the Early Miocene with the choking of the subduction zone with a Chinese continental fragment originally rifted from the Eurasian margin (Fig. 2). The collision caused the development of the Borneo–Palawan foreland basin into which prograded the major Baram delta system. Another Mio-Pliocene growth-faulted delta was positioned on the southern margin of Borneo in the present location of the Mahakam delta, filling the Kutei Basin.

Following inversion, many of the basins collapsed and subsided again to accommodate further Late Miocene and Plio-Pleistocene sediments.

The typical SE Asian geological history described above allows the development of several common petroleum play types involving clastic, carbonate and even basement reservoirs (Table 1).

Origin of petroleum

By far the majority of petroleum in SE Asia originates from non-marine kerogen, either lacustrine freshwater algal or freshwater or brackish (mangrove) higher land plant (Table 2; Figs 4 & 5). Roughly equal oil equivalent volumes of the two main petroleum types occur, but the lacustrine source rocks are clearly responsible for a substantially greater proportion of oil than the gas-rich paralic systems. This preference for oil in the lacustrine systems, which in turn tends to be sequestered into the

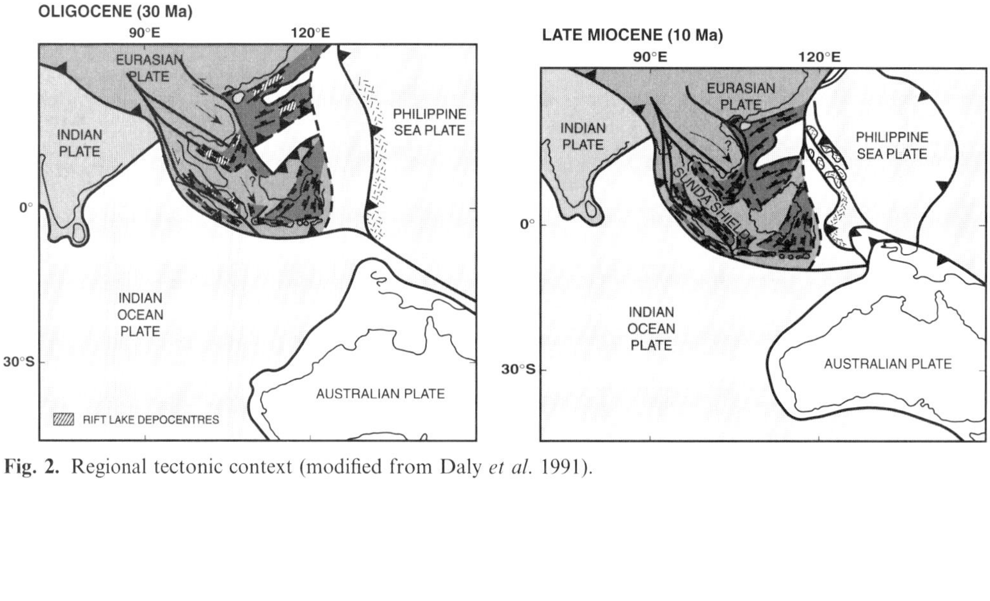

Fig. 2. Regional tectonic context (modified from Daly *et al.* 1991).

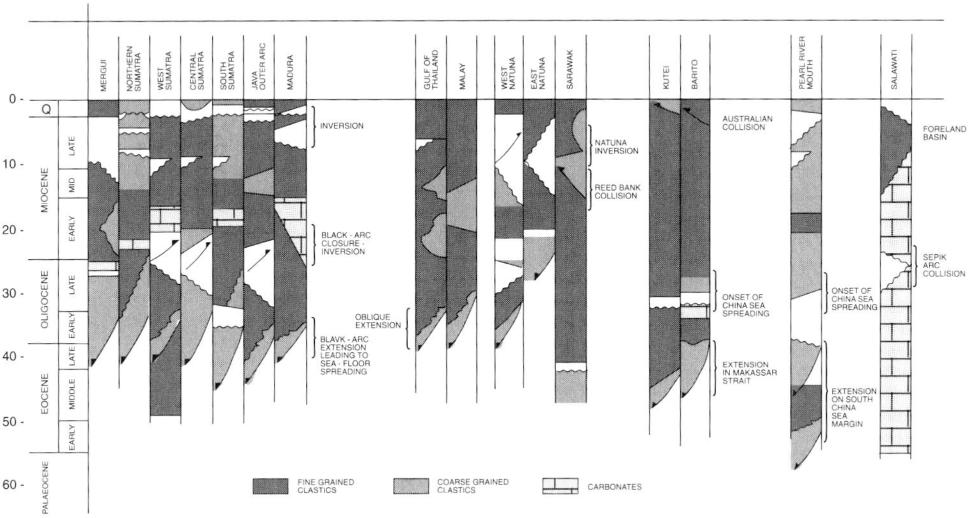

Fig. 3. Chronostratigraphic correlation of SE Asian basins (from Daly *et al.* 1991).

lower intervals of the basin fill, may help to explain the greater proportion of oil in Oligo-Miocene clastics as opposed to gas in the Miocene carbonates (Fig. 6). The preferential distribution of biogenic gas within the Miocene carbonates may be an additional but unquantified explanation (e.g. southern Vietnam; Fig. 7).

The only major exception is the North Sumatra basin where the source rock is believed to be deposited in a restricted marine environment but with a predominantly terrestrial land plant input (Robinson 1987). However it is acknowledged that the possible presence of a deeply buried, as yet unpenetrated lacustrine source rock, cannot be totally ruled out (Matchette–Downes 1994). The small remainder of petroleum is marine algal in origin, in both carbonate and clastic associations (Figs 4 & 5).

Table 1. *Typical play types in SE Asia*

Play	Reservoir	Seal	Trap	Examples
Mio-Pliocene turbidites	Slope turbidite fans	Hemipelagic mudstone	Stratigraphic, thrust anticline	Baram delta; Palawan
Miocene carbonate buildups	Carbonate buildup	Hemipelagic mudstone downlap onto buildups	Compactional drape, stratigraphic	Luconia; North Sumatra; East Natuna; PRMB
Oligo-Miocene 'post-rift' clastics	Paralic–fluvial, deltaic & shallow marine	Intraformational mudstone seals – multiple stacked pools common	Compressional anticlines and extensional tilt blocks	Baram delta; Kutei (Mahakam); Central Sumatra; Malay Basin; West Java
Palaeogene rift-fill clastics	Continental clastics	Intraformational seal, regional marine flooding mudstones	Compressional anticlines and extensional tilt blocks, stratigraphic	Bohai Basin
Pre-Tertiary basement	Fractured-weathered granite & limestone	Lacustrine shales	Extensional tilt blocks	Bohai Basin; Cuu Long Basin

Table 2. *Source types through oil-typing for major petroleum provinces in SE Asia*

Basin	Dominant source type	Age	Reference	Rec. Oil (10^6 bbls)	Rec. Gas (10^6 bbls)	Total Oil & Gas (10^6 bbls)
Bohai	Lacustrine mudstones	Palaeogene	In house BP	15 000	14 250	29 250
Pearl River Mouth	Lacustrine mudstones	Palaeogene	Cai (1987); In house	693	22	715
Gulf of Thailand	Lacustrine mudstones & paralic mudstones & coals	Palaeogene – Early Miocene	In house	452	2 339	2 751
Malay–Penyu	Lacustrine mudstones & paralic mudstones & coals	Palaeogene – Early Miocene	Creaney *et al.* (1994)	3 348	4 814	8 122
West Natuna	Floodplain lacustrine mudstones	Palaeogene	Pollock *et al.* (1984); In house	648	784	1 436
Nam Con Son	Paralic coals & coaly mudrocks	Oligo–Miocene	In house	258	626	884
Cuu Long	Brackish lacustrine mudstones	Oligocene	In house	859	216	1 075
Da Nang	Marine mudstones	Miocene	In house	0	363	363
Yinggehai	Paralic coals & coaly mudrocks	Oligo–Miocene	In house			
Palawan	Marine mudstones/ carbonates	Palaeogene	In house; Sofer (1984) but see Williams *et al.* (1992*a*)	554	941	1 491
Sabah	Paralic coals & coaly mudrocks	Miocene	In house	447	212	659
Baram	Paralic coals & coaly mudrocks	Miocene	James *et al.* (1984); In house	6 243	4 332	10 539
Luconia	Paralic coals & coaly mudrocks	Oligo–Miocene	In house	331	5 522	5 813
Balingian	Paralic coals & coaly mudrocks	Oligo–Miocene	In house	480	184	664
East Natuna	Paralic coals & coaly mudrocks	Oligo–Miocene	In house	70	3 593	3 663
Kutei	Paralic coals & coaly mudrocks	Miocene	Combaz & de Matharel (1978); Oudin & Picard (1982); Schoell *et al.* (1985)	3 610	7 780	11 434
North Sumatra	Marine mudstones	Miocene	Situmeang & Davies (1986); Kjellgren & Sugiharto (1989)	2 233	4 447	6 676
Central Sumatra	Lacustrine mudstones	Palaeogene	Katz (1991); Katz & Kelly (1987); Katz & Mertani (1989)	12 620	390	12 970
South Sumatra	Lacustrine mudstones & paralic coals & mudrocks	Palaeogene–Miocene	Sarjono & Sadjito (1989); Sardjito *et al.* (1991); Suseno *et al.* (1992)	2 282	1 099	3 822
West Java	Paralic coals & coaly mudrocks	Palaeogene	Noble *et al.* (1991)	2 334	1 175	3 469
East Java	Paralic coals & coaly mudrocks	Palaeogene	In house	435	1 198	1 590
Salawati	Marine mudstones	Miocene	Livingston (1992)	463	191	654
Sunda	Lacustrine mudstones	Palaeogene	Prayitno *et al.* (1992)	1 154	212	1 407

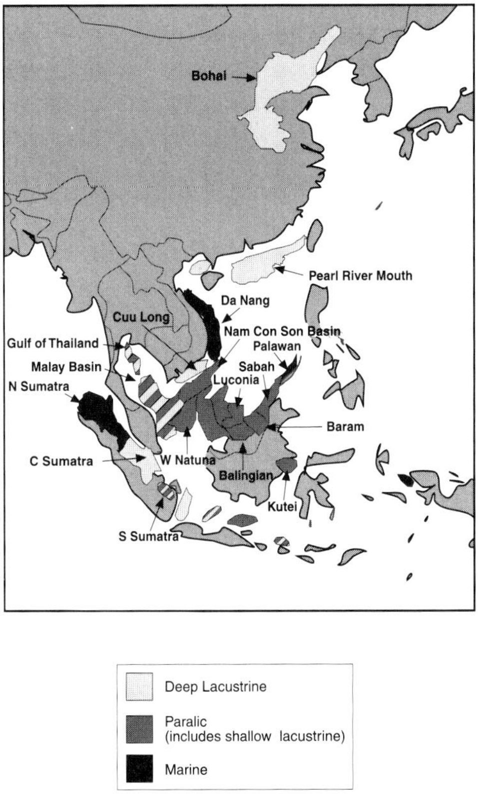

Fig. 4. Principal source types in main SE Asian petroleum provinces.

If Bohai is excluded, then by far the majority of the oil and gas in SE Asia is paralic in origin. It is also evident from Fig. 5 that paralic source rocks, although capable of expelling significant quantities of oil, appear to expel more gas than lacustrine sources. For example, the paralic-sourced Baram delta province contains more than 6×10^9 bbls oil, but also has nearly 25 tcf of gas.

SE Asia is characterized by waxy oil because of non-marine source rocks. This, together with the parameters listed in Table 3, which is modified after Robinson (1987), describe how paralic and lacustrine source rocks may be differentiated from marine sources and from each other.

Principal source depositional facies

It is striking in SE Asia that many large petroleum provinces have been explored without ever penetrating and sampling a significant source rock. Even where the source has been tagged it is commonly not until late in the exploration history. Thus information about the source rock responsible for the hydrocarbon frequently comes indirectly from the geochemical characteristics of the oils discussed above, rather than the source rocks themselves.

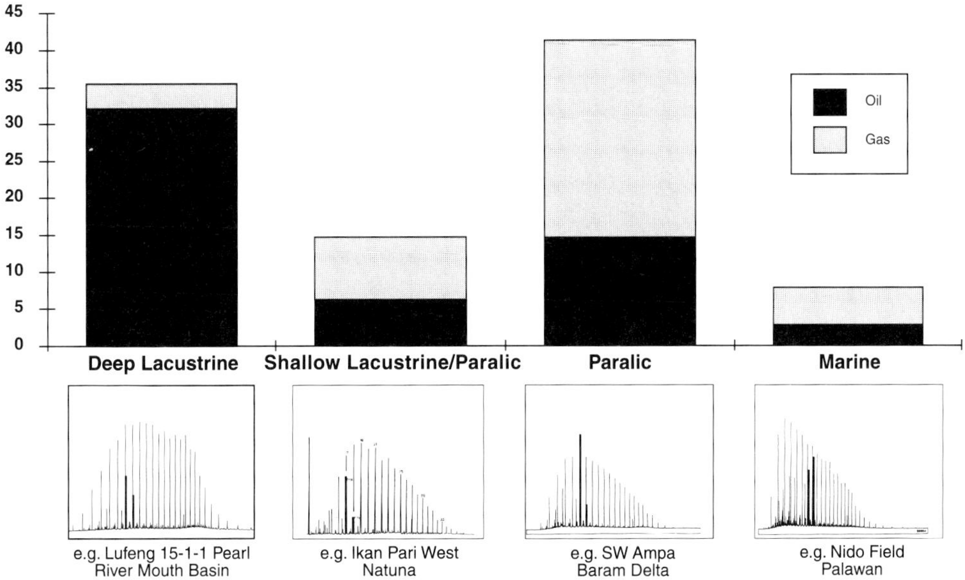

Fig. 5. Approximate volume of different petroleum (oil and gas) types, SE Asia.

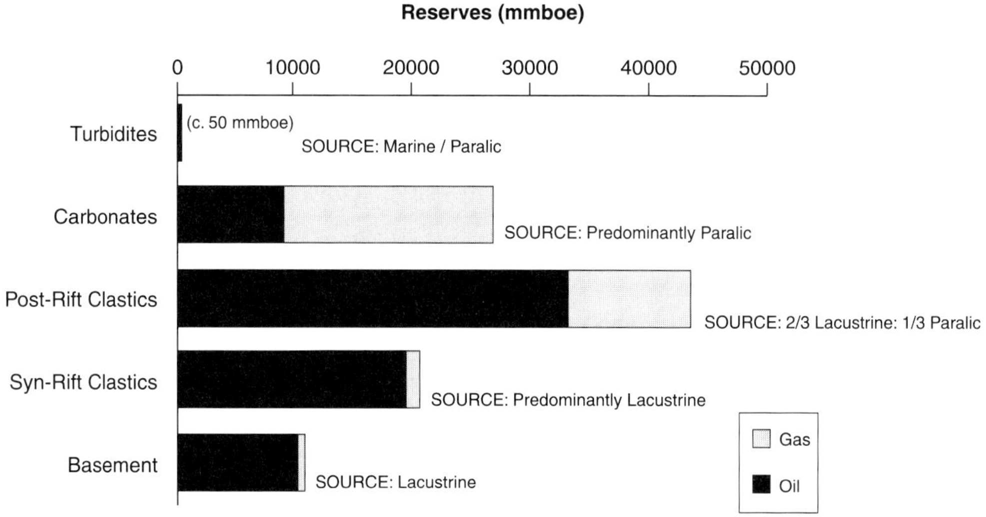

Fig. 6. Stratigraphic distribution of petroleum, SE Asia.

Lacustrine depositional systems

Two main types of lake can be defined with respect to tectonostratigraphic setting. These are:

- rift lakes, i.e. deeper, longer lived lakes enclosed by a hinterland, often mountainous, of exposed and eroding older basement;
- floodplain lakes, i.e. shallower, more ephemeral lakes within a floodplain or coastal plain setting.

While rift lakes are more likely to accumulate good, thick source rocks as discussed below, floodplain lakes can also form important source rocks.

Rift lakes. Freshwater algal source rocks are common in the Palaeogene rift depocentres of SE Asia. The common Eocene or Oligocene continent rift of many of the basins in the Palaeogene tropical latitudes of SE Asia created ideal conditions for the development of long-lived lakes in which organic-rich sediments could accumulate. The controls on lacustrine source rock accumulation that converged to produce some of the most prolific source rocks of the world in SE Asia include the following (Katz 1990; Sladen 1993).

(i) Rift basins with high rates of subsidence but low rates of clastic sediment supply. Such conditions occur early during basin evolution when fault-associated subsidence rates are high but sediment is derived from only local fault-block highs. These rifts are shielded from larger depositional systems of major rivers by the uplift of rift shoulders on the crests of tilt-blocks. The clastic supply will obviously be minimal if the catchment area bedrock is composed predominantly of limestone. In this relatively sediment-starved setting lake depths of several hundred metres can develop, with water columns stratified to include anoxic bottom waters.

(ii) Hydrologically closed or partially closed rift basins are optimal for lake development. Essentially 'a closed hole in the ground' is required for the standing body of water to develop. Rift basins are ideally suited to this factor, particularly where long rifts are split up into individual grabens or half-grabens by transfer zones or faults. A common method for identifying potential lake sites is by mapping closed isopach thicks in possible lacustrine syn-rift sequences.

(iii) High rates of precipitation are common between the latitudes of 15° N and 15° S. Thus the Palaeogene Asian rifts were well supplied with water to allow the development of deep, stratified lakes.

(iv) Organic productivity is controlled by a number of factors including nutrient supply, light, salinity and temperature. Tropical lakes with high surface run-off and higher surface temperatures are likely to fulfil these attributes.

(v) Preservation of organic matter is controlled by chemistry, organic flux, lake dynamics and biological agents. Clearly, deeper rift lakes with stratified water columns will

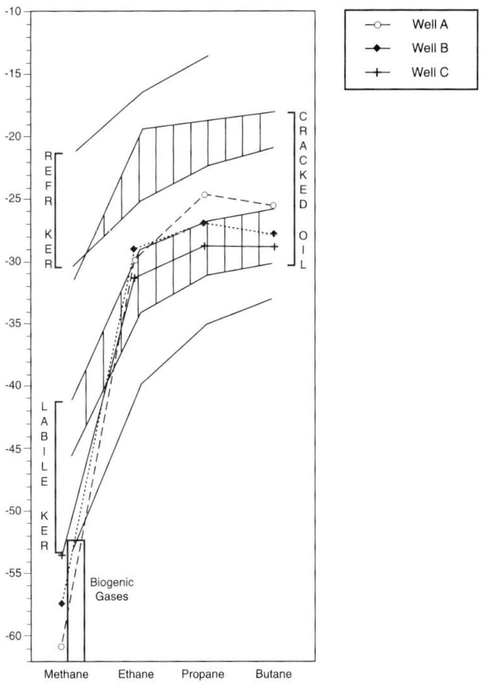

Fig. 7. Carbon isotope data for Miocene carbonate-reservoired gases from southern Vietnam.

have a better chance of producing conditions in which the preservation of organic material is optimum.

Several major SE Asia petroleum systems including Bohai and Central Sumatra (Williams *et al.* 1985) have proven rift lacustrine systems with very rich source rocks deposited during the earlier, Palaeogene syn-rift filling of the basin (Table 4). The common feature is the deposition of intervals at least several 10's of metres thick of mudrocks characterized by algal-rich laminates or 'lamalginities'. These pelagic or hemipelagic deep lacustrine mudstones typically contain *Botrococcus* and *Pediastrum*, the main petroleum kerogen generators (e.g. Williams *et al.* 1985, 1992b; Prayitno *et al.* 1992). Shallower water lake margin facies also have source potential, although less than the deep facies (Williams *et al.* 1985; Table 4).

The main source rock in the Cuu Long Basin, offshore south Vietnam (Figs 4 & 8) was deposited in a lacustrine depositional setting. The waxy nature of the Bach Ho oil's together with moderately low Pristane/Phytane ratios suggest that the oils has been sourced from a rock containing predominantly algal/bacterial debris. The biomarkers (dominance of extended tricyclic diterpanes and minor presence of the C_{30} 4-methyl steranes) suggest a brackish lacustrine environment (Fig. 8). One oil sample shows a slightly higher Pristane/Phytane ratio (3.0) suggesting a greater higher land plant input. The penetrated source rock deposited during the early post-rift is proven by pyrolysis data (unpublished Geochem Group data; Table 4, Fig. 8) in Bach Ho and a number of other wells. It contains a floral/faunal assemblage that indicates deposition in a brackish-water swamp facies with many back-mangrove elements and only limited indications of short-lived lakes (S. Lowe personal observations 1994). Most significant is the occasional presence of microforam test linings, indicating some limited connection to marine environments. In seismic data, the prospective source rock interval comprises a 'stripey' high-amplitude seismic facies in lenticular units with a classic post-rift 'Texas long-horn' geometry. An analogue is drawn with the present-day Lake Maracaibo, envisaging a tectonically influenced lagoonal system, with some connection to the sea. A rift-fill type lacustrine source

Table 3. *General characteristics of major oil types in SE Asia (modified after Robinson 1987)*

Attribute	Lacustrine	Paralic	Marine (carbonate)	Marine (clastic)
API gravity	20–39	34–50	20–34	20–34
Wt % sulphur	<0.2	<0.2	>1.0	>0.2
Wax (%)	>10	>10	<10	<10
Pr/Ph	1.5–3.0	>3.0	0.8–1.5	<3.0
Hop/Ster + Hop	na	>83%	<83%	<83%
T_m/T_s	1.50–0.20	6.0–1.0	3.0–1.0	3.0–1.0
C_{34} Resins	Low	Very high	Low–absent	Low–absent
Steranes	Equal dist.	Bias to C29	Equal dist.	Equal dist.
C_{30} 4-Methyl steranes	High	Low–absent	Low–absent	Low–absent

Table 4. *Significant lacustrine source rocks penetrated by wells in SE Asia*

Basin	Formation	Age	Facies	Thickness and quality	Reference
Central Sumatra	Pematang	Palaeogene	Deep lacustrine, well laminated, red-brown to black, organic-rich shales; silty/sandy & algal-rich laminae. Shallow lacustrine, light grey to grey-black shales with humic matter & thin coals. Common algae include *Pediastrum* & *Botryococcus*	Up to 580 m thick, best source in upper third. TOC 1–10%. P2 5100 kg t^{-1}. HI 500–1000	Williams *et al.* (1985); Prayitno *et al.* (1992); Katz & Kelly (1987); Longley *et al.* (1990)
Sunda	Banuwati	Early Oligocene	Lacustrine, dark-brown to grey-black shales, logs display high res, low density, but high GR	Up to 90 m thick in basin margin wells – may be thicker in depocentre. TOC 3–6% and P2 16–45 kg t^{-1}	Prayitno *et al.* (1992)
West Natuna	Barat	Late Oligocene	Lacustrine, grey to light-brown shale, with vitrinite, amorphous & liptinitic. *Pediastrum* & *Botryococcus* recognised as oil source	Tens of metres thick	Pollock *et al.* (1984)
Gulf of Thailand		Late Oligocene to Early Miocene	Lacustrine, organic-rich shales with minor interbedded deltaic sands. At base of interval of massive organic shales with common *Pediastrum* & *Botryococcus*	400–700 m thick. P2 average 20 kg t^{-1}	In house data; Williams *et al.* (1992*b*)
Bohai	Shahejie	Late Eocene	Lacustrine mudrocks	Interval up to 500 m thick gross, TOCs 3–11%, P2s 2–50 kg t^{-1}, HIs 400 to >1000	In house data
Cuu Long	Cau	Late Oligocene	Brackish lacustrine mudrocks	50–280 m gross thickness, average P2s 4–14 kg t^{-1} and HIs 367–590	This report and in house data
Pearl River Mouth Basin	Wenchang	Late Eocene	Lacustrine	P2s 8–19 kg t^{-1}. HI average 660	In house data

rock may also occur in the Cuu Long Basin but remains unpenetrated by the drill bit. This Cuu Long Basin example is obviously transitional between a rift lake and the flood plain lakes described below.

Floodplain lakes. Whilst they form less obvious and less easily mappable features, floodplain lakes nevertheless contribute significant oil and gas reserves to the SE Asia total through the West Natuna and possibly Malay Basins

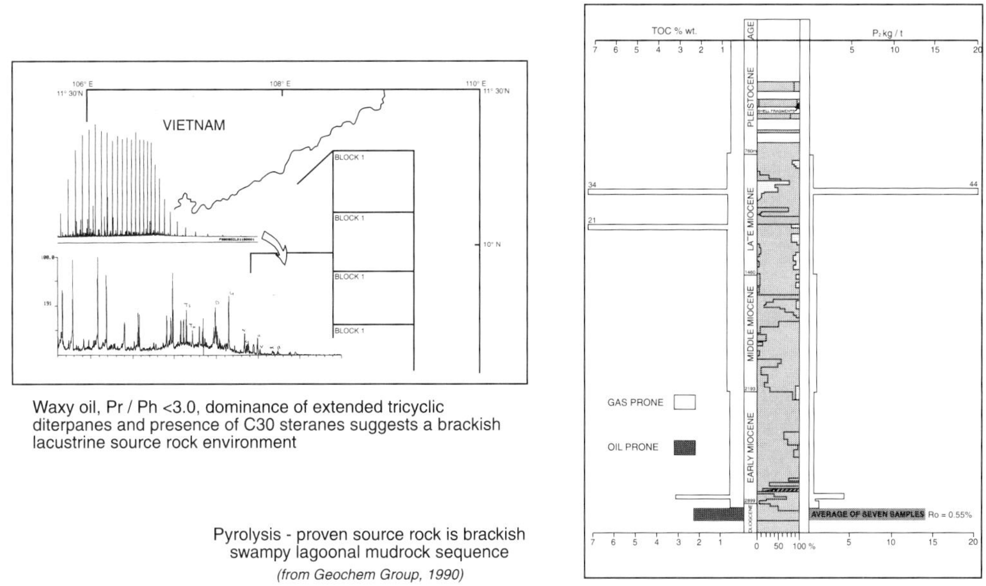

Fig. 8. Source rocks in the Cuu Long Basin.

(Table 2). Floodplain lakes develop on fluvial or coastal plains as ephemeral standing bodies of water and are typically less than 50 m in depth. A marine connection could incur a tidal or lagoonal (brackish) nature to the lake. An underlying tectonic control through locally higher rates of subsidence is possible, but the general characteristic is that at any one time in the history of the sequence no depositional significant topography was apparent. This makes them more difficult to identify in seismic data than rift lakes, although some of the controls on source rock development may be shared.

The oils in the West Natuna Basin are waxy indicating a non-marine source but display a range in Pristane/Phytane ratios (from 2 to 4) suggesting varying mixtures of algal and higher land plant kerogen (figs 4 & 5). Limited source rock data are available, and it is not known whether this mixing occurred at the depositional site or during oil migration, but it is possible that the oils record a floodplain setting in which both higher land plant and algal material are both preserved.

Paralic depositional systems

Paralic depositional environments include fluvial, coastal (delta plain), shoreface, interdistributary bay, estuarine, delta front and prodelta facies (Fig. 9). Source rocks, formed mainly by kerogens derived from higher land plants, but with a contribution of algal material, can accumulate anywhere in this range of facies. Lithologies are typically either shales or coals (Table 5).

In present-day SE Asia the distribution of peats is essentially controlled by climate with thicker peats developing in ever-wet regions. Peats accumulate over the upper coastal plain as far as the intertidal zone where they are replaced by beach ridges or mangrove swamps (Thompson *et al.* 1985).

Coals are known source rocks for oil and gas. While commonly capable of expelling gas, coals are less commonly oil prone. Particular circumstances appear to be required for the development of an oil-prone coal and these have been summarized by Fleet & Scott (1994) and include the following.

(i) Plant evolution – only since late Cretaceous times with the evolution of the angiosperms have plants had sufficient foliage to yield significant amounts of waxy lipid rich detritus.

(ii) Climate – these lipid rich plants favour the warm ever wet tropics.

(iii) Preservation – preferential preservation of potentially oil prone detritus takes place in the lower coastal plain environments under ever wet tropical conditions. This may be because bacterial reworking of the more lignitic material *in situ* is favoured in alkaline conditions which in

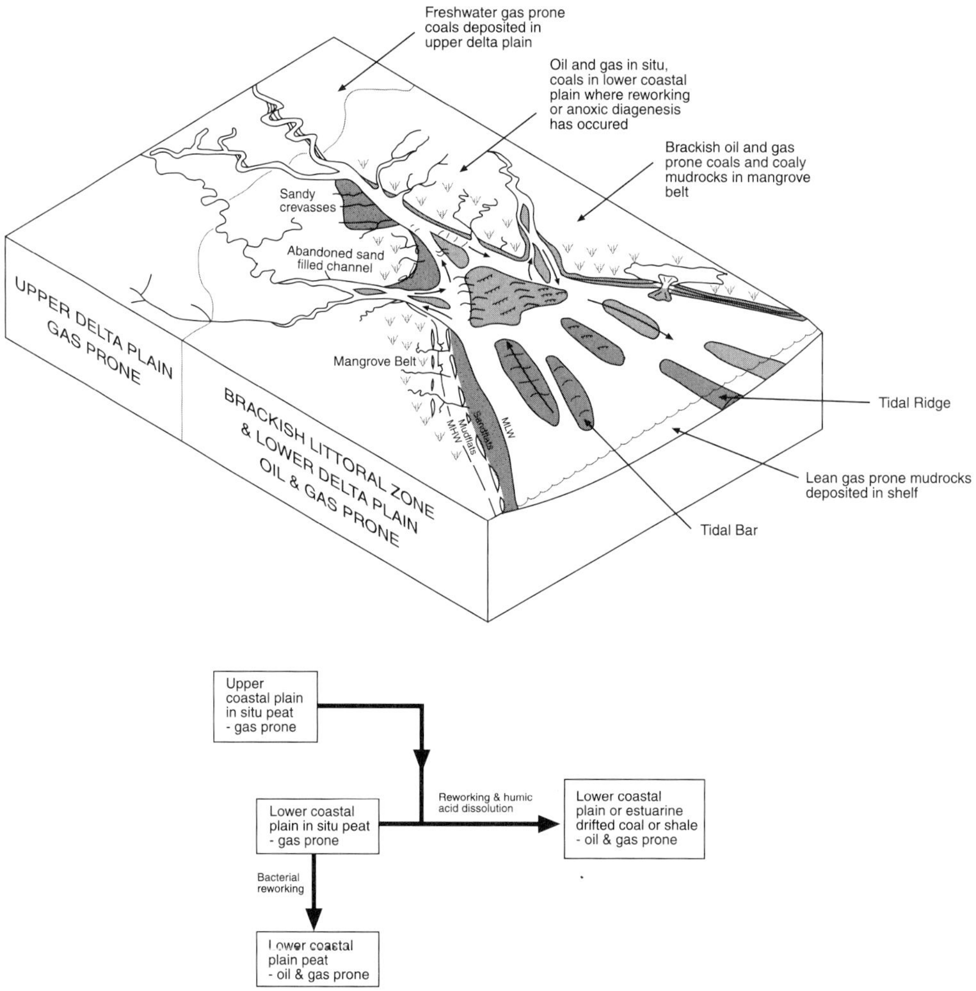

Fig. 9. (**a**) Source facies model for transgressive paralic systems (modified after a drawing by A. Reynolds). (**b**) Processes effecting source lithofacies in paralic environments.

turn are promoted in areas close to marine influence. Alternatively transport mechanisms have been suggested whereby humic acids are dissolved on erosion, while water-insoluble detritus such as plant cuticle, plant waxes, resins and humified woody fragments are carried downstream in suspension or as bedload. These particulate debris are generally deposited in intertidal or subtidal lower shoreface or estuarine or interdistributary bay settings as distributary rivers lose power as they enter the sea. In contrast, soluble humic acid components are carried to greater depths where they re-precipitate as vitrinite or are oxidized. The process of separation of soluble and insoluble organic matter results in the accumulation of liptinitic allochthonous or drifted coals and carbonaceous shales in estuarine or shallow marine environments along the coastal margin (Thompson *et al.* 1985; Fig. 9a). These processes are summarized in Fig. 9b.

Therefore for coals and indeed shales, a simple zone in the paralic realm from the lower coastal plain to the subtidal part of the littoral environment can be pinpointed as the area of highest probability of oil-prone source development (Fig. 9). Upper coastal plain coals tend to be more gas prone and prodelta to marine carbonaceous shales are likely to be gas prone, except where there is restriction of the marine basin.

Microscopy and geochemistry have been used separately and in an integrated fashion to

Table 5. *Significant paralic source rocks penetrated by wells in SE Asia*

Basin	Formation	Age	Facies	Thickness and quality	Reference
Kutei	Balikpapan	Mid-Miocene	Delta plain coals/lignites & carbonaceous shales. Higher land plant debris in shales important	Shales have TOCs typically 2.5% TOC, up to 8%	Combaz & de Matharel (1978); Oudin & Picard (1982); Hoffman *et al.* (1984); Fukasawa *et al.* (1987); Burrus *et al.* (1992)
West Java (Ardjuna)	Talang Akar	Oligocene	Lower delta plain coals and marine-influenced interdistributary shales	Coals have TOCs > 60–70%, P2s 218–288 $kg\,t^{-1}$, and HIs 260–420 Floodplain shales have little potential: P2 > 5 $kg\,t^{-1}$ Interdistributary bay shales have TOCs 2–21%, P2s 2–25 $kg\,t^{-1}$ and HIs 164–435	Noble *et al.* (1991)
East Java	Ngimbang	Late Eocene	Fluvio-deltaic coals & coaly mudrocks	Interval 40–90 m thick, carbonaceous shales have TOCs 0.2–17%, coals 26–82%	Phillips *et al.* (1991)
Balingian	Cycles I, II & III	Oligo-Miocene	Lower coastal plain coals, interdistributary bay/ floodplain mudrocks & drifted coals	Gross interval of 10s m, coals up to 10 m thick. Coals have TOCs > 40–80%, P2s 40–194 $kg\,t^{-1}$, and HIs 388–406. Floodplain shales have little potential: P2 < 5 $kg\,t^{-1}$. Interdistributary bay shales have TOCs 1–21%, P2s 3–90 $kg\,t^{-1}$ and HIs 164–435	This report

describe the oil potential of coals. While primary maceral content controlled by the type of plant material deposited exerts some influence on source rock type and quality, the relationship between maceral type and source quality is by no means systematic (e.g. Powell & Borham 1994). For example some coals rich in vitrinite and other refractory kerogen appear to be nonetheless oil prone. These coals are believed to represent kerogen that has been reworked by bacteria and the oil proneness is the result of the bacterial input. Instead, hydrogen indices and gas chromatograph data appear to give a much better estimate of the oil prone potential of a coal.

Several examples of oil-prone coals have been documented in SE Asia. Best described are the source rocks in the Kutei Basin (Mahakam Delta; Combaz & de Matharel 1978; Durand & Oudin 1979; Hoffman *et al.* 1984; Oudin & Picard 1982; Schoell *et al.* 1985). A full range of Miocene deltaic sediments is present with relatively high TOCs throughout. The best source rocks are concentrated in the coal-rich delta-plain facies. Another well documented example of oil-prone coals is from the Ardjuna sub-basin of West Java (Noble *et al.* 1991). Autochthonous delta plain coals there also combine highest P_2 and HI values, but Noble *et al.* (1991) report that marine-influenced interdistributory bay shales in this basin also have some potential for oil with HIs greater than 250.

Geochemical data, illustrating another example of oil prone coals obtained for Oligo-Miocene core from the Balingian Province, are reported here (Figs 10, 11, 12 & 13). The results of the Rock-Eval and PGC analysis are listed in Table 6 with the associated vitrinite reflectance data. The data are plotted on histograms in Figs 12 and 13 and illustrate the control exerted by depositional lithofacies on source rock quality. If a petroleum yield P_2 of $5\,kg^{-1}$ is taken as a cut-off for a good source then only non-marine sediments in the Oligo-Miocene of Sarawak have good source rock potential (Fig. 12). Marine or marginal marine shales are lean. Some floodplain/bay shales, which may have had a marine influence as interpreted from associated facies with marine indications, have P_2 values up to $50\,kg^{-1}$. However it is the floodplain shales with no marine influence and particularly the coals that apparently have greatest source potential (Fig. 12). The HI and GOGI values for the samples with P_2 yields greater than $5\,kg\,t^{-1}$ display a fairly large variability of HI values within all of the depositional environments, although a greater number of coal samples show oil potential (Fig. 13). In all these cases of oil prone coals a lower coastal plain environment is apparent from the sedimentological evidence.

Mudrocks. In the examples above it is clear that coaly mudrocks often have substantial source rock potential. These coastal plain and estuarine shales also have the potential to expel

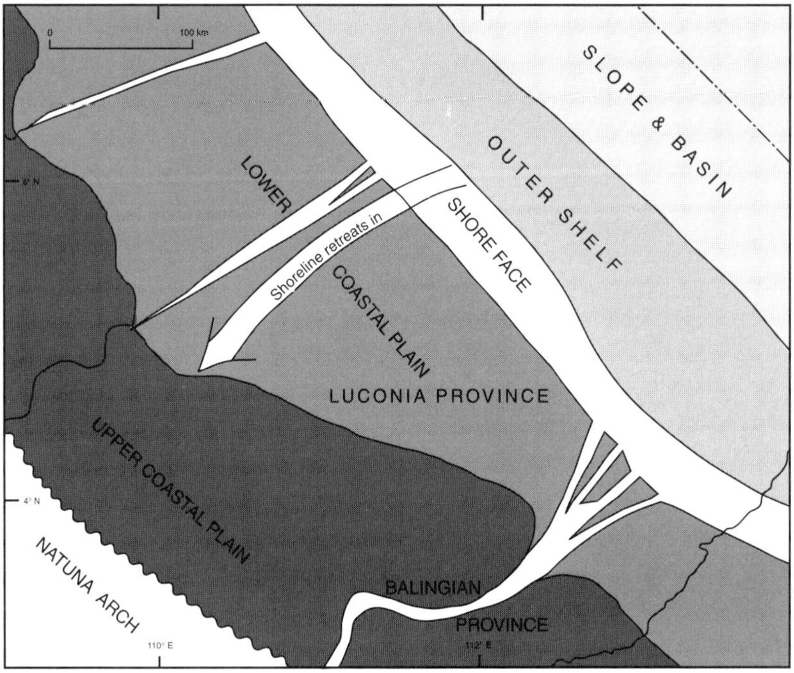

Fig. 10. Balingian Province: cycle 1 (Late Oligocene) paleogeography.

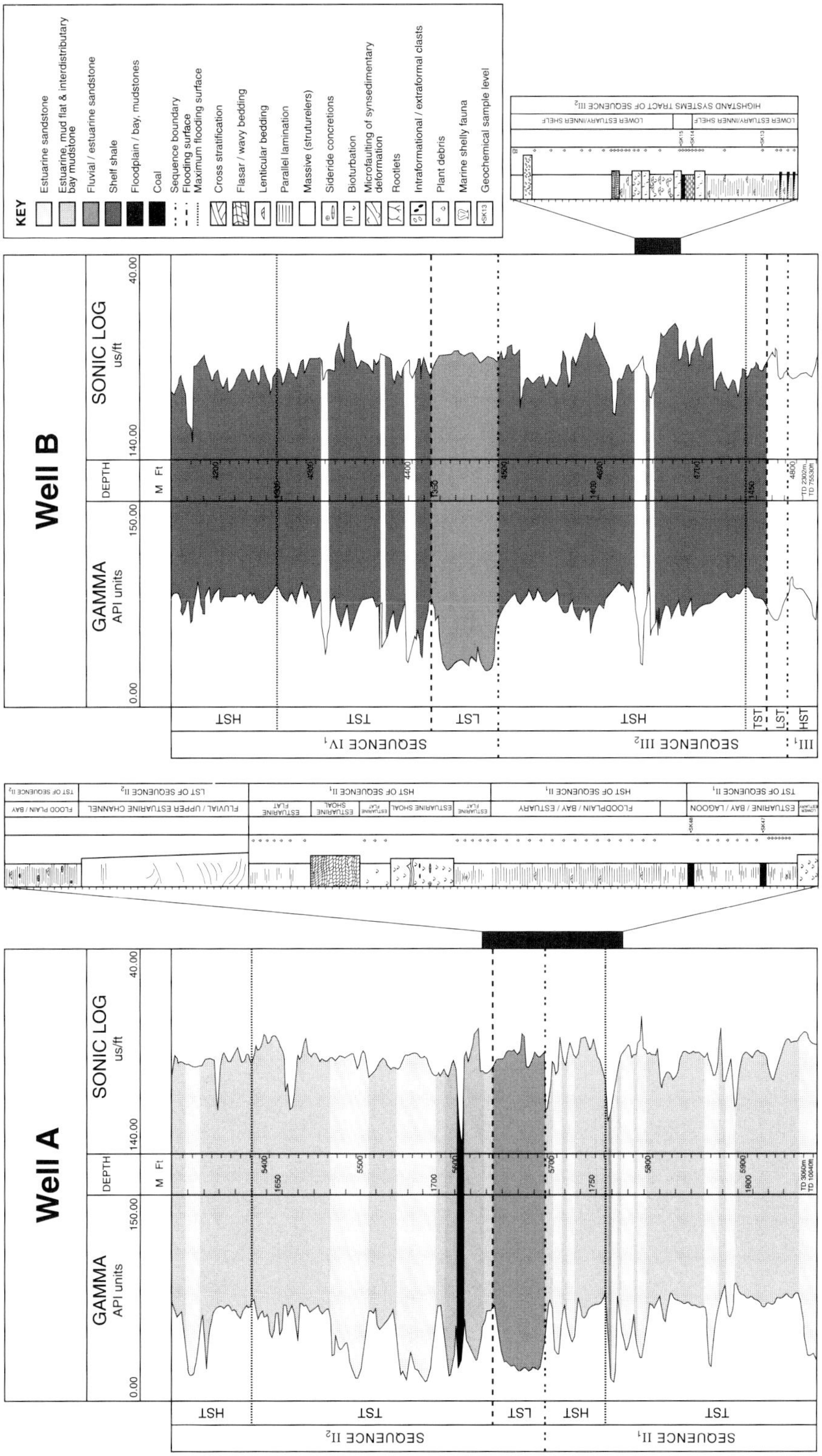

Fig. 11. Balingian Province: core and well log interpretations of two wells A and B.

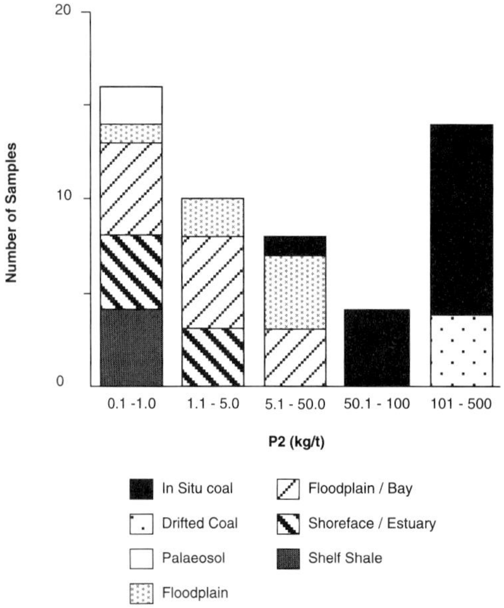

Fig. 12. Balingian Province: P_2 petroleum yield versus depositional facies.

oil as well as gas. The potential of shales has also been developed by depositional models involving highly prolific mangrove systems (the 'mangrove model'; Risk & Rhodes, 1985; Brown 1989). The model is based on the high organic productivity of mangrove systems and their propensity to produce oil-prone waxy detritus. The envisaged lithofacies is an intertidal or subtidal mudflat or lower estuary in which mangrove shales are deposited. The mangrove material would be supplemented by debris from freshwater higher land plants reworked from peats on the coastal plain (see above).

The mangrove model has appeal in a predictive sequence stratigraphic context because mangrove-rich fluvio-tidal settings often develop during transgressive systems tracts when fluvio-tidal systems are likely to occupy basins and embayments and are more likely to be restricted. The restriction and high rates of aggradation may promote mangrove shale accumulation (Fig. 9).

Some of oil-prone shales described above may well have a high mangrove content. However, the appropriate palynological data is not available to fully assess this notion. Brown (1989) provides a summary of uncited data from Indonesia. Apart from these data, there is little direct information to indicate just how important to oil and gas source development the mangrove shales are. However, it is possible to roughly gauge their significance by a more

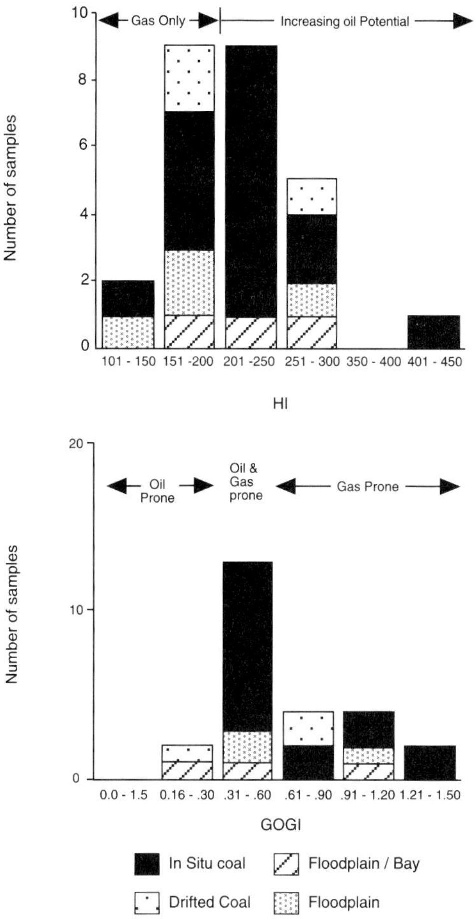

Fig. 13. Balingian Provinces: HI & GOGI versus depositional facies.

detailed interpretation of the properties of oils from the various fluvio-deltaic provinces in SE Asia. As discussed by Brown (1989) the properties of the rocks outlined in Table 6 should convert in some way to distinguishing characteristics of the oil products.

Oils from the Balingian and Baram delta have gas chromatograms that are characterized by relatively high Pristane/Phytane ratios and sterane and triterpane distributions that are characterized by high C_{29} sterane and bicadinane contents. These characteristics are typical of a coastal plain source rock that contains predominantly higher land plant angiosperm input.

However, when a number of carbonaceous shales and coals were analyzed by GC-MS from the Balingian province (Fig. 14) the following minor but possibly significant differences were obtained for the extracts:

Table 6. *Geochemical parameters measured in a series of Indonesian coals and mangrove shales (from Brown, 1989)*

Parameter	Coal	Mangrove shale
TOC (wt %)	>45%	1–5%
HI	140–280	150–240
Hydrocarbon/extractable matter	12.5–48.5%	45–65%
Saturates	3.5–7.5%	17–34%
Asphaltenes	20–57%	10–20%
Saturates/aromatics	0.17–0.56	0.61–0.99
Pr/Ph ratio	Mean 4.68	Mean 3.84
18α Oleanane	Moderately common	Abundant
Diasteranes	Very rare	Common
Steranes/17α Hopane	0.01–0.06	0.06–0.10

- Coals – $T_m > T_s$, moderate Oleanane, moderate C_{31} hopanes; and
- Coaly mudrocks – $T_m \approx T_s$, high Oleanane, low V_{31} hopanes.

A number of oils from the Balingian Province, the Baram Delta and NW Sabah were correlated with these carbonaceous mudstones and coal extracts as illustrated in Fig. 14. These correlations suggest that the Balingian oils are potentially predominantly coaly sourced whereas the source lithology is more mudstone/carbonaceous mudstone dominated for the Baram delta and NW Sabah oils.

A review of Mahakam Delta data using a similar approach also suggests that carbonaceous mudrocks may be more important volumetrically in sourcing the oils than the coals.

The correlation between the Early Miocene coaly mudrocks and oils with a typical SE Asian paralic signature within the Southern Nam Con Son Basin, offshore south Vietnam, is illustrated in Fig. 15. The striking similarity between the traces show the close tie between source and oil.

To summarize, two sub-classes of paralic type oils can be defined on the basis of the subtle but possibly significant lithological differences as described above. In reality, basins with paralic source rocks are likely to include both coal and coaly mudrock source rocks. Nevertheless, the discrimination of oils does suggest the possibility that the majority of paralic oil in SE Asia is possibly derived from coaly mudrocks, rather than coals.

Maturation, expulsion, migration and phase

In addition to the controls exerted by the starting source rock, the typical trends of SE Asian oil and gas distribution illustrated in Figs 5 and 6 can be explained by combining simple theoretical principles of expulsion and migration with empirical observations. These include the observed trends of oil type and phase distribution and distance between fields and kitchens.

Oil type distribution

The preponderance of lacustrine-generated oil in the lower part of the stratigraphic section of basins with a lacustrine source rock is obviously a result of the source rock itself being sequestered to that part of the stratigraphy in the syn-rift or early post-rift (Fig. 6). Migration of oil out of the stratigraphic interval of origin is often impeded by the many intraformational seals present in the lacustrine/fluvio-deltaic succession. Instead higher Miocene parts of the sedimentary fill of many basins rely on source rocks in the Miocene. As described above, these tend to be paralic source rocks. A greater marine influence in the higher, younger petroleum systems is a result of marine sources being restricted to this interval.

Phase distribution

The proportionally greater amounts of gas in the younger stratigraphic intervals is a product of the nature of the source rock in this interval and of preferential migration of gas (Fig. 6). The greater relative abundance of gas derived from paralic source rocks as opposed to the more oil-prone lacustrine source rocks is illustrated in Fig. 5 and helps explain why gas dominates post-rift clastic and carbonate plays (Fig. 6). Coals and coaly mudrocks are predicted from kinetic modelling (Pepper & Corvi 1995) to expel a greater proportion of gas in comparison to lacustrine algal source rocks (Fig. 16). This is because coaly kerogen tends to adsorb expelled oil in the early stages of hydrocarbon generation.

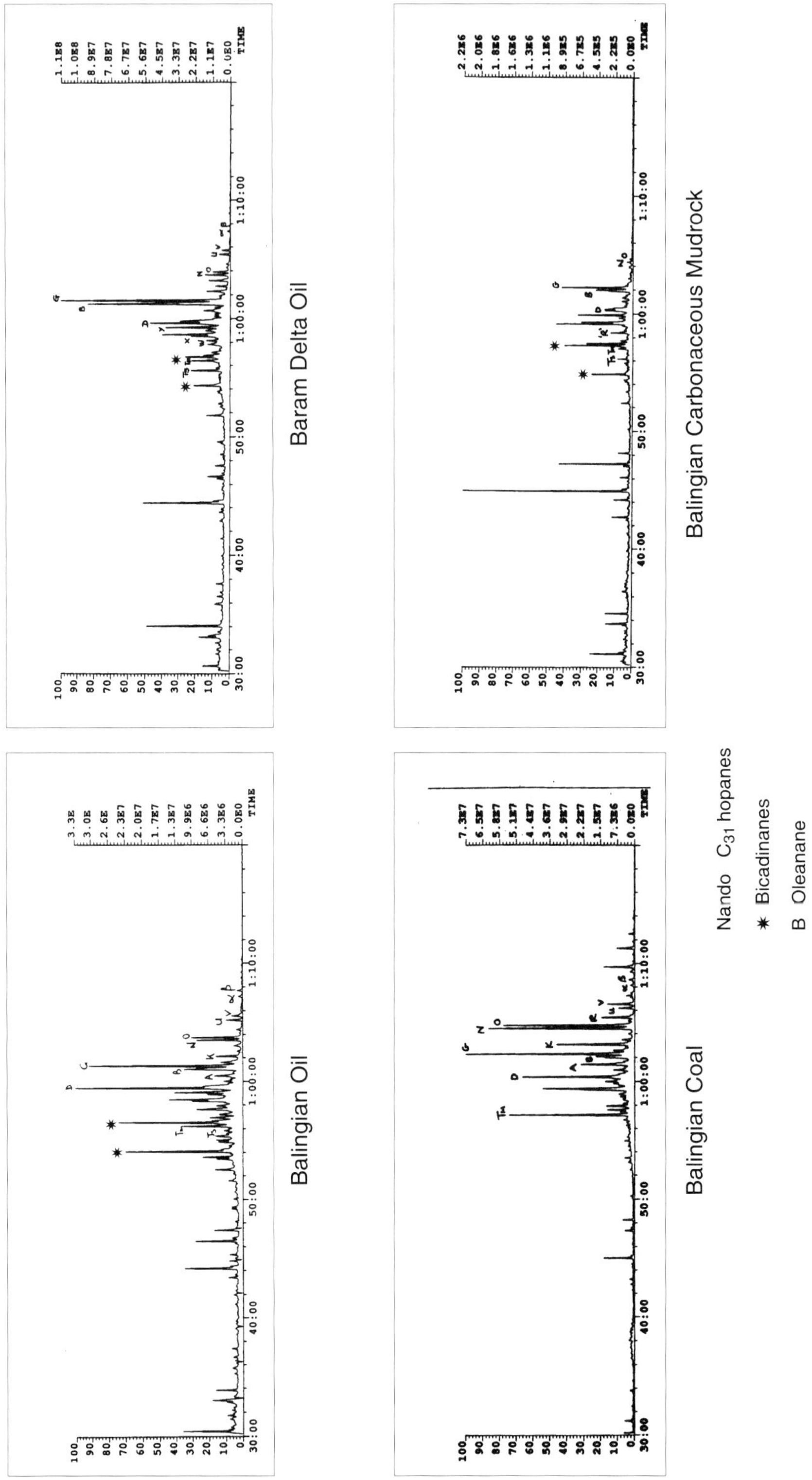

Fig. 14. Comparison of hopane distributions of Balingian mudstones/coals and oils with Baram Delta oils.

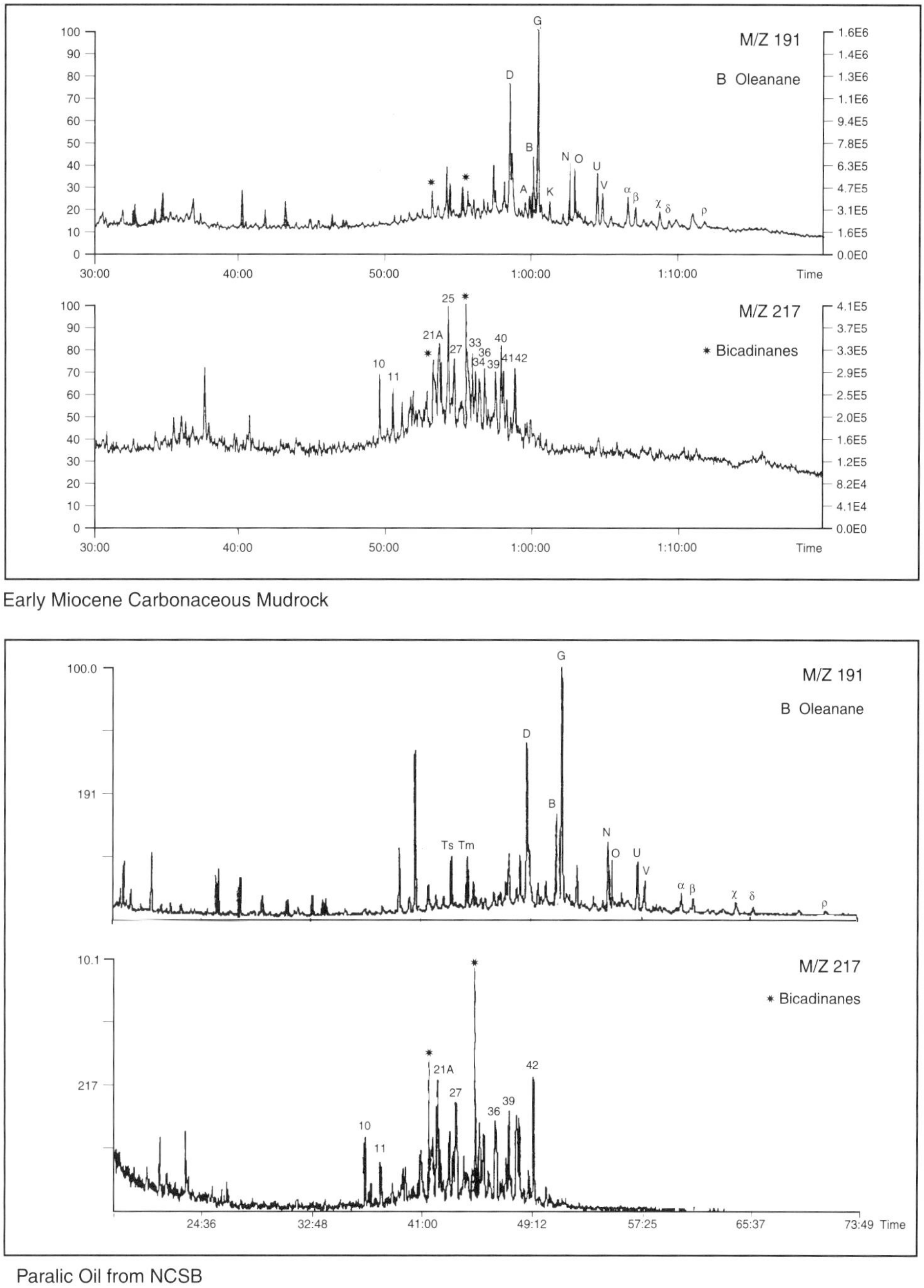

Fig. 15. Correlation of Early Miocene carbonaceous mudrock with a paralic oil from the Nam Con Son Basin.

This adsorbed oil is held in the source rock for sufficiently long to be cracked to gas as the temperature rises with deeper burial. Thus a proportionately greater amount of the kerogen in a paralic source, even it is oil prone in the first place, is converted to gas. The kinetic models (Fig. 16) indicate that volumetrically significant quantities of gas are expelled from paralic source

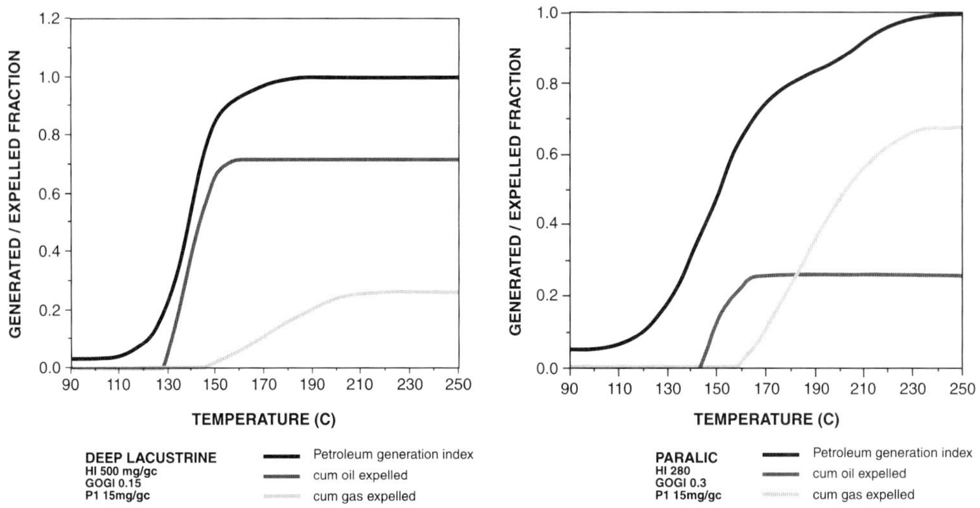

Fig. 16. Kinetics of SE Asian source rocks.

rocks over the temperature range 160–230°C. In contrast, most algal kerogen in lacustrine source rocks is expelled early as low GOR liquid phase petroleum (Fig. 16). These kinetic models are obviously consistent with the SE Asian field evidence (Fig. 5). Thus early analysis of oil samples in an exploration program in an unexplored basin is essential to obtain information on the source type (Table 3). Basins with lacustrine source rocks tend to be oilier than basins with paralic sources rocks – particularly if the paralic source rocks are buried to depths where temperatures are greater than approximately 160°C.

Gas is more mobile in the subsurface because of its lower viscosity and higher buoyancy than oil. For example, in typical subsurface conditions around 3 km depth gas is more than 100 times less viscous than oil. Hence seals tend to offer less resistance to gas than oil and gas tends to migrate further both laterally and particularly vertically than oil from the source. This helps to explain why gas tends to concentrate in younger stratigraphic plays in SE Asia (Fig. 6).

A further mechanism responsible at least in part for concentrating gas in the younger stratigraphic intervals of SE Asian basins is the tardiness of seal emplacement or trap configuration relative to the timing of oil charge. The best example of this is in northern Vietnam (Da Nang) where well 119-CVX-1X encountered oil-stained carbonate in a present-day gas reservoir; a late seal on the carbonate buildup is considered to have missed the oil charge.

A greater proportion of gas in shallower reservoirs is also a function of pressure and temperature. Quite simply, a saturated oil at 3500 m may evolve into a large volume of gas with a separate oil leg at shallower depths. If the trap has insufficient volume to accommodate both the gas volume and the oil, then the oil will be preferentially lost through gas flushing. This may explain why many SE Asian carbonate buildups (e.g. Luconia Province, Sarawak) have lean gas condensate accumulations underlain by thin oil rims.

Finally, the volumetric significance of biogenic gas supply to Miocene and younger reservoirs is poorly understood, but may be a significant factor (e.g. Fig. 7).

Migration distances from kitchen to trap

Typically major petroleum provinces in SE Asia are characterized by a close association of oil and gas pools to within 20 km of the source kitchen. Outside this 'halo' of accumulations, valid traps tend to be severely underfilled or dry (e.g. Figs 17 & 18). This is because lateral migration is largely an inefficient process because of the lack of continuous carrier beds and in some cases the lack of a volumetrically significant petroleum charge. The lack of continuous carriers is typically a function of poor gross permeability controlled by paralic clastic sandbodies stratigraphically interconnected only at scales of about 10 km or less and by complex basin structures with high degrees of faulting.

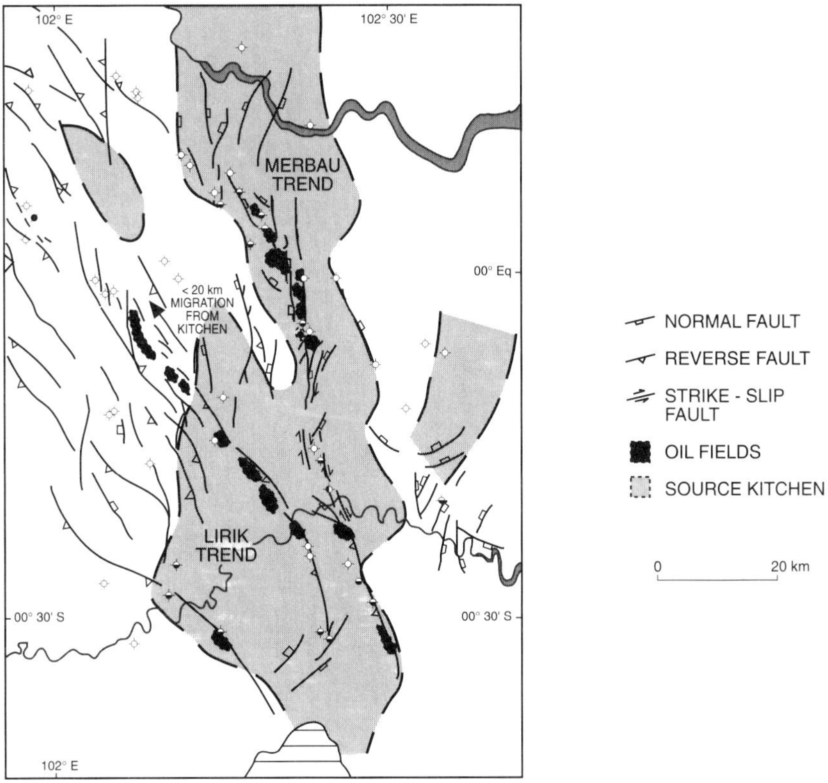

Fig. 17. Southern Central Sumatra Basin: oil field locations versus kitchen distribution (from Robinson & Kamal 1988).

Despite the high heat flows in SE Asia (sometimes over 100 mW m^{-2}), the emplacement of oil and gas into reservoirs stratigraphically separate from the source interval and well above source rocks within the oil and gas windows requires at least some vertical migration. Vertical migration may occur up along fault zones, especially when the faults are seismically active. However, for substantial vertical migration to occur, hard formation overpressure (>80% overburden pressure) is required in the kitchen. Hard overpressure is where pore pressure increases in carrier beds or reservoirs towards equivalence with the fracture pressure of the enveloping mudstones. In these conditions any hydrocarbon accumulation, however small, is likely to break its topseal and leak upwards. High overpressure also accelerates the process of capillary leakage through seals which are not being fractured. These processes of capillary leakage and/or hydraulic fracturing and upward hydrocarbon leakage allows vertical migration until the top of hard overpressure, marked by a good seal and/or an extensive carrier bed.

There is some evidence from well and seismic velocity data that many SE Asian basins are overpressured, e.g. Malay, Pearl River Mouth, Bohai, Baram and Kutei. Many basins experience high rates of post-rift Miocene and Plio-Pleistocene sedimentation and become overpressured through compaction disequilibrium. The Baram and Mahakam (Kutei) deltas are good examples of that (Fig. 1; Oudin & Picard 1982; Schaar 1976). In the Baram and Mahakam deltas the presence of an overpressured zone above the window of oil expulsion is considered to facilitate the migration of large volumes of hydrocarbon into the shallower reservoirs (e.g. Oudin & Picard 1982; Burrus *et al.* 1992; Fig. 19).

These are exceptions to the 20 km 'rule of thumb' outlined above for lateral migration in SE Asia. One case is the emplacement of oil into the Luihua Miocene carbonate buildup some 50 km from the lacustrine source kitchen in the Pearl River Mouth Basin in offshore China (Fig. 20; Tyrrell & Christian 1992). Immediate oil migration out of the Huizhou Sag was

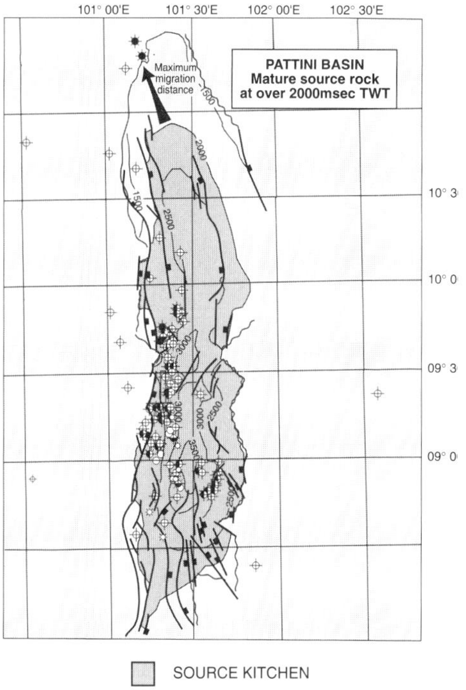

Fig. 18. Gulf of Thailand: oil and gas locations versus source kitchen distribution.

probably vertical as discussed above. Thereafter lateral migration up the Dongsha Massif was facilitated by an extensive, continuous, low pressure sandstone deposited on top of a break-up unconformity. Extensive sandbodies of this kind with connected lengths of more than 50 km are unusual in SE Asia. In general, long distance migration greater than 20 km should be considered as higher risk, with the risk only reduced when a favourable carrier bed can be mapped, much as Amoco Orient did before drilling Luihua (Tyrrell & Christian 1992). A second example are the oil fields on the eastern flank of the Malay Basin which are believed by Creaney *et al.* (1994) to have been charged as a result of long range migration.

Conclusons

Most SE Asian Tertiary petroleum has been derived from coaly paralic source rocks, although the larger proportion of oil is sourced from lacustrine algal mudrocks. Prolific lacustrine source rocks occur in rift-fill megasequences, but floodplain lake intervals in the post-rift are also important. Paralic source rocks include both coals and coaly mudrocks in an oil-prone zone developed in the paralic realm between the lower coastal plain and

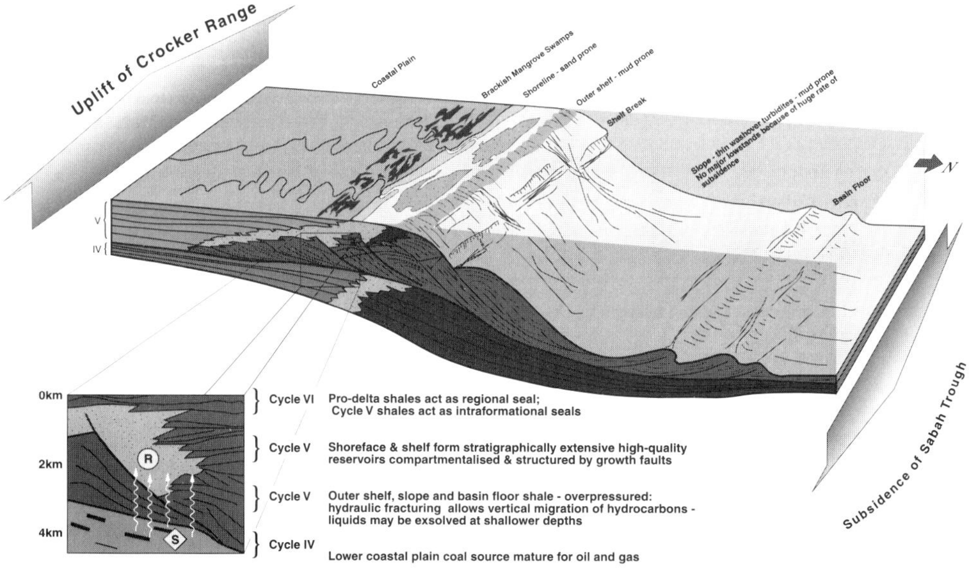

Fig. 19. Petroleum migration habitat in the Baram Delta.

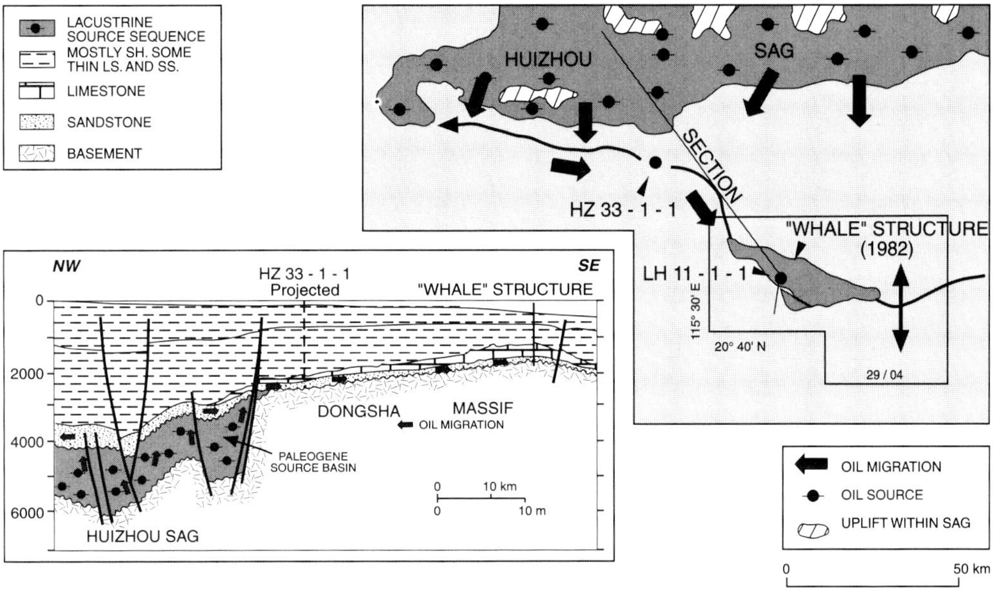

Fig. 20. Pearl River Mouth Basin: long-distance migration to Liuhua (from Tyrell & Christian 1992).

lower estuary/delta front facies, perhaps involving a mangrove system. Although coals can expel oil in certain circumstances, it is concluded that possibly the majority of SE Asian higher land plant sourced oils come from coaly mudrocks.

Early analysis of oil samples in an exploration program in an unexplored basin is essential to obtain information on the source type. Basins with lacustrine source rocks tend to be oilier than basins with paralic source rocks – particularly if the paralic source rocks are buried to depths where temperatures are greater than approximately 160°C. The stratigraphic position and kinetic expulsion behaviour of the different source types, migration, seal timing, pressure–temperature–depth relationships, and a biogenic gas component converge to make the younger plays more gas prone.

Vertical migration is considered to be common in overpressured basins, while lateral migration is believed to commonly restricted to distances of 20 km or less from the kitchen.

The authors would like to thank Petronas for granting permission to publish the source rock data for several Sarawak wells. Elsevier Science, the American Association of Petroleum Geologists, the Indonesian Petroleum Association and the Geochem Group for their permission to reprint several figures. T. Reynolds for his permission to publish a modified version of an unpublished figure. We also thank numerous colleagues particularly P. Kahn and S. Duppenbecker for their contributions, the two referees and A. Pepper for their constructive comments on the draft text. Finally BP Exploration Operating Company Limited for permission to publish.

References

Brown, S. 1989. The 'mangrove model' can it be applied to hydrocarbon exploration in Indonesia? *In*: *Proceedings of the 18th Annual Convention of the Indonesian Petroleum Association*. Indonesian Petroleum Association, Jakarta, 383–401.

Burrus, J., Brosse, E. *et al.* 1992. Basin modelling in the Mahakam Delta based on the integrated 2D model TEMISPACK. *In*: *Proceedings of the 21st Annual Convention of the Indonesian Petroleum Association*. Indonesian Petroleum Association, Jakarta, 23–43.

Cai, D. G. 1987. A preliminary analysis of hydrocarbon exploration of Dongsha Massif. *In*: *Collection of Papers from the International Petroleum Geological Convention, Northern South Sea Continental Shelf, China*. China Oil Magazine, Hong Kong, 435–455.

Combaz, A. & de Matharel, M. 1978. Organic sedimentation and genesis of petroleum in Mahakam Delta. *American Association of Petroleum Geologists Bulletin*, **62**, 1684–1695.

Creaney, S., Hussein, A. H. *et al.* 1994. Source facies and oil families of the Malay Basin, Malaysia. *American Association of Petroleum Geologists Bulletin*, **78**, 1139.

DALY, M. C., COOPER, M. A. *et al.* 1991. Cenozoic plate tectonics and basin evolution in Indonesia. *Marine and Petroleum Geology*, **8**, 2–21.

DURAND, B. & OUDIN, J. L. 1979. An example of hydrocarbon migration in a deltaic sequence: The Mahakam Delta. *In*: *Proceedings of the 16th Annual Convention of the Indonesian Petroleum Association*. Indonesian Petroleum Association, Jakarta, 123–139.

FLEET, A. J. & SCOTT, A. C. 1994. Coal and coal bearing strata as oil prone source rocks: an overview. *In*: SCOTT, A. C. & FLEET, A. J. (eds) *Coal and Coal-Bearing Strata as Oil Prone Source Rocks?* Geological Society, London, Special Publications, **77**, 1–88.

FUKASAWA, H., SUNARYO, R. & NAPITUPULU, P. H. 1987. Hydrocarbon Generation and Migration in the Sangatta Area, Kutei Basin. *In*: *Proceedings of Indonesian Petroleum Association*. Indonesian Petroleum Association, Jakarta, 123–139.

HOFFMANN, C. F., MACKENZIE, A. S. *et al.* 1984. A biological marker study of coals, shales and oils from the Mahakam Delta, Kalimantan, Indonesia. *Chemical Geology*, **42**, 1–23.

JAMES, D. M. D., CARTWRIGHT, J. A. *et al.* 1984. The oil and gas resources of Brunei. *In*: JAMES, D. M. D. (ed.) *Geology and Hydrocarbon Resource of Negara Brunei Darussalam*. Muzium Brunei, 103–139.

KATZ, B. J. 1990. Controls on distribution of lacustrine source rocks through time and space. *In*: KATZ, B. J. (ed.) *Lacustrine Basin Exploration: Case Studies and Modern Analogues*. American Association of Petroleum Geologists, Tulsa, Oklahoma, Memoirs, **50**, 61–76.

——1991. Controls on lacustrine source rock development: a model for Indonesia. *In*: *Proceedings of the 20th Annual Convention of the Indonesian Petroleum Association*. Indonesian Petroleum Association, Jakarta, 587–619.

—— & KELLY, P. A. 1987. Central Sumatra and the East African Rift Lake Sediments: An Organic Geochemical Comparison. *In*: *Proceedings of the 16th Annual Convention of the Indonesian Petroleum Association*. Indonesian Petroleum Association, Jakarta, 259–289.

—— & MERTANI, B. 1989. Central Sumatra – a geochemical paradox. *In*: *Proceedings of the 18th Annual Convention of the Indonesian Petroleum Association*. Indonesian Petroleum Association, Jakarta, 403–425.

KJELLGREN, G. M. & SUGIHARTO, H. 1989. Oil geochemistry: a clue to the hydrocarbon history and prospectivity of the southeastern North Sumatra Basin. *In*: *Proceedings of the 18th Annual Convention of the Indonesian Petroleum Association*. Indonesian Petroleum Association, Jakarta, 363–384.

LIVINGSTONE, H. J. 1992. Hydrocarbon source and migration, Salawati Basin, Irian Jaya (extended abstract) *Eastern Indonesian Exploration Symposium, Indonesian Petroleum Club, Jakarta*, 6.

LONGLEY, I. M., BARRACLOUGH, R. *et al.* 1990. Pematang lacustrine petroleum source rocks from the Malacca Strait PSC, central Sumatra, Indonesia. *In*: *Proceedings of the 19th Annual Convention of the Indonesian Petroleum Association*. Indonesian Petroleum Association, Jakarta, 279–297.

MATCHETTE-DOWNES, C. J., FALLICK, A. E. *et al.* 1994. A maturity and palaeoenvironmental assessment of condensates and oils from the North Sumatra Basin, Indonesia. *In*: SCOTT, A. C. & FLEET, A. J. (eds) *Coal and Coal-Bearing Strata as Oil Prone Source Rocks?* Geological Society, London, Special Publications, **77**, 139–148.

NOBLE, R. A., WU, C. H. & ATKINSON, C. D. 1991. Petroleum generation and migration from Talang Akar coals and shales offshore N.W. Java, Indonesia. *Organic Geochemistry*, **17**, 363–374.

OUDIN, J. L. & PICARD, P. F. 1982. Genesis of hydrocarbons in the Mahakam Delta and the relationship between their distribution and the over pressured zones. *In*: *Proceedings of the 11th Annual Convention of the Indonesian Petroleum Association*. Indonesian Petroleum Association, Jakarta, 181–202.

PEPPER, A. S. & CORVI, P. J. 1995. Simple kinetic models of petroleum formation: part III: modelling an open system. *Marine and Petroleum Geology*, **12**, 417–452.

PHILLIPS, T. L., NOBLE, R. A. & SINARTIO, F. F. 1991. Origin of hydrocarbons, Kangean Block Northern Platform, offshore N.E. Java Sea. *In*: *Proceedings of the 20th Annual Convention of the Indonesian Petroleum Association*. Indonesian Petroleum Association, Jakarta, 637–661.

POLLOCK, R. E., HAYES, J. B. *et al.* 1984. The petroleum geology of the KH Field, Kakap Indonesia. *In*: *Proceedings of the 13th Annual Convention of the Indonesian Petroleum Association*. Indonesian Petroleum Association, Jakarta, 405–423.

POWELL, T. G. & BORHAM, C. J. 1994. Terrestrially sourced oils: where do they exist and what are our limits to knowledge? – a geochemical perspective. *In*: SCOTT, A. C. & FLEET, A. J. (eds) *Coal and Coal-Bearing Strata as Oil Prone Source Rocks?* Geological Society, London, Special Publications, **77**, 11–29.

PRAYITNO, W., ARMON, J. W. & HARYONO, S. 1992. The implications of basin modelling for exploration – Sunda Basin case history, offshore southeast Sumatra. *In*: *Proceedings of the 21st Annual Convention of the Indonesian Petroleum Association*. Indonesian Petroleum Association, Jakarta, 379–416.

RISK, M. J. & RHODES, E. G. 1985. From mangroves to petroleum precursors; an example from tropical northeast Australia. *American Association of Petroleum Geologists Bulletin*, **69**, 1230–1240.

ROBINSON, K. M. 1987. An overview of source rocks and oils in Indonesia. *In*: *Proceedings of the 16th Annual Convention of the Indonesian Petroleum Association*. Indonesian Petroleum Association, Jakarta, 97–122.

—— & KAMAL, A. 1988. Hydrocarbon generation, migration and entrapment in the Kampar Block, Central Sumatra. *In*: *Proceedings of the 17th Annual Convention of the Indonesian Petroleum Association*. Indonesian Petroleum Association, Jakarta, 211–256.

SARDJITO, FADIANTO, E. *et al.* 1991. Hydrocarbon prospect of pre-tertiary basement in the Kuang Area South Sumatra. *In*: *Proceedings of the 20th Annual Convention of the Indonesian Petroleum Association*. Indonesian Petroleum Association, Jakarta, 255–277.

SARJONO, S. & SARDJITO 1989. Hydrocarbon source rock identification in the South Palembang subbasin. *In*: *Proceedings of the 18th Annual Convention of the Indonesian Petroleum Association*. Indonesian Petroleum Association, Jakarta, 427–467.

SCHAAR, G. 1976. The occurrence of hydrocarbons in overpressured reservoirs of the Baram Delta (offshore Sarawak, Malaysia). *In*: *Proceedings of the 5th Annual Convention of the Indonesian Petroleum Association*. Indonesian Petroleum Association, Jakarta, 163–169.

SCHOELL, M., DURAND, B. & OUDIN, J. L. 1985. Migration of oil and gas in isotope and biomarker studies. *In*: *Proceedings of the 14th Annual Convention of the Indonesian Petroleum Association*. Indonesian Petroleum Association, Jakarta, 49–56.

SITUMEANG, S. & DAVIES, P. R. 1986. A geochemical study of Asamera's Block A production sharing contract area, north Sumatra. *In*: *Proceedings of the 15th Annual Convention of the Indonesian Petroleum Association*. Indonesian Petroleum Association, Jakarta, 321–340.

SLADEN, C. P. 1993. Key elements during the search for hydrocarbons in lake basins. *In*: GIERLOWSKI-KORDESH, E. & KELTS, K. (eds) *Global Record of Lacustrine Basins*. Cambridge University Press, **3**, 3–17.

SOFER, Z. 1984. Stable carbon isotope compositions of crude oils: application to source depositional environments and petroleum alteration. *American Association of Petroleum Geologists Bulletin*, **68**, 31–49.

SUSENO, P. H., ZAKARIA, *et al.* 1992. Contribution of Lahat Formation as hydrocarbon source rock in South Palembang Area, South Sumatra, Indonesia. *In*: *Proceedings of the 21st Annual Convention of the Indonesian Petroleum Association*. Indonesian Petroleum Association, Jakarta, 325–337.

THOMPSON, S., MORLEY, R. J. *et al.* 1985. Facies recognition of some tertiary coals applied to prediction of oil source rock occurrence. *Marine and Petroleum Geology*, **3**, 288–297.

TYRRELL, W. W. & CHRISTIAN, H. E. 1992. Exploration history of the Liuhua 11-1 Field, Pearl River Mouth Basin, China. *American Association of Petroleum Geologists Bulletin*, **76**, 1209–1223.

WILLIAMS, H. H., KELLEY, P. A. *et al.* 1985. The Paleogene rift basin source rocks of central Sumatra. *In*: *Proceedings of the 14th Annual Convention of the Indonesian Petroleum Association*. Indonesian Petroleum Association, Jakarta, 55–90.

——, REYES, N. & EUBANK, R. T. 1992*a*. Geochemistry of Palawan Oils, Philippines: source implications. *In*: *Proceedings of the 9th Offshore Southeast Asia Conference*. Offshore Southeast Asia Pte Ltd, Singapore, 509–523.

——, FOWLER, M. & EUBANK, R. T. 1992*b*. Geochemical characteristics of Paleogene and Cretaceous hydrocarbon source basins of South East Asia. *In*: *Proceedings of the 9th Offshore Southeast Asia Conference*. Offshore Southeast Asia Pte Ltd, Singapore, 429–460.

Exploring the lake basins of east and southeast Asia

CHRIS SLADEN

BP Exploration Operating Company Limited, Villa A15, An Phu, Thu Duc, Ho Chi Minh City, Vietnam

Abstract: SE Asia contains a large number of late Mesozoic and early Tertiary lake basins, that together contain a significant proportion of the world oil and gas reserves, oil shale reserves, and coal reserves. Exploration and production of oil and gas from these basins is likely to continue throughout the twenty-first century. As exploration for oil and gas continues, and fewer large fields remain to be found, it will become increasingly vital to focus expenditure on low risk, small volume opportunities in a manner that generates a return on investments. To achieve this, there needs to be a much more detailed understanding of lake basins and the sequences that fill them. This has to be combined with innovative technology to reduce well costs and increase flow rates per well, reductions in operating expenditures and greater flexibility in fiscal terms.

Importance of lake basins for oil and gas in SE Asia

SE Asia has vast resources of oil and gas that owe their origins to sediments deposited in lake basins. In terms of worldwide production in 1994, lacustrine-derived crude oil in SE Asia is estimated to have contributed *c.* 6% of the world's oil production. China, for example, which is the world's oldest producer of petroleum, has about 90% of its oil production coming from petroleum generated by lacustrine source-rocks. In 1994, lake basins in the northeastern provinces of China produced at least 900×10^6 barrels of crude oil sourced from lacustrine source-rocks and around 250×10^9 cubic feet of lacustrine-derived gas (Table 1). The Daqing Field in northeast China has now produced more than 400×10^6 barrels of lacustrine-derived crude oil every year for the last 19 years. Indonesia has also been producing oil and gas derived from lacustrine source-rocks since the late 1890s from the Central Sumatra basin. Prolific producing areas such as the northeast China and Central Sumatra basins have combined proven and produced reserves of lacustrine hydrocarbons of at least 20×10^9 barrels of oil and 5×10^{12} cubic feet of gas. Other basins in Indonesia, and also in Thailand and Vietnam are now producing significant volumes of lacustrine crude oil and gas albeit relatively recently, mostly in the last 15 years. All of these producing areas are now firmly established and will continue to be important lacustrine crude producers for the next 20–30 years and probably throughout most of the twenty-first century.

Emerging areas for production of lacustrine crude oil and gas include the Pearl River Mouth basin offshore southern China and the Cuu Long basin offshore southern Vietnam. These are just two examples of numerous areas which have many billions of barrels of potential for lacustrine-derived crude oil and gas. With rapid economic growth in many countries in SE Asia resulting in primary energy consumption growing by 5–10% per year, and expected to continue at these rates for many years, the demand for oil and gas is expected to steadily increase, perhaps tripling in the next 15 years. At current production rates, most countries only have oil reserves sufficient to produce for the next 10–25 years. Many SE Asian countries that are currently oil and gas net exporters are therefore liable to become net importers sometime during the next 15 years unless significant new reserves can be found and developed. In this context, exploration for oil and gas in numerous lake basins throughout SE Asia is likely to continue, and have a key role for many years. This paper examines key issues for exploring these basins.

Figures 1 and 2 identify the lake basins covered in this paper and highlight some of the key fields containing lacustrine sourced oil and gas, together with other areas that may have future potential for lacustrine-derived oil and gas. All the examples of lake basins significant for oil and gas that are covered in this paper appear to be either late Mesozoic or early Tertiary in age and appear to have developed within extensional tectonic regimes. They extend from Mongolia, China and Korea in the north to Vietnam, Cambodia, Thailand, Malaysia and Indonesia in the south.

In addition to the numerous lacustrine sourced oil and gas accumulations, important reserves of algal-rich lacustrine mudstones ('oil shales')

From Fraser, A. J., Matthews, S. J. & Murphy, R. W. (eds), 1997, *Petroleum Geology of Southeast Asia*, Geological Society Special Publication No. 126, pp. 49–76

Table 1. *Examples of production of lacustrine-derived crude oil and gas from various fields and regions in China during 1994*

Oilfield/region	Oil production (million barrels)	Gas production (billion cubic feet)	Gas-oil ratio (cubic feet/barrel)
Daqing	415	82	200
Shengli	230	46	200
Liaohe	110	62	560
Zhongyuan	36	42	1150
North China	34	11	325
Dagang	31	14	450
Jilin-Songliao	24	7	290
Jiangsu-Zhenwu	7	0.8	115
Jianghan	6	2.8	460
	893 (total)	267 (total)	300 (average)

formed in many late Mesozoic and early Tertiary lake basins (Figs 1 & 2). For example, early Tertiary lacustrine oil shales are mined at outcrop at Maoming in SE China, and Fushun in NE China. Many millions of barrels of oil can be distilled from these lacustrine oil shales (see Cao Zhengyuan 1984; Fan Lang 1986). Oil shales are also found at outcrop in Thailand, Mongolia and Vietnam (see Stokes 1988; Gibling 1988; Pentilla 1993; Traynor & Sladen 1995, 1997). In Thailand, various schemes for commercial extraction have been considered in recent times for use as an energy source for power generation, for production of cement, and for distilling to produce oil (e.g. see Surapol Thanomsap & Suparerk Sitahirun 1992).

Coals (although not the focus of this paper) were often deposited in and around floodplain lakes, on lacustrine deltas, and as part of lake margin sequences around the edges of many of the late Mesozoic and early Tertiary lakes. Examples can be found in numerous extensional basins in Thailand, Indonesia, China and Mongolia (see Li Sitian *et al.* 1984; Gibling *et al.* 1985; Gorst 1986; Robinson 1987; Atkinson 1989; Sahat Muenlek 1992; Santichai Jitapunkal 1992; Yongyuth Uk-kakimapan 1992; Traynor & Sladen 1995). These coals are both important source-rocks for hydrocarbons when deeply buried, whilst when at outcrop, they constitute very important fuel sources which are mined at numerous localities both for local use and export.

Problems when establishing key issues for exploring lake basins

Given the importance of the late Mesozoic and early Tertiary lake basins as sources for oil and gas, it is appropriate to look at key issues when exploring them, some of the lessons learned, and also identify sets of working assumptions and principles to help constrain exploration strategies. This will help ensure efficient use of exploration funds, and lower the finding and development costs for oil and gas.

As a general statement, the state of the art in exploration techniques, strategy and philosophy is less sophisticated in lacustrine sequences when compared to exploration practices in marine clastic and carbonate sequences. Again, but only as a general statement, there is less worldwide history and experience to study past successes and failures in lacustrine basins, and we are less able to look back in detail at the industry track record in SE Asia for finding, developing and producing lacustrine derived oil and gas. For example, data analysis that allows us to study finding rates and field sizes, which are commonplace for play fairways and sequences in areas such as the North Sea, are simply not available for most SE Asia basins.

To make our difficulties in studying the industry track record even more complex, exploration in SE Asia is often characterized by relatively poor and incomplete datasets. Data collection and the maintenance of detailed records has not been comprehensive and has often had low priority; Governments and geological survey departments in many countries have often not had sufficient funds or experience to carry out key studies, analytical equipment has often been scarce, and the technology available has been limited. All too frequently, precious data, information and samples have either not been acquired or later lost. The database is thus not always as good as it could be and this complicates compiling and analysing the industry track record.

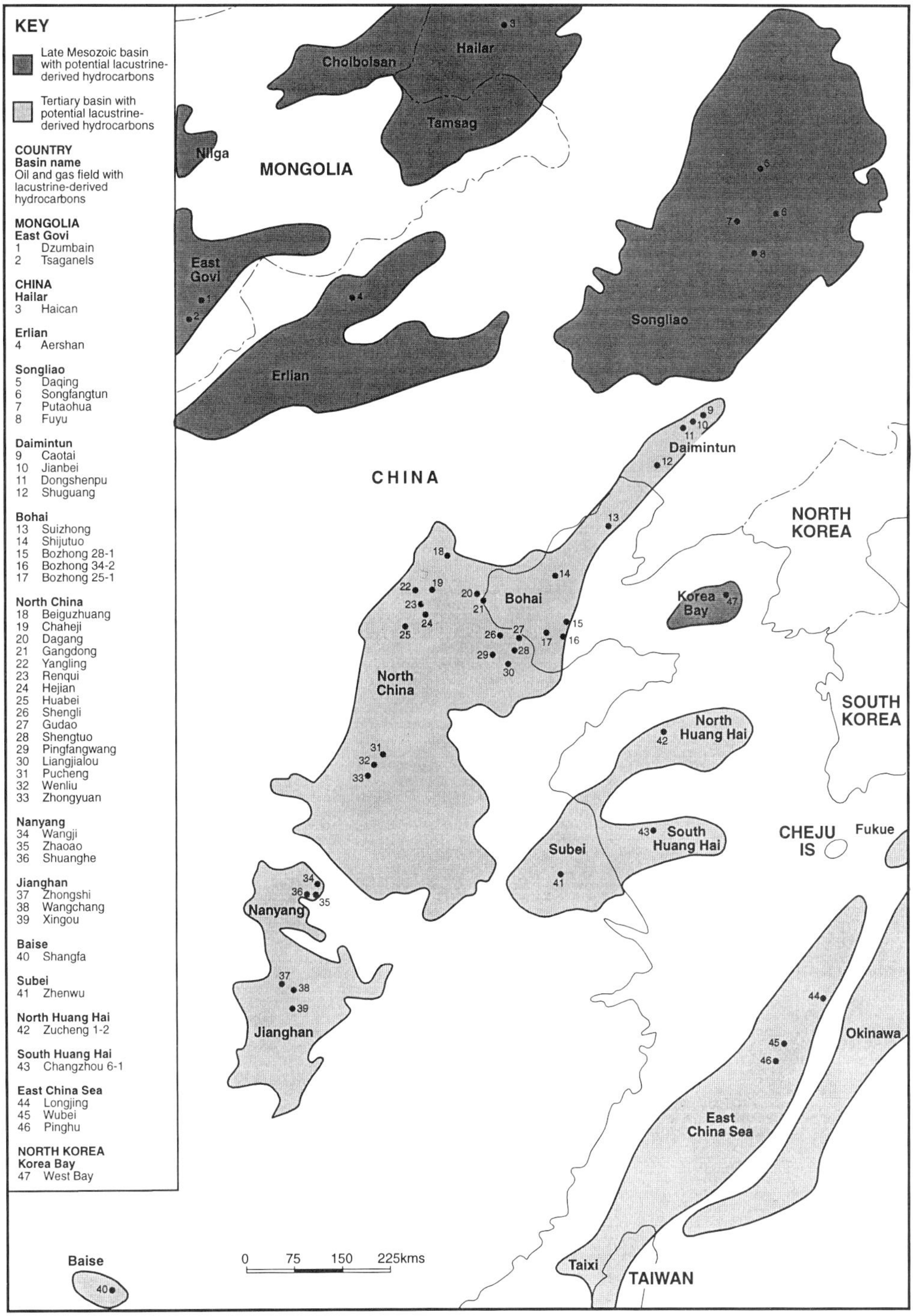

Fig. 1. Late Mesozoic and early Tertiary basins with potential for lake sequences and key fields containing lacustrine-derived oil and gas.

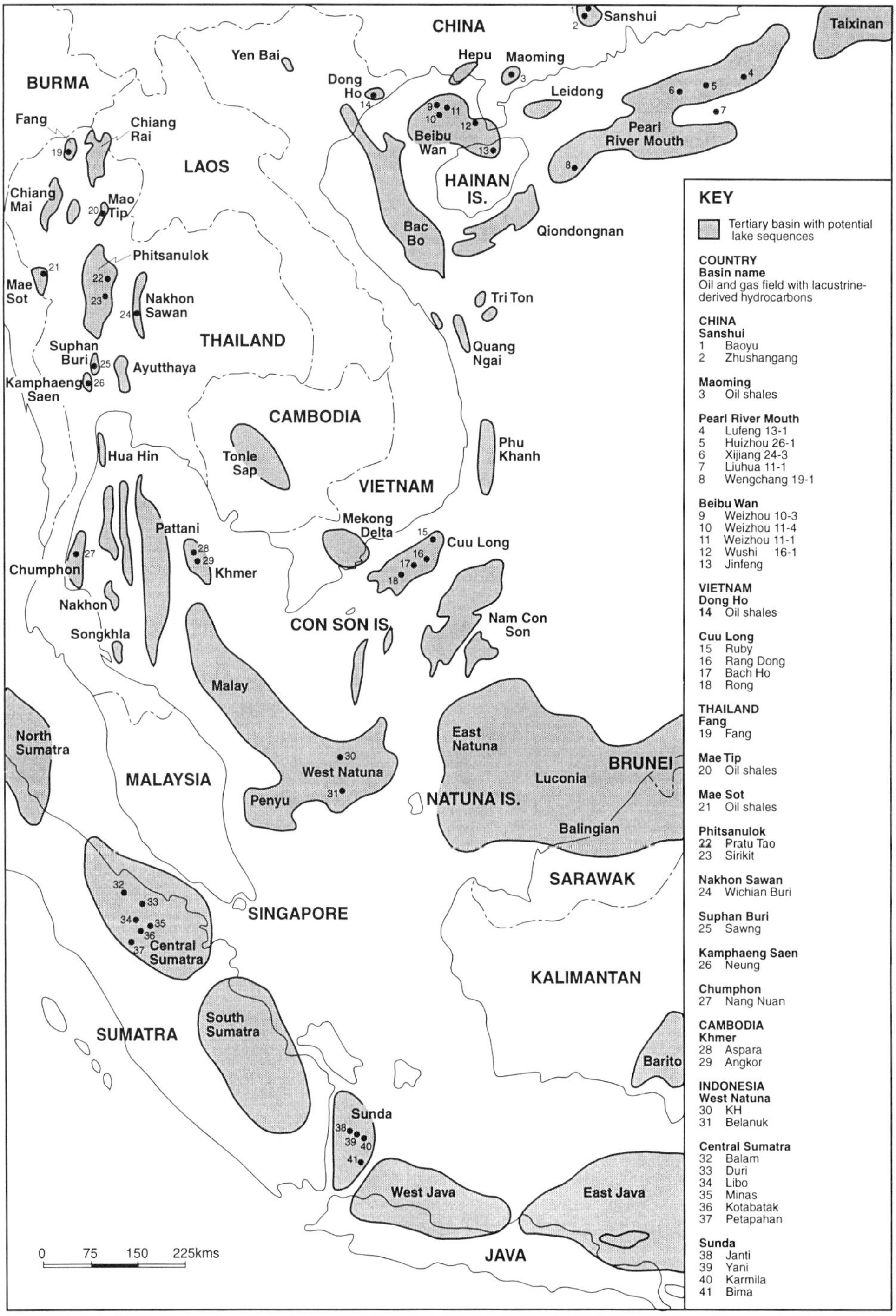

Fig. 2. Early Tertiary basins with potential for lake sequences and key fields containing lacustrine-derived oil and gas.

Overview of lake basins as hydrocarbon systems

The lacustrine basins of SE Asia that are important for hydrocarbons all appear to owe their origins to the development of a large 'closed hole' or depocentre, formed by extension resulting from rifting and strike-slip tectonics. A variety of grabens (rifts), half-grabens and strike-slip basins, developed during the late Mesozoic (especially the early and mid-Cretaceous) and early Tertiary (especially the Eocene and Oligocene, and also the Miocene in some areas). Minor basic volcanic activity was a common feature of the early basin development. Many of the extensional basins also experienced mild compression or transpression events during their evolution, and this led to partial inversion, and complex fault histories. In some instances, inversion and compression were sufficiently severe to completely destroy the extension-related architecture and create new types of basin. Numerous authors have documented and discussed these extensional processes in SE Asia and the reader is referred to these for details e.g. Lloyd 1978; Lao Qiuyuan & Gao Wenxu 1984; Liu Hefu 1986; Watson *et al.* 1987; Derksen & McLean-Hodgson 1988; Gibling 1988; O'Leary & Hill 1989; Li Desheng 1990; Daly *et al.* 1991; Bal *et al.* 1992; Surawit Pradidtan & Dook 1992; Traynor & Sladen 1995.

The lake systems that developed in these extensional basins were usually long-lasting features, existing for more than 1 Ma, frequently 5–10 Ma and sometimes as long as 30 Ma (Fig. 3). Given the typical high sedimentation rates in lakes, which can easily be 1–5 mm a^{-1} for mudstone and much higher for sandstone, many kilometres of lacustrine sequences can accumulate in only a few million years if subsidence outpaces (or equals) sedimentation rates (Table 2). Many basins contain between 3 and 8 km of non-marine sedimentary section of which lake and lake-related facies form the major part.

The lakes frequently occupied large areas or sometimes all of a basin, covering many thousands, or sometimes tens of thousands of square kilometres. Within these basins, a hydrocarbon system often developed with potential hydrocarbon source-rocks accumulating in low-energy anoxic parts of the lakes. During burial these lacustrine source-rocks (together with coals deposited in alluvial, deltaic and lake margin settings) frequently generated hydrocarbons that migrated and became reservoired in other lake basin facies such as deltas and fans that fed into the lake, or turbidites developed in the lake (Fig. 3). Hydrocarbons also escaped from these sequences to become reservoired in other older and younger sequences.

Three key factors for establishing and maintaining lacustrine hydrocarbon systems

When we examine the evolution of numerous late Mesozoic and early Tertiary lake basins in SE Asia, three factors consistently recur and point to these having been critical to developing and maintaining long-lived lakes and mud-prone hydrocarbon source-rocks. These factors comprise prolonged phases of subsidence rate outpacing sedimentation rate, a suitable climate for organic matter to flourish and widespread carbonate-rich hinterlands causing limited coarse clastic input.

The whole of SE Asia was an important area of extension during the late Mesozoic and early Tertiary (see Liu Hefu 1986; Ma Xingyuan &

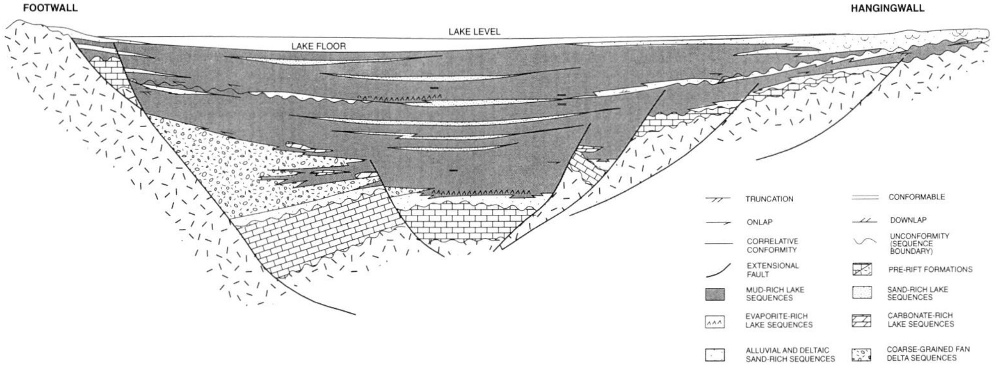

Fig. 3. Typical arrangement of facies and sequences developed in extensional lake basins in the Late Mesozoic and early Tertiary in SE Asia (modified from Sladen 1994).

Table 2. *Thickness of lacustrine mudstone sequences in some basins in SE Asia (many of these mudstones are potential hydrocarbon source rocks and also important seals)*

Basin	Location	Thickness of lake mudstones (m)	Age of lake sequences
East Govi	Mongolia	150	Cretaceous
Hailar	China	200	Cretaceous
Songliao	China	800	Cretaceous
Erlian	China	800	Cretaceous
Jianghan	China	2000	Tertiary
Bohai	China	2000	Tertiary
Nanyang-Biyang	China	1900	Tertiary
Changwei	China	2200	Tertiary
Baise	China	1000	Tertiary
Damintun	China	1000	Tertiary
Pearl River Mouth	China	150	Tertiary
Beibu Wan	China	400	Tertiary
Huang Hai	China	600	Tertiary
Cuu Long	Vietnam	300	Tertiary
Central Sumatra	Indonesia	600	Tertiary
Sunda	Indonesia	800	Tertiary
Chumphon	Thailand	500	Tertiary
Fang	Thailand	700	Tertiary
Suphanburi	Thailand	440	Tertiary
Kra	Thailand	500	Tertiary
Hua Hin	Thailand	700	Tertiary
Phitsanulok	Thailand	400	Tertiary

Wu Daning 1987; Watson *et al.* 1987; Tian Zaiyi 1990; Daly *et al.* 1991). Extension was in part related to subduction of the Pacific Plate, and part related to opening of the South China Sea and East Vietnam Sea. Elsewhere other important events dominated by strike-slip extension included the impact of India into Eurasia which led to the extrusion of Indochina, and strike-slip movements along the Australia Plate and Indian Ocean Plate margins. The prolonged periods of extension were conducive to the establishment and maintenance of long-lived lake sites where subsidence outpaced sedimentation.

Many strike-slip zones are still active today and these illustrate the way in which large deep basins capable of containing lakes can develop due to extension. Excellent present-day analogues include lakes such as Er Hai (near Dali) and Dian Chi (near Kunming) forming along splays of the Red River fault system in southwest China (see Ma Xingyuan & Wu Daning 1987 for details). Elsewhere, the Mongolia–China border region contains lakes such as Burr Nurr and Hulun Nurr which have formed in a basin developed along strike-slip faults that are translating the distant effects of the India–Eurasia collision (see Liu Hefu 1986; Traynor & Sladen 1995 for details).

Within the lake basins, the sequences can frequently be subdivided into three phases which reflect the development, acme and then decline in extension and rifting (Watson *et al.* 1987; Sladen 1994). During the early stages of extension and rift development, the sequences are usually a clastic alluvial fill which is lithologically very variable. Lake sites tend to be shallow and temporary with little source-rock potential. Later on, as subsidence outpaces sedimentation, stable long-lived lake sites tend to develop if a suitable closed basin form is maintained. A lake may be present for many millions of years. Many kilometres of lacustrine sediments may accumulate including thick mudstone source-rocks and seals (caprocks) in the lake, and various potential reservoirs in turbidites, deltas and lake margin facies (Fig. 3). Following this period, as sedimentation begins to outpace subsidence, the lake is gradually infilled becoming progressively shallower and smaller. As considerable rift topography may exist after extension ceases, this can enable lakes to persist into the post-rift period before being finally infilled and eliminated, often by alluvial and deltaic systems.

In addition to prolonged periods of extension during the late Mesozoic and early Tertiary, the climate in SE Asia also appears to have been particularly suitable to supporting the existence of lakes for long periods. The prevailing climatic conditions were hot and humid, tropical to subtropical. These conditions encouraged organic matter to flourish in the lakes, and also created lush vegetation in the hinterlands and on alluvial plains and deltas around the lakes. Key products were not only significant *in situ* organic matter in the lake, particularly algae, but also significant detrital organic matter input. The consequences were abundant hydrocarbon source-rocks in the form of lake mudstones and lake margin coals.

Humid climates were no doubt promoted by the close proximity of the warm oceans, which lay predominantly to the east and south. The fauna and flora of some basins show that they experienced marine influxes, implying low altitudes and temporary links to the oceans via narrow sea straits. These marine influxes are often recorded by the presence of marine bivalves and gastropods, forams, dinoflagellates, diatoms and glauconite. Many lakes may have been similar in style to the present-day Lake Maracaibo in Venezuela which maintains narrow and tenuous saltwater straits that are only a few kilometres wide but connect the lake with the ocean. The fauna and flora reflect the complex environments which are neither truly marine, nor truly lacustrine (see Sarmiento & Kirby 1962 for a detailed description).

Rainfall and temperature were no doubt also sensitive to altitude, not just proximity to oceans. The continual collision and rearrangement of tectonic plates would have affected atmostpheric circulation patterns and shifted pressure cells. In the early Tertiary for example, the India–Eurasia collision and uplift of Tibet would have cut off the northward penetration of humid southwesterly airstreams and caused the climatic extremes in the continental interiors to intensify, leading to increased aridity. Whilst the humid airstream of the SE Monsoon would have still penetrated inland and brought rainfall, it is clear from the presence in some lake basins of red bed sequences and evaporites (which include gypsum, ischelite and glauberite), that arid steppe and desert climatic belts existed in some inland areas. In the more humid areas, hinterlands to the lakes appear to have been densely vegetated and richly forested. These forests included both coniferous trees in cooler more mountainous areas, and broad-leafed deciduous trees and hardwoods, particularly on the calcium-rich soils of carbonate hinterlands.

The hinterland geology during lake development is also known to frequently be a key factor in controlling sediment input and sediment composition during lake development and evolution (see Sladen 1994 for details). In SE Asia, the widespread existence of carbonate rocks in the hinterlands, with limestones and dolomites varying from Sinian to Palaeozoic or early Mesozoic in age, is considered to have been a key factor in reducing coarse clastic infill of the lakes. The reduced availability of coarse clastics allowed more finer grained mud-rich lake sites to develop. Many of the resulting lake mudstone sequences were also organic-rich and became the key hydrocarbon source-rocks and seals (caprocks) that we encounter today.

Evaluating lake basin architecture and controls on the development of lake sequences

Understanding the architecture of a lake basin is of fundamental importance because it is critical to controlling positions of lakes and the development of source, reservoir and seal within the lake basins. The architecture helps to control hydrocarbon migration pathways and determine trap types.

Controls on lake sequences and sequence stratigraphy

Much work has been done on understanding the controls on lake sequences in recent years but in contrast to marine clastic and carbonate sequences, our understanding is still relatively poor. Although the three factors of subsidence outpacing sedimentation rate, climate and hinterland geology consistently recur during the evolution of lake basins in SE Asia, there are at least eleven important controls on the kind of lake sequences that may develop:

(1) structural evolution; subsidence v. sedimentation rate;
(2) basin size, lake size and shape;
(3) catchment area and topography;
(4) catchment bedrock, composition and geology;
(5) temperate or tropical climates;
(6) humid or arid climates;
(7) elevation, proximity and connectivity to oceans;
(8) open or closed lakes;
(9) fresh, brackish or saline lakes; water chemistry;
(10) shallow or deep lakes;
(11) volcanic and hydrothermal activity.

These controls and their relationships are described in detail in Sladen (1994) and various other authors have addressed controls on lake sequences (e.g. Chen Changming *et al.* 1984; Surawit Pradidtan 1989; Barron 1990; Katz 1990, 1991; Lambaise 1990; Gierlowski-Kordesch & Kelts 1994) so this will not be covered again here. Simply note that the controls need to be understood in detail to fully appreciate lake facies and lake sequence development so as to reliably predict in the subsurface and consequently reduce exploration risk. Consider for example, the complexity in the systems tracts shown in Figs 4 and 5 for the (relatively) simple reconstructions of humid subtropical and semi-arid subtropical lakes in an extensional regime.

The complex issues of controls on systems tracts in lake basins is in fact made more difficult because our techniques of sequence-stratigraphy in lake basins are currently less well advanced than, for example, in the marine realm. Many sequence-stratigraphy principles have been developed in the marine realm and these principles relate to sequences on carbonate platforms or on clastic dominated passive continental margins. These are simply inappropriate when applied to the confined lake basins of SE Asia which formed in areas of rapid extension and subsidence, often surrounded by mountains, fed by small alluvial systems tracts and often having no marine connection.

Using palaeotopographic analysis

Palaeotopographic analysis (a form of palaeogeographic analysis) is a particularly useful method for reconstructing lake basins to help understand lake sequences and lake facies distribution. The technique is described in more detail in Sladen (1994). It involves re-creating the lake basin, its architecture, its climate, its catchment and hinterland topography, elevation and geology, identifying the key (syn-depositional) faults controlling basin architecture and active during lake development, whilst also removing subsequent structural modifications. Isopaching of lacustrine sequences together with the use of seismic facies analysis, and detailed well data evaluation are all important parts of the process. By using all available data, and by applying a detailed analysis of the controls on lake facies, a palaeotopographic analysis gives the best possible predictions of facies variations in the subsurface.

The need for good data

Understanding lake basin architecture first and foremost requires good datasets. Good quality regional seismic grids, a free and easy exchange of data and information and early

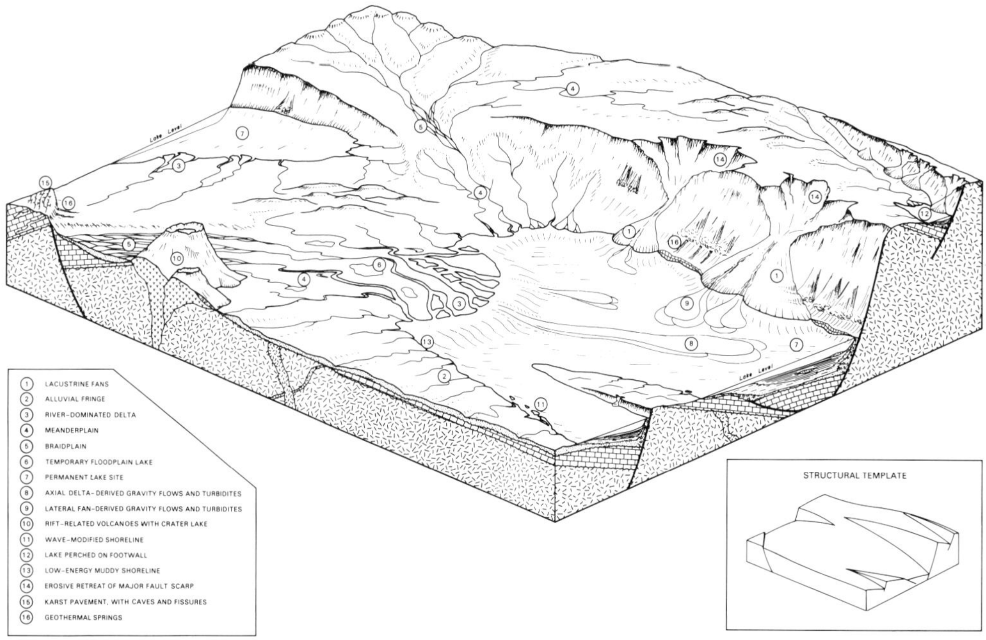

Fig. 4. Reconstruction of a syn-rift lacustrine basin with humid sub-tropical climate.

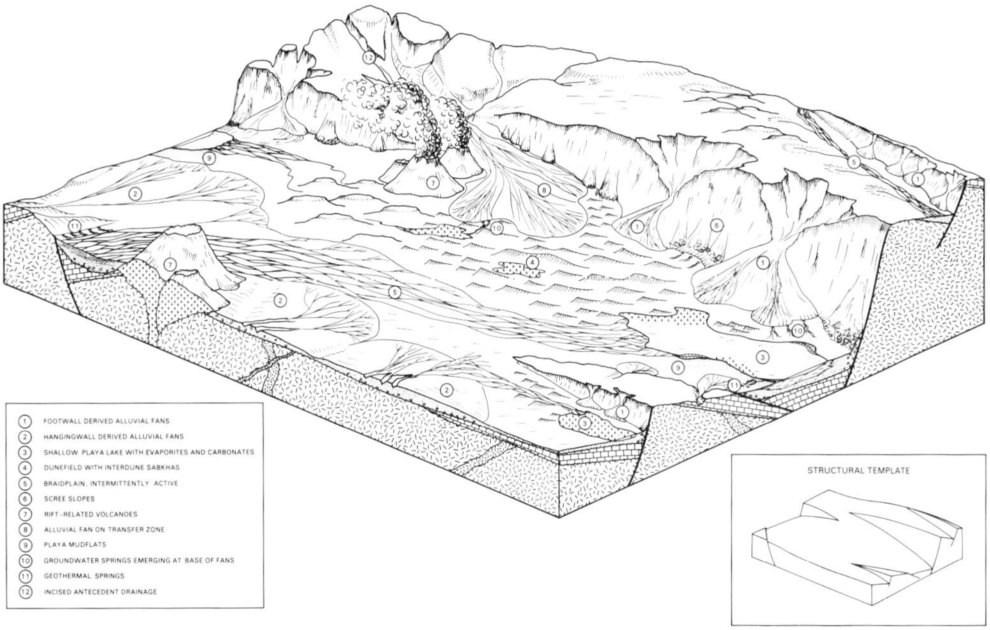

Fig. 5. Reconstruction of a syn-rift lacustrine basin with a semi-arid sub-tropical climate.

deep stratigraphic tests all help the process. So often throughout SE Asia, Government policies have severely restricted easy access and exchange of data and prohibited data trading or data release. Policies that encourage and promote data access undoubtedly lead to more effective exploration, best use of exploration dollars and ultimately more cost efficient and timely exploration successes.

Many basins in SE Asia unfortunately continue to be explored without regional seismic grids and tielines or without aeromagnetic and gravity data that are critical to a proper understanding of basin architecture. Often, understanding of a basin will end at a country's borders and key elements of basin architecture consequently remain unanswered. Programmes of data collection that cross international borders are needed, but in SE Asia these remain so difficult to agree and organize.

Benefits of early stratigraphic wells

The oil and gas companies that explore for hydrocarbons have perhaps been slow to recognize the importance of basin architecture. Also, they all too often fail to establish the full stratigraphy of lake basins as early as possible during an exploration programme by drilling deep stratigraphic tests towards the basin centres. Instead, most companies choose to drill many exploratory wells on the large structural highs that surround the basins in the belief that these will contain most of the hydrocarbons. Frequently they fail ever fully to understand the lake sequences and stratigraphy responsible for generating the hydrocarbons. Source-rocks responsible for generating the hydrocarbons in many of the hydrocarbon-producing basins in SE Asia have never actually been penetrated in wells; source-rock characteristics are simply inferred from geochemical analyses of the oils and gases.

Many oil and gas companies will argue that a deep stratigraphic test would be extremely costly and time consuming. In addition, the exploration periods available in many cases are very short and time does not permit the drilling and evaluation of such wells. The answer is surely for many companies to share the cost of early deep stratigraphic tests, and also for licencing authorities to recognize the logic of early deep stratigraphic tests, and encourage a free exchange of data and information. Otherwise vast amounts of exploration effort will continue to be poorly focused in areas that ultimately have little hope of success.

Focusing on the most effective trap styles in lake basins

Importance of structural traps

In most extensional basins (or structurally modified extensional basins), the bulk of reserves and the largest fields can be expected to be found in the large structural traps which are related to large fault blocks. Drilling results from the lake basins of SE Asia appear to confirm this, although the picture is complex to analyse, due not only to the exploration history and records that have been kept, but also to the different styles of exploration and drilling between onshore and offshore areas. Onshore, for example, the much lower drilling and production costs leads to a greater array of small structural traps being drilled, and often a variety of stratigraphic plays also being tested. Many of the smaller structures and relatively high risk stratigraphic plays often remain untested offshore due to the high cost of failed wells, and poor economics of developing and producing small fields offshore.

A preference for dip closure

Of the structural traps, whatever their sizes, those with four-way dip closure are, not surprisingly, found to be lowest risk and the proportion of fields and proven reserves in four-way dip and faulted four-way dip closures is higher than any others. Rollover anticlines, drape closures and gentle inversion anticlines usually represent the best dip-closed traps and they frequently also give rise to stacked hydrocarbon pay zones (Fig. 6). Although rare, diapirs of lacustrine mudstones and evaporites developed in some basins and these can also create four-way dip closed traps (e.g. in the Jianghan basin in China). Compression episodes during the development of many SE Asia lake basins often caused inversion or partial inversion of depocentres and the generation of compressional four-way dip closed anticlines (e.g. in the Central Sumatra and West Natuna basins in Indonesia). These traps usually formed in response to slight re-arrangements in plate motions, and the locking or squeezing of the fault systems during plate collisions.

Again, not surprisingly, risk increases with the extent of cross-fault seal required to make a trap. Whilst there may be vast numbers of fault traps in the extensional basins, traps which require cross-fault seal on one or more faults represent relatively high risk closures compared to four-way dip closures (Fig. 6). The presence of dip-reversal in the hanging-wall of major faults and on synthetic faults may improve the chances of trapping by inhibiting leakage of hydrocarbons and creating an overall four-way dip closed structural form, albeit faulted. Many strike-slip lake basins have flower structures and if the need for fault closure is limited these can also form low risk trap types dominated by dip closure.

Due to fault leakage, traps without an overall four-way dip geometry are likely, as a general rule to be relatively small, and higher risk, in comparison to well-defined four-way dip closures. Fluid contacts are defined by the point of seal failure, not by the mapped structural spill point. Fault closures on antithetic faults in the main hanging-wall of half-grabens are particularly high risk because these usually lack dip-reversal (Fig. 6). This allows most hydrocarbons to escape up-dip, and any volumes of trapped hydrocarbons are usually very small.

As exploration in the lake basins continues, and large, simple, low risk structures no longer exist to be drilled, more and more attention will need to be paid to the complex fault traps often with higher risk, smaller reserves potential and low reserves density. Accurate subsurface imaging of these complex traps will be essential to resolve the fault geometries, and often thin payzones. 3D seismic imaging will become critical to avoiding wrongly located wells and ensuring optimum development schemes.

Stratigraphic traps

Stratigraphic traps frequently develop in lake basins because of the complexity of many potential reservoir facies and because tectonic and climatically induced changes in lake level can cause dramatic shifts in facies belts. Many of the stratigraphic traps that are formed are available to receive a hydrocarbon charge from source-rocks toward the basin axis, as soon as these source-rocks begin to generate hydrocarbons. Numerous stratigraphic traps may develop where there are low depositional slopes, particularly on the hangingwall side of half-graben or at the 'ends' of graben.

Common stratigraphic trapping mechanisms (refer Fig. 7) include the following.

(a) Basinward pinch-out of lacustrine reservoir and basinward appearance of seal. For example, lacustrine reservoirs of this type include lacustrine fans, deltas, shoreline and shallow lake facies. Often they may have undergone some structural tilting and dip reversal, for example the Binnan Field, Zhaoao Field and Shaunghe Field in China (see Ma Li *et al.* 1982; Li Chunju *et al.* 1984).

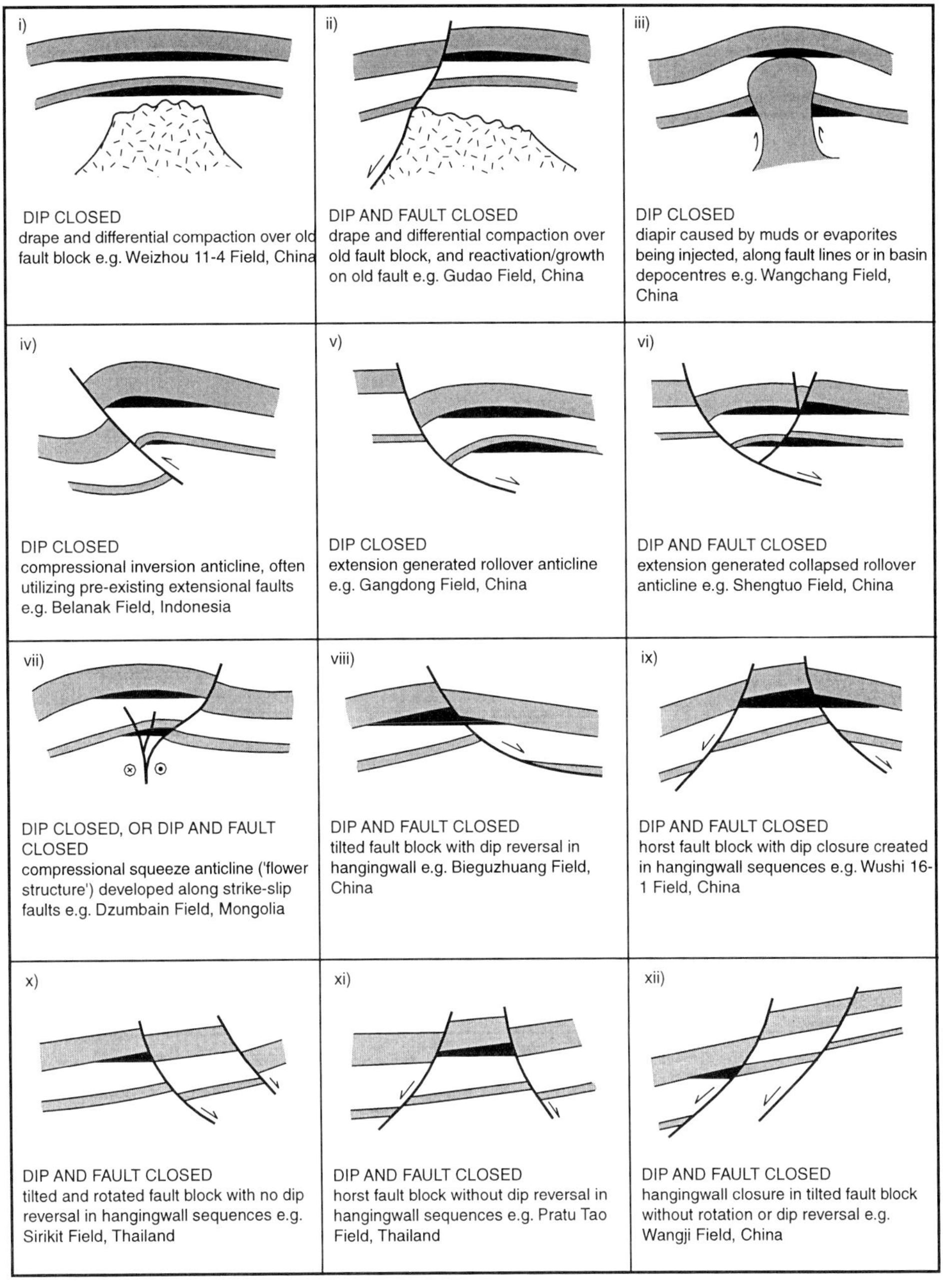

Fig. 6. Common structural trapping mechanisms in and around extensional lake basins in SE Asia.

(b) Basinward appearance of lacustrine reservoirs such as axial lacustrine turbidites developed in depositional hollows, for example the Ying 6 Field and, Liangjialou Field in China (see Ma Li *et al.* 1982; Hu Jianyi *et al.* 1984).

(c) Unconformities (up-dip truncation) within lacustrine sequences. These tend to be best developed around basin margins and frequently involve shallow lake or shoreline reservoirs. The traps are commonly formed due to episodes of tilting and erosion, or from changes in climate, which lead to changes in lake level. Examples include the Dhangxi Field, Shuguang Field and Jinjiazhuang Field

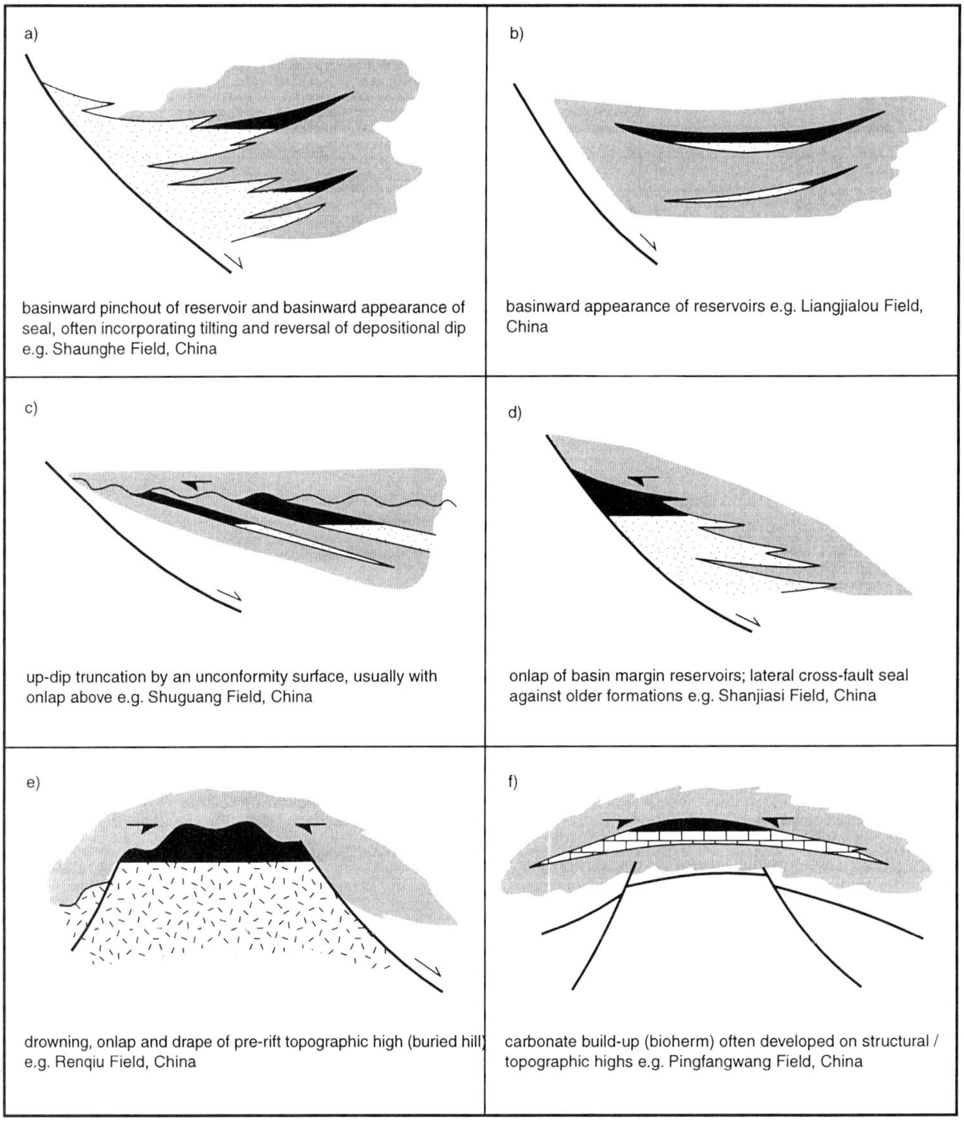

Fig. 7. Common stratigraphic trapping mechanisms in and around extensional lake basins in SE Asia.

in China (see Ma Li *et al.* 1982; Hu Jianyi *et al.* 1984).

(d) Simple onlap caused by expansion of the area of lake deposition and submergence of marginal fans, deltas and shallow lake reservoir facies by fine-grained lake sediments, for example the Shanjiasi Field and Zhongshi Field in China (see Hu Jianyi *et al.* 1984).

(e) Buried-hill traps are a type of unconformity trap that can usually also be mapped as a closed structural high. They are often created when lakes expand and drown pre-existing 'basement' fault block topography, causing onlap and drape of the basement blocks with lacustrine fines. This enables hydrocarbons to become trapped in pre-rift basement reservoirs, for example, in China examples include the Renqiu Field, Nanmeng Field, Balizhuang Field, Shangfa Field, Bozhong 25-1, and Wei 11-1 Field (see Zhai Guangming & Zha Quanheng 1982; Fei Qi & Wang Xie-Pei 1984; Zhang Qiming & Su Houxi 1989; Li Desheng & Lou Ming 1990).

(f) Palaeogeomorphic traps may develop where lacustrine carbonates (bioherms) develop on offshore shoals and highs within the lake and become encased in lacustrine muds, for example in China the Pingfangwang Field and Qi 26 Field (see Ma Li *et al.* 1982; Hu Jianyi *et al.* 1984).

Determining field characteristics and commercial value

The impact of well flow rates and onshore v. offshore developments

As general rule, flow rates from wells producing from lacustrine reservoirs are low throughout SE Asia. Oil flow rates are typically measured in hundreds of barrels per day for each well, as opposed to thousands; stabilized individual flow rates greater than 2000 barrels of oil per day are scarce, and flow rates greater than 5000 barrels of oil per day from lacustrine reservoirs are almost unheard of. The widespread problem of low flow rates is rooted in high wax, high viscosity and low gas-to-oil ratios (GORs), which are typical features of oils derived from lacustrine source-rocks, combined with poor reservoir quality and continuity. These widespread problems are further compounded by often low reservoir pressures in fields, particularly around the basin margins where pressures are frequently either at or close to hydrostatic.

Due to the fact that low GORs are typical of lacustrine derived oils, the amount of associated gas produced with the oil is also usually low (Table 1). Therefore, not only are oil flow rates typically low, but there is also little associated gas production to add extra value to a development programme and little gas is available for use in improving oil recoveries by gas re-injection or gas lift. Solution gas drive in the reservoirs is typically weak.

The low flow rates of oil (and associated gas) per well cause difficult economics particularly for offshore prospects as a very large number of production wells are required at considerable cost to be able to achieve a field flowrate that can justify the high infrastructure and operating costs of an offshore field development. Minimizing offshore development costs and operating expenditure is therefore essential to commercial success. In contrast, many onshore developments are possible due to the much lower onshore drilling and development costs.

In SE Asia, the overall production profile thus includes many small onshore fields producing oil from lacustrine reservoirs at low flow rates from numerous wells whereas offshore only development of large fields with high flow rates is practical. For example, well flow rates of only a few hundred barrels of oil per day or sometimes even a few tens of barrels per day are viable onshore Thailand and onshore China, and fields as small as 1 to 2 $\times 10^6$ barrels can be developed. In contrast, offshore developments are rarely viable unless individual well flow rates can be stabilized at more than 2500 barrels oil per day and oil reserves are greater than 20 $\times 10^6$ barrels.

Many additional offshore developments from small accumulations may be achievable in the future by a combination of innovative technology to reduce well costs and increase flow-rates per well, reductions in offshore operating expenditures and by greater flexibility in fiscal terms to enable a commercially viable development. Clearly this topic would benefit from both industry-led initiatives, and changes in Government policies and legislation.

Reservoir performance and recovery factors

Production from lacustrine reservoirs is often from multiple pays which are poorly interconnected, laterally inextensive and complex shapes (e.g. see Xu Shice & Wang Hengjian 1981; Yinan Qui *et al.* 1987). This inherent complexity causes considerable complications when appraising, developing and producing lacustrine-derived hydrocarbons trapped within lacustrine reservoirs. Due to the reservoir complexity, discoveries often require extensive appraisal drilling to delineate them, identify the most appropriate development scheme and confirm commerciality. Complex and expensive multiple completion technology is often necessary. Individual reservoirs are normally below the limit of seismic resolution and intra-reservoir correlation is typically highly complex. This makes development drilling and production planning a difficult process which must incorporate a number of risk assessments. Many development wells in the past have simply failed to find the expected reservoir due to reservoir heterogeneity being more complex and unpredictable than expected. This all adds to development costs.

Porosity and permeability characteristics in the reservoirs are very variable and there is a distinct lack of uniform, laterally extensive, homogenous reservoirs. These characteristics result in low recovery of hydrocarbons during primary depletion, with much hydrocarbon being left behind caught up in complex, poorly accessible, reservoir bodies. The combination of low pressures, low GORs, high wax crude and laterally inextensive, complex reservoirs frequently causes a rapid decline in production from wells when they are put onstream. Many wells may need pumping to maintain flow rates, and the inherent difficulty in predicting production performance leads to complex planning issues. Long term production tests are often needed to confirm a commercially viable development. Due to the complexity, numerous additional infill production wells may be required to access the hydrocarbons and add reserves. This can improve recovery factors but it is commercially difficult offshore due to the higher well costs.

Attempts to increase the recovery of oil and gas using secondary methods, such as waterflooding of the reservoirs to maintain pressure and push hydrocarbons toward producing wells is also fraught with difficulties. The reservoir complexity and permeability variability cause numerous parts of the reservoir to be by-passed during waterflooding, again leaving behind oil and gas in stranded pockets of hydrocarbons. As a consequence, waterflooding may easily cause more reservoir damage than improvement in production. If limited investment funds are available, the choice between drilling more infill production wells or installing waterflooding is often particularly difficult.

Value and price of lacustrine-derived crude oil and gas

The typical properties of lacustrine-derived crude oil and gas are important factors in determining what it can be used for and therefore its price on international markets. Lacustrine-derived crude oil typically has a very low sulphur content (<0.2 wt%) ultimately due to formation from organic matter in source-rocks accumulating in lake waters that are relatively sulphate-poor compared to marine source-rocks. This low sulphur content makes refining of the crude, and use of the crude, easier compared to a higher sulphur content crude. For example, many lacustrine-derived crudes are so low in sulphur content that they can be burnt directly for oil-fired power generation, whilst still meeting strict pollution controls. Consequently, they can command a higher price on international markets compared to other crude oils. Frequently they command a price premium of US$0.5–1.0 per barrel compared to the Middle East crudes which are sulphur-rich with often >1 wt% sulphur.

Unlike oil, gas price and gas use is much more controlled by local demand and proximity to markets. Nevertheless, lacustrine-derived gas shares the similar characteristic of lacustrine-derived oil of typically being low in sulphur. In the case of gas, this means low in hydrogen sulphide (H_2S), and low in sulphur-rich organic gas molecules such as mercaptans. This makes lacustrine-derived gas production, transportation and ultimately its use less technologically demanding, and hence lower cost and higher value compared to sulphur-rich gases.

Liquid yields (LPGs) from lacustrine-derived gases are typically high because the source-rocks are inherently oil-prone, and expel predominantly liquids. The LPGs consequently often represent a significant part of the value and economic viability of oil and gas development projects.

Wax content and pour point

Wax contents of 10–35% are common in oils derived from lacustrine source-rocks throughout SE Asia, and occasionally wax content even reaches 45%. The high wax content results in notably high-temperature pour points, causing them to solidify close to room temperatures. Many oils have pour points that are higher than 20°C and some are even higher at 30°C or more. The yield of gasoline and other light products (distillates) during refining is relatively low compared to, say, North Sea crudes. Consequently, during refining, SE Asian lacustrine crudes are often blended with other crudes (typically from the Middle East) to improve the proportion of light products available.

The high temperature pour points cause a variety of oil production problems. Offshore, it is often necessary to add a (costly) pour point depressant to avoid oil solidifying in subsea flowlines, pipelines or in tankers when it is being shipped. Small diameter flowlines are particularly prone to wax solidifying within them because they cool to ambient temperatures more quickly than large diameter flowlines and pipework. In the cooler northern latitudes, for example in northern China and Mongolia, where surface temperatures fall below zero for many months, heated flowlines and insulated tanks are commonly used to keep the oil warm and stop it solidifying during production and transportation. Associated gas is frequently used to heat water-filled pipes, which then warm oil flowlines. Heated pipework and insulation are also necessary to prevent wax solidifying during periods when wells and surface facilities are shut-in, perhaps for repair or maintenance checks. In extreme cases, where the oil solidifies at surface due to a lack of heated flowline technology and pour point depressants, the oil has in the past been transported in solid form simply by shovelling it! In other cases, entire fields have been shutdown during extremely cold winter months and only produced during warmer summer months.

Downhole, wax also causes production problems which can be costly to deal with. Wax build-up commonly occurs around perforations into the producing formation, reducing permeability into the wellbore. Wax build-up on the downhole tubing also reduces well productivity. To overcome these problems, frequent workovers of producing wells are required, and downhole scrapers may be added to stop wax building-up on sucker rods in pumping wells.

Understanding reservoir distribution, heterogeneity and quality

Lacustrine fans and deltaic reservoirs

Reservoirs may occur where lacustrine fans deposit a thick sequence of sandstones and conglomerates. In the extensional lake basins in SE Asia, fan reservoirs were commonly located along major fault scarps, with sediment typically entering transverse to the rift axis (Figs 3 & 4). Fan positions were controlled by antecedent drainage systems, transfer zones or transverse faults.

Lacustrine fan sequences can attain great thicknesses (100–1500 m) with individual sandstone and conglomerate reservoirs perhaps 10–100 m thick (Table 3). These offer the opportunity for large hydrocarbon columns. Whilst entire fans may constitute potential reservoir, they are characterized by extremely rapid basinward net-to-gross changes because coarse sediment is usually rapidly dumped on entering the lake. In proximal areas, the proportion of potential reservoir may approach 1.0 but decrease rapidly to <0.2 in under 2 km. Most of the fan may have net to gross of between 0.4 and 0.8. At the same time that the proportion of reservoir decreases rapidly at the lake margin, sandbody thickness and vertical connectivity also usually decrease markedly as beds wedge out into lake fines (Fig. 3).

Shoreline facies that form where the fans enter the lake tend to be only an extremely narrow strip which is difficult to locate in the subsurface using traditional seismic and well analysis techniques. In between fans, and on inactive fans, shallow water biogenic carbonates can develop (Fig. 3). These are laterally impersistent reservoirs in both dip and strike section (typically only a few kilometres), and only a few metres thick.

Lacustrine deltas contain the most important lacustrine reservoirs, with the majority of hydrocarbons found to date reservoired in various types of lacustrine delta sandbodies. Deltas commonly develop entering from the ends of lakes in elongate extensional basins where rivers have a relatively large catchment and drain axially (Fig. 4). Deltas may often also be found where rivers drained the hanging wall of a half-graben. Lacustrine deltas are typically river-dominated because waves are usually very small and tides non-existent in all but a few extremely large lakes (Sladen 1994). Even in the giant lake that existed in the Cretaceous Songliao Basin in northeast China, the deltas were river-dominated highly constructive forms (see Xu Shice & Wang Hengjian 1981). Arcuate delta fronts and birdsfoot deltas are common. These morphologies are important to consider in subsurface reservoir modelling and intra-reservoir correlation.

The principal deltaic reservoirs are distributary channel and mouth bar sandstones (Table 3). The proportion of reservoir is typically low (<0.5) and sandbodies are thin (<10 m) creating poorly connected, thin hydrocarbon columns. Complex and subtle stratigraphic trapping can develop in

Table 3. *Typical properties of individual lacustrine reservoir bodies in SE Asia (data assimilated from numerous sources covering various lake basins)*

Reservoir	Typical thickness (m)	Typical features
Meander channel sandstones	4–15	Fining-up
Distributary (meandering) sandstones	3–9	Fining-up
Distributary (straight) sandstones	2–6	Fining-up or uniform, laterally inextensive
Shoreline sandstones	1–5	Large lateral changes in position, fine-grained
Delta front sandstones	0.5–3	Some coarsening-up
Interdistributary sandstones	1–2	Crevasse splays of limited areal extent
Turbidite/gravity flow sandstones	1–5	Argillaceous and matrix rich
Carbonates	0.5–4	Fine-grained, algal, laterally impersistent
Fan sandstones and conglomerates	10–100+	Areally inextensive

deltaic sequences because of the irregular reservoir shapes (for example, see Ma Li 1985; Yang Wanli 1985). The channel sandbodies may have elongate shapes (shoestrings), typically only 100–1000 m wide and separated by intraformational seals comprising overbank floodplain and shallow lake fines (Yinan Qui *et al.* 1987). Crevasse splay deposits also form potential reservoirs but are usually poor quality thin reservoirs of limited areal extent, typically only 1–3 m thick, poorly sorted and incorporating considerable fines.

Considerable thicknesses of lacustrine delta sequences, perhaps more than 1000 m, may be present where the position of deltas is constrained by the basin architecture and multiple thin hydrocarbon zones are often present. Mudstone seals may develop in overbank areas on the delta plain and in interdistributary bays. These commonly only form local seals and discontinuous permeability barriers within the deltaic reservoirs leading to complex production problems. The Daqing Field in northeast China, in which there are many thousands of oil-bearing sandbodies, shows examples of many different phenomena that can occur (see Jin Yusun *et al.* 1985; Yinan Qui *et al.* 1987).

Clinoforms associated with lacustrine fans and deltas may be visible on seismic sections. These can help predict the presence, extent and quality of reservoirs, although individual reservoir bodies are usually below the limits of seismic resolution. Lacustrine deltas tend to create extremely gentle sigmoidal forms, often with vertically compressed topset/foreset/bottomset relationships where the deltas are building into very shallow lakes. Frequently these clinoforms may be unresolvable on seismic; squash plots of seismic data may be helpful when trying to locate them. An absence of clinoforms should not be taken to be indicative of the absence of a lacustrine delta. Lacustrine fans building out into a lake can also develop topset, foreset and bottomset clinoform relationships. Unfortunately on seismic sections, the clinoforms are always difficult to see because the fans typically develop adjacent to faults and faulted margins. Consequently seismic diffractions from the faults often obscure the clinoforms, and the diffractions are easily mistaken for stratigraphic reflections. Careful seismic processing is necessary to remove the effects due to diffractions.

Shoreline and shallow lake reservoirs

Shoreline and shallow lake reservoirs may be formed in low energy settings where the depositional slope is usually very low and many potential reservoir sandstones are thus very fine-grained and argillaceous. In most lake basins in SE Asia, there were normally only small waves with limited fetch and no significant tidal currents. The depth to which the substrate was reworked and scoured was small, often only a metre or two, and as a consequence 'offshore' sand bars and sand banks of the type that frequently form good quality reservoirs in a marine environment, are often minor or absent.

The shoreface zones of lakes are usually narrow with a rapid basinward transition to low energy silt and clay deposition in only a few tens of metres (Sladen 1994). Non-reservoir sequences are commonplace because the shoreline and shallow lake facies may be entirely mud and silt. Frequent vertical lithological variations occur because small lake-level fluctuations combined with the low depositional slopes shift facies belts large distances. Sequences may include, for example, variably coloured mudstones, shoreface siltstones and sandy siltstones, thin-bedded evaporites, algal micrites, bioturbated mudstones, palaeosols and carbonaceous beds.

Although thick, stacked, shoreface sandstone reservoirs are rare, barrier bar sandstones are sometimes developed (e.g. Chen Changming *et al.* 1984). These reservoirs are typically very thin (1–5 m), but can comprise laterally extensive sheet-like sandbodies (Table 3). The precise position of a lake shoreline at any one time is particularly difficult to locate in the subsurface, and most shoreline and shallow lake reservoir sandstones are too thin or lack diagnostic criteria, to be recognizable on seismic sections.

Carbonate rocks can be found in some shallow water lake sequences where there was reduced clastic input, often forming as small build-ups in shoals on structural highs within well-oxygenated parts of alkaline lakes. These may create paleomorphological traps if later drowned and encased in lake muds (Fig. 7). However, potential carbonate reservoirs are usually thin, with both low porosity and low permeability due to their predominantly fine-grained, usually algal micritic nature (Table 3). Facies include charophytes, oncolites, oolitic banks, shell blankets and coquinas of gastropods and bivalves, ostracodes, polychaetes such as *Serpula*, together with opalline silica and chert (for examples, see Chen Changming *et al.* 1984; Traynor & Sladen 1995). Shoaling and sub-aerial exposure of the carbonates leading to meteoric and vadose diagenesis, and to carbonate dissolution, is desirable in most lacustrine carbonates to enhance reservoir quality.

Deep lake reservoirs

The main potential reservoirs in the deep lake setting are lacustrine gravity flows (turbidites) and the distal portions of axial deltas (Figs 3 & 4). Turbidites are typically thin, poor-quality reservoirs, which are difficult to develop and produce hydrocarbons from, particularly in offshore fields as these require high flow rates to achieve commerciality. The proportion of potential reservoir is typically very low (<0.3), and the turbidites are characteristically thin-bedded (typically <1 m thick and normally <5 m), poorly sorted and matrix rich (Table 3). Overpressuring which is common in the deep axis of many lake basins may help preserve reservoir quality (for example, see Li Shaoguang & Wang Dexing 1985).

The main sites for gravity flows and turbidites are the relatively deepwater parts of lakes in front of lacustrine fans and deltas. (Deepwater is considered to be anything >20 m). Turbidite reservoirs when present have usually formed from flows spreading into and filling the topographic lows in the lake floor; accurate prediction therefore requires careful reconstruction of the syn-depositional lake floor topography. Turbidites may be generated by storms, by seismic shocks, by rapid lake level fluctuations, by increases in discharge into the lake, or simply because the high sedimentation rates characteristic of lakes creates oversteepening and unstable pore pressure regimes which result in slumping and slope failure. Large subaqueous fan systems analogous to submarine fans are however rare, because the area of deep lake suitable for their formation is typically quite small.

Alternative reservoirs

Much better alternatives to the rather mediocre quality of lacustrine and lake margin reservoirs can be looked for. A variety of alternative reservoirs may occur in the older (pre-rift) sequences, or in younger (post-rift) sequences. The widespread success throughout SE Asia of the 'buried hill' play where reservoir occurs in a topographic remnant of older pre-rift sequences, and the buried hill is onlapped and covered by lacustrine mudrocks has been a key factor in the development of the industry in many basins throughout the region (Figs 3, 4 & 7).

The most common reservoir in the buried hills has been fractured and kartstified limestones and dolomites, mostly Palaeozoic and Sinian (Precambrian) in age but occasionally Triassic in age. There are numerous examples of oil and gas fields of this type in China including onshore the Renqiu, Yang Ling, Shangfa, Zhuagxi, Yihezhuang, Luilubei and Hejian Fields, and offshore the Bozhong 28-1 Field in the Bohai, and the Weizhou 11-1 Field in the Beibu Wan (see Li Desheng 1981; Yan Dunshi & Zhai Guangming 1981; Hu Jianyi *et al.* 1984; Zu Jiaqi 1985; Li Desheng & Lou Ming 1990). Thailand also has examples of oil and gas fields of this type, for example the Nang Nuan Field (see Surawit Pradidtan & Dook 1992).

The carbonate reservoirs in the buried hills are characterized by low porosity, which is often exclusively present in fractures. Karstification may have improved porosity and permeability in the upper parts of the carbonate buried hills but usually it is the fractures that are most important. Fracture porosity is usually only in the 1–5% range but permeability, which is largely dependent on the interconnectedness of the open fractures can be very high. Stabilised oil flow rates of 5000 to 10 000 barrels per day have been achieved although there is a tendency for wells to begin flowing water quite soon after production start-up. To improve drainage and production of these fractured reservoirs, keys to success are horizontal well technology combined with an ability to image the fractures in the wellbore and predict the fracture patterns.

The 'buried hill' play concept is in fact, no different to the Brent tilted fault-block play concept of the North Sea where Cretaceous mudstones form the seal lying uncomformably on the Middle Jurassic Brent sandstones in eroded remnants of tilted fault blocks. In SE Asia, however, the noteworthy feature is the variety of potential reservoirs that can occur in the buried hills. Apart from the successes in fractured and karstified limestones and dolomites mentioned earlier, other sedimentary reservoirs include sandstones and conglomerates. For example, the Bozhong 25-1 Field in the Bohai basin, offshore China is reservoired in a Mesozoic sandstone buried hill. Perhaps more surprising though are many reservoirs that include fractured and weathered granites and various other igneous rocks including volcanics.

Although granitic rocks are normally considered to represent non-reservoir, vast reserves of oil and gas have been found by drilling deep into granite buried hill fault blocks in SE Asia. Perhaps the most famous at the present time are the fractured granites of the Cuu Long basin, offshore south Vietnam. In this basin, oil fields such as Bach Ho, Rang Dong, Rong and Ruby, together probably contain at least one billion barrels of oil reserves and one trillion cubic feet of associated gas reservoired in fractured and weathered granites and granodiorites.

Hydrocarbon columns of many hundreds of metres are present in these 'basement' rocks, and in the Bach Ho Field there is a column of lacustrine-sourced oil in excess of 1000 m (Areshev *et al.* 1992; Tran Canh *et al.* 1994). The variability in these fractured reservoirs is such that the upper parts of the reservoir may in fact be tight ineffective reservoir whilst better reservoir quality exists deeper within the granites. Many hydrocarbon accumulations may actually have been missed in the past in numerous basins throughout SE Asia because exploration wells stopped drilling as soon as granitic rocks were encountered in the belief that it was economic basement and no hydrocarbons would be reservoired there.

The effective porosity of most fractured igneous reservoirs is extremely low, typically 1–2%. However, similar to the fractured carbonates, they are able to deliver high flow rates when fractures are interconnected creating good permeability. Porosity and permeability of fresh granite may also have been enhanced by weathering or hydrothermal activity. The Bach Ho Field, for example, currently produces oil at around 100 000 barrels per day and associated gas at around $100 \times 10^6 \, ft^3$ per day from fractured, weathered and hydrothermally altered granite. Stabilized oil flow rates of 5000 barrels per day are common in vertical wells and flow rates of 10 000 barrels of oil per day have been achieved during testing (e.g. Areshev *et al.* 1992).

In addition to fractured granite reservoirs, fractured volcanic rocks can form equally good reservoirs. For example, fractured vesicular andesites, basalts and tuffs form important reservoir targets in the Erlian Basin in north-central China (see Wang Tonghe 1986; Tang Jieting 1988). Volcanic rocks were often extruded whilst the lake basins were developing and can occur in amongst the lacustrine mudstones. These have the advantage of close proximity to lacustrine source-rocks and examples of oil bearing lavas are known from the onshore Thailand basins and the offshore Pearl River Mouth basins (see Chen Sizhong 1985; Chunxiu Wang & Yuxiao Sun 1994). In the Bohai basin, offshore China, the Shijutuo Field produces from fractured Mesozoic volcanics (Hu Jianyi *et al.* 1984).

Elsewhere, even fractured metamorphic rocks have been found to be effective reservoirs for lacustrine sourced oil and gas. For example in the early Tertiary Damintun lake basin in northeast China, a number of fields have been found in fractured Archaen metamorphic rocks that include granulites, amphibolites, migmatites, quartzites and slates. In addition to the main oil field, Dongshengpu, a number of other smaller accumulations have now been established including Jiangbei, Caotai and Biantai (see Tong Xiaoguang & Huang Zuan 1991).

In younger (post-rift) sequences that formed after the lakes were infilled and eliminated, there is scope for a variety of alternative sedimentary reservoirs. Where marine transgression flooded into and eliminated the lake basins, a variety of marine sandstone reservoirs may be present (Fig. 8). For example, these developed after many of the lake basins in the Pearl River Mouth area and Yingehai area were flooded by the sea and then replaced by passive continental margins during the Oligocene and Miocene (e.g. Ke Ru & Pigott 1986; Chen *et al.* 1993). These shallow-marine sandstone reservoirs usually have much better reservoir quality compared to lacustrine reservoirs due to a combination of better sorting, greater mineralogical maturity, less clay matrix, considerable lateral extent (which may be 30 km or more), and relatively shallow burial. Large carbonate reefs also developed in some offshore areas that were starved of clastic input and these have excellent reservoir potential with both very high porosity and permeability. Classic examples are present offshore in the Pearl River Mouth basin and offshore central Vietnam (for details see Chen Sizhong & Hu Pingzhong 1989; Tran Canh *et al.* 1994).

Elsewhere, large-scale tectonic rearrangement of plates created giant rivers that eliminated many lake basins. These giant rivers fed enormous deltas that entered the oceans and then built out across continental margins towards the deep oceans. These systems tracts include a variety of alluvial, deltaic, marine shelf and deep marine reservoirs. This occurred for example in many Tertiary basins along the eastern China coast, in a number of Indonesian basins, the onshore central Thailand basins and the basins offshore southern Vietnam. Modern-day equivalents of these giant rivers are the Mekong River and Red River systems in Vietnam, the Pearl River and Yellow River systems in China, and the Chao Praya river system in the Gulf of Thailand.

Locating lacustrine hydrocarbon source-rocks, assessing their quality and identifying hydrocarbon migration pathways

Controls on source-rock development

Understanding the controls on lacustrine source-rock presence and quality is a critical component in understanding exploration risk. Defining

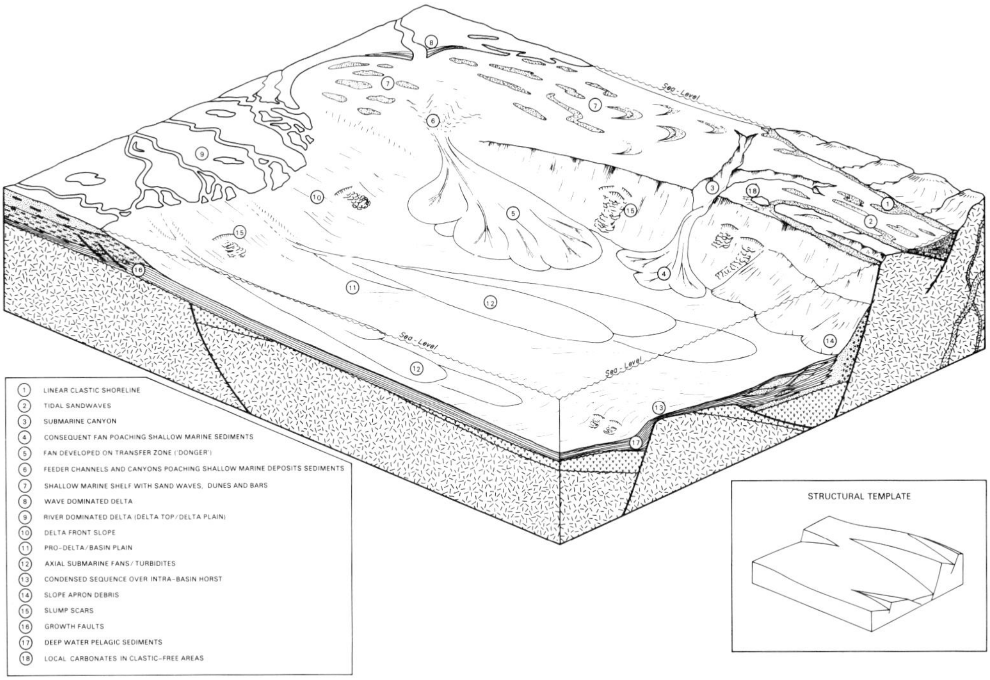

Fig. 8. Reconstruction of a syn-rift marine basin with humid sub-tropical climate.

adequate source-rock presence is a first priority in the frontier basins and underexplored basins in SE Asia. The three major controls on accumulation of a thick lacustrine sequence of mudstones that may be a good hydrocarbon source-rock appear to have been as follows.

Tectonics. Formation of a large, 'closed hole' or depocentre capable of containing a lake is required, accompanied by prolonged periods of subsidence outpacing or equaling sedimentation. Thick sequences of potential lacustrine source-rock commonly develop in these depocentres. In long-lived lake sites, where there is prolonged subsidence, significant thicknesses of fine-grained clastics may accumulate, often 500 m or more (Table 2).

Climate. Subtropical to tropical humid climates and warm temperatures encouraged organic matter to flourish in many lakes in SE Asia; humid climates were common and enabled vegetation to flourish in the hinterland, providing additional detrital organic input. Although humid climates may cause considerable sediment input to the lake basin, they also provide continual recharge of nutrients into the lakes, plentiful water to maintain the lake and help inhibit hypersalinity. Whilst high salinities may actually aid organic preservation due to salinity stratification, the advent of hypersalinity normally causes organic sterility and consequently poor source-rock potential. The lack of seasonal turnover in low latitude lakes no doubt assisted the preservation of organic matter throughout the region. Most lakes in SE Asia appear to have been fresh to mildly alkaline/saline although there are exceptions, for example the early Tertiary hypersaline lakes of the Jianghan basin in China (see Fu Jiamo *et al.* 1988).

Nature of clastic input. Hinterlands conducive to providing mostly silt and mud-grade sediment result in lakes dominated by fine-grained sequences which are more likely to preserve organic matter. Carbonate-rich hinterlands (limestones and dolomites) produce virtually no clastics, and usually liberate only small amounts of clay and silt-grade sediment during weathering. Examples of basins surrounded by widespread carbonate hinterlands during the early Tertiary include areas such as the North China basin onshore China, the South Yellow Sea basins offshore China, the Chumphon basin offshore Thailand, and the Li Basin onshore northern Thailand (see Hong Ye *et al.* 1985; Liu Hefu 1986; Surawit Pradidtan 1989; Yongyuth

Uk-kakimapan 1992). Areas of reduced coarse clastic supply may also develop where basins on the margins of the main rifts trap coarse clastics entering the rift system and prevent them being transported to the main lake depocentres. Early Tertiary lake basins in the Pearl River Mouth basins offshore China exhibit this characteristic (see Chunxiu Wang & Yuxiao Sun 1994), and a similar feature has been observed in the early Tertiary Suphanburi lake basin, onshore Thailand (see Bidston & Daniels 1992). Careful sequence stratigraphy studies and palaeotopographic analysis can help identify these areas.

Identifying source-rock intervals in the subsurface

Lake sediments that might be source-rocks frequently appear on seismic sections with a high degree of continuity and conformity of reflections. Lateral continuity of reflections develops because sedimentation over large parts of many lakes is dominated by rather uniform suspension fallout. Rising lake level is often seen as onlap in all directions as the lake expands and this is more likely to indicate a lake accumulating significant mudstones that may have source-rock potential.

In thick lake mudstone sequences, the frequency of seismic reflections may vary considerably. Monotonous mudstone/siltstone sequences may possess few lithological contrasts and be almost reflection-free on seismic sections. However, there are often changes in organic matter or calcareous content, or the presence of evaporites, biogenic oozes and thin turbidites that cause significant velocity variation, creating laterally extensive seismic reflections of differing amplitudes. These contribute to a distinctive set of 'stripey' seismic facies seen in many of the lake basins (for details and examples see Wang Xie-Pei *et al.* 1985; Flint *et al.* 1988; Burri 1989; Longley *et al.* 1990; Sladen 1994).

Maps of seismic interval velocities can also help to detect, or infer the presence of thick lake mudstones that may be source-rock sequences because the high organic contents create anomalously low velocities and densities. This technique has been proven very effective in the Pearl River Mouth, Bohai and South Yellow Sea basins. Overpressured sequences may also have relatively low seismic interval velocities. Overpressuring is common in very thick lacustrine mudrocks due to very rapid deposition impeding porewater expulsion and due to the onset of hydrocarbon generation. In some lake basins, re-establishing pressure equilibrium can result in the formation of mud diapirs, and in some cases this may be assisted by lacustrine evaporites if also present, e.g. see Wang Xie-Pei *et al.* (1985).

Characteristic features of source-rocks

Lacustrine mudrocks are typically grey, black and brown laminated mudstones with sometimes thin carbonates, interbedded evaporites, incipient siderite cementation and siltstone or very fine sandstone stringers. Mudrock deposition rates can be in excess of 1 mm/year and may reach 5 mm/year. Fossils often include fish, turtles, insects, freshwater diatoms, thin-shelled bivalves, gastropods and ostracodes. Highly fissile mudstones (often with preserved fish) may be present; other mudstones have waxy laminae which may be slightly plastic and flexible. The most oil prone mudstones (oil shales) generally have a dark brown colour and waxy texture.

Organic matter in the source-rocks can be divided into that produced in the lake (most notably algae) and that growing around the lake and in the hinterlands, which was carried into the lakes by rivers and during floods (typically plant debris). Throughout SE Asia, the algal source-rock component is dominated by *Botryoccocus* and *Pediastrum* and these green algae represent the main petroleum generators (e.g. see Robinson 1987; Brassell *et al.* 1988; Williams *et al.* 1992; Sladen 1994).

Palynofacies studies usually show that only one or two algal species flourished in the lakes at any one time. This is a situation which compares closely with the distribution of algae in many present-day lakes. *Botryococcus* develops as colonial planktonic algae and is resistant to a wide variety of environmental conditions, withstanding a pH range of 4.5–8. *Botryococcus* may form such a dense growth over the lake surface that it exhausts the oxygen supply and thick oily gelatinous and highly decay-resistant algal deposits may accumulate.

The abundance of algae in the source-rocks most probably reflects a succession of algal blooms over a number of years. As in modern lakes, algal blooms were probably initiated in a variety of ways depending on ecological and climatic circumstances. For instance, nutrient enrichment resulting from fluvial input after recent rain, alterations in water temperature, and seasonal changes causing seed populations to be activated.

In shallow lake and lake margin environments, source-rocks contain less algae but include more spores and pollen. Aquatic ferns, bullrushes and charophytes are frequently present together with wood and leaf debris which have often been reworked by bacteria. Pristane/phytane ratios of these facies tend to be higher than for algal source-rocks, reflecting the different mixture of organic matter and the greater oxidation of organic matter typical of shallow lake and lake margin environments. In more arid regions, shallow lakes were often hypersaline playas. These sequences may include reddened mudstones developed on exposed mudflats, evaporites and stagnant ponds with high algal productivity and salinity stratification aiding organic preservation. In contrast to humid areas, detrital organic input was minor comprising only scrub vegetation and dwarf shrub debris.

Lacustrine source-rocks are normally strongly oil-prone. The percentage of kerogen that is oil prone is typically 50–100%. Geochemical analyses and kinetic modelling of the source-rocks indicates that during burial and maturation, oil is the predominant phase expelled. The product is typically an oil with a low gas-oil ratio. In contrast, the other main source-rock in SE Asia comprising coals and coaly mudrocks generates mostly gas, although there is some oil potential (see various papers in this volume).

Given suitable subsidence and sedimentation rates, lake basins in SE Asia often accumulated potential source-rocks that may be many hundreds, or even thousands, of metres thick (Table 2). Analyses frequently show that much of this is of rather mediocre quality, interspersed with organically rich horizons. These variations can occur both on a millimetre scale, perhaps reflecting seasonal climatic variations (varves), and on a much larger scale, for example the development of oil-shale sequences perhaps 1–50 m thick and reflecting pulses of increased faulting and subsidence, reduced clastic input and high algal productivity.

Geochemical analysis of thick mudstones often identifies considerable thicknesses with mediocre quality. These typically have total organic carbon (TOC) contents of 0.5–1.5% and production capabilities (S2 values) of 3–10 $kg\,t^{-1}$. The low organic content is often due to dilution of organic matter by the high clastic sedimentation rates common in lakes. In contrast, oil shale sequences have considerably higher TOC contents, typically *c*. 20%, and production capabilities of up to 100 $kg\,t^{-1}$ or more. They usually have abundant algal organic matter and are so rich that they are often combustible. The density of these oil shales is very low due to the high organic matter content and this creates a distinctive response on wireline density and velocity logs. Gas chromatogram analyses and hydrogen indices (typically 400–900) indicate a clear potential to generate oil. Pristane/phytane ratios are normally very low (typically <2.5) but higher values may occur where there is more land plant kerogen.

Migration pathways

It is clear from drilling results that long-distance migration to traps more than 20 km from the lake source-rock depocentres is rare for most basins in SE Asia, and even less likely where the depocentre is significantly overpressured which tends to result in mostly short-distance vertical migration. Gas, being more mobile in the subsurface due to lower viscosity and higher buoyancy tends to migrate further than oil. Thus gas and lighter oils may be found trapped further from the lacustrine source-rock depocentres.

Given that the organic matter in the source-rocks is often diluted and spread through many hundreds of metres of section and there are also areal variations in organic matter content and type, the timing at which source-rocks become mature is frequently complex. Migration directions may shift significantly with time as these thick, mediocre-quality source intervals become more deeply buried and pass through different stages of maturity. For example, see Yang Wanli *et al.* (1985) for the sorts of complexity that occur in and around the Daqing Field in China, and Bal *et al.* (1992) for the complexity in and around the Sirikit Field in Thailand. Rarely is there a single short burst of hydrocarbon generation and expulsion which can be easily modelled and predicted.

In a simple lacustrine half-graben, the source-rocks usually occur toward the basin axis depocentre. The asymmetry of the basin results in the bulk of hydrocarbons first filling axial traps often involving turbidites and the distal parts of deltas, and then migrating up-dip toward the hanging wall (Fig. 9). Migration distances toward the footwall side of the basin from nearby 'deep lake' source kitchens can be short when source-rocks are developed in the adjacent basin axis. However, the asymmetry of most extensional basins causes relatively little of the hydrocarbon charge to move toward the footwall. There is often further complication on the footwall side because the drainage area of mature source-rocks is usually small, and can be almost negligible if sand-rich fans are well-developed and occupy most of the potential drainage area (consider Figs 3 & 9).

In the lake depocentres, source-rocks are more likely to become mature due to the greater subsidence and burial occurring in the axis. In

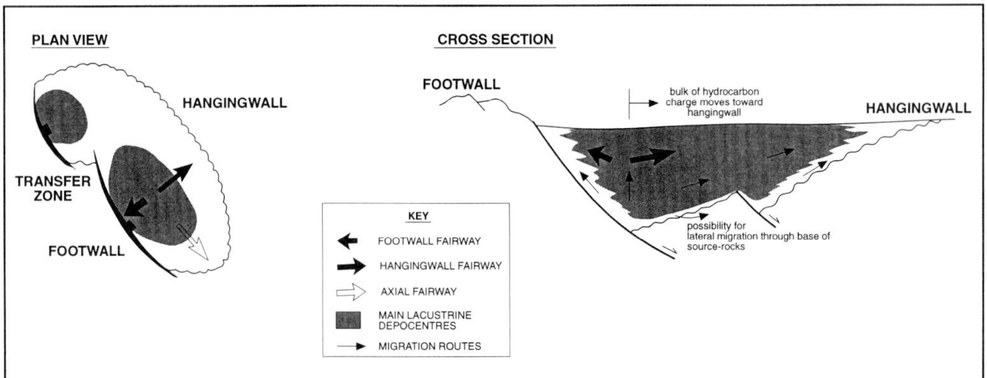

FAIRWAY	COMMON ADVANTAGES	COMMON DISADVANTAGES
FOOTWALL	• Short migration distances from nearby source kitchens • Potential for very thick, high net-to-gross fan reservoirs	• Difficulty in sealing fan and fault scarp reservoirs • Drainage area of mature source kitchen very small
HANGINGWALL	• Usually the preferred migration direction for hydrocarbons • Large number of synthetic or antithetic fault traps • Strongly interbedded reservoirs and seals helps trapping • Good opportunitues for stratigraphic trapping	• Migration usually poorly focussed toward traps • Ofen long distance migration with high migration losses • Laterally extensive, thick seals usually absent • Reservoirs rarely thick or good quality
AXIAL	• Thick sequences of mature source-rocks more likely • Migration distance from mature source to trap relatively small • Seals usually thickest, most uniform and laterally extensive • Source, reservoir and seal can be interbedded	• Structural traps usually small and scarce • Reservoirs often absent in basin centre • Reservoirs if present, typically thin and poor quality • Complex petroleum geology due to overpressure

Fig. 9. Summary of the main advantages and disadvantages of the fairways commonly developed in extensional lake basins.

the basin axis, the distances from hydrocarbon source 'kitchens' to lacustrine reservoirs such as turbidites are usually very short and there are few migration problems because source, reservoir and seal are interbedded. Migration losses are likely to be relatively low (i.e. high 'migration efficiency').

As mentioned earlier, a relatively large hydrocarbon charge is usually directed towards the hangingwall in extensional basins. However, the distance laterally between traps and source kitchens is relatively large and the gentle dips are not favourable toward focusing of hydrocarbons. Migration pathways therefore tend to be broad and losses of hydrocarbons during migration are often high (Fig. 9). Seal potential increases basinward as the facies become finer-grained but no regional seals are likely because of the strong interbedding of lithologies and facies; this adds further complexity to migration pathways.

Lateral migration of hydrocarbons through the base of source-rock sequences that developed in the depocentres is a critical phenomena for the success of the 'buried hill' traps where reservoir occurs in the pre-rift section. This style of migration is usually assisted by the presence of the early rift fill (basal) sequence which normally includes coarse-grained and permeable facies. After migrating out of the base of the source-rock sequences, hydrocarbons are then able to move updip through the early rift-fill and into pre-rift reservoirs, while the lacustrine mudstones (including the source-rocks) can also form a top seal to the reservoirs (Figs 3 & 9).

Optimal trap locations for hydrocarbon charge

Traps immediately adjacent to, or above the lacustrine source-rocks are considered to be optimally placed for hydrocarbon charge. Lowest risk is attached to four-way dip closed traps close to these depocentres and risk increases with lateral distance away from these depocentres.

Long-range up-dip lateral migration of petroleum from lacustrine depocentres to the large footwall highs and fault blocks that surrounded the lake basins has frequently been thought by many exploration companies to be an ideal exploration play concept. The large size of these structures has lured many exploration companies to drill these features in SE Asia. However, drilling results indicate this to be a largely mistaken concept with relatively high risk. Lateral hydrocarbon migration is frequently limited and poorly focused towards these distant footwall structures and instead hydrocarbons tend to focus on transfer zones or structural highs within the lake basin or move toward nearby parts of the hanging-wall side of the basin. In addition, connectivity and lateral extent of sandbodies in the lacustrine and lake margin alluvial facies is usually low and consequently regional lacustrine carrier beds for hydrocarbons for distances of >15 km are rare.

Hydrocarbons as a general rule therefore appear to have difficulty moving away from the immediate vicinity of the lake basin source-rocks once they become mature. Various authors have presented case studies that suggest lateral migration of lacustrine hydrocarbons is restricted to less than 20 to 30 km from the mature lacustrine source (e.g. Wang Shangwen *et al.* 1982; Hu Chaoyuan 1987; Derksen & McLean-Hodgson 1988; Niwat Chinbunchorn *et al.* 1989). But there are always exceptions to the rule, and one case appears to be the lacustrine sourced oil found in the Liuhua Field in the Pearl River Mouth basin. Here, once oil could escape vertically from the lake basin to the north, it was able to travel laterally southward for perhaps 40–60 km (Fig. 2). This appears to have been made possible because of a laterally extensive, highly porous and permeable shallow marine shelf sandstone capped by a regional seal of marine mudstone, both of these deposited after the lake basin had been infilled and flooded by the sea.

Lateral and vertical variations in oil gravity, wax content and gas–oil ratio

Lacustrine-derived hydrocarbons frequently have a very distinctive set of characteristics originating from the types of organic matter that accumulated in the lakes. Oils are typically low to medium gravity (20–40° API).

Gas–oil ratios (GORs) for lacustrine oils throughout SE Asia are usually low. Each barrel of oil typically contains less than 1000 scf of associated gas, and lower GORs of 100–500 scf/bbl are commonplace (Table 1). The primary reason for limited gas being present is the overwhelming oil-prone nature of the lacustrine source-rocks. In many instances, the oilfields do not have any gas cap present. In addition, many of the lacustrine source-rock sequences have never been sufficiently deeply buried to reach temperatures more likely to cause significant gas generation, or for oil to be cracked to gas. The source-rocks are frequently only marginally mature to mature for oil generation.

Changes in oil saturation, wax content, oil gravity and pour point may be apparent within oilfields; in many multilayered reservoirs there may be downward increases in GOR, together with lighter oils, related to increasing temperatures and pressures causing more gas to be dissolved in the oil. Laterally, toward the margins of large fields there may be decreases in oil saturation and wax content and lower viscosities and pour points (for example, see Yang Wanli 1985). During production, as reservoir pressure is depleted, complex secondary gas caps may form in the surface, and this is compounded by the inherently complex lacustrine reservoirs.

A typical characteristic of lacustrine oils is low sulphur contents, and low hydrogen sulfide contents in gases. This results from the low sulphate content of lake waters and lake sediments in which the source-rocks formed. Anoxic lacustrine shales are often rich in siderite ($FeCO_3$) whilst pyrite (FeS_2) content is very low (see Davison 1988 for details of the geochemical reactions that occur). Generally sulphur in lacustrine oils is <0.2 wt%. but there are a few instances of sulphur-rich oils where basins contained hypersaline lakes, the best known being oils from the evaporite-rich Jianghan basin in China where up to 12% sulphur occurs (see Powell 1986).

Finding a good seal (caprock)

Continuity and thickness

Potential seals in the form of mudstones generally improve in thickness and lateral continuity toward the lacustrine depocentres, so it again becomes crucial to undertake detailed palaeotopographic analysis to identify areas of good seal. Because the lake basin axis or depocentre attracts fine-grained sedimentation, caprocks in this area are usually thickest, more uniform and most widespread. The lacustrine seals thin and interdigitate towards the lake margins (Figs 3, 4 & 9).

The thick mudstone sequences that often contain hydrocarbon source-rocks are usually also the best seals and many of the principles and assumptions used in identifying and locating source-rocks can also be applied to identifying and locating seals. Seismic techniques include studying reflection continuity and conformity, seismic amplitudes, and interval velocities.

The presence of widespread thick lake mudstones frequently affects geothermal gradients because the mudstones have good insulation properties. This has the effect of a blanket, making it difficult for heat to escape from the basin. Geothermal gradients often show a marked increase when penetrating thick widespread mudrocks that form a good regional lacustrine seal (blanket).

Seals formed in the lake margin environments are particularly difficult to identify and quantify. Lateral continuity and thickness are usually very poor in these settings. Small lake level fluctuations move facies belts large distances laterally, and in the transition zone between rivers and deltas that enter the lakes, distributary channels destroy lateral continuity. Many sequences in this setting are silt dominated and form 'waste rock' which is neither good seal, nor good reservoir, and possesses poor source-rock potential. Seals tend to have only local sealing capability and create discontinuous permeability barriers. Mapping onlap surfaces that may indicate rising lake level and reflect a period of more widespread lake deposition are critical to identifying the better seals in these lake margin environments.

Problems with fans

Traps involving the proximal parts of alluvial or lacustrine fans frequently fail due to a lack of effective seal deposited on top of them. The seal is typically poor because the proximal parts of the fans lie close to the basin edges and coarse grained facies tend to locally persist even when there is a significant lake level rise that is sufficient to drown the fan. Additionally, cross-fault seal is required and this often means seal against pre-rift 'basement' formations (Figs 4 & 7). These formations may be fractured and weathered creating poor lateral seal.

It is rare in lake basins to drown the entire proximal parts of a fan system and cover the fan apex up to the fault scarps and uplands so as to drown the coarse sediment entry points from which the fans emerged. Seals, if deposited in these areas, are likely to be extremely thin and be easily broken or disrupted by basin margin fault movements. Thin seals may also be cut-out in places by remnants of active fan channels. Good seals can develop across the fans but it typically requires a dramatic lake level rise combined with a basinwide reorganization of the drainage systems to enable the fans to be choked, drowned and finally covered by thick lake mudstones (Fig. 7).

Traps involving the distal portions of lacustrine fans struggle to succeed because the fans lack any significant seal in an up-dip direction from where the fans built out into the lake. However, continued fault movement and growth frequently cause post-depositional rotation of the distal parts of the fans developed along the faulted basin margins. This rotation can create traps that comprise fan sandstones which pinch out basinward into sealing lake mudstones (Fig. 7). Locating this pinchout edge in the subsurface is particularly difficult without well calibrated, good quality, seismic data, and ultimately it may be necessary to drill many wells to accurately image the reservoir.

Problems with faults

Problems with faults disrupting seal continuity, and the difficulty of creating effective fault traps, are common on both faulted footwall and hanging-wall margins of extensional lake basins throughout SE Asia (see also p. 58 and Fig. 6). Lack of basinward counterdip (dip-reversal) across synthetic fault blocks is a common problem toward the main footwall side of half-grabens, whilst toward the hanging-wall side of half graben, antithetic fault blocks usually step down toward the basin centre without any counterdips. These 'domino' fault configurations are not conducive to creating large or easily sealed traps, especially in sand-rich sequences. For this reason, they are often known as 'deadly dominos' because of the high risk of failure to trap hydrocarbons.

Many faults in the lake basin sequences of SE Asia appear at best to be geologically temporary seals. The high frequency of present-day earthquakes in SE Asia that are due to extension and strike-slip tectonics is in fact a good reminder of the constant jostling of fault blocks and the difficulty of maintaining the integrity of fault traps for long periods. Smearing of clays along fault planes may help to increase the volume of trapped hydrocarbons in complex fault structures, for example in the Sirikit Field in Thailand (see Bal *et al.* 1992). Although clay-smear seals are genetically more likely to be associated with the larger faults, much ultimately relies on the juxtaposition of cross-fault sealing lithologies. Not surprisingly this is overall more effective in the lower net-to-gross sequences to be found more basinward in the lakes further away from the alluvial systems and deltas that drained into the lakes and basinward of the lake margin setting.

To find thicker and more laterally continuous seals capable of sealing faulted traps, it is therefore desirable to move toward the lake centre or basin axis (Figs 3 & 9). Here, it is common to find more effective fault traps and lower risk targets which comprise the old eroded fault blocks (buried hills) which have then been drowned and encased in thick lacustrine mudstones. The mudstones which encase the buried hills form not only thick and laterally extensive seals, but also usually include the lacustrine source-rocks. Migration distances for hydrocarbons are thus short, abundant mudstone helps focus hydrocarbons towards the structures thereby reducing migration losses, and opportunities for reservoir exist in both the early basin infill and in the pre-rift formations in the buried hills.

Concluding remarks

Common themes

The late Mesozoic and early Tertiary lake basins of SE Asia all owe their origin to extensional tectonics. They extend from Mongolia, China and Korea in the north to Vietnam, Cambodia, Thailand, Malaysia and Indonesia in the south. Lake basin evolution was frequently characterized by prolonged phases of subsidence outpacing sedimentation enabling a lake to exist, a suitable sub-tropical to tropical climate enabling organic matter to flourish, and a widespread carbonate-rich hinterland which caused a reduced coarse clastic input into the lakes.

Hydrocarbons that have been generated from lacustrine mudstone source-rocks possess a distinct set of characteristics. Oils are characterized by low to medium gravity (20–40° API), low gas to oil ratios (frequently <500 scf/bbl), high wax content (often 10–30%), high temperature pour points (20–30°C or more) and very low sulphur contents (<0.2 wt%). Due to the inherent complexity of lacustrine reservoirs, and the oil properties, oil flow rates from wells are typically measured in hundreds of barrels per day, as opposed to thousands, and recovery factors are low. To find better quality reservoir it is important to look for reservoirs in older (pre-rift), and younger (post-rift) sequences.

The success of the 'buried hill' where reservoir occurs in a topographic remnant of older pre-rift sequences, and the buried hill is onlapped and covered by lacustrine mudrocks has been a key factor in development of the hydrocarbon industry in many basins. Important pre-rift reservoirs include fractured and karstified carbonates, and fractured, weathered and hydrothermally altered granites.

Numerous structural and stratigraphic trap configurations have proven to be successful in the lake basins. Four-way dip closed traps are preferable to fault traps which have difficulty sealing large hydrocarbon volumes. Proximity to mature source-rocks is essential as it is clear that oil and gas generated from lacustrine source-rocks has poor migration efficiency. Hydrocarbons tend, as a general rule, to only migrate short distances laterally, typically 20–30 km.

Future trends

Interest in the late Mesozoic and early Tertiary lake basins of SE Asia is destined to continue for many decades given their importance in generating and reservoiring enormous volumes of good quality lacustrine-derived oil and gas, together with the presence in many basins of coals and oil shales which can be mined as energy sources. As exploration for oil and gas continues, and fewer large fields remain to be found, it will become increasingly vital to focus expenditure on low risk, small volume opportunities in a manner that generates a return on investments.

An essential part of any future focusing of exploration will be an increasingly accurate risk and reward assessment involving a detailed understanding of the lake basins and their lacustrine sequences. This will need to include identifying the low risk traps, understanding and quantifying reservoir complexity and well deliverability, assessing seal qualities, reliably predicting the lacustrine source-rock types and hydrocarbon phase, and modelling in detail the migration pathways. Building sophisticated reservoir models together with identifying and applying innovative technology will be essential to cost efficient production and effective development of the inevitably complex lacustrine reservoirs. Good quality datasets, accurately prepared, carefully kept and painstakingly evaluated, will be essential parts of this process. These are likely to be key challenges for the oil and gas finders and producers of the future; hopefully this paper has shared some useful thoughts and ideas with which to face the challenges.

Considerable assistance leading to the preparation of this paper has been provided throughout my career by BP Exploration. However, views expressed in this paper represent those of the author, not necessarily his employer. Mention must go to Vu Bich Hang and Nguyen Thi Bang Tam in Ho Chi Minh City for their tireless assistance with the manuscript and diagrams.

References

ARESHEV, E. G., TRAN LE DONG, NGO THUONG SAN & SHNIP, O. A. 1992. Reservoirs in fractured basement on the continental shelf of southern Vietnam. *Journal of Petroleum Geology*, **15**, 451–464.

ATKINSON, C. M. 1989. Coal and oil shale in Tertiary intermontane basins of Indonesia and eastern Australia. *International Symposium on Intermontane Basins: Geology and Resources, Chiang Mai University, Thailand. Proceedings*, 77–78.

BAL, A. A., BURGISSER, H. M., HARRIS, D. K., HERBER, M. A., RIGBY, S. M., THUMPRASERTWONG, S. & WINKLER, F. J. 1992. The Tertiary Phitsanulok lacustrine basin, Thailand. *National Conference on 'Geologic resources of Thailand: potential for future development', 17–24 November, Bangkok, Thailand. Proceedings*, 247–258.

BARRON, E. J. 1990. Climate and lacustrine source prediction. *In*: KATZ, B. J. (ed.) *Lacustrine Basin Exploration – Case Studies and Modern Analogs*. American Association of Petroleum Geologists Memoir, **50**, 1–18.

BIDSTON, B. J. & DANIELS, J. S. 1992. Oil from the ancient lakes of Thailand. *National Conference on 'Geologic resources of Thailand: Potential for future development' 17–24 November, Bangkok, Thailand. Proceedings*, 584–599.

BRASSELL, S. C., SHENG GUOYING, FU JIAMO & EGLINTON, G. 1988. Biological markers in lacustrine Chinese oil shales. *In*: FLEET, A. J., KELTS, K. & TALBOT, M. R. (eds) *Lacustrine Petroleum Source Rocks*. Geological Society, London, Special Publications, **40**, 299–380.

BURRI, P. 1989. Hydrocarbon potential of Tertiary intermontane basins in Thailand. *International Symposium on Intermontane Basins: Geology and Resources, Chiang Mai University, Thailand. Proceedings*, 3–12.

CAO ZHENGYAN 1984. Oil shale in China. *Oil & Enterprise*, **10**, 44–45.

CHEN CHANGMING, HUANG JIAKUAN, CHEN JINGSHAN & TIAN XINGYOU 1984. Depositional models of Tertiary rift basins, eastern China, and their application to petroleum prediction. *Sedimentary Geology*, **40**, 73–88.

CHEN SIZHONG 1985. There will be more large discoveries in the eastern part of the Pearl River Mouth Basin. *China Oil*, Autumn, 17–19.

—— & HU PINGZHONG 1984. Tertiary reef complexes in the Zhujiangkou (Pearl River Mouth) Basin and their significance for hydrocarbon exploration. *China Earth Sciences*, **1**, 21–30.

CHEN, P. P. H., ZHI YUONG CHEN & QI MIN ZHANG 1993. Sequence stratigraphy and continental margin development of the northwestern shelf of the South China Sea. *Bulletin of American Association of Petroleum Geologists*, **77**, 842–862.

CHUNXIU WANG & YUXIAO SUN 1994. Development of Paleogene depressions and deposition of lacustrine source rocks in the Pearl River Mouth Basin, northern margin of the South China Sea. *Bulletin of American Association of Petroleum Geologists*, **78**, 1711–1728.

DALY, M. C., COOPER, M. A., WILSON, I., SMITH, D. G. & HOOPER, B. G. D. 1991. Cenozoic plate tectonics and basin evolution in Indonesia. *Marine and Petroleum Geology*, **8**, 2–21.

DAVISON, W. 1988. Interactions of iron, carbon and sulphur in marine and lacustrine sediments. *In*: FLEET, A. J., KELTS, K. & TALBOT, M. R. (eds) *Lacustrine Petroleum Source Rocks*. Geological Society, London, Special Publications, **40**, 131–137.

DERKSEN, S. J. & MCLEAN-HODGSON, J. 1988. Hydrocarbon potential and structural style of continental rifts: examples from East Africa and Southeast Asia. *7th Offshore SE Asia Conference, Singapore, Proceedings*, 120–134.

FAN LANG 1986. The broad perspective of oil-bearing shale industries as viewed from Maoming. *China Oil*, Autumn, 57–59.

FEI QI & WANG XIE PEI 1984. Significant role of structural features in Renqui buried-hill oil field in eastern China. *Bulletin of American Association of Petroleum Geologists*, **68**, 971–982.

FLINT, S., STEWART, D. J., HYDE, T., GEVERS, E. C. A., DUBRULE, O. R. F. & VAN RIESSEN, E. D. 1988. Aspects of reservoir geology and production behaviour of Sirikit oil field, Thailand: an integrated study using well and 3D Seismic data. *Bulletin of American Association of Petroleum Geologists*, **72**, 1254–1269.

FU JIAMO, SHENG GUOYING & LIU DEHAN 1988. Organic geochemical characteristics of major types of terrestrial petroleum source rocks in China. *In*: FLEET, A. J., KELTS, K. & TALBOT, M. R. (eds) *Lacustrine Petroleum Source Rocks*. Geological Society, London, Special Publications, **40**, 279–289.

GIBLING, M. R. 1988. Cenozoic lacustrine basins of southeast Asia, their tectonic setting, depositional environment and hydrocarbon potential. *In*: FLEET, A. J., KELTS, K. & TALBOT, M. R. (eds) *Lacustrine Petroleum Source Rocks*. Geological Society, London, Special Publications, **40**, 341–351.

——, YONGYUT UKAKIMAPHAN & SUTHEP SRISUK 1985. Oil shale and coal in intermontane basins of Thailand. *Bulletin of American Association of Petroleum Geologists*, **69**, 760–766.

GIERLOWSKI-KORDESCH, E. & KELTS, K. 1994. Introduction. *In*: GIERLOWSKI-KORDESCH, E. & KELTS, K. (eds) *Global geological record of lake basins*, **1**. Cambridge University Press, xvii–xxxiii.

GORST, I. 1986. Onshore key to increased output. *Petroleum Economist*, December, 451–457.

HONG YE, SHEDLOCK, K. M., HELLINGER, S. J. & SCLATER, J. G. 1985. The North China Basin: an example of a Cenozoic rifted intraplate basin. *Tectonics*, **4**, 153–169.

HU CHAOYUAN 1989. Oil-source control on oil/gas field distribution: an example from the Jianghan Basin, Hubei Province and its application to Chinese offshore basins. *China Earth Sciences*, **1**, 87–92.

HU JIANYI, FAN CHENGLONG, ZHANG JINQUAN, LIU SHUXUAN, XU SHUBAO & TONG XIAOGANG 1984. Stratigraphic–lithologic oil and gas pools in continental basins, China. *International Petroleum Geology Symposium, Sept. 1984, Beijing, China. Proceedings.*

JIN YUSUN, LIU DINGZENG & LUO CHANGYAN 1985. Development of Daqing oil field by waterflooding. *Journal of Petroleum Technology*, **37**, 269–275.

KATZ, B. J. 1990. Controls on distribution of lacustrine source rocks through time and space. *In*: KATZ, B. J. (ed.) *Lacustrine Basin Exploration –Case Studies and Modern Analogs*. American Association of Petroleum Geologists Memoirs, **50**,' 61–76.

——1991. Controls on lacustrine source rock development: a model for Indonesia. *Proceedings Indonesian Petroleum Association, 20th Convention*, 587–619.

KE RU & PIGOTT, J. D. 1986. Episodic rifting and subsidence in the South China Sea. *Bulletin of American Association of Petroleum Geologists*, **70**, 1136–1155.

LAMBAISE, J. J. 1990. A model for tectonic control of lacustrine stratigraphic sequences in continental rift basins. *In*: KATZ, B. J. (ed.) *Lacustrine Basin Exploration – Case Studies and Modern Analogs*. American Association of Petroleum Geologists Memoirs, **50**, 256–286.

LAO QUIYUAN & GAO WENXUE 1984. The characteristics of Cenozoic sedimentary basins in the North China platform. *Sedimentary Geology*, **40**, 89–103.

LI CHUNJU, ZHU SHUIAN & ZHU SHAOBI. 1984. Sedimentary system and oil-bearing features of the small intermontane rift basin of Biyang. *International Petroleum Geology Symposium September 1984, Beijing, China, Proceedings.*

LI DESHENG 1981. Geological structure and hydrocarbon occurrence of the Bohai Gulf oil and gas basin (China). *In*: MASON, J. F. (ed.) *Petroleum Geology of China*. Pennwell Books, Tulsa, Oklahoma, 180–192.

——1984. Geological evolution of petroliferous basins on continental shelf of China. *Bulletin of American Association of Petroleum Geologists*, **68**, 993–1003.

——1990. Recent advances in the petroleum geology of China. *Journal of Petroleum Geology*, **13**, 7–18.

—— & LOU MING 1990. Hydrocarbon accumulation in Meso-Cenozoic lacustrine remnant petroliferous depressions and basins, southeastern China *In*: KATZ, B. J. (ed.) *Lacustrine Basin Exploration – Case Studies and Modern Analogs*. American Association of Petroleum Geologists Memoirs, **50**, 327–333.

LI SHAOGUANG & WANG DEXING 1985. Relationship between diagenesis and porosity of sandstones on the southern and northern flanks of Bei Dagang structural zone (in Chinese). *Acta Petrolei Sinica*, **3**, 13–21.

LI SITIAN, LI BAOFANG, YANG SHIGONG, HUANG JIAFU & LI ZHEN 1984. Sedimentation and tectonic evolution of Late Mesozoic faulted coal basins in north-eastern China. *Special Publications of the International Association Sedimentologists*, **7**, 387–406.

LIU HEFU 1986. Geodynamic scenario and structural styles of Mesozoic and Cenozoic basins in China. *Bulletin of American Association of Petroleum Geologists*, **70**, 377–395.

LLOYD, A. R. 1978. Geological evolution of the South China Sea. *SEAPEX Proceedings*, **4**, 95–137.

LONGLEY, I. M., BARRACLOUGH, R., BRIDDEN, M. A. & BROWN, S. 1990. Pemetang lacustrine petroleum source rocks from the Malacca Strait PSC, Central Sumatra, Indonesia. *Proceeding of the Indonesian Petroleum Association 19th Convention*, 279–297.

MA LI 1985. Subtle oil pools in Xingshugang delta, Songliao Basin. *Bulletin of American Association of Petroleum Geologists*, **69**, 1123–1132.

——, GE TAISHENG, ZHAO XUEPING, ZIE TAIJUN, GE RONG & DANG ZHENRONG 1982. Oil basins and subtle traps in the eastern part of China. *In*: HALBOUTY, M. T. (ed.) *The Deliberate Search for the Subtle Trap*. American Association of Petroleum Geologists Memoirs, **32**, 287–315.

MA XINGYUAN & WU DANING 1987. Cenozoic extensional tectonics in China. *Tectonophysics*, **133**, 243–255.

NIWAT CHINBUCHORN, SURAWIT PRADIDTAN & NARES SATTAYARAK 1989. Petroleum potential of Tertiary intermontane basins in Thailand. *International Symposium on Intermontane Basins: Geology and Resources, Chiang Mai University, Thailand. Proceedings*, 29–42.

O'LEARY, J. & HILL, G. S. 1989. Tertiary basin development in the southern central plains, Thailand. *International Sympsium on Intermontane Basins: Geology and Resources, Chiang Mai University, Thailand. Proceedings*, 254–264.

PENTILLA, W. C. 1993. The recoverable oil and gas resources of Mongolia. *Journal of Petroleum Geology*, **17**, 89–98.

POWELL, T. G. 1986. Petroleum geochemistry and depositional setting of lacustrine source rocks. *Marine and Petroleum Geology*, **3**, 200–219.

ROBINSON, K. M. 1987. An overview of source rocks and oils in Indonesia. *Proceedings of the Indonesian Petroleum Association, 16th Convention*, 97–122.

SAHAT MUENLEK 1992. Coal geology of Mae Than basin Amphoe Mae Tha, Lampang. *National Conference on 'Geologic resources of Thailand: potential for future development', 17–24 November, Bangkok, Thailand, Proceedings*, 112–121.

SANTICHAI JITAPUNKUL 1992. Geology of Ban Pa Kha coal deposit, Li basin, Lamphun. *National Conference on 'Geologic resources of Thailand: potential for future development', 17–24 November, Bangkok, Thailand, Proceedings*, 259–272.

SARMIENTO, R. & KIRBY, R. A. 1962. Recent sediments of Lake Maracaibo. *Journal of Sedimentary Petrology*, **32**, 698–724.

SLADEN, C. P. 1994. Key elements during the search for hydrocarbons in lake systems. *In*: GIERLOWSKI-KORDESCH, E. & KELTS, K. (eds) *Global geological record of lake basins*, **1**. Cambridge University Press, 3–17.

STOKES, R. B. 1988. Structural control of Neogene sedimentation in the Mae Sot basin (Thai-Burmese border): implications for oil-shale reserves. *Journal of Petroleum Geology*, **11**, 341–346.

SURAPOL THANOMSAP & SUPARERK SITAHIRUN 1992. The Mae Sot oil shale. *National Conference on 'Geologic resources of Thailand: potential for future development', 17–24 November, Bangkok, Thailand, Proceedings*, 676–691.

SURAWIT PRADIDTAN 1989. Characteristics and controls of lacustrine deposits of some of some Tertiary basins in Thailand. *International Symposium on Intermontane Basins: Geology and Resources, Chiang Mai University, Thailand, Proceedings*, 133–145.

—— & DOOK, R. 1992. Petroleum geology of the northern part of the Gulf of Thailand. *National Conference on 'Geologic resources of Thailand: potential for future development', 17–24 November, Bangkok, Thailand, Proceedings*, 235–246.

TANG JIETING 1988. Characteristics of andesite reservoir in north Alsen, Eren Basin (in Chinese, English abstract). *Oil & Gas Geology*, **9**, 389–400.

TIAN ZAIYI 1990. The formation and distribution of Mesozoic–Cenozoic sedimentary basins in China. *Journal of Petroleum Geology*, 13 19–34.

TONG XIAOGUANG & HUANG ZUAN 1991. Buried-hill discoveries of the Damintum depression in North China. *Bulletin of American Association of Petroleum Geologists*, **75**, 780–794.

TRAN CANH, DO VAN HA, CARSTENS, H. & BERSTAD, S. 1994. Vietnam – attractive plays in a new geological province. *Oil and Gas Journal*, March 14, 78–83.

TRAYNOR, J. J. & SLADEN, C. P. 1995. Tectonic and stratigraphic evolution of the Mongolian Peoples Republic and its influence on hydrocarbon geology and potential. *Marine and Petroleum Geology*, **12**, 35–52.

——1997. Seepages in Vietnam – onshore and offshore examples. *Marine and Petroleum Geology*, **4**, 345–361.

WANG SHANGWEN, HU WENHAI & TAN SHIDIAN Habitat of oil and gas fields in China. *Oil & Gas Journal*, June 14, 119–128.

WANG TONGHE 1986. A preliminary study on the Erlian Basin characteristics for structural petroleum geology (in Chinese, English abstract). *Experimental Petroleum Geology*, **8**, 313–324.

WANG XIE-PEI, FEI QI & ZHANG JIA-HUA 1985. Cenozoic diapiric traps in Eastern China. *Bulletin of American Association of Petroleum Geologists*, **69**, 2098–2109.

WATSON, M. P., HAYWARD, A. B., PARKINSON, D. N. & ZHANG, Z. M. 1987. Plate tectonic history, basin development and petroleum source rock deposition onshore China. *Marine and Petroleum Geology*, **4**, 205–225.

WILLIAMS, H. H., FOWLER, M. & EUBANK, R. T. 1992. Geochemical characteristics of Palaeogene and Cretaceous hydrocarbon source basins of southeast Asia. *9th Offshore SE Asia Conference, Singapore, Proceedings*, 429–459.

XU SHICE & WANG HENGJIAN 1981. Deltaic deposits of a large lake basin. *In*: MASON, J. F. (ed.) *Petroleum Geology in China*. Pennwell Books, Tulsa, Oklahoma, 202–213.

YAN DUNSHI & ZHAI GUANGMING 1981. Exploration practice in and prospects of the buried-hill oil fields in North China. *In*: MASON, J. F. (ed.) *Petroleum Geology in China*. Pennwell Books, Tulsa, Oklahoma, 92–100.

YANG WANLI 1985. Daqing oil field, People's Republic of China: a giant field with oil of non-marine origin. *Bulletin of American Association of Petroleum Geologists*, **69**, 1101–1111.

——, LI YONGKANG & GAO RUIQI 1985. Formation and evolution of nonmarine petroleum in Songliao Basin, China. *Bulletin of American Association of Petroleum Geologists*, **69**, 1112–1122.

YINAN QUI, PEIHUA XUE & JINGSUI XIAO 1987. Fluvial sandstone bodies as hydrocarbon reservoirs in lake basins. *In*: ETHRIDGE, F. G., FLORES, R. M. & HARVEY, M. D. (eds) *Recent Developments in Fluvial Sedimentology*. Society of Economic Paleontologists and Mineralogists, Special Publication, **39**, 329–342.

YONGYUTH UK KAKIMAPAN 1992. Geological variation of Ban Pu, Bang Hong and Mae Long coal deposits, Lam Phun. *National Conference on 'Geologic resources of Thailand: potential for future development', 17–24 November, Bangkok, Thailand, Proceedings*, 574–583.

ZHAI GUANGMING & ZHA QUANHENG 1982. Buried-hill oil and gas pools in the North China Basin. *In*: HALBOUTY, M. T. (ed.) *The Deliberate Search for the Subtle Trap*. American Association of Petroleum Geologists Memoirs, **32**, 317–335.

ZHANG QIMING & SU HOUXI 1989. Hydrocarbon accumulations in the Beibuwan (Beibu Gulf) Basin. *China Earth Sciences*, **1**, 31–42.

ZU JIAQI 1985. Summary of oil and gas development in Beibu Gulf. *China Oil*, Autumn, 92–96.

Exploration in the Gulf of Thailand in deltaic reservoirs, related to the Bongkot Field

COEN T. A. M. LEO

Statoil, Run-Vest, PO Box 300, 4001, Stavanger, Norway

Abstract: Statoil has been involved in exploration in the Gulf of Thailand since 1985, following an invitation by PTTEP (National Exploration and Production Oil Company of Thailand) to assist in the evaluation and possible development of the Bongkot gas and condensate Field. Statoil and PTTEP jointly studied the field in 1985 and 1986.

In 1990 PTTEP farmed 60% out to a group consisting of Total, British Gas and Statoil. Development work on the field started in 1990 with the acquisition of 3D seismic data. The field came on stream in June 1993, and production increased from the original planned $150\,10^6\,ft^2$/day to $350\,10^6\,ft^2$/day at present.

In the joint PTTEP/Statoil study the Oligocene to Middle Miocene reservoirs were described as fluvial point bar sandstones, deposited in a meandering river system located on a delta top or coastal plain. Based on available data it was not possible to map the individual reservoirs. 22 wells found intermingled gas- and water-bearing sandstones without obvious relation to structure. No field fluid contacts were found, and the pressure regime was thought to be hydrostatic.

Statoil simulated a reservoir model consisting of numerous pointbar deposits on a muddy delta plain. The pointbars were assumed to be isolated sandbodies which favoured stratigraphic trapping. Structural mapping of existing 2D and old 3D data sets was thought to give reliable structural maps, although it was difficult to correlate single events. Stratigraphic traps were interpreted as the main traps in the field because no relationship was found between structural closure and gas pools.

Since 1990 when field development started with Total as operator around 2500 km^2 of 3D data has been acquired, and approximately 20 exploration wells and 45 development wells have been drilled. These data resulted in a much more accurate description of the field, especially through the use of high-resolution 3D seismic data. The well information allowed the identification of high GR intervals with a continuous seismic expression. Also individual sand reservoirs could be correlated. It became evident from the detailed 3D seismic interpretation that structural traps form the main trapping mechanism, although a minor part of the reserves are found in isolated channel sands, which could be detected from amplitude maps. It has been difficult to identify stratigraphic traps but it is hoped that ongoing seismic attribute analyses will help.

In 1993 Statoil farmed in to licence B10/32, operated by Ampolex. The Bongkot model of isolated fluvial channels, with traps formed by several faults with or without structural closure was accepted as an attractive play type. Not withstanding this, the first wildcat well, Karawake-1 was a structural test. The well failed to find hydrocarbons because it intersected a series of overlapping alluvial fans; these lack internal seal and would have precluded hydrocarbon migration and entrapment. Ampolex made a reinterpretation of the block based on the existing 2D seismic, and concentrated on seismic facies mapping. The mapping resulted in two prospect areas where fluvial facies could be recognized above a lacustrine source rock; and also the results of the Karawake-1 well could be better explained. A second well should prove the model.

In 1985 Statoil was invited by PTTEP to assist in the evaluation and assessment of the reserves and possible field development of the Bongkot Gas and Condensate Field in the Gulf of Thailand. Statoil and PTTEP jointly studied the field in 1985 and 1986. After the study PTTEP acquired 100% of the licence from the company Texas Pacific, who discovered Bongkot and drilled 23 wells in the field. In 1990 PTTEP farmed 60% out to a group consisting of Total, British Gas and Statoil to develop the field with Total as operator. Statoil's share of 10% in the Bongkot Concession was its first licence interest in SE Asia. In 1993 Statoil farmed in to block B10/32, operated by Ampolex, and later assumed operatorship when Ampolex withdrew from the area.

The first part of the paper summarizes the exploration history, structure and stratigraphy of The Gulf of Thailand. The second part covers the 1986 Bongkot evaluation by Statoil, and how the ideas on the geology, structure, trapping

From Fraser, A. J., Matthews, S. J. & Murphy, R. W. (eds), 1997, *Petroleum Geology of Southeast Asia*, Geological Society Special Publication No. 126, pp. 77–87.

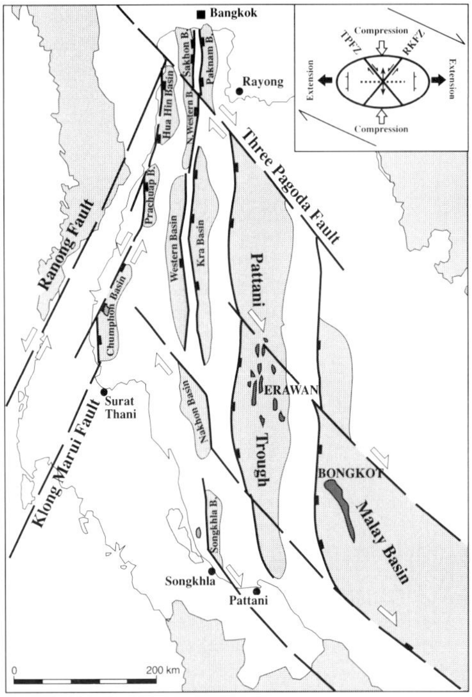

Fig. 1. Simplified structural map of the Gulf of Thailand (after Polachan and Sattayarak 1991).

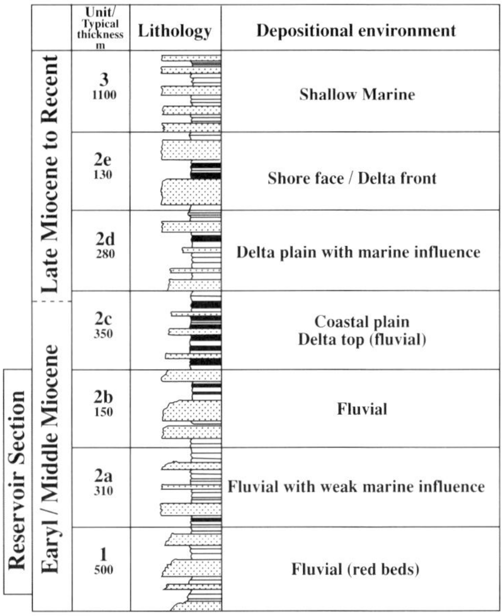

Fig. 3. Lithostratigraphic column of the Bongkot Field.

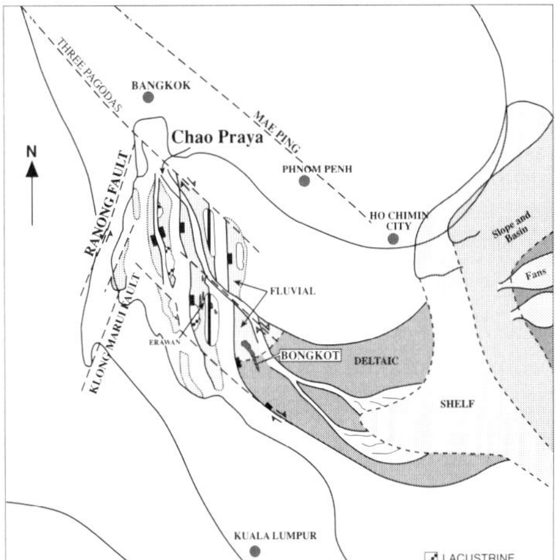

Fig. 2. Depositional environment during the Oligocene and Miocene.

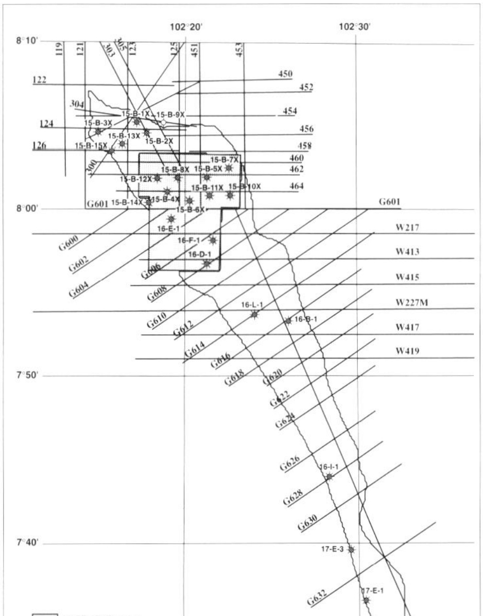

Fig. 4. Bongkot Field map showing the seismic and well database of 1986.

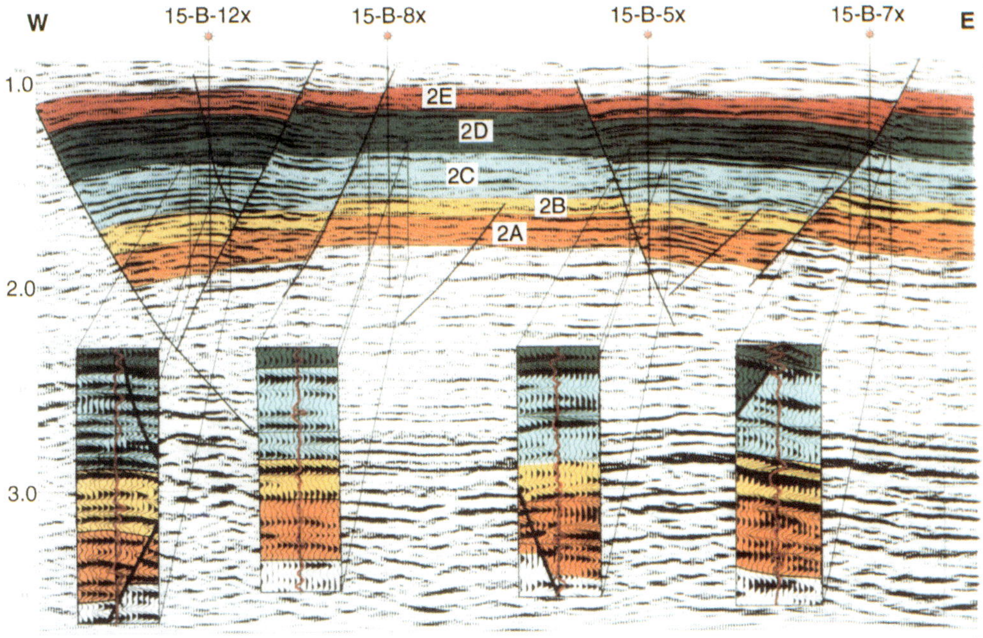

Fig. 5. Seismic crossection. Note the varied character of the well seismic tie at top formation 2B, due to lithology changes and gas presence. Line location on Fig. 6.

and reservoir evolved during the subsequent delineation and development of the Bongkot Field. The third part of the paper covers the exploration in Block B10/32.

Exploration history

In 1967 the Government of Thailand announced the delineation of 19 exploration blocks offshore in the Gulf of Thailand (Yongyut & Chatchawan 1992). Six foreign companies were awarded licences in 1968. The first well, Surat-1 was drilled by Conoco in 1971. Since then many exploration rounds have taken place, and the fourteenth round was announced in November 1995.

In the Gulf of Thailand two major gas complexes are in production, Unocal's Erawan Field complex and Total's Bongkot Field. Additional smaller gas fields, the Pailin and Tantawan discoveries, are expected to come on stream within a few years.

The Bongkot concession in the Malay Basin belonged originally to Blocks 15, 16 and 17, which were awarded to BP in 1967. In 1975 BP farmed out its interest to Texas Pacific who discovered the Bongkot Gas Field. Texas Pacific failed to negotiate a gas sales agreement, and in 1987 sold their entire interest to PTTEP. The present partner group of Total, BG, Statoil and PTTEP was formed in 1990 to develop the Bongkot Gas Field.

Tectonic history

The Gulf of Thailand developed during the Early Tertiary and lies near the intersection of two major strike-slip fault systems (Fig. 1). The NW trending Three Pagodas Fault runs from the Myanmar border and appears to extend into the northern part of the Gulf. The NE–SW-trending Ranong Fault cuts across peninsular Thailand and may extend into the northwestern part of the Gulf. These major fault systems bound a series of north–south-trending grabens and half grabens in the northern part of the Gulf. Polachan & Sattayarak (1991) described the Tertiary basins as transtensional pull-apart basins, resulting from the Eocene–Oligocene progressive northward collision of the Indian Plate with the Eurasian Plate. Right lateral strike slip movement along the Three Pagodas Fault zone and left lateral movement

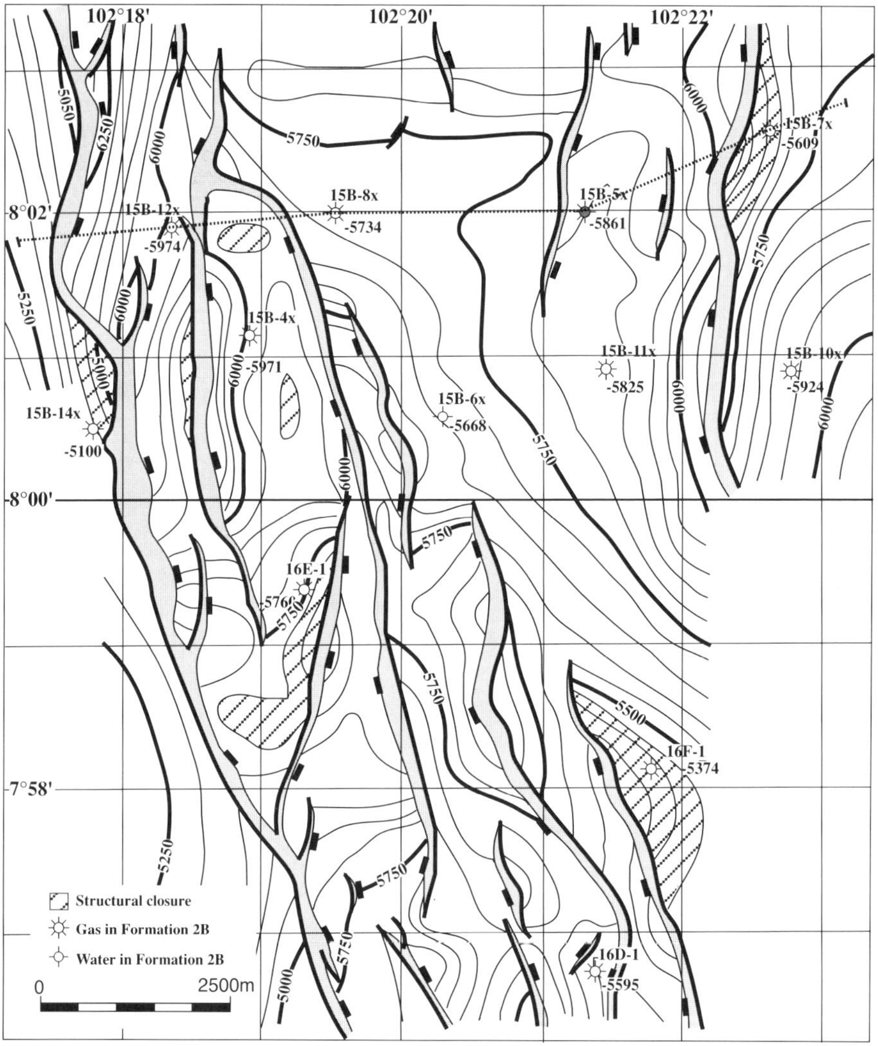

Fig. 6. Structural depth map of top reservoir 2B, 1977 3D area. Gas was present in formation 2B in all wells irrespective of structural closure. Only well 15-B-6x was dry (contours in feet).

along the conjugate NE Ranong fault system was the driving force behind the formation of the N-trending northern grabens and the NW-trending Malay Basin. The Gulf of Thailand falls into two regions, the western half with numerous smaller basins and the eastern half with two large basins, the Pattani Trough and the Malay Basin.

Stratigraphic overview

The Tertiary section overlies a basement consisting of Mesozoic to Carboniferous carbonates or Mesozoic granites (Pradidtan & Dook, 1992). Beneath the Malay Basin the basement is deeper than 3500 m and has not been penetrated. The Tertiary section can be separated into synrift and postrift sequences.

The Oligocene–Miocene synrift sequences were deposited in isolated grabens, and do not correlate, or only partly correlate, across intervening highs. In the north western basins and also in the Pattani Trough the sediments were deposited in alluvial/fluvial and lacustrine environments, and lack marine influences. In the more open and larger Malay Basin synrift deposition occurred in a fluvial to deltaic environment

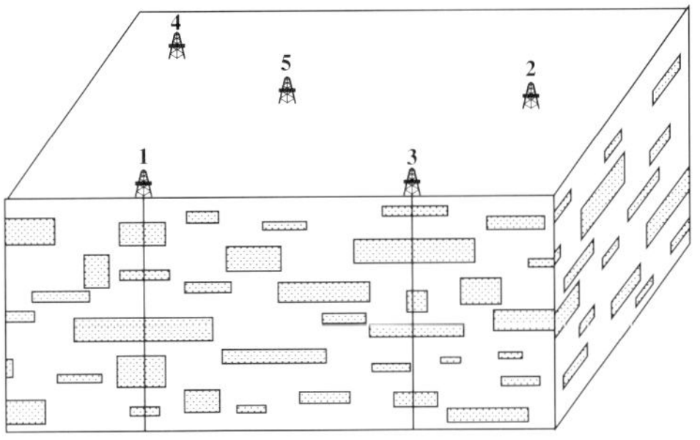

Fig. 7. Bongkot sand distribution sketch model of the 1986 reservoir study. The statistical sand distribution is influenced by formation thickness, net/gross, net pay, number of sands and gas filled sands in the drilled wells.

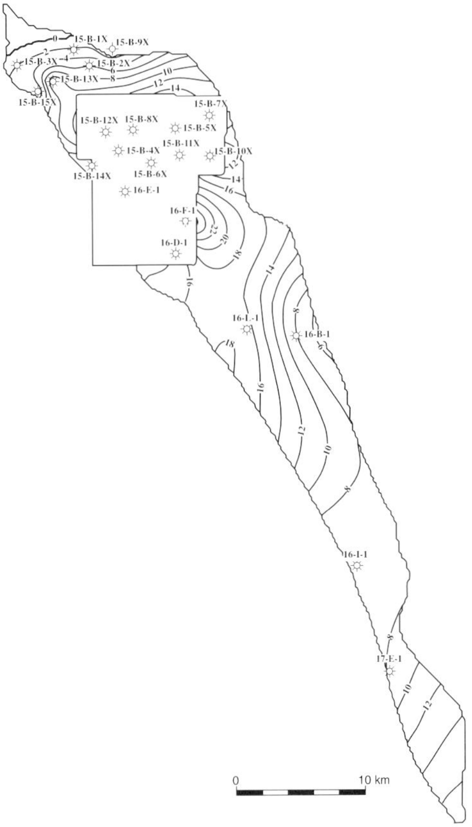

Fig. 8. Net gas column map of formation 1, 2a and 2b, used in 1986 to calculate reserves in the 2D area (contours in metres).

and marine highstand intervals are recognized (Fig. 2). The palaeo Chao Praya river delta system was probably the main clastic source, resulting in progradational sequences deposited from northwest to southeast. The delta system discharged into the South China Sea.

The Late Miocene post-rift sequence was deposited during a regional transgressive event which covered the Gulf of Thailand and paralic sediments were followed by more brackish deltaic marine to open marine sediments. The Bongkot Field is located in the northern Malay Basin, with typical fluvial and deltaic deposits in which more marine equivalents of flooding surfaces are recognized.

Statoil 1986 study

Statoil (1986) divided the prospective Miocene sequence into three intervals, based on lithostratigraphy (Fig. 3).

The oldest unit, Formation 1 consisted of red shales and fining-upward sandstones and was defined as fluvial meandering river deposits with a low sand/shale ratio. Thickness is more than 500 m.

Formation 2 represented a deltaic sequence and formed the main reservoir. The sequence was subdivided into five intervals, each characterized by a typical log pattern and lithology. The various interpreted environments ranged from meandering river systems on the lower delta plain (2A, 2C), fluvial meandering on upper delta plain and alluvial plain (2B), lower delta plain to delta front (2D), and coastal (2E). The overall trend was regressive. Thickness is between 800 and 1300 m.

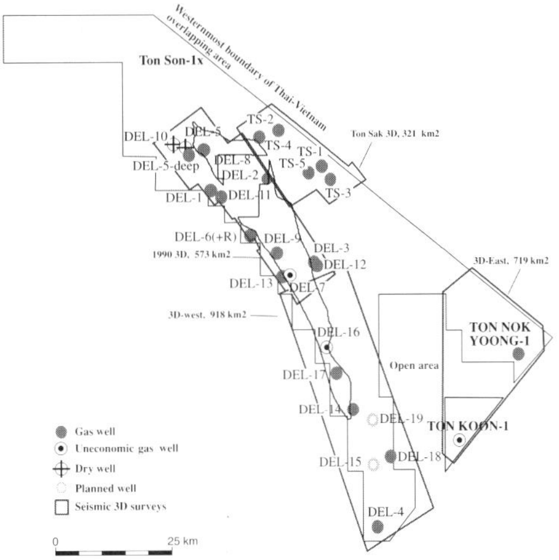

Fig. 9. Bongkot license map showing the seismic and well database of 1995.

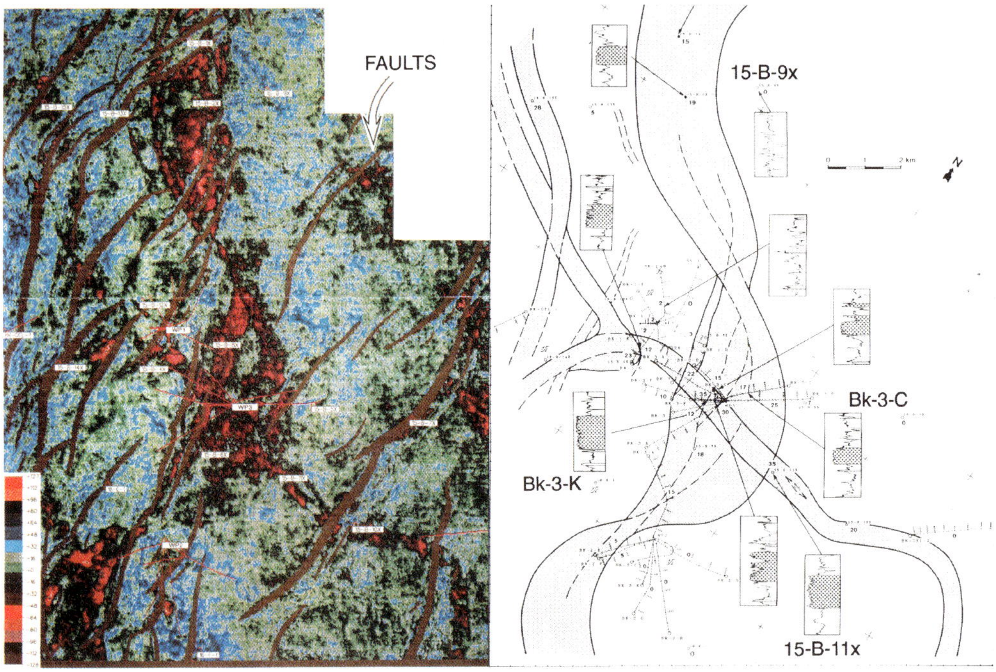

Fig. 10. Amplitude attribute map and interpretation along a horizon slice in formation 2D. The map shows several channels in a NW/SE direction, parallel to the axis of the Malay basin. Sand is found in wells inside the channels (well BK-3-C), shale in wells outside the channel (well 15-B-9x). Where channels cross each other, thick stacked sands are encountered (well BK-3-K) (after Total, internal report 1995).

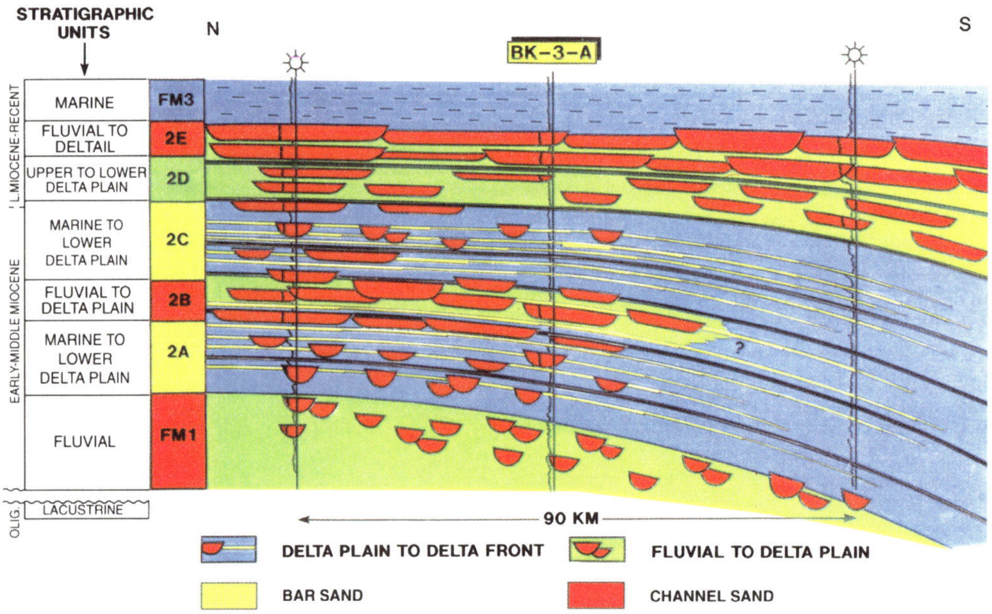

Fig. 11. Stratigraphic units and their depositional environment (after Total, internal report 1995).

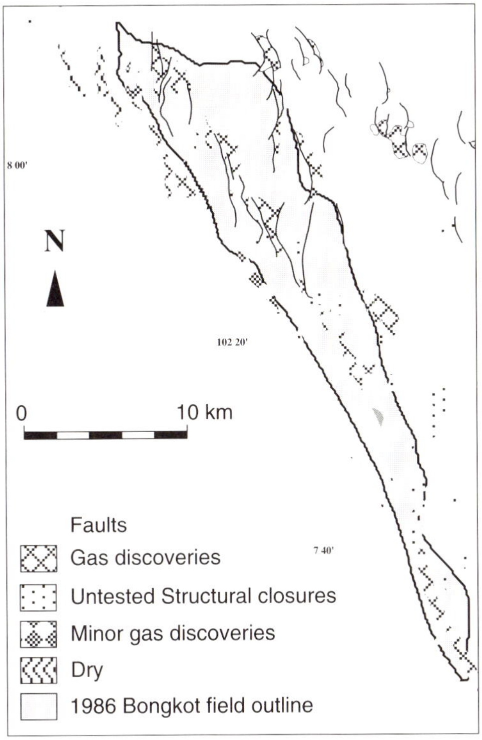

Fig. 12. Structural framework and closure map. In 1995 The Bongkot Field consist of many smaller structural closures, in contrast to the large field outline of 1986.

The youngest unit, Formation 3 is formed by marine clays, rich in foraminifera, alternating with some marine sands. Such open marine conditions have dominated the Gulf of Thailand up to the present time.

Most of the gas reservoirs penetrated by the 23 exploration wells drilled before 1986 were found in sands of Formations 2b, 2a and Formation 1 with mostly fluvial sand reservoirs. The sands could not be correlated from well to well and reservoirs were thought to be isolated sand bodies. Gas sands alternated randomly with water wet sands, and gas water contacts were seldom found. A common field contact was absent. Formation 2c was regarded as unproductive due to thin sands and Formation 2d and 2e were excluded from the evaluation, as only limited Net Pay was encountered.

Seismic structural interpretation was based on the first 3D survey shot in Asia in 1977, and on a coarse 2D survey covering the whole concession (Fig. 4). The interpretation concentrated on the

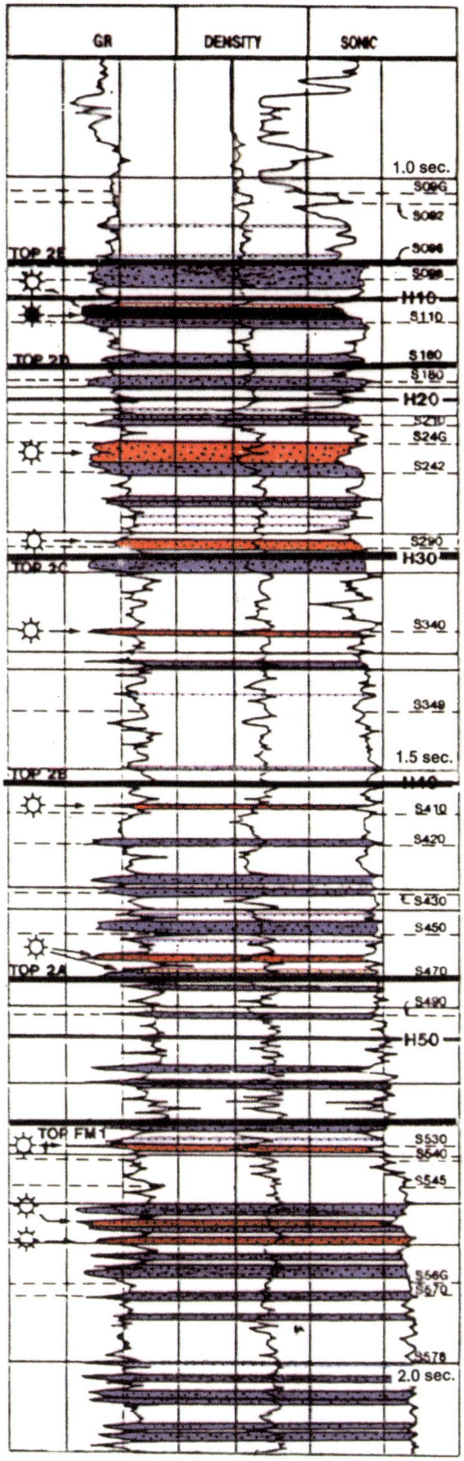

Fig. 13. Well 15-B-2x, showing an irregular alternation of gas- and water-filled reservoirs.

3D area, where structural maps were prepared for each formation top. Seismic horizon correlation was difficult due to the discontinuous character of the individual beds. The varying character of picked seismic horizons is demonstrated in Fig. 5. The Bongkot Field was characterized by normal and antithetic faults in the Miocene section; the deeper Oligocene had not been mapped. Most normal faults were dipping in an eastward direction, down into the basin. A detailed depth map of the 3D area (Fig. 6) shows several structural closures, but many wells were found to be hydrocarbon bearing outside of closure.

The main trapping system was concluded to be stratigraphic. Gas was trapped in isolated pointbars, and could be found all over the structure. Reserves in the 1977 3D area were calculated using a statistical model reflecting an unpredictable reservoir distribution. The model was based on the assumption of circular isolated pointbar reservoirs (Fig. 7), where sand distribution was derived from published sizes of meandering river systems, and vertical sand density was estimated from well data. Gas volumes were first calculated for individual wells and interpolated in between. Outside the 3D area net pay maps were constructed for reserve calculations (Fig. 8). As faults or structural closures where not considered to delimit the field, delineation of the field in the Bongkot concession (3D and 2D area) was a large problem, as most wells contained at least 5–10 m of gas reservoirs. An arbitrary but large area around the 22 wells had been chosen. This outline still marks the Bongkot Field on most license and scout maps.

Bongkot from 1990 onwards

In 1990 the development of the Bongkot Field started with the acquisition of a new 3D survey of 573 km^2, and a 2D survey with a line spacing of 500 m covered the rest of the concession (Cacheux *et al.* 1992) (Fig. 9). The surveys offered high density data with excellent resolution over the prospective zone from 1000 m to 2500 m depth. Based on the interpretation of the 3D survey, 13 exploration wells and approximately 40 development wells have been drilled. More 3D surveys were acquired later, and the wealth of information has led to a revised interpretation of the Bongkot Field.

A reinterpretation and correlation of old and new well information resulted in the recognition of high GR clay and coal markers which could be correlated over larger distances. A few of these markers could be recognized on seismic sections and made seismic horizon correlation more reliable. Because of the high quality of the 3D survey and consistent interpretation, depositional systems and individual reservoirs could be recognized on seismic amplitude maps and tied back into the wells.

Figure 10 is an amplitude map example from a time window along a seismic horizon. Several larger channel complexes (Fm 2D, depth around 1400 m) are visible, which were tied into nearby wells. The figure demonstrates the importance of amplitude maps for individual sand correlation.

The revised stratigraphic column and depositional system is shown in Fig. 11. The sedimentary facies vary from fluvial to delta plain, delta front and offshore marine. The depositional

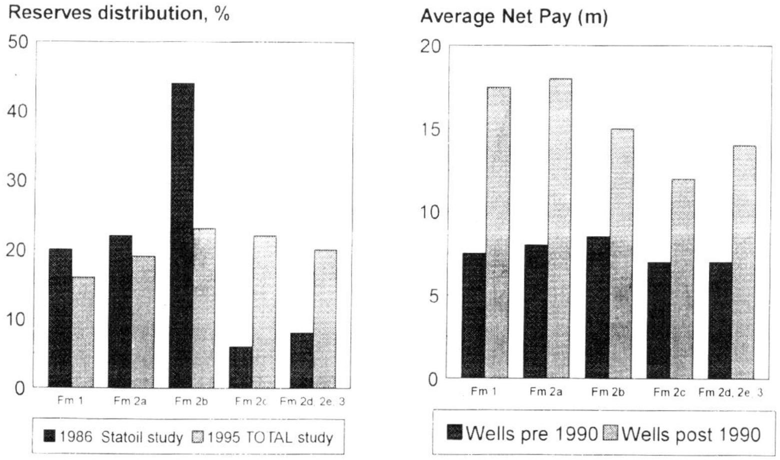

Fig. 14. Reserve distribution and net pay per formation in the Bongkot Field.

facies become more distal towards the south, where sands become thinner and less abundant, especially in Formation 2B. Reservoirs are represented by pointbars in high-sinuosity river systems, more elongated sand bodies in low sinuosity distributary systems and oval deltaic mouthbars. Gas was found in all types of reservoirs, not only in Formations 1 to 2b as assumed in the 1986 Statoil study. Also Formations 2c, 2d and 2e (Fig. 14) contained considerable amounts of gas resulting in a thicker reservoir section. Consequently the simplistic pointbar model of 1986 to calculate reserves was no longer applicable.

The new 3D and 2D seismic structural maps confirmed that the Bongkot Field is located on the north western flank of the Malay Basin, where a major basin fault zone with probably a lateral offset component, controls a normal en-echelon fault system in the Tertiary sequence. The normal faults have a typical throw of 50–100 m. Also it was found that most of the gas was trapped within structural closures. The maps indicated that the Bongkot field actually consists of many individual small structures and delineation of the field is a matter of drilling each of these (Fig. 12). Stratigraphic traps do exist in the Bongkot Field, which are often combinations of isolated reservoir bodies in a structural nose or fault trap. However, stratigraphic traps do not contain stacked reservoirs giving high net pay in one well, but occur more erratically. At present, a reliable methodology to define and delineate the stratigraphic traps is lacking and they therefore contribute little to the Bongkot reserves.

Trap limitation factors appear to be leakage across faults, when sands are juxtaposed and seal integrity, when vertical relief of the structure becomes too large. The apparent erratic gas- or water-filled sand distribution (Fig. 13) found in many of the older wells is likely to be due to leakage across juxtaposed sands, and possibly charge. However, the recent accurate depth and amplitude maps have allowed complex well paths to be designed for exploration and development wells (Gouadain & Duval 1994), which has improved the well efficiency and net pay results (Fig. 14). The average net gas pay in wells drilled before 1990 was around 35 m, but in the post-1990 wells this has improved to around 75 m.

The exploration and development efforts in Bongkot can be summarized as follows:

- 3D seismic structural and attribute maps improved the description of exploration and development well targets and reservoirs;
- optimized 3D well paths resulted in higher net pay results;
- well-defined structures allow for optimized development planning;
- more reliable reserve figures, allowing a higher production rate.

Licence B10/32

The Bongkot trapping model was Statoil's reference when farming in to licence B10/32 in 1993. The prospective area in licence B10/32 consisted of a small Tertiary sub basin west of the Pattani Trough, separated by the Insea High (Fig. 15). Miocene fluvial sands were expected with some alluvial influence, and a potential lacustrine source rock was recognized on a 2D seismic grid in the Oligocene section. Many smaller normal faults existed on the western flank of the basin, where small structural closures were mapped. Oil and gas were expected to be trapped in the fault closures, with an upside potential for stratigraphically trapped hydrocarbons. The Karawake-1 well failed to prove any hydrocarbons or shows, probably due to the abundance of Miocene alluvial reservoir rock. A seismic facies analysis study (Fig. 16) incorporating the well results

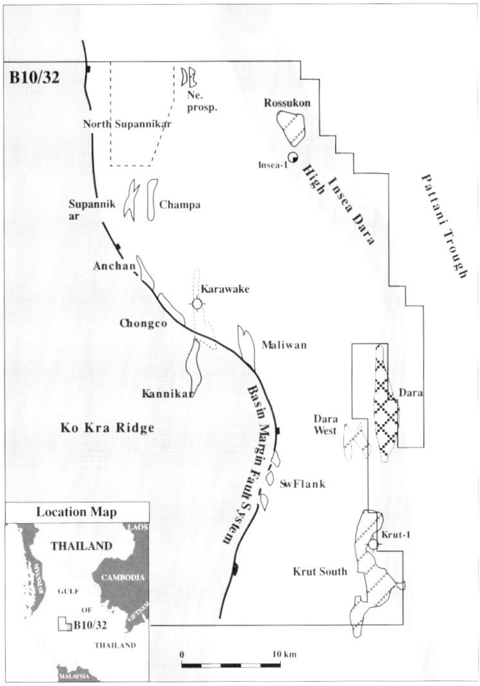

Fig. 15. Location and prospect map of licence B10/32. The detailed prospect map covers the southeastern part of the licence.

showed the existence of alluvial fans on the western flank of the basin, and the eastward continuation into fluvial sediments, possibly intermingled with lacustrine sediments. To be able to recognize the fans the seismic lines had to be condensed and structural interpretation was neglected during interpretation (Fig 17). Two prospects were defined based on the reinterpretation in the eastern part of the basin (Fig. 15), where fluvial sediments are expected on top of a lacustrine oil source rock. This model will hopefully be tested in a future exploration well.

Conclusion

It is concluded that detailed seismic facies analysis in the Gulf of Thailand is essential for

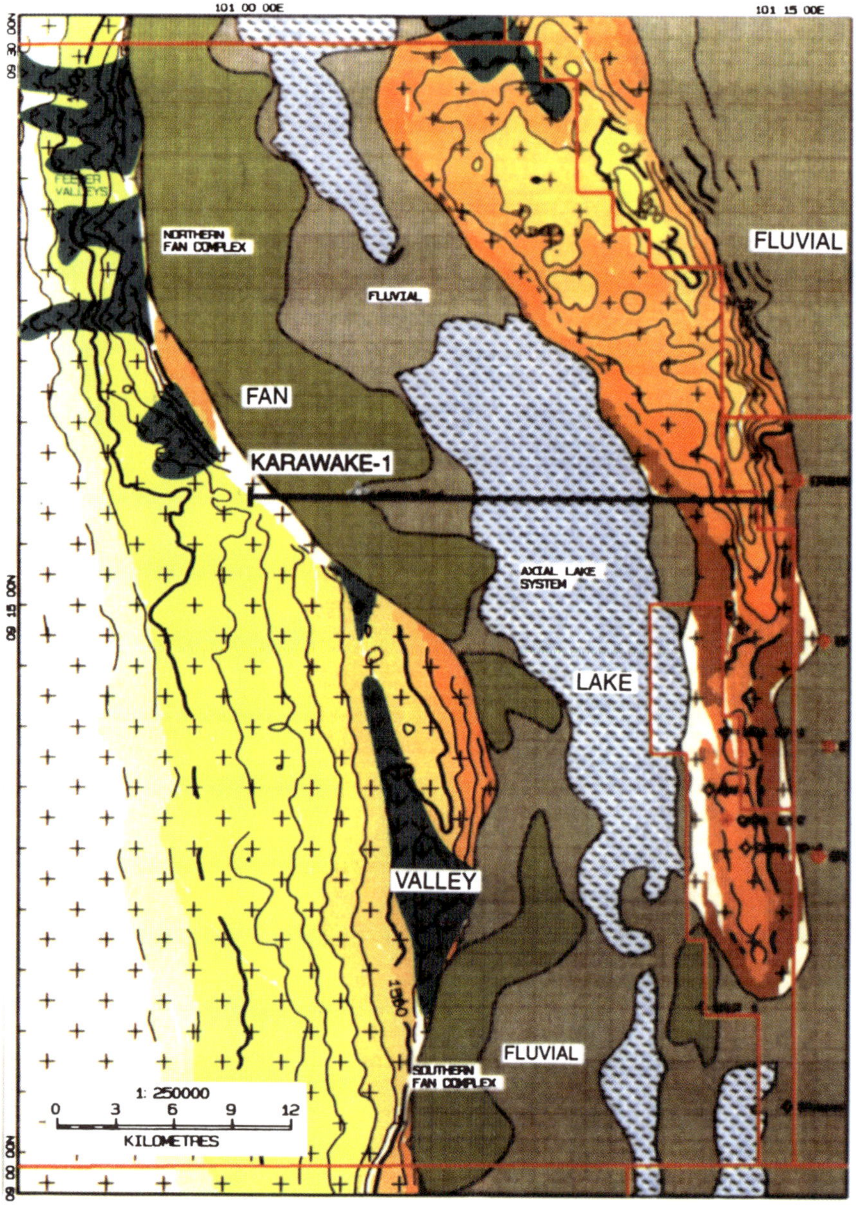

Fig. 16. Licence B10/32, paleofacies map of the Oligocene interval (Ampolex 1995 Internal report).

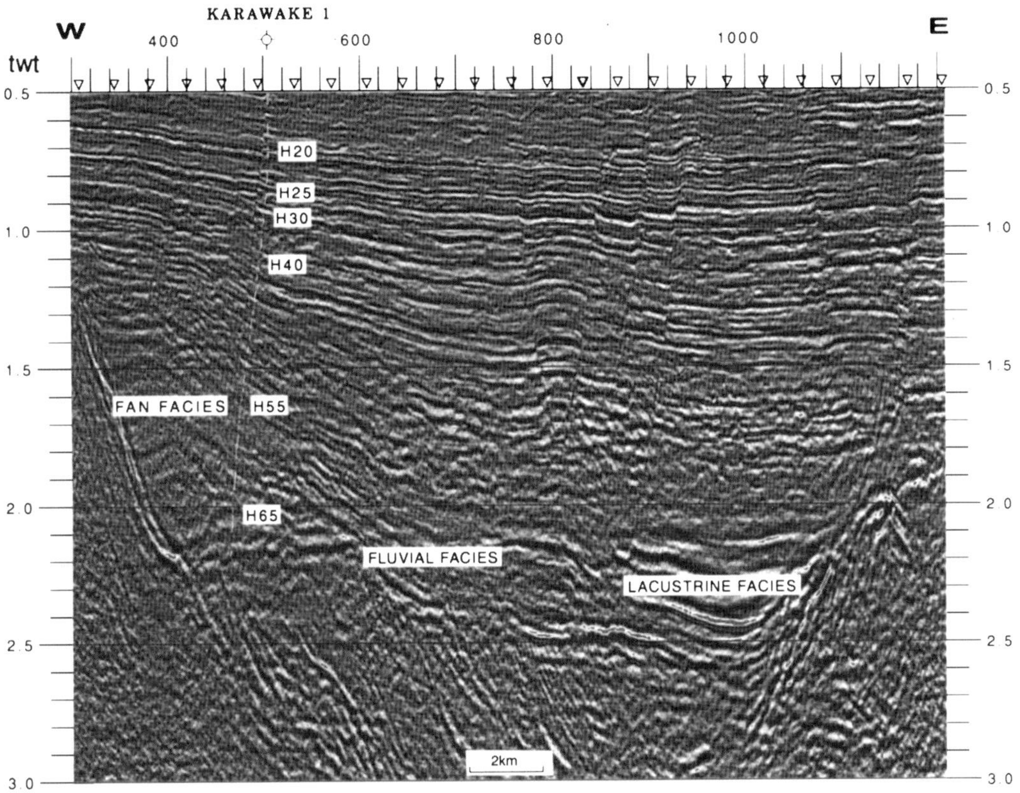

Fig. 17. Licence B10/32, condensed seismic section across the basin through the well Karawake-1, with interpreted fan facies (dull character), fluvial facies (discontinuous events) and lacustrine facies (high energy continuous events). Seismic markers H20 to H40 are intra Miocene, H55 is near top Oligocene and H65 is an unconformity level. Line location on Fig. 16.

exploration. For regional basin analysis 2D data should be used both for facies and structural mapping – before any decision on a well location is made. High resolution 3D is needed for delineation of single structures and attribute analyses for reservoir mapping.

The author acknowledges with gratitude the permission granted by Total, operator of the Bongkot Gas Field and partners British Gas and the National Oil Company of Thailand PTTEP, for the release of data and information. Much of the present work and ideas presented here originate from Total's work on the Bongkot Field.

Acknowledgement is also due to Statoil's former joint venture partner Ampolex, the previous operator in licence B10/32. Their work on seismic facies recognition improved geological understanding of the block substantially.

The views presented in this paper are those of Statoil and do not necessarily concur with those of the individual partners.

References

CACHEUX, A., FRAISSE, C. & BERNET-ROLLANDE, O. 1992. Bongkot Field – 3D seismic survey acquisition, processing, methodology of interpretation. *Proceedings of a National Conference on Geologic Resources of Thailand*, 000–000.

GOUADAIN, J. & DUVAL, B. 1994. Exploration and development in the Bongkot Field area (Gulf of Thailand): an integrated approach. *Proceedings of the 14th World Petroleum Congress*. John Wiley & Sons, 000–000.

POLACHAN, S. & SATTAYARAK, N. 1991. Development of Cenozoic Basins in Thailand. *Marine and Petroleum Geology*, **8**, 84–87.

PRADIDTAN, S. & DOOK, R. 1992. Petroleum Geology of the Northern Part of The Gulf of Thailand. *Proceedings of a National Conference on Geologic Resources of Thailand*, 000–000.

YONGYUT TRANGCOTCHASAN & CHATCHAWAN GLADNAKA 1992. Petroleum Exploration History in Thailand. *Proceedings of a National Conference on Geologic Resources of Thailand*, 000–000.

Structure, stratigraphy and petroleum geology of the SE Nam Con Son Basin, offshore Vietnam

S. J. MATTHEWS[1], A. J. FRASER[1], S. LOWE[2], S. P. TODD[3] & F. J. PEEL[4]

[1] *BP Exploration, BP Plaza, 200 Westlake Park Boulevard, Houston, Texas, 77079, USA*
[2] *BP Exploration, Chertsey Road, Sunbury-on-Thames, Middlesex TW16 7LN, UK*
[3] *BP Exploration Operating Company Ltd, Farburn Industrial Estate, Dyce, Aberdeen AB2 0PB, UK*
[4] *BHP Petroleum, GPO Box 1911R, Melbourne, Victoria, 3001, Australia*

Abstract: New well and seismic data acquired during recent exploration of the SE Nam Con Son Basin, offshore Vietnam, have been evaluated to assess the tectonostratigraphic evolution.

The offshore Vietnamese region has evolved in response to the complex relative motions of Indochina, Peninsular Malaysia, Borneo and the East Vietnam/South China Sea during the Cenozoic. On a regional scale these motions have been accommodated by strike-slip fault development, crustal extension and contraction. Rift pulses occurred in the SE Nam Con Son Basin from the Palaeogene to the earliest Late Miocene in response to the interaction of East Vietnam/South China Sea rift propagation and regional transtensional shear to the west of the evolving ocean basin. The structural evolution was complicated by mild contractional deformation during the Middle Miocene which was broadly synchronous with development of the largest inversion structures in the nearby West Natuna Basin.

The oldest dated Tertiary rocks in the SE Nam Con Son Basin are fluvio-deltaic sediments of Late Oligocene age which have been penetrated by several wells. Early to Mid-Miocene depositional environments ranged from non-marine to outer shelf, with a predominantly clastic basin-fill. The thickest Lower to Middle Miocene occurs in N–S- to NE–SW-trending half grabens.

A regionally recognized truncational unconformity of late Mid-Miocene age has resulted partly from the combined effects of the mild inversion and by the erosion of uplifted footwalls. During the Late Miocene there was rapid and widespread deepening of depositional environments, synchronous with net-extensional fault reactivation on many previously developed rift structures. The stratigraphic response to this increased bathymetry was the growth of isolated carbonate build-ups on pre-existing structural highs, progradation of the palaeo-Mekong delta system and associated deep-marine submarine channel development and erosion.

This pulsed structural and stratigraphic evolution has resulted in deposition of source, reservoir and seals, and produced a variety of potential trapping styles.

Nam Con Son Basin structural setting

The location of the Nam Con Son Basin (Fig. 1) at the junction of the East Vietnam/South China Sea, Red River Fault, Malay/Natuna Basin is indicative of its relevance to understanding SE Asian basin development and deformation. Many of the published interpretations of Cenozoic structural development onshore SE Asia (Fig. 1) suggest that the development of basin-defining faults and subsequent deformation of the basin-fill may be controlled by regional strike-slip displacements linked to the northward drive of India into Eurasia (Tapponnier *et al.* 1982, 1986; Peltzer & Tapponnier 1988). Offshore to the east of Vietnam, structural geometries displayed on seismic data (Holloway 1982; Pautot *et al.* 1990) are more typical of extensional basins where the main sub-basin defining faults have not experienced major strike-slip displacements. The Nam Con Son Basin is therefore located in the region where two fundamentally different structural styles may be expected to overlap, and therefore provides an opportunity to evaluate the relative importance and interplay between regional strike-slip and more orthogonal displacements during basin development and deformation.

The India–Eurasia collision and East Vietnam/South China Sea spreading history are only two of the several tectonic processes which have interacted to produce the present-day configuration of SE Asia. Additional processes, which may be linked to these collisional and spreading episodes, are subduction roll-back along the Sunda–Java trench and rotation (clockwise and/or anticlockwise) of large continental and oceanic blocks. Numerous plate tectonic models,

From Fraser, A. J., Matthews, S. J. & Murphy, R. W. (eds), 1997, *Petroleum Geology of Southeast Asia*, Geological Society Special Publication No. 126, pp. 89–106.

and regional constraints on plate motions, have been published for this region and include Daly *et al.* (1991), Fuller *et al.* (1991), Gower (1990), Hutchison (1989), Jolivet *et al.* (1989), Lee & Lawyer (1994), Le Pichon *et al.* (1992), Packham (1993), Rangin *et al.* (1990), Roberts (1988) and Tapponnier *et al.* (1986).

Stratigraphic framework

Increased understanding of the SE Nam Con Son Basin stratigraphy is based on recent interpretation of new seismic, palaeontological, wireline log and core data, and builds on earlier work by Matthews & Todd (1993) and Breare (1993). This recent work has provided the basis for refinement of the composite sequence stratigraphy (Figs 2 & 3). These composite sequences are defined by seismic boundaries which result from the interaction of pulses of structural motion and eustatic fluctuation, and which are typically recognised on the sub-basin scale. They are labelled numerically upwards (larger numbers are younger), with a prefix indicating Tertiary (T) and Quaternary (Q) composite sequences respectively. The characteristics of each of these composite sequences are summarised below.

Composite Sequence T10 (?Eocene–Oligocene)

Composite sequence T10 is the oldest rift-fill the top of which is defined seismically by an onlap surface. It is interpreted to have been deposited in approximately E–W-trending half grabens based on interpretation of seismic data. It has not been penetrated by wells and is interpreted to be restricted areally to the lower part of isopach thicks (Figs 4 & 5) where it onlaps 'basement', ('basement' in this area is typically a highly variable pre-Tertiary assemblage of granitic, volcanic and low-grade meta-sedimentary units which have been penetrated by several wells located on structural highs). The age and lithological characteristics of T10 are unknown presently, a partly Eocene–Oligocene fluvio-lacustrine sedimentary fill is possible based on comparison with potential analogues in the SE Asian region (Sladen 1993).

Composite Sequence T20 (Late Oligocene)

The T20 composite sequence has been penetrated by wells located on structural highs (Figs 4 & 5), and is characterized by high

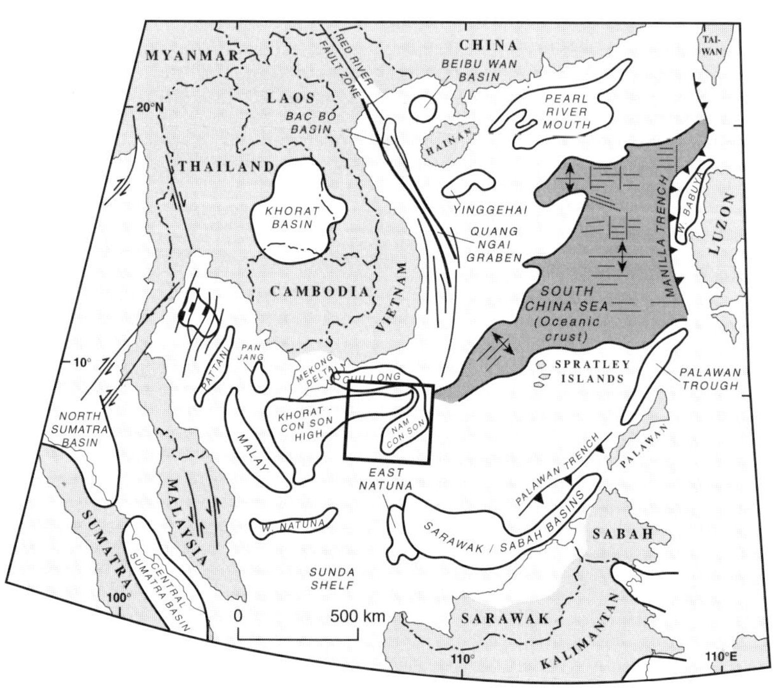

Fig. 1. Location of major sedimentary basins within SE Asia. Major basins are highlighted in the lighter tone. The location of Fig. 4 is shown as an inset.

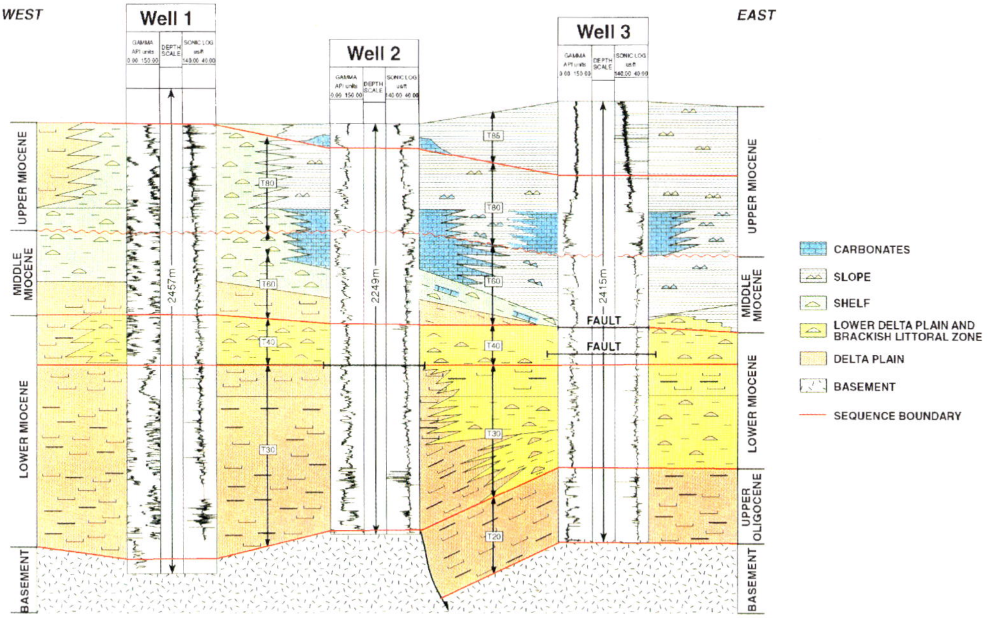

Fig. 3. Correlation of three wells from the SE Nam Con Son Basin.

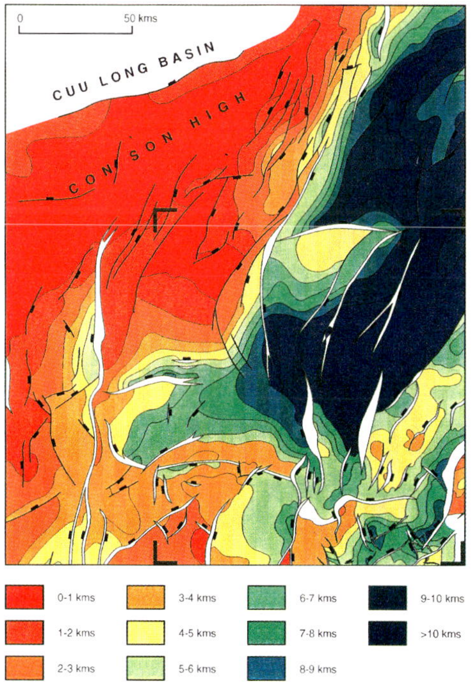

Fig. 4. Simplified structure of the Nam Con Son Basin at top pre-Tertiary ('basement'), showing the locations of Figs 5, 12, 13, 14 and 16 as an inset.

net : gross fluvial sandstones which are overlain by coals, mudstones and thinner sandstone units. Palynofacies and sedimentological data indicate that these sediments were deposited in backstepping fluvio-deltaic facies with minimal tidal influence. A westerly increase in net:gross and red beds suggests a regional source to the west and shoreline to the east. T20 is dated as Late Oligocene (*F. trilobata* palynological zone) and is equivalent to the lithostratigraphically defined Cau Formation (Bat *et al.* 1993).

Composite Sequence T30 (Early Miocene)

The lowermost sediments of the T30 composite sequence are thin transgressive sandstones and marine mudstones which are interpreted to be Early Miocene in age (NN2 nannofossil zone), suggesting that a time gap of up to 4 Ma may be present on drilled structural highs between T20 and T30. These transgressive sediments are overlain by interbedded sandstones and mudstones characterised by coarsening and fining-up cycles, and low net:gross values (10–30%). The seismic expression is typically a relatively conformable package of discontinuous, parallel and variable amplitude events. Palaeontological and sedimentary analysis suggests that T30 was

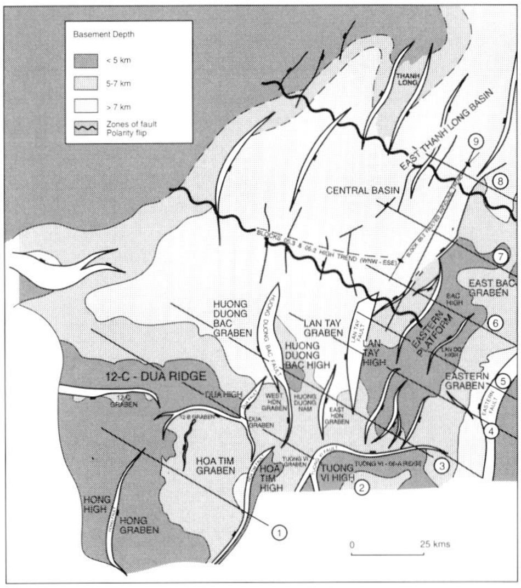

Fig. 5. SE Nam Con Son Basin structural elements and structural cross-section locations.

deposited predominantly in paralic environments; fluvial, estuarine and brackish shoreface facies are interpreted to exist with only occasional intercalation of more open marine sediments. It appears that both subsidence and sediment supply rates were high throughout most of T30 time. The top of T30 is an onlap surface coincident with a shift from predominantly fluvio-deltaic paralic facies below to shelfal marine clastics and carbonates above. This marks the onset of N–S- to NE–SW-orientated rifting, with resulting fault block rotation and subsidence. Lithostratigraphically this composite sequence is equivalent to the Dua Formation (Bat *et al.* 1993).

Composite Sequence T40 (late Early to early Mid-Miocene)

The top of composite sequence T40 is locally defined by erosional truncation of footwall highs with hangingwall dip slopes onlapped by younger stratigraphic units. Seismic data demonstrates that the magnitude of footwall erosion reduces in the east, suggesting that here it may be represented by a submarine unconformity or hiatus. Nannofossils and foraminifera date T40 as late Early Miocene to early Mid-Miocene. Lithostratigraphically this composite sequence is equivalent to the Thong Formation (Bat *et al.* 1993). T40 was deposited during a phase of rifting on faults orientated N–S- to NE–SW. Depositional environments are interpreted to have been controlled by the bathymetric variation resulting from this extensional phase. Subsidence rates appear to have outpaced sediment supply during the lower part of T40, when shelf clastics were being deposited in the east (Figs 4 & 5) synchronous with shelf carbonate deposition further west (Figs 4 & 5). A reversal of relative rates is apparent from the upper part of T40 where marginal marine to non-marine clastics have been encountered. More basinal depositional environments existed in some areas to the east (Figs 4 & 5) where backstepping carbonate shelfal sediments pass basinward into mudstones. Some well sections can be subdivided lithostratigraphically into lower clastic, middle carbonate and upper clastic components, but the boundaries between these are interpreted to be diachronous (Fig. 3).

Composite Sequence T60 (Mid-Miocene)

Depositional environments during early T60 times were predominantly shelfal with both clastic and carbonate units preserved, the upper part of this composite sequence is characterised by the preservation of deeper water clastic sediments. There is an overall back-stepping of facies belts observed from T40 to T60, and an increased variation in facies, palaeoenvironments and thickness. The uplift and/or magnitude of erosion on some footwall highs has resulted in the absence of T60, either through non-deposition or erosion. The top of T60 is the most prominent unconformity in the Nam Con Son basin. It is characterised by local erosional truncation, and by a basinward shift of facies in deep-water settings. Structural discordance is locally developed, and is observed best in the south east where it is interpreted to result from footwall uplift and mild inversion. The unconformity was enhanced in deeper marine settings by submarine channeling which eroded underlying sections, and may itself have been further modified by subsequent younger channelling events. Nannofossils and foraminifera date T60 as Mid-Miocene. Lithostratigraphically this composite sequence is equivalent to the Mang Cau Formation (Bat *et al.* 1993).

Composite Sequence T80 (early Late Miocene)

There is considerable variation regionally in thickness and facies in T80, mainly as a result of carbonate build-up growth which was controlled by the pre-existing topography on T60 and older composite sequences. The lower part of T80 is characterized by widespread shelf carbonates. Shallow water clastics have been encountered but are confined to the southwest. Deep water mudstones and some thin pelagic drapes of foraminiferal oozes are preserved between carbonate build-ups. An intra-T80 relative sea-level lowstand temporarily disrupted areas of carbonate deposition. Carbonate deposition was re-established only patchily afterwards, in the form of isolated buildups and reef pinnacles. The shelf clastics in the west prograded basinward in the upper part of T80 and supplied sediment through a system of submarine canyons into the deep-water settings. The top of composite sequence T80 contains a major channeling event which marks a basinward shift in deeper water settings. Nannofossils and foraminifera date T80 as the early part of the Late Miocene. Lithostratigraphically this composite sequence is equivalent to the lower part of the Nam Con Son Formation (Bat *et al.* 1993).

Composite Sequence T85 (Late Miocene)

T85 consists of shelf and slope clastics and some carbonates to the west. Condensed deepwater mudstones (with local sandy turbidites) surrounded by carbonate build-ups occur to the east. The top of this composite sequence is coincident with the maximum basinward progradation of the shelf break close to the Miocene–Pliocene boundary. Nannofossils and foraminifera date T85 as Late Miocene. Lithostratigraphically this composite sequence is equivalent to the upper part of the Nam Con Son Formation (Bat *et al.* 1993).

Composite Sequence T90 (Early Pliocene to early Late Pliocene)

The top of composite sequence T90 marks a major basinward shift in facies in the Pliocene. In the east, T90 comprises shelf clastics and minor deepwater mudstones. Further basinward, condensed slope mudstones sit on drowned carbonate build-ups and thicken into depositional lows which are interpreted to be filled partly by turbiditic sands. The top of T90 marks a major intra-Pliocene basinward shift in facies. Nannofossils and foraminifera date T90 as as Early Pliocene to early Late Pliocene. Lithostratigraphically this composite sequence and those above it are equivalent to the Bien Dong Formation (Bat *et al.* 1993).

Composite sequence T100 (Late Pliocene)

During T100 times the western shelfal depositional environments migrated eastward, with development of a number of successive prograding and backstepping units. T100 consists of shelf clastics in the west and slope deposits throughout most of the eastern parts, but sediments of this age are notably absent above some highs in the east. The top of T100 is a sequence boundary coincident with the base Pleistocene relative sea-level lowstand. Nannofossils and foraminifera date T100 as Late Pliocene.

Composite sequence Q10 (Pleistocene)

Q10 can be sub-divided into several units which reflect the spasmodic progradation of the shelf towards the east with intermittent exposure and backflooding. Submarine fans in the east correspond with phases of erosion and sediment by-pass in the west. The sea-bed is the top of Q10. Nannofossils date Q10 as Pleistocene and have enabled further subdivision into eight sub-zones which can be used in sand body identification and correlation.

Structural geometry

The Nam Con Son Basin developed during the Tertiary by complex rifting of a poorly known basement. Several wells have penetrated pre-Tertiary granites, but the detailed compositional and structural variations in the basement are not known. The trend of such deep fabrics, with respect to Tertiary extensional and compressional stress orientations, is likely to have partly controlled the Tertiary structural–geometrical evolution.

Three main fault trends are recognized; N–S, E–W and NE–SW (Fig. 4). In the west of the basin (Fig. 4) N–S faults are developed most strongly. The E–W-trending system is best imaged in the south and east, with the NE–SW-trending system dominant in the east (Fig. 5). A simplified interpretation of basement depth provides a guide to the displacements on the major faults (Fig. 5). Note that all of these faults define footwall structural highs and hanging-wall structural lows.

A consistent feature of many of the major faults is the rapid loss of displacement, toward tip lines, along the strike of the fault. It is typical for a fault with an opposite dip sense to be developed near to the tip zones of these major faults. This results in numerous polarity flips along the strike of the basin (Fig. 5). There is usually a zone of overlap between major faults of opposed dip, but it does appear that there is alignment of fault tip zones along an approximately E–W to ESE–WNW direction.

The scale and importance of polarity flips, and the relationships between the large faults, footwall highs and anticlines is illustrated by eight subparallel structural cross-sections orientated NW–SE and one cross-section oriented NE–SW. Cross-sections 1 to 4 are illustrated in Fig. 6 cross-sections 5 to 8 are shown in Fig. 7 and cross-section 9 in Fig. 10. The cross-sections are based on seismic lines (see e.g. Figs 8 & 9) spaced at intervals varying from approximately 12–26 km, their location is shown in Fig. 5.

Tectonostratigraphic model

The basic stratigraphic and structural observations have been integrated to produce a tectonostratigraphic model for the south east Nam Con Son Basin. This builds on an earlier interpretation (Matthews & Todd 1993) which was based on older seismic data.

Palaeogene

Prior to deposition of the oldest Palaeogene sediments, the Nam Con Son Basin area had experienced subaerial erosion of a structurally and compositionally complex basement terrain

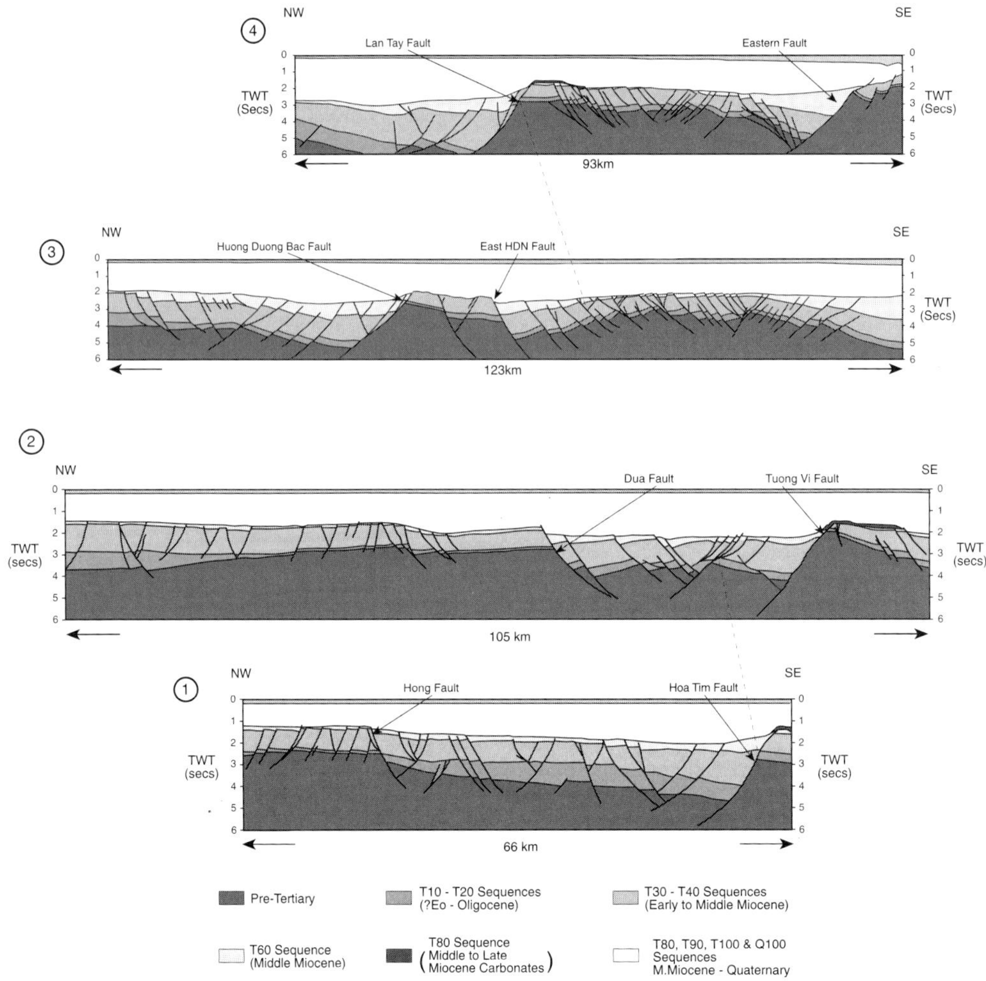

Fig. 6. Structural cross-sections 1–4, located on Fig. 5.

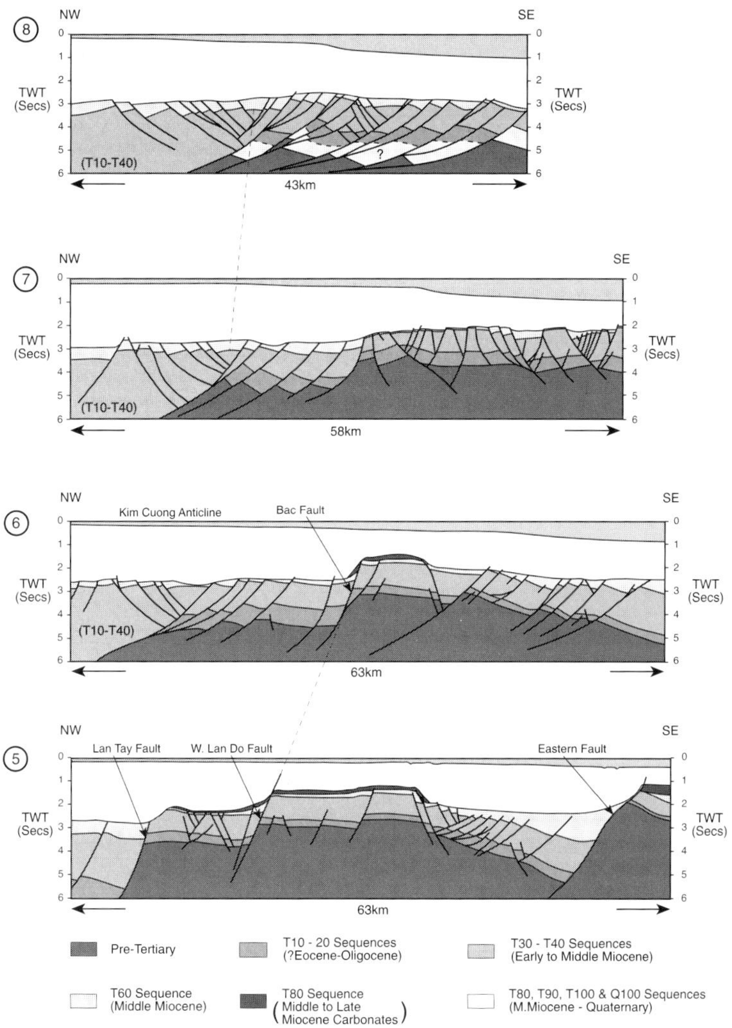

Fig. 7. Structural cross-sections 5–8, located on Fig. 5.

intruded by granites. This terrain was peneplaned in the SE part of the present-day basin, as the depositional topography at the base Palaeogene is negligible.

The oldest rift-fill stratigraphy in the SE Nam Con Son Basin has not been penetrated by wells and is preserved within seismically defined E–W-trending half grabens (Fig. 11), and displays a characteristic wedge-shaped rift-fill geometry. It is predicted that this early rift-fill was deposited in fluvial and/or lacustrine environments, based on regional analogues (Sladen 1993). Well penetrations on the structural highs demonstrate that subsidence was sufficiently regional during the Late Oligocene to allow accumulation and preservation of fluvial to coastal plain facies. It is possible that this was driven thermally in response to pre-Late Oligocene crustal extension in the SW area of the present-day East Vietnam/South China Sea. A combination of seismic and biostratigraphic evidence suggests therefore that the T10 sequence is a rift-fill in its lower part, and that the upper part (Upper Oligocene) was deposited during a phase of relatively minor fault motion in the SE Nam Con Son Basin. A speculative interpretation of the Oligocene distribution and depositional environments is shown in Fig. 12.

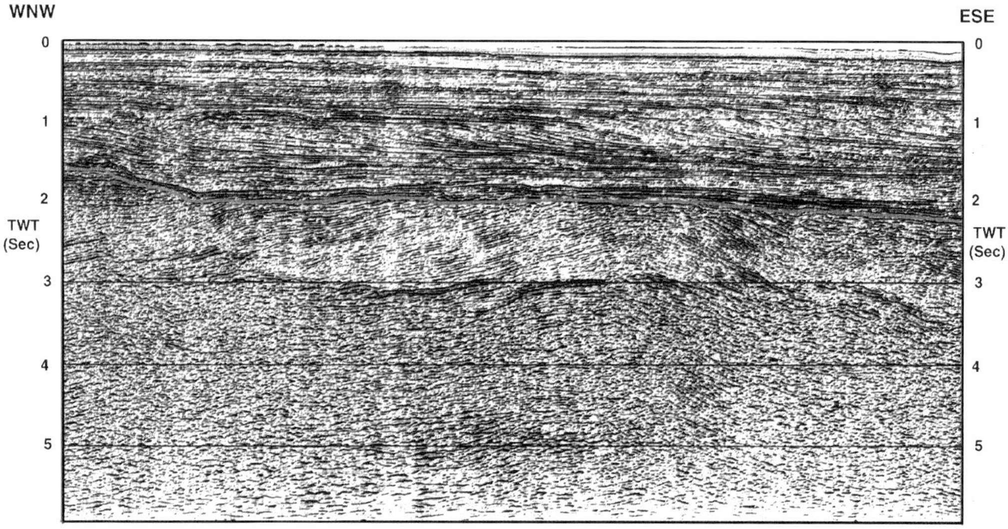

Fig. 8. Seismic example, forming the basis for part of structural cross-section 4 (Fig. 6).

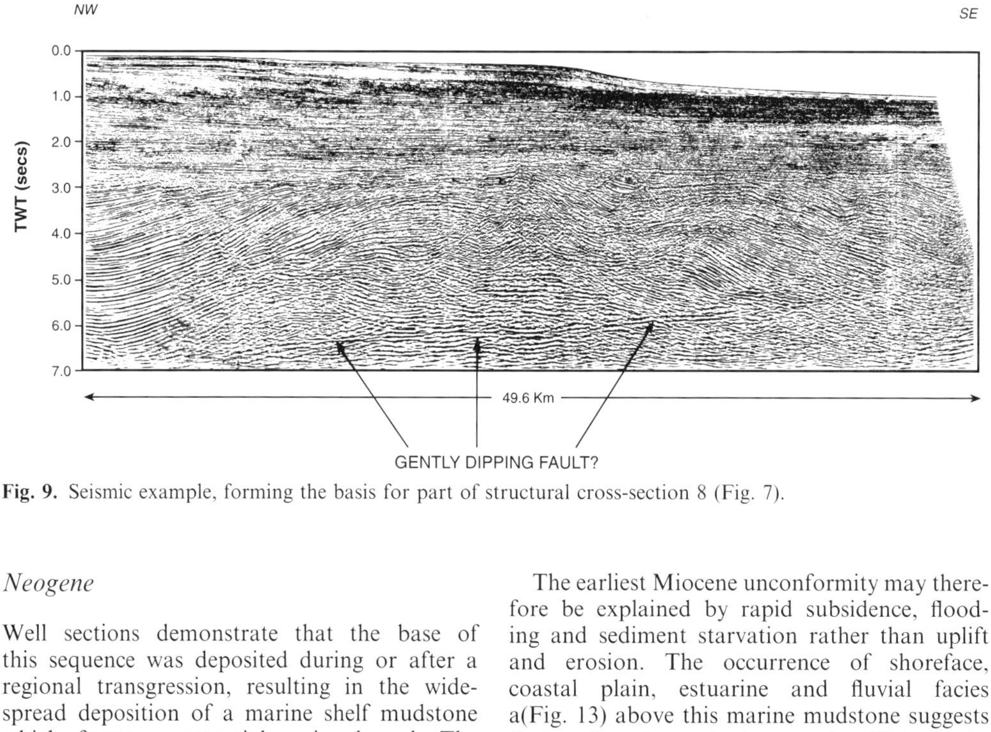

Fig. 9. Seismic example, forming the basis for part of structural cross-section 8 (Fig. 7).

Neogene

Well sections demonstrate that the base of this sequence was deposited during or after a regional transgression, resulting in the widespread deposition of a marine shelf mudstone which forms a potential regional seal. The occurrence of this marine mudstone immediately overlying the top T20 unconformity suggests that the unconformity may result from sediment starvation due to a major earliest Miocene flooding event, during which the shoreline retreated towards the west and NW.

The earliest Miocene unconformity may therefore be explained by rapid subsidence, flooding and sediment starvation rather than uplift and erosion. The occurrence of shoreface, coastal plain, estuarine and fluvial facies a(Fig. 13) above this marine mudstone suggests that sediment supply increased sufficiently to move the facies belts in an easterly direction, after the earliest Miocene flooding. Seismic geometries demonstrate that there is a perceptible thickening of the T30 sequence into some of the N–S- and NE–SW-orientated faults which may have started to propagate into the SE Nam Con Son Basin during the Early Miocene.

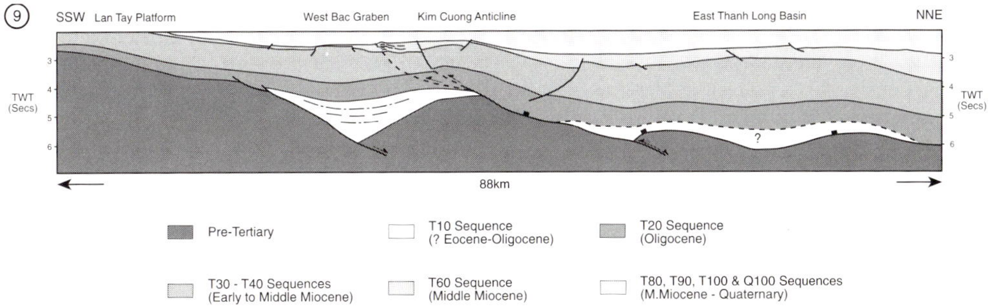

Fig. 10. Structural cross-section 9 located on Fig. 5.

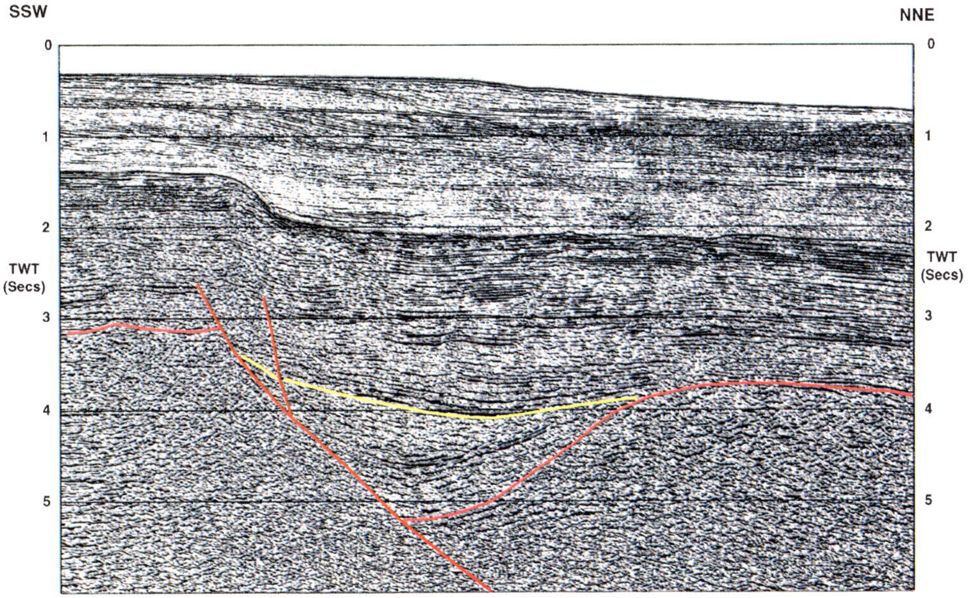

Fig. 11. Seismic example of deep rift geometry.

The uppermost part of the T30 sequence is characterized by a more marine shoreface facies, with the top of this sequence defined by an onlap surface.

A reduction in clastic sediment supply to parts of the basin is suggested by the onset of carbonate platform growth during T40 times, a sedimentary evolution compatible with west to north west movement of the shoreline. The interpretation of regressive cycles in well sections in the upper part of this sequence may result from genuine uplift of the footwalls to the active faults. This mechanism is supported by the observation locally of footwall erosion at the top of the T40 sequence.

The structural and stratigraphic characteristics of the base T60 sequence are similar to those which define the base T40 sequence; onlap onto rotated fault blocks in response to a further pulse of fault movement. A progressive west to north west shift in facies belts may have continued from T40 to T60 times, and is evidenced by the initiation of carbonate platform growth to the west and deeper marine clastic deposition to the east (Fig. 14). These facies interpretations suggest that the structural rotation which may be the key control on development of the T40/T60 boundary probably punctuated ongoing regional subsidence.

The top of the T60 sequence is defined by a regional unconformity. This boundary is best expressed as an angular unconformity across the top of anticlines. These anticlines may have formed initially during extension or transtension (involving no uplift) and locally have been accentuated during compression or transpression (involving uplift) (Fig. 15). The example seismic lines and cross-sections demonstrate the probability that anticlines formed dominantly during local extension with the development of the 'faulted anticline trend'. A minor contractional component to anticline formation is suggested by the structural geometry and strong erosional truncation of some anticlines which is incompatible with only hanging-wall extensional slip on bounding faults. This contractional pulse most likely occurred during the Mid-Miocene (mainly during late T60 times but contraction may have started in latest T40 times), and probably punctuated ongoing regional subsidence. Although the localized inversion interpreted to exist in the SE Nam Con Son Basin is very mild, it is relevant to note that the strongest inversions documented in the nearby West Natuna Basin (Fig. 1) (Ginger *et al.* 1993; Daines 1985), and in the Indonesian East Java Sea Basin (Matthews & Bransden 1995) occurred during the late Mid-Miocene. There is therefore a strong regional basis for interpretation of contraction during this time.

Extensional reactivation of several pre-existing faults occurred after development of the late Mid-Miocene regional unconformity. These are largely coincident with the margins of Mid- to Late Miocene carbonate build-ups (Fig. 16), suggesting that the main control on the youngest carbonate build-up distribution was the 'late' displacement on these faults. It is suggested that this most recent local extensional pulse interrupted ongoing regional subsidence, in a similar way to the preceding pulses of movement.

Thermal subsidence is interpreted to be the control on basin development during the Pliocene, Pleistocene and Quaternary. During the earliest part of this stage, major submarine channel and canyon systems were developed and filled with deep marine clastics. Commonly, the position of the new major downcutting geometries is coincident with the major anticlinal crests causing modification of any previously developed unconformity.

Petroleum geology

Source, charge and trap timing

Exploration of the Nam Con Son Basin during the 1970s and 1980s proved the presence of an oil prone source rock in the drainage area of the Dai Hung field (Fig. 5) and the Dua field (Fig. 5). Thermogenic gas and gas/condensate accumulations were discovered to the west. Exploration further east was restricted by the relatively deep water.

The geochemical analysis of oil samples has shown a dominance of kerogen from higher land plants, suggesting paralic carbonaceous mudstones and lower coastal plain coals as the main source facies. Other possible source facies include lacustrine sediments, deposited either in floodplain lakes or in earlier Palaeogene fault-block defined half-grabens, and marine mudstones deposited in a restricted anoxic environment in the Miocene depocentres.

Structural cross-sections demonstrate the large variation in thickness observed from depocentre to structural high. If the Oligocene to Lower Miocene section includes source rocks in the depocentres then it is probable that most of the liquid hydrocarbons would have been expelled during the Mid-Miocene (assuming an average heat flow of approximately 60 mW m^2). The onset of expulsion would have been synchronous with rifting but predated the last phases of structuring resulting from inversion and/or rift reactivation, and was also prior to deposition of Late Miocene carbonate build ups. In contrast, if Oligocene to Lower Miocene source rocks exist on the structural highs then oil generation and expulsion in these areas was probably initiated during burial by the Pliocene to Recent prograding deltaic system. The few laterally extensive carrier beds in the basin contribute to the development of overpressure and may result in the predominance of vertical rather than lateral migration as the principal charge mechanism.

Mild inversion will have modified the geometry of the drainage cells such that migration directions after inversion are different to those which existed earlier. From the Late Miocene to present-day the depocentres are predicted to have been expelling gas which may have flushed deeper oil pools to remigrate oil into younger stratigraphy or structures. Remigration may have been assisted by rift reactivation in the Late Miocene.

The source rock systems of the south east Nam Con Son Basin appear to be typical of SE Asia (Todd *et al.* this volume). The dominance of paralic coaly mudrocks in the sources rather than lacustrine (floodplain or rift lake) is probably one of the main controls on the predominance of gas in the SE Nam Con Son Basin. Biogenic gas is also known to contribute to discoveries in this area.

The Indochina region has experienced widespread extrusion of volcanics during the late Middle Miocene to present-day. This

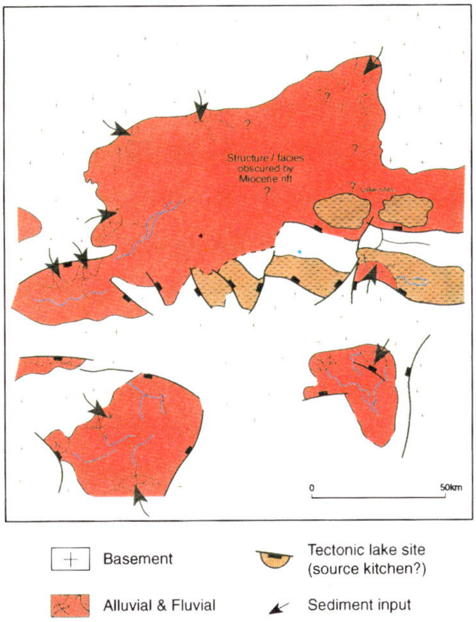

Fig. 12. Interpretation of gross depositional environments for the T10 and T20 composite sequences.

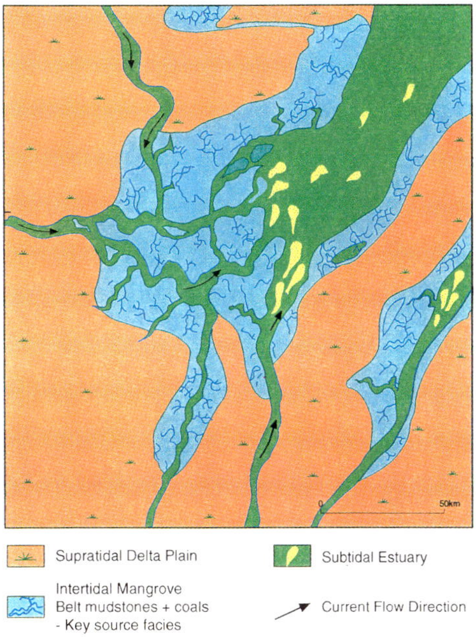

Fig. 13. Interpretation of gross depositional environments for the T30 composite sequence.

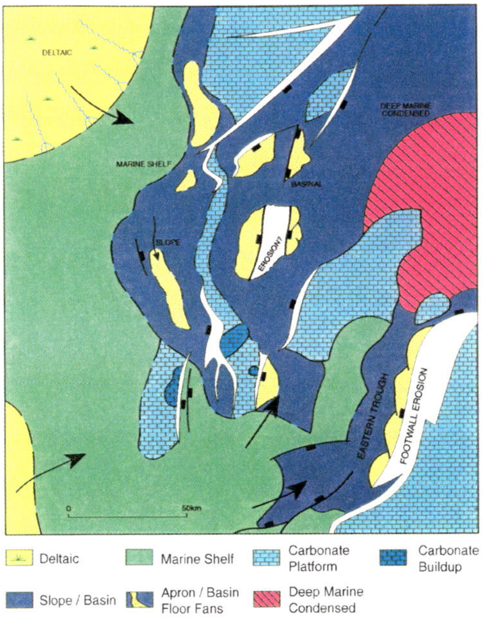

Fig. 14. Interpretation of gross depositional environments for the T60 composite sequence.

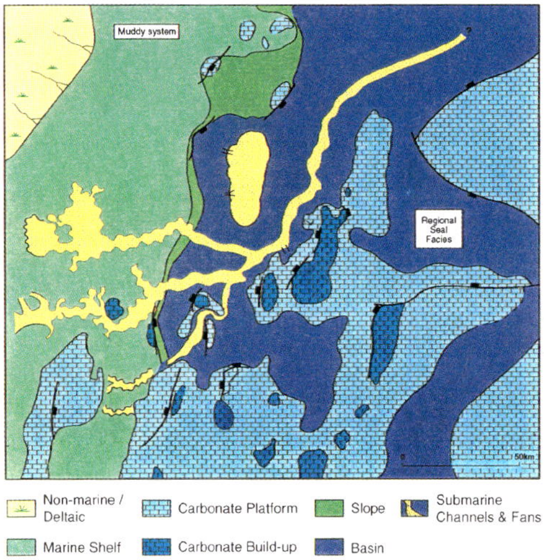

Fig. 16. Interpretation of gross depositional environments for the T80–T85 composite sequences.

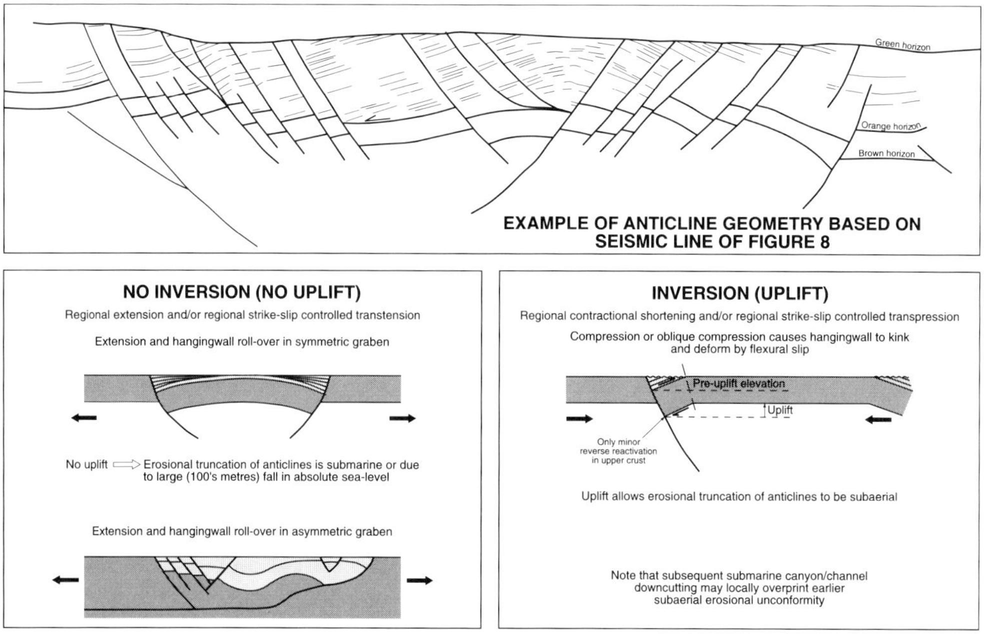

Fig. 15. Anticline geometry and evaluation of mechanisms of formation.

igneous activity is interpreted to have contributed to the carbon dioxide known to exist in some of the Nam Con Son Basin gas accumulations.

Seal

There are two regional seals in the south east Nam Con Son Basin, the Lower Miocene shelf mudstones which seal the underlying Upper Oligocene clastics and the deep-marine mudstones which overlie the regional late Mid-Miocene unconformity.

There are several potential sub-regional mudstones seals demonstrated by wells located on the structural highs. It is predicted that these units are likely to thicken from the highs into the hangingwall depocentres. They are interpreted to have been deposited during rises in relative sea-level. Transgressions linked to the rifting pulses during the Mid-Miocene may also have resulted in deposition of mud-dominated packages in a variety of shelfal environments. Thickness of seal may be related directly to fault block rotation with thinnest seal deposition on the structural culminations.

The Miocene clastics appear typical of many SE Asian basins in the presence of multiple seals, of only sub-regional significance, which formed as marine flooding surfaces in the paralic succession. Regional subsidence during the Late Miocene, following the latest phases of gentle inversion and extensional reactivation, resulted in deposition of a mudstone facies on top of the unconformity. Late Miocene carbonate build-ups are overlain by deep water mudstones varying in age from Late Miocene on the flanks to Pleistocene on the crests.

Clastic plays

Miocene sandstone reservoirs are known to exist in the Dai Hung and Dua oil and gas accumulations. It is probable that the better reservoir characteristics will be encountered in fluvial channel, shoreface and shelf sandstones (within T10–T60 composite sequences), particularly when deposited in lowstands above sequence boundary unconformities. In the brackish littoral zone and inner shelf the sandstones may be more heterogeneous and offer lower reservoir effectiveness. Plio-Pleistocene turbidite or slump related reservoirs may exist on the slope and/or base of slope as channel levee complexes and depositional fan lobes.

Carbonate plays

Middle Miocene to Pliocene platform and build-up carbonate reservoirs are known in the Nam Con Son Basin. These carbonates were deposited in shallow open marine to possibly restricted marine back-reef environments. In the south east Nam Con Son Basin, spectacular build-ups of this age form the reservoir for the Lan Tay and Lan Do gas accumulations.

Basement potential

Fractured basement is an important reservoir offshore Vietnam, particularly in the Cuu Long basin (Fig. 1). There is a speculative possibility that fractured granitic basement may prove to be a viable reservoir in parts of the SE Nam Con Son Basin also.

Trapping styles

Structural traps which contain potential T10–T60 sequence clastic reservoirs are dominantly three-way dip and fault, and two-way dip and two fault, closures. These structures formed during the pulsed, superimposed rift evolution from the Palaeogene to the Mid-Miocene. Bedding in the hanging wall of some of the faults which define these traps frequently dips away from the fault, thereby creating a possible four-way dip closure. Traps of this type show a variety of orientations from E–W (mostly early, probable Palaeogene structures) to N–S and NE–SW (mostly later, Mid-Miocene structures). Fault blocks of this type are frequently draped by younger mudstone seals.

The other principle structural trapping style is fault-segmented four-way dip anticlines. These occur in hanging walls of the main faults, and have formed during the rifting phases in response to hangingwall motion and fracturing over a variably dipping fault surface. In many cases these anticlines appear to have been tightened, and inverted, by mild contraction during the Mid-Miocene. The gentle inversion is often expressed on a wavelength of tens of kilometres such that the three-way dip and two-way dip structures are contained within a sub-regional four-way dip closure.

Fault seal effectiveness is a key consideration for trap integrity in the three-way dip and two-way dip structures (Knott *et al.* this volume). SE Asia contains many oil fields which are controlled by sealing faults, often these are seals because of the large proportion of mudstone in the hanging wall and because the fault walls have experienced some invasion by clay. A major concern with faulted traps in the Nam Con Son Basin is that the proportion of mudstone may be too low to provide good fault seals. For these traps to be effective a fortuitous juxtaposition of hanging-wall sealing facies against footwall reservoir facies is necessary, or very effective clay smearing and/or cataclastic gouge generation within the fault zone.

Late Miocene carbonate traps often have a four-way dip drape seal of deep water mudstones which overlie the build ups. More extensive stratigraphic traps can be proposed to involve the flank portions of the carbonates

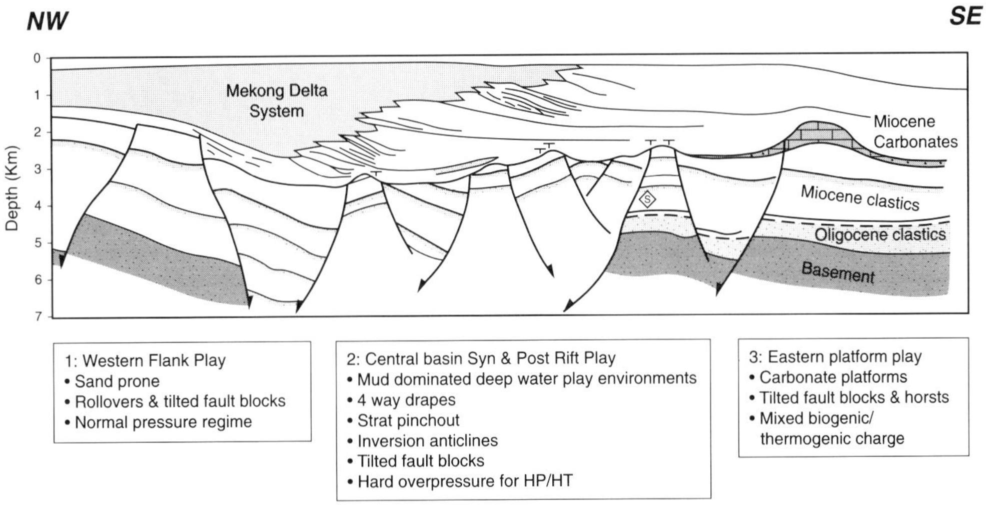

Fig. 17. Petroleum geology summary cross-section.

build ups where a facies change to mudstone or tight carbonate is predicted to exist, but these have not been proven to date. The petroleum geology is summarised on the schematic cross-section of Fig.17.

Discussion

The main aim of this paper is to describe the petroleum geology of the SE Nam Con Son Basin in the context of the basin-scale structural and stratigraphic development. A secondary objective is to use the new seismic and well data to help constrain the alternative models for regional structural and tectonic controls on basin development, especially to assist understanding of the relative importance of strike-slip and orthogonal fault displacements offshore south Vietnam.

Palaeogene regional structural evolution: observations and implications

Observation of approximately E–W-orientated faults and corresponding sub-basins prior to the Late Oligocene suggests that the SE Nam Con Son Basin was extended locally in an approximately N–S direction at that time.

There are several possible driving mechanisms which may have operated independently or combined to produce the structural geometries which evolved prior to the Late Oligocene. The first of these is East Vietnam/South China Sea extension, which could have propagated as far west as the Nam Con Son Basin in pre-Late Oligocene times. Palaeogene crustal extension, linked directly to South China Sea development, is proposed in many publications (Taylor & Hayes 1980, 1982; Holloway 1982; Ru & Pigott 1986; Pautot *et al.* 1986, 1990; Briais *et al.* 1989; Su *et al.* 1989; Yu 1990; Pigott & Ru 1994; Tongkul 1994). Extension during the Palaeogene phase of South China Sea development is most reasonably interpreted to have been N–S directed, producing a dominant E–W fault trend. The coincidence of the trend of the deeper pre-Late Oligocene half grabens in the SE Nam Con Son Basin with those which evolved during the Palaeogene phase of South China Sea evolution, suggests that the two may be linked and have evolved in response to the same driving mechanism.

One difficulty with this model is the observed offset between the South China Sea E–W-trending extensional fault systems and the SE Nam Con Son Basin examples. This offset may be explained by the presence of a crustal-scale transfer zone in the offshore region subparallel to the east coast of Vietnam.

A second possibility is that the pre-Late Oligocene half-grabens of the SE Nam Con Son Basin may be a response to crustal-scale strike-slip displacement on major faults within the Vietnamese region. The development of major strike-slip fault zones in this area is frequently linked to the collision of India with Eurasia, and the subsequent lateral displacement of major crustal blocks to the east of the northward-driving Indian continent (Tapponnier *et al.* 1982, 1986, 1990; Peltzer & Tapponnier 1988; Scharer *et al.* 1990; Polachan *et al.* 1991; Harrison *et al.* 1992; Le Pichon *et al.* 1992). If the observed Palaeogene rifts are linked to Palaeogene strike-slip then the E–W orientation of the deep half grabens can be reconciled with the SE–NW trend of known major strike-slip faults (Fig. 1), by interpreting the SE Nam Con Son Basin to have been positioned within a sinistral (left-lateral) shear zone during the Palaeogene. However, there is no direct evidence in our existing seismic data-set for the existence of large Palaeogene strike-slip displacements in the SE Nam Con Son Basin. Minor transfer faults with an approximate N–S trend are interpreted to segment the major E–W-trending rift-defining faults.

Neogene regional structural evolution: observations and implications

The Early to Mid-Miocene structural history of the SE Nam Con Son Basin was most likely one of pulsed movement on faults orientated approximately orthogonal to the Oligocene structures. This local extension was apparently synchronous with regional subsidence on a wavelength of several hundred kilometres.

Early to Mid-Miocene fault development may have been controlled by the most recent propagation of the East Vietnam/South China Sea towards the SW, but our current interpretation of timing of the main phase of pulsed faulting (approximately 23–11 Ma) is partly later than the generally accepted cessation of South China Sea rift propagation at approximately 17–15 Ma (Taylor & Hayes 1982). It is therefore a possibility that the driving mechanism for the Early to Mid-Miocene fault propagation may be a combination of extension linked to propagation of South China Sea structures and strike-slip shear. If so, and assuming that any strike-slip is the result of motion on regional-scale faults such as the SE-NW trending structures of the Vietnamese region (Fig. 1), then a regional dextral (right-lateral) shear sense is most plausible for the Early to Mid-Miocene history.

Our current understanding of fault displacement patterns on the major faults in the SE Nam Con Son Basin which have experienced Mid-Miocene movement, is that all of them reduce in displacement to tip lines. This fault geometry has produced numerous polarity flips, suggesting that the amount of strike-slip motion on individual mapped faults is low. It is possible that greater strike-slip movement has been concentrated on principal displacement zones outside the interpreted area.

Recent modelling of oblique extension by McClay & White (1995) has demonstrated that if the extension direction is orientated oblique to a pre-existing structural heterogeneity then a distinct fault geometry develops in map-view, characterized by relatively short faults and polarity flips. It is observed that a strong heterogeneity, in the form of E–W-orientated faults are present at depth in the SE Nam Con Son Basin, and these may have produced a similar effect to the models of McClay & White (1995).

The Mid-Miocene was also the time of mild inversion in some parts of the basin. There is seismic evidence to suggest that this mild inversion may have been controlled partly by deep reactivation of the older E–W-trending fault system, and that the resultant anticlinal uplifts are locally developed almost orthogonal to the trend of the major Middle Miocene faults. If correct, a simple two-dimensional consideration of inversion and extension is inappropriate and an oversimplification of a structurally complex three-dimensional evolution. This apparent contrast in the directions of local extension and local contraction during the Mid-Miocene may explain the absence of demonstrable net-reverse stratigraphic offsets on the major Mid-Miocene faults. It is possible that these faults are more likely to have accommodated a minor amount of lateral or strike-slip displacement during the contractional pulse.

The preferred model for map-view and cross-sectional development of structures in the SE Nam Con Son Basin is illustrated in Fig. 18. Note that if a model of partial control on basin formation and deformation by regional strike-slip is applicable to both the Palaeogene and Neogene then the shear sense appears to have switched from sinistral (left-lateral) during the Palaeogene to dextral (right-lateral) during the Early to Mid-Miocene.

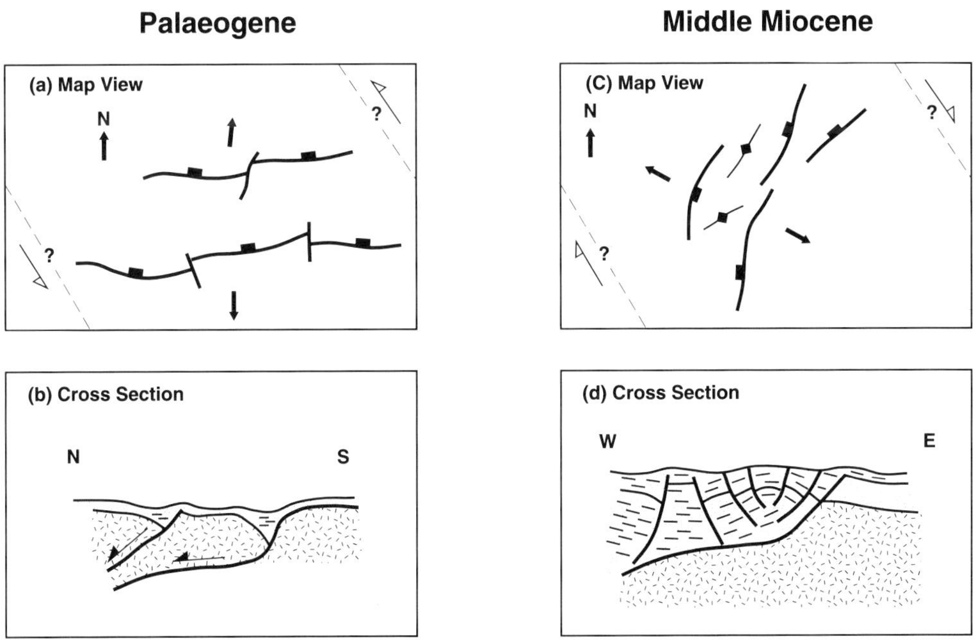

Fig. 18. Model for development of structures observed in the SE Nam Con Son Basin.

Conclusions

Seismic and well data has allowed assessment of models for the structural and stratigraphic history of the SE Nam Con Son Basin. Tertiary basin development commenced during the Palaeogene with the formation of rift basins within a compositionally and structurally varied basement which is known to be partly granitic. The Palaeogene rifts are orientated along approximately E–W trends, and are interpreted to have been filled with fluvial and/or lacustrine sediments of pre-Late Oligocene age. Upper Oligocene fluvio-deltaic clastic sediments were deposited regionally across the crests of the older rift-margin footwall highs. The Palaeogene structural and stratigraphic history has therefore combined to produce a proven clastic footwall play in the Upper Oligocene, and a potential deeper rift-fill play. This deeper play has numerous analogues around the margins of the East Vietnam/South China Sea.

Miocene rifting along N–S to NE–SW trends was synchronous with deposition of dominantly clastic sediments in environments ranging typically from inner to outer shelf, but with occasional fluvio-deltaic intercalations. Platform carbonates were deposited locally during the Mid-Miocene. This synchronous structural and stratigraphic development resulted in the formation of proven clastic and carbonate footwall play systems in the Mid-Miocene. The prospectivity of these footwall plays increases if there is effective four-way dip closure formed by the hanging-wall sedimentary section dipping away from the fault.

The Miocene rifting was accompanied by the formation of large extensionally-driven faulted hangingwall anticlines, which locally have been further deformed by obliquely-orientated mild contractional tightening. Middle Miocene clastic, and possibly carbonate, plays exist therefore in several large anticlinal structures which are four-way dip closed, but which are highly segmented internally into numerous smaller, but proven, three-way dip and fault plays. As with the larger footwall plays the prospectivity of these anticlines, and of internal fault-blocks within the anticlines, is increased if there is effective four-way closure resulting from the structural dip of the hanging wall or subsequent depositional drape of a sealing unit.

Late Miocene and younger carbonate build-ups nucleated on structural highs and sealed by deep marine muds are proven gas plays. Potential also exists in deep-marine clastic sediments of Late Miocene to recent age. In 1995 the basin is still at an early stage of what promises to be long and exciting exploration history.

The authors would like to thank the directors of BP Exploration and Statoil for permission to publish this paper. The technical evaluation of this area since 1991 has involved the substantial contribution of many staff from partners; they are thanked for their considerable and important input to these studies. All projects of this scope are completed by a team of geoscientists, and our thanks are extended to all of our many colleagues in BP and Statoil who worked within our joint team between 1992 and 1995. D. G. Roberts, J. Hossack (who generated an original version of part of Fig. 15) and D. Pegrum (who generated an original version of Fig. 1) are thanked for technical guidance throughout the project. D. G. Roberts and C. Sladen are thanked for their detailed helpful comments on an earlier version of this paper.

References

BAT, D., QUYNH, H., QUE, P. H. & DONG, T. L. 1993. Tertiary stratigraphy of continental shelf of Vietnam. *In*: *Proceedings of the 1st international seminar on the stratigraphy of the southern continental shelf of Vietnam*, January 1993.

BREARE, M. J. 1993. A sequence stratigraphic approach for defining the Tertiary and Quaternary of the Nam Con Son Basin, offshore Vietnam. *In*: *Proceedings of the 1st international seminar on the stratigraphy of the southern continental shelf of Vietnam*, January 1993.

BRIAIS, A., TAPPONNIER, P. & PAUTOT, G. 1989. Constraints of sea beam data on crustal fabrics and seafloor spreading in the South China Sea. *Earth and Planetary Science Letters*, **95**, 307–320.

DAINES, S. R. 1985. Structural history of the W. Natuna basin and the tectonic evolution of the Sunda region. *Proceedings Indonesian Petroleum Association*, **14**, 39–61.

DALY, M. C., COOPER, M. A., WILSON, I., SMITH, D. G. & HOOPER, B. G. D. 1991. Cenozoic plate tectonics and basin evolution in Indonesia. *Marine and Petroleum Geology*, **8**, 2–21.

FULLER, M., HASTON, R., LIN, J-L., RICHTER, B., SCHMIDTKE, E. & ALMASCO, J. 1991. Tertiary paleomagnetism of regions around the South China Sea. *Journal of Southeast Asian Earth Sciences*, **6**, 161–184.

GINGER, D. C., ARDJAKUSUMAH, W. O., HEDLEY, R. J. & POTHECARY, J. 1993. Inversion history of the west Natuna basin: examples from the Cumi-Cumi PSC. *Proceedings Indonesian Petroleum Association, Twenty second Annual Convention, October 1993*. 635–658.

GOWER, R. J. W. 1990. Early Tertiary plate reconstructions for the South China Sea region: constraints from northwest Borneo. *Journal of Southeast Asian Earth Sciences*, **4**, 29–35.

HARRISON, T. M., WENJI, C., LELOUP, P. H., RYERSON, F. J. & TAPPONNIER, P. 1992. An Early Miocene transition in deformation regime within the Red River Fault Zone, Yunnan, and its significance for Indo-Asian tectonics. *Journal of Geophysical Research*, **97**, 7159–7182.

HOLLOWAY, N. H. 1982. North Palawan Block, Philippines – its relation to Asian mainland and role in evolution of South China Sea. *AAPG Bulletin*, **66**, 1355–1383.

HUTCHISON, C. 1989. *Geological Evolution of South-East Asia*. Oxford Monographs on Geology and Geophysics, **13**, Clarendon Press, Oxford.

JOLIVET, L., HUCHON, P. & RANGIN, C. 1989. Tectonic setting of western Pacific marginal basins. *Tectonophysics*, **160**, 23–47.

KNOTT, S. D., BEACH, A., BROCKBANK, P. J., BROWN, J. L. & WELBON, A.I. 1997. Fault seal analysis of south east Asian basins with examples from west Java. *This volume*.

LEE, T. Y. & LAWVER, L. A. 1994. Cenozoic plate reconstruction of the South China Sea region. *Tectonophysics*, **235**, 149–180.

LE PICHON, X., FOURNIER, M. & JOLIVET, L. 1992. Kinematics, topography, shortening, and extrusion in the India-Eurasia collision. *Tectonics*, **11**, 1085–1098.

MATTHEWS, S. J. & TODD, S. P. 1993. A tectonostratigraphic model for the southern Nam Con Son basin, offshore Vietnam. *In*: *Proceedings of the 1st international seminar on the stratigraphy of the southern continental shelf of Vietnam*, January 1993.

—— & BRANSDEN, P. J. E. 1995. Late Cretaceous and Cenozoic tectono-stratigraphic development of the East Java Sea basin, Indonesia. *Marine and Petroleum Geology*, **12**, 499–510.

MCCLAY, K. R. & WHITE, M. J. 1995. Analogue modelling of orthogonal and oblique rifting. *Marine and Petroleum Geology*, **12**, 137–151.

PACKHAM, G. H. 1993. Plate tectonics and the development of sedimentary basins of the dextral regime in western southeast Asia. *Journal of Southeast Asian Earth Sciences*, **8**, 497–511.

PAUTOT, G., RANGIN, C., BRIAIS, A., TAPPONNIER, P., BEUZART, P., LERICOLAIS, G., MATHIEU, X., WU, J., HAN, S., LI, H., LU, Y. & ZHAO, J. 1986. Spreading direction in the central South China Sea. *Nature*, **321**, 150–154.

——, ——, ——, WU, J., HAN, S., LI, H., LU, Y. & ZHAO, J. 1990. The axial ridge of the South China Sea: a seabeam and geophysical survey. *Oceanologica Acta*, **13**, 129–143.

PELTZER, G. & TAPPONNIER, P. 1988. Formation and evolution of strike-slip faults, rifts, and basins during the India–Asia collision: an experimental approach. *Journal of Geophysical Research*, **93**, B12, 15 085–15 117.

PIGOTT, J. D. & RU, K. 1994. Basin superposition on the northern margin of the South China Sea. *Tectonophysics*, **235**, 27–50.

POLACHAN, S., PRADIDTAN, S., TONGTAOW, C., JANMAHA, S., INTARAWIJITR, K. & SANGSUWAN, C. 1991. Development of Cenozoic basins in Thailand. *Marine and Petroleum Geology*, **8**, 84–97.

RANGIN, C., JOLIVET, L., PUBELLIER, M. & THE TETHYS PACIFIC WORKING GROUP 1990. A simple model for the tectonic evolution of southeast Asia and Indonesia region for the past 43 m.y. *Bulletin de la Société Géologique de France*, **8**, 889–905.

ROBERTS, D. G. 1988. Basin evolution and hydrocarbon exploration in the South China Sea. *In*: WAGNER, H. C., WAGNER, L. C., WANG, F. F. H. & WONG, F. L. (eds) *Petroleum resources of China and related subjects*. Circum-Pacific Council for Energy and Mineral Resources Earth Science Series, **10**, Houston, Texas, 157–177.

RU, K. & PIGOTT, J. D. 1986. Episodic rifting and subsidence in the South China Sea. *AAPG Bulletin*, **70**, 1136–1155.

SCHARER, U., TAPPONNIER, P., LACASSIN, R., LELOUP, P. H., DALAI, Z. & SHAOCHENG, J. 1990. Intraplate tectonics in Asia: a precise age for large-scale Miocene movement along the Ailao Shan – Red River shear zone, China. *Earth and Planetary Science Letters*, **97**, 65–77.

SLADEN, C. P. 1993. Lake sequences in Tertiary hydrocarbon basins of Vietnam. *In*: Proceedings of the 1st international seminar on the stratigraphy of the southern continental shelf of Vietnam, January 1993.

SU, D., WHITE, N. & MCKENZIE, D. 1989. Extension and subsidence of the Pearl River Mouth basin, northern South China Sea. *Basin Research*, **2**, 205–222.

TAPPONNIER, P., LACASSIN, R., LELOUP, P. H., SCHARER, U., DALAI, Z., HAIWEI, W., XIAOHAN, L., SHAOCHENG, J., LIANSHANG, Z. & JIAYOU, Z. 1990. The Ailao Shan/Red River metamorphic belt: Tertiary left-lateral shear between Indochina and South China. *Nature*, **343**, 431–437.

——, PELTZER, G., LE DAIN, A. Y., ARMIJO, R. & COBBOLD, P. 1982. Propagating extrusion tectonics in Asia: new insights from simple experiments with plasticine. *Geology*, **10**, 611–616.

—— & ARMIJO, R. 1986. On the mechanics of the collision between India and Asia. *In*: COWARD, M. P. & RIES, A. C. (eds) Collision Tectonics. Gelogical Society Special Publication **19**, 115–157.

TAYLOR, B. & HAYES, D. E. 1980. The tectonic evolution of the South China basin. *In*: HAYES, D. E. (ed.) *Tectonic and geologic evolution of southeast Asian seas and islands, Part 1*. American Geophysical Union Geophysical Monograph, **22**, 87–104

TAYLOR, B. & HAYES, D. E. 1982. Origin and history of the South China Sea basin. *In*: HAYES, D. E. (ed.) *Tectonic and geologic evolution of southeast Asian seas and islands, Part 2*. American Geophysical Union Geophysical Monographs, **27**, 23–56.

TODD, S. P., DUNN, M. E. & BARWISE, A. J. G. 1997. Characterizing petroleum charge systems in the Tertiary of SE Asia. *This volume*.

TONGKUL, F. 1994. The geology of northern Sabah, Malaysia: its relationship to the opening of the South China Sea basin. *Tectonophysics*, **235**, 131–137.

YU, H. S. 1990. The Pearl River Mouth basin: a rift basin and its geodynamic relationship with the southeastern Eurasian margin. *Tectonophysics*, **183**, 177–186.

Predicting reservoir quality during exploration: lithic grains, porosity and permeability in Tertiary clastic rocks of the South China Sea basin

R. H. WORDEN[1], M. J. MAYALL[2] & I. J. EVANS[2,3]

[1] *School of Geosciences, The Queen's University of Belfast, Belfast BT7 1NN, UK*
[2] *BP Exploration Operating Company Limited, Uxbridge One, 1 Harefield Road, Uxbridge, Middlesex UB8 1PD, UK*
[3] *Present address: BP Exploration Operating Company Limited, Sherwood House, Blackhill Road, Holton Heath Trading Park, Poole, Dorset BH16 6LS, UK*

Abstract: Tertiary sandstones in the Gulf of Thailand, Malay and South China Sea basins have been supplied with sediment rich in pelitic metamorphic rock fragments, weathered basic igneous rock fragments and micaceous rock fragments by similar drainage systems. We have analysed data from the Gulf of Thailand and Malay basins to predict reservoir quality in the South China Sea basin. These Tertiary sandstones contain 20–40% lithic rock fragments that are rich in clay minerals. There are good correlations between both porosity and permeability and the total clay mineral content. They undergo rapid loss of porosity with depth at a rate commensurate with sandstones bearing 20–40% ductile grains. The relationship between porosity and permeability is also strongly influenced by the presence of ductile grains. The rapid loss of permeability with porosity is a result of ductile grains being squeezed between the rigid grains, blocking pore-throats and leaving isolated, ineffective porosity. Mineral cements seem to be unimportant modifiers of porosity or permeability. The rapid loss of porosity with depth and the decrease in permeability with porosity leads to low permeabilities at shallow depths relative to many other hydrocarbon provinces (e.g. North Sea, Gulf of Mexico, etc.) which have lower ductile grain contents. We have used modelled compaction curves for sandstones with 20 and 40% ductile grains to predict a porosity value for any given depth. We have also used sandstones of 100 and 300 μm grain size to predict permeability for a given porosity. The combination of these predictions can be used to predict permeability as a function of depth of burial. The choice of ductile content and grain size will depend upon the conditions of deposition. The main positive reservoir quality anomaly may result from overpressure generation before or during the final phase of burial.

Reservoir quality is one of a number of critical parameters in petroleum systems. Porosity is required to store the hydrocarbons in the reservoir whilst permeability is needed to allow them to flow out of the rocks. A theoretical framework to predict *ab initio* reservoir quality in sandstones does not exist largely due to the very large number of variables that influence the evolution of porosity and permeability during deposition and diagenesis. Nevertheless, in many under-developed petroleum provinces, it is important to attempt to establish reservoir quality at the prospect scale because it can critically affect the relative economic viability of prospects or play systems. Under a given set of economic conditions, it is possible to define minimum petroleum (oil or gas) flow-rates required to make a prospect profitable. For sandstones of a given thickness and for defined drawdown pressures (i.e. pressure difference between formation pressure and well-bore pressure), flow-rates depend largely upon permeability. A key and perennial problem is how to estimate permeability at the prospect scale in frontier regions.

In the South China Sea basin (Fig. 1) there are no direct evidence or data that are openly available from the area itself upon which to base a simple predictive model. Unlike the North Sea, where there is a huge database and where the variation of porosity and permeability with depth have been frequently described (see for example the Brent Group of the North Sea: papers in Morton *et al.* 1992), the South China Sea basin Tertiary clastic plays have no well-established reservoir quality predictive schemes. However, it is important to be able to estimate reservoir quality to be able to rank different areas and prospects for their economic viability.

We have adopted a simple and pragmatic approach to facilitate prediction of reservoir quality in the South China Sea basin. First of all,

From Fraser, A. J., Matthews, S. J. & Murphy, R. W. (eds), 1997, *Petroleum Geology of Southeast Asia*, Geological Society Special Publication No. 126, pp. 107–115.

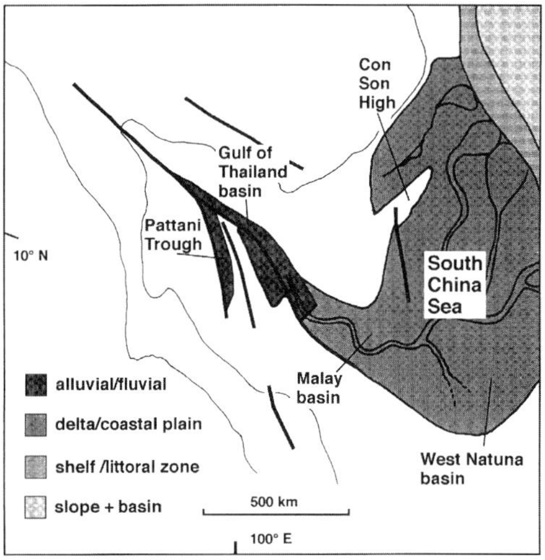

Fig 1. Late Oligocene palaeogeography of the South China Sea after Fraser (pers. comm.). Note that sands in the South China Sea probably have the same general source and transport pattern as the sands in the Gulf of Thailand and the northern Malay Basins.

we have identified that the Oligocene and Miocene sediments of the South China Sea basin were probably sourced by the same major fluvial system that sourced the Oligocene and Miocene sandstones of the Gulf of Thailand and Malay basins (see Fig. 1; Wood 1980; Du Bois 1985). These three neighbouring basins are rift basins that originated following compressional tectonics and have been subject to extensive subsidence and burial since the early Oligocene (Du Bois 1985). All three basins have experienced similar, high geothermal gradients during their burial (approximately 40°C per km; Du Bois 1985). It is therefore likely that the general controls on reservoir quality in the Gulf of Thailand and Malay basins are similar to those in the South China Sea basin.

Consequently, we will review the limited available literature from these basins and discuss the general nature of the Tertiary clastic sediments, likely controls on reservoir quality, porosity–depth relationships, porosity–permeability relationships and ultimately propose a scheme to predict permeability as a function of burial depth. We will also compare our derived predictive relationships to modelled theoretical depth–porosity–permeability patterns. Data available in the literature include sandstone compositions, general depositional environments, porosity–depth patterns and porosity–permeability patterns.

Controls on reservoir quality

Tertiary sandstone compositions: general controls on reservoir quality

The drainage system that sourced and transported the Tertiary sediments in the Malay and Gulf of Thailand basins typically led to lithic rich sandstones (Fig. 2). These sandstones also have fairly low feldspar contents. The figure includes mono- and polycrystalline quartz grains under the grouping of quartzose grains. Thus the lithic grains represent polymineralic rock fragments. In thermodynamic terms the sediment supplied to these basins is mineralogically immature; this does not auger well for reservoir quality. The lithic grain content typically falls between 20 and 40% of the total although individual samples fall outside these limits. Lithic grains are potentially composed of many different minerals although the weathering process will tend to produce clay minerals from ferro-magnesian minerals and feldspars. Thus is it likely that the lithic fragments typically will be rich in clay minerals.

Malay Basin sediment data have been collated to show that the clay mineral content has a strong bearing on reservoir quality (Barr *et al.* 1992; Hill *et al.* 1992). The data in Fig. 3 were taken from a restricted depth interval to minimize differential compaction and diagenetic

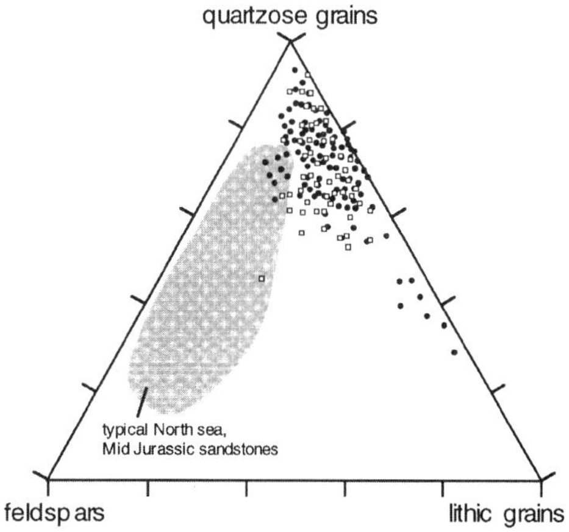

Fig. 2. Sandstone ternary composition diagram illustrating abundance of lithic grains in South China Sea sandstones (after Shing 1992 and Trevena & Clark 1986). The sandstones are rich in lithic grains in comparison to data from other important petroleum-bearing basins. Typical North Sea Jurassic sandstone compositions are shown on the diagram for comparison (after data in Morton *et al.* 1992).

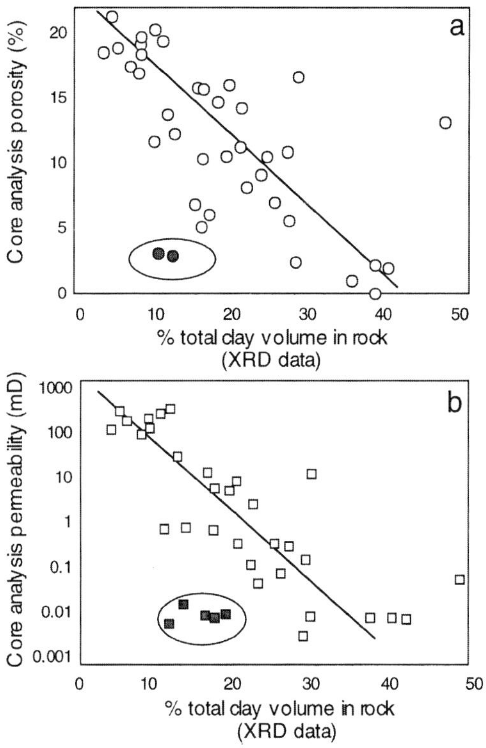

Fig. 3. Comparison of the clay mineral content to core (**a**) porosity and (**b**) permeability (after Hill *et al.* 1992 and Barr *et al.* 1992). The carbonate cemented samples are anomalous and are shown high-lighted. The carbonate cement represents an entirely different control on reservoir quality. Reservoir quality appears to be controlled by the overall content of clay minerals in the rock although localised carbonate cement also reduces reservoir quality. Note that XRD data do not differentiate clay minerals that were initially fine-grained matrix deposited at the same time as the sand grains, diagenetic mineral cements or clay rich sand (lithic) grains.

effects. Both porosity and permeability are strongly correlated with the overall clay mineral content. The clay-contents were estimated from semi-quantitative XRD analysis. This technique does not discriminate between clay-rich lithic fragments, clay matrix, detrital micas or diagenetic clays but it does give an approximation of the total *ductile* grain content of these rocks. There is another subordinate reservoir quality control: carbonate cement. However, this represents a net pay problem rather than a reservoir quality problem. When it is present it almost totally obliterates porosity.

Thus the sandstones are rich in lithic sand grains that probably contain a high proportion of clay minerals and the clay mineral volume appears to be the main control on porosity and permeability.

Porosity–depth patterns

Core porosity data are available over a limited depth range (1500–3000 m) (Fig. 4). The porosity data show an extensive range although there is a general decrease in porosity with increasing depth of burial. It is theoretically possible to force a regression line through the data although this is generally not recommended for predicting outside the range of the data. It is better to build an understanding of the controls on the porosity–depth pattern and then to predict how those controls will evolve or vary with depth of burial.

Porosity in sandstones is controlled by two processes: compaction and cementation (dissolution usually leads to redistributed porosity rather than a net porosity-gain). The former is the result of increasing effective stress during burial that initially leads to the development of an approximately close-packed fabric. The latter is due to the gradual attainment of thermodynamic equilibrium or the changing chemical and physical environment of sandstones as they are buried and leads to the growth of mineral cements. The nature of compactional porosity loss depends critically upon the physical nature of the composite sand grains that is (largely) controlled by the mineralogy of the sand grains. Some minerals have a high shear stress (i.e. they are rigid), others have a low shear stress (i.e. they are ductile). Clay minerals and lithic fragments rich in clay minerals behave in a ductile fashion during burial and compaction. Most other minerals (commonly quartz and feldspars in sandstones) behave in a rigid manner. Thus the way in which compaction affects the porosity of sandstones depends on the proportions of rigid and ductile grains in the sandstone. Sandstones with high ductile grain contents lead to lower porosities than low ductile grain contents at a given depth of burial.

The way in which the proportion of ductile grains affects porosity has been defined experimentally by Atkinson & Bransby (1978) and Kurkjy (1988). Compactional porosity-loss curves constructed from the experimental data are superimposed on Fig. 4. The modelled data agree closely with some of the compacted lithic sand data in Pittman & Larese (1991). There is close correspondence between the core porosity data and the porosity loss curves suggesting that compactional porosity is more important than

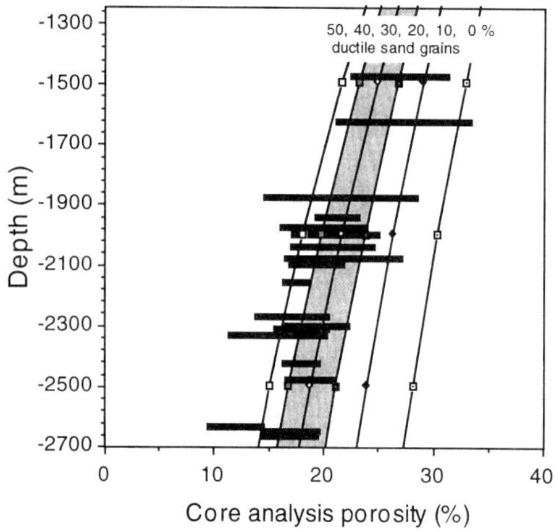

Fig. 4. Porosity–depth plot (after Trevena & Clark 1986) compared to modelled trends. Porosity loss with depth has been modelled as a function of ductile sand grain content (after Atkinson & Bransby 1978; Kurkjy 1988). No account has been taken of the effects of overpressure or the growth of mineral cements. The measured data show that porosity decreases rapidly with increasing depth, commensurate with a sandstone in which the key porosity-loss mechanism is ductile grain compaction. The data generally lie between the modelled 20 and 40% ductile grain content curves (shaded). This agrees with the petrographic data (Fig. 2), which show that most sandstones have between 20 and 40% lithic grains. An optimistic estimate of porosity may be made using the 20% ductile curve, a pessimistic estimate would follow the 40% ductile grain curve.

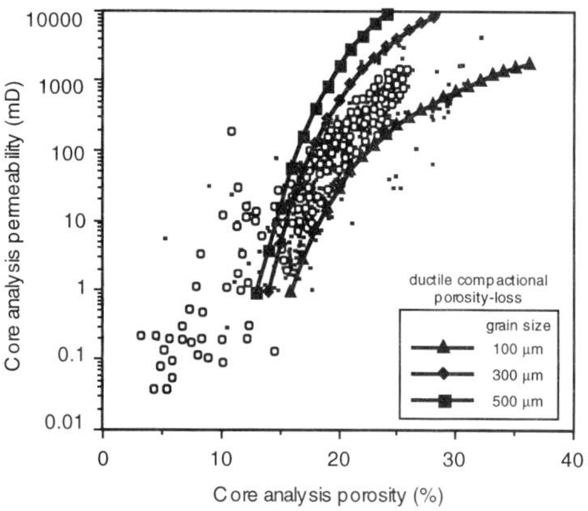

Fig. 5. Porosity–permeability data (after Ibrahim & Madon 1990 and Bowers *et al.* 1994) compared to modelled trends for compactional porosity loss in ductile (i.e. clay-rich) sandstones of different grain size (after Cade *et al.* 1994 and Evans *et al.* 1994). The data show a rapid decrease in permeability over a limited porosity range commensurate with sandstones for which reservoir quality is largely controlled by the compactional porosity loss of ductile grains. Permeability may be predicted from porosity using the modelled 300 and 100 μm curves for ductile grain compaction.

cementation in these rocks. The data are generally enclosed by the curves possibly reflecting that the sandstones have a variable ductile grain content although the majority (approximately 80% of the data) of the data falls within the 20 and 40% ductile grain content curves. Thus there is good agreement between the petrographic and XRD data and the core porosity data (considering the experimental compaction curves). We can conclude that these sandstones contain abundant lithic fragments rich in clay minerals and that these lithic fragments control porosity through ductile grain compaction.

Porosity–permeability patterns

Core analysis data are commonly used to predict permeability from porosity. Core analysis data from the Gulf of Thailand and Malay Basins fall in the same area on a porosity–permeability diagram (Fig. 5) suggesting that permeability in the sandstones in these basins is controlled by the same critical parameter. Analysis of the ϕ–k trend shows that these sandstones have relatively low permeability for a given porosity in comparison to, for example, Brent Group sandstones from the North Sea (e.g. Morton *et al.* 1992) or Tertiary sandstones from the Gulf of Mexico. The sandstones from SE Asian basins are intrinsically poorer hydrocarbon producers than sandstones in many other hydrocarbon provinces.

We have already shown that the ductile grain content is a key control on porosity (Fig. 4). It is also likely that ductile grains will exert a critical control upon permeability. Permeability results from pore spaces being connected so that, in the extreme, a rock containing isolated pores will be impermeable. A critical parameter is the average pore throat radius, high values equate to high permeability for a given porosity and vice versa. The different porosity-loss mechanisms discussed earlier will have different effects upon permeability because they affect the pore connectedness, i.e. pore-throat radii, differently. The effects of the different porosity-loss mechanisms upon permeability have been modelled theoretically by Bryant *et al.* (1993), Evans *et al.* (1993) and Cade *et al.* (1994) using coordinates from a random sphere pack, network

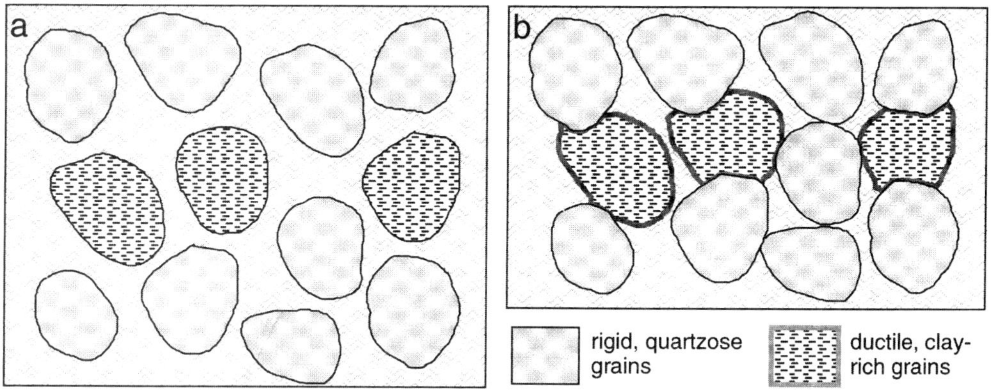

Fig. 6. Schematic diagram of the mechanism of ductile sand grain compaction for a sandstone with about 25% ductile sand grain content (nominal grain size of 200 μm). (**a**) The rock at deposition has typical sandstone porosities of about 40–45%. (**b**) Upon burial to about 2500 m, the porosity has been reduced to about 20% and the permeability to about 50–100 mD. The clay-rich ductile grains have deformed around the rigid quartzose grains as the effective stress of the system has increased during burial. The ductile grains may have recrystallised (probably *in situ* due to the profound insolubility of the component aluminium) leading to neomorphic clay crystals on the surfaces of the ductile grains and possibly ineffective microporosity. The equivalent porosity and permeability for a compacted quartz arenite would be about 28% and up to 10 000 mD respectively.

modelling and idealised porosity-loss processes. They concluded that ductile-grain compaction should have a profoundly different effect upon permeability than rigid-grain compaction. Ductile grain compaction reduces permeability by blocking pore throats as the ductile grains deform and 'flow' between the rigid grains as effective stress is applied to the rock (Fig. 6). The result is a growing number of isolated pores during burial leading to relatively low permeability for a given porosity. Initial grain size also plays an important role in controlling permeability (Beard & Weyl 1973). We have therefore superimposed three curves (100, 300 and 500 μm grain size sandstones) of the modelled effects of ductile grain compaction upon the core analysis data from the Malay and Gulf of Thailand basins (Fig. 5) following Evans *et al.* (1993) and Cade *et al.* (1994). The majority of the data are enclosed between the 100 and 300 μm ductile-grain compaction curves. There are a number of very low permeability and porosity data that fall outside the scope of the modelled relationships; these may represent extreme cementation (by carbonate cement) as indicated by the XRD data (Fig. 3). The close agreement between the data and the modelled 100 and 300 μm curves in Fig. 5 shows that the evolution of permeability as well as porosity is probably controlled by ductile grain compaction.

Predicting reservoir quality in the South China Sea basin

Palaeogeographic considerations show that the drainage system that supplied the Malay and Gulf of Thailand basins also supplied the South China Sea basin (Fig. 1). Thus it is likely that the controls on reservoir quality are the same in the three basins. We have shown that (i) Tertiary sandstones from the Malay and Gulf of Thailand basins are rich in lithic sand grains (20–40% on average), (ii) that the total clay mineral content has a profound effect upon porosity and permeability, (iii) that porosity loss during burial is largely controlled by compaction reflecting sandstones with 20–40% ductile grains and (iv) that permeability is also largely controlled by ductile grain compaction. We are thus in a good position to predict porosity and permeability at the prospect scale based on depth-converted seismic data.

Predicting porosity

Figure 4 clearly shows that porosity can vary at a given depth of burial. All the signs are that the variability is due to different ductile grain contents within the sandstones. It would therefore be incorrect to force one regression line through porosity–depth data. This is especially important when trying to predict outside the range of the data. We have shown that the sandstones seem to fall within the 20 and 40% ductile grain content curves (confirmed by the petrographic data, Fig. 2). We have therefore used the 20 and 40% ductile compaction curves to predict an optimistic and a pessimistic case respectively. These curves may be used to predict porosity at greater burial depths than the data (>2700 m) although it must be appreciated that other porosity-affecting processes may come into play at depth. Examples of this are temperature-sensitive quartz cementation, illite precipitation or the development of overpressure. No attempt has been made to incorporate these processes into the prediction because the theoretical basis for predicting if, when, or to what extent these diagenetic processes may occur is yet imperfect.

Predicting permeability

A commonly used method for predicting permeability is to force an exponential regression curve through core analysis data. We could do that for the data from the Malay and Gulf of Thailand basins. However, we have a detailed understanding of the controls on permeability and we have modelled curves that closely fit the spread of the data (Fig. 5). The vast majority of the core data that represent net pay fall between the 100 and 300 μm grain size ductile compaction curves. Consequently we will use the former as the pessimistic case and the latter as the optimistic case respectively to predict permeability from a given porosity value. This range realistically reflects the range of reservoir quality and reflects the uncertainty in the prediction.

Thus we can predict porosity as a function of depth and permeability as a function of porosity so that we can now predict permeability as a function of depth. However, we have selected optimistic and pessimistic predictive curves that produce a matrix of permeability prediction with depth. The best case will be represented by the 20% ductile compaction curve and the 300 μm sandstone. The worst case will be represented by the 40% ductile compaction curve and the 100 μm sandstone. These best and worst case predictions have been superimposed upon permeability–depth data from the Gulf of Thailand basin (Fig. 7a). The predictive curves enclose

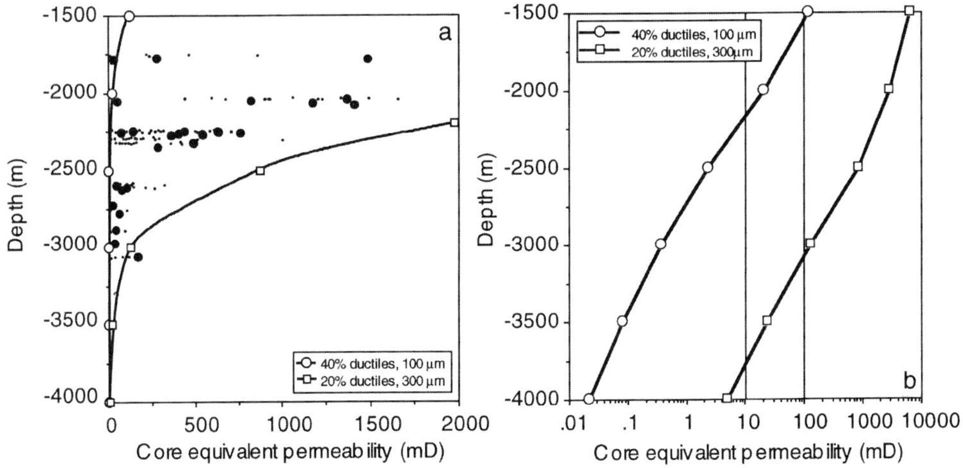

Fig. 7. (**a**) Permeability–depth data (after Bowers *et al.* 1994) showing rapid curtailment of permeability with depth. The depth–porosity modelled curves have been combined with the porosity–permeability curves to enable the prediction of permeability from depth. Curves have been defined for 300 μm sands with a 20% ductile content and 100 μm sands with a 40% ductile content to represent a best and worse case respectively. The modelled curves comfortably enclose the data. The curves can also be used to predict beyond the range of the data (i.e. to greater depths). (**b**) Modelled permeability–depth relationship (as in Fig. 6a) using a logarithmic permeability scale. Arbitrary cut-off permeabilities of 10 and 100 mD have been added to reveal the economic basement for the two types of sandstone. For the better sand requiring a low core permeability (as in a gas field), basement is at about 3700 m, for a higher core permeability (as in an oil prospect), basement is at 3000 m. For the poorer sand, basement is at 2200 m for gas and 1500 m for oil. The equations of these curves (or any intermediary case) may be used to compute average reservoir quality as a function of depth (for a given sedimentological model, overpressure model) and be used to devise the depth of economic basement.

practically all the data confirming that the predictive approach is probably correct. The shape of the best case curve reflects the limit of the best permeability data although the best case is a slight overestimate. The agreement between the predictive curves and measured data therefore gives us confidence to use this approach to predict permeability at the prospect level.

Some general insights into the value of permeability prediction can be gained by examining Fig. 7b that contains the predictive curves on a logarithmic permeability axis. Gas fields require lower permeability to be economic than oil fields so that we have arbitrarily selected 10 and 100 mD respectively as the minimum permeabilities required to achieve economic flowrates (the specific critical permeabilities will depend upon the licence agreement and global oil/gas price). These permeability limits have been added to Fig. 7b. In the gas case, economic basement is between 2200 and 3800 m. In the oil case, economic basement is between 1500 and 3000 m. The greater depths can be used as a guide to the maximum depth to which exploration wells should be drilled depending on the anticipated hydrocarbon charge. The two depths for each case represent variations in grain size and ductile grain content of the sand. These are probably controlled by sedimentological factors at the time of deposition about which it might be possible to be more predictive using sediment-source and depositional models for the basin.

Predicting the effects of the anomalous preservation of porosity

There are two commonly cited reasons for anomalously good reservoir quality: early oil charge and overpressure generation before, or during, compaction. The former is thought to operate because oil emplacement inhibits inorganic diagenetic process therefore preventing mineral cementation (discussed at length by Worden *et al.* 1997, with references to earlier work). In the South China Sea basin, mineral cements are not the main cause of porosity loss (see earlier discussion and Figs 2, 3, 4 and 5). It is not clear that oil emplacement would necessarily prevent compaction although the elevated capillary pressure associated with the steepened hydrostatic pressure in hydrocarbon columns may locally reduce effective stress and inhibit compaction. Regional overpressure generation is a more likely cause of anomalous porosity-preservation. Predicting regional overpressure is outside the scope of this paper although the effects of overpressure generation could be converted into enhanced porosity, and therefore permeability, due to the reduction of effective stress using the predictive curves on Figs 4 and 5. Overpressures as low as 1000 psi may improve the permeability by up to an order of magnitude because compactional porosity loss is the main control on permeability.

Global perspective

Economic basement for oil at 3000 m (Fig. 7b) is much shallower than many hydrocarbon provinces. The general reason for such a shallow economic basement is that the primary sediment was mineralogically immature probably due to abundant ferro-magnesian mineral-rich rocks in the provenance area (e.g. metapelites or metabasites) as well as rapid uplift and erosion limiting the time available for weathering and decomposition of sand grains composed of ferro-magnesian mineral. Other rift basins in the proximity of continental collision zones (where metamorphic rocks provide the sediment) are likely to contain sandstones of similar poor reservoir quality potential.

Other ductile lithic-rich sandstones show similar behaviour to SE Asian sandstones. They typically display rapid loss of porosity with increasing depth (e.g. Smosna 1989; McBride *et al.* 1991). They also display a steep porosity–permeability trajectory commensurate with the predicted ductile compaction trends and data in Fig. 5 (e.g. Ranganathan & Tye 1986; Bloch *et al.* 1990, Sousa *et al.* 1995). However, care must be taken in comparing the SE Asian sandstones with any other province because their diagenetic history appears to be particularly simple.

In comparison to ductile sand provinces, the Jurassic North Sea basin sub-arkosic sandstones have intrinsically better prospects of retaining good reservoir quality because they are probably second-cycle sediments (double the opportunity to develop mineralogical maturity). The lower heat flow in the North Sea basin may also contribute to the better reservoir quality potential. Thus whilst burial to 3000 m reduces permeability to a maximum of 100 mD in the SE Asian basins discussed thus far, permeabilities of the order of 1000 mD are present in North Sea sandstones at 4000 m burial (Bjørlykke *et al.* 1992; Giles *et al.* 1992). Indeed, many passive margin or failed rift basins where the sediment is sourced from slowly eroding and weathering hinterlands, have similar good permeabilities as found in the Jurassic reservoirs of the North Sea basin.

Conclusions

(1) We have developed a scheme to predict porosity and permeability in the frontier South China Sea basin by making the assumption that the same drainage system carrying the same sand compositional type that filled the Gulf of Thailand and Malay basins also filled the South China Sea basin. The database used to make the prediction contains a very limited quantity of sedimentological and petrophysical data from Tertiary sandstones from the Gulf of Thailand and Malay basins.

(2) The Tertiary sandstones of these basins contains 20–40% clay-rich lithic fragments. For a given depth interval, bulk rock XRD data show that there is good correlation between reservoir quality and clay mineral content.

(3) Porosity decreases with increasing depth commensurate with sandstones that contain 20–40% ductile lithic grains. It appears that ductile grain compaction is the dominant porosity-loss mechanism. Mineral cementation may be relatively insignificant.

(4) Permeability declines steeply with decreasing porosity due to ductile grain compaction. The ductile grains become deformed between rigid grains and tend to block pore throats leaving unconnected porosity. The analytical data correspond to the ductile compaction of sandstones between 100 and 300 μm grain size.

(5) Permeability may be predicted as a function of burial depth during exploration by combining porosity–depth curves with porosity–permeability curves (Fig. 7b). Predicted permeability–depth curves agree well with analytical data. Permeability declines steeply with increasing depth leading to a predicted shallow economic basement in the South China basin (relative to other basins containing lithic-poor sandstones).

(6) The most likely cause of anomalously good reservoir quality in the South China basin is the development of 'hard' overpressure pressure during burial. Even moderate overpressures may lead to good permeabilities because compaction is the chief cause of reservoir quality destruction.

(7) The prediction of permeability is not an exact science but is as good as the sedimentological model of the sandstone, the depth conversion of the seismic data and the overpressure model. It should be used in conjunction with all the other tools available to the explorer to rank prospects and fairways and should not be expected to give exact permeabilities.

C. Cade and J. Gluyas are credited for the development of the compaction curves for quartzose and ductile sandstones. A full account of this work will be published by them in due course. D. Swarbrick is thanked for drawing to our attention some of the publications cited in this paper. Anonymous reviewers are thanked for a careful examination of the manuscript.

References

ATKINSON, J. H. & BRANSBY, P. L. 1978. *The mechanics of soils. An introduction to critical state soil mechanics.* McGraw-Hill, England.

BARR, D. C., KENNAIRD, A. & ALTUNBAY, M. 1992. Understanding reservoir behaviour through an integrated geological and engineering study. *Geological Society of Malaysia Bulletin*, **32**, 45–57.

BEARD, D. C. & WEYL, P. K. 1973. Influence of texture on porosity and permeability of unconsolidated sand. *American Association of Petroleum Geologists Bulletin*, **42**, 578–587.

BJØRLYKKE, K., NEDKVITNE, T., RAMM, M. & SAIGAL, G. C. 1992. Diagenetic processes in the Brent Group (Middle Jurassic) reservoirs of the North Sea: an overview. *In*: MORTON, A. C., HASZELDINE, R. S., GILES, M. R. & BROWN, S. (eds) *Geology of the Brent Group.* Geological Society, London, Special Publications, **61**, 263–287.

BLOCH, S., MCGOWEN, J. H., DUNCAN, J. R. & BRIZZOLARA, D. W. 1990. Porosity prediction, prior to drilling, in sandstones of the Kekiktuk Formation (Mississippian), North Slope of Alaska. *American Association of Petroleum Geologists Bulletin*, **74**, 1371–1385.

BOWERS, M. C, EHRLICH, R. & CLARK, R. A. 1994. Determination of petrographic factors controlling permeability using petrographic image analysis and core data, Satun Field, Pattani Basin, Gulf of Thailand. *Marine and Petroleum Geology*, **11**, 148–156.

BRYANT, S., CADE, C. A. & MELLOR, D. 1993. Permeability prediction from geologic models. *American Association of Petroleum Geologists Bulletin*, **77**, 1338–1350.

CADE, C. A., EVANS, I. J. & BRYANT, S. L. 1994. Analysis of permeability controls: a new approach. *Clay Minerals*, **29**, 491–501.

DU BOIS, E. P. 1985. Review of principal hydrocarbon-bearing basins around the South China Sea. *Geological Society of Malaysia Bulletin*, **18**, 167–209.

EVANS, I. J., BRYANT, S. L. & CADE, C. A. 1993. Modelling the effects of diagenetic cements on sandstone permeability. *In*: PARNELL, J., RUFFELL, A. H. & MOLES, N. R. (eds) *Contributions to an international conference on fluid evolution, migration and interaction in rocks*, 212–215.

GILES, M. R., STEVENSON, S., MARTIN, S. V., CANNON, S. J. C., HAMILTON, P. J., MARSHALL, J. D. & SAMWAYS, G. M. 1992. The reservoir properties and diagenesis of the Brent Group: a regional perspective. *In*: MORTON, A. C., HASZELDINE, R. S., GILES, M. R. & BROWN, S. (eds) *Geology of the Brent Group.* Geological Society, London, Special Publications, **61**, 289–327.

HILL, J. A., SOO, D. K. Y. & VERRIAH, T. 1992. Clay mineralogy in subsurface sandstones of Malaysia and the effects on petrophysical properties. *Geological Society of Malaysia Bulletin*, **32**, 15–43.

IBRAHIM, N. A. & MADON, M. 1990. Depositional environments, diagenesis and porosity of reservoir sandstones in the Malong Field, offshore West Malaysia. *Geological Society of Malaysia Bulletin*, **27**, 27–55.

KURKJY, K. A. 1988. *Experimental compaction studies of lithic sands*. MSc thesis, University of Miami.

MCBRIDE, E. F., DIGGS, T. N. & WILSON, J. C. 1991 Compaction of Wilcox and Carrizo sandstones (Paleocene-Eocene) to 4420 m, Texas Gulf Coast. *Journal of Sedimentary Petrology*, **61**, 73–85.

MORTON, A. C., HASZELDINE, R. S., GILES, M. R. & BROWN, S. (eds) *Geology of the Brent Group*. Geological Society, London, Special Publication, **61**.

PITTMAN, E. D. & LARESE, R. E. 1991. Compaction of lithic sands: experimental results and application. *American Association of Petroleum Geologists Bulletin*, **75**, 1279–1299.

RANAGATHAN, V. & TYE, R. S. 1986. Petrography, diagenesis and facies controls on porosity in Shannon sandstone, Hartzog field, Wyoming. *American Association of Petroleum Geologists Bulletin*, **70**, 56–69.

SHING, C. Y. 1992. Petrographic and diagenetic studies of the reservoir sandstone of the Malay Basin. *Geological Society of Malaysia Bulletin*, **32**, 261–283.

SMOSNA, R. 1989. Compaction law for Cretaceous sandstones of Alaska's North Slope. *Journal of Sedimentary Petrology*, **59**, 572–584.

SOUSA DE, R. S., ROS DE, L. F. & MORAD, S. 1995. Dolomite diagenesis and porosity preservation in lithic reservoirs: Carmopolis Member, Sergipe-Alagoas basin, northeastern Brazil. *American Association of Petroleum Geologists Bulletin*, **79**, 725–748.

TREVENA, A. S. & CLARK, R. A. 1986. Diagenesis of sandstone reservoirs of Pattani Basin, Gulf of Thailand. *American Association of Petroleum Geologists Bulletin*, **70**, 299–308.

WOOD, P. W. J. 1980. Hydrocarbon plays in Tertiary S. E. Asia basins. *Oil and Gas Journal*, **78** (29), 90–96.

WORDEN, R. H, OXTOBY, N. H. & SMALLEY, P. C. 1997. Does oil emplacement stop quartz cementation in sandstones? *Petroleum Geoscience*, **3**, in press.

Miocene carbonate buildups offshore Socialist Republic of Vietnam

M. J. MAYALL, A. BENT & D. M. ROBERTS

BP Exploration, Chertsey Road, Sunbury-on-Thames, Middlesex TW16 7LN, UK

Abstract: A variety of carbonate accumulations can be identified offshore Vietnam. The best reservoirs are developed in large, fault-controlled, buildups which have undergone extensive leaching during emergence. More moderate reservoir quality is present in platform facies which extend over large areas and in small buildups usually developed on footwall crests. This paper describes the setting, facies, diagenesis and reservoir quality of the various carbonate types.

A variety of carbonate buildups, ranging in age from Middle Miocene to Lower Pliocene, can be observed offshore Vietnam (Matthews *et al.* this volume figs 6–8). This paper presents a brief review in terms of their setting, facies, diagenesis and reservoir quality.

Five main types of carbonate accumulations can be recognized (Fig. 1):

Platforms, broad areas of essentially flat carbonates;
Small buildups, typically 1–2 km across;
Large buildups, typically 5–10 km across;
Pinnacles; more uncommon but have a higher thickness to area ratio then the big buildups;
Pelagic carbonates, developed on structural highs.

Platforms

The platforms comprise intervals 50–200 m thick, of horizontal, strong, continuous parallel seismic reflectors. Locally minor mounding is present. The platform facies covers large areas and underlies the other types of carbonate accumulation. It is absent in deeper water areas and towards the palaeoshorelines becomes a mixed carbonate–clastic sequence.

The facies are dominated by foraminiferal packstones and wackestones. In one example penetrated by a well the platform can be divided into three broad intervals:

(1) a lower interval of bioclastic wackestones and packstones deposited in >50 m water depth, separated by thin deeper water calcareous shales from
(2) a middle interval of bioclastic packstones deposited in 10–50 m water depth, separated by a thin deeper water calcareous shale from
(3) an upper interval comprising bioclastic wackestones and mudstones deposited in >50 m water depth with mixed reworked shallower water fauna presumably derived from a small buildup a few kilometres from the well location.

Reservoir quality in the platform facies is generally moderate, porosity is mostly <15% and permeability <10 mD. Porosity has been excluded largely by compaction, giving a rapid decrease in reservoir quality with depth. Cements are also present in the form of an early marine isopachous cement and a blocky pore filling spar. Fluid inclusion analysis of the blocky cements indicates that they precipitated, at least in part, at relatively high temperatures from the late Pliocene to present. A few intervals of better-quality reservoir are present and higher permeabilities may be present due to fracturing associated with faults or gentle compression.

Potential traps in the platform carbonates are formed by small four-way dip or fault closure.

Small buildups

The small buildups are typically 1–2 km across and 100–200 m thick. On seismic surveys they show a mounded external form with weakly reflecting to slightly mounded internal character. They are mostly developed on the footwall crests of tilted fault blocks and in some places show some evidence of growth during the faulting. There are also some small buildups which appear to be rooted in the underlying platform facies without any prominent structural control. Facies data on these small buildups is limited but they appear to comprise bioclastic packstones. Reservoir quality data is also limited with moderate porosities but with a few higher porosity zones appears to be typical. As with the platform facies there may be a strong depth control on the reservoir quality. Potential traps are formed by three- or four-way dip drape over the buildups.

From Fraser, A. J., Matthews, S. J. & Murphy, R. W. (eds), 1997, *Petroleum Geology of Southeast Asia*, Geological Society Special Publication No. 126, pp. 117–120.

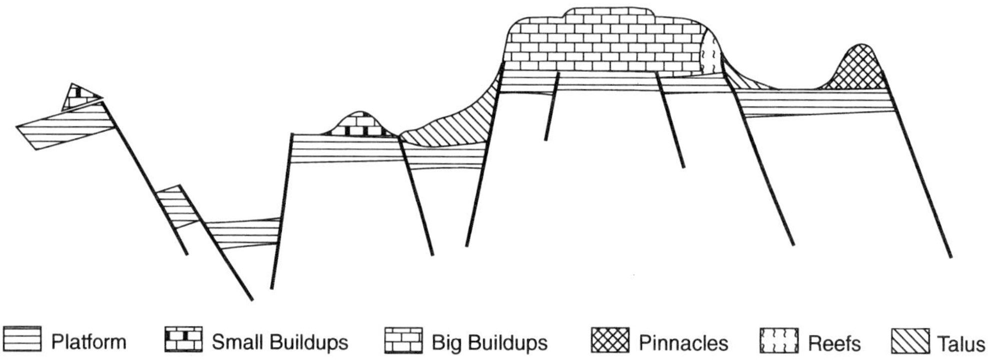

Fig. 1. Schematic summary of the carbonate types offshore Vietnam.

Pinnacles

Pinnacles are less common and are separated here from the large buildups by their smaller areal extent. The pinnacles are typically 300 m thick but only 2–5 km across. They form prominent features although the internal seismic character is difficult to resolve. They mostly seem to comprise chaotic or mounded seismic facies. No facies or reservoir quality information on the pinnacles is available. Potential traps would be formed by four-way drape closure.

Large buildups

The large buildups are the most spectacular carbonate accumulations in the area with a prominent vertical relief. They are typically 5–10 km across and ≥300 m thick. The margins of the buildups are usually sharp and steep, being defined by large deeper faults. In a few cases, however, they are more gradational and have less-steep margins. Well-defined talus accumulations are developed around the buildups particularly along the fault controlled margins. The top of the buildups is generally quite sharp on seismic data (Fig. 2) (although there are locally difficulties in picking top carbonate) and they often show a well-defined back-stepping pattern with the area of the carbonate accumulation becoming smaller with time. The interior of the big buildups are dominated by variable strong to weak semi-continuous seismic reflectors with a few more conspicuous continuous reflectors. Along the buildup margins rather narrow (<0.5 km) discontinuous zones of chaotic-mounded facies with convex compaction over them are interpreted as coral–algal framestone reefs. The buildups are underlain by the platform facies described above with the top of the platform facies often showing extensive karstification indicated by the presence of large open fissures.

Data on the facies in the buildup is entirely from the buildup interior. A wide range of facies are present but bioclastic (coral, red algae and benthic forams) packstones dominate with wackestone, grainstones and local framestones also present. The facies are arranged in variably well-defined shallowing-upwards sequences 5–20 m thick. Evidence for repeated exposure and karstification is in the form of large solution fissures, 1–10 cm wide and at least several metres deep (as seen in cores). The fissures are occasionally open but are mostly filled by carbonate sediment and often have a very thin iron-stained rim to them. In some areas of the buildup interior the karstic surfaces are locally stacked while other areas may contain only a few exposure surfaces. This strongly suggests that areas of the large buildups were repeatedly emergent while other areas remained mostly below sea level. In some cases these emergent areas may be controlled by deeper faults within the buildup interior.

The overall shallow-water facies are punctuated by several major deepening events across the buildups which resulted in the deposition of carbonate mudstones with pelagic foraminifera. It is these deeper-water events which cause the more conspicuous continuous seismic reflectors across the buildups.

Shallow-water carbonates with exposure surfaces continue almost to the top of the buildup but appear to be abruptly overlain by pelagic carbonates.

The large buildups contain excellent reservoir quality. Porosities are 20–40% and permeabilities 200–2000 mD. Porosity is dominated by

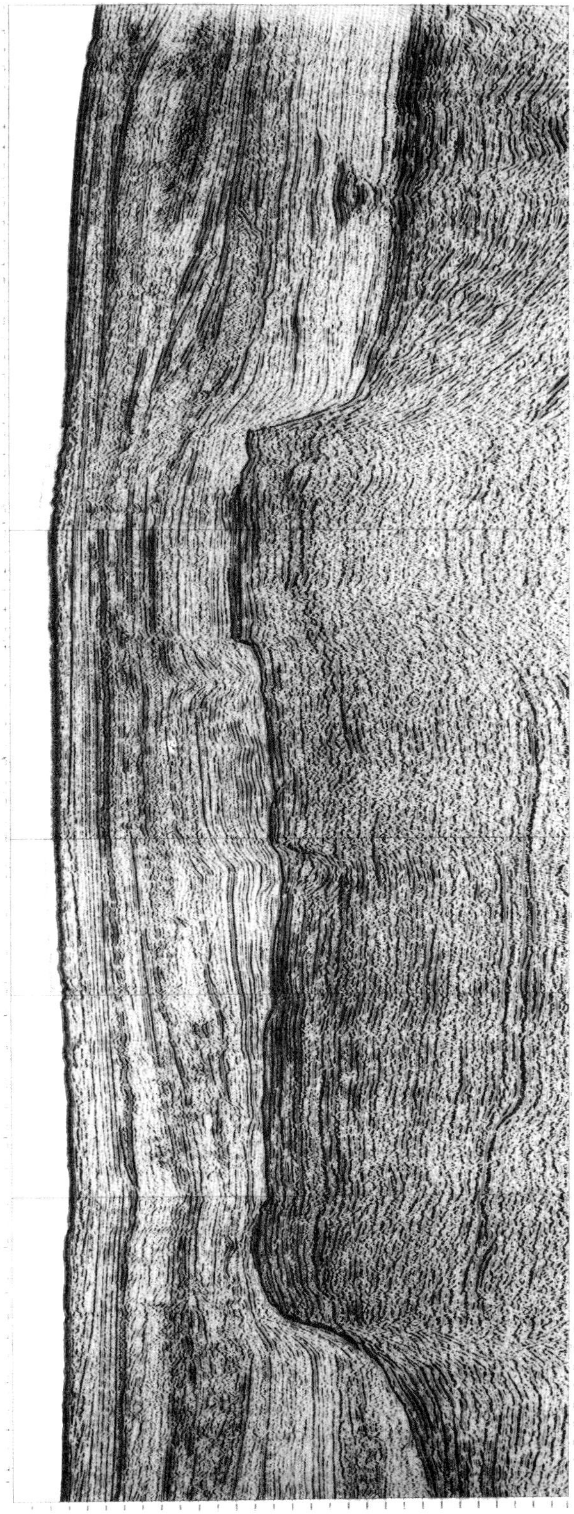

Fig. 2. Seismic line across large buildup. The buildup is approximately 12 km long with a vertical relief of about 400 m. Note the steep fault controlled margins, the talus cones, especially on the left hand end and the prominent back-stepping geometry.

leaching of bioclasts, presumably during emergence. Pore occlusion by blocky spar cements is very localised and is usually most obvious at the karstic surfaces. Dolomitization is extensive and very limited isotopic data suggests that it is of mixed water origin.

The large buildups form the best potential reservoirs in the area, traps would be formed by four-way drape closure. Potential reservoir heterogeneities are the finer grained deep water events, the vertical arrangements of the facies in shallowing-upwards sequences, cementation and/or open fissures associated with the karst surfaces and perhaps more cemented reef frameworks at the buildup margins.

Pelagic carbonates

Pelagic carbonates, generally less then 5 m thick, are developed at the crest of the big buildups or over other structural highs. They comprise bioclastic packstones-wackestones composed almost entirely of pelagic foraminifera. Porosity may be high (>30%) but permeabilities are low. The pores are mostly intraskeletal and poorly connected.

Conclusions

(1) Platform carbonates and small buildups form widespread carbonates and low relief mounds dominated by bioclastic packstones and wackestones. Reservoir quality is moderate but there may be a few higher porosity and permeability intervals.

(2) The large buildups form thick accumulations with fault-controlled margins. Excellent reservoir quality is present due to extensive leaching during exposure.

References

MATTHEWS, S. J., FRASER, A. J., LOWE, S., TODD, S. P. & PEEL, F. J. 1997. Structure, stratigraphy and petroleum geology of the southeast Nam Con Son Basin, offshore Vietnam. *This volume*.

Exploration history of the offshore Southeast Sumatra PSC, Java Sea, Indonesia

A. WIGHT[1], H. FRIESTAD[2], I. ANDERSON[1], P. WICAKSONO[1] & C. H. REMINTON[3]

[1] *Maxus Southeast Sumatra, Inc., PO Box 2759, 7th Fl., Jakarta Stock Exchange Bldg., Jl. Jend. Sudirman Kav. 52/53, Jakarta 12190, Indonesia*
[2] *Maxus International Inc., 717 North Harwood Street, Dallas, Texas 75201, USA*
[3] *Pertamina-BPPKA (Indonesian State Oil Company), Jl. Merdeka Timur 1A, Jakarta Pusat, Indonesia*

Abstract: The Offshore Southeast Sumatra Production Sharing Contract (PSC) is situated in the Java Sea, just north of Indonesia's capital, Jakarta. It was the first to produce offshore oil in Indonesia and has so far produced over 0.8×10^9 BOE. The crude is low sulphur, waxy and sourced by Oligocene lacustrine shales. Over 200 exploration wells have been drilled from 1970 to the present, with 21 commercial accumulations discovered. Reservoirs are Oligocene to early Miocene in age with around 25% of the reserves in carbonates and 75% in fluvial–alluvial clastics. The main basins, Sunda and Asri, are retro-arc, N–S-oriented, asymmetric half-grabens situated between the stable Sunda Shield and the active volcanic arc on Java island.

This paper describes the acquisition of the area and traces its history from the initial discovery at Cinta-1 in 1970, through the oil price rise and fall of the 1980s, to the discovery of a new productive basin in 1987. Exploration is not only reliant on technological advances and available budgets, so other factors relevant to the history of exploration are also included.

The Offshore SE Sumatra Production Sharing Contract (SES PSC) is situated in the west Java Sea adjacent to the coasts of west Java and southeast Sumatra in Indonesia (Fig. 1). It has been one of the prime oil producing areas of Indonesia for the last 25 years, having produced a cumulative 800×10^6 barrels of oil at an average of around 90 000 BOPD. The area is favourably located near the capital city and port of Jakarta. It lies in shallow waters with relatively calm sea conditions, facilitating jackup drilling for the mainly shallow, Tertiary age reservoirs. Twenty one commercial accumulations (Fig. 2) have so far been discovered, with just over 200 exploratory wells, an overall 10% success rate. The pace of exploration, apart from the overriding dictates of oil price, has also been variously influenced by contract terms, production incentives, technological advances and company strategies.

After a brief summary of the hydrocarbon system and geological history, this paper turns to more recent events; the history of exploration. After the initial negotiations for the PSC, themselves forming an historic landmark in formulating a production-sharing agreement, exploration in the Offshore Southeast Sumatra PSC began in 1969 with an extensive seismic survey covering the 123 995 km^2 (47 875 square miles) original area. The PSC area has since undergone five relinquishments and a 20 year extension (Fig. 3). Drilling began in January 1970 and the first commercial discovery was made by the ninth well, Cinta-1, in August 1970. It was rapidly placed on production, using a temporary caisson, in only two months, to become Indonesia's first ever offshore production.

This was followed by the ups and downs of further discoveries and concomitant high exploration and development activity, then reduced drilling in the mid-1970s, but re-building to a climax in exploration drilling with the high oil prices of the early 1980s. Thereafter, the cycle was completed by the equally dramatic fall in oil price and consequent reduction in exploration to an all-time low. These swings in activity occurred in parallel with the corporate fortunes of IIAPCO, as its main shareholder, Natomas Co., merged with Diamond Shamrock, later to be restructured to Maxus SE Sumatra, and which has recently become a unit of the major Argentinian company, YPF.

The events leading up to the relatively late discovery of the second largest field in the whole PSC, Widuri, in the under-explored basin to the NE of Sunda are then recalled. These began during an interlude when drilling had virtually

From Fraser, A. J., Matthews, S. J. & Murphy, R. W. (eds), 1997, *Petroleum Geology of Southeast Asia*, Geological Society Special Publication No. 126, pp. 121–142.

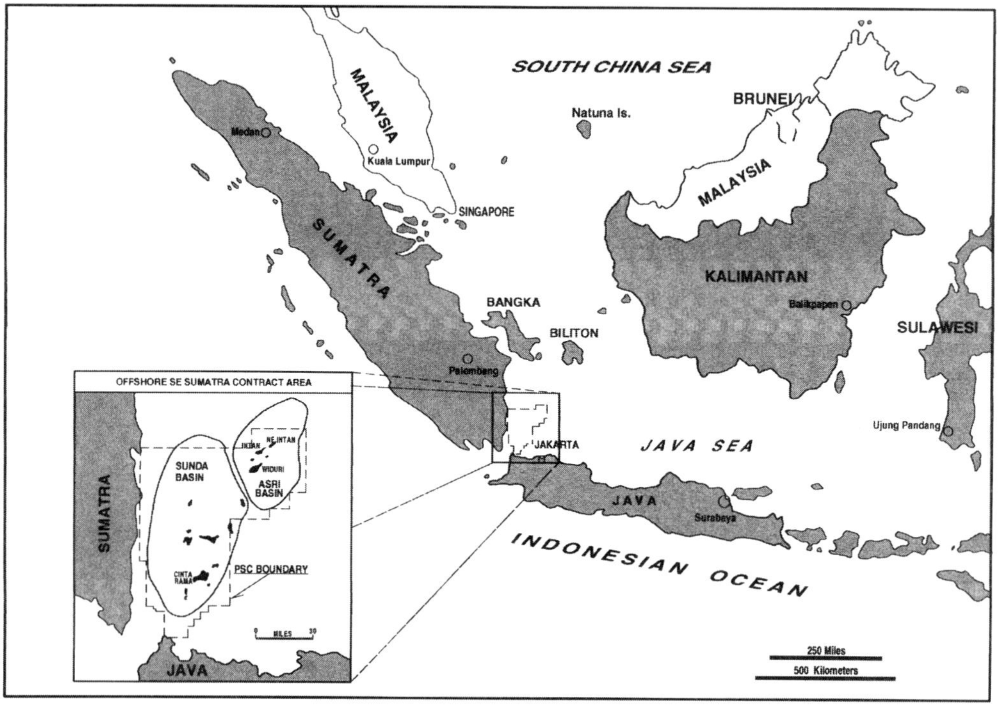

Fig. 1. Location map. The Offshore SE Sumatra PSC is located in the western Java Sea, adjacent to the SE coast of Sumatra, NW of Java and the capital city and port of Jakarta. Water depths are all less than 60 m (200 ft) and weather largely devoid of major storms.

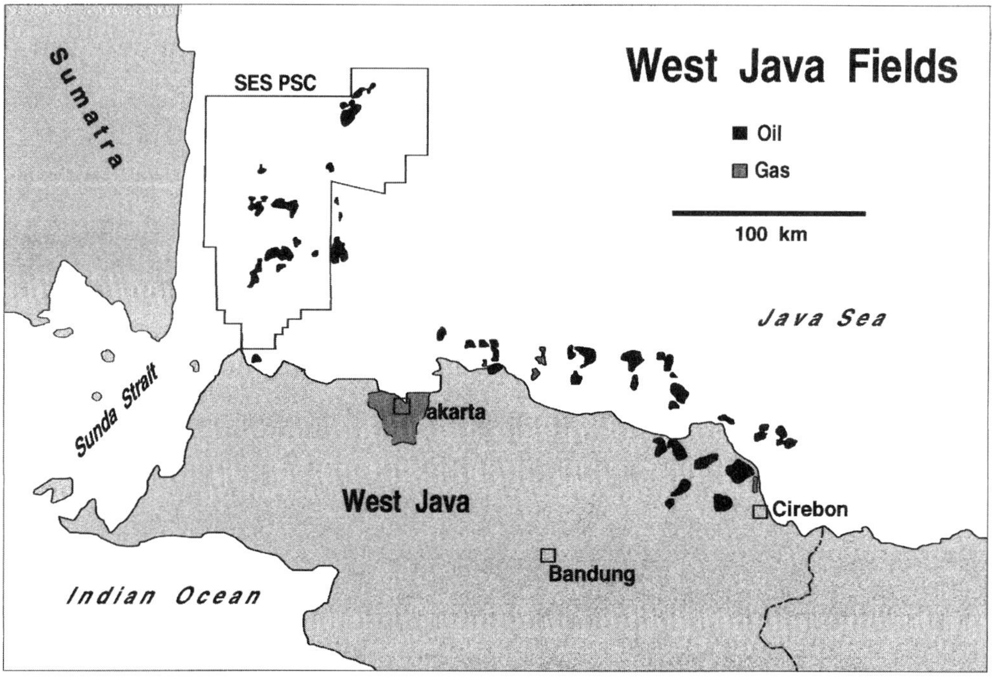

Fig. 2. West Java fields. Numerous offshore and onshore oil fields, but few gas fields, are located in the SE Sumatra PSC and adjacent ONWJ PSC (ARCO) to the east, as well as onshore (PERTAMINA).

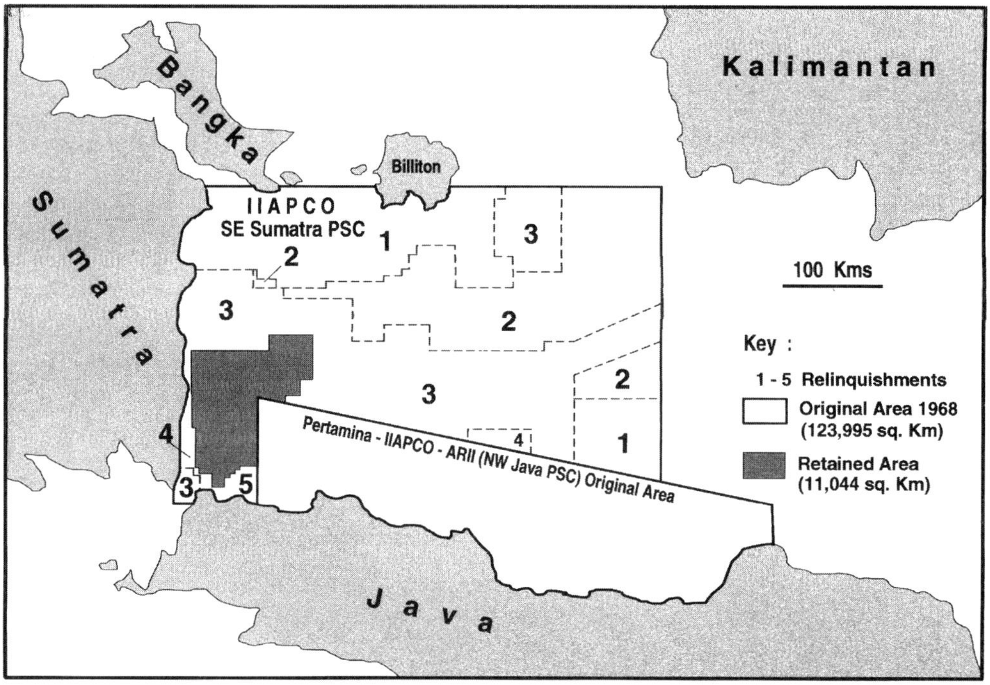

Fig. 3. PSC relinquishments. Five relinquishments between 1971 and 1994 have reduced the area from an original 124 000–11 000 km^2.

ceased. A Sunda basin re-evaluation and mapping programme covering all the remote parts of the PSC, to define a final relinquishment outline were followed by a 'final' exploration campaign which resulted in the discovery of Intan field in the Asri basin. This dramatically reversed the declining fortunes of the expiring exploration programme and resulted in the acquisition of 3D seismic data, the formation of exploitation Asset Management Teams, with successful infill drilling and the achievement of record production rates. The closing discussion attempts to list several important facets of both the human and technical sides of exploration, a case history which it is hoped will assist other explorers in future, perhaps hoping that 'history repeats itself'?

Basin summary

The Sunda and Asri Basins (Fig. 4) are part of a series of Tertiary half-grabens developed on the Asian continental margin that have occupied a retro-arc setting since early Neogene times. These basins extend from NW Sumatra to west Java and lie approximately 200 km. north of the present-day volcanic arc of Java and 400 km. north of the subduction trench which extends east–west for the length of Java and beyond (Fig. 5). The basins are bounded by roughly north–south boundary faults on the east and maximum thickening occurred mainly during the Oligocene, (Talang Akar Fm, Zelda Mbr), establishing the asymmetric half-graben tectonic style typical of the Sunda and Asri Basins today (Figs. 6, 7). These depocentres were succeeded by broader Neogene ones, representing sag phase development. The north–south faulting may be a younger manifestation of the major basement shear system attributed to the early Tertiary (53–40 Ma) collision of India with Eurasia (Tapponier *et al.* 1982), which simultaneously gave rise to major NW-oriented shear zones e.g. the Sumatran, Thai-Burmese and Red River faults. This NW trend also parallels the present subduction zone off Sumatra and the major right-lateral Sumatra fault zone and is clearly a fundamental regional crustal feature. In Java this orientation swings round to become roughly east-west and is also well developed within the Sunda Basin in particular.

In common with the neighbouring Palaeogene depocentres, both Sunda and Asri Basins

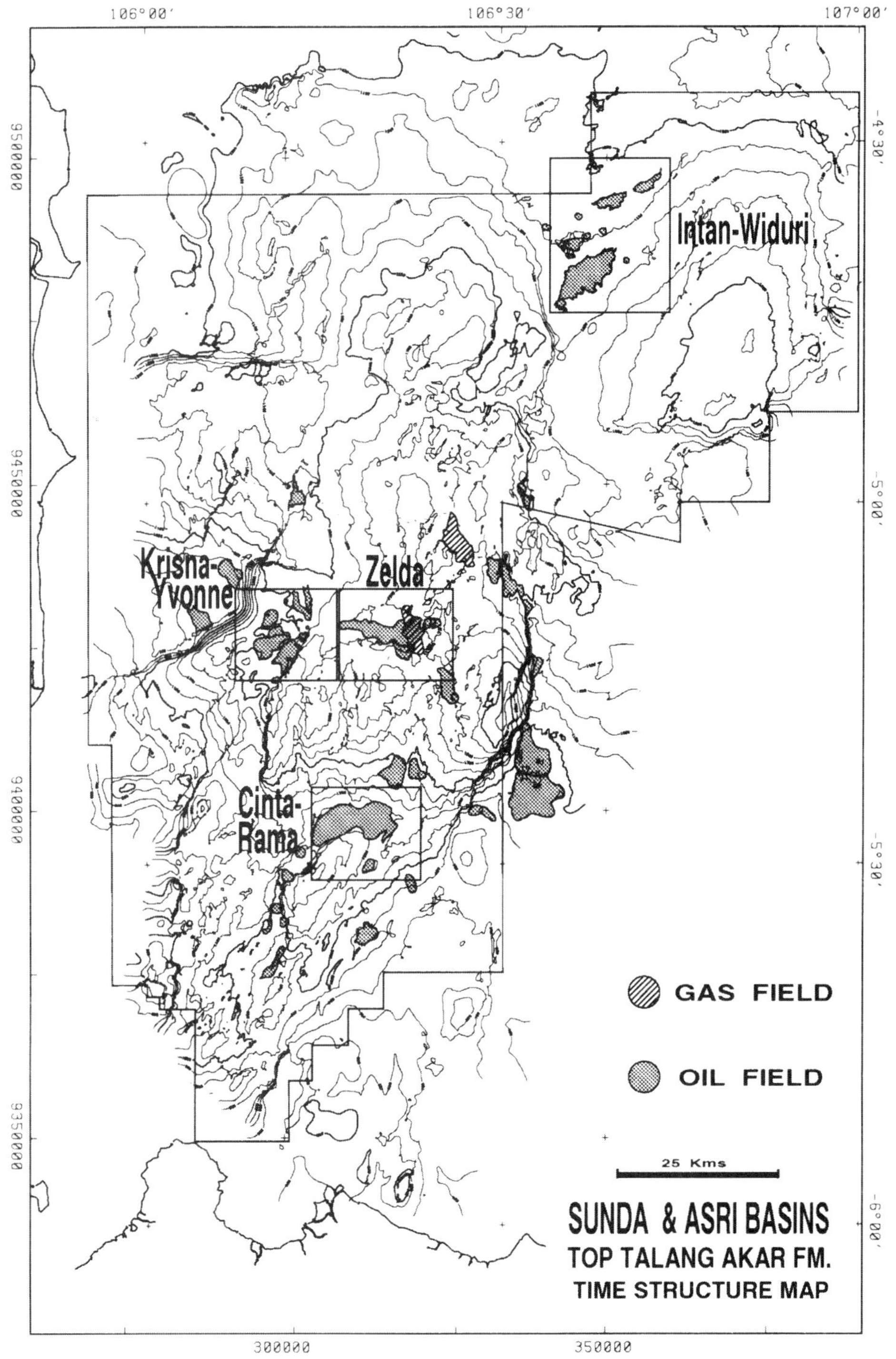

Fig. 4. SE Sumatra PSC structure: Top Talang Akar Fm. The major depocentres, the Sunda and Asri Basins, contain up to 5000 m of Tertiary sediments. Smaller, shallower, sub-basins lie around the main Sunda Basin depocentre. The major fields (shown in squares) are located on inter-basin highs and horsts or large hanging-wall and footwall noses around the Sunda Basin. The Asri Basin fields found so far are located only in the NW part of the hanging-wall dip slope. (Maxus Exploration Dept. & Ridwan Syah, Maxus SE Sumatra Inc.).

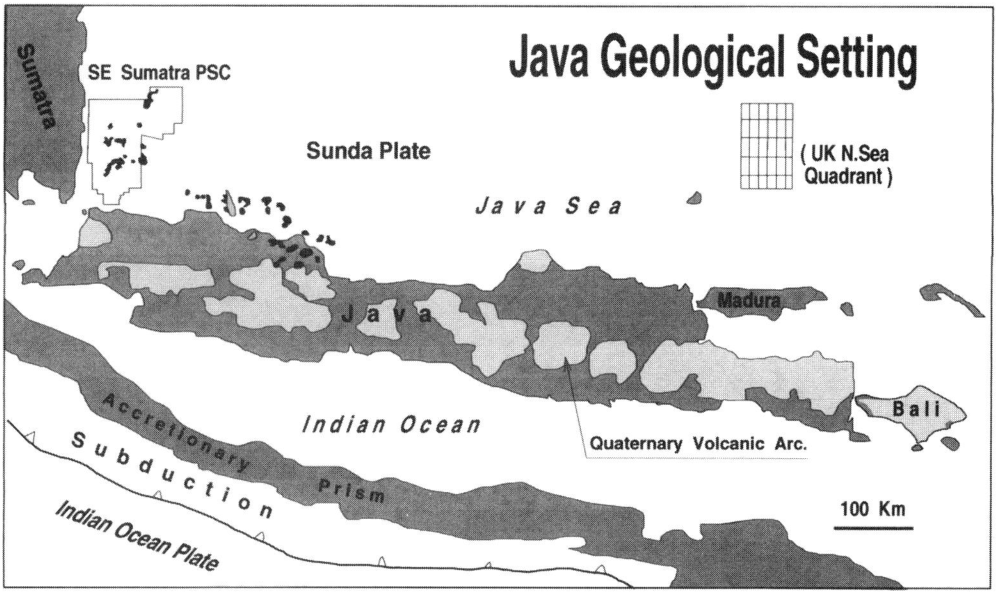

Fig. 5. Geological setting of Java. The retro-arc position of the PSC is shown, north of the present-day volcanic arc on Java. This in turn lies north of the subduction trench south of Java, where the Indian Ocean plate is being subducted northwards below Java and accounts for higher temperatures, maturing Palaeogene, oil-prone source rocks below 3000 m subsea. A North Sea quadrant is shown for scale comparison.

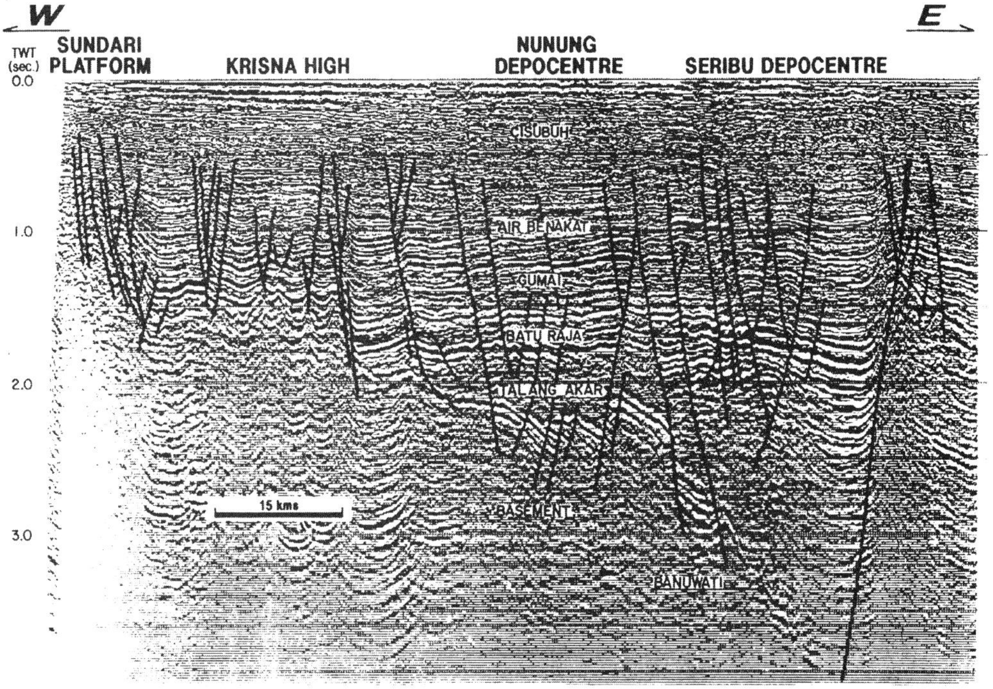

Fig. 6. Sunda Basin regional seismic line. The asymmetry of the Sunda Basin generative depocentre is shown, together with the Zelda horst which separates it from a smaller depocentre to the west. The Krisna high is a major hydrocarbon focus and is downfaulted relative to the western basin margin, where basement is only around 600 m subsea. The principal formations shown range in age from Palaeogene to Recent.

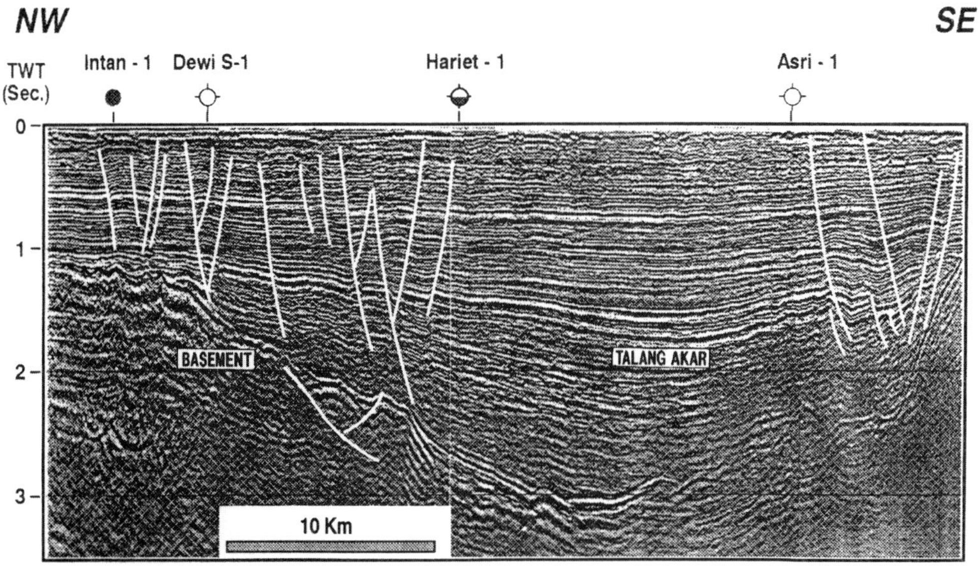

Fig. 7. Asri Basin regional composite seismic line. The Asri Basin is more symmetrical than the Sunda Basin, especially in the deep Palaeogene section. The Widuri and Intan oilfields are located on the hanging wall in the NW corner. To the east, near the Asri-1 well, a zone of poor data corresponds to poorly-bedded debris flows and fans in the Talang Akar Fm, contrasting with the higher-amplitude reflectors to the west, where thick sandstones are interbedded with dense shales. The Talang Akar Fm thins westwards to only a few hundreds of metres in the Intan field area. The basin has less faulting than Sunda and faults are mostly down-to-basin, with the notable exception of the Intan area. (Dennis Kucinskas, Maxus SE Sumatra Inc.).

contain a thick sequence of non-marine Palaeogene syn-rift sediments. They reach a maximum thickness of around 2500 m in the Sunda Basin and are overlain by thick Neogene post-rift sediments. The oldest Tertiary sediments proven from drilling are of early Oligocene age, but sediments as old as Eocene may be present in the deepest parts of the basin. The Palaeogene sequence contains thick lacustrine shales in the Banuwati Fm, constituting excellent oil-prone source rocks that are particularly well developed due to the high degree of basin restriction. The largely Neogene development of volcanic arcs in Sumatra and Java coincided with relatively recent elevation of heat flow (which could also be attributed to crustal thinning). Whatever the cause, this has resulted in high levels of maturation in the retro-arc areas. The Banuwati Fm. lacustrine shales attain thicknesses of several hundred feet and contain Type 1 kerogens which have generated abundant oil but as yet little gas. Maturity may have been attained as far back as early Miocene and the present oil window is estimated to lie around 3000 m subsea (Wicaksono *et al.* 1992). Lateral migration pathways are several tens of kilometres long and are thought to be along the weathered basement/sediment interface, via channel sandstone systems and karstified pipes which riddle the carbonates. Vertical migration nearer to the basin deep is via the numerous faults which have been active since Oligocene times and frequently reactivated (Fig. 8).

The main reservoir rocks in both Sunda and Asri Basins are Oligocene to early Miocene alluvial, fluvial and marginal marine, deltaic sandstones of the Talang Akar Fm, and, in the Sunda Basin alone, shallow marine carbonates of the overlying Batu Raja Fm (Fig. 9). Typical composite logs of the three largest fields show the vertically stacked sandstone reservoirs and thick carbonate pay zones (Fig. 10). Regional seals are provided by shales at the top of and directly above the Talang Akar Fm and for the Batu Raja Fm by thick mudstones of the succeeding Gumai Fm. The close interplay of abundant fault and permeable migration pathways, high quality reservoir, source and seal facies in a region of high heat flow have thus been responsible for the accumulation of large reserves of oil within the Sunda and Asri Basins. The Sunda basin petroleum system was accurately described by Bushnell & Atmawan (1986)

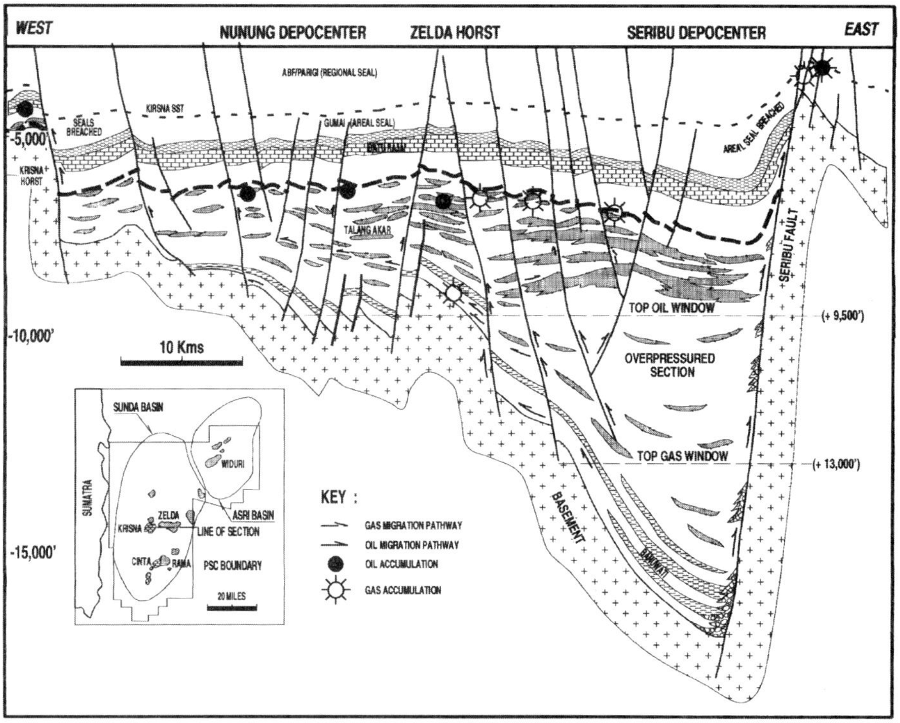

Fig. 8. Sunda Basin migration pathways. Migration from the (partly overpressured) Sunda depocentre is mainly up faults adjacent to the deep basin. Lateral migration is predominant further west, via multiple, stacked sandstones of the Talang Akar Fm and eventually into the younger Batu Raja Fm carbonates. A regional seal is provided by the overlying Gumai Fm shales.

Contract history

'Billings cowboys'

The contract for the offshore Southeast Sumatra Area was signed on 6 September 1968, between IIAPCO (which later became Diamond Shamrock, now Maxus Energy) and the government of Indonesia. It was the culmination of four years of dogged perseverence, (often in the face of potential danger), faith in the prospect of finding offshore basins and skill in developing personal relationships and mutual trust with the officials of the fledgling Pertamina organization. The original founder members, Don Todd, Larry Barker and Pat McDonough were all independents from Billings, Montana. Don Todd's first trip to Jakarta in June, 1964, was originally inspired by Pat McDonough's interest in an article in the *Oil and Gas Journal* on Indonesia (Todd 1971). Conditions at the time were singularly uninspiring: there was political unrest, likely to become worse, the geological predictions were that basement of the granitic Sunda Shield would be at shallow depths immediately offshore and that even if reservoirs were present, they would have a high volcaniclastic content. The founders of IIAPCO did not agree, based firstly on the lack of data, the favourable coincidence of trends with those of the oilfields onshore Sumatra and Java and lastly that the bathymetry indicated a depression offshore.

After an interruption due to the attempted communist coup in 1965, Todd returned to continue negotiating a contract. Despite several difficulties, however, a contract was finally hammered out, which was destined to become a model throughout the world for production sharing (the original ones were profit sharing) and IIAPCO's good relations with Pertamina officials led to the negotiation of the first offshore contract, for the Sinclair-IIAPCO, ONWJ block, in 1966. Its competitive offer of spending \$22.5 million in 10 years with a bonus of \$2.5 million gave it the approval of the government

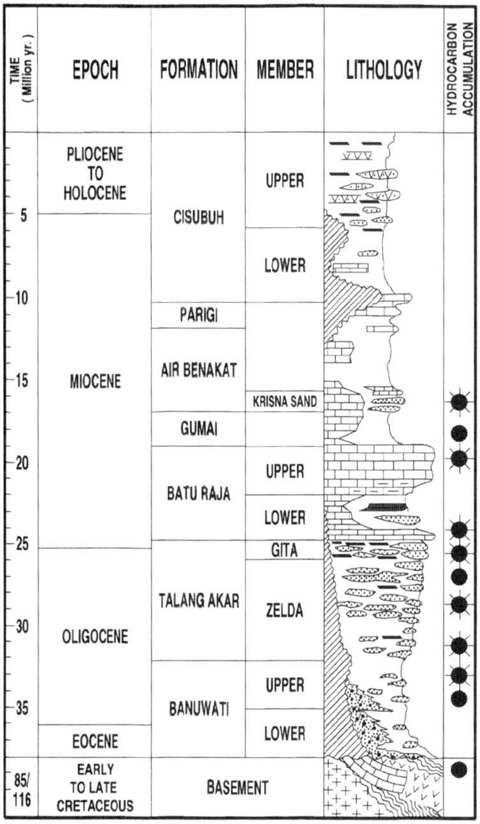

Fig. 9. Stratigraphy of the Sunda Basin. Both Sunda and Asri Basins contain up to ±5000 m of Tertiary sediments, ranging in age from Oligocene or perhaps older, to Recent. Basement comprises Cretaceous metamorphic, igneous and sedimentary rocks. The Palaeogene sequence (Banuwati Fm) contains fluvial and lacustrine oil source rocks. Above, the Talang Akar Fm (Zelda Mbr) contains coals and some source shales but mainly the porous alluvial and fluvial reservoirs. The Neogene sequence begins with the uppermost Talang Akar Fm carbonates. The Gumai Fm shale is an excellent regional seal, broken only at depocentre boundary or basin-marginal faults.

of Indonesia. Many of the early contracts signed with Indonesia were with small independent oil companies since many of the major companies were still concerned with political risk and precedents which might affect their assets in other countries. One of IIAPCO's guiding principles, however, was that it was better to have an area of good geology and uncertain politics than good politics and poor geology (Todd 1971). Seismic data from the NW Java area also assisted IIAPCO in choosing the SE Sumatra area. The favourable logistics of offshore acreage obviated the political and logistical complications found onshore, and proximity to Jakarta's port facilities and the relatively calm seas also contributed to its selection. The Offshore Southeast Sumatra contract was signed in September 1968, after discoveries had been made in mid-Miocene clastics in the ONWJ area. An early review of the exploration status was given by Todd & Pulunggono (1971). The early IIAPCO partnership (shared 35/65% with Pertamina) comprised Natomas Co., Reading & Bates drilling company, (which initially provided drilling service in lieu of funds) Carver-Dodge and Warrior; of these, only Warrior remains in the present group.

Early success (1968–1972)

In February 1969, a 14 400 km seismic program was initiated, which resulted in ten drillable prospects. The first exploratory well, Anda-1, was spudded in the extreme north of the block, in January 1970, using the Reading & Bates rig 'Dickson M. Saunders'. The location of the well was positioned using satellite, the first use of this technology in the industry. The second well, Banuwati-1, was drilled in the central portion of the Sunda Basin, tested some hydrocarbons and although non-commercial, it encountered a thick section of petroleum generative rocks, giving encouragement that oil was going to be found in this basin. The next six wells tested structural prospects in widely separated areas to gain a better understanding of the geology of the huge contract area. Well names were in alphabetical sequence and at first named after the daughters of Pertamina friends.

The first commercial discovery occurred on the ninth well, Cinta-1, which was drilled on the largest structural closure mapped in the basin. (Figs. 11, 12). The well should have been the fifth drilled but the rig had to have the legs extended for the water depth and drilling was deferred. Cinta-1 encountered oil shows in Batu Raja Fm carbonates, and over 61 m of oil shows, from 1006 m, in the Talang Akar Fm, a sandstone–shale–coal sequence (Fig. 10, centre). Tests of the three main zones from several thick pay sandstones in the Talang Akar Fm. yielded over 6000 BOPD, which proved to be a commercial rate. Since a barge and caisson facility had been prepared before drilling, production was quickly achieved, and Cinta-1 was then tested for an extended period into a barge, which confirmed a very large reservoir, of

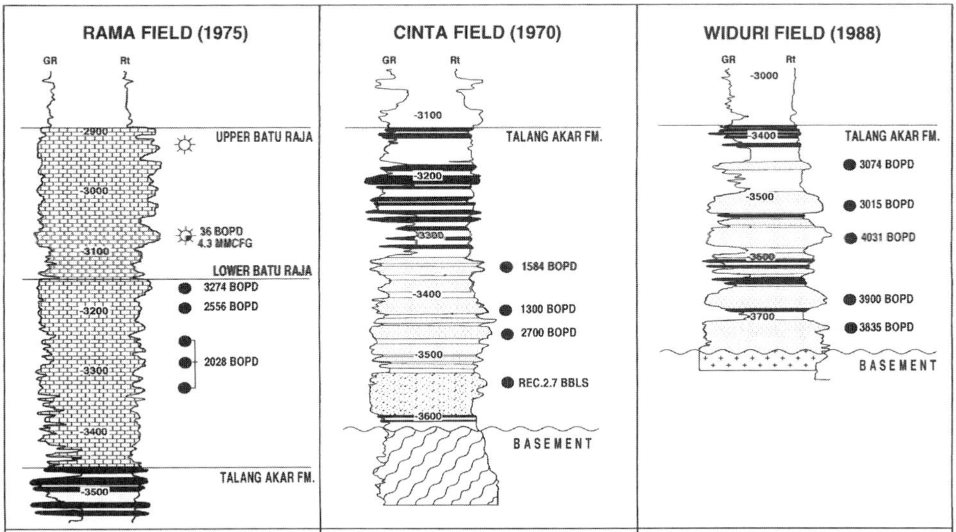

RAMA FIELD (1975)	CINTA FIELD (1970)	WIDURI FIELD (1988)
VERTICAL OIL COLUMN : 242 feet	VERTICAL OIL COLUMN : 350 feet	VERTICAL OIL COLUMN : 316 feet
AREA / AV. OIL PAY : 3470 acres / 67 feet	AREA / AV. OIL PAY : 4293 acres / 87 feet	AREA / AV. OIL PAY : 6196 acres / 51 feet
AV. Sw / AV. POR. : 41.3% / 24.5%	AV. Sw / AV. POR. : 22% / 25%	AV. Sw / AV. POR. : 20.1% / 28.5%
AV. K : 4500 md.	AV. K : 505 md.	AV. K : 7020 md.

Fig. 10. Oilfields type logs. These examples, from the three largest fields in the PSC, show the pay zones, lithologies, thicknesses and reservoir data. Reserves in clastics are three times as large as in the carbonates so far, and all reservoirs are either Oligocene or early Miocene.

stacked, alluvial sandstones. A dedication ceremony was held on 23 October 1970, to commemorate the first offshore oil production in Indonesia. Less than one year later, on 10 September 1971, commercial production began from Cinta field, which was the first permanent offshore oil production from Indonesia. Production soon reached 50 000 BOPD after the second platform was put on stream in December 1971. The initial price of Cinta crude was $2.60/barrel.

Exploration continued in the area from 1971–1972 with the discovery of oil at the Kitty, Zelda, and Gita structures. These fields were subsequently developed, each initially with single platforms. The principal type of prospect drilled during this time was structural, based on the extensive seismic data which had been acquired. The Kitty field was drilled on a prominent horst block, southwest of the recently discovered Cinta field and encountered a thick pay section in a highly fractured Batu Raja limestone. The lost circulation zones were part of the best producing intervals, which also contained good matrix porosity in the reefal limestone. This was also the first significant production from the Batu Raja Fm in Indonesia.

Growing pains (1972–1979)

In 1972, a technical agreement was signed between IIAPCO and Shell, which was to explore the Asri Basin and area to the east in exchange for a loan and an interest in anything found in this area. Over 25 000 km of seismic were shot and ten wells drilled in this area over the next three years. A large part of the eastern area was relinquished in September 1974. The main play in the Asri Basin was stratigraphic truncation, interpreted to occur in the uppermost part of the thick Talang Akar sequence, below the 29 Ma unconformity. This potentially regional-scale trap was identified on the northwest flank (the hanging wall) of the basin. The play was tested in 1973 by the Umi-1 well, which although encountering a cumulative 500 ft of porous sandstones, had no indications at all of oil shows. In 1974, a second stratigraphic test, Dewi South-1, encountered a cumulative 244 m of water–wet sandstones. It was not until 1987 that the basin was to be first proven oil productive by the drilling of a 'last' prospect, Intan-1, the history of which will be discussed later. It was not until 15 years after Umi-1 was

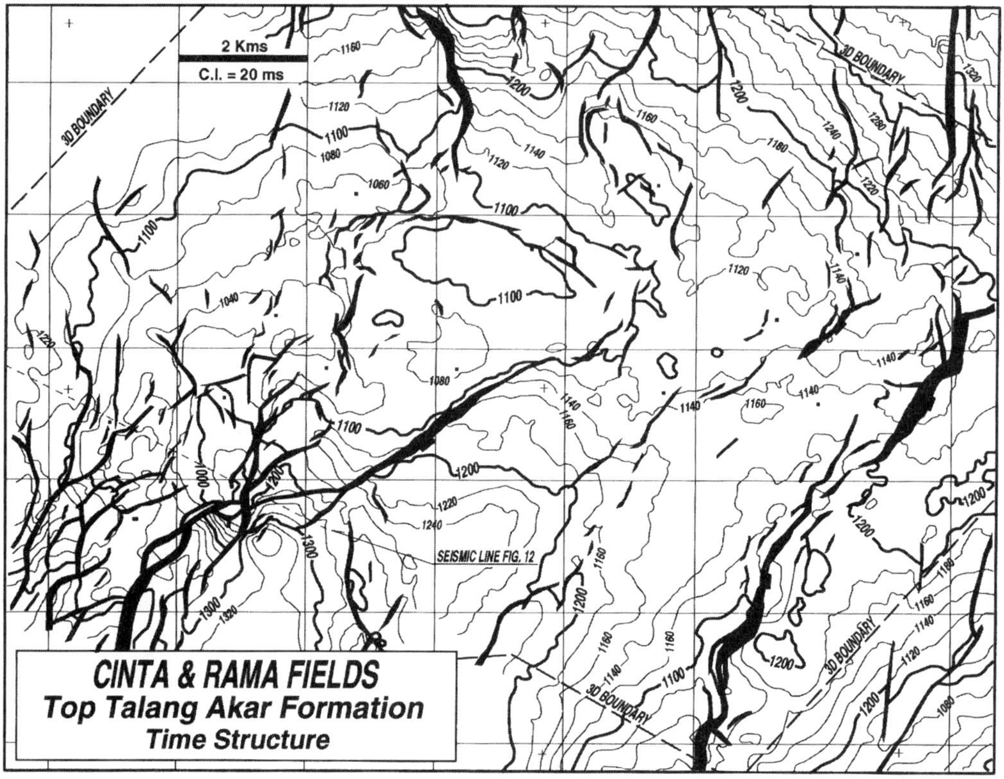

Fig. 11. Cinta–Rama composite structure. The complex Cinta trap is controlled to the south by a NE-oriented fault system bounding a small depocentre. The Cinta map is at top Talang Akar Fm reservoir level. To the east, the large Rama field lies structurally higher and contains carbonate reservoirs, primarily of the Upper Batu Raja (immediately above mapped horizon) which is stratigraphically controlled by westerly shale-out, although the overall trap for both fields is provided by an east-west arch 18 km long. (Jeremy Dyer, Independent Consultant/Maxus SE Sumatra Inc.).

drilled that a different stratigraphic trap, Widuri field, was found in this area. Persistence and new ideas surely had paid off. It is ironic that the structural play at Intan had ultimately led back to the discovery of a stratigraphic trap at Widuri.

By 1973, a well established exploration staff and a continuous drilling program by IIAPCO, yielded further successful results. Nora-1 was mapped as a discrete buildup on seismic, which was confirmed by drilling. The well tested over 5000 BOPD and encouraged the company to explore for additional reef buildups in the contract area. By 1974, a new interpretation of the depositional environment of the Cinta reservoirs indicated that they were a series of braided, stacked channel and point bar sands. The limits of the field were estimated to be much larger than had been developed at the time. A major extension program of five delineation wells was proposed. This successful program created the need for five additional platforms, which increased production significantly in 1975–1976. Also, the discovery of Rama field, in 1974, (mainly Upper Batu Raja-GUR 100 MMBO) resulted from one of these extension wells (Cinta-7, renamed Rama-A1). A detailed seismic program followed by delineation drilling was undertaken at Rama field to define the limits of this very complex field, which has been described in detail by Ardila & Kuswinda (1982) and Tonkin *et al.* (1992).

In 1976, oil exploration by companies operating in Indonesia was severely curtailed, due to the change in the contract terms from a 65/35% Pertamina/Contractor split, to 85/15%. IIAPCO, however, continued development drilling at Rama field, installing five platforms. To encourage oil exploration, which had virtually stopped, the government of Indonesia granted new incentives in March 1977. IIAPCO quickly responded with the discovery of significant oil at

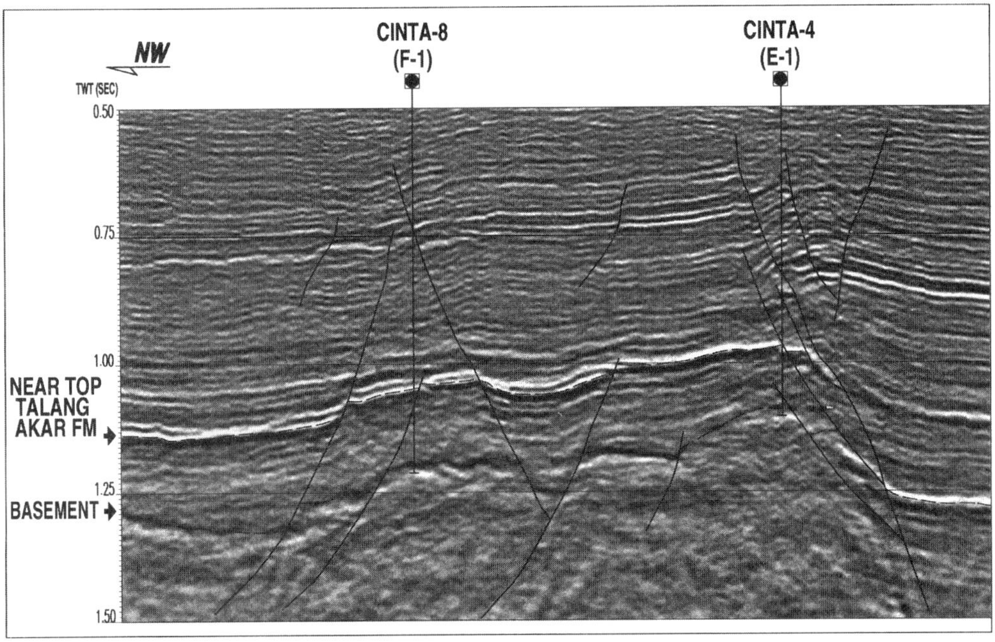

Fig. 12. The Cinta structure is a narrow, horst feature with stacked alluvial and channel sandstone reservoirs. Conversely, the adjacent Rama field to the east is a broad feature, with carbonate reservoirs which shale-out towards the eastern edge of Cinta field. (Jeremy Dyer, Independent Consultant/Maxus SE Sumatra Inc.). See Fig. 11 for location.

Selatan field in 1978 and the commerciality of the undeveloped reserves at Gita field was confirmed. The Selatan prospect was based on a simple geological re-interpretation of one of the early wells, which had mistakenly been drilled on the downthrown side of a major fault due to a navigation error. The fault throw had placed the very porous, reefal Batu Raja section in the water leg with some poor oil shows. Selatan-7 tested the upthrown block and found this porous reefal limestone full of oil and 76 m higher than the downthrown well. This field was put on stream in only ten months.

With significant production increases, new incentives, a substantial increase in oil prices, and renewed confidence in the exploration potential of the area, IIAPCO management changed their strategy in 1979. They brought in a large team of geologists and geophysicists to intensify exploration in order to maximize the potential of the contract area.

Boom years (1979–1984)

The combination of aggressive management strategy and the serendipitous oil price rise (from \$9/bbl to \$21/bbl between 1978 and 1979) encouraged the delineation of earlier subcommercial finds and the revival of old play concepts. This resulted in several 'rediscoveries', the most notable being that of Krisna field, one zone of which had been discovered by Krisna-1 in 1976. That well encountered 8 m of net pay in a 'reefal' Upper Batu Raja carbonate and, on test, flowed 35° API oil at a rate of 2686 BOPD. Although the step-out, Krisna-2, appeared to indicate downdip shale-out and consequently areally restricted reservoir, cleaner Upper Batu Raja carbonates were interpreted from the high amplitudes on E–W-oriented lines (and not seen on the N–S lines over Krisna-1 and -2). An earlier play concept, the onlap of the underlying Talang Akar Fm, was also invoked to give a secondary objective, enabling a third well to be proposed, downdip to the dry hole.

The Krisna-3 well, drilled in 1979, instead encountered a second carbonate buildup, the Lower Batu Raja Member, which encircles and partly covers (Figs. 13, 14) the flanks of the underlying andesitic basement high (Ardila 1982). It is a very prolific reservoir having produced 54 MMBO, from three Krisna platforms and one at Yvonne 'A'. During the early Miocene, the Lower Batu Raja carbonate buildup consisted of a coral atoll (now tight)

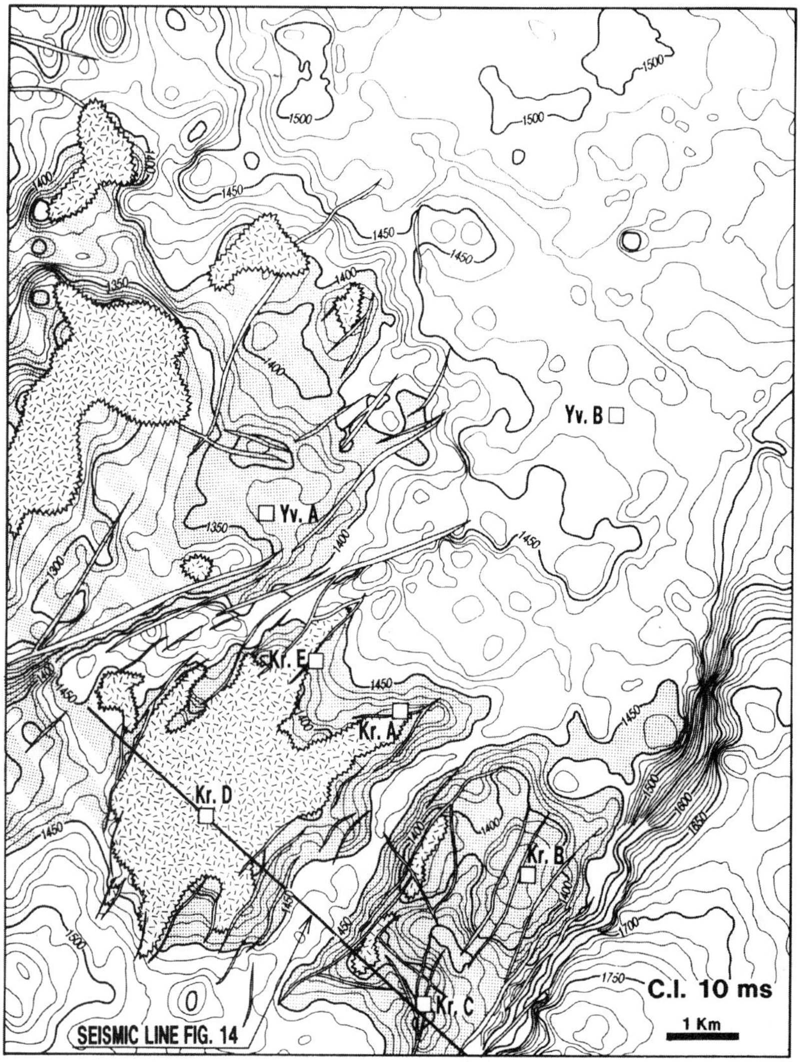

Fig. 13. Krisna–Yvonne structure, top lower Batu Raja. A large andesitic basement high which was progressively onlapped by the Talang Akar Fm and Batu Raja Fm sediments. The main reservoir in Krisna field is the Lower Batu Raja carbonate, which flanks the 'bald' areas or is thin where it overlies the deeper parts of the structure. (Mary W. Haddock, Maxus SE Sumatra Inc.).

surrounding a large back-reef largely composed of packstone-wackestone, which was subsequently exposed to leaching (Wight & Hardian 1982). Freshwater phreatic diagenesis generated secondary porosity lenses, the good interconnection of which was enhanced by karstification. These properties, together with the circular distribution and moderate drilling depths, facilitated rapid and efficient drainage of the reservoir. Production was further enhanced by a successful water injection programme (McChesney *et al.* 1992).

During 1980 and 1981, the Lower Batu Raja was also found in the adjacent Yvonne 'A' field, and the old play, the onlapping Talang Akar clastics, was finally vindicated with the Krisna-10 and Yvonne 'B' discoveries, the latter a stratigraphic trap. Also, the potential of the originally discovered Upper Batu carbonate was confirmed due to an interesting quirk, in the Krisna-26 well, which penetrated a pinnacle of Gumai Fm atop the crest of the whole structure. This dolomitized carbonate has a high gamma-ray response, so high calculated water saturations at

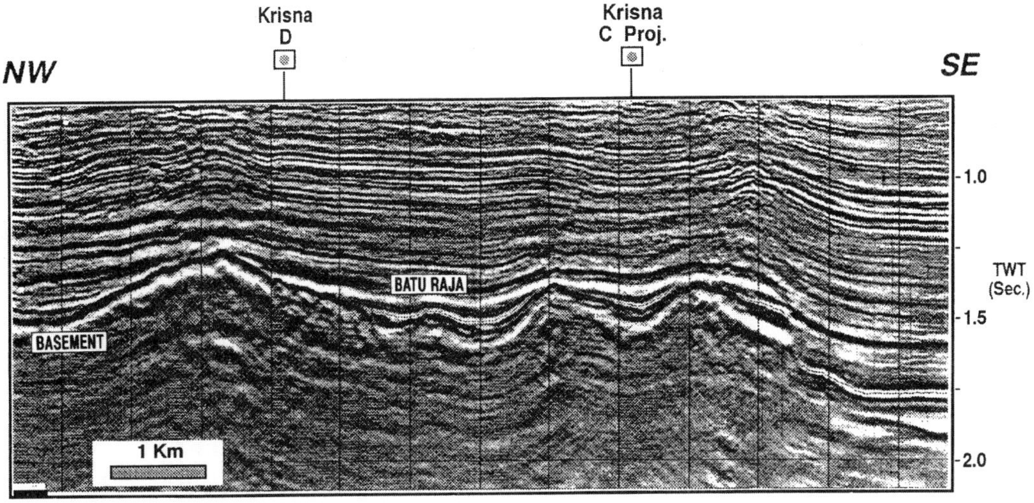

Fig. 14. Krisna field seismic line. The Krisna high comprises a series of andesitic basement closures, onlapped successively by the Talang Akar Fm and Lower Batu Raja Mbr in the east and eventually by the Upper Batu Raja in the west. Trapping is structural–stratigraphic, with mainly carbonate reservoirs (on this line) but clastic reservoirs onlap further down-dip, to the north. (Mary W. Haddock, Maxus SE Sumatra Inc.). See Fig. 13 for location.

first appeared not to warrant testing. Fortunately, one of the first spectral gamma-ray tools was run in the logging programme and the corrected curve enabled determination of 51 m of probable pay, which tested a cumulative 5418 BOPD. This success was responsible for the development of the Upper Batu Raja/Gumai reservoirs from two more platforms, with a current cumulative production of 23 MMBO. Thus the entire complex Krisna–Yvonne accumulation was found to comprise three carbonate zones and one clastic zone, exploited from seven platforms, the combined GUR of which puts the whole accumulation in the Giant class.

The importance of the Krisna-3 discovery at that time, however, was to catalyse not only acceleration of delineation and development drilling there, but also speeded exploration drilling, with the 'rediscovery' of the nearby Farida accumulation, by Farida-4 (1981). In 1973 a single, thin sandstone reservoir with oil shows had been found in Farida-1, but this literally proved to be the 'thin end of the wedge' as increasing thicknesses of sandstones were encountered farther east, trapped in a series of down-to-margin, fault-dependent accumulations (Fig. 15). Other smaller structures now became economically viable, and were drilled in rapid succession, yielding discoveries at Sundari (1981), Wanda (1982, the first stratigraphically-trapped sandstone accumulation found) and Karmila (1983). Karmila, although identified in 1979, was not drilled until later due to its small area (741 acres), but has turned out to be the most prolific field in terms of recovery per unit area.

The rig count reached an unprecedented high of ten during these years, due to development of these and other fields, (eg. Zelda downthrown block) and exploration of a multi-prospect, deep Sunda basin play as well as many other smaller structures around the periphery of the basin. The number of exploration wells peaked at 31 in 1983 but reserve additions began to decline as the largest structures were all drilled (Fig. 16). The 1979 'renaissance' thus resulted in a timely production high averaging 112 000 BOPD in 1981, coinciding with peak oil prices. (This figure was slightly lower than the record high of 113 000 BOPD in 1977). Over 90 exploration wells were drilled from 1979–1984, resulting in discovery of ten commercial accumulations with a success rate of 10.5% (1 in 9.5). Gross recoverable reserves of these are now estimated to be 198 MMBO. In addition, many sub-commercial discoveries were made, including some with large (*c.* 100 MM BOIP in-place reserves, such as Nurbani field, near Krisna (Djuanda 1985).

Declining fortune (1984–1987)

Other strategic considerations influencing the ten rig programme were a 1991 relinquishment and maximization of extraction in time for the PSC expiry in 1998. The discovery rate could not

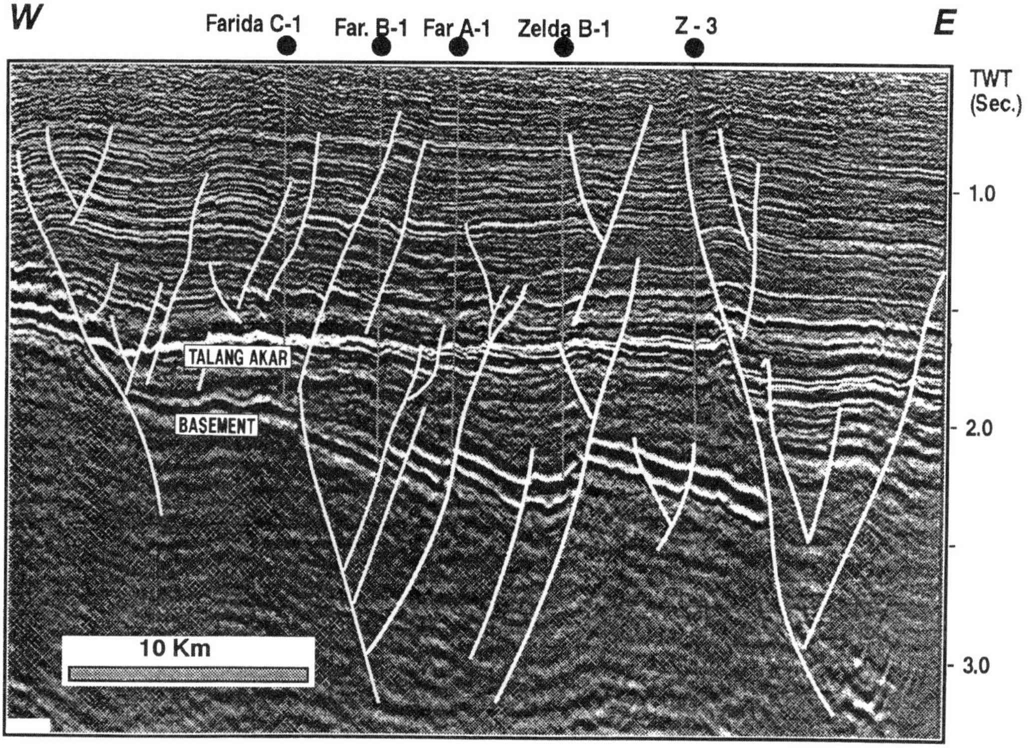

Fig. 15. Zelda–Farida fields seismic line. This line shows the complex, syndepositional and rejuvenated faulting of the Zelda horst in the east and the down-to-margin fault traps of the Farida field, in the west. Reservoirs are stacked sandstones in the Zelda and Gita members of the Talang Akar Fm. (Dave Carter, Maxus SE Sumatra Inc.).

be maintained, however, and coupled with the oil price tumble from a high of around $32/bbl (WTI yearly average) down to $15/bbl between 1984 and 1986, the recession throughout the oil industry affected exploration so severely that only seven exploration wells were drilled between 1985 and 1987. During this respite a major Sunda Basin re-evaluation was undertaken, in order to determine the remaining potential.

Also at this time another mapping project covered all the remote areas around the edges of the Sunda Basin, the Asri Basin, (where there had been no activity since 1983), and the distant Vera–Zaitun Basin, a separate part of the PSC far to the east. This measured pace of exploration was to yield dramatic results, although not as predicted by the conclusions of the mature Sunda Basin re-evaluation. There, several diverse plays were mooted, including downthrown fault plays at the extreme ends of the Seribu fault (one of which led to the Retno discovery, yet to be developed) two near the Karmila field and two plays in areas where migration was thought to be possible but highly unlikely (as was proven) and one each in the Hera sub-basin and the Asri Basin. The Asri Basin, lying adjacent to, but offset to the NE of Sunda, had been regarded as having very low potential, primarily on account of ten dry holes, either directly within, or immediately on the flanks of the basin.

Rejuvenation

The reason for the pessimism on the Asri Basin was the lack of any real oil shows in all wells (Fig. 17a), the lack of source rocks, the low temperature gradient, and the lack of structures in the basin. Detailed source rock analyses run on all the wells drilled in and around the basin indicated insufficient source rock for the generation of oil, although none of them penetrated stratigraphically deep enough to encounter the Banuwati Fm, now known to be the primary source sequence. Two wells drilled in 1983 further seemed to confirm the lack of hydrocarbons

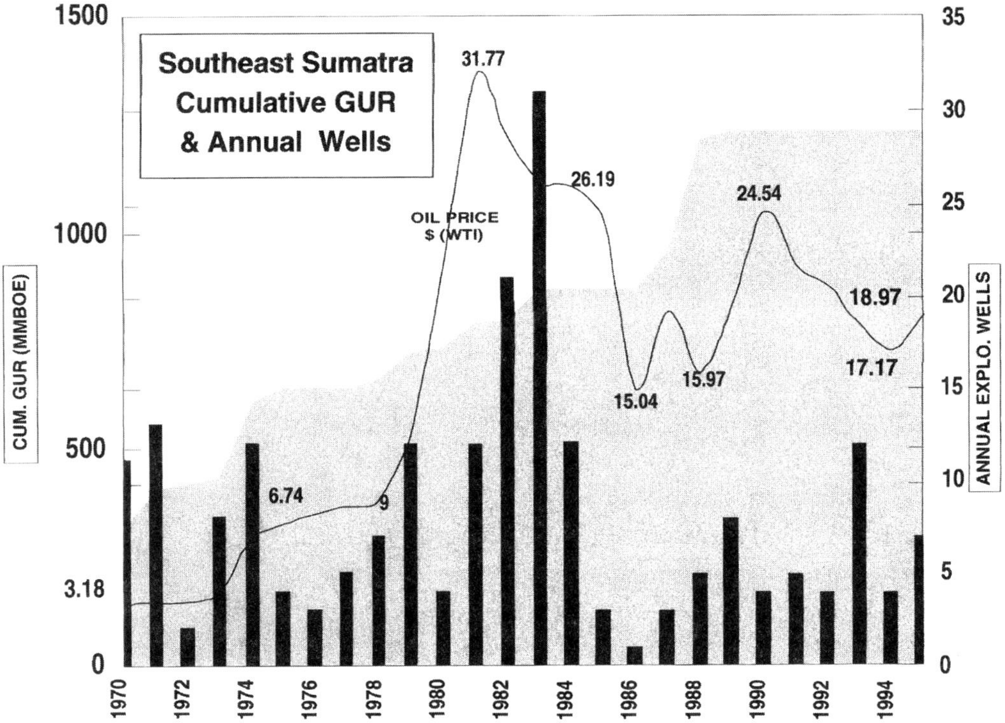

Fig. 16. SE Sumatra cumulative GUR and annual wells. The currently estimated GUR exceeds 1.2 billion barrels of oil, the bulk of which was discovered in three major phases – Cinta field in 1970, Rama (1974) and Intan–Widuri (1987 & 1988). Annual exploration drilling reached a peak in 1983, with 31 wells. The oil price (WTI) is shown, with a steady rise in the 1970s until the peak in 1981 and subsequent drop in the mid-1980s, when drilling virtually ceased.

in the basin, and one of these, Saleha-1, was even unique in the PSC in having no sandstone at all in the Talang Akar. The location is now mapped very close to the OWC of Widuri field, and in fact lies within the overall contact at 1121.6 m (3680 ft) subsea. With hindsight, after the Intan-1 discovery, this shaly sequence was identified as a possible stratigraphic seal to the west of the next prospect, at Widuri.

Mapping of the Asri Basin during 1986, using poorly positioned 1973 vintage seismic, revealed a large closure updip to the Dewi South-1 well. Evaluation of all the dry holes demonstrated that none of them showed conclusive evidence of lack of generation and migration, despite the monotonous lack of shows. Previously, it had been assumed that the central-eastern basin deep might not contain Talang Akar source rocks because of the sand-rich, tight fan/debris flows encountered in the Asri-1 well and thought to be the cause of a zone of poor seismic data, which can still be seen to the east on recent seismic (Fig. 7). Source rock data, although not revealing the presence of oil-prone rocks in the wells, at least showed that geothermal gradients were theoretically sufficient for generation at greater depths. Apart from these geological aspects, the human factor once again proved crucial. The perseverance of the exploration team in proposing the Intan-1 prospect against long odds, and just as important, for keeping the concept alive, is largely responsible for its ultimate success.

Intan-1, drilled in late 1987, was located on an 11 700 (max.) acre, fault-dependent closure bounded by a down-to-margin fault (Figs. 17b, 18 & 19). It encountered 23 m of net oil pay in two upper Talang Akar sandstones and flow tested a cumulative 5845 BOPD, establishing the Asri Basin as a major oil-bearing area and proving to be the turning point in the history of the moribund exploration of the PSC. The NE Intan field, in a separate fault block, was soon added, by the Intan-4 well, which lies only 610 m NW of the Dewi South-1 well, which did not have a single show.

These discoveries in turn led in the following year, to the Widuri-1 discovery on a similar, but

(a)

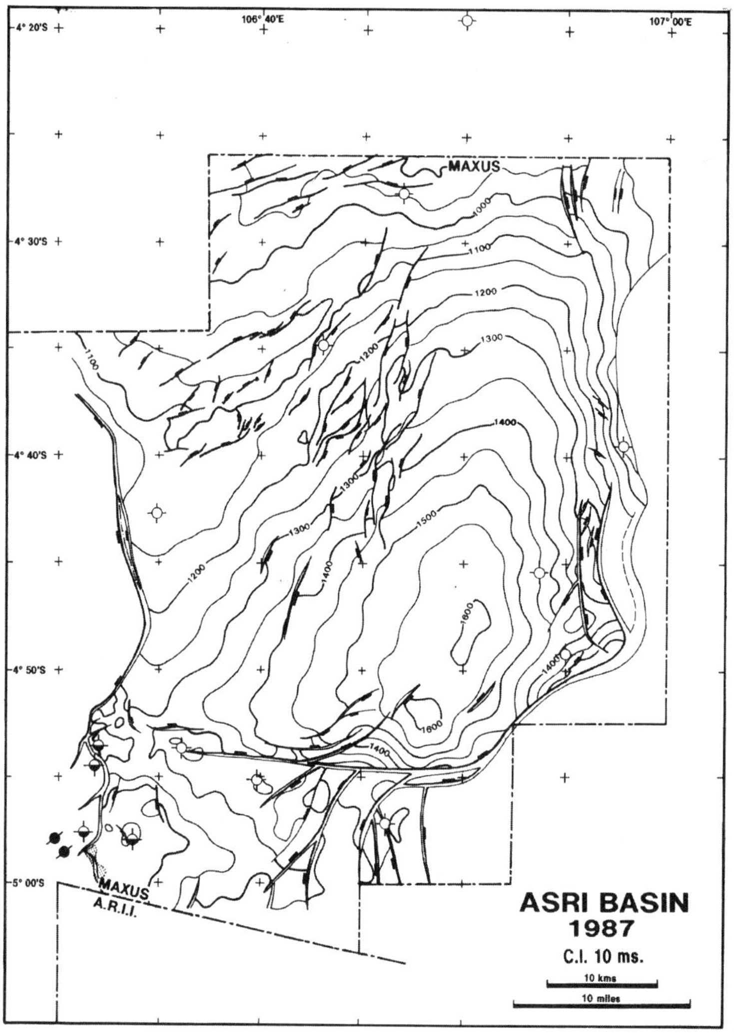

Fig. 17. (**a**) Asri Basin, 1987. Prior to the Intan-1 discovery, there were six dry holes in the basin, one to the north and three on the southern, upthrown flank. The lack of shows and reservoir rocks had virtually condemned the basin until a re-appraisal of its potential in 1985/6. (**b**) Asri Basin, 1995. There are currently four producing fields, one commercial discovery and several undeveloped accumulations, with two pre-1987 dry holes very close to producing area boundaries.

much smaller, faulted structure (Fig. 20). The trap in the critical NW direction, as at Intan, is provided by a young, NE-oriented fault, but the closure is greatly enlarged by the sub-regional westerly shale-out and NW–SE orientation of the channeling. Although it was postulated that the maximum closure might be stratigraphically enhanced (originally estimated reserves ranged from 11 to 63 MMBO); it was never imagined that there may be a giant field so close to a dry hole. This was one of those fortunate examples where all expectations were vastly exceeded and the GUR, after the drilling of 83 wells, is currently estimated as 260 MMBO, the second largest field after Cinta in the PSC. Production is from six principal sandstones in the uppermost Zelda and the Gita Members of the Talang Akar Fm. (Fig. 10). These reservoirs have porosities averaging 25% and permeabilities of tens of Darcies, which coupled with a strong water drive, give a recovery factor in the range 45–50%. They were deposited in alluvial and fluvial to marine distributary channels, some of which are comparable to the present-day Mahakam

(b)

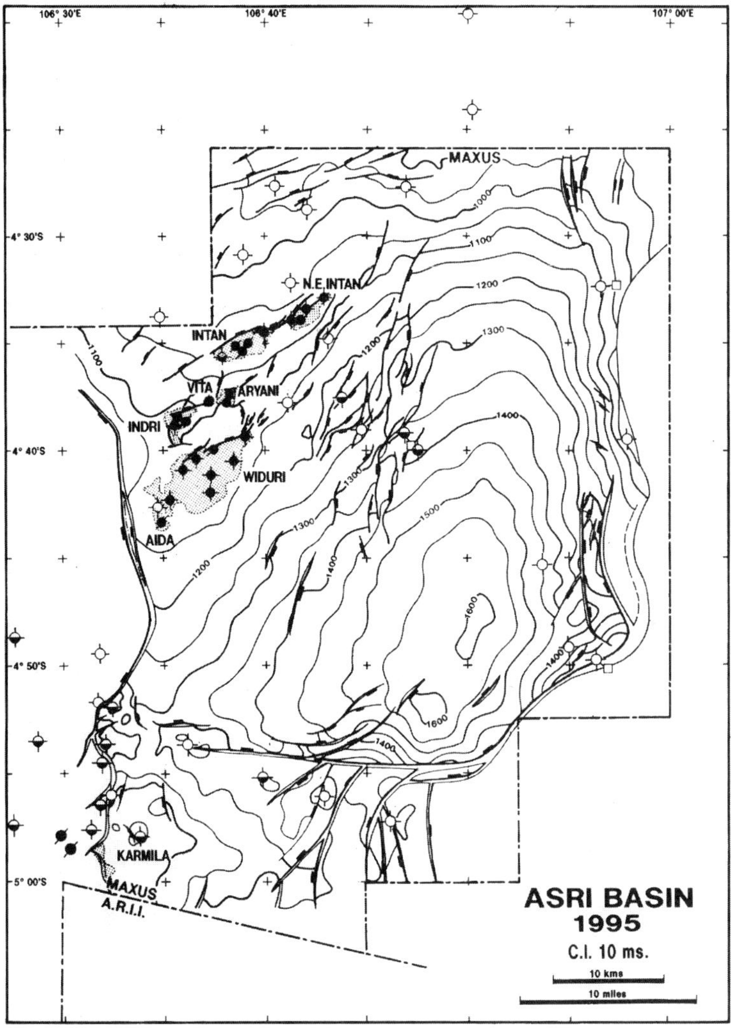

Fig. 17. Continued

delta of E. Kalimantan (Young *et al.* 1991). Other environments have now been identified using sequence stratigraphy, resulting in a continual improvement in drilling success in the area as a whole.

Several satellite accumulations have been discovered recently (Indri in 1989 and Aida in 1994) as well as several sub-commercial finds. Currently, exploration is being conducted in conjunction with the Widuri–Intan multidisciplinary team, a highly successful combination (Girgis *et al.* 1995) and sedimentary studies are being combined with seismic attributes and AVO work (Smith *et al.* 1995) to chase the many narrower sand-filled channels missed by the present well pattern (Fig. 21, from Smith *et al.* 1995). Although this course has good commercial viablity, the future, large exploration potential must surely lie in the eastern and central parts of the basin, in stratigraphic traps. Deep (but relatively inexpensive per metre) wells are thus being drilled to test the vast remaining area.

Conclusions and discussion

The key find which made the basin commercial, Cinta, was made first, in 1970. This funded the exploration programmes for several years.

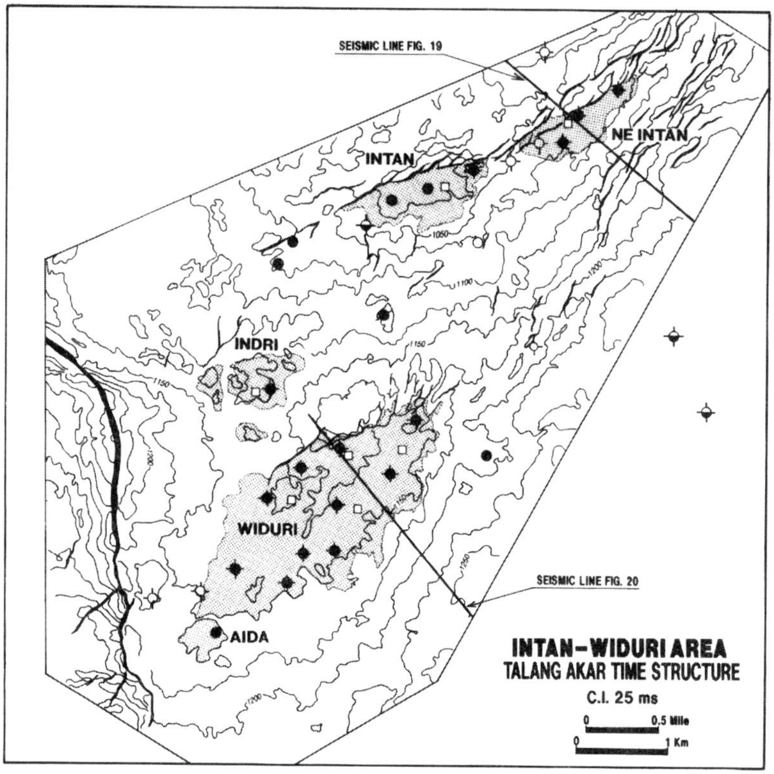

Fig. 18. Structure of Intan–Widuri area. This map, on an intra-Gita member (uppermost Talang Akar Fm) horizon, shows the NE-oriented faulting which is down to the basin margin at Intan, NE-Intan and Widuri. The large area of Widuri field is enlarged by stratigraphic trapping. (Romina Himawan/Pujianto/Bill Harmony, Maxus SE Sumatra Inc.).

Contrasting with the early finds in clastics, those found during 1973–1979 were all in carbonates. Exploration activity peaked from 1981 to 1983, but although several small accumulations were found, only one large field, Karmila, notable for its small areal extent and high productivity, was discovered.

It took 17 years for the potential of the Asri Basin to be recognized, due to a perceived lack of shows in many wells and lack of proven oil source rock. The Intan-1 discovery was drilled on a very large closure in the NW corner, using 1972 vintage seismic data. It rejuvenated activity in the PSC to this day. Although stratigraphic traps were discovered in 1982 (Wanda field being the first), the largest stratigraphic-structural trap, at Widuri, was drilled based on a much smaller structural closure and only after another structural look-alike (Intan) had been successful. In addition, the Saleha-1 well (1983), r are i – that it was devoid of sandstone, yet lies a mere 230 m from the Widuri field boundary, appeared at the time further to condemn the basin. The future potential of the Asri Basin is not likely to mirror that of Sunda, inasmuch as there is less structural closure and no major carbonate reservoirs; however it is hoped that the stratigraphic element may be successful. Additional potential, in both basins, may be identified in gas prospects. The infill and near-field potential has been successfully identified using new technology such as 3D seismic, however new oil will continue to be found in the minds of the explorationists with new and creative ideas and persistence.

There are many important lessons to be learned from the acquisition of the block, the formulation of a new company/Government contract and exploration progress. Firstly the perseverance of individuals, often in unfavourable circumstances and bucking supposed 'conventional wisdom', has been the most important. This human tenacity enabled the acquisition of the area in the first place and later in making

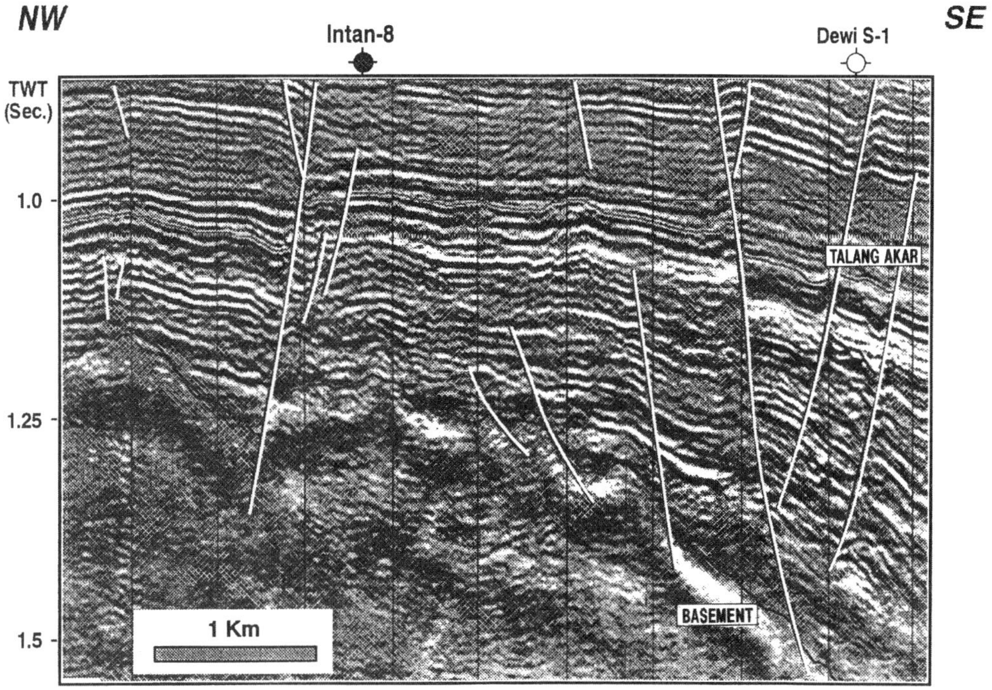

Fig. 19. Intan–Dewi S.-1 seismic line. The down-to-basin-margin, NE-oriented faulting forms the main trap at Intan field, but at NE Intan, lateral shale out of the NW-oriented channels enables some stratigraphic trapping. There are two principal producing zones, in the Gita Member (uppermost Talang Akar Fm). The field OWC is only 610 m NW of the dry hole at Dewi South-1. (Romina Himawan/Pujianto, Maxus SE Sumatra Inc.). See Fig. 18 for location.

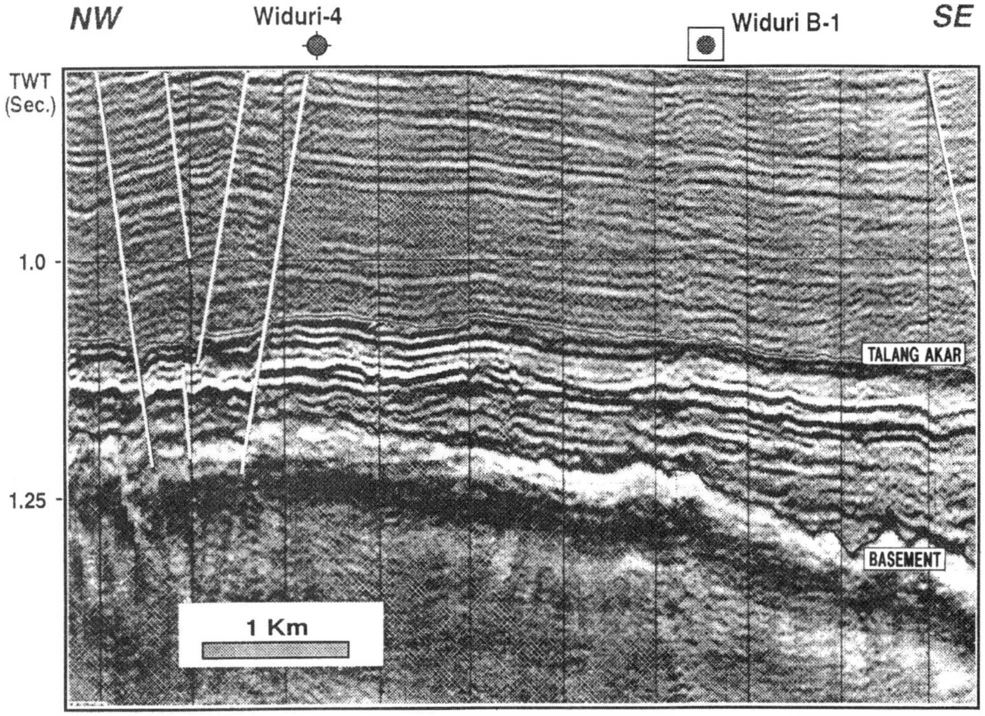

Fig. 20. Widuri field seismic line. The fault-trap style is similar to that of Intan and NE-Intan, but here the thinner Talang Akar Fm sequence shales out to the NW, providing a much larger area of oil accumulation. Six major zones in the uppermost Zelda Member and overlying Gita Member are productive. (Romina Himawan/Pujianto, Maxus SE Sumatra Inc.). See Fig. 18 for location.

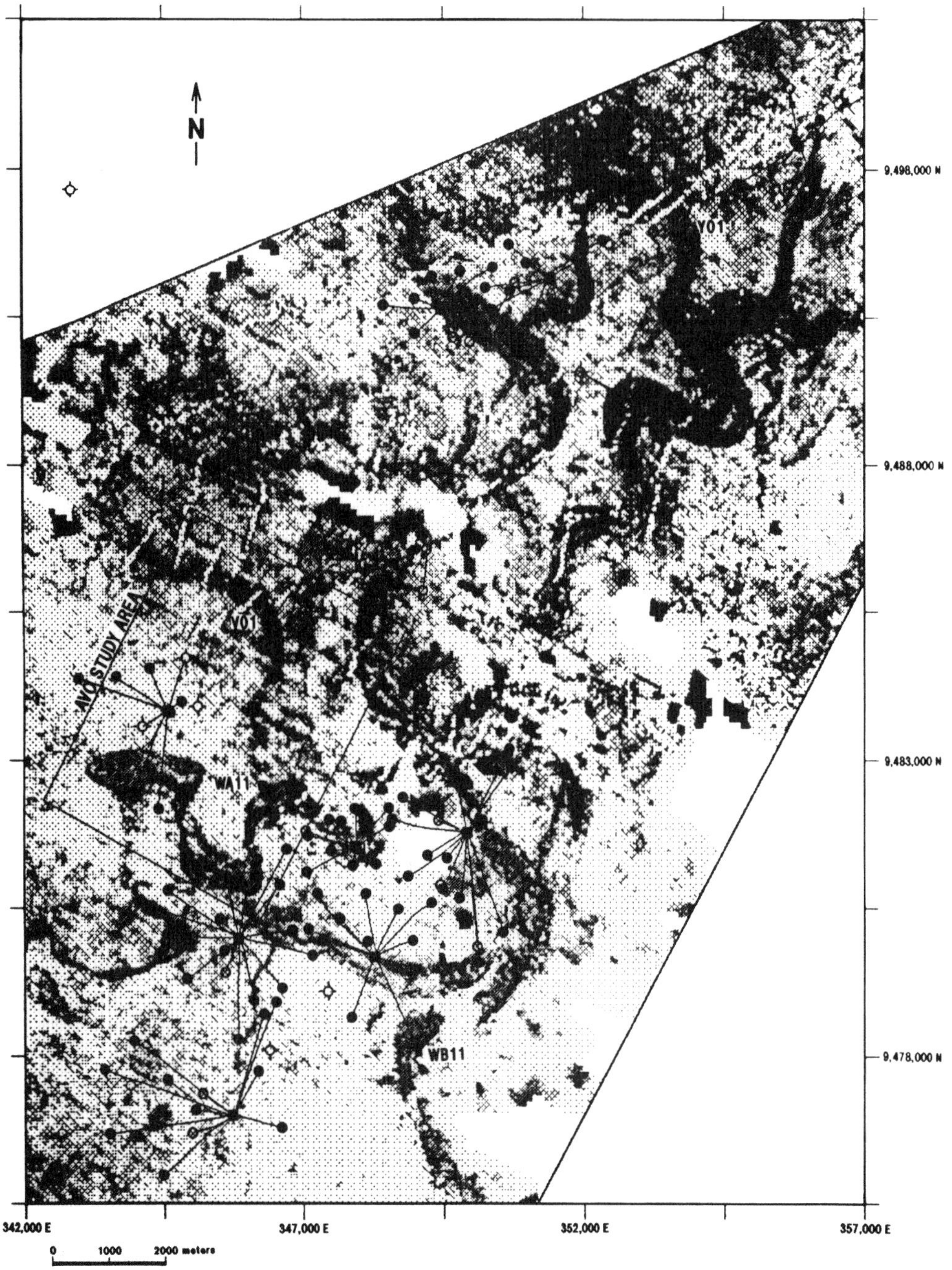

Fig. 21. This attribute map on an intra-Talang Akar sequence boundary shows how the channel thalwegs can be interpreted. AVO is being used to identify which channels are active and sand-filled rather than abandoned/shale-filled, as channel-fill densities are similar. (Steven W. Smith, Maxus SE Sumatra Inc.).

possible the discovery of a new oil-bearing basin. It does not seem likely in this competetive age, that a company could negotiate a block almost as large as Cambodia in area, where no basin was thought to exist. However, this is what Hunt achieved in North Yemen and Oxy in Ecuador, as recently as the 1980s. There remain many large areas of the world still improperly evaluated and, as Todd and co. noticed, there always seems to be insufficient hard data to provide

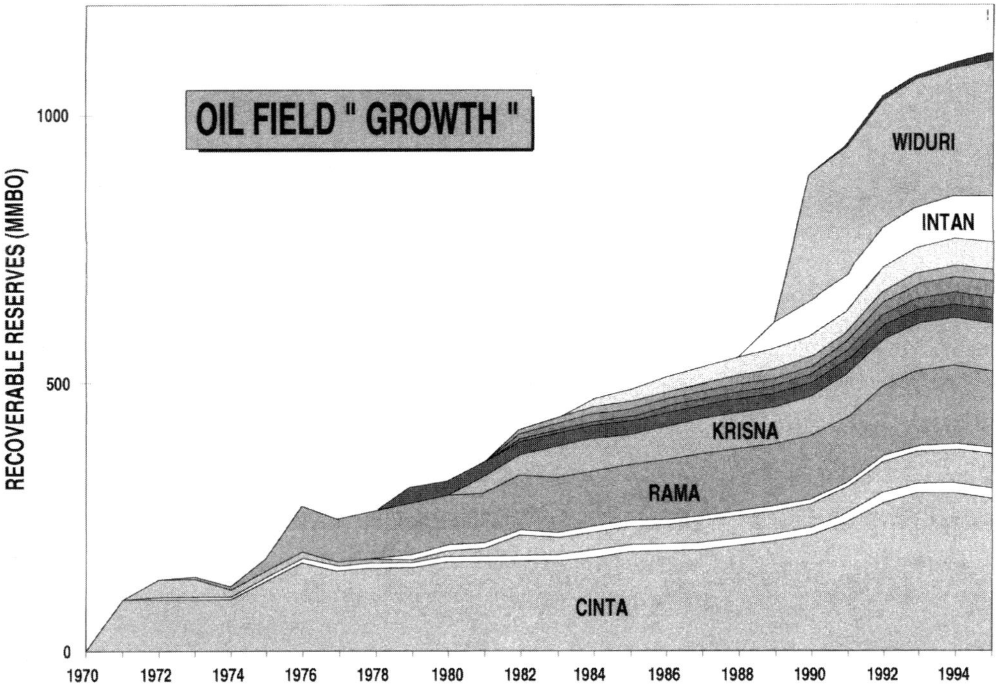

Fig. 22. Reserve 'growth'. A consistent increase in the annually revised field reserve estimates reflects the good performance of all fields except two in the PSC. Infill drilling and realization of better reservoir parameters and drive mechanisms from production history have contributed to this apparent 'growth'.

positive evidence to managements in the short times available for bids.

Although technology has improved the way in which we operate in many ways, all the major fields were discovered with old vintages of seismic, often mis-positioned and with less resolution. Modern logging has also given us more detail, but in our basins at least, conventional pay zones can usually be identified by observant wellsite geologists using mudlog data (drilling breaks, gas and oil shows) and modest log suites with gamma-induction and density-neutron logs; above all, the sidewall core is an indispensable tool. Indeed, there are instances, especially in karstified and /or fractured carbonates, where the logs positively cannot identify pay, thus conventional coring and testing should be *de rigeur*.

Geological models based on dry wells apparently indicating non-migration (e.g. Dewi South-1) are to be studiously avoided, as indeed are any wells in which post-drill evaluation cannot unequivocally assess the various risk factors. All wells have a high percentage of poor cement jobs, (as evidenced by the 'California water shut-off test', a state regulation which enables perforation above supposedly cemented-off zones) and high invasion, both during and after drilling, even with good mud cake. These factors, combined with the poor flow properties of viscous and waxy oils, can lead to very slow recovery of the well bore giving spurious non-recovery on what are almost always short-term tests. At field scale, 'small can be beautiful', as shown by the tiny Karmila field and the originally-mapped relatively small Widuri feature. Oil has continually been found downdip to existing fields where the boundaries appeared to be defined, eg. shaled-out carbonates (Krisna Lower Batu Raja), downfaulted block of Zelda horst, Yvonne 'B' Talang Akar sandstones downdip to the younger Yvonne 'A' carbonate reservoirs and Rama carbonates, downdip to Cinta.

The cumulative addition of reserves (Fig. 16) has been sporadic, largely due to: (a) discovery of a previously unsuspected carbonate reservoir at Rama, (b) the oil price rise of 1979 and unsuspected bonus of the Krisna carbonates, (c) the high-risk Asri Basin finds opening a new basin to exploration. Conservatism in initial estimation of reserves has partly contributed, in all fields but Duma and Titi, to a continual upward revision in field sizes (Fig. 22). Whereas this is a welcome occurrence, it occasionally inhibited exploration plays, where estimates of recovery factors in particular could lead to condemnation especially of prospects, as the economic climate makes us risk averse. In 1986, it was thought that the PSC had limited horizons, but that prediction failed. Will this be the case for the future, have we used up all our luck? Certainly not; we hazard the guess that Mr Micawber's philosophy that 'something will turn up' will again prove true and confidently predict that the continuous evolution of new (and old) technology will enable what was once considered impossible to be repeated.

The authors are grateful for the managements of Pertamina and Maxus SE Sumatra for encouraging and permitting the publication of this paper. In addition we thank the many geoscientists who provided their interpretations and especially for their help in preparing workstation outputs at a particularly busy time; D. Kucinskas, Pujianto, R. Himawan, D. Carter, B. Harmony, J. Dyer, R. Syah and M. Welker-Haddock. Thanks also go to T. Budiyento for his data on Widuri and to P. Lunt of Rocktech for printing the stratigraphic column with Mincom software. B. Lodwick kindly gave his time to review the text.

References

ARDILA, L. E. 1982. The Krisna High: Its geologic setting and related hydrocarbon accumulations. *SEAPEX Offshore S.E. Asia '82 Conference, Singapore.*

—— & KUSWINDA, I. 1982. The Rama field: An oil accumulation in Miocene carbonates, W. Java Sea. *In: Ascope/CCOP Workshop, Surabaya, Tech. Paper TP/2*, 241–381.

BUSHNELL, D. C. & ATMAWAN, D. T. 1986. A model for hydrocarbon accumulation in the Sunda Basin, West Java Sea. *Proceedings of the Indonesian Petroleum Association Annual Convention*, 48–75.

DJUANDA, H. 1985. Facies distribution in the Nurbani carbonate build-up, Sunda Basin. *Proceedings of the Indonesian Petroleum Association Annual Convention*, 507–533.

GIRGIS, J., SNEIDER, R. M. & BUDIYENTO, T. 1995. Multi-disciplinary teams – what is the "right structure"? Based on ten years of MDTs in Indonesia. *Proceedings of the Indonesian Petroleum Association Annual Convention*, 545–564

MCCHESNEY, D., RUSMANTORO, A., SMITH, M. G. & MURSID, S. 1992. The Krisna lower Batu Raja waterflood: an updated case history. *Proceedings of the Indonesian Petroleum Association Annual Convention*, 403–430.

SMITH, S., DANAHEY, L., HIMAWAN, R., HARMONY, W., GILMORE, L., ARMON, J. & SMITH, W. 1995. An example of 3D AVO for lithology discrimination in Widuri field, Asri Basin, Indonesia. *Proceedings of the Indonesian Petroleum Association Annual Convention*, 397–424.

TAPPONIER, P., PELTZER, G., LE DAIN, A. Y., ARMIJO, R. & COBBOLD, P. 1982. Propagating extrusion tectonics in Asia, new insights from simple experiments with plasticene. *Geology* **10**, 611–616.

TODD, D. & PULUNGGONO, A. 1971. The Sunda basinal area, an important new Indonesian oil province. *American Association of Petroleum Geologists Convention Abstracts.*

TONKIN, P., TEMANSJA, A. & PARK, R. K. 1992. Reef complex lithofacies and reservoir, Rama field, Sunda basin, Southeast Sumatra, Indonesia. *In: Carbonate rocks and reservoirs of Indonesia – a core workshop.* Indonesian Petroleum Association Core Workshop Notes, 5 October 1992. 7-1–7-14.

WICAKSONO, P., ARMON, J. W. & HARYONO, S. 1992. The implications of basin modelling for exploration-Sunda Basin case study, Offshore Southeast Sumatra. *Proceedings of the Indonesian Petroleum Association Annual Convention*, 379–415.

WIGHT, A. W. R. & HARDIAN, D. 1982. Importance of diagenesis in carbonate exploration and production, Lower Batu Raja carbonates, Krisna Field, Java Sea. *Proceedings of the Indonesian Petroleum Association Annual Convention*, 211–235.

YOUNG, R., HARMONY, W. E., JUNIARTO, G. & BUDIYENTO, T. 1991. Widuri field offshore Southeast Sumatra, sandbody geometries and the reservoir model. *Proceedings of the Indonesian Petroleum Association Annual Convention*, 385–418.

Sumatra/Java oil families

CRAIG SCHIEFELBEIN[1] & NICK CAMERON[2]

[1] *GeoMark Research, Inc., 9748 Whithorn Drive, Houston, TX 77095, USA*
[2] *Chalkyfold, Grove Lane, Chesham, Bucks HP5 3QQ, UK*

In areas where substantial oil production has been established, regional crude oil geochemical studies are excellent ways of identifying, evaluating and comparing the various petroleum systems that have contributed to reserves. The number and type of effective source rocks can be determined by establishing the number of compositionally distinct oil families. Inferences regarding source rock depositional environment are possible since the geochemical characteristics of crude oils reveal important information on paleoenvironmental conditions of source rock deposition and possible age.

In order to understand better the petroleum systems of Sumatra and Java, 122 crude oil samples (Fig. 1) have been geochemically analysed. The detailed analytical program included measurements of bulk parameters (API Gravity, %S, V and Ni content), stable carbon isotope composition of C15+ hydrocarbon fractions, whole oil gas chromatography, and terpane and sterane biomarker distributions using GC/MS. The analytical data were compared using multivariate statistical techniques such as cluster and principal component analyses (Fig. 2). With this approach, the majority of the oils were divided into two main petroleum systems, 'lacustrine' and 'terrigenous'. Both of these petroleum systems are supplied by Tertiary-aged source rocks. Two oils from the North Sumatra Basin exhibit marine affinities, whilst several oils from Central and South Sumatra exhibit intermediate geochemical characteristics suggestive of mixing. Further detailed examination of each petroleum system, sometimes using more specific geochemical criteria, typically resulted in the establishment of sub-groups of oils according to specific source environment, such as lacustrine fresh versus lacustrine saline.

The oil families

'Lacustrine' oils

'Lacustrine' oils originated from mainly algal kerogen (Type I) deposited in a lacustrine environment with variable but minor admixtures of terrigenous organic matter. 'Lacustrine' oils

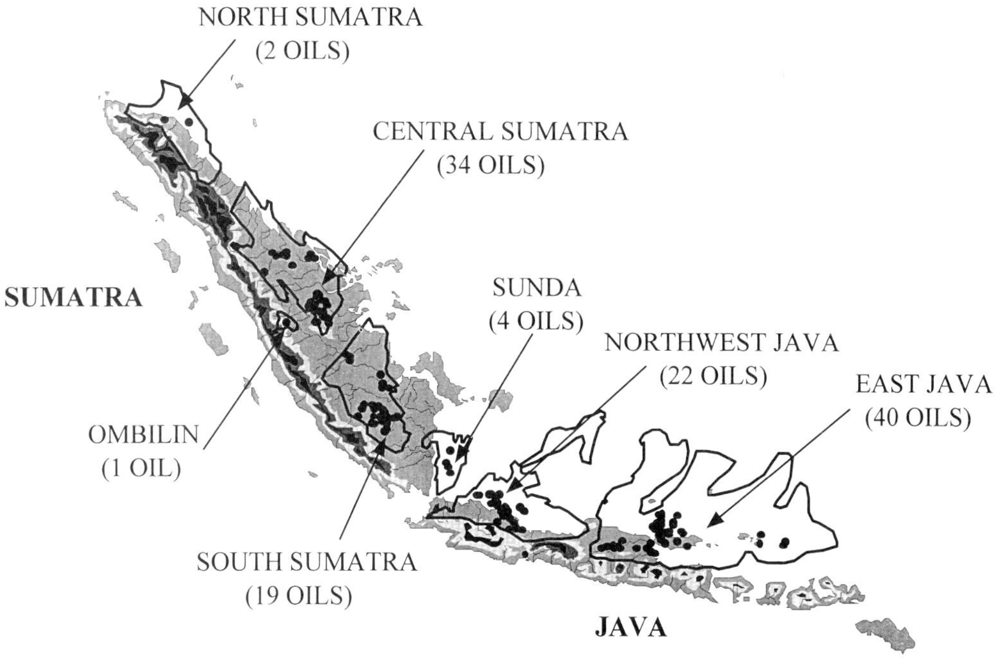

Fig. 1. Crude oil location map.

From Fraser, A. J., Matthews, S. J. & Murphy, R. W. (eds), 1997, *Petroleum Geology of Southeast Asia*, Geological Society Special Publication No. 126, pp. 143–146.

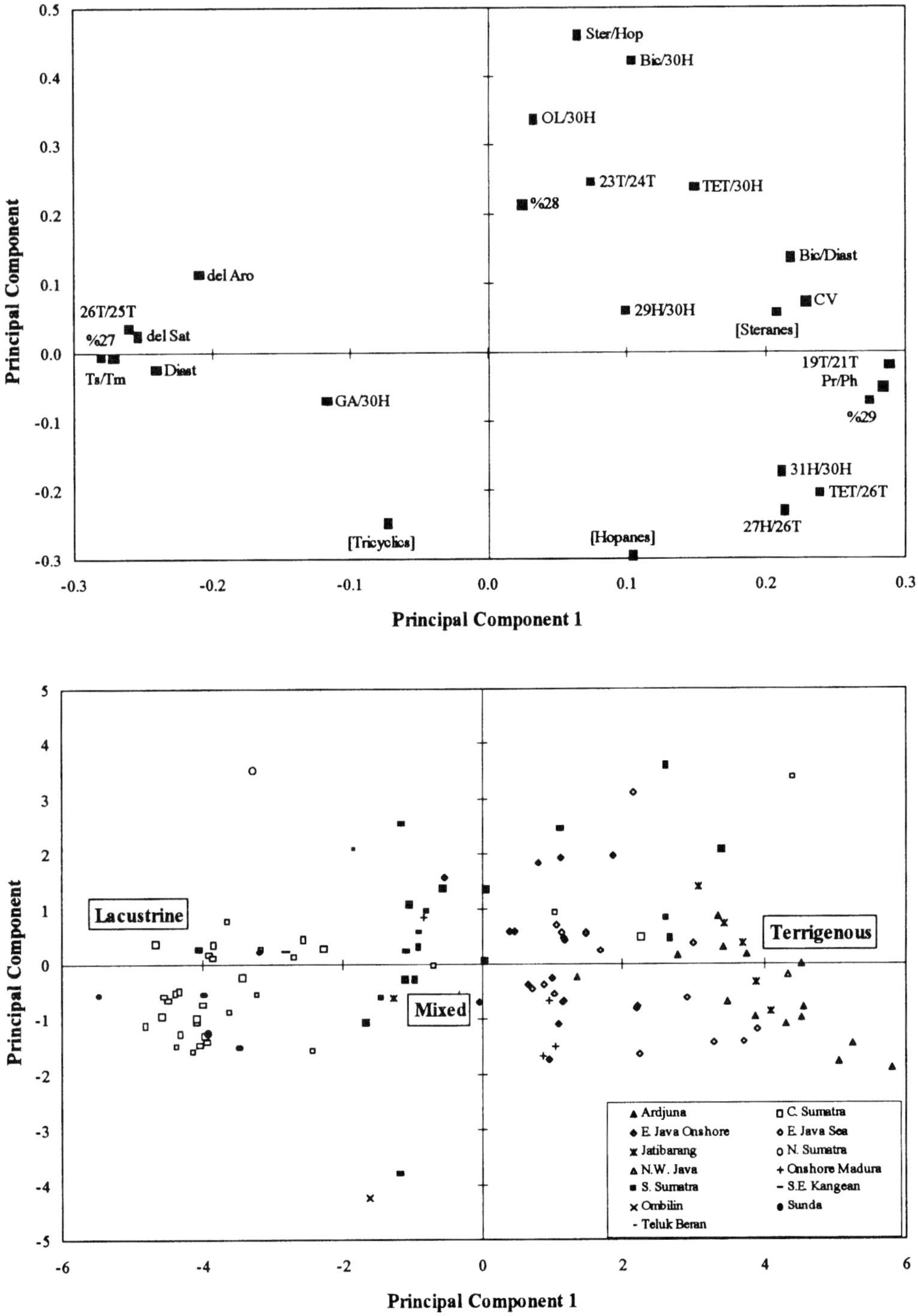

Fig. 2. Principal component loadings (top) and scores (bottom) plots.

are characterized by the absence of 24-*n*-propylsteranes, an abundance of hopanes relative to steranes, a C26-tricyclic/C25-tricyclic ratio >1, high T_s/T_m ratios, abundant extended tricyclic terpanes, low proportions of C31+ pentacyclic terpanes, and, in general, a heavy carbon isotopic signature. However, the isotopic signature should be treated with caution, as it can vary from −20 to −30‰ for lacustrine oils (ten Haven & Schiefelbein 1995). Bottryoccocane, a highly specific biomarker for fresh water algae *Botryococcus braunii* Race B (Derenne *et al.* 1988), was identified in several lacustrine oils.

Oil families with lacustrine affinities are present in Central and South Sumatra, the Sunda/Asri Basins, and in the East Java Kangean Sub-basin (Fig. 3). In Sumatra, these oils are related to the early Tertiary rift phase (Murphy this volume). Previously, pure lacustrine (algal-rich) oils were unknown from the South Sumatra Basin and the Kangean Sub-basin. 'Lacustrine' oils were not identified in the North West Java Basin, even though the rift succession is now known to contain lacustrine source rocks ('Deep Arjuna Study Team', this volume). The Sumatra 'lacustrine' oil family can be divided into three subgroups. Katz & Mertani (1989) propose that all variations seen in the geochemical characteristics of lacustrine oil from Central Sumatra are due to facies variations within the Pematang source rock, in turn related to different water chemistries in the different subbasins. Alternatively, the differences in oil chemistries may be related to facies variations within a single, long term lake system.

Oils with both 'lacustrine' and 'terrigenous' affinities

Oils with geochemical characteristics of both 'lacustrine' and 'terrigenous' source types are widespread in South Sumatra (Fig. 3). These oils may have been derived either from a mixed lacustrine/terrigenous kerogen assemblage deposited in the lake margins or from two different source rocks (or source rock facies). This oil type is similar to the Type II oil of Suseno *et al.* (1992). When examined from a geographical viewpoint, these oils appear to be mixed as they are bordered by pure 'lacustrine' oils to the northwest and by pure 'terrigenous' oils to the southeast. Interestingly, in the *X–Y* space of principal component analyses, these oils plot between the 'lacustrine' and 'terrigenous' oils.

'Terrigenous' oils

'Terrigenous' oils originated from mainly terrigenous kerogen (Type III) deposited in a fluvial deltaic environment with variable but minor

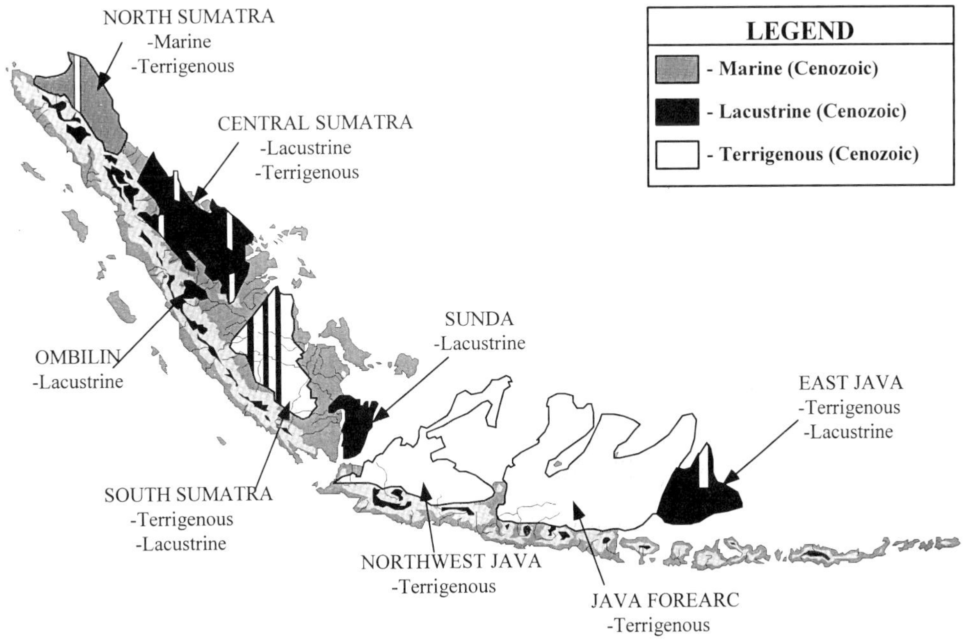

Fig. 3. Map illustrating location of the different petroleum systems present in Sumatra and Java.

admixtures of algal organic matter. The depositional environment could also be a lake, but the kerogen is mainly composed of Type III organic matter which often contains an abundance of dammar resins derived from trees of the *Dipterocarpus* family. 'Terrigenous' oils are characterized by the absence of 24-*n*-propylsteranes, high abundances of low molecular weight tricyclic terpanes and C24 tetracyclic terpanes, a predominance of C29 steranes, elevated pristane/phytane ratios, positive canonical variables, and abundant terrigenous triterpanes such as oleananes and bicadinanes. Oleanane and bicadinane are *angiosperm* biomarkers associated with a Tertiary (or late Cretaceous) source.

'Terrigenous' oils are occur primarily in South Sumatra and Java, with isolated occurrences in Central Sumatra (Fig. 3). The Java oils are derived from terrigenous to near-shore environments dominated by terrigenous input. However, two of the analysed oils from East Java contained appreciable quantities of C30 *n*-propyl steranes and 30-norhopanes, suggesting that a marginal marine source with carbonate affinities may also be present.

Oils from the North Sumatra Basin may have originated from terrigenous organic matter deposited in a marginal marine environment. Marine oils are distinguished by the presence of 24-*n*-propylsteranes, an abundance of steranes relative to hopanes, a predominance of C27 steranes, and low pristane/phytane ratios. The North Sumatra oils examined in this study contain terrigenous triterpanes such as oleananes and/or bicadinanes. Kirby *et al.* (1993) suggest that a lacustrine oil is also present in North Sumatra. Matchette-Downes *et al.* (1994) consider that the source is located in the Oligocene syn-rift succession. They present a case for a lacustrine succession dominated by terrigenous input. Kjellgren & Sugiharto (1989) identified a suite of terrigenous and marine oils, including one sourced from a carbonate. Subroto *et al.* (1992) also reported that a carbonate source rock exists in North Sumatra, but no sample from this putative source rock was represented in the oils included in the present study.

Conclusions

Though the Sumatra and Java regions are considered mature in exploration terms, the variety of oil types requires a greater range of source rock types than previously suspected. Most notable are the extended geographical range of the 'lacustrine' oils and the complexity of the source succession associated with the Oligo-Miocene transition from extension to thermal subsidence (Murphy this volume). Also evident is the variety of lacustrine source rocks present in the Central Sumatra Basin. Finally, now that the deep architecture of the North Sumatra Basin is known, there is a need to better understand where the source rocks are in the succession.

References

DEEP ARJUNA STUDY TEAM 1997. Early Tertiary rift evolution and its relationship to the Petroleum System in the offshore Northwest Java Basins, Indonesia. *This volume.*

DERENNE, S., LARGEAU, C., CASADEVALL, E., TEGELAAR, E. W. & DE LEEUW, J. W. 1988. Relationship between algal coals and resident cell wall polymers of extant algae as revealed by Py-GC-MS. *Fuel Processing Technology*, **20**, 93–101.

MATCHETTE-DOWNES, C. J., FALLICK, A. E., KARMAJAYA & ROWLAND, S. 1994. A maturity and palaeoenvironmental assessment of condensates and oils from North Sumatra Basin, Indonesia. *In*: SCOTT, A. C. & FLEET, A. J. (eds) *Coal and Coal-bearing Strata as Oil-prone Source Rocks*. Geological Society, London, Special Publications, **77**, 55–72.

KATZ, B. J. & MERTANI, B. 1989. Central Sumatra – a geochemical paradox. *Proceedings of the Indonesian Petroleum Association, 18th Annual Convention*, 403–435.

KIRBY, G. A., MORLEY, R. J., HUMPHREYS, B., MACHETTE-DOWNES, C. J., SARGINSON, M. J, LOTT, G. K., NICHOLSON, R. A., YULIHAUTO, B., WIDIASTUTI, R., KARMAJAYA, SUNDER, FIRTHS, F., SOFYAN, S. & WIJAYA, S. 1993. A re-evaluation of the regional geology and hydrocarbon prospectivity of the onshore central North Sumatra Basin. *Proceedings of the Indonesian Petroleum Association, 22nd Annual Convention*, 243–264.

KJELLGREN, G. M. & SUGIHARTO, H. 1989. Oil geochemistry: a clue to the hydrocarbon history and prospectivity of the southeaster North Sumatra Basin. *Proceedings of the Indonesian Petroleum Association, 18th Annual Convention*, 363–384.

MURPHY, R. W. 1997. Petroleum Systems of SE Asia. *This volume.*

SUBROTO, E. A., ALEXANDER, R., PRANJOTO, U. & KAGI, R. I. 1992. The use of 30-norhopane series, a novel carbonate biomarker in source rock to crude oil correlation in the North Sumatra. *Proceedings of the Indonesian Petroleum Association, 21st Annual Convention*, 145–163.

SUSENO, P. H., ZAKARIA, MUJAHIDIN, N. & SUBROTO, E. A. 1992. Contribution of Lahat Formation as hydrocarbon source rock in South Palembang area, South Sumatra, Indonesia. *Proceedings of the Indonesian Petroleum Association, 21st Annual Convention*, 325–337.

TEN HAVEN, H. L. & SCHIEFELBEIN, C. 1995. The Petroleum Systems of Indonesia. *Proceedings of the Indonesian Petroleum Association, 24th Annual Convention*, 443–459.

Early Tertiary basin formation and the development of Lacustrine and quasi-lacustrine/marine source rocks on the Sunda Shelf of SE Asia

J. M. COLE[1] & S. CRITTENDEN[2]

[1] *The Geochem Group Ltd, Chester Street, Saltney, Chester CH4 8RD, UK*
[2] *Waye Cottage, Chagford, Devon TQ13 8HN, UK*

Abstract: The Tertiary basins of the Sunda Shelf of SE Asia were formed in the ?Mid- to Late Eocene and accumulated thick intervals of syn-rift lacustrine and low salinity organic rich shales throughout the Late Palaeogene. Towards the end of the Oligocene and into the Early Miocene marine transgression occurred throughout the region. A change to deep open marine sedimentation followed by a general shallowing and deposition throughout the Neogene, of great thicknesses of marginal marine and paralic sediments. The syn-rift sediments are the most important hydrocarbon source rocks in this region, although precise dating of them is difficult due to the long-ranging nature of the palynomorphs and the paucity of the microfauna. However characteristics of the depositional environments and nature of the source rocks deposited within separate provincial basins that were wholly or partially landlocked can be assessed from a detailed examination of the isolated palynomorph and kerogen assemblages. A classification of lacustrine depositional environments within the region on this basis is presented.

The Sunda Shelf or Sundaland, is the collective term given to the geographical and geological entity of southern Indochina, the Malay Peninsula, Sumatra, the NW Java Sea and southern Borneo. Formed of fragments from Gondwanaland that separated, drifted northwards across 'proto-Tethys' and accreted on to the Laurasian continent throughout the Late Palaeozoic and Mesozoic, it had formed a stable continental block by the Early Tertiary. This stable block of accreted microcontinents that now forms a southeastern protrusion from Laurasia, bordered in part by subduction zones of oceanic crust, was an uplifted area throughout the Latest Cretaceous and Early Palaeocene. No definitively dated sediments of this age are recognized within the region. However by the Late Palaeocene peripheral intracratonic rifts had developed in the area of the Gulf of Tonkin. By the Mid- to Late Eocene such peripheral back-arc rifting was occurring further south and west around the bordering edge of the of Sundaland (Sumatra, Java and Kalimantan). Rifting on the eastern side (Thai–Malay Basin), in what has been called an 'interior and accreted zone rift margin' tectonic setting, commenced in the Oligocene (Kingston *et al.* 1983).

These rifts accumulated thick intervals of syn-rift lacustrine shale within semi-permanent lacustrine depositional environments. By analogy with present-day tropical lacustrine depositional environments (e.g. the East African rift valley lakes) and other well documented ones from the fossil record (e.g. in the Eocene interior Great Basin of the USA), it is known that such environments form prolific oil source beds. Such depositional environments occurred throughout the ?Mid–Late Eocene and Early to early Late Oligocene syn-rift (or interior fracture) phase of Sundaland. Sediments within these basins are typically characterised by thick intervals of dark grey or brown, slightly to moderately calcareous shale. Finely comminuted organic debris, that can be isolated for study from such rock by the dissolution of siliceous and carbonate rock matrix using strong acids, is often dominated by large amounts of hydrogen rich type I and type II kerogen. This material can be very autofluorescent when viewed at high-power magnification, using the appropriate equipment on the microscope stage. It may occur with an abundance of Chlorococcalean algae such as *Botryococcus* and *Pediastrum* together with other indeterminate algal/fungal/bacterial debris. This material can be highly oil prone.

By the Late Oligocene, rifting had ceased and these interior fracture basins had been largely filled. Thermal subsidence (or thermal sag) across the entire region lead, by the Early Miocene, to widespread marine transgression. This caused a termination of sedimentation in semi-permanent stratified lakes and a change to thick reefal carbonate sedimentation. In turn this was followed by deep marine shales and associated marginal marine deltaics in the Mid-to early Late Miocene of different parts of Sundaland. Renewed tectonic activity in the Late Miocene associated with the impinging of India into Laurasia led to uplift of Sundaland and renewed influence of shallow marginal marine and

From Fraser, A. J., Matthews, S. J. & Murphy, R. W. (eds), 1997, *Petroleum Geology of Southeast Asia*, Geological Society Special Publication No. 126, pp. 147–183.

continental sedimentation in which lacustrine depositional environments occurred. These lakes and those formed during the thermal sag stage, were much more ephemeral and shallow than those present earlier on during the Late Palaeogene, and are less important as oil source beds.

The present paper summarizes aspects of the sedimentology, geochemistry and palynofacies of lacustrine sediments throughout the Tertiary of SE Asia. A classification of deep and shallow lacustrine depositional environments is presented based on studied stratigraphical sections throughout the region from the Palaeogene and Neogene. Support is given for the Late Palaeogene syn-rift semipermanent lake sediments as those that are by far the most prolific hydrocarbon source beds of Sundaland.

Tectonic development

The Sunda Shelf or Sundaland on the south-eastern extremity of the Laurasian continent, is a regional promontory that juts out into, and divides the South China Sea to the east from the Indian Ocean to the west. Approximately half of

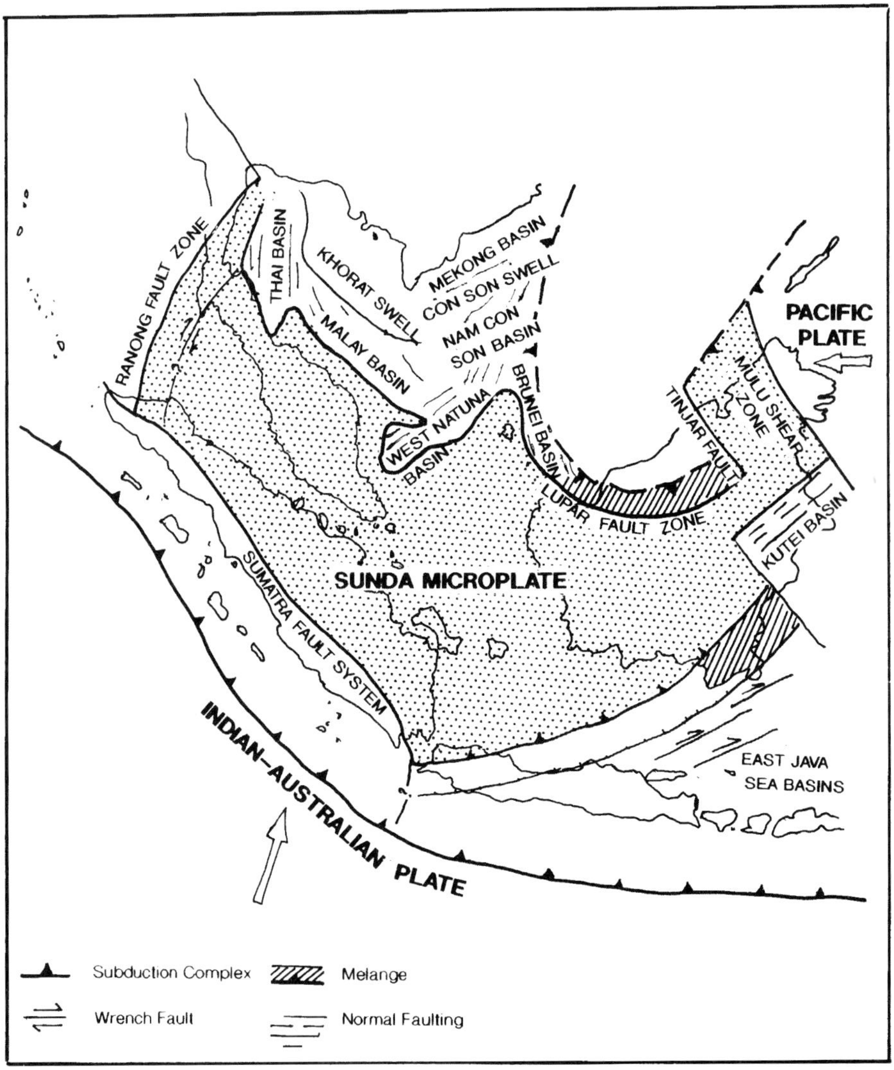

Fig. 1. The Sunda Microplate (after Davies 1984).

this shelfal area is occupied by the landmasses of Sumatra, Peninsula Malaysia/southern Thailand and western Borneo, whilst the remainder is covered by the shallow shelf seas of the Straits of Malacca, south Natuna Sea (southern S. China Sea) and the West Java Sea, (Figs 1 & 2). Almost all of this area was dry land during periods of glacial maxima, as evidenced by the presence of peat in the shallow marine surficial sediments. The equator runs through the middle of this region, which has been equatorially located since the Early Tertiary and the natural vegetation is tropical evergreen rainforest.

This region of SE Asia was formed by accretion of microcontinental fragments at various times from the Triassic onwards that parted from the Gondwanaland continent and travelled across the proto-Tethys ocean on the back of the 'conveyor belt' system of subducting oceanic crust, (Hamilton 1979; Audley-Charles 1988; Audley-Charles & Hallam 1988; Fig. 3). Fragments of Gondwanaland separated at various times and accreted on to Laurasia in successive 'slices'. Metcalfe (1988) has highlighted those microcontinental fragments, which do not coincide very exactly with the present-day islands of

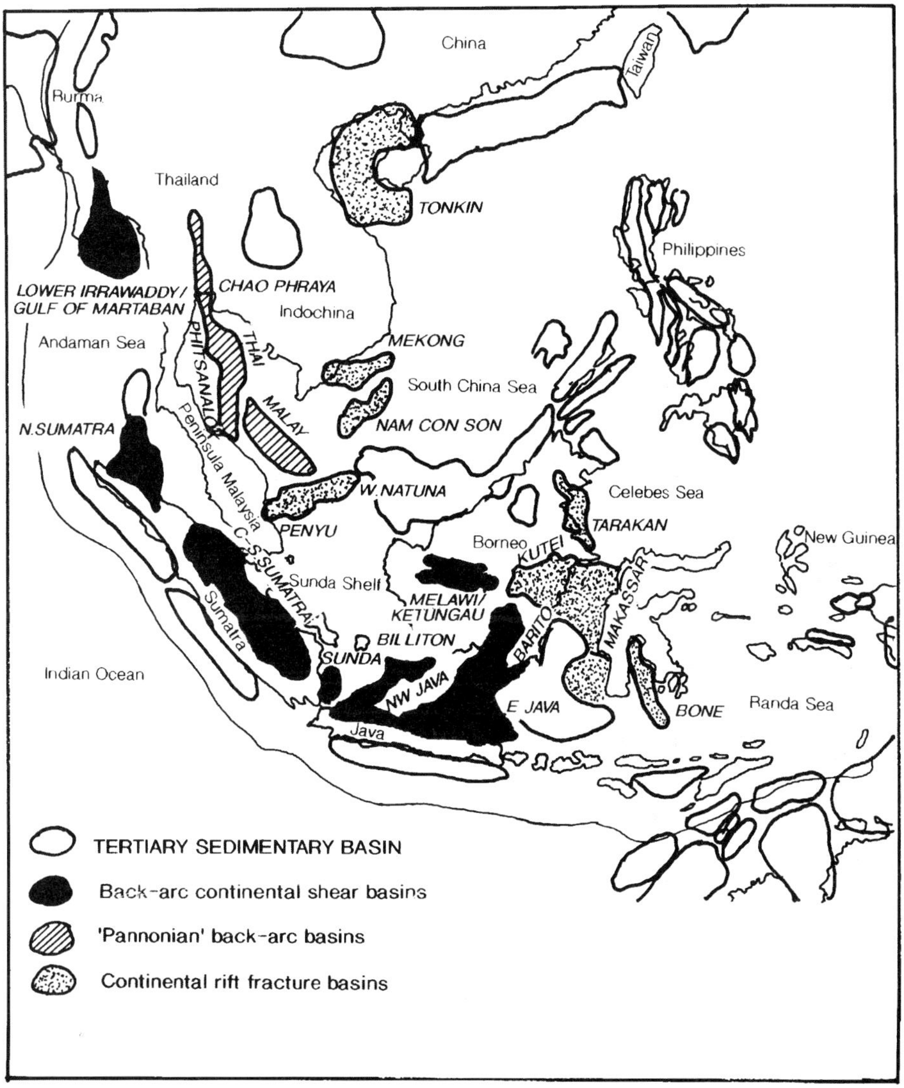

Fig. 2. Main Tertiary basins of SE Asia (after Hutchinson 1989).

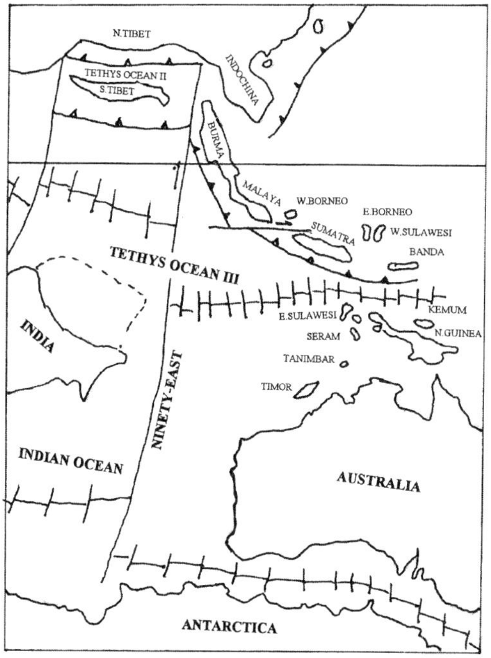

Fig. 3. Fragmentation of northern Gondwanaland to form the microcontinents of SE Asia (after Hamilton 1979; Audley-Charles 1988; Audley-Charles & Hallam 1988).

the west Indonesian and Malay archipelago (Fig. 4). For instance the microcontinental fragment of Sibumasu includes parts of Burma, Thailand Malaysia and Sumatra. As a very general rule the age of the microcontinental fragments (in terms of their suturing and accretion) young to the south from which direction continental drifting brought them into the region.

These accreted fragments have behaved as a single unified microcontinent (Sunda microplate, Fig. 1) since the Early Tertiary, though other microcontinental fragments arrived throughout the Tertiary that now form parts of eastern Borneo, east Java, Sulawesi and the Banda Arcs. This region is tectonically quite different and is usually referred to as Eastern Indonesia. These differences are also reflected in their present-day floras that have a much more antipodean character as observed by the early naturalists, the floral provinces being separated by the so called 'Wallace's line'.

The northward drift of microcontinental fragments from Gondwanaland and their accretion onto Laurasia were a rapid advance effect of plate motions that carry on to the present, as reflected by the general slower northward movement of the whole of the Gondwana continent that eventually closed the Tethys Ocean. This movement continues today as the Australian continent and Pacific plate continue to impinge into Asia at an extensive and complicated triple junction in the SE Asian region.

Some authorities (Audley-Charles 1987) believe the microcontinental fragments drifting from Gondwana throughout the Mesozoic formed closed 'arks' that bore rich tropical floras that harboured and nurtured the evolution of the first modern flowering plants. The angiosperms may have rapidly radiated out and evolved to colonise all the earth's land areas. It is known that island habitats often allow more rapid evolution, as species develop to occupy and exploit new environmental niches that may be created in the absence of already adapted colonizers. Examples in the pollen record of SE Asia exist of *in situ* evolution of important angiosperm plant groups that are endemic today, (see Morley 1977, 1990; Muller 1964, 1972, 1981; Muller & Caratini 1977). For example the mangrove palm *Nypa* first appears in the fossil record in the Palaeocene and a succession of species of its pollen of the genus *Spinizonocolpites* are known throughout the Tertiary to the present day. It is also known that the richness and diversity of plants in the tropical rainforest of this region surpasses that of Africa and the Americas due to a very broadly consistent equatorial position being maintained by this region that left it less effected by catastrophic environmental changes.

The accreted continental fragments of the Sunda microplate comprise rock suites of various age and type as old as Carboniferous. Some have been named (e.g. the Woyla terrane of Central Sumatra; see Cameron & Pulunggono 1984; Hartono & Tjokorosapoetro 1984; Pulunggono & Cameron 1984). Collectively they moulded together to form the basement rocks of SE Asian Sundaland that produced a framework substratum for the initiation of renewed basin formation and sedimentation in the Early Tertiary.

The separation of India from Africa/Antarctica and its collision with Laurasia is considered to have been the driving force for rifting and basin formation in SE Asia from the Late Palaeocene onwards. The eventual closure of proto-Tethys has now been superseded by the continued westward impingement of the Australian continent into the region with extensive east–west transcurrent faulting along the Sorong fault of Irian Jaya. This together with the oblique compression and subduction of the Pacific plate and the trapping of fragments of oceanic crust has produced the geological

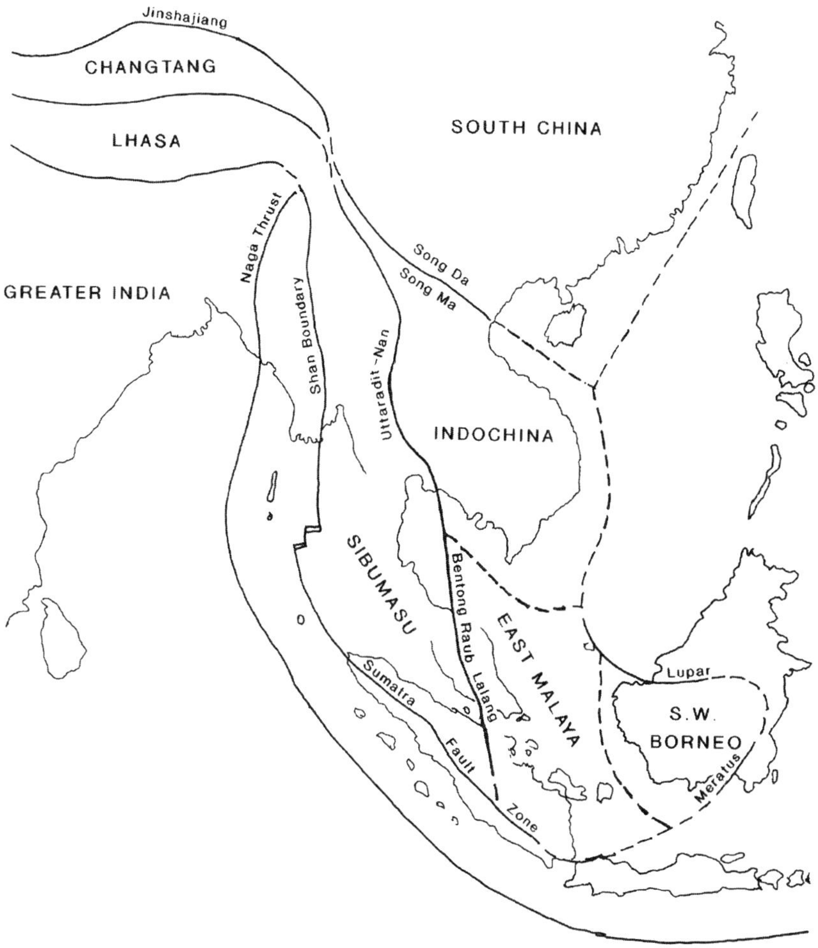

Fig. 4. Tectonostratigraphic terranes of SE Asia (after Metcalfe 1988).

complex regions of the Philippines and the Banda arcs, that are still tectonically highly active (Daly *et al.* 1987, 1991; Hall this volume). Numerous sedimentary basins occur within this region of Eastern Indonesia, but this is outside the scope of the present paper that deals only with Sundaland.

Subduction is now occurring along the western margin of Sundaland adjacent to Sumatra and Java (the Java trench) as India continues to impinge into Asia. This is only the latest phase in subduction zones that have successively 'stepped back' as subsequent accretion added further younger continental crust onto the core of Sundaland, (Hutchinson 1989). Subduction and accretion southwards and westwards against the other side (South China Sea – Borneo side) of Sundaland ceased in the Early Tertiary (Fig. 5).

A spreading centre may have been present in the South China Sea that was responsible for producing the old suture of the Lupar line across southern Borneo that forms the southeast edge of Sundaland, with this dead spreading centre now itself subducted under the westward moving Pacific plate. Later transcurrent movement along the Lupar line has further obscured this old subduction zone, and this together with the difficulty of field mapping has lead to the eastern tip of Sundaland not being very clearly defined, its position often dotted on to many published structural maps.

Until comparatively recently, before explorationists turned towards Eastern Indonesia, almost all petroleum exploration in the region has concentrated on the prolific hydrocarbon plays of Sundaland. To date almost all petroleum exploration (with the exception of some

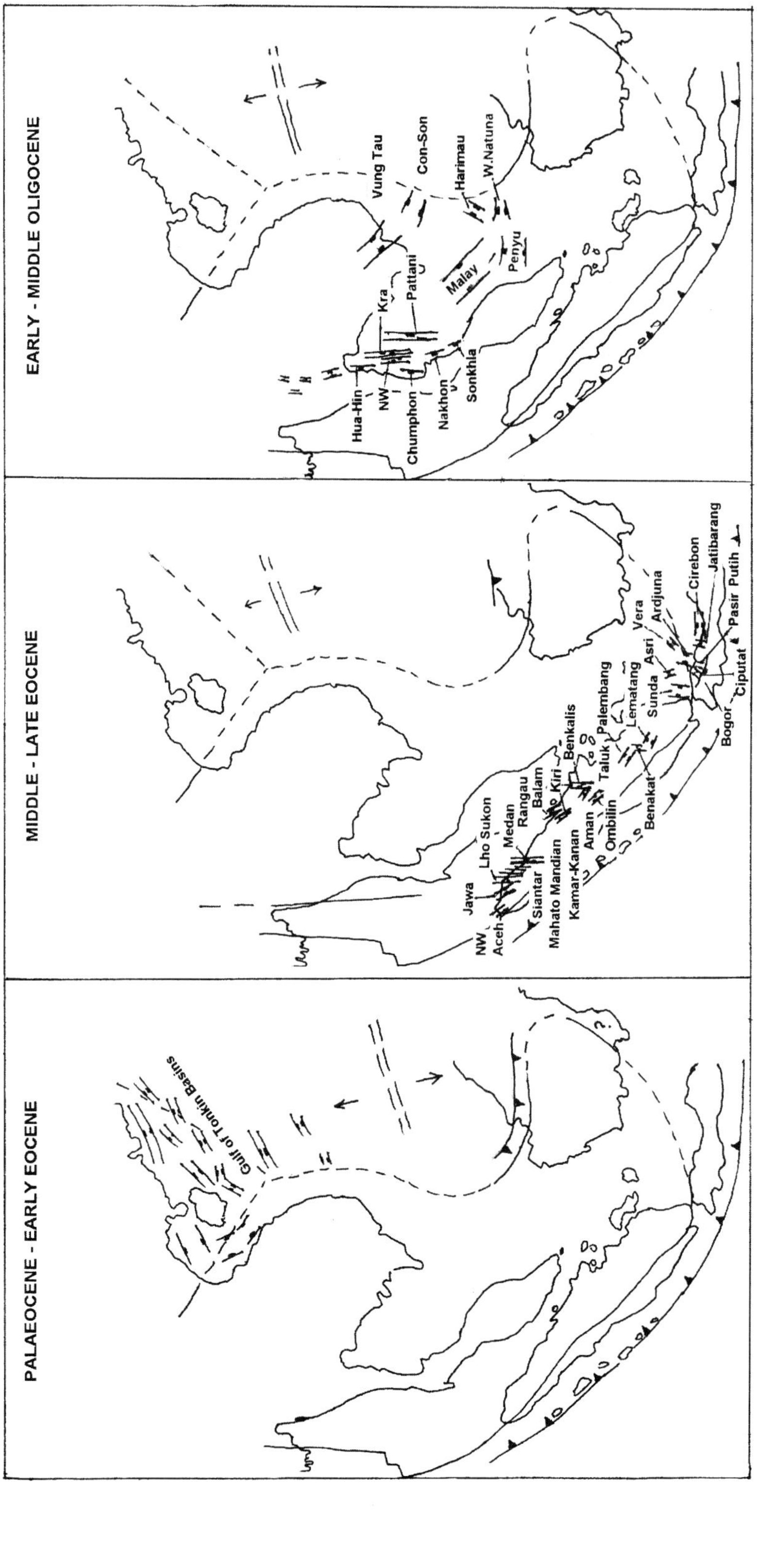

Fig. 5. 'Interior fracture' Palaeogene basin formation in SE Asia. **(a)** Palaeocene–Early Eocene; **(b)** Mid–Late Eocene; **(c)** Early–Mid-Eocene.

minor Cretaceous plays and fractured basement reservoirs, (Koning & Darmono 1984; Mulhadiono & Sutomo 1984; Sardjito *et al.* 1991) has been within the Tertiary (?Mid–Late Eocene to Pliocene) sedimentary pile.

Basin formation

It is evident that the stable cratonic block of Sundaland formed by the Early Tertiary, was effected by NE–SW-aligned rifting in the Gulf of Tonkin (offshore Vietnam and southern China) in the Palaeocene (Fig. 5a). This rifting may have been associated with a subduction zone in this region from a spreading centre in the south China Sea. A similar subduction zone with associated basin formation along the along the complex structural hinge of the Lupar Line of Borneo may subsequently have itself been subducted and no longer detectable.

A coincidence between the timing of collision of India and Laurasia and basin formation was noted by Daly *et al.* (1991). The 'hard' collision of India with Laurasia and change in motion of the Pacific Plate has been put at 42–44 Ma in the latest Mid-Eocene to early Late Eocene, (Murphy 1992), when intracratonic rift basins opened along the Sumatra–Java–southern Borneo back-arc (Fig. 5b).

This rifting may have occurred later further to the south at locations further away from the direction of origin of the compressive stress. The NW Java and Sunda Basins do not appear to have the same thickness of Palaeogene sediments as seen in Sumatra, particularly in the North and Central Sumatra basins (see Fig. 18). Within these latter two basins sediments have been dated as old as Late Eocene whilst in the Java Sea sedimentation may have commenced in the Oligocene. On the eastern side of Sundaland, crustal stretching as India went on past SE Asia and further extrusion within Eastern China and Indochina, resulted in rotation of Sundaland (Holcombe 1977; Clure 1991), more regional rifting with the opening of the South China Sea basin in the Early to Middle Oligocene, (White & Wing 1978; Fig. 5c).

Rifting around the periphery of the Sunda Shelf, western and southern margins (Sumatra, Java and Borneo) in the latest Mid-Eocene to early Late Eocene is known from the age of pollen dated from some of the earliest known Tertiary sediments in these basins such as Nangullan Formation of Central Java (Morley 1982). Early Palaeogene sediments (Palaeocene–early Mid-Eocene) are not definitely dated from the region. Older dated Tertiary sediments such as the Plateau sandstone of Sarawak (Muller 1968) are now open to doubt due the effects of possible recycled pollen (Morley pers. comm. 1992).

The age of the Upper and Lower Tanjung beds of the Barito Basin is given as latest Palaeocene to Eocene (Kusuma & Darin 1989), but this seems to be radically different from the age of syn-rift sediments in neighbouring basins. Further palynological work needs to be conducted in order to verify it as it does appear anomalous at present. The pollen genus *Proxapertites* has been used as a Palaeocene marker in other basins such as in the East Java Sea. However there are now indications that this taxon may in fact range well into the Eocene. Work by Satyana (1995) has suggested a Mid–Late Eocene age for the Tanjung Formation in the Barito basin that is more in keeping with the neighbouring basins. Biostratigraphical evidence provided in this latter paper is restricted to microfaunal dating of the latest Eocene unconformity at the top of the Tanjung Formation. Similar ages for the Tanjung Formation were given by Mason *et al.* (1993) and Rotinsulu *et al.* (1993).

A serious problem exists with the precise dating of the Early Tertiary sediments of SE Asia as they are generally devoid of microfaunas and are landward of the mangrove zone, the depositional environment where many of the important palynological age markers occur. Lacustrine algae are notoriously evolutionarily conservative with extremely long geological ranges (Cole 1992). Biostratigraphical age dating of this sequence is thus poor, (Courtney 1994). Further refinements to it will come from further work on establishing local biofacies zones (Morley 1991) that are related to third-order systems tracts. For example the work of Goodall *et al.* (1991) demonstrated that cyclical changes in the type of amorphous kerogen within a lacustrine sequence can be used to indicate shallowing events associated with eustatically driven base level changes. It is unlikely that radical improvements to the biostratigraphy will come from the recognition of new palynological marker taxa in the Early Tertiary of SE Asia.

Early Tertiary rifting occurred in a back-arc setting along the Sumatra–Java–Borneo margin of Sundaland, influenced by the active subduction zone of the Indian Ocean Plate against this margin of Sundaland. The basins were generally elongate fault-bounded grabens that accumulated sediments that were largely freshwater lacustrine in character. During this early syn-rift phase of sedimentation these lakes were quite localized (Fig. 5) in comparison to the whole area of Sundaland, but sufficiently large to have laid down very significant volumes of organic rich source rock shales. Only by the latest Oligocene/Early Miocene did extensive marine transgression occur. In this 'post-rift' phase, sedimentation took place over almost all of Sundaland, producing the classic 'steers head' basin infill geometry, and the development of extensive regional reservoirs.

Not all basins of Sundaland are back-arc (Murphy 1975, 1976; Woodside 1984; Hutchinson 1986, 1989; Pulunggono 1985), but some of the largest known hydrocarbon occurrences in the region are associated with hydrocarbon plays within such a structural setting. However Williams & Eubank (1995) remark that the Sumatra–Java back-arc may not actually have been such in the Early Tertiary.

The South China Sea basins on the eastern side of Sundaland are not formed immediately adjacent to a subduction zone. Subduction is believed to have occurred along this margin, but it had already ceased by the Eocene (Hutchinson 1989). The South China Sea is essentially a very large rifted re-entrant embayment within Sundaland that was initiated in the Mid-Oligocene (Fig. 5c). Alternative hypotheses of its origin have been given. It may have opened when compressive forces on Sundaland from the west were relaxed as India moved on past SE Asia (White & Wing 1978). Rifting on this eastern margin may therefore have been caused by clockwise rotation of Sundaland as the Indian continent passed the region by.

Downwarping along the old subduction zone of the Lupar line has been invoked by some authors, (Wongsosantiko & Wirojudo 1984). An element of 'extrusion tectonics', (Molnar & Tapponier 1975; Tapponier *et al.* 1982) induced by the collision and continued impingement of India into Laurasia, caused lateral movement along major faults through China and down into the Indochina/Malay peninsula. This may also have played a part in basin formation in the South China Sea (Daines 1985).

Such extrusion tectonics are believed to have had significant effects at second order sequence stratigraphical level in Late Palaeogene and Neogene basin development and sedimentation of SE Asia. The South China Sea basin is classified as a 'craton and accreted zone rift margin' type by Klemme (1980). Basin formation around the periphery of Sundaland on this stable foreland craton set the scene for the subsequent geological history of Tertiary petroliferous sediment fill (Beddoes 1980; Fletcher & Soeparjadi 1976; Najoan 1972; Soeparjadi *et al.* 1975).

Most of the major known productive syn-rift basins of Sumatra are located on land, unlike those of Java that occur mostly in the area of the Java Sea. The 'spine' of Sumatra is composed of the Barisan Mountains. Uplift of this mountain range largely came about with volcanic activity in the Pliocene. This was the same time as volcanic activity in Borneo that lead to the uplift of Mount Kinabalu and the Central Borneo Mountains. This uplift is registered in the palynological record with the in-coming of *Podocarpus imbricatus*, an upland pollen type that becomes an established element in most depositional environments and is an important marker taxon.

Palaeogene sedimentation

Rifting of Sundaland in the Early Tertiary produced a number of pronounced elongated, generally north–south-trending grabens along the Sumatra–Java back-arc. The newly formed isolated basins of various ages on the periphery of Sundaland initially accumulated continental red beds, alluvial fans and conglomerates of Late Eocene–?Early Oligocene age along the Sumatra–Java back-arc and ?'Mid'–Late Oligocene age in the South China Sea. These sediments tend to be thickest and best developed along the margins of the basins. Sedimentation within each of these grabens occurred in similar ways. Alluvial fans developed at either end and along the edges.

After the initial alluvial phase, they quickly became established as lakes with the succeeding sediments accumulating largely as fine grained lacustrine syn-rift sediments. Continued intermittent alluvial fan sedimentation followed by further lacustrine shales form important potential oil reservoirs that are enclosed by their source rock shales. Good examples of this, on a scale large enough to form discrete reservoirs are the Jani and Yanti fan-glomerates of the Sunda Basin (Bushnell & Atmawan 1986).

The South China Sea opening was a regional event that occurred subsequent to the development of interior sag basins along the Sumatra–Java back-arc, after India had completely passed Sundaland. The South China Sea can be regarded as a single very large extensional 'tear' feature across the eastern part of Sundaland, unlike the discrete smaller grabenal features of the Sumatra–Java back-arc. In terms of the early alluvial fan deposits, these tend to be thickest and best developed at the proximal end of this extensional basin, i.e. within the northern extremity, the Gulf of Thailand (Thai Basin). Stratigraphical nomenclature and recognition of distinct lithological units (Woollands & Haw

1976; ASCOPE 1981; Lian & Bradley 1986; Polachan *et al.* 1991) are very unclear in the Thai Basin partly because of this. Moving south from the Gulf of Thailand into the northern and then southern Malay basin, facies tend to become finer grained and more distal in aspect.

The lacustrine sequences formed during the syn-rift phase developed into prodigious source rocks. Based on characteristics of the organic material in a number of these depositional settings it has been possible to recognise a number of types of lacustrine environments, related to tectonic setting at second- and third-order sequence stratigraphical level. Further discussion of sequence stratigraphical units is beyond the scope of the present paper.

Study of the organic material from lacustrine source rocks reveals large quantities of highly hydrogen-rich amorphous material. Various types of amorphous debris can be distinguished. In addition a variety of pseudo-structured particles enclosed within the amorphous material can often be seen (Fig. 6). Some samples contain abundant lacustrine algae. Thick monotonous sequences occur in the back-arc basins of Sumatra and Java, where deep semi-permanent lakes were most persistently developed in the Late Palaeocene. In the South China Sea, fluctuations are to be seen in response to marine incursions and shallowing within the depositional environment (Armitage & Viotti 1977). The basin was intermittently open to marine incursion from the southeast, with marine influence diminish ing northwards into the Thai Basin. Such cyclical regressive–transgressive events that can be well registered in the lithofacies and palynofacies record and may represent third-order seismic sequences, *sensu* the Exxon model (Haq *et al.* 1987).

A model of Early Tertiary lacustrine source bed generation within syn-rift back-arc basins has been well established in the Central Sumatra Basin. In this basin amongst all of them, the conditions came together in the most optimal way as a hydrocarbon prospect. A large number of oil fields are known from this region, but two of the them are giants, the Duri and Minas fields that alone account for over 52% of the entire production for the whole of Indonesia, (Williams & Eubank 1995). Despite being the subject of some debate as to its relevance in the Sunda Basin where Late Oligocene/Early Miocene coals have recently been strongly proposed as the primary source, (Gordon 1985) this lacustrine model now provides an analogue of hydrocarbon generation around the periphery of the whole of Sundaland; the North Sumatra Basin, South Sumatra Basin, Sunda Basin, NW Java Basin and Barito Basin. All of these basins contain named dark grey to black organic rich source shales as those at or near the earliest known sedimentary accumulations within these respective Tertiary basins (ten Haven & Schiefelbein 1995).

In the case of the North Sumatra Basin a lacustrine source rock is not widely recognised. This may be due to the nature of the Miocene carbonate hydrocarbon play in this region (Graves & Weegar 1973; Kingstone 1978; Situmeang & Davies 1986; Abdullah & Jordan 1987; Kjellgren & Sugiharto 1989) the ?Palaeogene source rocks within this basin may not have been fully penetrated. Evidence exists that suggest a marine/lacustrine source rock for some of the hydrocarbons from this basin (Anderson *et al.* 1993; Kirby *et al.* 1993; Matchette-Downes *et al.* 1994). Organic-rich sediments laid down under conditions of anoxia from a lacustrine stratified water body have been documented in the Bampo Formation of the North Sumatra Basin by Tiltman (1987). Work by ten Haven & Schiefelbein (1995) has yielded a lacustrine oil from the North Sumatra Basin, but this was explained away as an anomaly caused by fractionation during migration.

Neogene sedimentation

By the Early Miocene rifting had ceased and the interior fracture basins had become completely filled. With continued subsidence the region became ripe for marine transgression and sedimentation over a much wider area, during the interior sag or post-rift phase giving a classical 'steers head' shape to the basins. Thick intervals of shallow marine limestones and clastics were laid down over much of the region. These sediments formed the reservoir units for the source rock kitchen within the deeper basinal syn-rift lacustrine sediments, but also contain ephemeral shallow water lacustrine sediments that are a component of their source rock systems.

The major known oil province of Kalimantan is associated with the Mahakam Delta and the various Tertiary precursors to the present day Mahakam River. This alluvial clastic sedimentary setting is different from the post-rift setting of most of the rest of Sundaland, where such major river systems did not develop. The Barito Basin of the most southeastern tip of Borneo however lies on the eastern periphery of Sundaland and like the back-arc basins of the Sumatra–Java arc, displays all the characteristics of interior fracture and 'steers head' geometry.

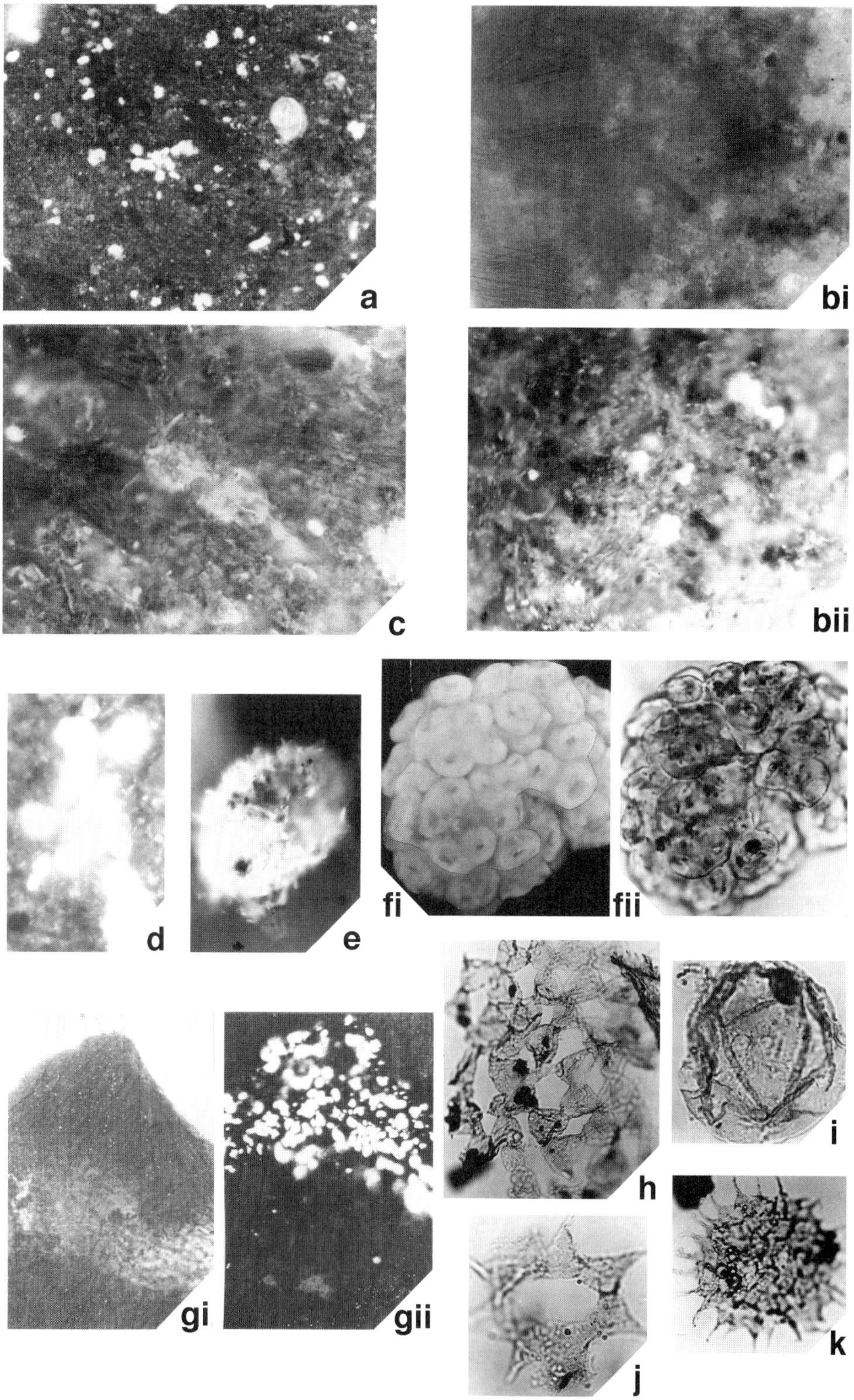
a
bi
c
bii
d
e
fi
fii
h
i
gi
gii
j
k

Tertiary hydrocarbon source rock systems occur in southeast Asia where good organic rich shales are developed in the Neogene, some within ephemeral shallow lacustrine systems of various types. However the semi-permanent syn-rift lacustrine source shales are by far the most prolific hydrocarbon source yet discovered in the region (Courtney & Williams 1994).

Geochemistry of lacustrine source rocks and oils

Lacustrine rocks are often characterized by high total organic carbon (TOC), with values in the range of 1–10% and high hydrocarbon generation potential (pyrolysate yield – S2) in the region of 10–25 $kg\,t^{-1}$. The hydrogen index (HI) which is given by the number of milligrams of pyrolysate yield per gram of TOC, is typically in the region of 400–900 $mg\,g^{-1}$ for lacustrine rocks. Values of this order are very high and therefore implicitly an indicator of organic material type, as type 1 kerogens only have HI values this high. An example of a typical lacustrine oil prone screening 'Rock-Eval' plot is given on Fig. 7a.

Pyrolysis-gas chromatography analysis of the pyrolysate yield generated from the kerogen during pyrolysis provides a detailed confirmation of the hydrocarbon prone nature of a sample. The Central Sumatra Basin Pematang shale has been compared to Palaeogene sediments from the East African rift valley by Katz & Kelly (1987). Pyrolysable material has been found to be dominated by a series of doublets of normal alkanes and alkenes (see Fig. 7d). Aromatic hydrocarbons are only represented in minor amounts with long chain hydrocarbons forming a large part of the pyrolysable material, similar to that found by Mann *et al.* (1987) in Oligocene source rocks from China. Larter & Douglas (1980) demonstrated the dominance of alkane–alkene doublets with a lack of aromatics in alginite pyrolysis-GC traces. Robinson (1987) noted these doublets and a virtual absence of aromatic and phenolic compounds in Indonesian crudes.

High wax content, a tendency towards low pristane/phytane ratio, low sulphur and bimodal n-alkane distribution are typical features of Duri and Minas type crudes. High wax oils are typically associated with large lacustrine basins, e.g. the Uinta Basin of the USA and the Mesozoic and Tertiary lake basins of China, that mostly contain 10–20% wax, (Powell & Snowdon 1983). This high wax content may be due to lipid concentration from higher plants, as their cellulosc, lignin and fats are easily degraded leaving long chain n-alkanes and wax esters that are relatively inert (Tissot & Welte 1984), but it seems more likely to be derived from phytoplankton that lived in the water column itself. Land derived kerogens such as vitrinite and inertinite are often present in very low abundance in the kerogen assemblage of lacustrine rocks, (Fig. 8d). Oils from the Central Sumatra Basin are characteristically waxy, a fact noted by Katz & Kelly (1987) who believed this waxy nature to result from non marine algal remains as opposed to the 'waxier' constituents of higher plants.

The paraffin distribution is an indicator of waxiness with lacustrine sediments showing an abundance of normal-paraffins greater than C22 in the C15+ n-paraffin distribution (Fig. 7c). Mann *et al.* (1987) found a marked bias to longer carbons chains with n-alkane distributions peaking at n-C27 and a strong odd over even dominance, in Chinese lacustrine crudes. Bimodal peaks in the n-alkane distribution in the ranges C15–C19 and C23–C33 in Indonesian lacustrine oils were found by Robinson (1987). Gelpi *et al.* (1970) demonstrated that many groups of algae contain these two groups of odd carbon numbered n-alkanes and n-alkenes. A highly specific C34 isoprenoid termed botryococcane derived from *Botryococcus* (Moldowan & Siefert 1980) has been found in the Brown Shale of the Central Sumatra Basin.

Pristane/phytane and pristane/nC17 ratios give an indication of depositional environment, although variations in land plant input into the system will have a primary effect on these. A dominance of pristane indicates oxidising environments, whilst dominant phytane is indicative of reducing conditions in the depositional environment. This is caused by the reduction of phytol residues from chlorophyll or bacterial cell walls under anoxic conditions forming

Fig. 6. (**a**) Type I amorphous organic material under flourescence, ×100; (**b**) Type I amorphous organic material, ×400 (i) transmitted light, (ii) under flourescence; (**c**) Amorphous organic material enclosing ?algal cyst debris, ×400; (**d**) Highly autoflourescent inclusions within amorphous organic material, ×1000; (**e**) Preserved ?algal palynomorph with highlighted detail of processes, ×1000; (**f**) *Botryococcus braunii*, ×1000 (i) under flourescence, (ii) transmitted light; (**g**) ?Crystalline autoflourescent mineral or kerogen, ×1000 (i) transmitted light, (ii) under flourescence; (**h**) *Pediastrum kajeites*, ×400; (**i**) *Bosedinia infragranulata*, ×500; (**j**) *Pediastrum sp.*, ×1000; (**k**) *Pediastrum bifidites*, ×400.

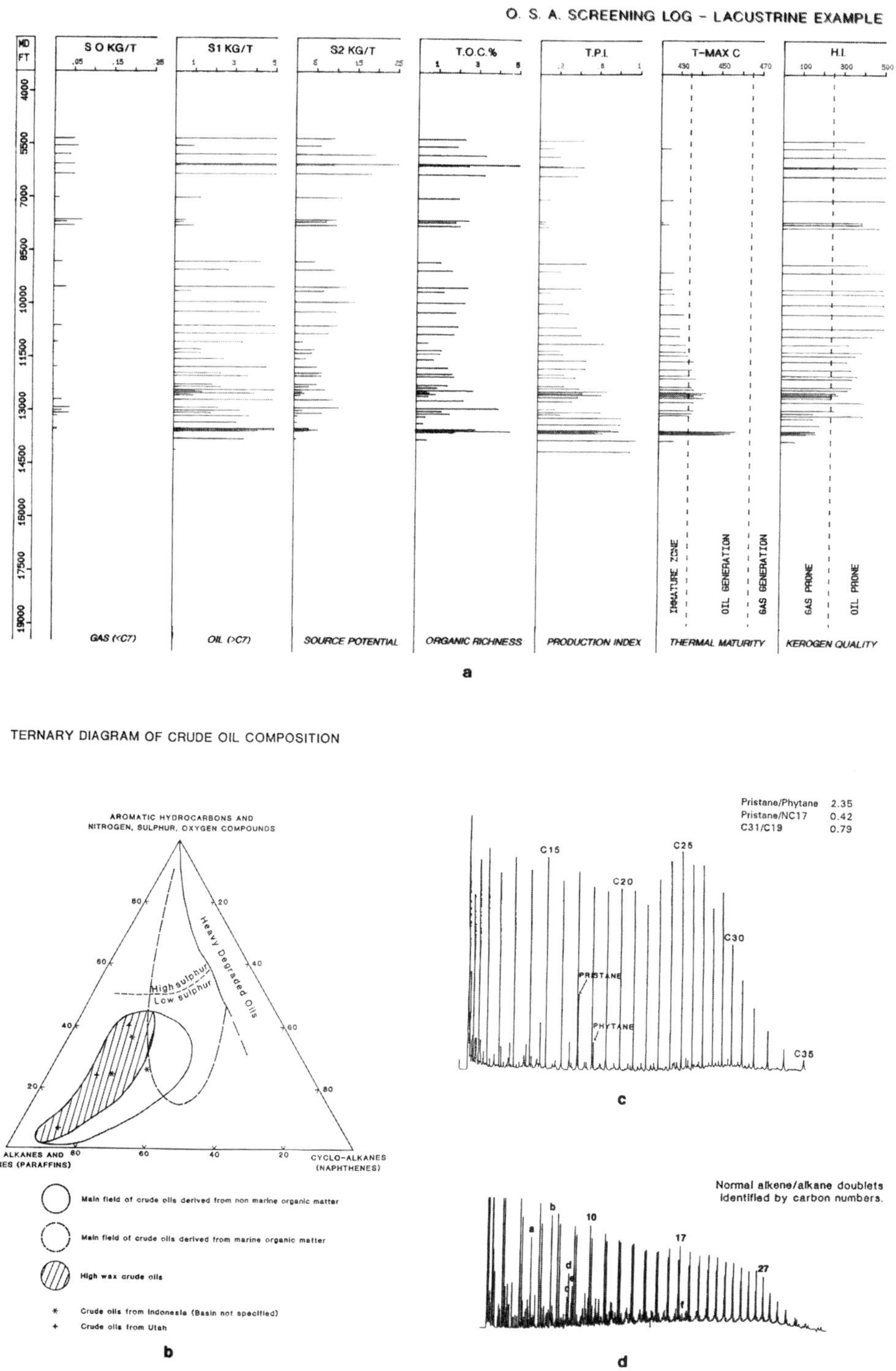

Fig. 7. Geochemistry of lacustrine source rocks and oils. (**a**) Oil-source analysis geochemical screening log; (**b**) ternary diagram of crude oil composition (adapted from Tissot *et al.* 1984); (**c**) lacustrine oil gas chromatogram (after Robinson 1987); (**d**) pyrolysis gas chromatogram, lacustrine source (after Horsfield *et al.* 1994).

phytane, whilst under oxidising conditions pristane forms from phytol by an oxidative pathway via phytanic and phytenic acids (Moldowan *et al.* 1985). Pristane concentration can be initially enhanced however by early aerobic decay. Indonesian lacustrine oils typically have a pristane–phytane ratio of less than 3.0 (Robinson 1987).

The sterane and terpane distribution of lacustrine oils can also be distinctive, with a dominance of pentacyclic 17 alpha hopanes from C27 to C35 plus moretanes and little else in the terpanes of Indonesian crudes (Robinson 1987), together with unusually high concentrations of C30 4-methylsteranes. The latter are commonly attributed to a dinoflagellate source (Wolff *et al.* 1986). Carotenoid tetra-terpanes (C40) can be abundant in algae and their corresponding saturated hydrocarbons turn up in lacustrine sediments. One pentacycloalkane or terpane gammacerane has been correlated with lacustrine rocks (Waples 1985) in being derived from a non marine protozoan *Tetrahymena pyriformis* though it also occurs in ferns (Hills *et al.* 1966). Steranes relative to hopanes tend to be low (< 0.2) in non-marine oil (Robinson 1987).

A summary of the bulk compositional characteristics of lacustrine sourced oils is given on Fig. 7b. Algal derived non-marine oils tend to have low to medium API gravity (20–35°), high wax ($C31/C19 > 0.4$) and low sulphur (< 0.2 wt%). Organo-metallic compounds tend to be moderate to low and nickel is prevalent over vanadium.

Lacustrine depositional systems

Lacustrine depositional systems are responsible for some of the giant oil provinces of the world (Swain 1956, 1966; Hutton 1984). Best known are those in mainland China such as the Daqing oil field in the Songliao Basin, (Fleet *et al.* 1988). Some of the first produced and best known oil in SE Asia comes from a similar giant lacustrine system of the Minas Field of Central Sumatra (Katz & Mertani 1989; Longley *et al.* 1990; Macgregor & McKenzie 1986; Mertosono & Nayoan 1974; Robinson & Kamal 1988; Williams *et al.* 1985; Moulds 1989). Crudes from the Minas Field are of a distinctive heavy waxy type and the presence of biomarkers such as Botryococcene have long been an indicator of freshwater lacustrine source beds. (Maxwell *et al.* 1968; Metzer & Casdevall 1983; Moldowan & Seifert 1980; Moldowan *et al.* 1985).

When large fields were discovered in the South Sumatra Basin (de Coster 1974; Harsa 1975; Hutapea 1981; Hasan & Soebandrio 1988; Sarjono & Sardjito 1989; Suria 1991), the Sunda and NW Java basins (Bushnell & Atmawan 1986; Fletcher & Bay 1975; Fainstein 1987; Fainstein & Promono 1986; Molina 1985; Heru Pramono *et al.* 1990; Reminton 1985; Soulisa & Sujanto 1979; Wahab & Martono 1985; Suherman & Syahbuddin 1986), in the Barito Basin of southeast Kalimantan (Siregar & Sunaryo 1980; Kusuma & Darin 1989) and the Southern Malay Basin (Pupilli 1973; Pollock *et al.* 1984) it was also found to be of a similar heavy, waxy character.

However the connection to a lacustrine source was not immediately clear, either because the source shales themselves had not been commonly penetrated by the drill-bit, or because the lacustrine origin of the source rocks themselves was not initially recognised. A lively debate continues over the lacustrine rocks versus coals as the major source rocks of Sundaland (Gordon 1985; Bushnell & Atmawan 1986; Ponto *et al.* 1988; Subroto *et al.* 1992; Bissada *et al.* 1992).

Lacustrine shales form very rich oil source beds due to the closed nature of lake systems that favour the preservation of organic material at the sediment–water interface. This organic material is derived from photosynthetic activity of the algal biomass of the photic zone within the surface waters.

Under tropical conditions of high non-seasonal (i.e. non-fluctuating) temperatures, semi-permanent enclosed aquatic systems become rapidly stratified or meromictic. Heated surface waters are less dense than cooler waters lower down, and the absence of profound diurnal and seasonal temperature fluctuations results in the surface water retaining their heat and lower density on a more or less permanent basis. This effectively isolates the upper and lower waters (epilimnion and hyolimnion respectively) across a boundary surface termed the pycnocline, (Demaison & Moore 1980). Meromictic waters create very good conditions for the preservation of organic material at the sediment water interface (Figs 9 & 10; Bradley 1966; Hutchinson 1957).

In some of the lacustrine sedimentary sequences of SE Asia intermittent marine influence is evident. Marine influence within a large tropical lake would not necessarily have upset the stratification. Any marine ingress would have been as a heavier layer well within the hypolimnion and would have contributed to further strengthening of the thermal stratification (Fig. 11a).

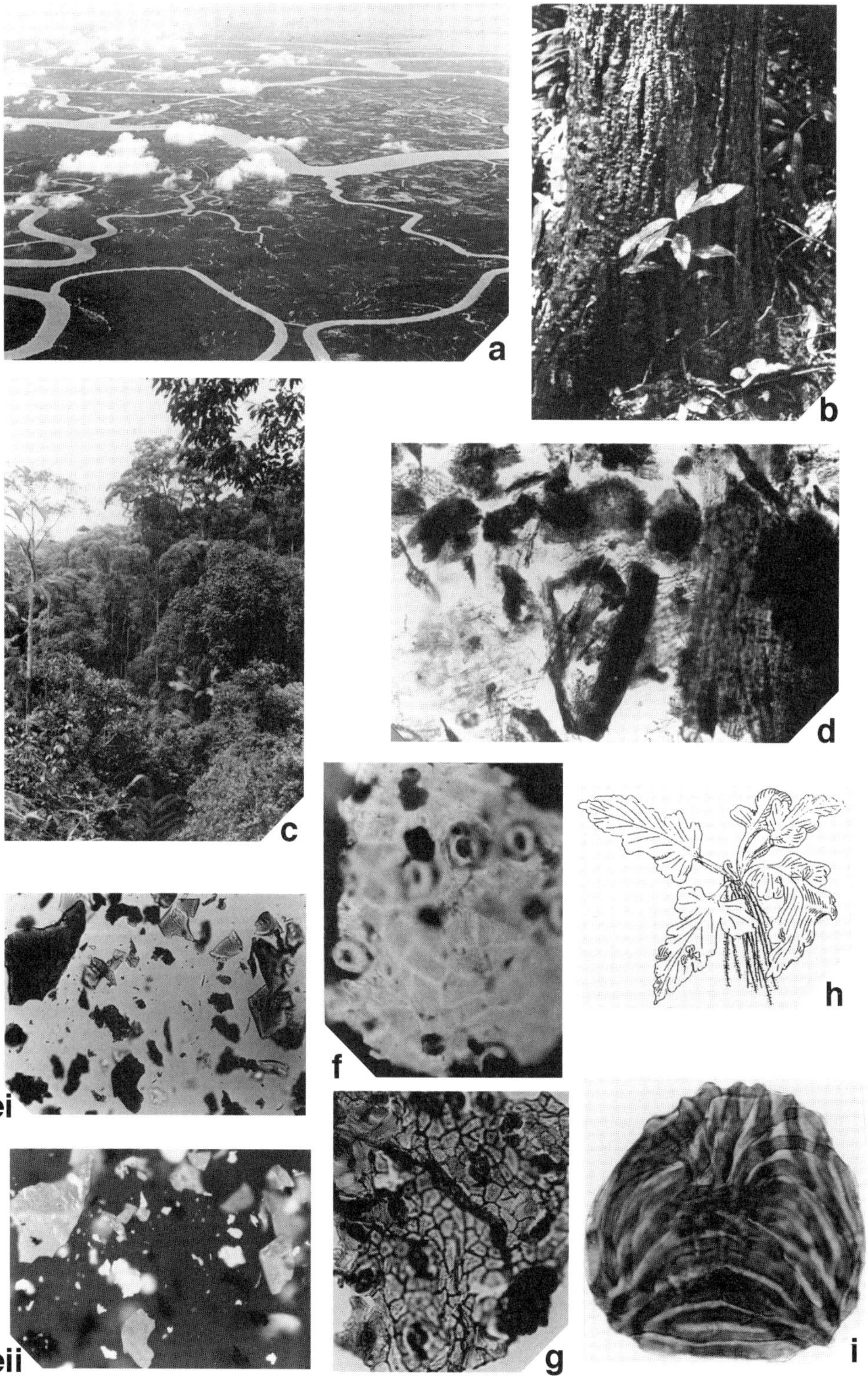
a
b
c
d
ei
f
h
eii
g
i

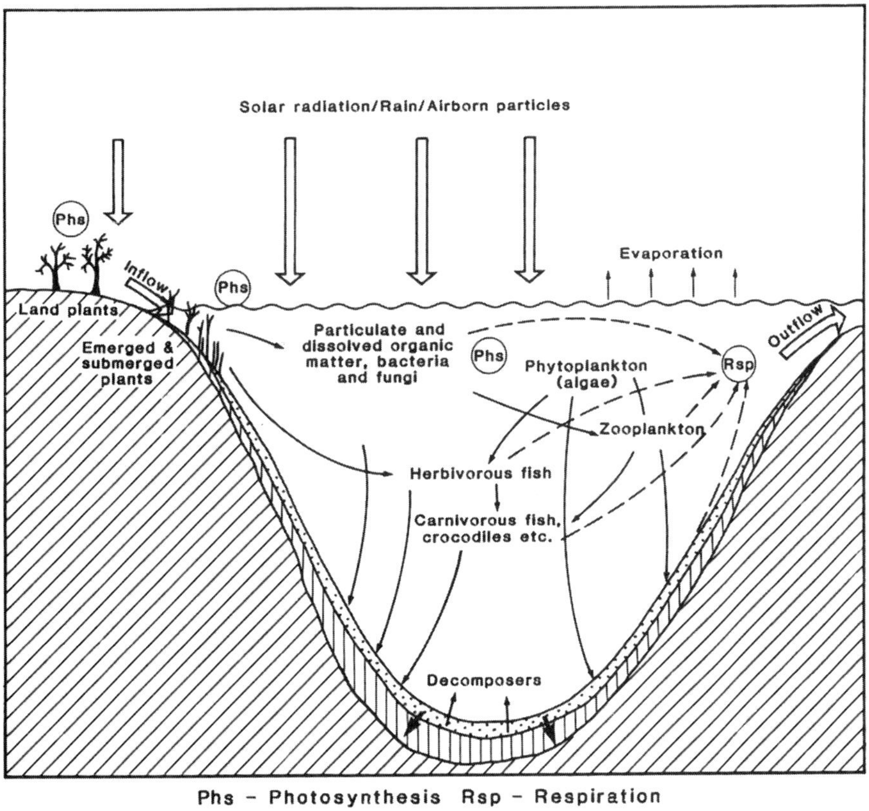

Fig. 9. The circulation of matter and energy in a tropical lake, (after Beadle 1974).

Under conditions of high diurnal rainfall within semi-enclosed basins, the lakes would have been dominantly freshwater, certainly as far as the microplankton in the photic zone were concerned (Fig. 11b). Very large or elongate basins that were open to the sea at one end may have remained essentially freshwater under the tropical conditions of very high regular rainfall and fluviatile input coupled with low evaporation rates associated with continuous high humidity (Fig. 11c). Some of the basins may have been far more open to the sea and thoroughly marine though even under such 'restricted' basin configuration with a silled margin, stratification can still allow anoxia to form in the lower part of the water column that is not readily circulated (Fig. 11d). It is evident from studying the syn-rift sequences of a number of the Sundaland basins that marine and quasi-marine influence effected them to a lesser or greater degree, without having a profound effect on the overall organic productivity of the system that remained dominantly freshwater.

With no winter to cause cooling and sinking of oxygenated surface waters that would otherwise disrupt the stratification, the only disturbance that can be caused is by the effect of the wind setting up a circulation accompanied by a 'seich' as wind generated low amplitude waves begin to oscillate from one side of the lake to the other (Fig. 12a). In a well-stratified lake, circulation within the epilimnion may cause a reverse circulation of the hypolimnion (Fig. 12b), which under extremely stormy conditions in

Fig. 8. (**a**) Aerial view of the Mekong Delta; (**b**) Green-black resin oozing from a Dipterocarpaceous tree, Pantai Forest, Malaysia; (**c**) Lowland evergreen rainforest at the water's edge, Taman Negara, Malaysia; (**d**) Vitrinite kerogen, ×100; (**e**) Resinite kerogen, ×400 (i) transmitted light, (ii) under flourescence; (**f**) Cutinite kerogen under flourescence, ×100; (**g**) Cutinite kerogen under transmitted light, ×100; (**h**) Outline drawing of the aquatic fern *Ceratopteris*; (**i**) *Magnastriatites howardi*, ×400.

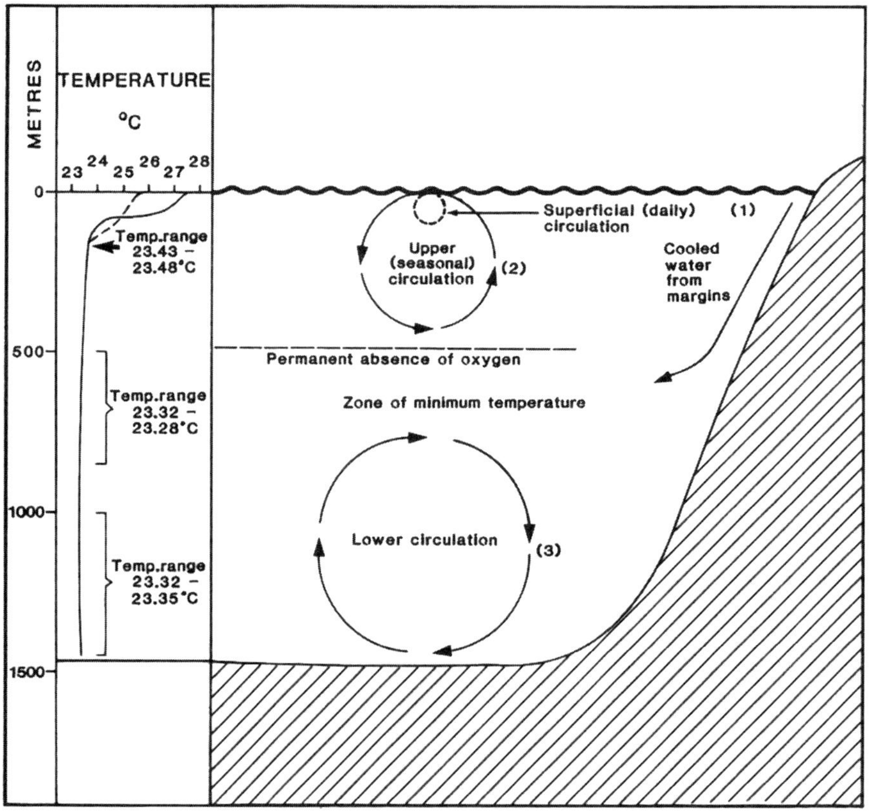

Fig. 10. Temperature profile in ake Tanganyika and suggested circulation (in Beadle 1974). (1) 'Epilimnion' down to 50–80 m, undergoes daily circulation, separated by thermocline from: (2) 'metalimnion' down to at least 200 m, seasonally circulated, separated by less marked thermocline from: (3) 'hypolimnion' free of oxygen, almost uniform temperature to the bottom apart from 0.1°C rise in lower half. N.B. These terms are not really equivalent to summer stratification of temperate lakes.

modern tropical lakes can cause upwelling of anoxic waters that will poison and kill all fish (Fig. 12c). Shallow-water bodies will remain regularly oxygenated throughout the water column, although high organic input under tropical environments into the bottom muds may be sufficient to cause anoxia within them only a few millimetres below the surface (Fig. 12d).

Almost all the organic productivity occurs in the epilimnion by planktonic algae. The surface waters tend to be quite nutrient starved and this together with temperature fluctuations makes them highly stressed habitats. Organisms that are capable of thriving in such conditions tend to be evolutionary conservative forms such as *Botryococcus* (see Fig. 6); (Traverse 1955; Tappan 1980; Wake & Hillen 1981; Cole 1987). Unlike marine continental shelf habitats characterized by upwelling currents that enrich surface waters with nutrients and have an abundance and diversity of planktonic taxa, stressed environments are characterized by very low species diversity, or even by monospecific assemblages (Chu 1942). However given the high input of solar energy into the system, such algae as can tolerate the conditions thrive prodigiously. *Botryococcus* secretes an oil droplet into its central cavity in order to assist buoyancy and it is this droplet that forms the precursor to eventual reservoired hydrocarbons (Brown *et al.* 1969; Gelpi *et al.* 1970).

The nutrient starvation of the epilimnion is contrasted with the great nutrient enrichment of the hypolimnion. Another contrast is provided by the highly reducing conditions of the latter in comparison to the oxygenation of surface waters. The large amount of decaying organic material quickly exhausts any oxygen in the system and aerobic decay that is many times

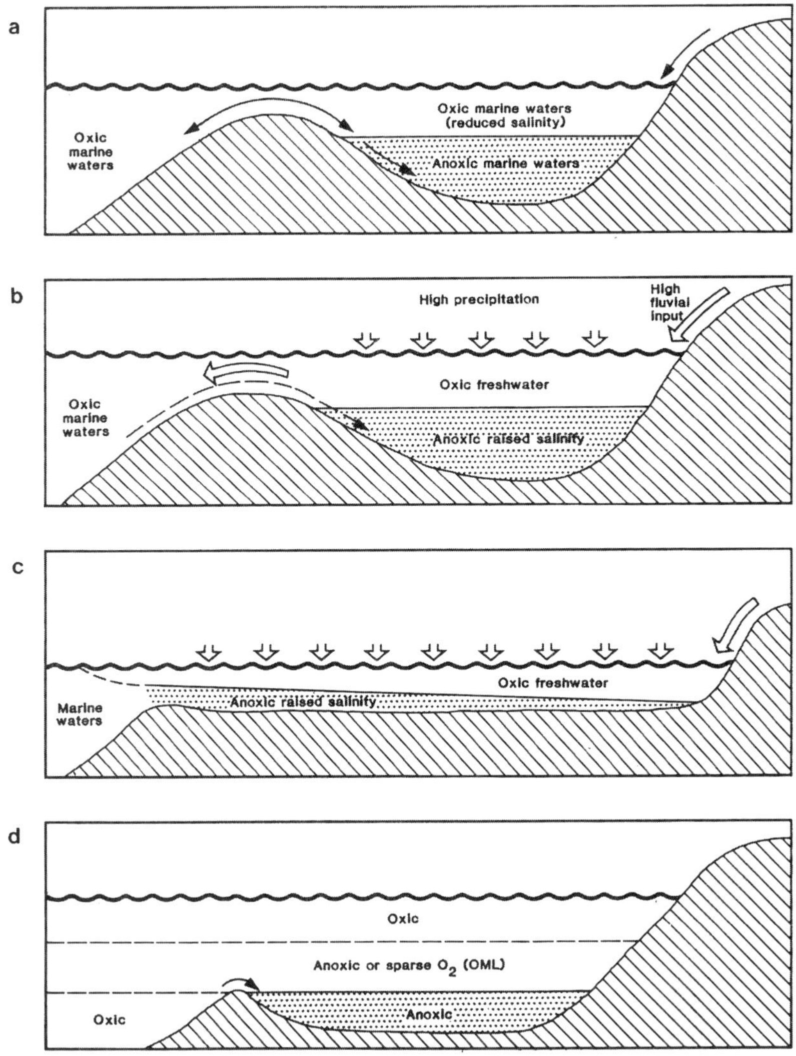

Fig. 11. Schematic hydrology dynamics in shallow restricted basins (adapted from Waples 1985).

more rapid and efficient than anaerobic, cannot take place. Anaerobic decay does not bring about a complete breakdown of material and the reducing acidic conditions act to preserve organic material that falls below the pycnocline (Hutchinson *et al.* 1932; Hutchinson 1957)

At the sediment–water interface conditions may be too hostile for bottom-feeding vertebrates and invertebrates and this, associated with incomplete putrefaction of material, results in a net gain in organic input into the bottom muds. Even despite the favourable conditions for preservation of organic material, it has been estimated that less than 1% of the organic productivity is eventually preserved within the sediment. For lacustrine systems this 1% incorporation of organic material is one of the highest of any sedimentary regime, sufficient to result in development of very rich source bed shales.

Any marine influence into a dominantly lacustrine basin may occur within a stratified water body as a heavier lower layer. The normally freshwater alga *Pediastrum* (Fig. 6h, j & k) may have been able to tolerate conditions of slight brackishness (Evitt 1963; Matsuoka & Hase 1977).

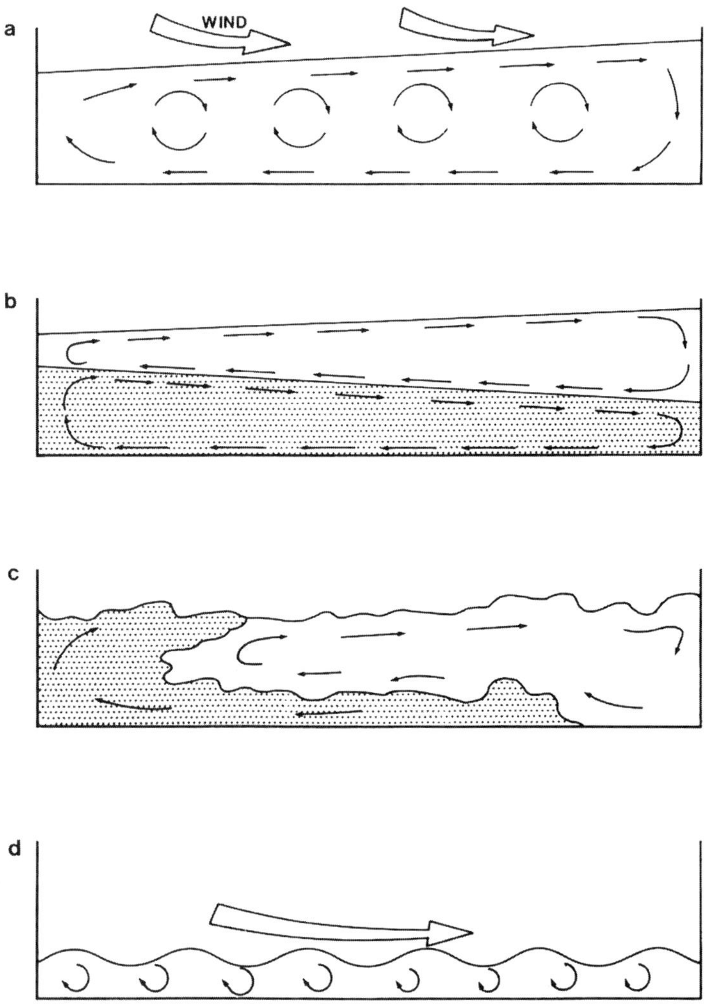

Fig. 12. Wind circulation of lake water (after Hutchinson 1957; Mortimer 1969)

Palynofacies analysis of lacustrine source rocks

Detailed micoscopical study of the organic material in the sediments reveals that it is comprised of a complex mixture of particles of various shapes and sizes (Fig. 6a). Much of this material is derived from algal and bacterial debris that does not produce a recognizable resting cyst type. It is often referred to as amorphous material. In the case of this material deposited under anoxic conditions it is highly auto-fluorescent indicating that it to be type 1 highly oil prone kerogen, or amorphous liptinite. Amorphous organic material (AOM) observed within many kerogen preparations from sedimentary sequences elsewhere in the world deposited under conditions of far less anoxia is only slightly auto-fluorescent or non-fluorescent.

This material must be studied under the fluorescent microscope in order to fully characterize it, as under normal transmitted light (Fig. 6bi) it only appears as a characterless 'mush' compared to the same field of view observed under fluorescence (Fig. 6bii). Detailed plotting of the abundance of this material together with other kerogen macerals and recognisable algal cyst taxa over a given stratigraphical interval is the technique of palynofacies analysis. Close study of this material can often reveal structures resembling acanthomorph acritarchs (Fig. 6c) though some are of unknown, possibly bacterial origin (Fig. 6d). Sometimes the bodies have a more clear outline with fine processes (Fig. 6e). Most of these forms have never been formally taxonomically described as they are not normally fossilized. Other observed features of the amorphous liptinite included larger opaque fragments of kerogen (Fig. 6gi) that have clusters of highly fluorescent ?mineral crystals attached (Fig. 6gii).

Sequences from the Sumatra–Java back-arc series of basins often consists of monotonous intervals of dark brown or black shales. The dominant palynoflora appears to be the freshwater lacustrine alga *Botryococcus* (Fig. 6). Other algal cysts that can occasionally be very abundant in these sequences are forms such as *Bosedinia* spp. (Cole, 1992; Fig. 6i).

Similar thick intervals of black and dark brown shale also occur within the sedimentary pile of the South China Sea Basins (Md Nazri Ramli 1988). Due to differences in basin origin and timing as discussed above, the age and characteristics of this material is not the same as that in the Sumatra–Java–Kalimantan back-arc. Regressive third-order seismic stratigraphical sequences that display a gradation up into shallow higher energy lake sediments with abundant *Pediastrum* spp. (Fig. 6h, j & k) succeeded by lake shore forested peatswamps have been observed and described below (see Fig. 14). More terrestrially influenced palynofacies are characterized by large amounts of land derived kerogen such as woody inertinite and vitrinite (Fig. 8d), and leaf derived cutinite (Fig. 8a & g).

In the ensuing section various examples of lacustrine palaeoenvironments within the SE Asian Tertiary are described and illustrated. Lithological and palynofacies features from a number of Tertiary lacustrine facies in the region are described and explained in terms of modern lake processes. The expected separation into deep and shallow lacustrine types is possible, following the same broad separation that can be recognized in modern lacustrine systems.

A number of palaeoenvironmental sub-types can be recognized for both deep and shallow lacustrine basins in the SE Asian Tertiary and examples of each are described below. A short paragraph on source potential is included with each modelled example based on the observed 'visual' geochemistry. Unfortunately no supporting analytical geochemical data are available to accompany any of the general qualitative observations made. It must be noted that in the presence of such data it is not uncommon to encounter contradiction between the visual and analytical data, possibly because of the bacterial contribution that may not be readily seen.

Palynofacies in a deep semi-permanent lacustrine setting

Intramontane lacustrine

Occurrence. This depositional facies has been encountered in the following syn-rift Oligocene formational units of Sundaland; Bampo Formation of the North Sumatra Basin; the Ombilin Basin Sangkarewang Formation; the Pematang Formation Brown Shale of the Central Sumatra Basin; the Gritsand Formation of the Lower Talang Akar of the South Sumatra Basin, the Banuwati Formation of the Sunda Basin, the 'Deltaic' Talang Akar of the NW Java Basin, the Lower Tanjung Formation of the Barito Basin; the Natuna Group (Tertiary I and II) of the Malay Basin and the Lama/Gabus and Barat Formations of the West Natuna Basin.

Lithology. Sediments are dominated by laminated dark-coloured mudstone, occasionally silty and sometimes with a high calcareous content. Sand intervals are sometimes evident as a gradational unit from the mudstones, via siltstones, or can occur abruptly within a mudstone interval.

Kerogen type. Kerogen types are dominated by fluorescent amorphous material (amorphous liptinite, Fig. 6a, b & c). Occasionally *Botryococcus* is sufficiently abundant to comprise up to 15% of the bulk kerogen assemblage. In some cases the occurrence of exinite (pollen and spores) can be very high (see Fig. 13b). Occasionally the percentage of humic macerals (vitrinite and inertinite) are raised and this is usually coincident with more arenaceous lithologies.

Palynology. Palynomorphs are distinctly freshwater, often with an abundance of montane bisaccate pollen. Freshwater pollen are usually abundant, sometimes overwhelmingly so, comprising up to 40% of the bulk kerogen assemblage. *Botryococcus* may be very common and in such cases is usually the only microplankton present (Fig. 13b). However, where *Botryococcus* is less abundant, other lacustrine algae such as *Pediastrum*, *Bosedinia* sp. (Cole 1992) and certain acanthomorph acritarchs are present in large numbers.

Palaeoenvironmental interpretation. Sediments within this kind of basin are generally fine grained, dark and laminated, indicating deposition under low-energy conditions, generally from suspension, at low sedimentation rates.

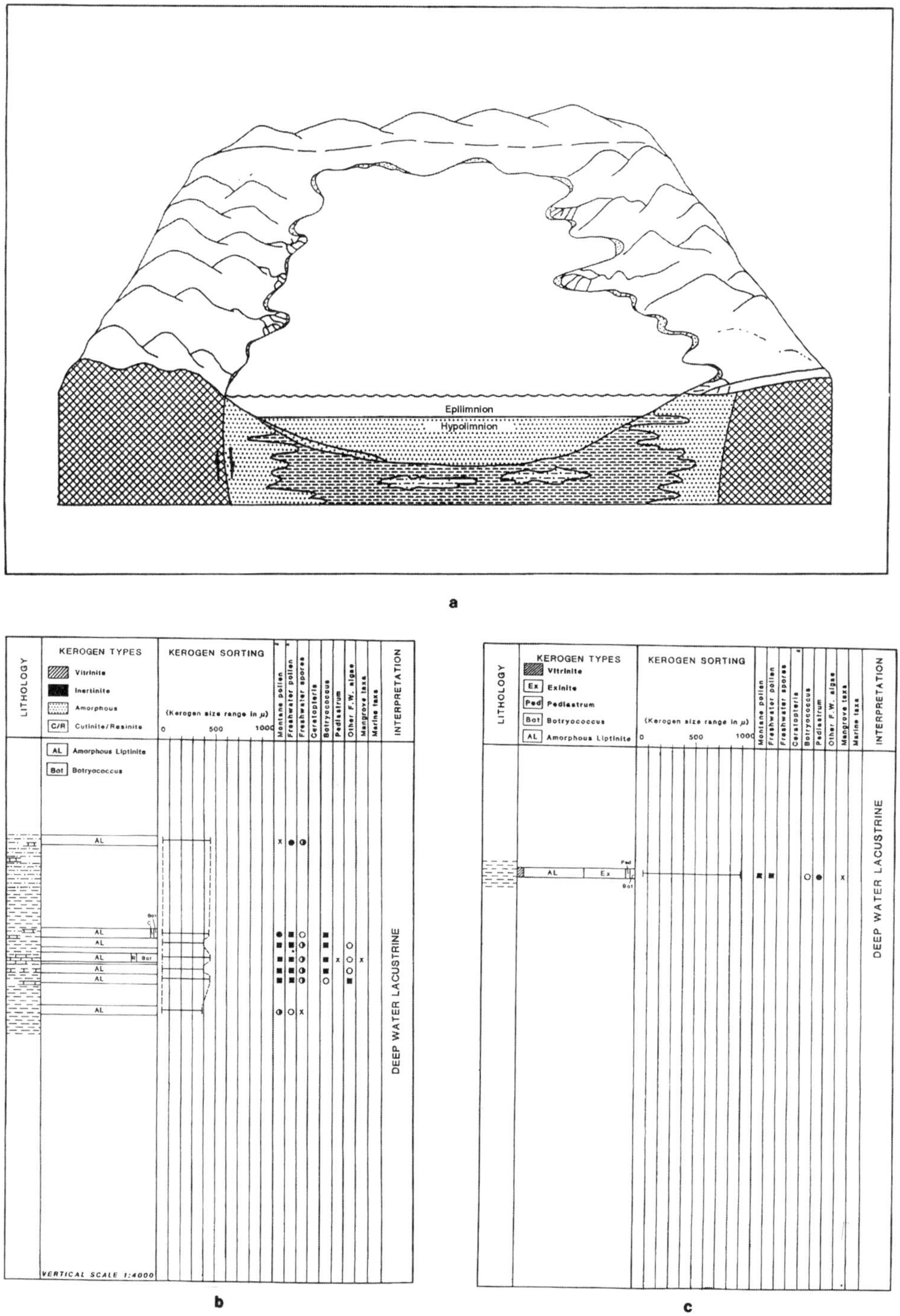

Fig. 13. Intramontane lacustrine. (**a**) Lake model; (**b**, **c**) palynofacies diagrams.

They contain a high organic component. Palynomorphs are almost all of freshwater origin, indicating a depositional site landward of any tidal marine influence.

The abundance of well-preserved amorphous liptinite suggests that the bottom waters were anoxic, fitting well with the meromictic deep lacustrine model. Under such circumstances any marine influence would not effect the overall salinity of the lake but would remain as a lower saline layer within the anoxic hypolimnion. This amorphous liptinite appears as small membranous particles of slightly different fluorescence intensities (see Fig. 6) and probably represents remains of phytoplankton that have been partially altered by anaerobic bacteria. Under oxidizing conditions this material would be completely destroyed.

Significant recognizable remains of lacustrine algae, e.g. *Bosedinia* sp., ?algal bodies and *Botryococcus* can also be abundant within this amorphous debris. This material is all formed within the lake basin itself by the phytoplankton community of the surface water. Humic land derived material (vitrinite and inertinite) is present in very low amounts or is totally absent in this deep lacustrine facies. A general lack of coarse clastic material also suggests a depositional site towards the centre of the basin.

A dominance of the alga *Botryococcus* in this realm may indicate that the surface waters are oligotrophic, i.e. lacking nutrients, consistent with the known ecological limits of modern *Botryococcus* dominance (Naumann 1931; Jarnefelt 1956; Wake & Hillen 1981). However, given the slow sedimentation rate and the ability of *Botryococcus* to fix carbon and secrete oil droplets, and the high preservation potential of cells that fall into the hypolimnion, the surface waters must be sufficiently productive to allow accumulation of organic rich sediments. The lack of circulation that is responsible for preservation of sedimented organic matter is thus also responsible for the lack of return of nutrients to the surface waters favouring a dominance of *Botryococcus*.

High proportions of montane bisaccate pollen that occur within this realm suggest the presence of high reliefs not far from the basin edge (Fig. 8c). The absence of extensive shallows adds to the comparatively low productivity of the lake, while high reliefs around the lake, limit the effect of the wind to set up surface waves and seiches (low frequency waves travelling in an oscillating fashion from one end of the lake to the other causing a corresponding oscillation in the slope of the pycnocline, see Fig. 12b). These effects would tend to cause greater recirculation of nutrients.

Steep basin margins hint at an active fault graben system that maintains the base level of the lake below the rate of sedimentary infill. Steep basin margins indicate that clastic influx in the form of alluvial fans that may be coalescent along the basin edge, are to be expected. The alluvial fans probably did not prograde towards the basin centre because subsidence matched or even exceeded sediment supply.

Sedimentary input into the lake via alluvial fans would have been comparatively intermittent with phases of high discharge of clastic debris, probably to a large extent seasonally controlled. Under these conditions there would be less scope for continued transportation of debris into the deeper waters as might be expected from a low angle coastal delta where river flow is more regular. However, such alluvial fans may cause intermittent covering of the deep lacustrine muds by turbidite flow. In the more distal regions within the basin centre, distal turbiditic material apparent as more silty horizons may occur within the deep water muds.

Towards the basin edge where coarser clastic material is laid down, the proportion of humic material, vitrinite and inertinite (land derived) is greater. In this shallower zone, other freshwater algae including *Pediastrum* can become quite abundant. The shallow lake waters may be more productive (eutrophic) allowing other lacustrine algae such as *Pediastrum* to flourish. Comparatively eutrophic conditions are favoured by modern *Pediastrum*, and a greater variety of alga without a dominance of *Botryococcus* can be expected. *Botryococcus*, although often still remaining common, is normally reduced in numbers compared with its predominance in the deeper-water facies. A greater variety and abundance of land-derived spores and pollen occurs in the shallower nearer-shore facies.

Source potential. The source potential of this facies is very high and is oil prone. Clear evidence that most of the organic material is derived from phytoplankton laid down within an anoxic hypolimnion, indicates a Type I kerogen source. Where exinite forms the dominant maceral the source is less rich (Type II), but it is still a very good oil source. Towards the basin edge a greater amount of humic material is to be found and influx of clastic material is greater. Given the steep shoreline slope, however, the accumulation of amorphous liptinite within the hypolimnion may continue very close to the shore but it is diluted by clastic input. The inclusion of humic material that will source mainly gas has the effect of reducing the source potential of the shallower-water deposits.

Reservoir rocks. Sands with reservoir potential are not normally laid down in the basin centre with the source rocks. However, sandy lacustrine fan units at the basin edge will form the nearest reservoir facies. Migration conduits to these sands are provided by coarse turbiditic detritus that occasionally extends into deeper water (distal turbidite fine sands and silts). Ma Li *et al.* (1982) noted that the largest oil accumulation of the Songliao Basin lies within the Heiyupao delta system along the northern part of the basin, wedging into the deeper lake facies. In the Dongying Basin alluvial sands are also noted to wedge into the source rocks and form traps up dip against faults. Intermittent episodes of vertical tectonic movement may effect intramontane rift grabens and may cause pulses of deepening/shallowing, with coarse clastics laid down over source muds allowing direct vertical migration into reservoirs. Ma Li *et al.* (1982) noted the occurrence of some oil reservoired in fractured mudstone of the deep lake facies of the Songliao Basin.

Lowland marine-influenced lacustrine

Occurrence. This depositional facies has been encountered in following Oligocene and Miocene formational units of Sundaland; Bampo Formation of the North Sumatra Basin; the Pematang Formation Brown Shale of the Central Sumatra Basin; the Gritsand Formation of the Lower Talang Akar of the South Sumatra Basin, the Banuwati Formation and Gita/Zelda Members of Talang Akar Formation of the Sunda Basin, the 'Deltaic' Talang Akar of the NW Java Basin, the Cycle I and II sediments of the Thai Basin; the Natuna Group (Tertiary I and II) and Trengganu/Bekok Groups (Tertiary III–VI) of the Malay Basin; and the Keras, Gabus, Barat and Arang Formations of the West Natuna Basin.

Lithology. Dark–medium grey mudstone, with intercalated sandstones, siltstones and occasional thin coals.

Kerogen types. Humic kerogen (vitrinite and inertinite) can comprise over 50% of the total kerogen assemblage. However, amorphous material can still be prominent and occasionally forms the dominant maceral. Much of this amorphous material falls into the amorphous liptinite category. *Botryococcus* can sometimes be a dominant part of the entire kerogen assemblage. Parts of the sequence have abundant cutinite (Fig. 8f & g) and resinite (Fig. 8ei & ii).

Palynology. The assemblages are nearly always dominated by freshwater fern spores, including the taxon *Magnastriatites howardi* (Fig. 8i) derived from the aquatic fern *Ceratopteris* (Fig. 8h) that often chokes slow moving lowland water courses. Montane bisaccate pollen are occasionally abundant and freshwater pollen are consistently present in moderate numbers. Other freshwater algae, such as *Pediastrum*, are generally fairly common over some intervals. Marine dinocysts and microforaminiferal linings are occasionally present.

Palaeoenvironmental interpretation. This lacustrine palaeoenvironment is thought to be 'lowland', showing evidence of consistent marine influence and the presence of extensive shallows and alluvial plains surrounding the deeper water lacustrine facies. One of the example sections discussed here is believed to show a regressive trend from deep lacustrine to freshwater peat swamp (Fig. 14a, b & c). Lithologies in the deeper lacustrine settings are dominated by medium grey mudstones with fairly low TOCs suggesting more rapid sedimentation than seen in the deep water intramontane lacustrine palaeoenvironment.

Kerogen types show comparatively high proportions of humic material that is well size sorted. Much of this humic material is composed of inertinite that probably comes from distributary channels associated with deltas on the basin edge.

The basin was not enclosed by high relief, allowing the wind to set up surface waves and seiches within the lake, helping more effectively to distribute nearshore material into the deeper water areas. The presence of an extensive alluvial plain is indicated by the abundance of pteridophyte spores from ferns that grew on the unstable water-logged ground where more mature lowland forest could not become established.

A stratified lake model is indicated by the good preservation of abundant *Botryococcus* in the deeper water facies, but the surface waters contained more recycled nutrients, allowing other 'eutrophic' algae such as *Pediastrum* to flourish. Marine taxa (dinocysts and microforaminiferal linings) can be quite common, particularly within the deeper water facies. This suggests that the lake was in direct connection with the sea, and that under the meromictic conditions seawater entered the lake along its floor causing raised salinities in the deeper water but allowing the surface lake waters to remain fresh.

The basin can still be called 'lacustrine' rather than 'marine' due to the expected very high positive fresh-water balance from abundant precipitation and fluvial input under tropical

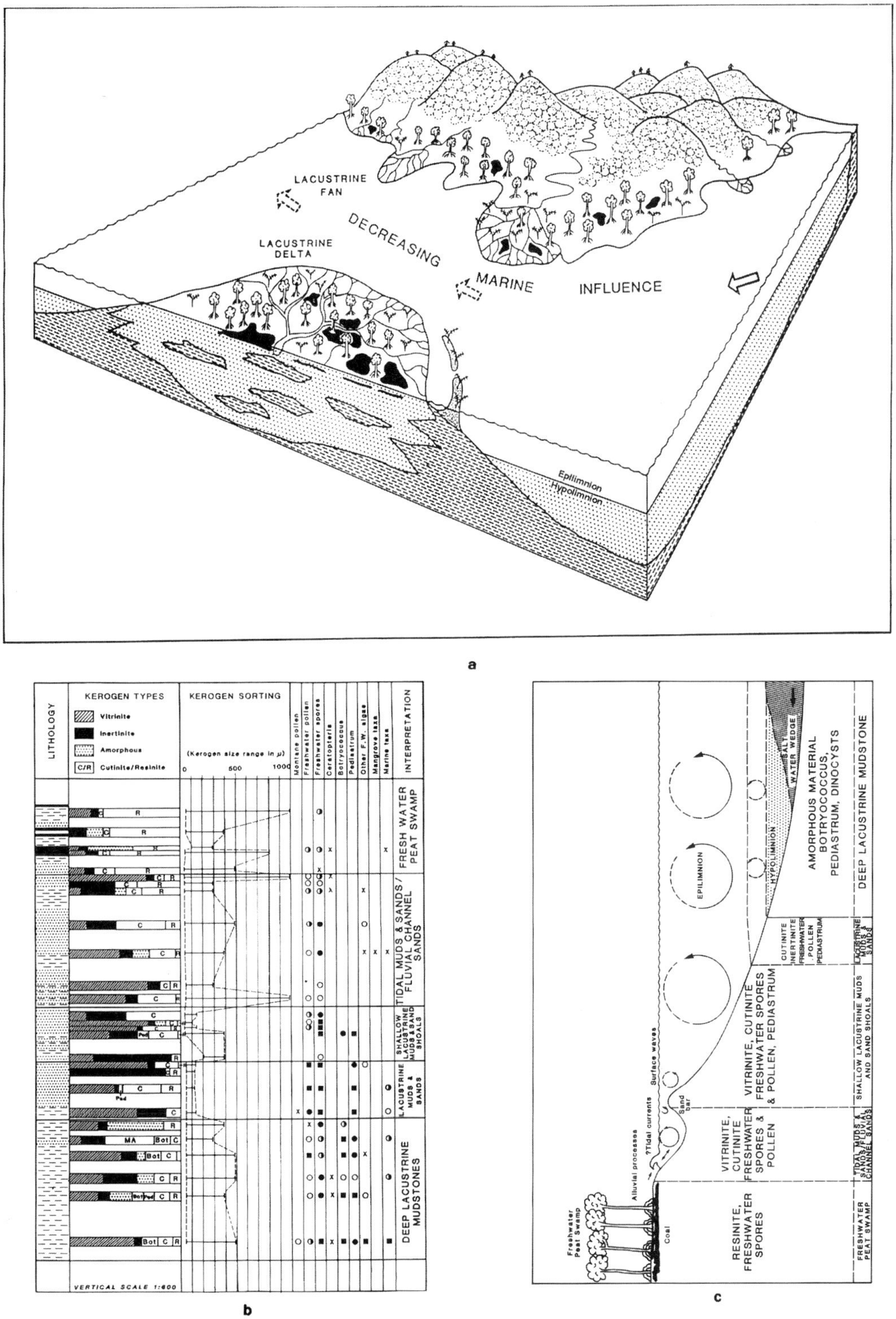

Fig. 14. Lowland marine influenced lacustrine. (**a**) Lake model; (**b**) palynofacies diagram; (**c**) depth zonation of palynofacies.

conditions, (Fig. 11c) and comparatively low evaporation due to high humidity. Montane pollen may occasionally be abundant in the lowland deep lacustrine facies, indicating that high reliefs may be present on the landward edge of the basin. In these circumstances the alluvial plain may be less well developed with consequently fewer fern spores.

The palynofacies sequence of Fig. 14 has been interpreted to show an uphole regressive trend. Comparatively deep-water facies with abundant freshwater algae, *Botryococcus*, *Pediastrum* and marine taxa, occur towards the base of the sequence together with common amorphous material. Much of this amorphous material falls into the amorphous liptinite category derived from phytoplankton. The top of the zone of abundant *Botryococcus* and amorphous liptinite may correspond to the highest sample laid down within the hypolimnion,

Samples above this zone contain higher proportions of often well-sorted humic material, indicating better circulation by currents, and contain larger numbers of freshwater pollen and spores with fewer marine taxa and arenaceous lithologies become more common. The evidence suggests deposition within the epilimnion. *Pediastrum* remains common, corresponding with more eutrophic waters in the epilimnion. Further upsection samples contain higher proportions of cutinite, often poorly sorted, with many freshwater pollen and spores indicating deposition within quiet low energy lagoonal environments adjacent to the strandline, or within overbank environments on the alluvial plain.

The highest samples are dominated by resinite, poorly size sorted due to lack of current activity within dominantly coaly lithologies. Palynomorphs are mainly freshwater pollen. This facies corresponds to a freshwater peat swamp developed on more stable land between distributary channels not far from the strandline of the lake.

Source potential. Similar to the intramontane lacustrine palaeoenvironment, Type I kerogens in the form of amorphous liptinite laid down in deeper water within an anoxic hypolimnion provide a good source rock. The presence of a wide alluvial plain along the strandline, and shallow eutrophic waters in the nearshore zone, provides scope for a '*Pediastrum*' facies to be well developed with abundant cutinite and spores deposited in more restricted areas. The abundance of these Type I and Type II kerogens makes for good oil-source facies but the overall organic richness (TOC) is lower than the deep anoxic facies.

Peat swamp environments that develop within the alluvial plain led to rich Type II resinite deposition that is oil prone. Tropical evergreen angiospermous trees (*Dipterocarpaceae*) of the rainforest produce prodigious quantities of resin unlike it's more common association with coniferous gymnospermous trees at higher latitudes. Sometimes on injured trees a thick green-black oleaginous coating of this material can be found (Fig. 8b). Lumps of hard yellow-orange, white-weathering resin can be found littering the forest floor, remaining once the wood of fallen trees has rotted. Resin is a common constituent of peat and therefore also, coal. Peat swamp is not of course always associated with lacustrine environments of this type, but given this particular low angle freshwater coastal plain it may be regarded as a closely associated facies. Lawton (1985) detected clearly related alternating cyclic coal deposits and lacustrine shales in seismic sequences, each representing lateral facies equivalents in the overall palaeoenvironment. The occasional abundance of humic macerals in the shallow-water facies indicates gas, in addition to oil potential.

Reservoir rocks. Reservoir rocks are not obvious in the deep-water source facies, although turbidite interbeds may locally be important. However, potential reservoirs are present in abundance in the shallow-water and alluvial-plain facies at the basin edge where large deltaic sand-bodies linked to alluvial channel sands are to be found. A regressive shallowing trend that tends to exhibit upwards coarsening produced by deltas prograding into the basin, is to be seen. This results in coarse sand overlying the deep water shale and is the most obvious aspect of the depositional environment that brings the source and reservoir facies into close proximity.

Palynofacies in a shallow lacustrine setting

Low-energy freshwater embayment

Occurrence. This depositional facies has been encountered in the Coal Zone and Lake Fill formations of the Pematang Formation of the Central Sumatra Basin; the Tanjung and Warukin formations of the Barito Basin.

Lithology. Medium grey mudstones dominate most of the sequence, with some sand dominated intervals. The sands are never massive but have argillaceous horizons and they grade vertically into the muds.

Kerogen type. Kerogen types are dominated by amorphous liptinite. Vertical changes in proportions of kerogen types can be quite abrupt, with cutinite, exinite and humic components (vitrinite and inertinite) occasionally dominant. Kerogen size sorting can also be quite variable, indicating comparatively abrupt changes in energy levels.

Palynology. The palynology records a general dominance of pteridophyte spores, but these also show marked vertical changes in abundance. Freshwater pollen are generally consistently abundant. Montane bisaccate pollen are very rare or absent. Freshwater algae, including *Botryococcus* and *Pediastrum* are intermittently abundant. Likewise marine taxa are generally consistently present, but of variable abundance.

Palaeoenvironmental interpretation. The variable but consistent presence of amorphous liptinite suggests that anaerobic conditions existed for most of the time on the lake floor. Fairly rapid nutrient recycling within the water column may have lead to generally eutrophic conditions resulting in oxygen depletion just below the surface. *Botryococcus* is only occasionally abundant indicating that stable, stratified conditions with oligotrophic surface waters within a deep lake did not exist for long periods. It is possible that deeper areas of pooled anoxic water may have existed, a pycnocline being partly established by the ingress of sea water. The occurrence of common *Botryococcus* does coincide with higher numbers of marine taxa (see Fig. 15b).

Mangrove taxa are rare or absent indicating that true tidal marine conditions did not exist. Raised percentages of humic debris or cutinite in some samples, coincident with coarser clastic deposits, indicates shallower near shore depositional sites. In one case (Fig. 15b*) the size sorting of the humic debris is very poor suggesting rapid dumping during flood stage from an alluvial channel. Generally, however, conditions are low energy with fine material laid down from a slowly circulating water column. In another case (Fig. 15b+) fern spores are extremely abundant, forming 50% of the total kerogen assemblage. This must indicate very peculiar hydrodynamic conditions in one very restricted part of the lake embayment. The general abundance of fern spores indicates the presence of a wide alluvial plain adjoining the lacustrine embayment, with no evidence of high relief in the viscinity as suggested by the absence of montane taxa.

Source potential. The presence of common amorphous liptinite indicates that these deposits would make good oil source rocks. Some samples laid down in the deepest 'pooled' areas where a thermocline develops would make the richest oil-source facies similar to that of the deep lacustrine basins. Most of the samples contain high proportions of land-derived kerogens. Those with common land-derived humic material in shallower water facies are partially gas prone. However, some of the land-derived kerogens are Type II, cutinite and exinite that can also be expected to be good oil sources. The principle limit on the source potential of this facies is the inconsistent and limited vertical extent of the anoxic facies.

Reservoir rocks. In this depositional environment reservoir deltaic sands at the basin-edge and alluvial-plain sand facies are developed in reasonably close proximity to the deeper-water source facies. Evidence of rapid delta progradation into the lacustrine embayment may indicate that sands occasionally extended right across the anoxic source muds. This would allow short distance migration and provide conduits to the shallower water reservoir facies. As a general rule in this type of depositional environment, anoxic mudstones and shallow water oxic sands are well intercalated providing for good gas/oil migration from source rock to reservoir.

High-energy freshwater embayment

Occurrence. This depositional facies has been encountered in the Meucampli and Bruksah formations of the North Sumatra Basin; the Coal Zone and Lake Fill formations of the Pematang Formation of the Central Sumatra Basin; the Tebidah Formation of the Melawi Basin; the Tanjung and Warukin formations of the Barito Basin and the Belut Formation of the West Natuna Basin.

Lithology. Fairly rapidly alternating carbonaceous silty sandstones and mudstones comprises the lithological sequence.

Kerogen type. Kerogen types are very consistent throughout the sequence, being dominated by humic land derived types, vitrinite and inertinite. Some intervals show greater proportions of amorphous material. Kerogen size sorting is moderate throughout.

Palynology. The most distinctive aspect of this type of sequence is the continuous and

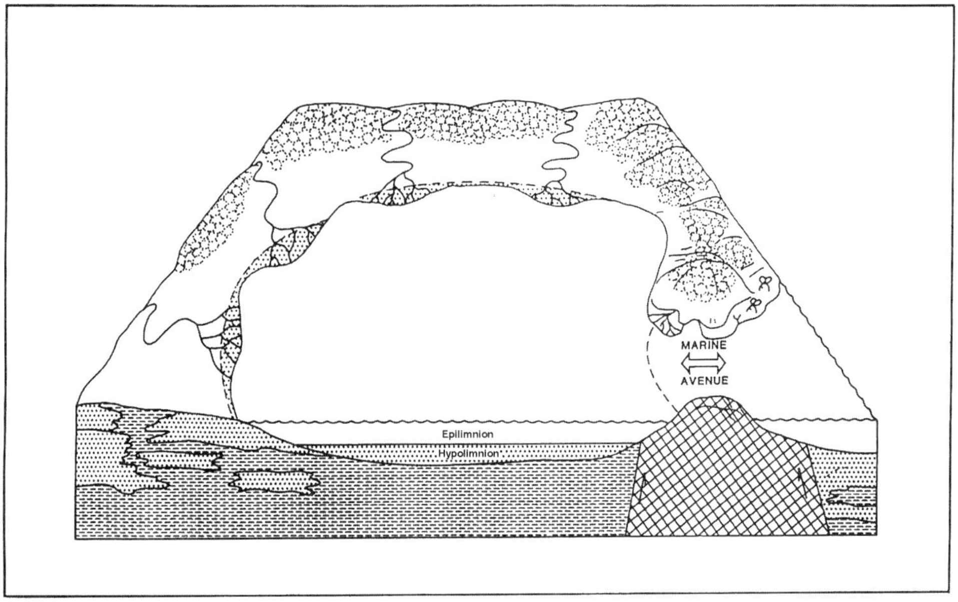

a

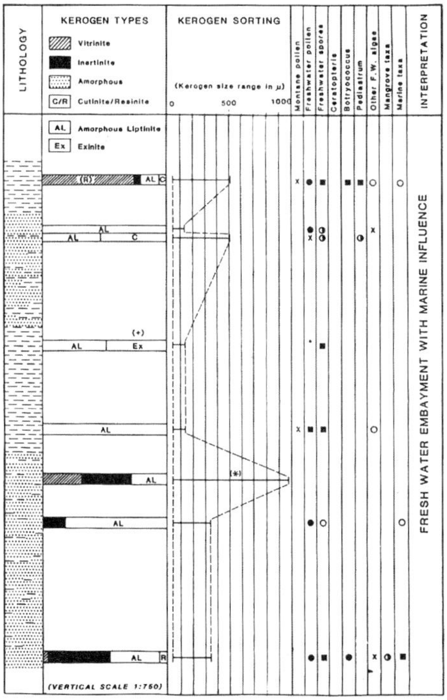

b

Fig. 15. Low-energy freshwater embayment. (**a**) Lake model; (**b**) palynofacies diagram.

overwhelming dominance of *Pediastrum* throughout. Freshwater spores are consistently present and to a lesser extent freshwater pollen. Marine taxa are consistently recorded. No *Botryococcus* or montane taxa are recorded.

Palaeoenvironmental interpretation. Rapidly alternating lithologies of sand and shale suggest rapid sedimentation rates. Palynofacies evidence of markedly unchanging character, suggests that despite rapid sedimentation, rapid subsidence maintained the base level at an approximately constant position. The abundant presence of *Pediastrum* throughout a comparatively thick sequence suggests eutrophic lake waters, as would be expected with such rapid alluvial input. No part of the lagoon or lake can have been very far from alluvial supply, as indicated by the consistent occurrence of poorly sorted humic macerals and pteridophyte spores in an arenaceous dominated sequence. A complex of unstable interconnecting lakes and ponds on a rapidly subsiding alluvial plain is indicated.

Consistently occurring marine taxa indicate tidal marine influences throughout the alluvial plain. The amorphous material is possibly partly marine in origin. There is no clear indication of permanent stratification within the water column, and the good preservation of *Pediastrum* may be a function of rapid burial of deposited organic material rather than the creation of anoxia. Tidal influences within the basin may have caused the lagoonal/ lacustrine waters to be slightly brackish but not sufficiently saline to inhibit the growth of *Pediastrum*. The alluvial plain must have been of comparatively large size as no apparent lateral facies changes are evident throughout the vertical sequence (Fig. 16c). This suggests subsidence on a more regional scale rather than locally within a rift/graben system. Figure 16b is an example of a similar sequence that becomes abruptly transgressive to inner shelf and then outer shelf as subsidence exceeded the rate of sedimentation.

Source potential. Amorphous material within this depositional environment is mostly non-fluorescent. The abundance of well preserved *Pediastrum* (Type I kerogen), together with amorphous material of possible phytoplankton origin indicates good oil source potential in this depositional environment. However, evidence of shallow water in a sequence generally dominated by coarse clastic material suggests that oil prone organic material is much less abundant than in the deep meromictic lake facies. The abundance of humic material throughout the sequence indicates good gas proneness. Palynofacies analysis may reveal that higher proportions of cutinite (Type II) can be expected in this facies, which would increase the oil-source potential.

Reservoir rocks. Reservoir rocks are intimately associated with source facies in this depositional environment. The sequence is coarse clastic-dominated with comparatively thin shale intervals. Gas/oil migration can be expected to be easily achieved under these conditions. Within a wide alluvial plain that accumulates an abundance of coarse sandstone there is good potential for reservoir development.

Fluvio-lacustrine overbank

Occurrence. This depositional facies has been encountered in the Sawatambung Formation of the Ombilin Basin; the Lake Fill Formation of the Pematang Formation of the Central Sumatra Basin; The Talang Akar Formation of the South Sumatra, NW Java and Sunda basins and throughout the Miocene sequences of the South China Sea Basins.

Lithology. A sequence of alternating sandstones and mudstones forms the lithological sequence of this facies type. Mudstone units are more persistent than sandstone units. Occasional coals are developed within the mudstones.

Kerogen type. Kerogen is all derived from land plants and is generally dominated by cutinite and vitrinite. Resinite and inertinite are consistently evident in all samples and the kerogen assemblage is generally moderate to poorly size sorted.

Palynology. Pediastrum is persistent and occasionally abundant throughout this facies, together with freshwater pollen and spores. The spore type *Magnastriatites howardi* (Fig. 8i) derived from the freshwater aquatic fern *Ceratopteris* (Fig. 8h) is present throughout and is most common coincident with abundant *Pediastrum*. Other freshwater fern spores are common. Very rare occurrences of *Botryococcus* are to be seen. Marine and mangrove taxa are extremely rare or absent.

Palaeoenvironmental interpretation. The absence of marine fossils indicates deposition above any tidal influences (Fig. 17b, lower two biofacies zones). Abundant poorly sorted cutinite and vitrinite in a freshwater setting of variable sandstone, mudstone and coal lithologies is typical of a low energy alluvial plain. Persistent occurrences of the alga *Pediastrum* indicates the presence of lacustrine facies within this setting. Lacustrine environments in such a setting would take the form of small ponds, some interconnected, in low-lying areas between small braided distributary channels (Fig. 8a, Mekong Delta).

These ponds would receive abundant supplies of nutrients allowing *Pediastrum* to thrive, but amorphous liptinite that is associated with anaerobic decay is not present, suggesting an absence of anoxia in the water column. A largely overbank organic debris of leaves from the local fern flora would be rapidly dumped within the ponds during probable frequent floods, leading to cutinite enrichment of the kerogen assemblage. The environment was probably too unstable for the growth of trees, hence the lack of *in situ* peat swamps and the lesser amount of humic material (derived from wood) in the kerogen assemblage. However, thin inpersistent coal seams may have developed, similar to those described by Cabrera & Saez (1987). The abundant supply of organic debris, together with comparatively high sedimentation rates and high aerobic activity, may have caused anoxia to develop in the sediment, ensuring the preservation of a comparatively large part of the organic debris.

Source potential. Oil-source potential is moderate in this depositional facies by virtue of the abundance of cutinite (Type II kerogen). This

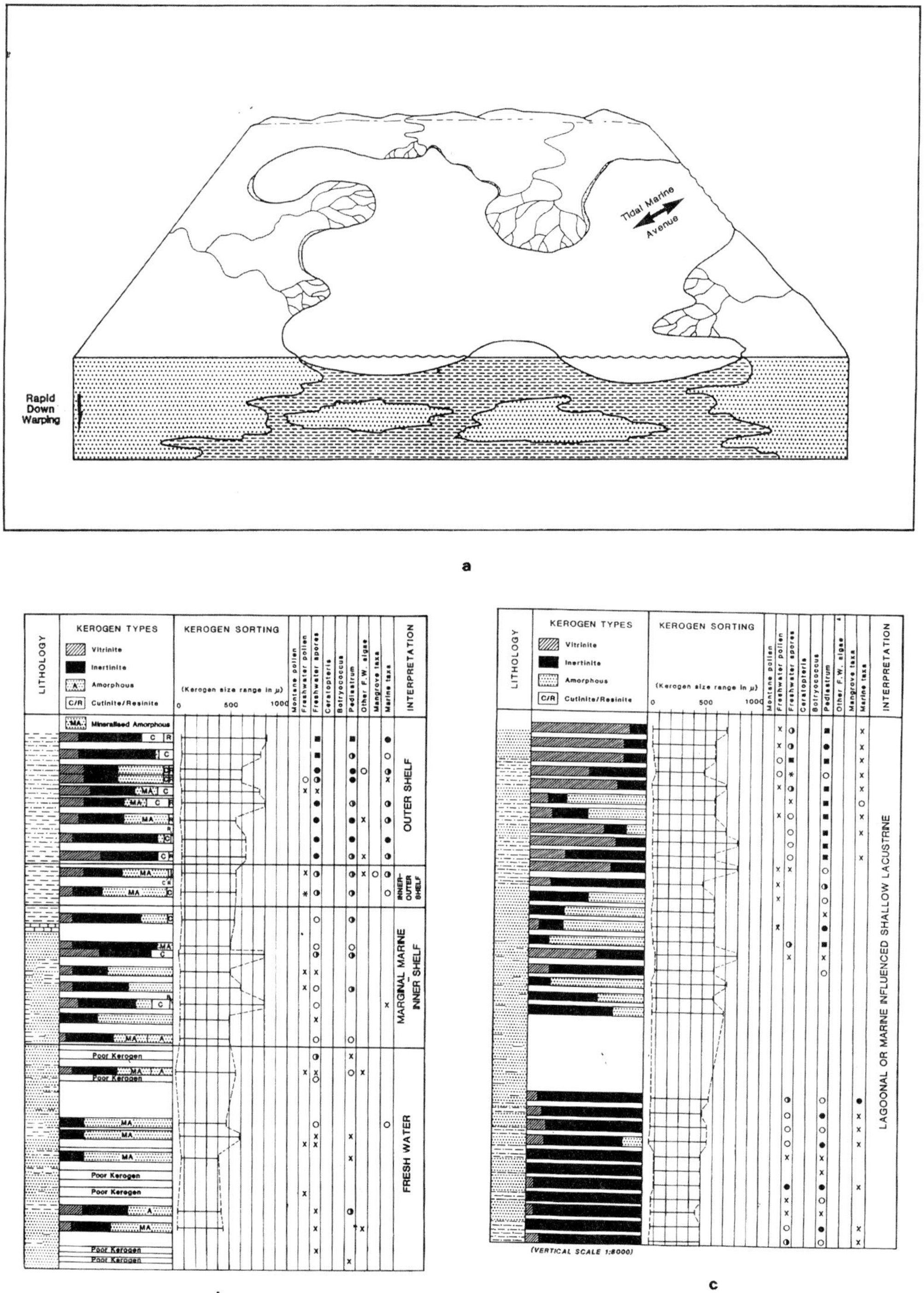

Fig. 16. High energy freshwater embayment. (**a**) Lake model; (**b**, **c**) palynofacies diagrams.

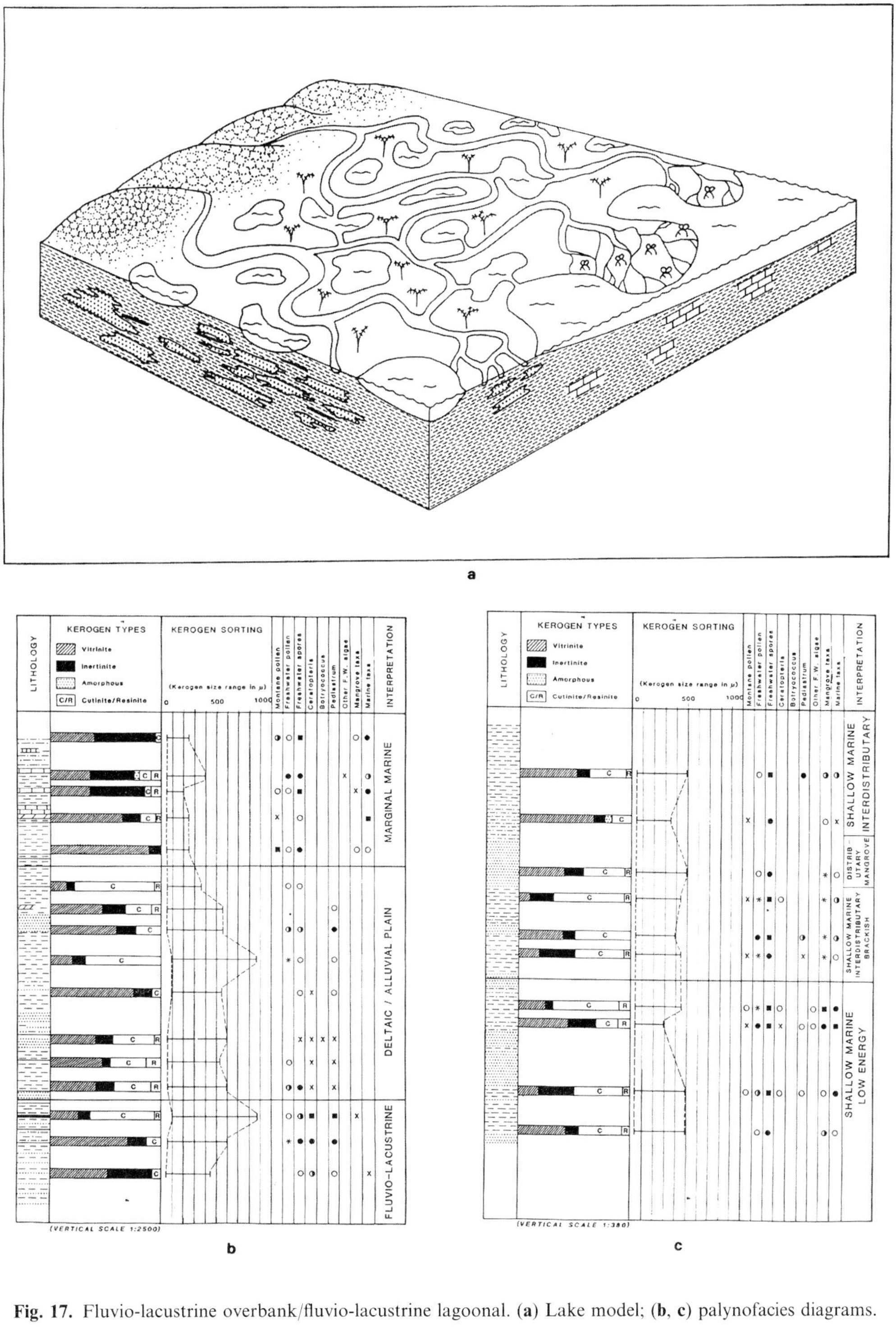

Fig. 17. Fluvio-lacustrine overbank/fluvio-lacustrine lagoonal. (**a**) Lake model; (**b**, **c**) palynofacies diagrams.

material is dispersed in thin shale units and comprises a large part of the thin coal seams. Occasional abundance's of *Pediastrum* (Type I kerogen) are to be seen but in the absence of amorphous liptinite, Type I kerogen is volumetrically comparatively small.

Reservoir rocks. The pond and overbank lake muds and coals comprising the source rock are laid down in close juxtaposition to channel and point-bar sands that form the main reservoir development. Large thick deltaic sands and sands associated with major distributary channels are absent. Efficient migration routes that drain the source facies are provided by the minor channel sands.

Fluvio-lacustrine lagoonal

Occurrence. This depositional facies has been encountered in the Sawatambung Formation of the Ombilin Basin; the Lake Fill Formation of the Pematang Formation of the Central Sumatra Basin; The Talang Akar Formation of the South Sumatra, NW Java and Sunda basins and throughout the Miocene sequences of the South China Sea Basins.

Lithology. A lithological sequence comprising alternating mudstones and sandstones occurs. Some of the individual sandstone and mudstone units can be comparatively persistent.

Kerogen type. Similar to the freshwater overbank setting, the kerogen assemblage is dominated by land derived cutinite and vitrinite and is moderately to poorly size sorted. Rare to moderate percentages of resinite and inertinite are also recorded.

Palynology. Abundant and persistent records of mangrove pollen (*Spinizonocolpites echinatus*, derived from Nypa together with persistent records of marine taxa (dinocysts and microforaminiferal linings) are to be found. Freshwater pollen and spores are present in moderate to common numbers throughout the sequence. Montane taxa occur as a rare component. Intermittent records of *Pediastrum* are to be seen, rarely it is abundant.

Palaeoenvironmental interpretation. The palaeoenvironment represented by this palynofacies is similar to that of the fluvio-lacustrine overbank, but with greater brackish/marine influence (mangrove taxa and marine palynomorphs), nearer to the strandline (Fig. 17c). Some records of *Pediastrum* in this palaeoenvironment may be reworked from upstream. However, this is not necessarily invariably the case. Persistent records of *Pediastrum* within this palynofacies indicates the development of pools and ponds subject to slightly increased salinity from tidal affects that this alga may have been able to tolerate.

Input of organic material is high in this palaeoenvironment and current energies are reduced by the dense vegetation and low angle high sinuosity streams, resulting in moderately sorted vitrinite and cutinite dominated kerogen assemblages. Rapid rates of decomposition would have created anoxic conditions within the shallow water muds. Thicker sand units probably represent sediments closely associated with large distributary channels.

Source potential. As with the freshwater overbank depositional environment, the consistent and general abundance of cutinite (Type II kerogen) in a brackish/marine lagoonal environment makes this a 'moderate' oil prone source rock. However, organic material is not concentrated into thin coal seams and thicker channel sands are developed. The overall TOC content of the sequence is, therefore, comparatively low. In addition, *Pediastrum* is present in comparatively low numbers, so Type I oil prone kerogen is sparse. The common occurrence of humic material, particularly vitrinite, is indicative of good gas generation potential.

Reservoir rocks. As with the freshwater overbank type, a fluvio-lacustrine setting results in a good intercalation of organic-rich muds and high porosity/permeability sands. In this case the sands are comparatively well developed, so apart from providing good migration routes they may form potential reservoirs in their own right. In this strandline environment the possibility of migration routes via channel sands, to shallow marine sands, and offshore bars may be an important consideration.

Conclusions

Stratified or meromictic lake systems, by analogy with modern tropical lake systems in the African rift valleys can be demonstrated to be prolific oil source beds. That such a system existed within the Central Sumatra Basin has been convincingly demonstrated in the literature produced over the last 10 years. Few can now dispute that the lacustrine syn-rift depositional

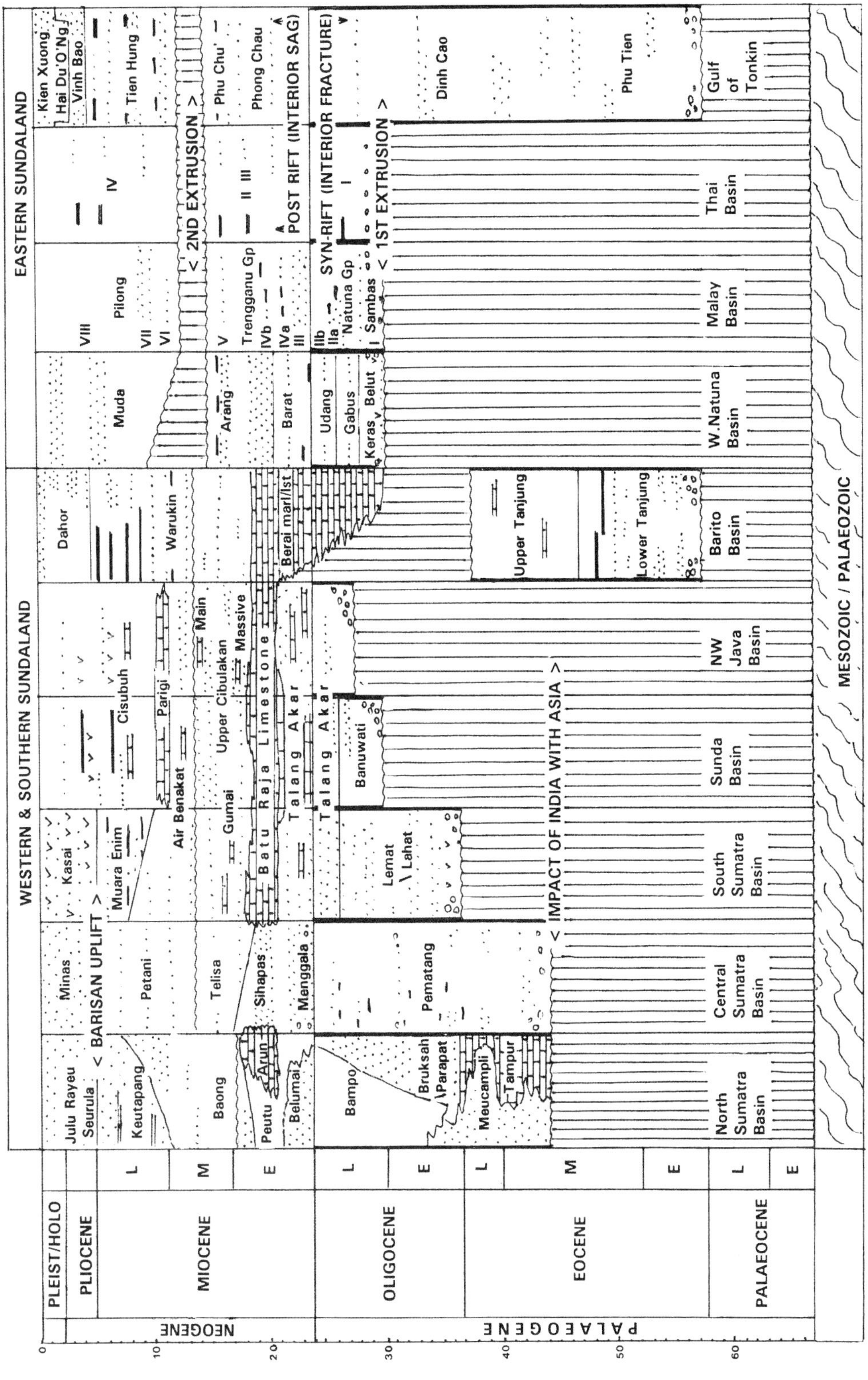

Fig. 18. Schematic stratigraphy of Sundaland.

system is well established as a model of hydrocarbon source rock for the best known and most drilled area of this basin. The existence of interior fracture basins that accumulated thick intervals of lacustrine sediments can be shown from palynofacies analysis to have been present in all of the basins formed on Sundaland. This system has now been widely adopted in the adjoining NW Java and Sunda basins and has also been found to be applicable to the Barito Basin of southeastern Borneo. These sedimentary sequences and their relationship to the principle tectonic events are summarized on Fig. 18.

Some of the interior fracture basins of Sundaland accumulated quasi lacustrine sediments that have a palynofacies that includes marine influence. For example Late Eocene to Oligocene productive source rocks are evident within the North Sumatra basinal area. Marine influence from the Andaman Sea is known from the presence of microfaunas in this sequence. Marine influence within a partially enclosed restricted basin does not necessarily upset stratification within the water column, indeed such stratification may be enhanced.

A major rifting phase in the Oligocene that effectively caused opening and basin formation on the eastern side of Sundaland lead to the establishment of the Gulf of Thailand/Malay basinal system of the South China Sea. This was not a land locked interior fracture setting but was open to the sea towards the southeast. Marine influences are present that diminish northwards into the Gulf of Thailand. However extensive organic rich sediments were also laid down in this region throughout the Oligocene and Early Miocene from water bodies that were essentially lacustrine in character.

While a variety of other shallow-water lacustrine facies and delta top/shallow marine source rocks can be demonstrated to occur throughout the Palaeogene and Neogene of SE Asia, the thick syn-rift lacustrine and quasi-lacustrine (marine influenced) organic rich shales may be by far the most prolific source for hydrocarbons that account for the great majority of the oil in this region.

Help and discussion with numerous colleagues over the years is gratefully acknowledged. Improvements to the manuscript came from the suggestions of two anonymous reviewers. Special advice and assistance with aspects of the geochemistry came from P. Walko whose efforts led to some significant improvements to the text.

W. Bryan assisted with the typing. The directors of the Geochem Group Ltd gave permission for publication. A special thanks to Jessica and Edward.

References

ABDULLAH, M. & JORDAN, C. F. 1987. The geology of the Arun Field Miocene reef complex. *Proceedings of the Indonesian Petroleum Association 16th Annual Convention*, 65–96.

ADNAN, A., SUKOWITONO & SUPRIYANTO 1991. Jatibarang sub basin – A half graben model in the onshore of Northwest Java. *Proceedings of the Indonesian Petroleum Association 20th Annual Convention*, 279–297.

ANDERSON, B. L., BON, J. & WAHONO, H. E. 1993. Reassessment of the Miocene stratigraphy, paleogeography and petroleum geochemistry of the Langsa Block in the offshore North Sumatra Basin. *Proceedings of the Indonesian Petroleum Association 22nd Annual Convention*, 169–189.

ARMITAGE, J. H. & VIOTTI, C. 1977. Stratigraphic nomenclature-southern end Malay Basin. *Proceedings of the Indonesian Petroleum Association 6th Annual Convention*, 69–94.

ASCOPE 1981. *Tertiary sedimentary basins of the Gulf of Thailand and South China Sea: stratigraphy, structure and hydrocarbon occurrences.* The ASCOPE Secretariat, Jakarta, Indonesia.

AUDLEY-CHARLES, M. G. 1987. Dispersal of Gondwanaland: Relevance to evolution of the angiosperms. *In*: WHITMORE, T. C. (ed.) *Biogeographical evolution of the Malay archipelago.* Clarendon Press, Oxford, 5–25.

——1988. Evolution of the southern margin of Tethys (North Australia region) from Early Permian to Late Cretaceous. *In*: AUDLEY-CHARLES, M. G. & HALLAM, A. (eds) *Gondwana and Tethys.* Geological Society, London, Special Publication **37**, 79–100.

—— & HALLAM, A. 1988. Introduction. *In*: AUDLEY-CHARLES, M. G. & HALLAM, A. (eds) *Gondwana and Tethys.* Geological Society, London, Special Publication **37**, 1–4.

BEDDOES, L. R. 1980. Hydrocarbon plays in Tertiary basins of Southeast Asia. *Offshore South East Asia Conference SEAPEX* Feb. 1980, 000–000.

BISSADA, K. K., ELROD, L. W., DARNELL, L. M., SZYMCZYK, H. M. & TROSTLE, J. L. 1992. Geochemical inversion – a modern approach to inferring source-rock identity from characterisation of accumulated oil and gas. *Proceedings of the Indonesian Petroleum Association 21st Annual Convention*, 165–199.

BRADLEY, W. H. 1966. Tropical Lakes, copropel and oil shale. *Geological Society of America Bulletin*, **77**, 1333–1338.

BROWN, A. C., KNIGHTS, B. A. & CONWAY, E. 1969. Hydrocarbon content and its relationship to physiological state in the green alga *Botryococcus braunii. Phytochemistry*, **8**, 543–547.

BUSHNELL, D. C. & ATMAWAN, D. T. 1986. A model for hydrocarbon accumulation in Sunda Basin, West Java. *Proceedings of the Indonesian Petroleum Association 15th Annual Convention*, 47–55.

CABRERA, L. L. & SAEZ, A. 1987. Coal deposition in carbonate rich shallow lacustrine systems: the Calaf and Mequinenza sequences (Oligocene, eastern Ebro Basin, NE Spain). *Journal of the Geological Society, London*, 451–461.

CAMERON, N. R. & PULUNGGONO, A. 1984. Sumatran microplates, their characteristics and their roles in the evolution of the Central and South Sumatra basins. *Proceedings of the Indonesian Petroleum Association 13th Annual Convention*, 121–143.

CHU, S. P. 1942. The influence of the mineral composition of the medium on the growth of the planktonic algae. *Journal of Ecology*, **30**, 284–325.

CLURE, J. 1991. Spreading centres and their affect on oil generation in the Sunda region. *Proceedings of the Indonesian Petroleum Association 20th Annual Convention*, 37–49.

COLE, J. M. 1987. Some fresh/brackish water depositional environments in the S.E. Asian Tertiary with emphasis on coal bearing and lacustrine deposits and their source rock potential. *Proceedings of the Indonesian Petroleum Association 16th Annual Convention*, 429–449.

——1992. Freshwater dinoflagellate cysts and acritarchs from Neogene and Oligocene sediments of the South China Sea and adjacent areas. *In*: HEAD, M. J. & WRENN, J. H. (eds) *Neogene and Quaternary Dinoflagellate cysts and Acritarchs*. American Association of Stratigraphic Palynologists Foundation, Dallas, 181–192.

COURTNEY, S. 1994. Sequence stratigraphy applied to the hydrocarbon productive basins of Western Indonesia (abs). *AAPG International Conference and Exhibition 23 August, 1994, Kuala Lumpur*.

—— & WILLIAMS, H. H. 1994. The Palaeogene rifts of western Indonesia (abs). *AAPG Annual Meeting "Analogs for the World", 15 June 1994, Denver*.

DE COSTER, G. L. 1974. The geology of the central and south Sumatra basins. *Proceedings of the Indonesian Petroleum Association 3rd Annual Convention*, 77–110.

DAINES, S. 1985. Structural history of the West Natuna Basin and the tectonic evolution of the Sunda region. *Proceedings of the Indonesian Petroleum Association 14th Annual Convention*, 39–62.

DALY, M. C., HOOPER, B. G. D. & SMITH, D. G. 1987. Tertiary plate tectonics and basin evolution in *Proceedings of the Indonesian Petroleum Association 16th Annual Convention*, 399–411.

——, COOPER, M. A., WILSON, I., SMITH, D. G. & HOOPER, B. G. D. 1991. Cenozoic plate tectonics and basin evolution in Indonesia. *Marine and Petroleum Geology*, **8**, 2–21.

DAVIES, P. R. 1984. Tertiary structural evolution and related hydrocarbon occurrences, North Sumatra Basin. *Proceedings of the Indonesian Petroleum Association 13th Annual Convention*, 19–53.

DEMAISON, G. J. & MOORE, G. T. 1980. Anoxic environments and oil source bed genesis. *American Association of Petroleum Geologists Bulletin*, **64**(8), 1179–1209.

EUBANK, R. T. & MAKKI, A. C. 1981. Structural geology of the Central Sumatra Back-arc Basin. *Proceedings of the Indonesian Petroleum Association 10th Annual Convention*, 153–196.

EVITT, W. R. 1963. Occurrence of the freshwater alga Pediastrum in Cretaceous marine sediments. *American Journal of Science*, **261**, 890–893.

FAINSTEIN, R. 1987. Exploration of the North Seribu area, Northwest Java Sea. *Proceedings of the Indonesian Petroleum Association 16th Annual Convention*, 191–214.

—— & HERU PRAMONO 1986. Structure and stratigraphy of AVS Field, Java Sea. *Proceedings of the Indonesian Petroleum Association 15th Annual Convention*, 19–45.

FLEET, A. J., KELTS, K. & TALBOT, M. R. (eds) 1988. *Lacustrine Petroleum Source Rocks*. Geological Society Special Publications, **40**.

FLETCHER, G. L. & BAY, K. W. 1975. Geochemical evaluation – NW Java Basin. *Proceedings of the Indonesian Petroleum Association 4th Annual Convention*, 211–241.

—— & SOEPARJADI, R. A. 1976. Indonesia's Tertiary basins – The land of plenty. *Southeast Asian Petroleum Exploration Society, Program, Offshore South East Asia Conference* Feb. 1976, Paper 8, 1–54.

GELPI, E., SCHNEIDER, J. & MANN, J. 1970. Hydrocarbons of geochemical significance in microscopic algae. *Phytochemistry*, **9**, 603–612.

GOODALL, G. S., COLES, G. P. & WHITAKER, M. F. 1992. An integrated palynological, palynofacies and micropalaeontological study of the pre-salt formations of the south Gabon subbasin and the Congo Basin. *In*: CURNELLE, R. (ed.) *Geologie Africaine*. Compte-rendu des Colloques de geologie de Libreville, 6–8 mai 1991. Bulletin du Centres Recherches Exploration Production Elf Aquitaine Memoires, **13**, F-31360 Boussens, 365–399.

GORDON, T. L. 1985. Talang Akar coals – Ardjuna sub basin oil source. *Proceedings of the Indonesian Petroleum Association 14th Annual Convention*, 91–120.

GRAVES, R. R. & WEEGAR, A. A. 1973. Geology of the Arun gas field (north Sumatra). *Proceedings of the Indonesian Petroleum Association 2nd Annual Convention*, 23–51.

HALL, R. 1997. Cenozoic plate reconstructions of SE Asia. *This volume*.

HAMILTON, W. 1979. *Tectonics of the Indonesian region*. US Geological Survey Professional Paper **1078**.

HAQ, B. U., HARDENBOL, J. & VAIL, P. R. 1987. Chronology of fluctuating sealevel cycles since the Triassic. *Science*, **235**, 1156–1167.

HARSA, A. E. 1975. Some of the factors which influence oil occurrences in the south and central Sumatra basins. *Regional Conference on Geology and Mineral Resources of southeast Asia, Jakarta, Indonesia*.

HARTONO, M. H. S. & TJOKOROSAPOETRO, S. 1984. Preliminary account and reconstruction of Indonesian terranes. *Proceedings of the Indonesian Petroleum Association 13th Annual Convention*, 185–226.

HASAN, M. M. & SOEBANDRIO, D. S. 1988. The petroleum geology of Tanjung Laban Field, South Sumatra. *Proceedings of the Indonesian Petroleum Association 17th Annual Convention*, 257–274.

HERU PRAMONO, WU, C. H. & NOBLE, R. A. 1990. A new oil kitchen and petroleum bearing subbasin in the offshore northwest Java area. *Proceedings of the Indonesian Petroleum Association 19th Annual Convention*, 394–422.

HILLS, I. R., WHITEHEAD, E. V., ANDERS, D. E., CUMMINS, J. J. & ROBINSON, W. E. 1966. An optically active triterpane, gammacerane in Green River, Colorado, oil shale bitumen. *Journal of the Chemical Society, Chemical Communications*, 752–754.

HOLCOMBE, C. J. 1977. How rigid are the lithospheric plates? Fault and shear zones in southeast Asia. *Journal of the Geological Society, London*, **134**, 325–342.

HORSFIELD, B., CURRY, D. J. & 9 OTHERS 1994. Organic geochemistry of freshwater and alkaline lacustrine sediments in the Green River Formation of the Washakie Basin, Wyoming, U.S.A. *Advances in Organic Geochemistry 1993: Organic Geochem*istry, **22**(3–5), 415–440.

HUTAPEA, O. M. 1981. The prolific Talang Akar Formation in Raja Field, South Sumatra. *Proceedings of the Indonesian Petroleum Association 10th Annual Convention*, 251–267.

HUTCHISON, C. S. 1986. Tertiary basins of S.E. Asia – their disparate tectonic origins and eustatic stratigraphical similarities. GEOSEA V. Proc. Vol. 1 *Geological Society of Malaysia Bulletin*, **19**, 109–122.

—— 1989. *Geological evolution of South-East Asia*. Oxford Monographs on Geology and Geophysics, **13**. Oxford Science Publications.

HUTCHINSON, G. E. 1957. *A treatise on limnology. Vol 1. Geography, Physics and Chemistry*. John Wiley, New York.

——, PICKFORD, G. E. & SCHUURMAN, J. F. M. 1932. A contribution to the hydrobiology of pans and other inland waters of South Africa. *Archives in Hydrobiology*, **24**, 1–54.

HUTTON, A. C. 1984. Geology of oil shale deposits within the Narrows Graben Queensland Australia: Discussion. *American Association of Petroleum Geologist Bulletin*, **68**, 1055–1057.

JARNEFELT, H. 1956. *Zur limnologie einiger Gewasser Finnlands XVI. Mit besonderer Berucksichtigung des Planktons sus mal. elain-ja Kasirt. Sevr. van elain. Julk. Annuls Zool. Soc. Vanamo* **17**(7).

KATZ, B. J. & KELLY, P. A. 1987. Central Sumatra and the East African rift lake sediments: An organic geochemical comparison. *Proceedings of the Indonesian Petroleum Association 16th Annual Convention*, 259–290.

—— & MERTANI, B. 1989. Central Sumatra – a geochemical paradox. *Proceedings of the Indonesian Petroleum Association 18th Annual Convention*, 403–425.

KINGSTON, D. R., DISHROON, C. P. & WILLIAMS, P. A. 1983. Global basin classification system. *American Association of Petroleum Geologist Bulletin*, **67**, 2175–2193.

KINGSTONE, J. 1978. Oil and gas generation, migration and accumulation in the North Sumatra Basin. *Proceedings of the Indonesian Petroleum Association 7th Annual Convention*, 75–104.

KIRBY, G. A., MORLEY, R. J. & 12 OTHERS 1993. A re-evaluation of the regional geology and hydrocarbon prospectivity of the onshore central North Sumatra Basin. *Proceedings of the Indonesian Petroleum Association 22nd Annual Convention*, 243–264.

KJELLGREN, G. M. & SUGIHARTO, H. 1989. Oil geochemistry: A clue to the hydrocarbon history and prospectivity of the southeastern North Sumatra Basin, Indonesia. *Proceedings of the Indonesian Petroleum Association 18th Annual Convention*, 363–384.

KLEMME, H. D. 1980. Petroleum basins classification and characteristics. *Journal of Petroleum Geology*, **3**, 187–207.

KONING, T. & DARMONO, F. X. 1984. The geology of the Beruk northeast field, Central Sumatra: oil production from Pre-Tertiary basement rocks. *Proceedings of the Indonesian Petroleum Association 13th Annual Convention*, 385–406.

KUSUMA, I. & DARIN, T. 1989. The hydrocarbon potential of the Lower Tanjung Formation, Barito BASIN, S.E. Kalimantan. *Proceedings of the Indonesian Petroleum Association 18th Annual Convention*, 107–138

LARTER, S. R. & DOUGLAS, A. G. 1980. A pyrolysis-gas chromatographic method for kerogen typing. *In*: DOUGLAS, A. G. & MAXWELL, J. R. (eds) *Advances in organic geochemistry* 1979. Pergamon Press, New York, 579–583.

LAWTON, D. C. 1985. Seismic facies analysis of delta plain coals from Camrose, Alberta and lacustrine coals from Pictou coalfield, Nova Scotia. *American Association of Petroleum Geologists Bulletin*, **69**, 2120–2129.

LIAN, H. M. & BRADLEY, K. 1986. Exploration and development of natural gas, Pattani Basin, Gulf of Thailand. *4th Circum-Pacific Energy Mineral Resource Conference, Singapore*.

LONGLEY, I. M., BARRACLOUGH, R., BRIDDEN, M. A. & BROWN, S. 1990. Pematang lacustrine petroleum source rocks from the Malacca Strait PSC, Central Sumatra, Indonesia. *Proceedings of the Indonesian Petroleum Association 15th Annual Convention*, 279–297.

MA LI, GE TAISHING, ZHAO XUEPING, ZIE TAIJUN, GE RONG & DANG ZHENRONG 1982. Oil basins and subtle traps in the eastern part of China. *In*: The deliberate search for the subtle trap. *American Association of Petroleum Geologists Bulletin*, **69**, 1123–1132.

MACGREGOR, D. S. & MCKENZIE, A. G. 1986. Quantification of oil generation and migration in the Malacca Strait region, Central Sumatra. *Proceedings of the Indonesian Petroleum Association 15th Annual Convention*, 305–320.

MANN, A. L., GOODWIN, N. S. & LOWE, S. 1987. Geochemical characteristics of lacustrine source rocks: a combined palynological/molecular study of Tertiary sequence offshore China. *Proceedings of the Indonesian Petroleum Association 16th Annual Convention*, 241–258.

MASON, A. D. M., HAEBIG, J. C. & MCADOO, R. L. 1993. A fresh look at the North Barito Basin,

Kalimantan. *Proceedings of the Indonesian Petroleum Association 22nd Annual Convention*, 589–606.

MATCHETTE-DOWNES, C. J., FALLICK, A. E., KARMAJAYA & ROWLAND, S. 1994. A maturity and palaeoenvironmental assessment of condensates and oils from the North Sumatra Basin, Indonesia. *In*: SCOTT, A. C. & FLEET, A. J. (eds) *Coal and Coal-bearing Strata as Oil-Prone Source Rocks*. Geological Society, London, Special Publications, **77**, 331–371.

MATSUOKA, K. & HASE, K. 1977. Fossil Pediastrum from the Pleistocene Hamamatsu Formation around Lake Hamana, Central Japan. *Translational Proceedings of the Palaeontological Society of Japan*, **104**, 432–441.

MAXWELL, J. R., DOUGLAS, A. G., EGLINTON, G. & MCCORMICK, A. 1968. The botryococcenes – hydrocarbons of novel structure from the alga *Botryococcus braunii* Kitzing. *Phytochemistry*, **7**, 2157–2171.

MD NAZRI RAMLI 1988. Stratigraphy and palaeofacies development of Carigali's operating areas in the Malay Basin, South China Sea. *Geological Society of Malaysia Bulletin*, **22**, 153–187.

MERTOSONO, S. & NAYOAN, G. A. S. 1974. The Tertiary basinal area of Central Sumatra. *Proceedings of the Indonesian Petroleum Association 3rd Annual Convention*, 63–76.

METCALFE, I. 1988. Origin and assembly of south-east Asian continental terranes. *In*: AUDLEY-CHARLES, M. G. & HALLAM, A. (eds) *Gondwana and Tethys*. Geological Society, London, Special Publications, **37**, 101–118.

METZER, P. & CASDEVALL, E. 1983. Structure de trois nouveaux 'botryococcenes' systhetises pas une souche de Botryococcus braunii cultivee en laboratoire. *Tetrahedron Letters*, **24**, 4013–4016.

MOLDOWAN, J. M. & SIEFERT, W. K. 1980. First discovery of botryococcane in petroleum. *Journal of the Chemical Society, Chemical Communications*, 912–914.

——, —— & GALLEGOS, E. J. 1985. Relationship between petroleum composition and depositional environment of petroleum source rocks. *American Association of Petroleum Geologists Bulletin*, **69**, 1255–1268.

MOLINA, J. 1985. Petroleum geochemistry of the Sunda Basin. *Proceedings of the Indonesian Petroleum Association 14th Annual Convention*, 143–179.

MOLNAR, P. & TAPPONIER, P. 1975. Cenozoic tectonics of Asia: effects of a continental collision. *Science*, **189**, 419–426.

MORLEY, R. J. 1977. Palynology of Tertiary and Quaternary sediments in southeast Asia. *Proceedings of the Indonesian Petroleum Association 6th Annual Convention*, 255–276.

—— 1990. Tertiary stratigraphic palynology in southeast Asia. *Geological Society of Malaysia Petroleum Geology Seminar, Kuala Lumpur, 27–28 November, 1990*.

—— 1982. Fossil pollen attributable to *Alangium* LAMARCK (Alangiaceae) from the Tertiary of Malesia. *Reviews of Palaeobotany and Palynology*, **36**, 65–94.

—— 1991. Tertiary stratigraphic palynology in Southeast Asia: current status and new directions. *Geological Society of Malaysia Bulletin*, **28**, 1–36

——1989. Development of the Benkalis Depression, Central Sumatra and its subsequent deformation – a model for other Sumatran grabens. *Proceedings of the Indonesian Petroleum Association 16th Annual Convention*, 217–247.

MULHADIONO & SUTOMO, J. A. 1984. The determination of economic basement of rock formation in exploring the Langkat – Medan area, North Sumatra Basin. *Proceedings of the Indonesian Petroleum Association 13th Annual Convention*, 75–107.

MULLER, J. 1964. A palynological contribution to the history of the mangrove vegetation in Borneo. *In*: CRANWELL L. M. (ed.) *Ancient Pacific Floras*. University of Hawaii Press, 33–42.

——1968. Palynology of the Pedawan and Plateau Sandstone Formations (Cretaceous–Eocene) in Sarawak, Malaysia. *Micropalaeontology*, **14**, 1–37.

—— 1972. Palynological evidence for change in geomorphology, climate and vegetation in the Mio-Pliocene of Malesia. *In*: ASHTON, P. S. & ASHTON, M. (eds) *The Quaternary Era in Malesia*, Geography Department, University of Hull, Miscellaneous Series, **13**, 6–16.

—— 1981. Fossil pollen records of extant angiosperms. *The Botanical Review*, **47**, 1–142

—— & CARATINI, C. 1977. Pollen of *Rhizophora (Rhizophoraceae)* as a guide fossil. *Pollen et Spores* **19**, 361–388.

MURPHY, R. W. 1975. Tertiary Basins of Southeast Asia. *Southeast Asian Petroleum Exploration Society Proceedings*, **2**, 1–36.

—— 1976. Pre-Tertiary framework of Southeast Asia. *Southeast Asian Petroleum Exploration Society*, Program of offshore South East Asia Conference, February 1976, Paper 3, 1–2.

—— 1992. *The Petroleum Geology of Southeast Asia*. OGCI course notes.

NAJOAN, G. A. S. 1972. Correlation of Tertiary lithostratigraphic units in the Java Sea and adjacent areas. *Proceedings of the Indonesian Petroleum Association 1st Annual Convention*, 11–16.

NAUMANN, E. 1931. Limnologische Terminologie. *In*: ABBERHALDEN, E. (ed.) *Handbuch der biologischen Arbeitsmethoden*. Abt IX, Teil 8. Urban & Scharzenberg, Berlin & Wien.

POLACHAN, S., PRADIDTAN, S. TONGTAOW, C., JANMAHA, S., INTARAWIJITR, K. & SANGSUWAN, C. 1991. Development of Cenozoic basins in Thailand. *Marine & Petroleum Geology*, **8** 84–97.

POLLOCK, R. E., HAYES, J. B., WILLIAMS, K. P. & YOUNG, R. A. 1984. The petroleum geology of the KH field, Kakap Indonesia. *Proceedings of the Indonesian Petroleum Association 13th Annual Convention*, 407–424.

PONTO, C. V., WU, C. H., PRANOTO, A. & STINSON, W. H. 1988. Improved interpretation of the

Talang Akar depositional environment as an aid to hydrocarbon exploration in the ARII offshore northwest Java contract area. *Proceedings of the Indonesian Petroleum Association 17th Annual Convention*, 397–422.

POWELL, T. G. & SNOWDON, L. R. 1983. A composite hydrocarbon generation model. Erdol Undkohle-*Erdgas-Petrochemic vereinigt mit Brennstoff chemie*, **36**, 163–170.

PULUNGGONO, A. 1985. The changing pattern of ideas on Sundaland within the last hundred years. Its implications to oil exploration. *Proceedings of the Indonesian Petroleum Association 14th Annual Convention*, 347–378.

—— & CAMERON, N. R. 1984. Sumatran microplates, their characteristics and their role in the evolution of the Central and South Sumatra basins. *Proceedings of the Indonesian Petroleum* Association 13th Annual Convention, 121–143.

PUPILLI, M. 1973. Geological evolution of South China Sea area tentative reconstruction from borderland geology and well data. *Proceedings of the Indonesian Petroleum Association 2nd Annual Convention*, 223–241.

REMINTON, C. H. 1985. A Hydrocarbon generation analysis in Northwest Java Basin using Lopatin's method. *Proceedings of the Indonesian Petroleum Association 14th Annual Convention*, 121–141.

ROBINSON, K. M. 1987. An overview of source rocks and oils in Indonesia. *Proceedings of the Indonesian Petroleum Association 16th Annual Convention*, 97–122.

—— & KAMAL, A. 1988. Hydrocarbon generation, migration and entrapment in the Kampar block, Central Sumatra. *Proceedings of the Indonesian Petroleum Association 17th Annual Convention*, 211–256.

ROTINSULU, L. F., SARDJONO, S. & HERIYANTO, N. 1993. The hydrocarbon generation and trapping mechanism within the northern part of Barito Basin, South Kalimantan. *Proceedings of the Indonesian Petroleum Association 22nd Annual Convention*, 607–633.

SARDJITO, FADIANTO, E., DJUMLATI & HANSEN, S. 1991. Hydrocarbon prospect of Pre-Tertiary basement in Kuang area, South Sumatra. *Proceedings of the Indonesian Petroleum Association 20th Annual Convention*, 255–277.

SARJONO, S. & SARDJITO 1989. Hydrocarbon source rock identification in the South Palembang subbasin. *Proceedings of the Indonesian Petroleum Association 18th Annual Convention*, 427–467.

SATYANA, A. H. 1995. Palaeogene unconformities in the Barito Basin, southeast Kalimantan: A concept for the solution of the "Barito dilemma" and a key to the search for Palaeogene structures. *Proceedings of the Indonesian Petroleum Association 24th Annual Convention*, 263–275.

SIREGAR, M. S. & SUNARYO, R. 1980. Depositional environment and hydrocarbon prospects, Tanjung Formation, Barito Basin, Kalimantan. *Proceedings of the Indonesian Petroleum Association 9th Annual Convention*, 379–400.

SITUMEANG, S. & DAVIES, P. R. 1986. A geochemical study of Asamera's Block 'A' production sharing contract area, North Sumatra Basin. *Proceedings of the Indonesian Petroleum Association 15th Annual Convention*, 321–340.

SOEPARJADI, R. A., NAYOAN, G. A. S., BEDDOES, L. R. & JAMES, W. V. 1975. Exploration play concepts in Indonesia. *Proceedings of the Ninth World Petroleum Congress*, **3**, 51–64.

SOSROMIHARDJO, S. P. C. 1988. Structural analysis of the North Sumatra Basin with emphasis on synthetic aperture radar data. *Proceedings of the Indonesian Petroleum Association 17th Annual Convention*, 187–209.

SOULISA, B. & SUJANTO, F. X. 1979. Hydrocarbon occurrences in the Kandanghaur-Cemara area north west Java. *Proceedings of the Indonesian Petroleum Association 8th Annual Convention*, 223–245.

SUBROTO, E. A., ALEXANDER, R., PRANJOTO, U. & KAGI, R. I. 1992. The use of 30-Norhopane series, a novel carbonate biomarker, in source rock to crude oil correlation in the North Sumatra Basin, Indonesia. *Proceedings of the Indonesian Petroleum Association 21st Annual Convention*, 145–163.

SUHERMAN, T. & SYAHBUDDIN, A. 1986. Exploration history of the MB Field coastal area of northwest Java. *Proceedings of the Indonesian Petroleum Association 15th Annual Convention*, 101–122.

SURIA, C. 1991. Development strategy in the BZZ Field and the importance of detailed depositional model studies in the reservoir characterisation of Talang Akar channel sands. *Proceedings of the Indonesian Petroleum Association 20th Annual Convention*, 419–451.

SWAIN, F. M. 1956. Stratigraphy of lake deposits in Central and Northern Minnesota. *American Association of Petroleum Geologists Bulletin*, **40**, 600–653.

——1966. Bottom sediments of Lake Nicaragua and Lake Managua, Western Nicaragua. *Journal of Sedimentary Petrology*, **36**, 522–540.

TAPPAN, H. 1980. *The palaeobiology of plant protists.* W. H. Freeman & Co. San Francisco.

TAPPONIER, P., PELTZER, G., LE DAIN, A. Y., ARMIJO, R. & COBBOLD, P. 1982. Propagating extrusion tectonics in Asia: New insights from simple experiments with plastercine. *Geology*, **10**, 611–616.

TEN HAVEN, H. L. & SCHIEFELBEIN, C. 1995. The petroleum systems of Indonesia. *Proceedings of the Indonesian Petroleum Association 24th Annual Convention*, 443–459.

TILTMAN, C. J. 1987. Tertiary sedimentation and tectonics in the North Sumatra Basin, Indonesia. *British Sedimentological Research Group, Aberdeen, December 1987*, Abstracts 136.

TISSOT, B. P. & WELTE, P. H. 1984. *Petroleum formation and occurrence.* 2nd edition. Springer-Verlag.

TRAVERSE, A. 1955. Occurrence of the oil forming alga Botryococcus in lignites and other Tertiary sediments. *Micropalaeontology*, **1**, 343–350.

WAHAB, A. & MARTONO, D. 1985. Application of oil geochemistry for hydrocarbon exploration in Northwest Java. *Proceedings of the Indonesian Petroleum Association 14th Annual Convention*, 657–682.

WAKE, L. V. & HILLEN, L. W. 1981. Nature and hydrocarbon content of blooms of the alga Botryococcus braunii occurring in Australian freshwater lakes. *Australian Journal of Marine and Freshwater Research*, **32**, 353–367.

WAPLES, D. W. 1985. *Geochemistry in petroleum exploration*. International Human Resources Corp Reidel Publ.

WHITE, J. M. & WING, R. S. 1978. Structural development of the South China Sea with particular reference to Indonesia. *Proceedings of the Indonesian Petroleum Association 7th Annual Convention*, 159–177.

WILLIAMS, H. H. & EUBANK, R. T. 1995. Hydrocarbon habitat in the rift graben of the central Sumatra Basin, Indonesia. *In*: LAMBIASE, J. I. (ed.) *Hydrocarbon Habitat in Rift Basins*. Geological Society, London, Special Publication, **80**, 331–371.

——, KELLEY, P. A., JANKS, J. S. & CHRISTENSEN, R. M. 1985. The Palaeogene rift basin source rocks of Central Sumatra. *Proceedings of the Indonesian Petroleum Association 14th Annual Convention*, 57–90.

WOLFE, G. A., LAMB, N. A. & MAXWELL, J. R. 1986. The origin and fate of 4 – methyl steroid hydrocarbons I. 4 – methyl steranes. *Geochimica et Cosmochimica Acta*, **50**, 335–342.

WONGSOSANTIKO, A. & WIROJUDO, G. T. 1984. Tertiary tectonic evolution and related hydrocarbon potential in the Natuna area. *Proceedings of the Indonesian Petroleum Association 13th Annual Convention*, 161–183.

WOODSIDE, P. R. 1984. A look at the petroleum geology of Indonesia. *Oil and Gas journal* Feb. 1984, 78–82.

WOOLLANDS, M. A. & HAW, H. 1976. Tertiary Stratigraphy and Sedimentation in the Gulf Of Thailand. *SEAPEX Program Offshore S.E. Asia Conference*, Paper 7, 1–22.

Fault seal analysis of SE Asian basins with examples from West Java

ALASTAIR BEACH, J. LAWSON BROWN, PAUL J. BROCKBANK, STEVEN D. KNOTT, JEAN E. McCALLUM & ALASTAIR I. WELBON

Alastair Beach Associates Ltd, 11 Royal Exchange Square, Glasgow G1 3AJ, UK

Abstract: A regional fault seal analysis study of SE Asian basins was carried out in order to define trends for predicting sealing faults. Map and well data were analysed and relationships were obtained between fault and reservoir parameters and fault seal. The parameters found to be most useful for discriminating between sealing and non-sealing faults in the region were fault strike, fault mode (oblique slip faults are most likely to be sealing), fault throw (faults with high throws are most likely to be sealing), depth (shallow faults are most likely to be sealing) and thickness–throw. Other parameters measured, some of which are useful for fault seal prediction in some basins, are: fault type (bounding faults, within-field faults, across-field faults), net-to-gross ratio, average reservoir porosity, average reservoir permeability, differences in hydrocarbon contact elevation across faults, hydrocarbon column height held back by sealing faults, and reservoir pressure and temperature. Data from over 150 faults from fields within West Java are presented to illustrate the main results of the analysis.

Fault seal is just one of the critical risks that need to be assessed during exploration, appraisal, development and production of hydrocarbons from the prolific clastic plays of the SE Asian region (Villarroel 1985). A fault seal analysis study of SE Asian basins was initiated in order to understand the controls on fault seal in the explored part of the region, and to obtain insight into the processes responsible for fault seal that may also be applicable to the fault seal related traps identified in other, similar and underexplored parts. Areas analysed in the regional study included East Java, West Java, South, Central and North Sumatra, and the West Natuna and Malay basins. Results from West Java are presented here. Data were collated from IPA Atlases and other industry proprietary databases.

The main objectives of this study were to obtain systematic relationships between the key parameters related to fault seal in the area, and from these relationships make predictions regarding the likely controls on fault seal. In a semi-quantitative way, these relationships may be used in the frontier exploration acreage in SE Asia where fault seal is perceived to be a significant risk for trap integrity.

Geological setting

The basins of SE Asia lie within a zone of distributed deformation between the Sunda forearc and subduction zone to the south, and the Indo China continental block to the north (Davies 1984; Fig. 1). The area has been affected by collision tectonics from Cretaceous through to early Tertiary time with the development of accretionary complexes and back-arc igneous intrusives. This basement complex was stretched at various places and various times throughout early to mid-Tertiary time in response to back-arc extension (Davies 1984). Local basin inversions followed in mid- to late Tertiary time (Davies 1984). Some areas are still undergoing uplift and shortening, notably regions of Sumatra and Java adjacent to the Sunda arc. Other areas are currently undergoing subsidence in the back arc region, notably the West Natuna and Malay basins (Davies 1984). Reviews of the regional and sub-regional geology of each of the areas studied can be found in this volume (McCarthy & Elders, Matthews *et al.*, Wight *et al.*, Todd *et al.*, Sladen, this volume) and the reader is referred to these articles for further information.

Methodology

For each basin, maps of individual fields were analysed systematically, measuring key fault seal parameters such as: strike, mode, type, throw, depth, reservoir interval thickness, interval net-to-gross ratio, porosity, permeability, hydrocarbon contact differences across faults, hydrocarbon column heights held back by faults, and reservoir pressure and temperature.

In most cases the data were reasonably reliable, except for net-to-gross ratio data where reservoir

From Fraser, A. J., Matthews, S. J. & Murphy, R. W. (eds), 1997, *Petroleum Geology of Southeast Asia*, Geological Society Special Publication No. 126, pp. 185–194.

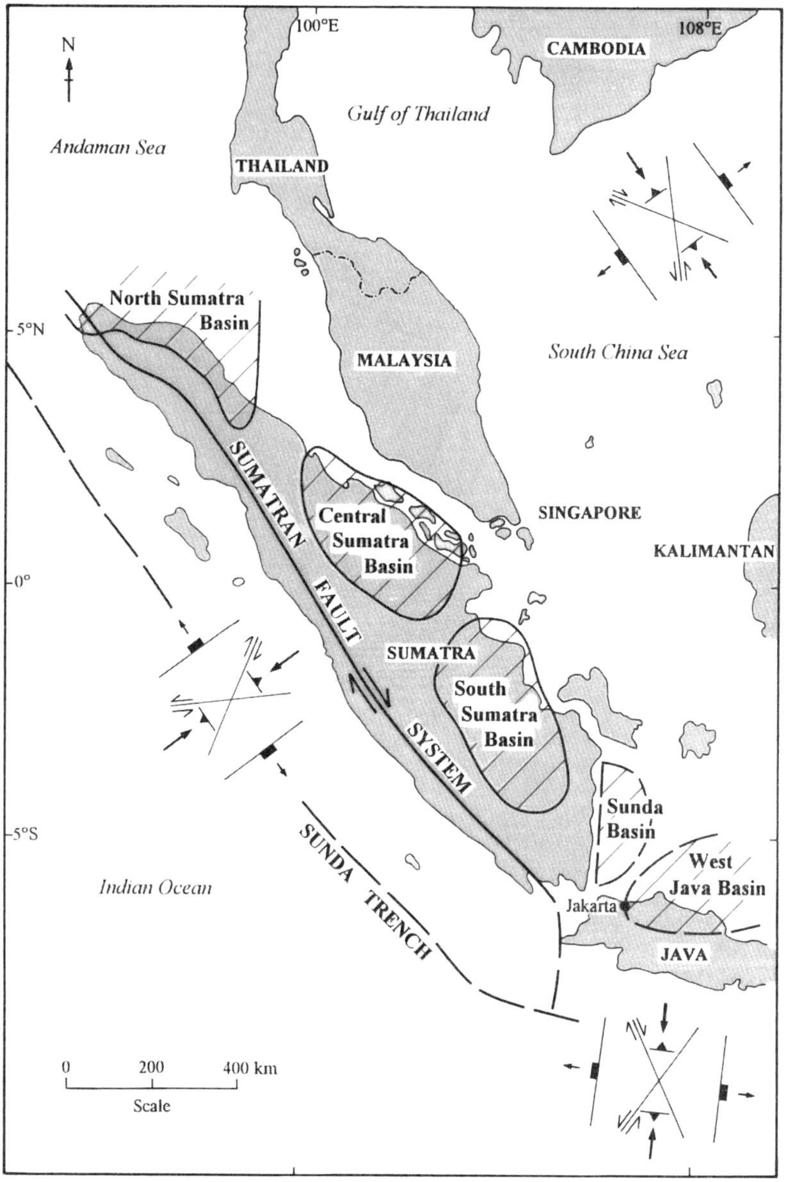

Fig. 1. Geological setting and fault kinematics (after Villarroel 1985).

stratigraphies were lacking in detail, and top and base of intervals were referred to hydrocarbon-bearing zones (i.e. 'pay') and not the entire section of interest. Where possible, stratigraphic templates were created from the limited well data available; however, these cases are in the minority. The results of the net-to-gross analysis were, accordingly, limited.

The classification of faults into sealing and non-sealing categories was based on various types of evidence, including differences in fluid contacts, phase differences across faults, cross sections, well penetrations, and map data. If there was any uncertainty in the evidence, a question mark was placed against the sealing or non sealing category. This led to four classes of fault in the database: sealing, ?sealing, non-sealing and ?non-sealing. Where it was not possible to make any comment on the sealing capability of a fault, a question mark was placed against that fault in the database without stating whether it was sealing or non-sealing.

Fault strike data were obtained from maps of fields and orientations were divided into 30° classes, from 0° to 180° east of north. Fault mode data were obtained from maps and were based on a kinematic model for the SE Asia basins (Fig. 1) (Davies 1984) using a pure shear strain ellipse. This ellipse was oriented separately for each individual basin to align correctly with the prevailing maximum horizontal compression direction. Faults were divided on the basis of this model into extension (e), left oblique slip extension (los/e), right oblique slip extension (ros/e), reverse (r), left oblique slip reverse (los/r), and right oblique slip reverse (ros/r) classes. Previous studies in the SE Asia region (Villarroel 1985) have considered oblique slip to be an important control on fault seal and hence the fault mode plots may allow the identification and quantification of the importance of this parameter. Fault type data were collated for maps and were divided into the following classes: 'across' – where faults completely cut across the field; 'bounding' – faults that form an essential trapping element to the closure; 'within' – faults that are contained within the field limits or cut across only part of the field, with hydrocarbons able to pass around the fault.

Fault throw data were collated from maps and cross sections. The maximum fault throw along each fault was plotted. Depth data were obtained from maps and cross sections. The maximum depth, at the deepest cut off against the point of maximum throw, was measured whether this was a normal or reverse fault. Reservoir thickness data are the most unreliable of the parameters measured for the SE Asia region. The original data sources tended to supply data for thickness of pay (i.e. hydrocarbon-bearing reservoirs), and top and base pay refer to the top of the closure and the deepest hydrocarbon contact respectively. The net pay thickness in the original data sources refers to the combined thickness of producing/hydrocarbon-bearing reservoirs. Because of the way the data were presented, the interval thickness used is not the true thickness of the stratigraphy under study, and the net thickness is not the total combined thickness of all sandstones. The data are therefore not in the most useful format because thicknesses are not true stratigraphic thicknesses. Net-to-gross ratio values will, therefore, be lower than the normal values used in fault seal studies (e.g. Knott 1993), which are based on the entire stratigraphy not simply the hydrocarbon-bearing zones. Porosity and permeability data were taken from the original datasets and averages for each field were calculated for the main reservoir sections.

Differences in the elevation of hydrocarbon contacts across faults and hydrocarbon column heights held back by sealing faults were taken from maps and cross sections. Each of these values were plotted against the maximum throw for the corresponding fault and separated into gas columns and oil columns. Pressure–depth and temperature–depth data were taken from the original source documents.

Results for the West Java study area

To illustrate some of the main trends and results of our fault seal analysis, data are presented from the West Java area which produced some of the clearest correlations and also some unexpected results. Each parameter will be treated either separately or in combinations, the latter based on experience from previous studies (Knott 1993) where plots of throw–thickness, for example, proved useful in discriminating between sealing and non-sealing faults.

Strike

Fault strike data (Fig. 2) display the main trends of faults within the West Java area, i.e. predominantly north–south and northeast–southwest. Sealing faults predominate in the class between 0° and 29° east of north. In the class between 30° and 59° east of north, the ratio of sealing to non sealing faults is close to one with slightly more sealing than non sealing faults. In the remaining classes there are insufficient data to make any firm conclusions except for the class between 150° and 180° east

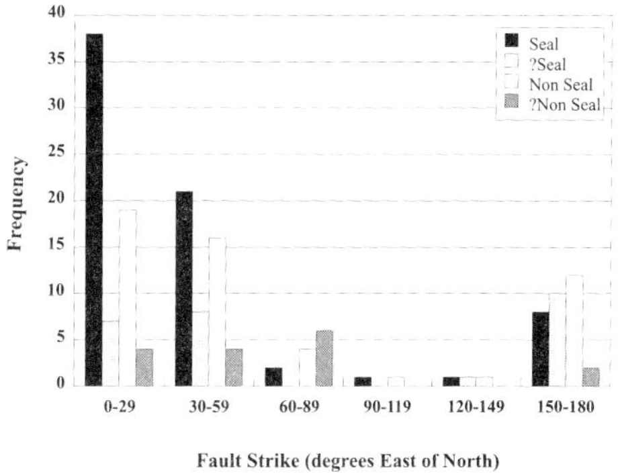

Fig. 2. Strike data.

of north where, again, the ratio of sealing to non-sealing is approximately equal to one, with a slight predominance of non-sealing faults.

The main conclusion from this dataset is that approximately north–south-trending faults are most likely to be sealing. This is a very useful first pass tool to differentiate sealing from non-sealing faults on the basis of orientation.

Fault mode

Fault mode data (Fig. 3) for the West Java area do not show major discrimination between sealing and non-sealing faults. Extensional faults of all modes are approximately evenly distributed between sealing and non-sealing faults. This is at odds with the analysis of fault mode from other SE Asian basins where mode does discriminate between sealing and non sealing faults.

Fault type

In the fault type dataset (Fig. 4), faults which cut across a field are generally found to be sealing. Most of the faults that are completely or partly contained within one side of the field, i.e. do not

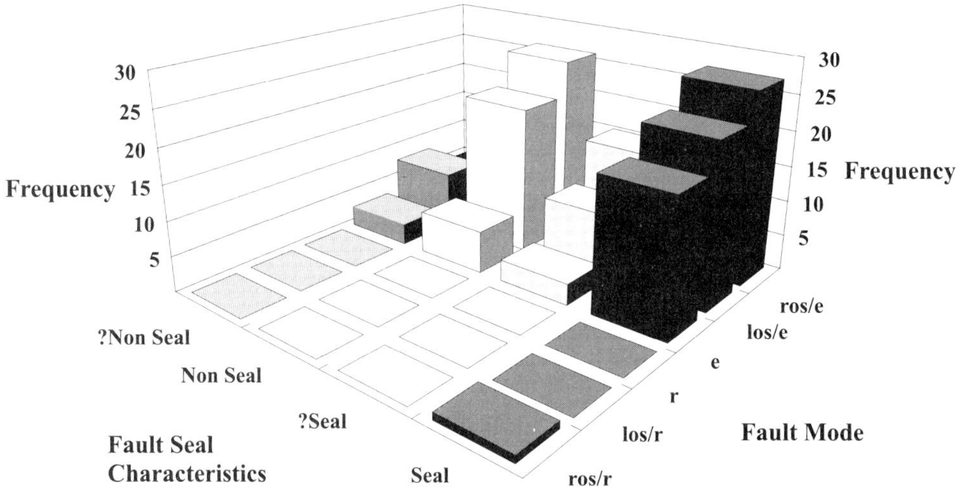

Fig. 3. Fault mode data.

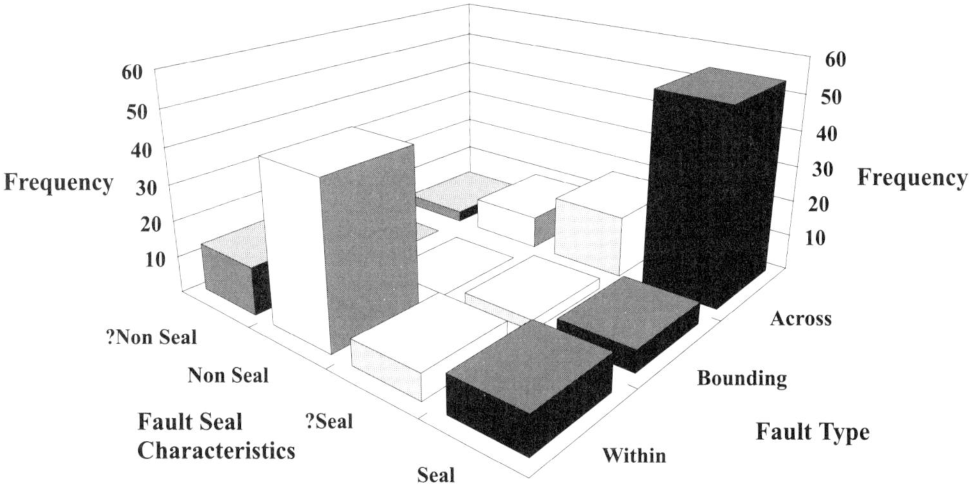

Fig. 4. Fault type data.

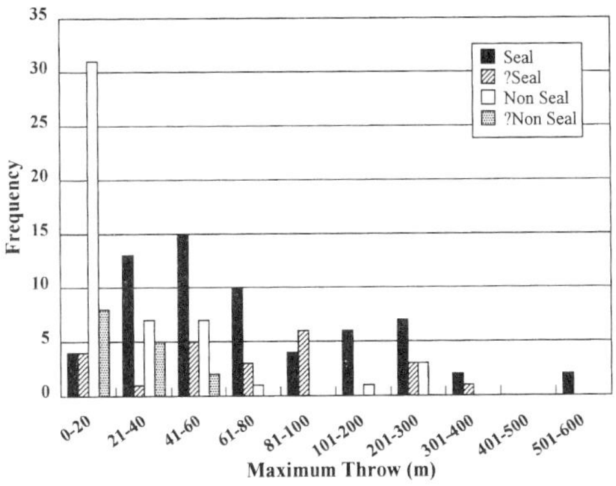

Fig. 5. Throw data.

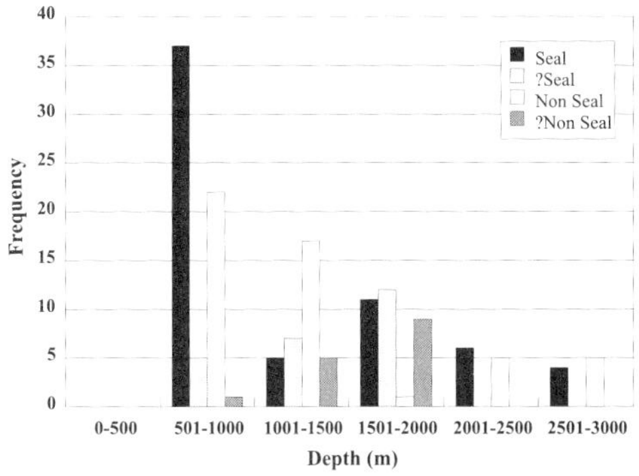

Fig. 6. Depth data.

cut right across the field, are non-sealing, in the sense that hydrocarbons were able to migrate around their tips. There are few bounding faults in the West Java dataset.

Throw

Fault throw data (Fig. 5) are a very clear discriminant of sealing and non sealing faults. Faults with throws less than 20 m are in the majority of cases non sealing. The likelihood of sealing appears to increase with fault throw. The ratio of sealing faults to non sealing faults increases with fault throw, although data values are few above throws of 100 m.

Depth

Sealing faults appear to be most common at shallow depths in the West Java dataset (Fig. 6). At depths shallower than 1000 m, sealing faults are by far the most common type. Between 1000 and 1500 m, the number of sealing faults decreases within a sample that is statistically significant. Explanations for this are outlined in the discussion.

Depth versus throw

A clear division is seen on the depth versus throw plot (Fig. 7), with non-sealing faults

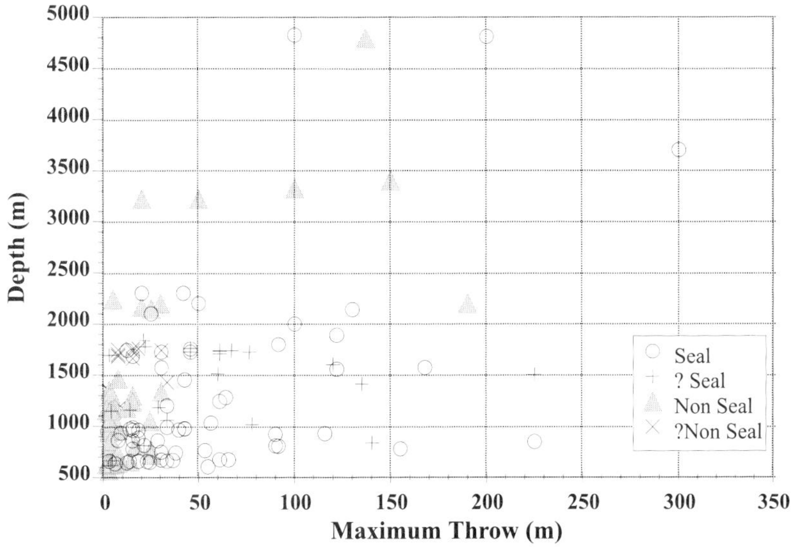

Fig. 7. Depth versus throw data.

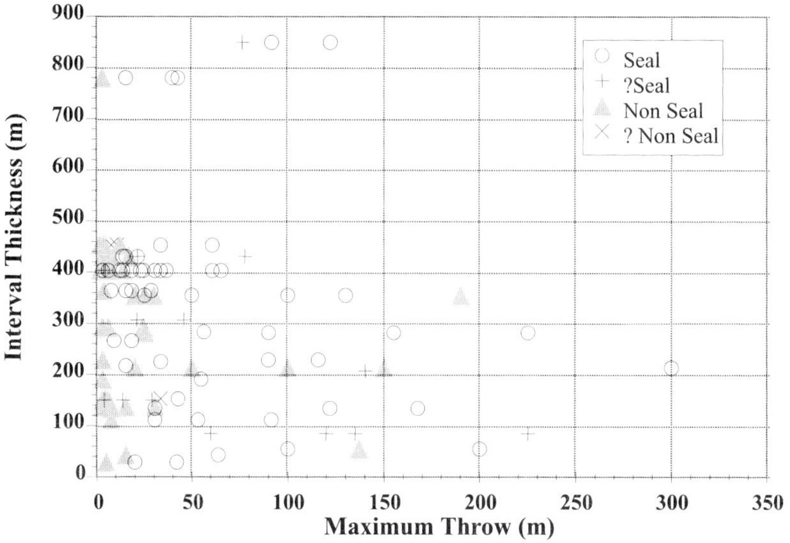

Fig. 8. Thickness versus throw data.

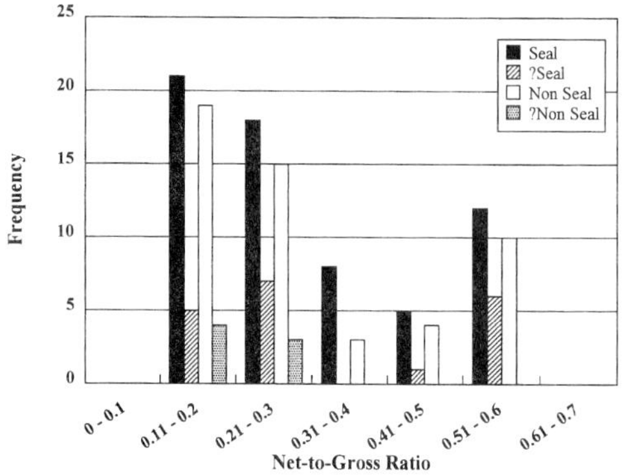

Fig. 9. Net-to-gross ratio data.

clustered mainly on the left hand side of the plot and sealing faults more broadly scattered across the right hand side. With increasing depth values the non-sealing faults appear to be more common at larger fault throws compared with shallower depths. The main control is probably throw in this case. A broad line can be drawn to divide the two main fault types and this plot could be used successfully to predict sealing faults within this basin.

Thickness versus throw

Separation of sealing and non-sealing faults is again evident on this plot (Fig. 8), which is dominantly controlled by the throw values. Even though there is some uncertainty as to the exact interval thickness from the data supplied (in most cases the net pay is given rather than the stratigraphic thickness), there is still a reasonable separation between the sealing and non-sealing types. A broad dividing line can be drawn between sealing and non-sealing faults, and this plot serves as a useful predictor for fault seal in this basin.

Net-to-gross ratio

Net-to-gross data (Fig. 9) have some uncertainty attached to them as mentioned previously. In this plot there is no discrimination at all between sealing and non-sealing faults. In most classes the ratio of sealing to non-sealing faults is approximately the same. No one class appears to have a predominance of sealing faults over any other class.

Porosity

In most porosity classes, the ratio of sealing faults to non sealing faults is similar (Fig. 10). However, above 25%, sealing faults predominate, particularly in the 26–30% class. This is statistically significant, but it is unclear why this should be. There may be a control on sealing by cataclasis in the higher porosity sandstones, which has been observed in outcrop and core from elsewhere (e.g. Antonellini & Aydin 1994).

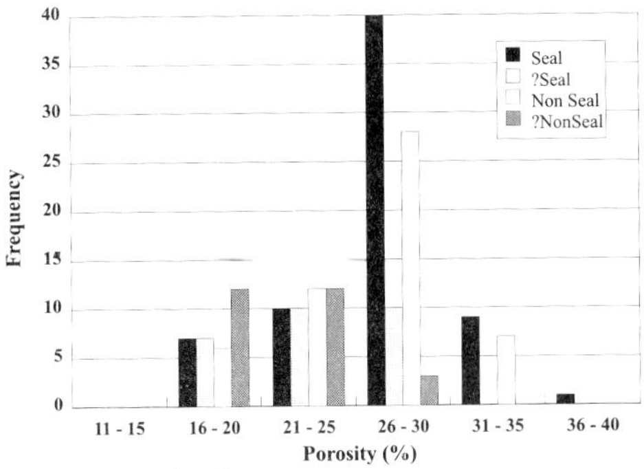

Fig. 10. Porosity data.

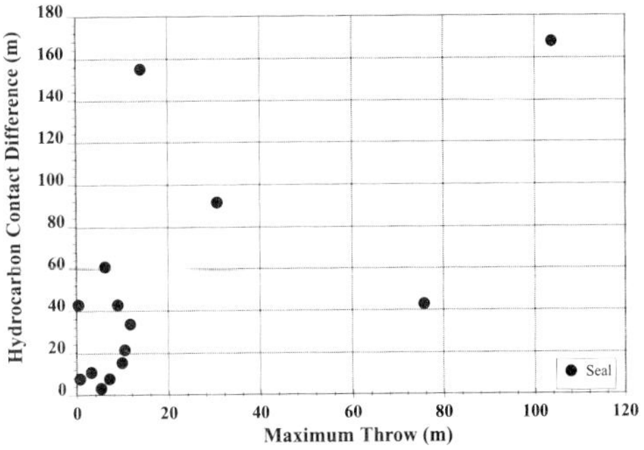

Fig. 12. Hydrocarbon contacts data.

Permeability

The range of permeability values collected for West Java was limited, and as Fig. 11 shows, it is difficult to define any discrimination on the basis of permeability. The cluster of data values between 0 and 1000 mD contains the largest number of sealing faults, but there are insufficient data values in the other classes to say whether or not this is due to the effect of lower permeabilities in the reservoir rock controlling seal capacity.

Hydrocarbon contacts

For the few data values available for differences in hydrocarbon contact elevation from the West Java area (Fig. 12), there is a broad scatter. Overall, however, there is a very general increase in the difference in contact elevation with increasing fault throw, although it is difficult to draw a trend confidently through these data.

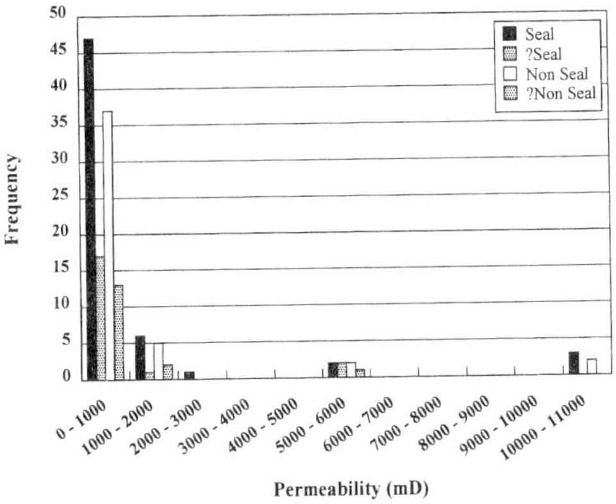

Fig. 11. Permeability data.

Hydrocarbon columns

Figure 13 shows that with increasing throw, there is a general increase in hydrocarbon column height, at least for oil. Above throws of about 200 m however, there is significant scatter in throw and column height values. There are insufficient data for gas columns.

Temperature versus depth

Sealing and non-sealing faults are evenly distributed along the general trend of increasing temperature with depth (Fig. 14) following the regional geothermal gradient. There appears to be no control on sealing properties based on reservoir temperature in the West Java area.

Trap type

Five main trap types have been designated for the fields in SE Asia: four-way dip closures (4wd); faulted four-way dip closures (f4wd); faulted three-way dip closures having one main

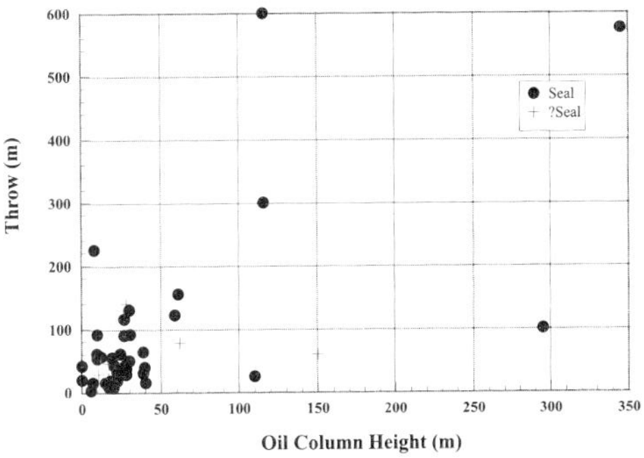

Fig. 13. Hydrocarbon columns data.

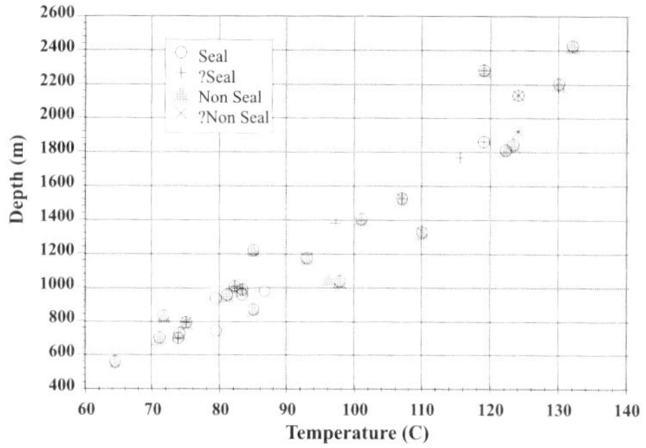

Fig. 14. Temperature versus depth data.

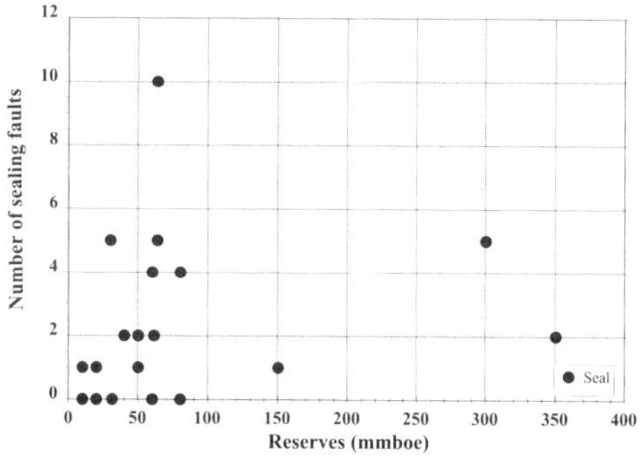

Fig. 17. Reserves and number of sealing faults.

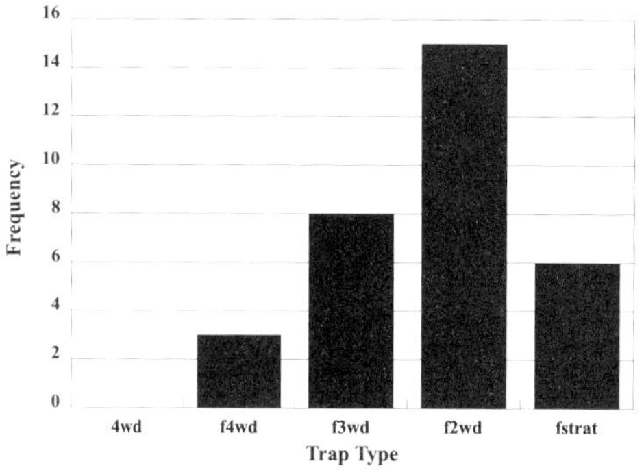

Fig. 15. Trap type data.

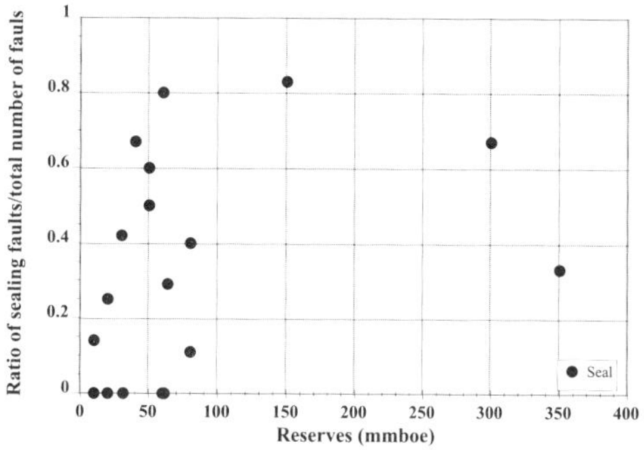

Fig. 18. Reserves and proportion of sealing faults.

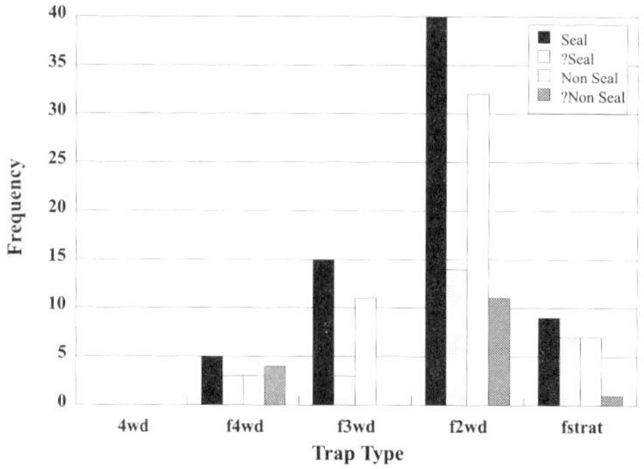

Fig. 16. Trap type and seal characteristics.

bounding fault (f3wd); faulted two-way dip closures having two main bounding faults (f2wd); combination faulted and stratigraphic traps (fstrat). For the West Java area, the dominant trap type is the faulted two-way dip closure (Fig. 15). The division of sealing and non-sealing faults among the trap types is fairly even (Fig. 16), no one trap type appearing to have a dominance of sealing faults over others.

Reserves

The number of sealing faults within a field plotted against field reserves (Fig. 17) shows only a very broad correlation for West Java, with two loosely defined populations. There is an increase, in general, in field reserves with number of sealing faults. This increase is steeper for fields that plot to the left hand side of the figure than those on the right hand. This relationship is also seen in the plot of reserves and the ratio of sealing faults to the total number of faults in a field (Fig. 18). The correspondence between the two plots suggests that the correlation between reserves and sealing faults is not simply because bigger field have more faults but, significantly, reserves appear to be related to the proportion of sealing faults in a field.

Discussion and conclusions

Faults analysed from the West Java dataset show a number of correlations between fault seal and the parameters measured. Some parameters, however, showed no correlation with fault seal. Fault strike is an important discriminant of fault seal, as it is in most SE Asian basins. North–south faults are most likely to be sealing. A related parameter, fault mode, is also a useful discriminant for fault seal. This is probably the result of greater net slip on oblique slip faults than dip slip faults for the same fault throw. Increased net slip across a fault leads to increased fault zone thickening and therefore a better development of fault gouge membrane seals (Villarroel 1985; Watts 1987; Antonellini & Aydin 1994; Knott 1994). In West Java itself, fault mode does not clearly discriminate between sealing and non-sealing faults. This is perhaps not surprising as oblique slip will not be strongly favoured in the West Java region as the direction of subduction and back arc extension is oriented sub-perpendicular to the structural grain. In cases such as Sumatra the marked obliquity of the subduction vector to the trend of the faults is more likely to produce a significant component of oblique slip on faults.

Faults were classified into three main types related to whether they cut across a field, bounded a field, or lay within a field closure. Most across-field faults were sealing for all the areas studied, including West Java. Although there are few bounding faults in the West Java dataset, most field bounding faults were sealing whether they were normal faults or reverse faults (it was not possible, with the available data, to separate out reverse reactivated faults from non-reactivated faults). Within-field faults were, in the main, non-sealing.

Fault throw proved to be a very important discriminant for separating sealing and non sealing faults. Faults with throws as small as 20 m were found to be sealing in most basins, and there is an increase in the likelihood of fault seal with increasing fault throw. This is probably a result of increased fault zone thickening with increasing fault throw as mentioned previously.

Depth data were also useful in separating sealing from non-sealing faults. In contrast to the North Sea, where the likelihood of fault seal increases with depth, in West Java sealing faults occurred mainly at shallow depths. This perhaps counter-intuitive result may be due to the existence of low capillary pressures at shallow depths enabling fault seal to occur when the same fault, at greater depths and pressures, would be unable to hold back a significant hydrocarbon column (e.g. Berg 1975; Watts 1987). Another possibility is that minor reverse reactivation of normal faults may have breached the membrane seals within fault zones. This process of fault reactivation is more likely to occur at deeper crustal levels and therefore leaky faults are more likely at greater depths (Sibson *et al.* 1975).

Combination plots of depth versus throw and stratigraphic thickness versus throw proved to be useful discriminants of sealing and non-sealing faults. Sealing faults lie to the right hand side of these plots and non sealing faults lie to the left hand side. At larger throws and greater stratigraphic thicknesses, the trend of the line of separation between sealing and non-sealing faults changes, probably due to variations in stratigraphic content in the upper part of the sections. Usually, sandier sections in the upper parts of the stratigraphies analysed lead to an increase in the number of non-sealing faults with increasing fault throw. This is probably due to breaching of top seal successions across faults and juxtaposition of the upper sandy section with deeper reservoirs leading to leakage across faults.

Net-to-gross ratio data are not available in the format that is most useful for fault seal analysis – only data for hydrocarbon-bearing sections were available. Porosity data did not span a large range of values in most areas, but where they did, it appeared that sealing faults were more common in the higher porosity reservoirs. This may be due to the formation of granulation seams in the high porosity sections, a common feature described from core and outcrop (Antonellini & Aydin 1994; Knott 1994). These features can hold back significant (50–100 m) hydrocarbon columns (Antonellini & Aydin 1994). Permeability data were similarly restricted in their range of values, and were undiscriminatory for sealing faults.

Differences in the elevation of hydrocarbon contacts were observed to increase broadly with increasing fault throw for both oil and gas. This may be due to increases in seal capacity of faults with increasing fault throw, or may be related to offset of fluid contacts by late faulting post-hydrocarbon emplacement. No clear correlation was found between hydrocarbon column heights and fault throw.

The most common trap types in the region were faulted two- and three-way dip closures. Both these trap types contained a large number of sealing faults. An explanation for this may be that increased reserves are associated with traps that have hydrocarbon columns held back by sealing faults that form an essential part of the trap closure. Whereas, in contrast, four-way dip closures have columns controlled by structural closure and spill. If correct, this interpretation indicates that fault-seal-related traps are more likely to contain greater volumes of hydrocarbons than four-way dip closures of a similar areal extent because of the faults' ability to hold back taller hydrocarbon columns than the structural spill point would indicate.

From these observations some straightforward rules of thumb can be obtained. In West Java, faults are most likely to be sealing if their throws are greater than 20 m, if they occur at depths shallower than 1000 m and if they occur in high porosity sandstones. Non-sealing faults are most commonly normal faults with throws less than 20 m that occur at depths greater than 1000 m.

Fault reactivation was difficult to assess from the data provided in this study. But the observation, as mentioned above, that non-sealing faults increase in number with increasing depth may be related to fault reactivation occurring at depth. Therefore, shallow structural traps are more likely to be sealing, and deeper structural traps have a greater likelihood of leaking.

Breaching and leakage are also likely to occur where the upper part of the stratigraphy is sand prone. Juxtaposition and shale smear potential should be carefully evaluated for main bounding structures.

Trap type classification for the basins studied revealed that the most common type of traps are those that rely on at least one fault for closure. In West Java the most common trap type was the faulted two-way dip closure – this trap type also contained the largest number of sealing faults in this area. In Sumatra the most common trap type was the faulted three-way dip closure. Most sealing faults in Sumatra occurred in both faulted four-way and three-way dip closures.

Plots of reserves against number and relative proportion of sealing faults in a field revealed a very broad correlation. A general increase in field reserves with number and proportion of sealing faults is observable. The implication of these plots is that reserves are related not simply to the number of faults in a field, but to the relative proportion of sealing faults. In small to moderate sized fields, this latter result may reflect the upside potential due to field extensions related to fault seal. This situation is now beginning to be exploited in other areas including the North Sea.

Thanks are due to A. Fraser, S. Matthews and S. Todd of BP. Thanks also to BP for permission to publish the data. M. Smith and B. Bruce kindly prepared the figures.

References

ANTONELLINI, M. & AYDIN, A. 1994. Effect of faulting on fluid flow in porous sandstones: petrophysical properties. *AAPG Bulletin*, **78**, 355–377.

BERG, R. R. 1975. Capillary pressure in stratigraphic traps. *AAPG Bulletin*, **59**, 939–956.

DAVIES, P. R. 1984. Tertiary structural evolution and related hydrocarbon occurrences, North Sumatra basin. *Proceedings of the Indonesian Petroleum Association, 13th Annual Convention*, 105–116.

KNOTT, S. D. 1993. Fault seal analysis in the North Sea. *AAPG Bulletin*, **5**, 778–792.

——1994. Fault zone thickness versus displacement: results from Permo-Triassic sandstone outcrops in NW England. *Journal of the Geological Society, London*, **151**, 17–25.

MATTHEWS, S. J., FRASER, A. J., LOWE, S. TODD, S. P. & PEEL, F. J. 1997. Structure, stratigraphy and petroleum geology of the SE Nam Con Son Basin, offshore Vietnam. *This volume*.

MCCARTHY, A. J. & ELDERS, C. F. 1997. Cenozoic deformation in Sumatra: oblique subduction and the development of Sumatran Fault System. *This volume*.

SIBSON, R. H., MOORE, J. MCM. & RANKIN, A. H. 1975. Seismic pumping – a hydrothermal fluid transport mechanism. *Journal of the Geological Society, London*, **131**, 653–659.

SLADEN, C. 1997. Exploring the lake basins of east and southeast Asia. *This volume*.

TODD , S. P., DUNN, M. E. & BARWISE, A. J. G. 1997. Characterizing petroleum charge systems in the Tertiary of SE Asia. *This volume*.

VILLARROEL, T. 1985. Observations on the sealing properties of faults in Sumatra. *Proceedings of the Indonesian Petroleum Association, 14th Annual Convention*, 105–116.

WATTS, N. L. 1987. Theoretical aspects of cap-rock and fault seals for single and two-phase hydrocarbon columns. *Marine and Petroleum Geology*, **4**, 274–307.

WIGHT, A., FRIESTAD, H., ANDERSON, I., WICAKSONO, P. & REMINTON, C. H. 1997. Exploration history of the offshore Southeast Sumatra Production Sharing Contract, Java Sea, Indonesia. *This volume*.

Permeability heterogeneity within the Jerudong Formation: an outcrop analogue for subsurface Miocene reservoirs in Brunei

D. J. PROSSER[1] & R. R. CARTER[2]

[1] *Z&S (Asia) Ltd, First Floor, 46 Ord Street, West Perth, Perth 6005, WA, Australia*
[2] *Brunei Shell Petroleum Co. Snd. Bhd. Seria 7082, Negara Brunei Darussalam*

Abstract: The Jerudong Formation (early Tortonian) cropping out at Punyit Beach (Negara Brunei Darussalam) comprises a heterolithic succession of deltaic strata. Probe-permeameter data have been used to evaluate the spatial nature of permeability heterogeneity within shoreface, lagoon and distributary channel environments. Strata at outcrop comprise the same lithologies as those at depth in the abandoned Jerudong Field, and provide an analogue for parts of the Jerudong Formation and equivalents in other Bruneian oilfields.

Identified permeability heterogeneities are associated with contrasts of up to three orders of magnitude, occurring at facies, genetic facies unit, bed-set, bed and laminar scales. Distributary-channel facies display the best outcrop permeability (>7000 mD). They include trough cross-stratified fills of relatively straight (often tidal) distributaries, and laterally accreted point bar sandstones deposited within meandering channels. These facies display upward decreasing permeability trends. Shoreface facies may also display high (>6000 mD) permeability, but are characterized by upward increasing permeability and are significantly more heterogeneous than channel facies. Lagoon fill successions are highly heterolithic, comprising sandstones, siltstones and mudstones with a wide range of permeabilities.

5 cm spaced outcrop permeability data are more variable, and display 'average' permeabilities several orders of magnitude larger than occur within similar facies encountered in the subsurface.

The spatial scale, frequency and magnitude of permeability variation within the Jerudong Formation indicate that facies packages observed at outcrop are of sufficient scale (4–10 m thickness) to provide potentially useful scaled elements for flow simulation.

A model using 0.25 m scale grid blocks has allowed assessment of intra-facies permeability variation (associated with bedding/cross-bed sets) upon recovery within different channel fills. The model predicts both trough cross-stratified and laterally accreted channel fills are heterogeneous with respect to waterflood, and preferential imbibition into their lower parts. Trough cross-stratified channels flood more rapidly than laterally accreted channel fills, where more fingering of the flood front occurs. Lagoon fill deposits display variable waterflood characteristics dependent upon sand content. Within both simulations, heterolithic lagoon fill facies appear to effectively compartmentalise modelled reservoir successions with respect to vertical cross flow.

The objectives of this study were to use a portable nitrogen probe-permeameter to obtain an outcrop permeability data-base that would allow investigation of the spatial nature of permeability heterogeneity within a section of the Jerudong Formation cropping out at Punyit Beach near Jerudong on the coast of Negara Brunei Darussalam. Experience gained from observing the detailed geology and production characteristics of many of the reservoirs in Brunei suggests that permeability heterogeneity is an important parameter for reservoir simulation studies in this setting. Simulation work is currently in progress on a number of field projects, and the work presented in this paper has formed the basis of an improved understanding of the controls upon, and magnitude of, permeability heterogeneity at a variety of scales pertinent to the reservoir engineer. The outcrop data at Jerudong form an excellent analogue for some of the subsurface reservoirs in Bruinei. For example, depositional facies observed at the studied outcrop are similar to those from Champion field well CP-194 (James 1984, fig. 17). Outcrop permeability data from the Jerudong Formation are compared with subsurface permeability data from offshore Bruneian oilfields.

Techniques

Outcrop permeability data were gathered with a portable nitrogen probe-permeameter. Permeameter gas flow rates measured upon core-plug samples of known permeability were used to construct calibration curves for conversion of permeameter gas flow rates to permeability.

From Fraser, A. J., Matthews, S. J. & Murphy, R. W. (eds), 1997, *Petroleum Geology of Southeast Asia*, Geological Society Special Publication No. 126, pp. 195–235.

Comparison of probe-permeameter flow rates with Hassler Sleeve permeabilities was carried out using a similar procedure to that outlined by Sutherland *et al.* (1993).

Prior to analysis, the outcrop surface was cleaned by careful scraping with a strong knife blade, to ensure removal of any surface weathering rind/iron oxide coatings. The typically friable nature of the sediments exposed at surface outcrop allowed easy preparation of the sample surface, which could be excavated to several cm depth where necessary, e.g. where weathering had lead to the development of a low permeability surface crust upon the rock. Probe permeameter data were gathered using a probe tip of 0.4 cm internal diameter, which corresponds to the estimated radius of investigation of the tool. The outcrop surfaces were remarkably clean and free from the thick crusts normally associated with lateritic weathering in a tropical climate, a factor probably largely attributable to their coastal location, and high rate of erosion within the cliffs in which they are exposed. The studied outcrops were located away from intertidal or spray zones, where repeated salt precipitation/dissolution on and within the strata is likely to result in their yielding spurious permeability data.

Sedimentology

The Early Tortonian Jerudong Formation at Punyit Beach is exposed within the steeply dipping eastern limb of the north–south-trending, thrust-faulted, Jerudong Anticline. Hydrocarbon seeps occur in the area, which has been a site of hydrocarbon exploration since the early 1900s. The Punyit Beach exposures comprise the same lithofacies as those encountered at depth within the abandoned Jerudong oilfield located some 5 km inland, and provide a potential analogue for parts of the Jerudong Formation and its equivalents in other Bruneian oilfields. Hydrocarbon wells were first drilled upon the Jerudong Anticline in the early part of the twentieth century, but the Jerudong Field was not discovered until 1955, and was exploited commercially by BSP until 1962.

A framework for the description of the Tertiary strata of NW Borneo was initially provided by the work of Leichti *et al.* (1960) and Wilford (1961), who divided the sediments into a number of very thick (kilometre scale) formations. Subsequently, the concept of 'sedimentary cycle recognition' and its application to fine scale subdivision of Tertiary strata (e.g. Ho 1978) greatly aided correlation at scales more

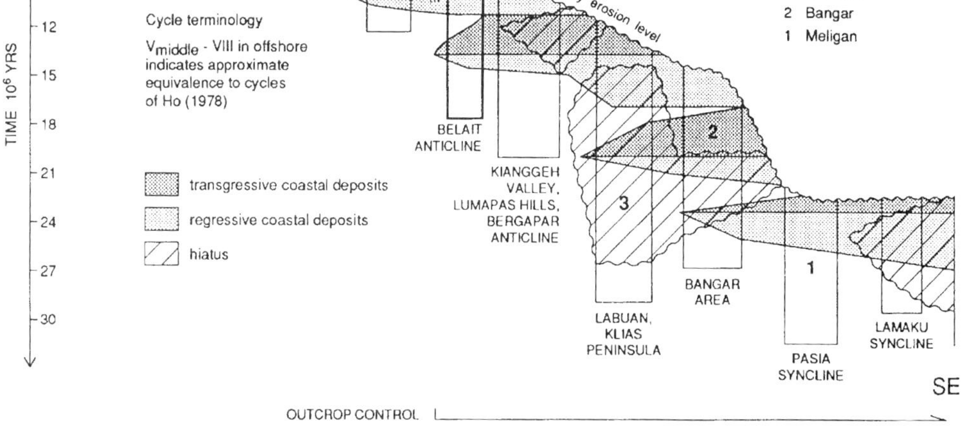

Fig. 1. Cycle stratigraphic terminology, onshore Brunei. Note the Belait cycle is not the equivalent of the Belait Formation. (From Potter *et al.* 1984).

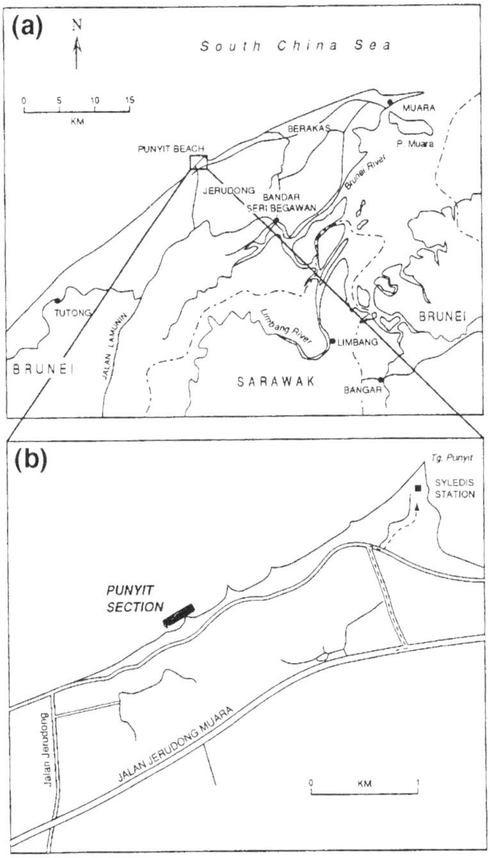

Fig. 2. Location map for outcrops at Jerudong.

Fig. 3. The southernmost outcrop at Punyit Beach, Jerudong.

suitable for reservoir studies (Potter *et al.* 1984). The successions of stacked sedimentary 'cycles' into which larger formations were divided (Fig. 1) are analogous to the para-sequence units now commonly used during sequence stratigraphic analysis of marine/paralic successions (Van Wagoner *et al.* 1990), and were constrained by micropalaeontological data (e.g. Eckert 1970). Ideally cycles comprise a basal succession of shale deposited during marine transgression from the north, and are overlain by a sandier succession of northerly prograding regressive shoreface/deltaic strata. Within this framework of cycles, the Jerudong Formation cropping out at Punyit Beach comprises part of the Middle Miocene Cycle V^L (see Fig. 6), as described by Potter *et al.* (1984). For a detailed discussion of the stratigraphy and exploration history of Brunei the reader is directed to James (1984).

The section of Jerudong Formation analysed forms the southernmost cliff exposures at Punyit Beach (Figs 2 and 3), and represent the stratigraphically oldest sediments exposed along the coast in the eastern limb of the Jerudong anticline. The studied section comprises some 56 m of strata and is overlain by a thick succession of predominantly fine-grained lithologies (mudstones, siltstones and fine sandstones), with intercalations of fine–medium-grained sandstone up to 10 m thick. The succession exposed along the coast at Jerudong has been described briefly by Potter *et al.* (1984).

A sedimentological log through the strata used for permeability data gathering is illustrated in Fig. 4. The strata are divided into eight 'units' defined by dominant lithofacies type present (Table 1). These 'units' simply comprise large scale groupings of similar facies types, and may themselves contain internal bounding surfaces of significance in terms of sequence analysis. This simple sub-division of strata has been used because of a known relationship between lithofacies and flow units during production. The studied lithofacies associations include distributary channel sandstones, shoreface sandstones, and heterolithic lagoon fills comprising sandstone, siltstone and mudstone.

Shoreface facies

Shoreface facies comprise heterolithic successions of mudstones, siltstones and fine–medium-grained sandstones (Fig. 5a) arranged in broadly coarsening and thickening upward sedimentary 'cycles'. In an ideal cycle, the lower parts of shoreface facies successions comprise fine-grained lithologies, e.g. black-grey mudstones with ripple laminated silt and fine sandstone lenses which pass upward into parallel and current ripple laminated siltstones and very fine sandstones with abundant plant debris. The upper parts of shoreface successions are characterized by more thickly

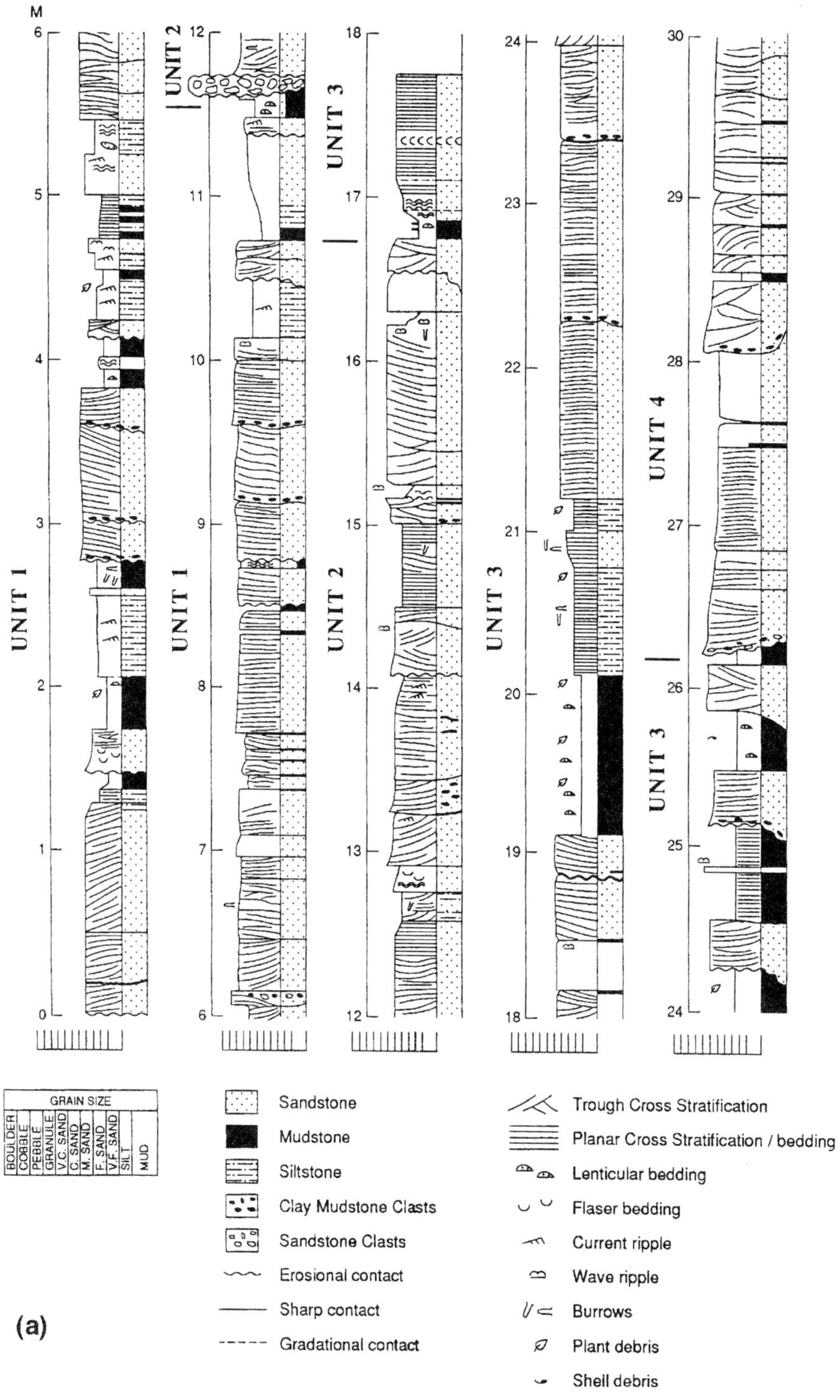

Fig. 4. Sedimentological logs for studied section.

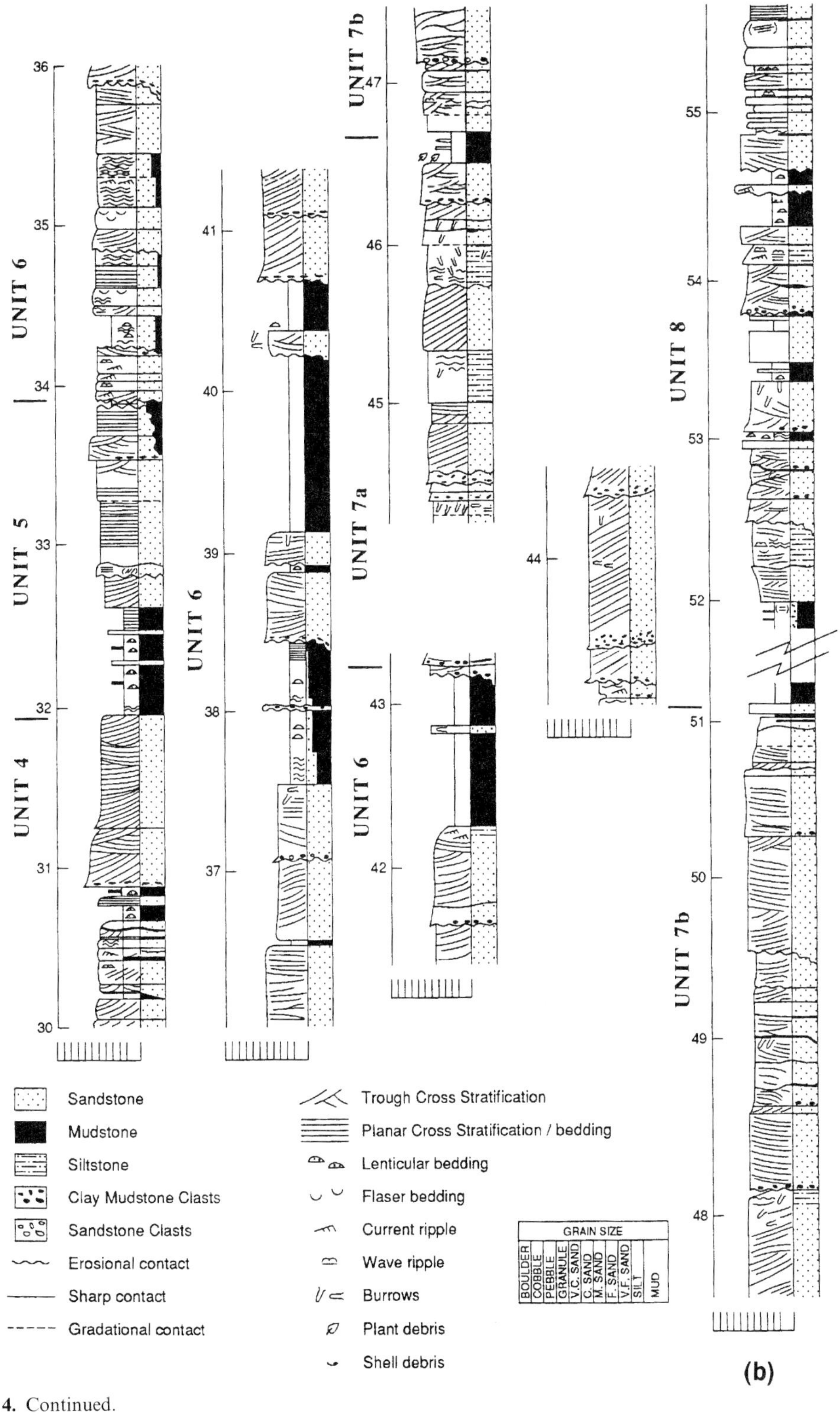

Fig. 4. Continued.

bedded (up to 40 cm), fine–medium-grained argillaceous sandstones. The sandstones often comprise tabular–lenticular (20–40 cm thick) bed-sets containing planar, low angle trough, or 'hummocky' cross-stratification fabric. Hummocky cross-stratification fabrics (HCS) are defined by gently curved, concave–convex-upward erosion surfaces; (Harms *et al.* 1975), and may form in shallow intertidal zones within metres of a shoreline, or to depths of up to 200 m on the outer shelf (Cheel & Leckie 1993).

Table 1 *Lithofacies units identified within the Jerudong Formation studied at Punyit Beach*

Height in section	Unit	Facies
0.0–11.75	Unit 1	Shoreface/lagoon
11.75–16.60	Unit 2	Tidal channel
16.60–26.2	Unit 3	Shoreface/lagoon
26.2–31.95	Unit 4	Distributary channel
31.95–33.85	Unit 5	Shoreface
33.85–43.00	Unit 6	Lagoon Fill
43.00–46.90	*Unit 7a*	*Distributary channel*
43.00–51.60	Unit 7	Distributary channel
46.90–51.60	*Unit 7b*	*Distributary channel*
51.60–51.85	Unit 8	Shoreface

Sandstone beds contain millimetre–centimetre-scale low-angle internal cross-lamination, fine upwards, and may contain both horizontal and inclined burrows, of which *Ophiomorpha* is common. Clay rip-up clasts overly scour surfaces at the base of individual sandstone bed-sets, or low angle internal scour surfaces within bed-sets in the upper sandstone dominated portion of the shoreface.

An upward increase in sand content, sandstone bed thickness, and inclination of low-angle trough cross-stratification fabrics within tabular–lenticular bedsets is interpreted as indicative of

Fig. 5. (**a**) Photograph illustrating the heterolithic nature of shoreface facies sediments (Unit 8 of studied section). (**b**) Photograph illustrating the heterolithic nature of sediments comprising Unit 6 within the studied section. Sandstones display marked lateral thickness variations and erosive bases which down-cut into underlying mudstones. The base of Unit 7 and tops of Unit 5 Unit 4 and Unit 3 are indicated by numbering. The succession is interpreted as the deposits of a coastal lagoon cut by minor distributary channels. (**c**) Small-scale, low-angle trough cross stratification sets within tidal distributary channel sandstones. Note, the high frequency of opposing foreset dip directions giving rise to 'herring-bone style' cross stratification. Foreset relief is due to weathering along concentrations of coaly or argillaceous matter. (**d**) Photograph illustrating large scale, low angle, laterally accreted bedforms within Unit 7b of the studied section. The tops of units 7a and 7b are also indicated. Numbers refer to height in metres above base of logged section illustrated in Fig. 4.

an upward transition from lower to middle/upper shoreface environments. Stacked coarsening upward shoreface deposits are often capped by fine grained siltstones and mudstones (which are in turn overlain by channelised sandstones) reflecting an upward passage from shoreface to lagoonal (and finally distributary channel) environments.

Channelized sandstone facies

Trough cross-stratified distributary channel fill. Trough cross-stratified distributary channel sandstones are medium–fine grained, contain millimetre–centimetre-scale scale internal lamination, and occur as beds 10–80 cm thick, which are arranged in erosively based fining-upward units of up to 2 m thickness (Fig. 5b). Fining-upward units are often stacked to form multistory channel fills. Within multistory channel fills, the bases of both individual fining upward sediment packages, and erosive trough cross-set boundaries defining co-sets within fining-upward units are typically overlain by gravel–pebble-grade 'lags'. Pebble lags are comprised of dark grey mudstone clasts, or less commonly fine-grained sandstone intraclasts. Trough cross-stratification sets are typically 1–2 m thick, and 3–10 m wide near the base of distributary channels, and bedforms decrease in scale upwards. The sandstones contain sparse vertical and horizontal *Ophiomorpha*, which are most common within slightly finer-grained, more argillaceous lithologies found towards the top of individual fining-upward units. Distributary channel deposits have markedly erosive bases which typically down-cut into underlying argillaceous lagoon fill sediments.

The erosive based, trough cross-stratified, fining-upward sandstone successions described are interpreted as representing the deposits of relatively straight distributary channels. The common occurrence of pelleted *Ophiomorpha* burrows (*Skolithos* ichnofacies association) within trough cross-stratified distributary channel sandstones generally supports deposition within a marginal marine setting.

Tidal distributary channel fill. Tidal distributary channel facies comprise fine–medium-grained, trough cross-stratified sandstones arranged in stacked, broadly fining upward units up to 1.5 m in thickness (Fig. 5c). Tidal distributary channel fill successions typically display an erosive base which in some cases is overlain by a pebble- grade conglomeratic lag. More rarely the lag comprises a pebble–cobble-grade breccia containing lithified sandstone clasts of similar character to the sandstone lithologies found throughout the studied succession. Erosional trough cross-set boundaries occur at vertical spacings of 15–70 cm within fining-upward sandstone units, and set size decreases upward within tidal distributary channel fills. Intraformational coal and clay clasts are also common near the bases of individual fining-upward units. The stratigraphically higher beds within these fining-upward units comprise very low angle ripple (wave and current) or planar laminated (millimetre–centimetre-scale) fine-grained sandstones, often penetrated by pelleted vertical burrows (*Ophiomorpha*) several centimetres long. Foreset lamination (centimetre scale) within trough cross-stratification sets are draped by thin concentrations of coaly/argillaceous matter, and may display opposing current directions, giving rise to 'herring-bone' style cross-stratification. The uppermost beds within tidal channel fill successions display wave rippled surfaces, penetrated by paired, sand-filled, U-shaped burrows (*Diplocraterion*). Wave ripple crests seen in successive laminations on these uppermost bedding plane surfaces often trend at angles which differ by up to 90° from one another, and complex patterns of interference ripples are also preserved. The assemblage of sedimentary structures, including clay/organic draped foresets (indicative of fines deposition by settling from suspension during periods of slack water), wave ripples, foreset bedding indicating deposition from flows of opposing direction, and burrows characteristic of the skolithos ichnofacies, all support a shallow water inter-tidal–subtidal depositional environment.

Laterally accreted distributary channel fill. In contrast with the grey argillaceous sandstones typical of trough cross-stratified distributary channel fills, one distributary channel fill within the studied section (Unit 7b, Fig. 4) comprises orange-white, medium grained, clean, well sorted, cross-stratified sandstones (Fig. 5d). The basal sandstones within this interval comprise some 50 cm of grey-white, fine–medium- grained sandstones, containing black-clay-draped foresets, clay drapes over current ripple forms, discontinuous coaly lamination, and scattered pebble-grade coal clasts. The basal sandstones pass upward into a succession of well-sorted orange-white erosively based sandstones, in which large scale, low angle (5–15°) wedge or sigmoidal shaped cross-stratification sets are evident (Fig. 5d). In places, low-angle cross-stratification sets are cross-cut by erosional scours up to 50 cm deep. Individual sandstone beds are 10–50 cm thick, and arranged in broadly fining-upward bed-sets up to 1.5 m in thickness.

The bases of individual fining-upward units are erosive, and overlain by thin lags of scattered (centimetre scale) black mudstone pebbles. Internally, individual sandstone beds display millimetre–centimetre-scale sub-parallel lamination, often in association with low-angle trough cross-stratification fabrics which indicate flow sub-parallel to the strike of the inclined bedforms. *Ophiomorpha* burrows are common within slightly finer-grained lithologies towards the top of fining-upward bed-sets, or towards the tops of individual beds. Current ripples often characterize the uppermost few cm of individual beds, and thin black clay partings (typically <2 cm thick) are present along many bed boundaries.

The predominance of large-scale, low-angle inclined, wedge-shaped cross-stratification sets within this facies is indicative of deposition upon a point bar surface within a meandering distributary channel. Friable grey sandstones with abundant clay drapes, coaly laminae and coal clasts near the base of this facies are interpreted as 'bottom-set beds' within the base of the channel, over which the point-bar migrated. The clean well sorted quartzitic nature of sandstones may indicate that point bars were deposited within 'proximal' distributary channels which perhaps transected and reworked more texturally and compositionally mature sediments within beach–foreshore environments.

Heterolithic lagoon fill

Lagoon-fill facies comprise heterolithic successions of interbedded mudstone, siltstone and sandstone (Fig. 5b). Dark grey-black mudstones rich in plant debris predominate within lagoon fill successions, and may contain thin (millimetre–centimetre scale) sandstone lenses, often with wave or current ripple laminations. Rare, small (<1 cm) thin-shelled bivalves occur within some darker (black) mudstones. In places mudstones are highly fractured, and some are heavily oil stained or display marked yellow sulphur staining.

Interbedded with the mudstone lithologies are very fine–medium-grained sandstones. The sandstones occur within erosively based, trough cross-stratified, fining-upward sediment packages typically 10–50 cm thick, but which may reach thicknesses of 1.5 m.

The thicker sandstone intercalations within lagoon fill successions (sandstone beds up to 1.5 m thick) have sharply erosive bases, and strongly erosive internal trough cross-set boundaries which are typically overlain by a gravel–pebble-grade 'lag' of dark grey mudstone clasts. The uppermost beds within fining upward sandstone packages often display current ripple lamination, and contain pelleted, mud-lined vertical and horizontal burrows (*Ophiomorpha*). Sandstone intervals display channel-like geometry, suggesting deposition from minor distributaries which transected a muddy lagoon and associated tidal flats. The thinner (10–20 cm) sandstone beds within lagoon fill successions are often 'sheet-like' in geometry, and display planar or low-angle internal trough cross-stratification, or fine-scale ripple lamination (both wave and current ripples are present). Ripple forms are frequently clay draped, giving rise to abundant internal wavy and flaser-bedded fabrics. These centimetre-scale laminated, heterolithic sandstone–mudstone bedsets (up to 25 cm thick) may contain as much as 50% mudstone laminae. They may reflect deposition from crevasse events from distributary channels into the surrounding lagoon, or onto a tidal mudflat. Heterolithic lagoon fill successions (e.g. Unit 5, Fig. 5b) often display an upward increase in sand content, and the scale at which sandstone and mudstone lithologies are interbedded.

The heterolithic sediments described are interpreted as having been deposited during the infilling of partially closed lagoons along a wave dominated coastline. The common occurrence of sandy wave ripples within muddy intervals, a very low diversity ichnofauna characteristic of the *Skolithos* ichnofacies (dominated by the vertical dwelling structure *Ophiomorpha* within channelized sandstones, with rare horizontal *Palaeophycus*) indicate relatively high wave energy within lagoons. Pemberton *et al.* (1992) suggest that the Skolithos ichnofacies is characteristically developed within slightly muddy–clean, well-sorted loose or shifting substrates, in environments characterized by high levels of wave or current activity.

Sedimentological model

A simple sedimentological framework within which the sediments cropping out along Punyit Beach could be described is illustrated in Fig. 6. The model envisages deposition within a wave dominated coastal setting, with recognized subenvironments including distributary channels (some of which are tidally influenced), lagoons transected by minor distributaries, and shoreface deposits.

A schematic representation of the studied strata exposed at Punyit Beach is illustrated in Fig. 7. These strata can be simply described as comprising a number of stacked, broadly coarsening upward parasequences documenting a vertical transition through shoreface, lagoon

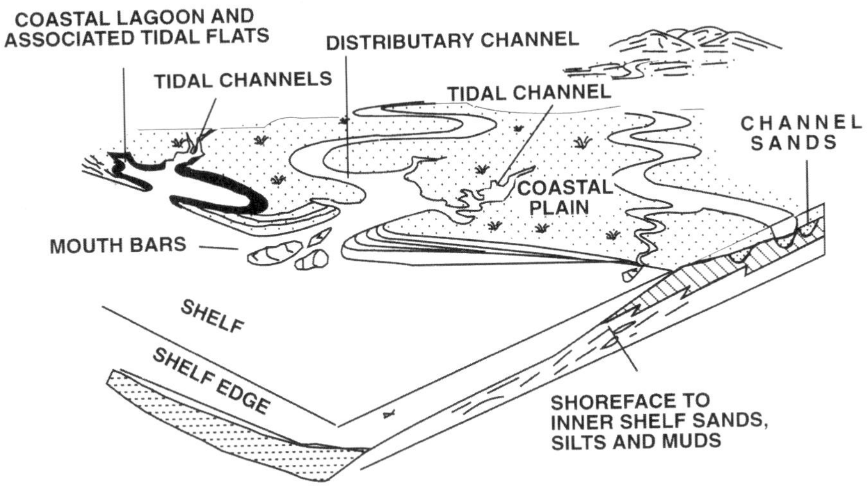

Fig. 6. Sedimentological model within which strata exposed at Punyit Beach may be described (adapted from Potter *et al.* 1984).

and channelised distributary facies. Within this simplistic model, a number of smaller 'lower order' parasequence sets occur as a result of the vertical stacking of a number of individual coarsening upward shoreface parasequences prior to their passing into more 'landward' lagoonal and distributary channel facies. A significant sequence boundary occurs at the contact between Units 7b and 8 (Fig. 7), where distributary channel sandstones are overlain by an erosive surface along which a thin veneer of well rounded, pebble–cobble-grade sandstone clasts, representing a transgressive lag, is developed. The transgressive lag is overlain by a broadly coarsening upward shoreface sedimentary package, which is then itself overlain by a thick succession of dark grey silty mudstones. In a sequence stratigraphic context, the erosional surface separating the distributary channel facies of Unit 7b (deposited within a lowstand systems tract?) from the transgressive lag and subsequent shoreface sediments of Unit 8 is best interpreted as a ravinement surface, superceded by a landward shift in facies belts. The coarsening-upward shoreface parasequence set overlying the transgressive lag might be interpreted as the deposits of a transgressive systems tract, with the final passage of Unit 8 into a thick succession of dark grey silty mudstones representing a maximum marine flooding surface.

The erosional nature of the surface marking the base of Unit 6 within the studied section (Fig. 7) may also represent an important surface in terms of sequence analysis. The base of Unit 6 (the deposits of a lagoon fill cut by distributary channels), down-cuts into underlying shoreface sediments, perhaps representing an erosional unconformity above which there is a *slight* basinward shift in facies type. However, in this case interpretation is subjective. i.e. the overall facies stacking pattern from Unit 5 through to Unit 7 is still broadly that of a large scale coarsening upward parasequence set, and it is unclear whether the surface at the base of Unit 6 represents a correlateable erosional unconformity, or just the effect of localized channel incision into a shoreface. The deepest erosional scours along this surface were infilled by mudstone dominated lithologies with thin sandstone interbeds, and the initial deposits overlying these scours were sandy in nature, comprising wavy bedded (and flaser-bedded) sandstones, perhaps indicative of tidal influence within this particular part of the lagoon fill succession. Subsequently, the lagoon fill again became dominated by more muddy lithologies, into which erosive based, minor distributary channel sandstones incised.

The lower part of the succession exposed at Punyit Beach (i.e. represented by units 1–5) comprises a number of stacked coarsening upward, progradational parasequences. However, the very limited outcrop studied invalidates interpretation of these strata in the context of whether they were deposited during highstand or lowstand progradation.

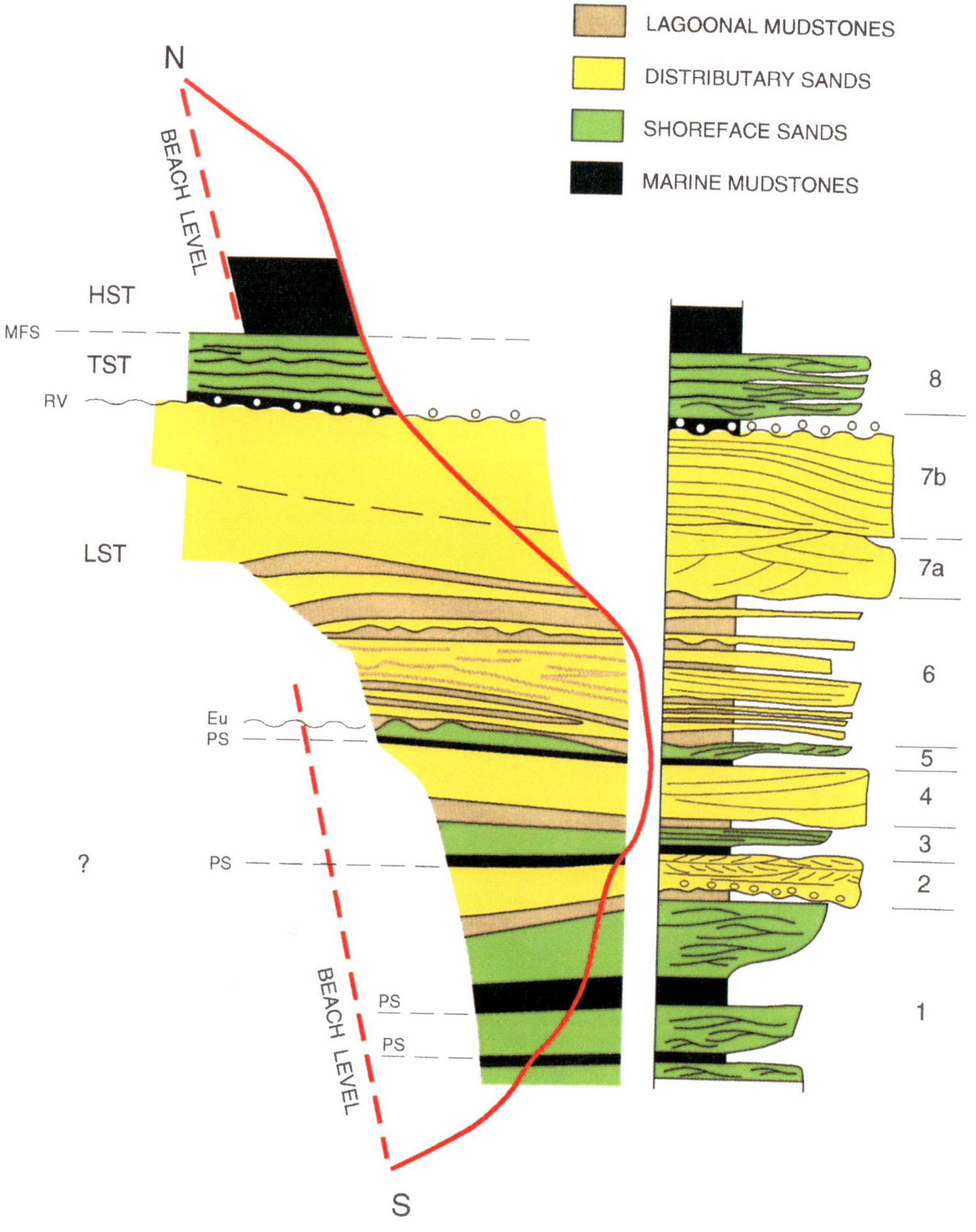

Fig 7. Schematic representation of strata exposed at Punyit Beach. Numbers refer to units cited in text.

Permeability data

A graphical representation of a 5 cm spaced permeability data traverse taken along a vertical transect (i.e a traverse oriented perpendicular to bedding) through the studied outcrop section is illustrated in Fig. 8 and summary statistics for these data are provided in Table 1. Permeability histograms for both vertical and lateral data traverses through the facies exposed at outcrop are illustrated in Figs 9 and 10. Permeability heterogeneities identified within outcrop are associated with permeability contrasts of up to three orders of magnitude, and occur at facies, facies unit (e.g. fining-upward or coarsening-upward sediment packages), bed-set, bed and

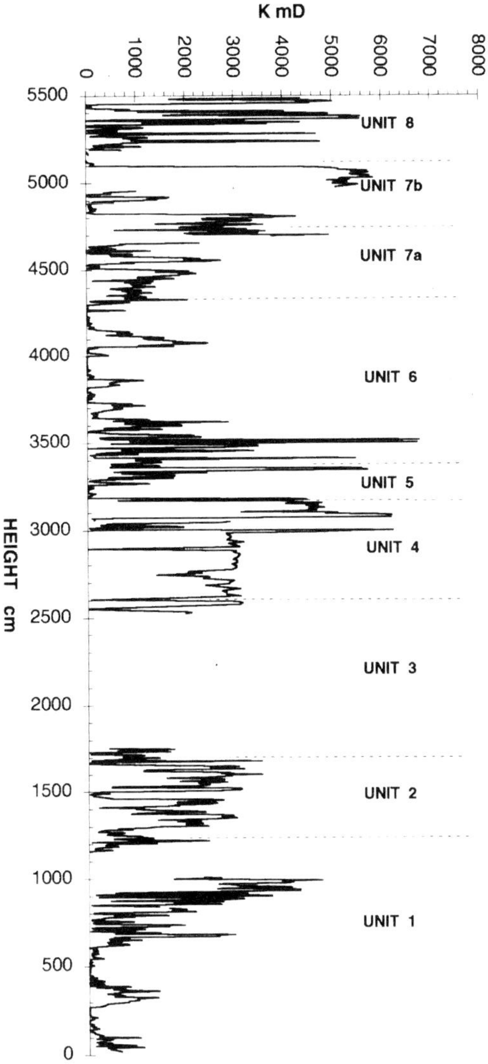

Fig. 8. Permeability data traverse vertically through sediments of studied outcrop section.

laminar scales. Permeability traverses through examples of subsurface Lower Miocene reservoir sections (approx. 2500 m depth) containing comparable sedimentary facies to those encountered in the Jerudong Formation at surface outcrop are illustrated in Fig. 11, and permeability data histograms are provided in Fig. 12.

Average permeabilities

Summary statistics for both outcrop data, and for similar facies encountered in the subsurface of offshore Brunei are provided in Tables 2–5.

The highest permeabilities measured at outcrop (up to 7235 mD) occur within the distributary channel facies. Shoreface facies also display permeabilities up to 5713 mD but are more heterogeneous, with lower 'average' permeability than distributary channels. Lagoon fill successions comprise significantly more heterolithic deposits, containing interbedded sandstone, siltstone and mudstone, with a wide range in permeability (<1 mD to >1000 mD).

Average permeabilities for similar facies types encountered at depths of up to 2500 m in the subsurface of offshore Brunei are several orders of magnitude smaller than those encountered at surface outcrop (Fig. 12). There is also a trend towards higher variability within the subsurface data set.

Variation within permeability data-sets.

The interquartile range (i.e. the range between the 25th and 75th data percentiles) provides a useful measure of variation or dispersion within data, being less influenced by extreme data values than the standard deviation. Interquartile ranges for outcrop data sets studied are summarised in Tables 2–5, and illustrated graphically in Figs 9 and 10. For some of the data illustrated the arithmetic mean is greater than the 75th percentile permeability value, indicating that data are strongly skewed, and that arithmetic mean is generally not a good 'average' permeability estimator. The geometric mean appears to provide a far better estimate of central tendency within the studied data.

The normalized standard deviation or coefficient of variation (C_V) provides an estimate of variability or 'heterogeneity' within permeability data (Goggin *et al.* 1988; Corbett & Jensen 1992*a*, *b*, 1993*a*). C_V is defined by

$$C_V = \frac{\text{Sample standard deviation}}{\text{Arithmetic mean}}. \qquad (1)$$

Permeability data are termed homogeneous if their C_V are less than 0.5; heterogeneous if their C_V lie between 0.5 and 1.0; and very heterogeneous if their C_V are greater than 1.0 (Corbett & Jensen 1992*b*). C_V determined for permeability traverses through studied outcrop sections are illustrated in Tables 2–4.

Channel-fill sandstones. Channel-fill sandstones display upward decreasing permeability and grain size (see Tables 2 and 3 for summary statistics). However, some channel fills may be stacked such that stratigraphically higher fining-upward sediment packages are of overall higher

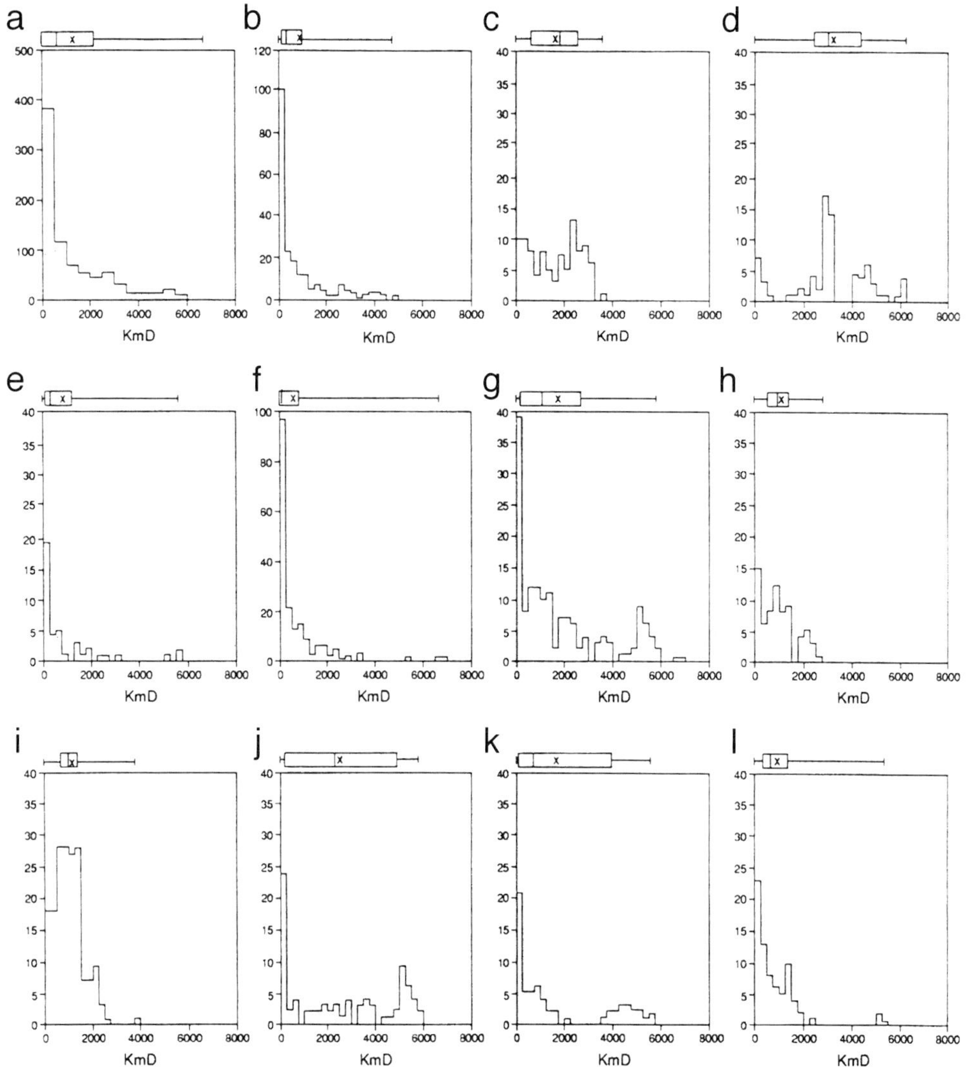

Fig. 9. Permeability histograms for vertical data transects through the studied outcrop section. (**a**) Total data set (5 cm sample spacing). (**b**) Unit 1, shoreface/lagoon (5 cm sample spacing). (**c**) Unit 2, tidal distributary channel (5 cm sample spacing). (**d**) Unit 4, distributary channel (5 cm sample spacing). (**e**) Unit 5, shoreface (5 cm sample spacing). (**f**) Unit 6, lagoon fill (5 cm sample spacing). (**g**) Unit 7 (total), distributary channel complex (5 cm sample spacing). (**h**) Unit 7a, trough cross-stratified distributary channel fill (5 cm sample spacing). (**i**) Unit 7a, trough cross-stratified distributary channel fill (1 cm sample spacing). (**j**) Unit 7b, laterally accreted distributary channel fill (5 cm sample spacing). (**k**) Unit 8, shoreface (5 cm sample spacing). (**l**) Unit 8, shoreface (0.5 cm sample spacing).

permeability than the lowest fining-upward packages, yielding an overall upward increase in permeability for the stacked channel-fill succession. Vertical data transects through trough cross-stratified channelized sandstone facies and tidal distributary channel facies are homogeneous–heterogeneous with respect to permeability variation ($C_v = 0.46$–0.74). Permeability appears homogeneous ($C_v = 0.2$–0.5) within lateral data traverses (Figs 13 and 14, Table 3) through these trough cross-stratified facies. The values of C_v indicated for trough cross-stratified distributary channel sandstones are comparable to those cited by Dryer *et al.*

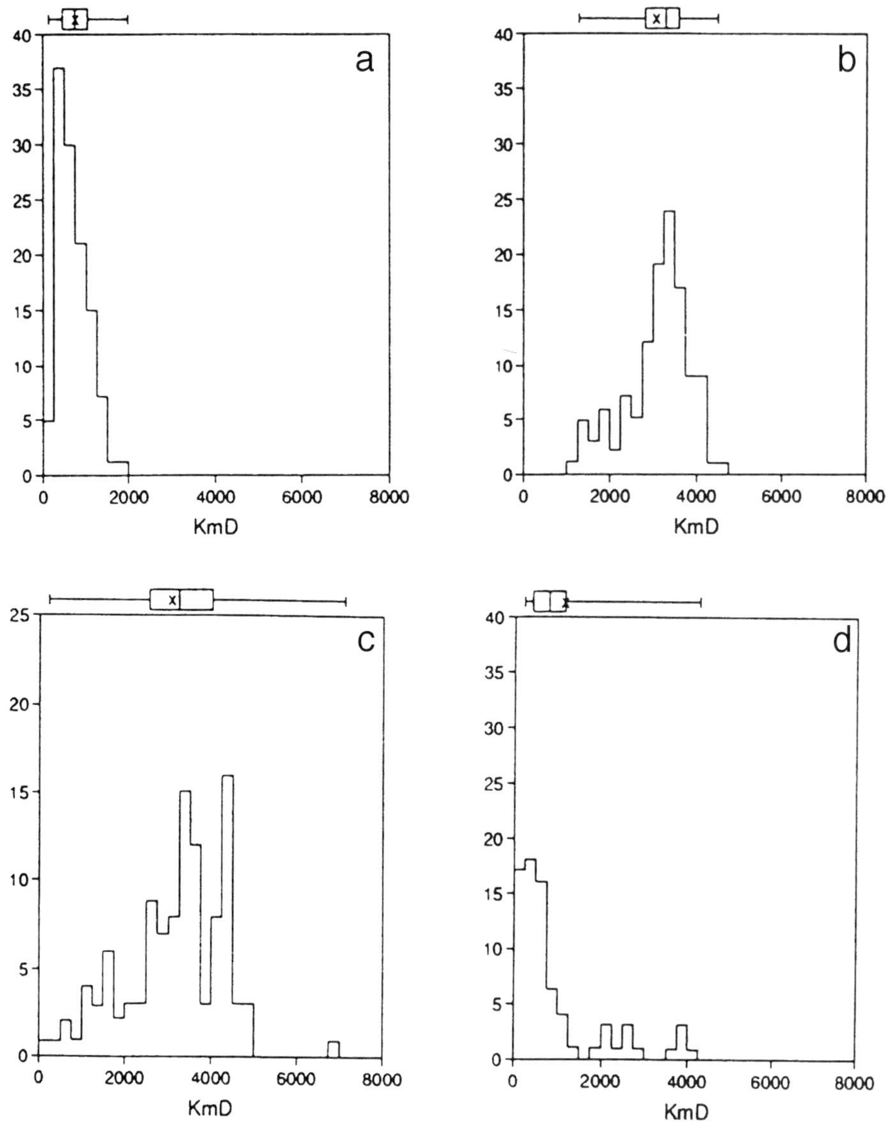

Fig. 10. Permeability histograms for lateral data transects through the studied outcrop section. (**a**) Unit 7a, trough cross stratified distributary channel fill. (**b**) Unit 7b, laterally accreted distributary channel fill. (**c**) Unit 2, tidal distributary channel. (**d**) Unit 8, shoreface.

(1990), for similar facies ($C_V = 0.29$–0.81) within the Scalby and Saltwick Formations (Jurassic, UK Yorkshire Coast). Vertical data transects through laterally accreted deposits of meandering distributary channels are heterogeneous with respect to permeability ($C_V = 0.9$). However, as is the case with trough cross-stratified channel fills, permeability appears relatively homogeneous within a lateral data traverse (Figs 14–16, Table 4) through laterally accreted distributary channel fill facies ($C_V = 0.24$). Values of C_V for 5 cm spaced vertical data traverses through thin (metre scale) minor distributary channels within lagoon fill successions (e.g. within Unit 6 of the studied outcrop section) range from 0.25 to 1.0 (Table 3).

Subsurface data sets for distributary channel facies display a distinct upward decrease in permeability (Fig. 11), which broadly corresponds to upward-fining grain-size profiles

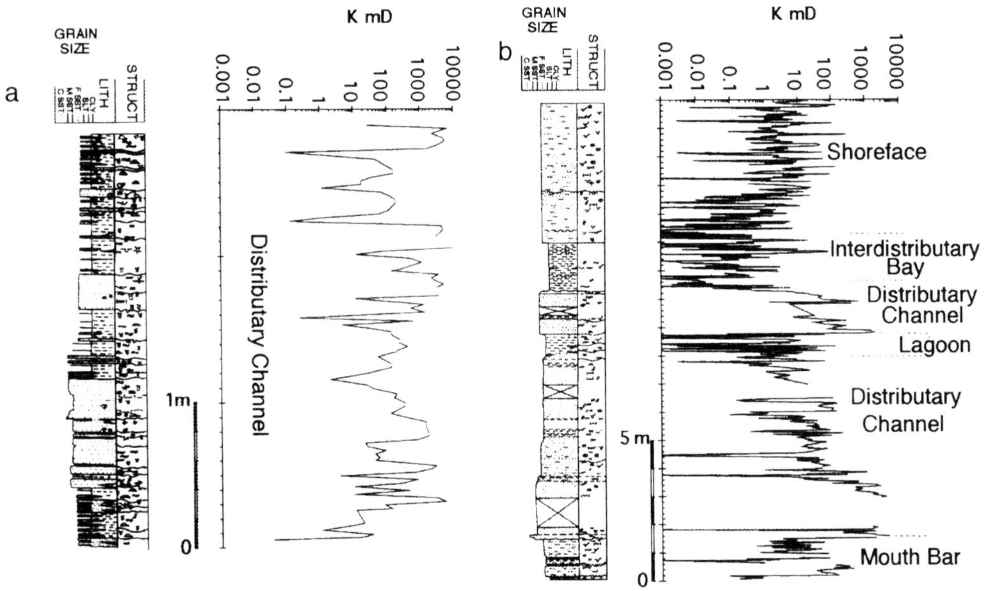

Fig. 11. (**a**) Core log and probe-permeameter data traverse (0.03 m spaced) through studied core interval from subsurface reservoir section A. (**b**) Core log and probe-permeameter data traverse (0.03 m spaced) through studied core interval from subsurface reservoir section B.

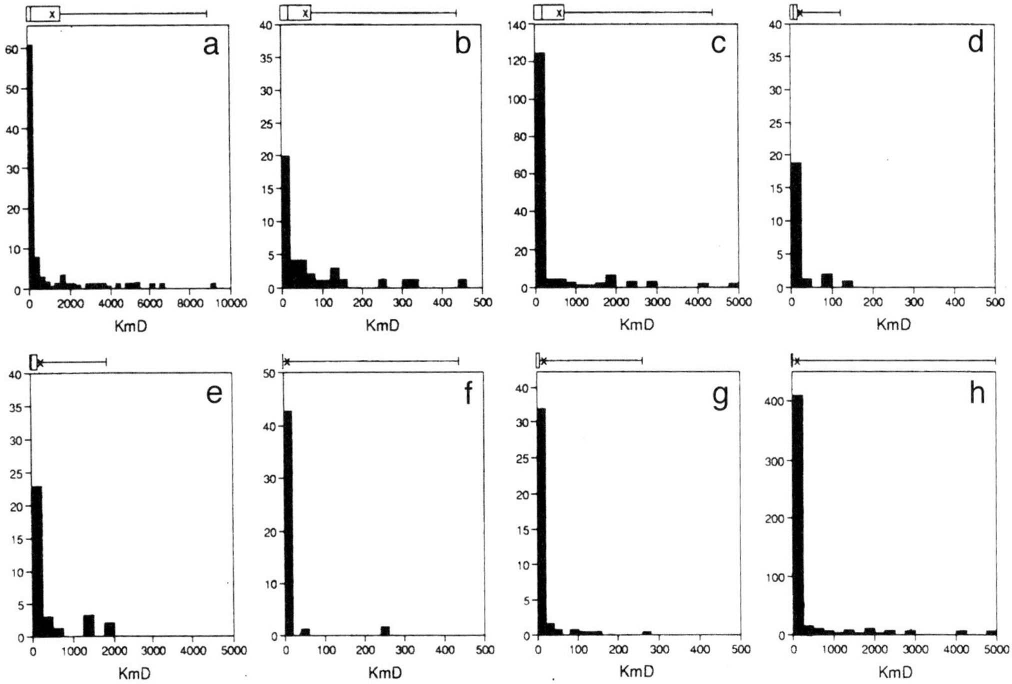

Fig. 12. Permeability histograms for subsurface reservoir sections containing similar lithofacies to those exposed along Punyit Beach. (**a**) Section B, total data set (all distributary channel). (**b**) Section A, prodelta–lower shoreface. (**c**) Section B, interdistributary bay. (**d**) Section B, marine-influenced distributary mouth bar. (**e**) Section A, distributary channel. (**f**) Section A, total data set. 5 cm spaced data.

Table 2. *Permeability data – summary statistics for 5 cm spaced vertical data traverse through studied outcrop section at Punyit Beach*

Height (m)	Facies	Minimum (mD)	Maximum (mD)	Arithmetic mean (mD)	Standard deviation (mD)	Harmonic mean (mD)	Geometric mean (mD)	Median (mD)	25th Percentile	75th Percentile	IQR	C_v	No. samples
0.0–11.75 Unit 1	Shoreface/ lagoon	0.001	4761.42	793.12	1137.75	0.04	139.84	278.23	51.98	947.92	895.94	1.44	203
11.75–16.60 Unit 2	Tidal channel	23.62	3553.67	1629.46	1018.97	504.15	1138.37	1750.31	579.79	2476.2	1896.41	0.63	97
16.60–26.2 Unit 3	Shoreface/ lagoon fill	INSUFFICIENT DATA											
26.2–31.95 Unit 4	Distributary channel	8.12	6233.88	3056.06	1999.86	334.20	2087.73	2982.88	2403.82	4322.99	1919.17	0.52	76
31.95–33.85 Unit 5	Shoreface	0.37	5712.73	961.38	1513.21	10.93	207.56	416.51	26.20	1262.25	1236.06	1.57	40
33.85–43.00 Unit 6	Lagoon fill	0.001	6771.58	641.95	1062.84	0.01	49.63	175.63	10.33	873.45	863.12	1.66	184
43.00–51.60 Unit 7	Distributary channel	0.001	5836.81	1806.27	1814.49	0.15	658.99	1113.35	239.11	2472.98	2233.87	1.01	155
43.00–46.90 Unit 7a	*Distributary channel*	*0.001*	*2734.71*	*962.93*	*708.44*	*0.07*	*475.70*	*881.72*	*461.91*	*1340.84*	*878.93*	*0.74*	*71*
46.90–51.60 Unit 7b	*Distributary channel*	*0.74*	*5836.81*	*2519.10*	*2136.08*	*29.34*	*868.00*	*2358.33*	*152.03*	*4943.41*	*4791.38*	*0.84*	*84*
51.60–51.85 Unit 8	Shoreface	0.37	5580.37	1669.80	1935.66	12.85	416.13	765.91	92.25	3942.46	3850.21	1.16	64
Total data set		0.001	6771.58	1326.33	1558.29	0.04	271.69	641.83	91.23	2242.51	2151.28	1.18	849

Table 3. *Permeability data – summary statistics for lateral data traverses through selected facies within outcrops at Punyit Beach*

Length (m)	Facies	Minimum (mD)	Maximum (mD)	Arithmetic mean (mD)	Standard deviation (mD)	Harmonic mean (mD)	Geometric mean (mD)	Median (mD)	25th Percentile	75th Percentile	IQR	C_v	No. samples	Sample interval (m)
6.45	Distributary channel (Unit 7a)	112.40	1981.94	690.40	344.79	520.08	605.82	650.10	412.27	892.06	479.79	0.50	117	0.05
6.05	Distributary channel (Unit 7b)	1245.65	4513.09	3114.91	732.09	2883.34	3010.81	3264.02	2784.24	3565.94	781.70	0.24	121	0.05
1.50	Shoreface Unit 8)	29.66	4116.03	927.73	1057.26	283.05	529.39	550.80	269.55	923.04	653.49	1.13	60	0.01–0.05 (nested)
5.55	Tidal distributary (Unit 2)	195.10	7234.58	3159.68	1192.62	2244.28	2832.21	3330.19	2577.44	4076.74	1499.30	0.28	111	0.05

Table 4. *Permeability data – summary statistics for vertical data traverse through channels within the lagoon-fill succession of Unit 6 within the studied outcrop section*

Channel	Thickness (cm)	Minimum (mD)	Maximum (mD)	Arithmetic mean (mD)	Standard deviation (mD)	Harmonic mean (mD)	Geometric mean (mD)	Median (mD)	C_v	No. samples
C1	190.00	0.37	2875.34	640.94	642.59	9.58	269.72	501.20	1.00	38
C2	65.00	2.58	146.44	393.53	358.07	22.85	174.03	469.61	0.91	13
C3	15.00	261.30	435.02	346.78	86.89	332.13	339.40	344.40	0.25	3
C4	170.00	15.50	2461.73	760.78	757.22	89.00	304.48	571.51	1.00	34

Table 5. *Permeability data – summary statistics for studied core intervals from subsurface reservoir sections A and B*

Facies	Minimum (mD)	Maximum (mD)	Arithmetic mean (mD)	Standard deviation (mD)	Harmonic mean (mD)	Geometric mean (mD)	Median (mD)	25th Percentile	75th Percentile	IQR	C_v	No. samples
Section B												
Shoreface/ Prodelta	0.01	261.00	11.15	30.56	0.01	0.88	2.00	0.30	6.93	6.63	2.74	145
Interdistributary bay	0.001	248.00	7.39	37.30	0.01	0.07	0.07	0.004	0.77	0.77	5.04	45
Distributary channel	1.36	1890.00	290.80	513.07	18.98	77.53	55.60	26.60	206.00	179.40	1.76	31
Lagoon	0.001	126.00	16.01	29.88	0.003	0.40	4.19	0.001	12.03	12.02	1.87	23
Distributary channel	0.001	4920.00	355.64	804.73	0.04	20.11	25.50	5.43	98.70	93.27	2.40	153
Distributary mouth bar	0.001	445.00	64.58	104.00	0.04	8.57	16.00	0.98	77.3	76.32	1.61	40
Total data set	0.001	4920.00	150.73	518.89	0.01	3.24	5.94	0.43	40.83	40.40	3.44	453
Section A												
Distributary channel	0.04	9070.00	1061.92	1768.20	1.92	147.36	165.00	32.95	1462.50	1429.55	1.67	103

Fig. 13. (**a**) Outcrop photograph illustrating position of lateral permeability traverse through trough cross stratified distributary channel deposits of unit 7a of the studied outcrop section (44 m above base of section shown in Fig. 4). (**b**) Lateral permeability traverse.

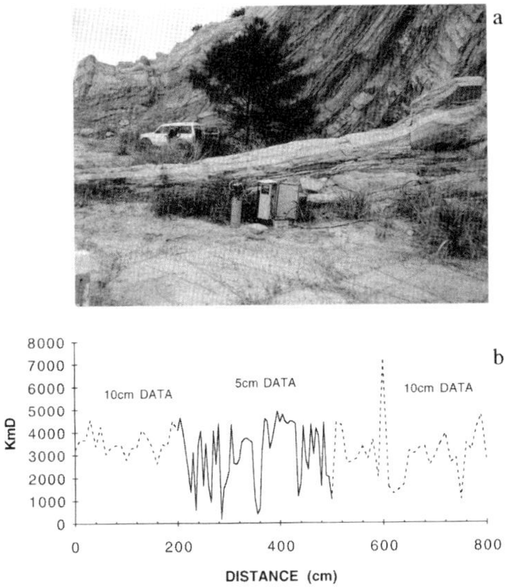

Fig. 14. (**a**) Outcrop photograph illustrating position of lateral permeability traverse through tidal distributary channel deposits of unit 2 of the studied outcrop section (15.5 m above base of section shown in Fig. 4). (**b**) Lateral permeability traverse.

within distributary channel fills. Permeabilities in excess of 1000 mD at the base of channels contrast with permeabilities of tens to hundreds of milliDarcies within higher parts of channel

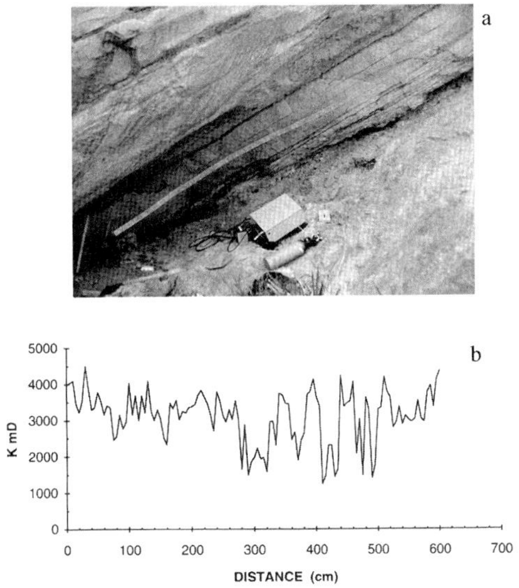

Fig. 15. (**a**) Outcrop photograph illustrating position of lateral permeability traverse through laterally accreted distributary channel fill deposits of unit 7b of the studied outcrop section (47.5 m above base of section shown in Fig. 4). (**b**) Lateral permeability traverse.

fills. The values of C_V for subsurface permeability data range from 1.67 to 2.40 indicating that distributary channels are very heterogeneous with respect to permeability (Table 5). These values of C_V are of comparable magnitude to those cited for Etive Formation tidal channels sampled at similar spacing ($C_V = 0.82$–1.99) within the Jurassic Brent Group of the UK North Sea (Daws & Prosser 1992; Corbett and Jensen 1992*a*).

Shoreface facies sandstones. Shoreface successions sampled at outcrop are characterised by distinct upward increase in grain size, bedform scale and permeability (Fig. 8). They are very heterogeneous with respect to permeability variation ($C_V = 1.16$–1.57 for vertical traverses). A 1.0 cm spaced bedding parallel permeability data traverse (Fig. 16) and 0.5 cm spaced traverse perpendicular to the bedding within outcrop Unit 8 yield C_V of 1.13 and 1.22 respectively, confirming a high degree of lateral permeability heterogeneity within this facies. The values of C_V for outcrop data traverses through shoreface sandstones are similar to those documented by Corbett & Jensen (1993*a*) for shoreface facies sandstones (C_V = up to 1.49) from the subsurface Rannoch Formation (Brent Group, UK North Sea).

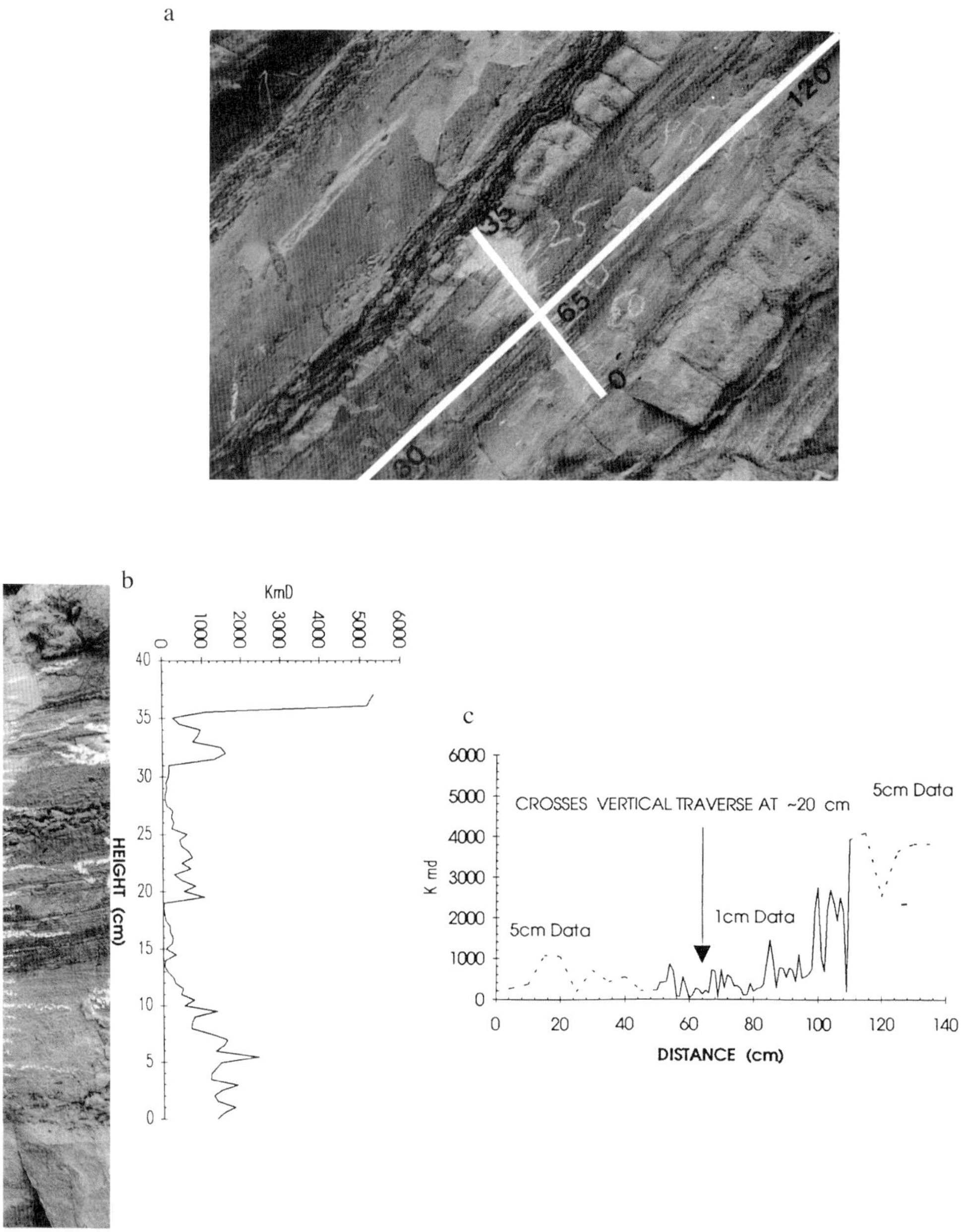

Fig. 16. Permeability data for 1 cm spaced bedding parallel and 0.5 cm bedding normal traverses through shoreface facies sediments of Unit 8 of the studied outcrop section. (**a**) Outcrop photograph. (**b**) Bedding- normal permeability traverse. (**c**) Bedding-parallel permeability traverse.

Shoreface facies sandstones sampled in the subsurface also display marked upward increase in permeability; e.g. from <0.1 mD near the base of this facies, to >10 mD towards the top of the facies (Fig. 11), corresponding to upward-coarsening grain-size trends. High values of C_V (2.74 for the example illustrated) indicate subsurface data (Table 5) are very much more heterogeneous with respect to permeability variation than outcrop data (C_V = 1.16–1.57).

Lagoon-fill successions. Permeability is very heterogeneous within lagoon fill successions sampled at surface outcrop ($C_V = 1.66$ for a traverse oriented perpendicular to bed boundaries, Table 2). Permeabilities recorded for lagoon-fill successions sampled in the subsurface range from <0.01 mD to several tens of mD (Table 5). Permeability heterogeneity is comparable for subsurface ($C_V = 1.87$) and surface outcrop data ($C_V = 1.66$, Unit 6).

Spatial variation within permeability data

Techniques. Geostatistical analysis of outcrop permeability data using semivariograms was carried out in order to evaluate the spatial nature of permeability variation. The construction of experimental semivariograms is summarized in the Appendix. Not all of the permeability data-sets gathered during this study were normally distributed. Permeability data were transformed to normal data distributions required for semivariogram modelling by using a normal scores data transformation, validated by use of a Lilliefor's test (Conover 1971) to test for normality at a 95% confidence level.

Experimental semivariograms have been constructed for both outcrop data, and for a sample of data from offshore Bruneian oilfields. Experimental semivariograms for the outcrop data are illustrated in Figs 17–21. Experimental semivariograms for 0.03 m spaced subsurface permeability data sets are illustrated in Fig. 22. Experimental semivariograms constructed for both outcrop and subsurface permeability data are complex, and their interpretation is not straightforward. All experimental semivariograms for the outcrop section show 'moderate' positive projected Y-axis intercepts (nugget effect), typically 20–50% of the semivariogram sill values. These nugget effects would imply that sampling a given site more than once will yield different results either because of low measurement precision, or as a result of the presence of permeability heterogeneity at a finer scale than the sampling interval (Journel & Huijbregts 1978; Hohn 1988).

Experimental semivariograms for some intervals modelled appear to show marked decreases in the calculated semivariogram function at regular multiples of a positive lag spacing. These 'hole effects' reflect periodicity within the permeability data set, and result in the semivariograms displaying a sinusoidal variation about their sill. Periodicities within successions of sedimentary strata also provide spatial data useful during the modeling of fluid flow in reservoirs, and may reflect the repeated stacking of strata containing similar permeability trends or structures. Within both the outcrop and subsurface permeability data, the observed spatial variability frequently approximates the scales at which sedimentary phenomena are observed, occurring at bed thickness, fining/coarsening upward unit or facies scale.

The results of geostatistical analyses of outcrop and subsurface permeability data are summarized in Tables 6 and 7

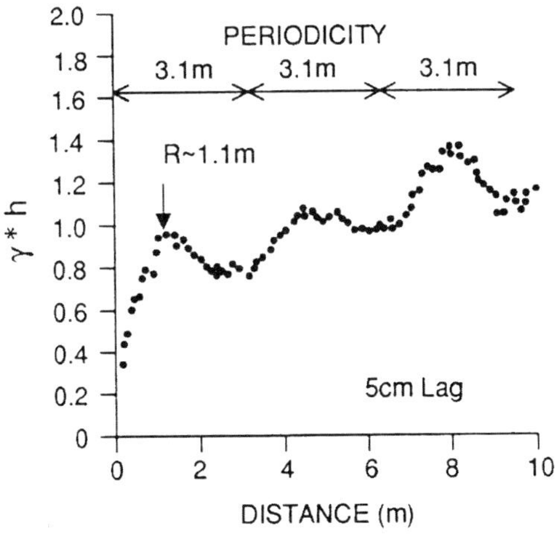

Fig. 17. Experimental semivariogram – total outcrop data set.

Experimental semivariograms of 'total' permeability data sets

A semivariogram for the 'total' permeability data-set from the studied outcrop section (Fig. 17) indicates a correlation length of (1.1 m) approximating average bed/bed-set thickness, and periodicity of 3.1 m, approximating the average thickness of genetic facies units present (e.g. large-scale coarsening-upward shoreface sediment packages, or fining-upward distributary channel facies). A semivariogram for the interbedded facies within the total studied interval from subsurface Section A (Fig. 22) displays a two-fold correlation structure with correlation lengths at 0.18 m and 1.35 m. These probably reflect the presence of both bedding scale heterogeneity and the stacking of genetic facies units (fining- or coarsening-upward units), and the results are comparable to those obtained for surface outcrop data-sets.

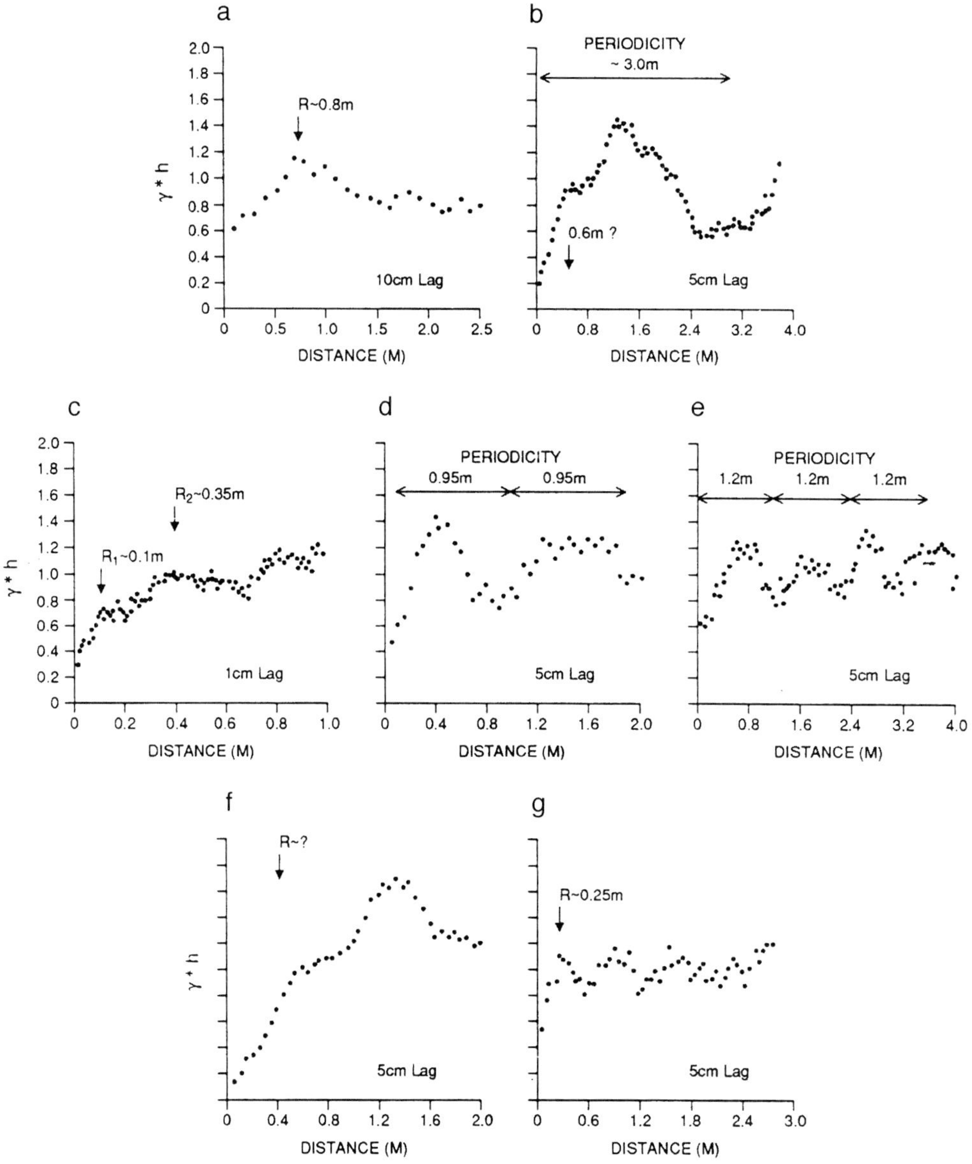

Fig. 18. Experimental semivariograms – trough cross stratified and laterally accreted distributary channels. (**a**) Outcrop Unit 4, vertical data traverse (10 cm sample lag). (**b**) Outcrop Unit 7 (total), vertical data traverse (5 cm sample lag). (**c**) Outcrop Unit 7a, vertical data traverse (1 cm sample lag). (**d**) Outcrop Unit 7a, vertical data traverse (5 cm sample lag). (**e**) Outcrop Unit 7a, lateral data traverse(5 cm sample lag). (**f**) Outcrop Unit 7b, vertical data traverse (5 cm sample lag). (**g**) Outcrop Unit 7b, lateral data traverse (5 cm sample lag).

Experimental semivariograms constructed for individual facies types

Channelized sandstone facies. Semivariograms for vertically oriented permeability data transects through channelized sandstone facies at surface outcrop (Figs 18 and 19) display short correlation lengths (0.1–0.8 m), corresponding to bedding or cross-set thickness. Periodicities (up to 3.0 m) within these semivariograms reflect stacking of genetic facies units (fining upward) with similar permeability correlation structures. Lateral data traverses through trough cross-stratified distributary channel sandstones indicate periodicities

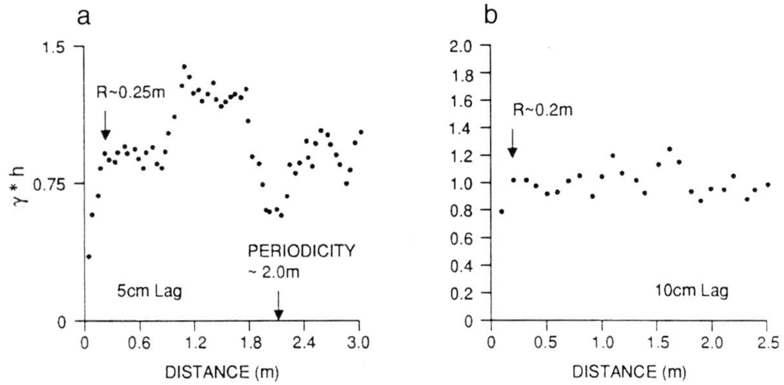

Fig. 19. Experimental semivariograms – tidal distributary channel. (**a**) Outcrop Unit 2, vertical data traverse (5 cm sample lag). (**b**) Outcrop Unit 2, lateral data traverse (10 cm sample lag).

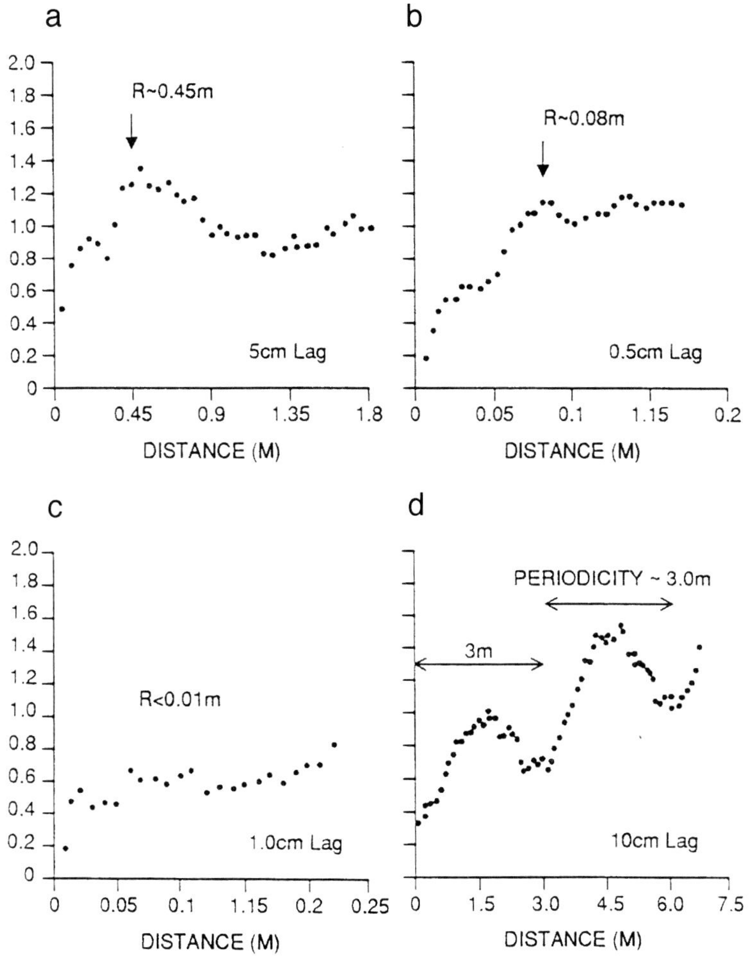

Fig. 20. Experimental semivariograms – shoreface facies. (**a**) Outcrop Unit 8, vertical data traverse (5 cm sample lag). (**b**) Outcrop Unit 8, vertical data traverse (5 cm sample lag). (**c**) Outcrop Unit 8, lateral data traverse (1 cm sample lag). (**d**) Outcrop Unit 1, vertical data traverse (10 cm sample lag).

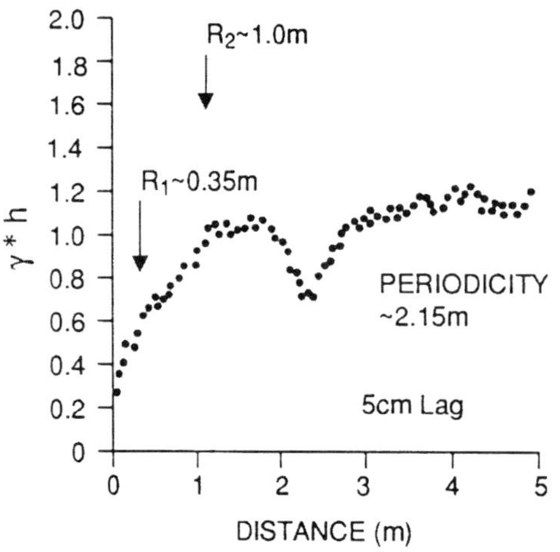

Fig. 21. Experimental semivariograms – lagoon fill facies Outcrop Unit 6 (5 cm sample lag).

(*c.* 1.2 m) corresponding to average trough bed widths (Fig. 18). Short (*c.* 0.25 m) correlation lengths within semivariograms for lateral data traverses through a point-bar deposit and trough cross-stratified tidal channel fill approximate foreset spacing within lateral accretion sets, and lateral spacing of clay draped cross bed boundaries within tidal trough cross sets (Figs 18 and 19).

Short correlation lengths (0.15–0.25 m) for permeability within deltaic distributary channels sampled in the subsurface (Fig. 22) are comparable to those obtained from similar facies at outcrop. The larger (i.e. 0.25 m scale) correlation lengths approximate the average frequency of bedding scale grain size contrasts within this facies (typically 0.15 m to >0.3 m), whereas *c.* 0.15 m correlation length recorded for permeability data from the deltaic distributary from subsurface Section A (Fig. 22) appears to reflect the presence of small scale 'within-bed' heterogeneity.

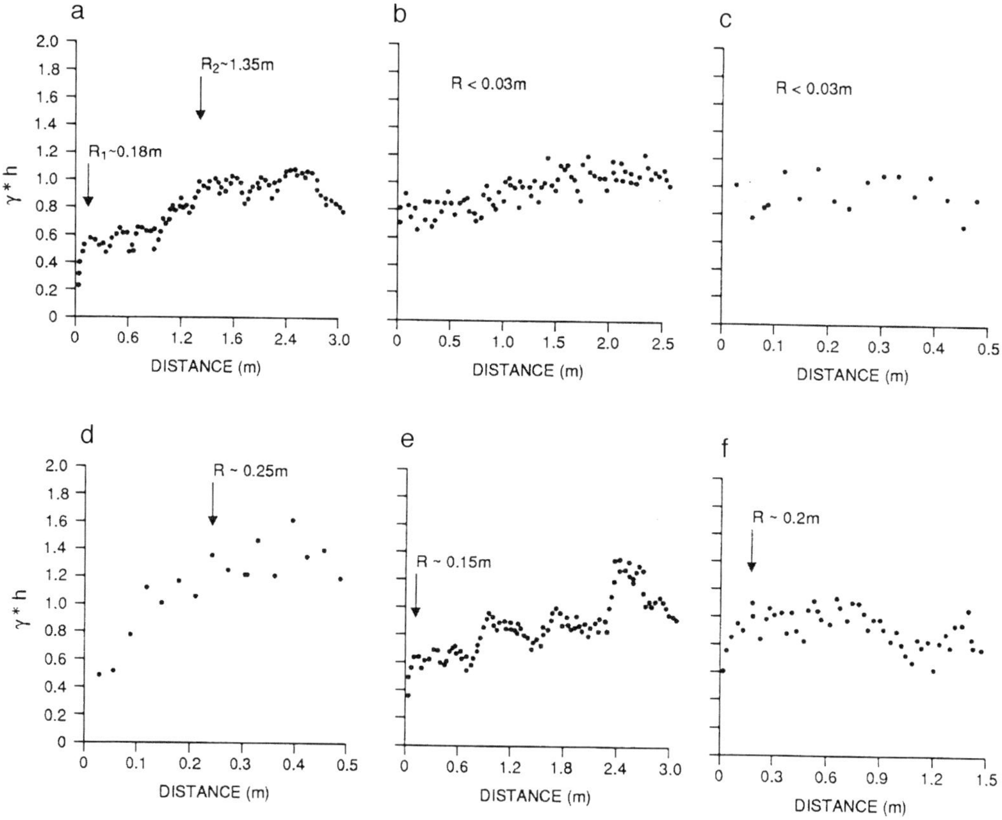

Fig. 22. Experimental semivariograms – subsurface reservoir sections (all plotted at 3 cm sample lag spacings). (**a**) Section B, total data set. (**b**) Section B, prodelta – lower shoreface. (**c**) Section B, interdistributary bay. (**d**) Section B, marine-influenced distributary mouth bar. (**e**) Section B, distributary channel. (**f**) Section A, total data set (all distributary channel).

Table 6. *Summary results of geostatistical analyses of outcrop permeability data*

Unit	Facies	Orientation	Range 1 (m)	Range 2 (m)	Periodicity (m)	Notes
Total	See below	Vertical 1.1 m			3.1	P = Average genetic facies unit thickness
Unit 1	Shoreface	Vertical 5 cm			3.0	P = Average genetic facies unit thickness
Unit 2	Tidal distributary	Vertical 5 cm	0.25			R1 = Genetic unit (fining-up) thickness
Unit 2	Tidal distributary	Lateral 5 cm	0.2			R1 = Lateral spacing of clay-draped foresets
Unit 3	Insufficient data					
Unit 4	Distributary	Vertical 5 cm	0.8			R1 = Coset/genetic unit thickness
Unit 5	Insufficient data					
Unit 6	Heterolithic lagoon fill	Vertical 5 cm	0.35	1.0	2.2	R1 = Average bed thickness in sandstone facies R2 = Average channelized sandstone thickness P = Average thickness grossly fining upward packages
Unit 7	Total distributary channel	Vertical 5 cm	0.6		3.0	R1 = Average bed-set thickness P = Large scale fining-up/coarsening-up units within stacked facies
Unit 7a	Trough cross stratified channel	Vertical 1 cm	0.1	0.35		R1 = Average foreset thickness R2 = Trough set thickness
Unit 7a	Trough cross stratified channel	Vertical 5 cm			0.95	P = Average genetic unit (fining-upward) thickness
Unit 7a	Trough cross stratified channel	Lateral 5 cm			1.2	P = Average trough set width
Unit 7	Laterally accreted channel fill	Vertical 5 cm			>2.0	P = Average genetic unit (fining-upward) thickness?
Unit 7	Laterally accreted channel fill	Lateral 5 cm	0.25			R1 = Bed thickness (foreset spacing) within lateral accretion sets
Unit 8	Shoreface	Vertical 0.5 cm	0.08			R1 = Bed thickness
Unit 8	Shoreface	Vertical 5 cm	0.45			R1 = HCS bed-set thickness
Unit 8	Shoreface	Lateral 1 cm	< 0.01			R1 = Laminar scale hetrogeneity

Table 7. *Summary results of geostatistical analyses of permeability data from subsurface reservoir sections A and B*

Unit	Facies	Orientation	Range 1 (m)	Range 2 (m)	Notes
Section B					
Total data	See below	Vertical 3.0 cm	0.18	1.35	R1 = Within bed heterogeneity? R2 = Genetic facies unit?
Unit 6	Prodelta/lower shoreface	Vertical 3.0 cm			Laminar scale (<0.03 m) heterogeneities
Unit 5	Interdistributary bay	Vertical 3.0 cm			Laminar scale (<0.03 m) heterogeneities
Unit 2	Deltaic distributary	Vertical 3.0 cm	0.15		R1 = Within bed heterogeneity?
Unit 1	Marine-influenced mouth bar	Vertical 3.0 cm	0.25		R1 = Average bed/bed-set thickness
Section A					
Total data set	Trough cross-stratified channel	Vertical 3.0 cm	0.2		R1 = Average bed thickness

Shoreface successions. Semivariograms for permeability traverses through shoreface sandstones at outcrop (Fig. 20) illustrate short correlation lengths (0.08–0.5 m), which approximate bed or bed-set thickness. Lateral permeability traverses through shoreface sandstones yield semivariograms displaying 'pure nugget' effect, indicating the presence of heterogeneity at a scale smaller than the sampling intervals used.

Semivariograms for prodelta–lower shoreface facies sandstones sampled in subsurface section A (Fig. 22) show 'pure nugget' effect (i.e. projected y-axis intercept is approximately 100% of total sill value). No correlation lengths can be inferred for these semivariograms, which imply the presence of a high frequency of permeability heterogeneity at a finer scale than the sampling interval.

Lagoon-fill successions. A vertical data traverse through a lagoon fill succession sampled at outcrop (Unit 6 Fig. 5b) yields a semivariogram with a two-fold 'nested' correlation structure (Fig. 21), reflecting the influence of both average sandstone bed thickness within sandier parts of the succession, and average thickness and spacing of channelized sandstone units within the heterolithic lagoon fill. A hole effect at approximately 2.2 m may reflect the scale of stacking of sandstone–mudstone sediment packages (effectively gross fining-upward packages) with upward decreasing permeability trends. Insufficient data were available to construct experimental semivariograms for lagoon fill successions encountered in the subsurface. However, a similar, highly heterolithic mudstone dominated facies interpreted as an interdistributary bay fill succession within subsurface data (subsurface section A), yields an experimental semivariogram displaying pure nugget effect (Fig. 22), indicating the presence of heterogeneity at a scale smaller than the 3 cm sampling interval.

Outcrop simulation study

Incorporation of probe permeameter and outcrop analogue data into a model for flow simulation

Although the magnitude of recorded probe-permeabilities and the degree of heterogeneity within surface and subsurface data-sets are substantially different, core and outcrop analogue studies have revealed complimentary information concerning the spatial scale, frequency and likely magnitude of permeability variation within the Jerudong Formation. A link clearly exists between the spatial scale at which permeability heterogeneities occur and the frequency of observed sedimentological features. e.g. average spacing of cross set boundaries, or the average scale of coarsening/fining-upward sediment packages. In addition, facies packages observed at outcrop are of sufficient thickness (4–10 m) to provide potentially useful scaled elements for flow simulation.

The Jerudong Formation outcrop permeability data have been used as a template for construction of two hypothetical reservoir segments for investigation of the likely influence of facies stacking patterns and inter/intra-facies permeability variation upon flow performance and hydrocarbon recovery during simple 2D reservoir simulation. Averaging of the permeability data traverses over 2 m intervals through the outcrop allowed construction of a 'coarse' grid-block model (Fig. 23) which captured gross permeability heterogeneity within individual facies. The geometric mean permeability was used to estimate K_h (Fig. 24a), whereas the harmonic mean was used as an estimator of K_v (Fig. 24b). Assignment of porosity–permeability values to individual cells within the model reservoir segment has resulted in a layer cake reservoir structure which assumes a high degree of lateral homogeneity. Although this has produced a somewhat unrealistic and over-simplified reservoir architecture (Fig. 24), it has enabled evaluation of the impact of vertical variations in permeability upon hydrocarbon recovery. During construction of this 'coarse grid' reservoir segment, gaps in the original data set (e.g. where unstable cliff faces preventing safe gathering of data) were 'infilled' simply by substituting data for similar facies encountered in more accessible parts of the outcrop. The coarse grid reservoir segment constructed comprised 7250 grid blocks of 2 m height, 4 m length and 2 m depth.

A second hypothetical reservoir segment was constructed through laterally accreted and trough cross stratified distributary channel fill facies of outcrop Units 7a and 7b (Fig. 25), in order to compare sweep efficiencies etc. within these two different styles of channel fill sandstone. In this detailed model 16 000 grid blocks of 0.25 m height, 0.25 m length, and 2 m depth, were arranged in a simple repeatable 40×20 block pattern to allow construction of a reservoir segment 100 m in length. Averaging of a single vertical data transect over 0.25 m windows provided input variables for K_h and K_v for a column of grid blocks, which were simply extrapolated along beds/foresets through the model (Fig. 26a, b). The results of geostatistical analyses of semivariograms were used to condition incorporation of permeability heterogeneities into the grid-block model at appropriate vertical and lateral frequencies.

Hg injection porosimetry was used to derive porosity for nine outcrop samples representative of lithologies from lower shoreface, upper shoreface, tidal distributary channel, trough cross-stratified distributary channel and laterally accreted distributary channel facies (Table 8). Hg injection porosimetry was carried out using a Carlo Erba Macropore Unit 120 for operation

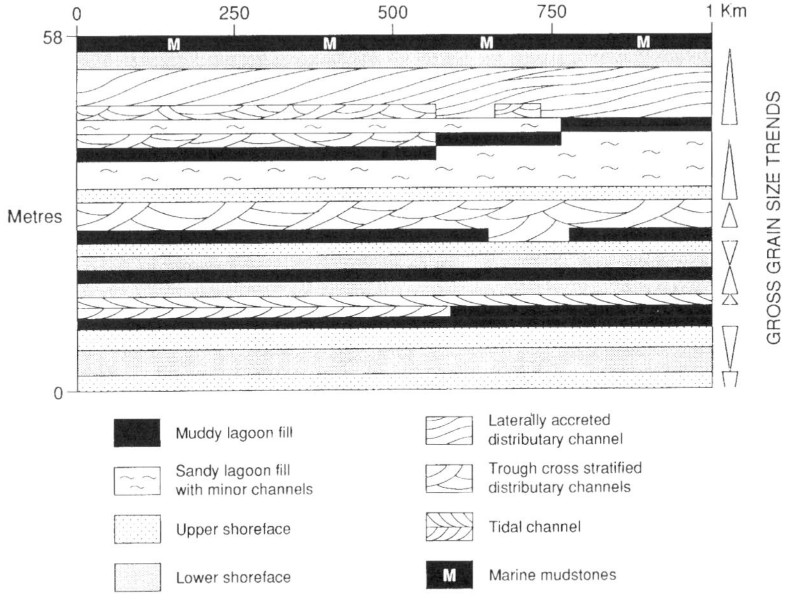

Fig. 23. Facies template used in construction of 'coarse grid' reservoir segment comprising 7250 blocks of 2 m height, 4 m length, and 2 m depth.

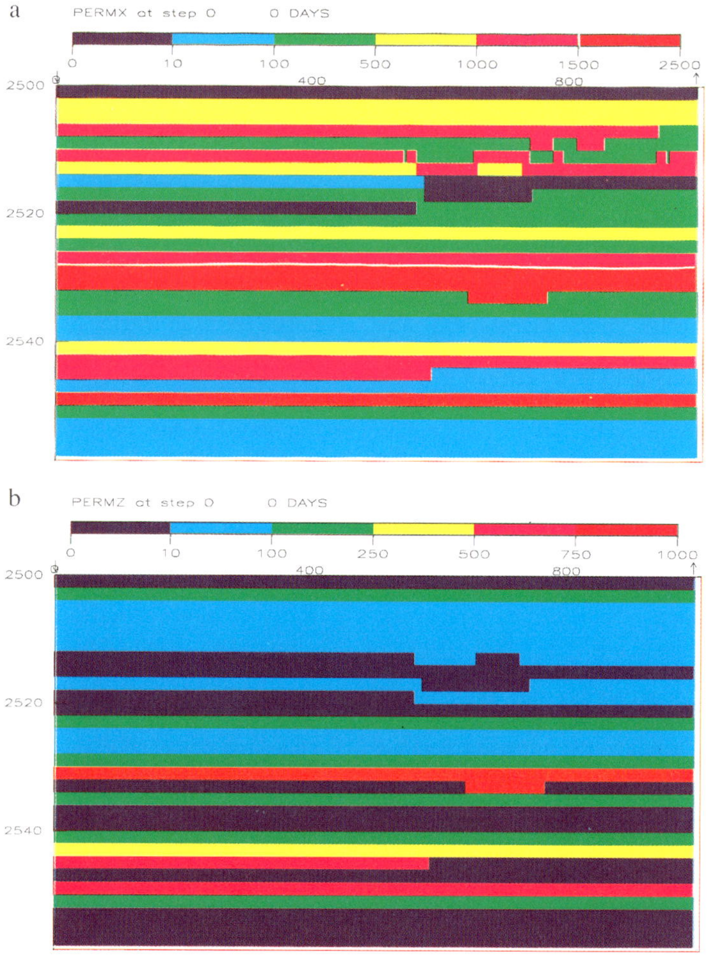

Fig. 24. (**a**) Coarse grid simulation model – horizontal permeability display. (**b**) Coarse grid simulation model – vertical permeability display. (**c**) Coarse grid simulation model – distribution of oil saturation 150 days into waterflood. (**d**) Coarse grid simulation model – distribution of oil saturation 550 days into waterflood.

within a 0–300 kPa (0–3 bar) range; and an automated Carlo Erba WS2000 system allowing incremental injection of Hg into sample chips of known permeability at pressures of up to 200 MPa (2000 bar). The system is suitable for quantification of volume changes as low as 0.1 mm^3, and allowed determination of microporosity with pore radius 2.0–7500 nm. Pore size distributions were determined from Hg injection data using the Washburn Equation (Washburn 1921), and Hg withdrawal efficiencies were determined using the method outlined by Wardlaw & Taylor (1976). Factors including fluid density, saturation, viscosity, wettability, interfacial tension, contact angle, displacement rate, and the geometry of the pore system will all effect withdrawal efficiency within the Hg–air–solid system and also the efficiency of hydrocarbon recovery. Capillary pressure curves for the Hg–air–solid system were converted to the oil–water–solid system using the methods outlined in Vavra *et al.* (1992) and Archer & Wall (1986).

Porosities were assigned to grid blocks using a correlation which integrated outcrop porosity–permeability data with subsurface data. For the subsurface data set, probe permeability data gathered at 1 cm spacing were averaged over 30 cm intervals (comparable to the resolution of the formation density logging tool), and compared with porosities interpreted from density logs (Fig. 27a). The porosities interpreted using

c

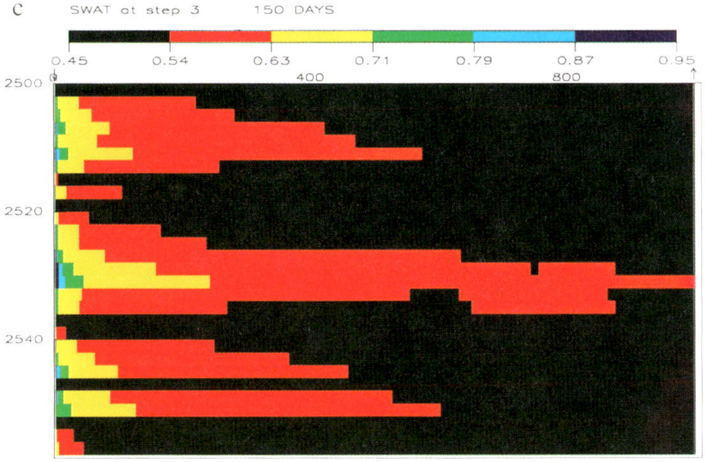

d

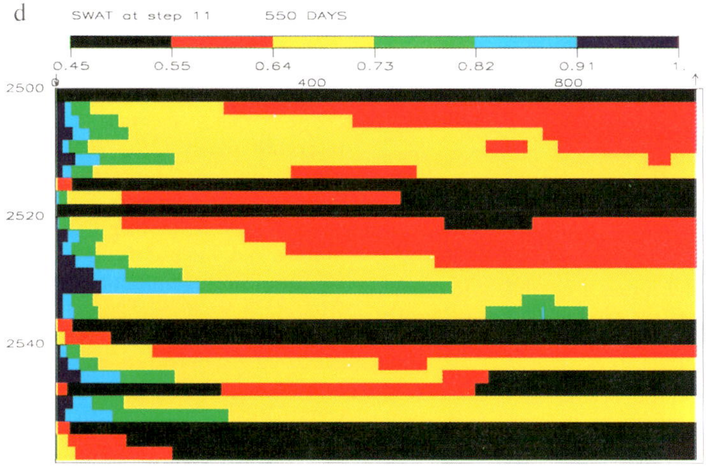

Fig. 24. continued.

the formation density log response were noted to be similar to those determined by conventional analyses of core plugs, but are representative of significantly larger volumes of investigation. Permeabilities recorded for outcrop samples are typically higher than those recorded for similar facies within subsurface reservoir sections (Fig. 27a). However, combination of the outcrop and subsurface data resulted in construction of a simple correlation which allowed porosity values to be assigned to the wide range of grid block permeabilities.

For simulation, the model reservoir segments were located at a depth of 2500 m (an intermediate depth for Bruneian reservoirs) and subject to a live oil simulation using an oil having viscosity of *c.* 2.65 MPa S^{-1}, and tank density of 930 kg m^{-3}. The modeled reservoir segments were located within the oil column, well away from the oil–water and gas–oil contacts input for the model scenario. Model rock curves for relative permeability, capillary pressure etc. were constructed using relationships developed by BSP following review of SCAL data from offshore Bruneian oilfields (Fig. 27b). Comparison of model curves with laboratory derived capillary pressure curves (water/oil system) for the outcrop samples used for porosity determination have indicated that model curves calculated for some facies (e.g. distributary channel sandstone facies) can illustrate reasonable correspondence with experimentally derived data over a significant range of water saturations (5–80% in the case of some 'clean' distributary channel facies). In these cases the model curves appear to represent reasonable 'average' relationships. However, having been

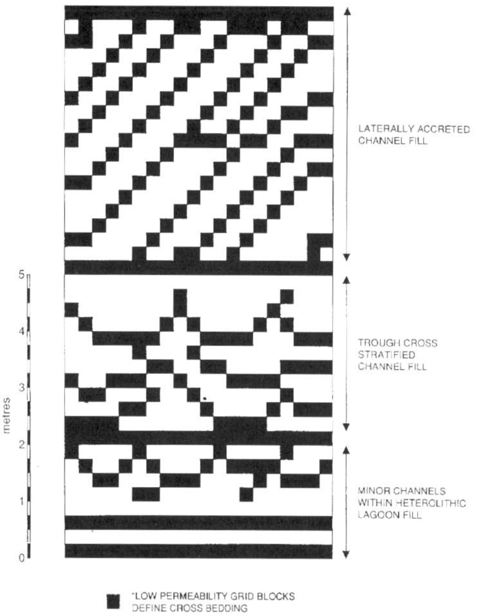

Fig. 25. Facies template used in construction of 'fine grid' reservoir segment comprising 7250 blocks of 2 m height, 4 m length, and 2 m depth.

originally derived via analysis of subsurface samples, the model curves are unlikely to predict rock capillary pressure characteristics of weathered surface outcrop samples with any great accuracy, other than in a relative context.

The model reservoir segments were set to an initial reservoir pressure of 260 bar, which represents some 10 bar overpressure relative to estimated hydrostatic pressure. Waterflood through the hypothetical reservoir segment was modelled using a simple injector–producer pair located at opposite ends of the model. Water was injected over the entire injection well interval intersecting the model, with production occurring over the entire production well interval. The top and bottom of the model were defined as no flow boundaries. The system was constrained during simulation by maintenance of production well pressures in excess of 200 bar. These constraints coupled with high permeabilities (typically >1000 mD) result in a relatively high flow rate during waterflood (typically around 1.5–3 m per day), so that viscous fluid forces dominate within the system, and capillary forces are subordinate.

Results of coarse grid simulation

The results of the coarse grid simulation are summarized in Figs 24c and d. The results illustrate four zones of preferential imbibition within the studied section:

(1) a succession of upper shoreface facies sandstones located near the base of the studied section (6–10 m in Fig. 4; 2548–2452 m Fig. 23);
(2) a tidal distributary channel (11.75–16.6 m Fig. 4; 2542–2546 m Fig. 23);
(3) a trough cross-stratified distributary channel (36.2–31.95 m Fig. 4; 2526–2532 m Fig. 23);
(4) a composite distributary channel fill unit comprising (a) a lower succession of trough cross-stratified distributary channel sandstones which incise into a heterolithic lagoon fill succession (43–46.9 m Fig. 4; 2512–2516 m Fig. 23) and (b) a laterally accreted distributary channel fill (46.9–51.6 m Fig. 4; 2506–2512 m Fig. 23).

The water–oil front advances most rapidly through the trough cross-stratified distributary channel facies located in the centre of the reservoir panel (36.2–31.95 m Fig. 4; 2526–2532 m Fig. 23). Horizontal and vertical permeabilities within this flow unit range from 1028–2472 mD, and 67–885 mD respectively, and are among the highest in the modelled reservoir section. Interpreted porosities for this unit range from 23% to 30%. The data for this particular distributary channel suggest that the basal parts of trough cross-stratified distributary channel sandstones are likely to waterflood more efficiently than the upper parts of trough cross-stratified distributary channel fills. However, this is not always the case. For example, the fluid saturation data suggest that for the uppermost distributary channel fill succession, the basal part of the trough cross-stratified distributary channel fill facies (43–46.9 m Fig. 4; 2512–2516 m Fig. 23) floods considerably slower than the upper parts of this facies. In this case, the difference in sweep is attributed to the fact that the base of this channel is heterolithic in places, and down-cuts into underlying mudstone dominated lagoonal facies – a feature which is reflected in permeability averages calculated over 2 m windows.

Laterally accreted distributary channel fill facies (46.9–51.6 m Fig. 4; 2506–2512 m Fig. 23) also appear to be heterogeneous with respect to waterflood performance. Figures 24c and d indicates that the basal–middle parts of these laterally accreted channel fill deposits are subject to most rapid waterflood. This is discussed further within the context of results of a fine grid waterflood simulation model.

Within shoreface facies, upper shoreface sandstones are subject to more rapid imbibition of water than the finer grained, more argillaceous lower shoreface deposits.

Heterolithic lagoon fill facies are generally least responsive to waterflood. However, if significant sandstone laminae (several centimetres scale) are present, lagoon fill deposits may display increased horizontal permeability, and waterflood morc successfully than their more mud-rich counterparts. Vertical permeability within lagoonal facies is always likely to be poor, and in some scenarios stacking of facies may effectively compartmentalise a succession, enabling multiple completions to optimise recovery from interbedded channel and shoreface facies. The coarse grid model thus allows some evaluation of the importance of facies stacking patterns upon waterflood performance. This appears particularly true of shoreface and distributary channel facies.

Facies stacking patterns within the Jerudong Formation at Punyit Beach reveal that coarsening upward parasequences may comprise both upward transitions from shoreface through lagoon fill to distributary channel environments; or transitions from shoreface directly into higher permeability channel environments.

Coarsening upward shoreface successions (K_h typically 0.01 mD at their base, but hundreds to thousands of milliDarcies near their top; geometric mean >100 mD) are sometimes seen to be truncated by an overlying distributary or tidal channel sandstone with significantly higher permeability (K_h typically 1000 mD at the channel base to 10 mD near the channel top; geometric mean *c.* 1000 mD). In this scenario, the better reservoir quality channel sandstone may act as a preferential site for water imbibition, and early water breakthrough. However, where coarsening upward shoreface successions are overlain by successions of impermeable lagoonal mudstones (0.001–1000 mD; geometric mean *c.* 10 mD), it may be possible to obtain more favourable sweep efficiencies and predictions of early water breakthrough during waterflood. In the coarse grid simulation model, a distributary channel at approximately 800 m lateral distance in the section (2526–2532 m Fig. 23) 'down-cuts' through a heterolithic succession of lagoonal sandstones and mudstones (K_h 100–500 mD, K_v 0–10 mD), into upper shoreface facies sandstones. In this case, juxtapositioning of tidal channel facies sandstones (K_h 1500–2500 mD, K_v 750–1000 mD) with upper shoreface facies sandstones (K_h 100–500 mD, K_v 100–250 mD) appears to result in preferential sweep of the upper shoreface facies where in contact with the tidal channel (Fig. 24d). This is largely due to gravitational draw-down of the water from the tidal channel into the upper shoreface facies, and suggests the potential for perhaps considerable cross-flow where these facies are juxtaposed.

Results of fine grid simulation.

Results of fine scale grid-block simulation through a model (Fig. 25) of laterally accreted and trough cross-stratified distributary channel fill deposit (Units 7a & 7b) and part of a heterolithic lagoon fill succession (Unit 6) are summarized in Fig. 26.

The model section is very heterogeneous with respect to waterflood performance. For the modelled conditions, the water–oil front advances most rapidly through the trough cross-stratified distributary channel fill deposits of Unit 7a (43.0–6.9 m Fig. 4; *c.* 2504.75–2507.75 m Fig. 25). The water–oil front advances considerably slower through laterally accreted channel fill deposits (Unit 7b 46.9–51.6 m Fig. 4; *c.* 2500–2504.75 m Fig. 25) and minor trough cross-stratified channels within heterolithic lagoon fills (Unit 6 33.85–43.0 m Fig. 4; 2508–2510 m Fig. 25).

The heterogeneous waterflood performance for channelized sandstone facies with upward decreasing permeability is similar to that described by Weber (1986), Xue Peihua (1986), Van De Graaf & Ealey (1989) and Werren *et al.* (1990). Heterogeneous waterflood performance is especially evident within laterally accreted distributary channel fill deposits (Figs 25 and 26c, d). In this facies, preferential imbibition of water (under conditions where viscous forces dominate) occurs along high porosity–permeability facies near the base of the channel fill. In addition, the model waterflood also clearly indicates fingering of the flood front up high porosity–permeability horizons within the inclined foreset beds characteristic of laterally accreted channel fill deposits. Xue Peihua (1986) has described very similar heterogeneous waterflooding of laterally accreted fluvial point bar deposits, recording early water breakthrough in the middle–lower point bar (where permeabilities are highest), with less than 50% of the total point bar being efficiently flooded.

The fine scale grid simulation also reveals trough cross-stratified distributary channel fills to be heterogeneous with respect to waterflood, and subject to preferential imbibition along higher porosity–permeability deposits towards their base.

Heterolithic lagoon fill deposits display variable waterflood characteristics. Where sandy

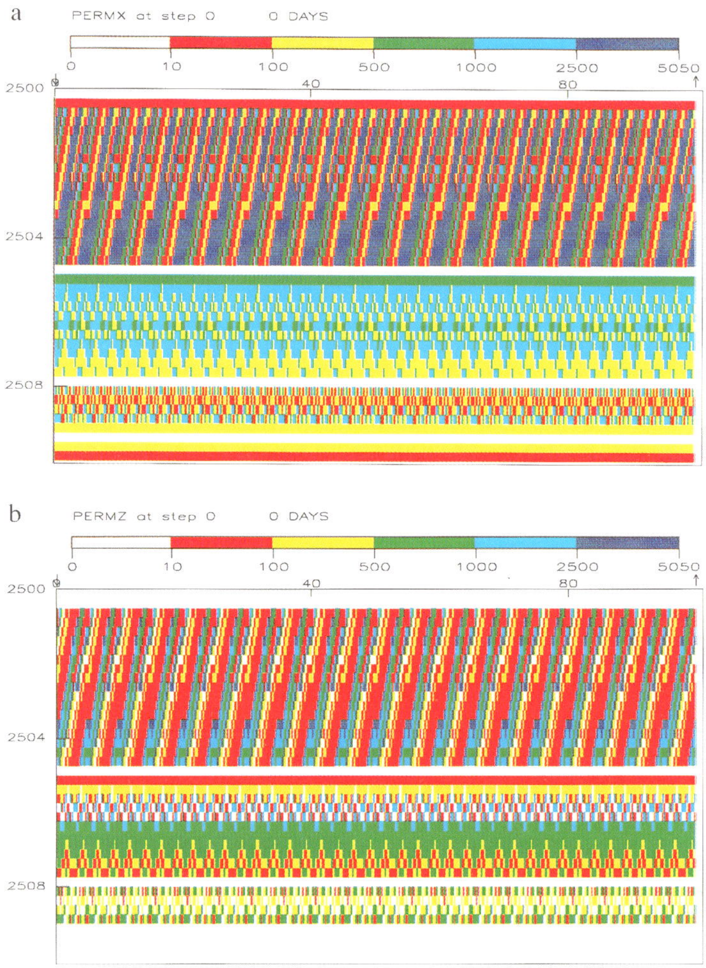

Fig. 26. (**a**) Fine grid simulation model – horizontal permeability display. (**b**) Fine grid simulation model – vertical permeability display. (**c**) Fine grid simulation model – distribution of oil saturation 40 days into waterflood. (**d**) Fine grid simulation model – distribution of oil saturation 80 days into waterflood.

lithologies are present, they may have significant horizontal permeability (hundreds of milliDarcies), and hence may be subject to reasonable horizontal sweep. Vertical permeability however is always low (rarely 10 mD) and vertical sweep efficiency is therefore poor. Lagoon fill facies again appear to effectively compartmentalise the modelled succession with respect to vertical cross-flow.

Discussion

Examination of strata at outcrop clearly allows identification of potential reservoir heterogeneities and their likely distribution. Although relatively thin in terms of many offshore Bruneian reservoir sections, the Jerudong Formation studied at outcrop comprises packages of heterolithic sandstone, siltstone and mudstone (tens of metres thick) within thicker mudstone-dominated successions, providing one basic scale at which flow units may be defined. These are identifiable on wireline logs, and may be correlateable on an interwell scale. The smaller scale 'Units' or facies associations used to construct a vertical subdivision of studied outcrop sections may also be identifiable on wireline logs. The detailed probe-permeameter traverse through the studied outcrop (Fig. 8) clearly indicates the subdivision of the succession into a number of sub-units having similar permeability characteristics, and these approximate the facies

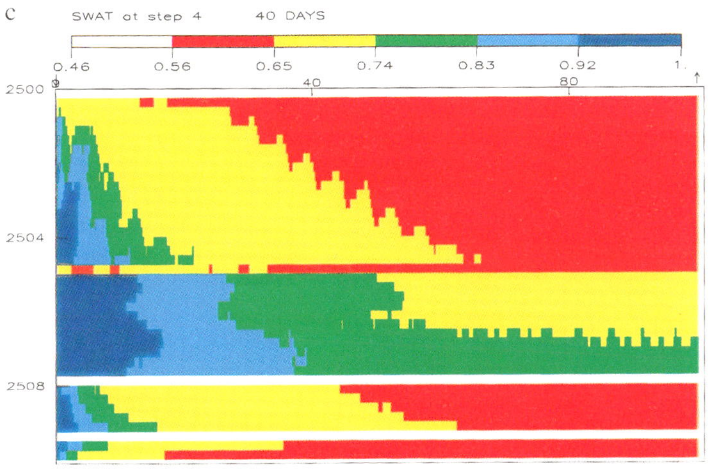

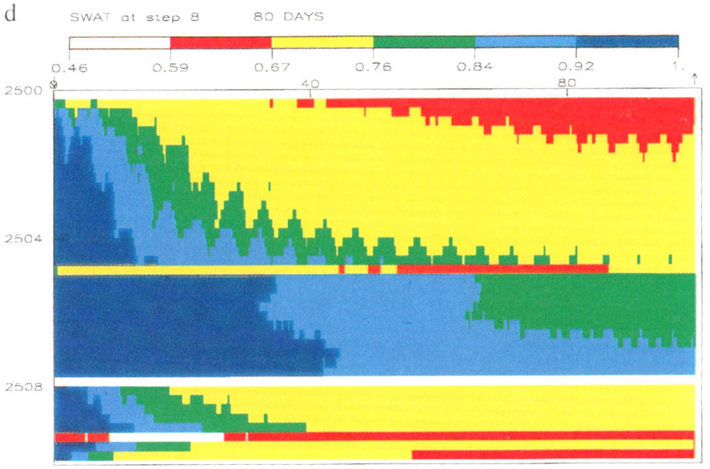

Fig. 26. continued.

divisions present. The facies units identified are generally of comparable thickness (4–10 m) to vertical grid block dimensions indicated by Weber (1993) to be normal for simulation studies, and are similar to those used in documented simulations by Wolcott & Chopra (1993) and Nybraten *et al.* (1990). They thus appear to provide potentially useful scaled reservoir elements for flow simulation.

The simple outcrop simulations illustrated predict heterogeneous waterflood performance, with preferential imbibition into channel base and upper shoreface facies. Differences in the capillary pressure (P_c) characteristics of different facies types are clearly indicated by the Hg injection capillary pressure curves (Figs 28 and 29). All the facies studied at outcrop are highly laminated, and a similar degree of capillary pressure contrast is predicted to occur at laminar scale. Laminar scale capillary pressure variations are indicated by step-like profiles in some capillary pressure curves (Figs 28 and 29). Capillary pressure contrasts of the scale observed both within and between outcrop samples will effect the mobility of oil within laminated and heterogeneous lithologies (Abrahms 1975; Kortekaas 1985; Kocberber & Collins 1990; Dawe *et al.* 1992; Hartkamp-Bakker 1993; Ringrose *et al.* 1993). In particular, these authors have shown that during production of hydrocarbons from heterogeneous porosity–permeability systems, capillary trapping of otherwise moveable hydrocarbons within large pores may occur if the ratio of viscous to capillary forces is low, and that recovery efficiency will also vary with waterflood rate, as well as and waterflood

Table 8. *Results of Hg injection porosity determination*

Sample	Grain size	Height in section (m)	Porosity %	K (mD)	Hg WE (%)	Pore sizes (μm)	
						Median	Mode
Laterally accreted channel fill (top)	Fine sst	50.9	34.61	3700	10	10	10
Laterally accreted channel fill (middle)	Fine sst	49.05	30.59	1534	53	4	6
Lterally accreted channel fill (base 2)	V. fine sst	47.3	27.35	575	36	10	10
Laterally accreted channel fill (base 1)	Fine sst	47	36.03	2500	15	0.16	10 (0.05)*
Trough cross stratified distributary (base)	Fine sst	43.45	20.83	1130	2	1.4	4
Trough cross stratified distributary (top)	V. fine sst	45.8	27.29	670	2	8	10
Tidal channel	Fine sst	14.25	28.25	1750	23	8	10
Lower shoreface	V. fine sst	54.8	21.66	1700	11	0.85	40 (0.8)*
Upper shoreface	Fine sst	8.6	24.99	2450	16	3.5	7

* Bimodal distribution (minor mode in brackets).

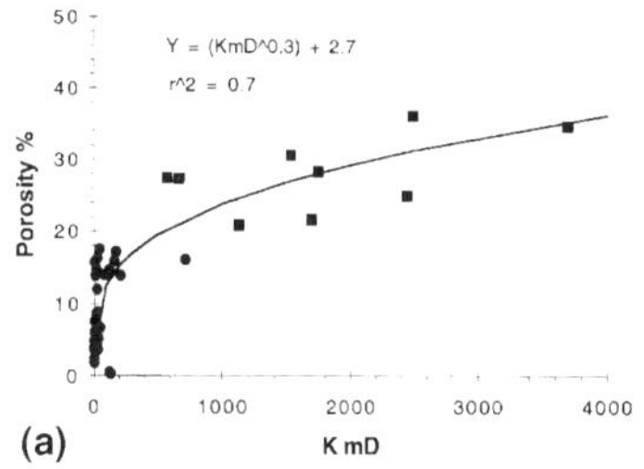

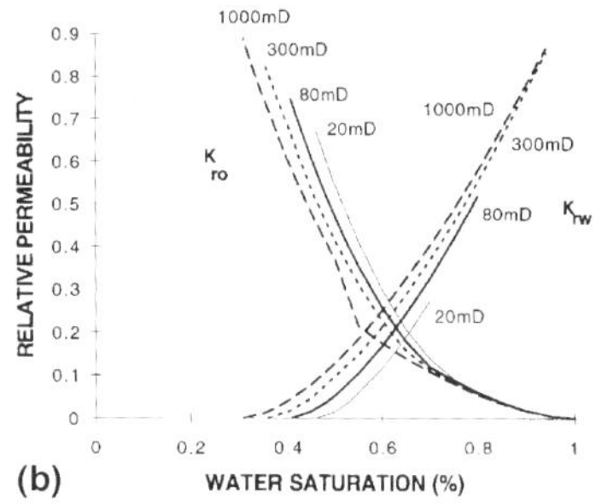

Fig. 27. (**a**) Porosity versus permeability correlation constructed using both outcrop and subsurface data. (**b**) Relative permeability curves for rocks of 20 mD, 80 mD, 300 mD and 1000 mD permeability. Curves constructed using relationships developed by BSP following review of SCAL data from offshore Bruneian Oilfields.

direction relative to sedimentary fabric. Ringrose *et al.* (1993) illustrated that for heterolithic sediments, during low rate waterflood parallel to lamination, viscous forces move oil from high permeability laminae, while capillary forces drive oil from the fine-grained laminae into coarse grained laminae by capillary imbibition of water. This process results in higher recovery than will occur for flow normal to laminar permeability heterogeneities, in which case impaired sweep efficiency reduces the effectiveness of the capillary forces that drive oil from fine grained laminae into coarse grained laminae, and capillary trapping of oil becomes likely. In the Jerudong Formation, the degree and scale of lamination and bedding fabrics varies between different facies, but inferences as to their possible effects upon hydrocarbon recovery efficiency can be made.

Marine shoreface successions are characterized by distinct upward coarsening grain-size trends, with an accompanying upward increase in bedform scale and permeability. Internally they display a wide variety of both laminar and bedding scale heterogeneities. They are characterized by low-angle, undulatory bedding surfaces in which gently curved concave–convex erosion surfaces truncate one another at low angles, creating a 'hummocky cross-stratified fabric' mimicked by millimetre-scale internal lamination. Thin, discontinuous mudstones whilst present within the studied shoreface sections, are certainly not abundant, and show little evidence of significant stacking/vertical overlapping. Their potential effects upon vertical sweep efficiency are interpreted as minimal (cf. Richardson *et al.* 1987). Van De Graaff & Ealey (1989) and Slatt & Galloway (1993) have shown that the presence

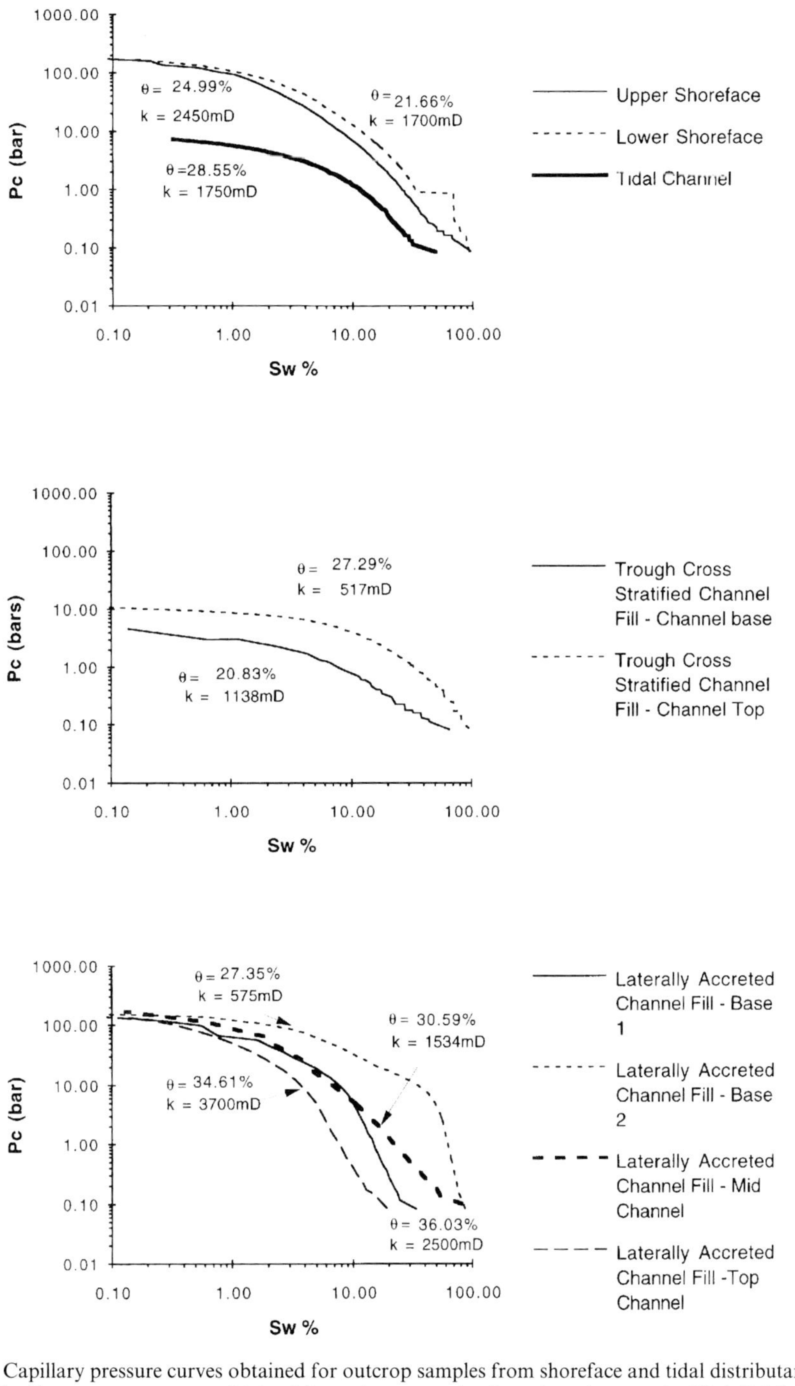

Fig. 28. (**a**) Capillary pressure curves obtained for outcrop samples from shoreface and tidal distributary channel facies. (**b**) Capillary pressure curves obtained for outcrop samples from trough cross-stratified distributary channel fill facies. (**c**) Capillary pressure curves obtained for outcrop samples from laterally accreted distributary channel fill facies.

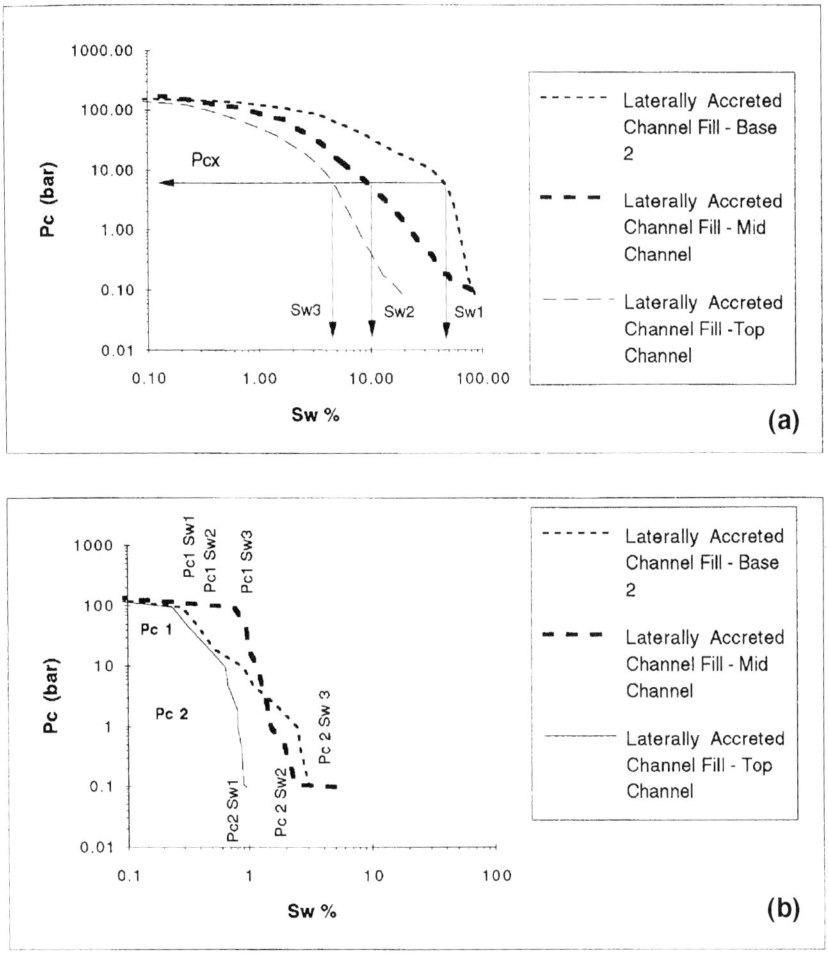

Fig. 29. Capillary pressure curves for samples collected from base, middle and upper parts of laterally accreted distributary channel fill. (**a**) Drainage curves (in which wetting phase is decreasing) illustrate that at capillary pressure P_{cX}, sandstones from the lower point bar facies are close to maximum water saturation when S_w within middle and upper point bar facies is considerably reduced. High S_w and hence lower relative permeability to oil within lower parts of point bars could result in their becoming a site of preferential water retention relative to the higher permeability middle and upper parts of point bar deposits. (**b**) Imbibition curves (in which wetting phase is increasing, as in waterflood) indicate that at P_{c1}, higher S_w, and hence relative permeability to water, within mid channel facies (effectively middle point bar) could result in their forming sites of preferential capillary imbibition,during waterflood. However at P_{c2}, S_w and hence relative permeability to water is higher within lower permeability facies from the base of laterally accreted distributary channel fills, which will become the preferential sites of capillary imbibition. These capillary effects are likely to be especially effective under low rate waterflood conditions where the ratio of viscous to capillary forces is low.

of a high permeability horizon at the top of coarsening upward sediment packages combined with gravitational draw-down of waterflood results in shoreface facies successions displaying better recovery factors than, for example, upward fining sediment packages of similar permeability range in the same production scenario. Figure 28a and Table 8 illustrate the differences in capillary pressure characteristics of upper and lower shoreface facies.

Detailed, laminar scale simulation studies of two-phase flow through highly laminated, hummocky cross-stratified shoreface sandstones, using sedimentological and probe-permeameter data sets from the Brent Group Rannoch Formation (UK North Sea) have been documented by Corbett & Jensen (1993*b*), and Corbett *et al.* (1992). Via combination of subfacies (i.e. laminar scale) pseudofunctions with models of bedform architecture, Corbett & Jensen (1993*b*) and Corbett *et al.* (1992) were able to realistically up-scale reservoir characteristics via a 'Geopseudo' approach which incorporates small-scale capillary effects of the type described by Ringrose *et al.* (1993). Corbett & Jensen (1993*b*), and Corbett *et al.* (1992) found that at subfacies scale, lamination fabrics similar to those observed within the Jerudong Formation are associated with order of magnitude permeability contrasts and large capillary pressure contrasts. These authors demonstrated that although laminae and subfacies constitute discontinuous elements at the interwell scale, shoreface facies sediments are still likely to behave as layered systems in which two-phase flow through hummocky cross-stratified bedforms will be strongly anisotropic, resulting in better recovery efficiency for horizontal flow than vertical flow. Similar effects are predicted for shoreface facies within the Jerudong Formation, where laminar scale permeability heterogeneities may be associated with permeability contrasts of up to three orders of magnitude (i.e. >1000 mD).

Facies stacking patterns observed at outcrop along Punyit Beach reveal that coarsening upward parasequences may comprise both upward transitions from shoreface through lagoon fill to distributary channel environments; or transitions from shoreface directly into distributary channel environments. The latter scenario has been described for paralic successions within the Brent Group (UK North Sea) by Daws & Prosser (1992), where a transition from middle shoreface environments within the Rannoch Formation to stacked tidal channel deposits within the Etive Formation is associated with permeability contrasts of up to three orders of magnitude. In this scenario, Hurst (1993) has illustrated that the presence of thin high permeability 'thief zones' at the base of the Etive Formation strongly influences waterflood performance of the Lower Brent Group (Etive and Rannoch Formation), and that prediction of water breakthrough timing can be difficult (mainly due to inadequate characterisation of permeability heterogeneity when defining flow units). The implication of these studies is that where coarsening upward shoreface successions of the type seen at outcrop (typically <0.01 mD at base, but 100–1000 mD near top; geometric mean *c.* 100 mD) are truncated by an overlying distributary or tidal channel sandstone with significantly higher permeability (typically 1000 mD at its base to 10 mD near its top, geometric mean *c.* 1000 mD), then the higher permeability channel sandstones form sites of preferential water imbibition and early water breakthrough. However, where coarsening upward shoreface successions are overlain by more impermeable successions, (e.g. heterolithic lagoonal facies having permeability of 0.001–1000 mD, geometric mean *c.* 10 mD) cross flow will be limited. In this scenario it may be possible to obtain better prediction of early water breakthrough during waterflood, and more favourable sweep efficiencies via use of multiple completions.

Channelized sandstone facies display highest overall permeability within both outcrop and in subsurface sections studied. Higher permeability within deltaic distributary channels compared to other deltaic facies has also been documented by Tyler *et al.* (1991), within the Cretaceous Ferron Delta (Utah, USA). Within the distributary channel facies, marked permeability heterogeneities essentially occur as a result of grain size and sorting contrasts at sedimentary boundaries which exist at a variety of different scales and frequencies.

Fining upward unit or coset scale (typically up to 2 m thick). Stacking of cosets is associated with permeability contrasts of two to three orders of magnitude.

Bed-set (e.g. trough cross-set) scale: trough cross sets are typically 0.5–2 m thick and 3–10 m wide. Set boundaries are typically associated with permeability contrasts of up to two orders of magnitude. Clay clasts at the base of erosive trough cross-stratification sets locally increase permeability contrasts.

Bed scale: typically 10–80 cm scale and associated with permeability contrasts of two to three orders of magnitude.

Laminar scale: millimetre–centimetre scale laminar heterogeneities are typically associated with permeability contrasts of two to three orders of magnitude. In particular, clay draping of foresets within tidally influenced distributary systems creates large permeability contrasts.

Channel fill facies are typically characterized by an upward decrease in grain size and permeability. The simulation models indicate the combined effects of gravity and the presence of high permeability zones at the base of channel deposits results in poor vertical sweep of channelised facies, with concentration of waterflood towards the lower parts of the channel. This agrees with the observations of Van De Graaf & Ealey (1989).

Both the coarse and fine grid simulation models indicate that hydrocarbon production will be affected by the presence of sedimentary structure (bedding, lamination etc.) within individual fining-upward unit sedimentary packages. Two-phase fluid flow through trough cross bedded sandstones, similar to those seen at outcrop in the Jerudong Formation, is influenced by capillary effects, and hydrocarbon recovery will vary with the direction of waterflood relative to lamination fabric (Kortekaas 1985). The vertical and lateral stacking and overlapping of trough cross-stratification sets within trough cross-stratified distributary channel fills produces a heterogeneous deposit in which laminar scale sub-facies of low and high capillary pressure characteristics are frequently juxtaposed. The significant porosity–permeability contrasts associated with these juxtaposed heterogeneities indicate that capillary trapping of oil may be possible under waterflood conditions where capillary forces dominate over viscous forces. Porosity–permeability contrasts of this type are further enhanced by draping of individual foreset lamina by concentrations of coaly/argillaceous matter within tidally influenced distributary channel fills.

Similarly, although the laterally accreted distributary channel sandstones of Unit 7b display overall high permeabilities (typically >5000 mD), they do contain internal permeability contrasts of up to two orders of magnitude within sandstone beds (associated with 0.5–1 cm scale lamination), and up to three orders of magnitude at inclined bounding surfaces between beds and lateral accretion bed-sets. These heterogeneities are generated by grain size-sorting contrasts, together with the development of thin (often discontinuous) clay laminae along inclined set/foreset boundaries. The fine scale waterflood simulation indicates the latter could form 'barriers' to fluid flow. Laterally accreted channel fills are also clearly heterogeneous with respect to their internal capillary pressure characteristics, and this may also effect hydrocarbon recovery in this facies (Fig. 29). Similar variations in capillary pressure characteristics have been documented for point bar facies by Hartkamp-Bakker (1993) and Kocberber & Collins (1990).

Lagoonal-fill successions are comprised predominantly of low-permeability mudstones and silty mudstones with thin (centimetre scale) typically discontinuous rippled sandstone laminae. The reservoir potential of lagoon fill successions is thus low, except where the lagoon-fill is dissected by deltaic distributary channels. In addition, minor distributary channels (up to 2 m thick) within lagoon fill successions may provide connection with larger 'trunk' distributary systems. Sharma *et al.* (1990) investigating the nature of heterogeneities within barrier island deposits of the Bell Creek Field (Lower Cretaceous Muddy Formation, Powder River Basin, USA) found that the thickest occurrence of lagoonal facies coincided with areas of poorest production following waterflood. However, these authors were able to demonstrate that the geometrical relationship between barrier island facies (shoreface and foreshore sandstones) and lagoonal facies was responsible for the location of highest cumulative waterflood production within the barrier shoreface deposits of the Bell Creek field. This arose as a result of low-permeability lagoonal facies effectively barring the oil bank being swept by the advancing waterflood. Knowledge of geometrical relationships between similar facies within the subsurface Jerudong Formation or equivalent successions could exploit these observations during the development of waterflood strategies.

Conclusions

(1) The Jerudong Formation (early Tortonian) cropping out at Punyit Beach on the coast of Negara Brunei Darussalam comprises a heterolithic succession of deltaic strata containing equivalent lithologies to those encountered at depth in the now abandoned Jerudong Field, and other Bruneian oilfields.

(2) Identified permeability heterogeneities are associated with permeability contrasts of up to three orders of magnitude (10^3 mD). They occur at facies, genetic facies unit, bed-set, bed and laminar scales. Distributary channel facies display the best outcrop permeability (>7000 mD), and include trough cross-stratified fills of relatively straight (often tidal) distributaries, and laterally accreted point bar sandstones deposited within meandering channels. These facies typically (though not always) display an upward decreasing permeability trend. Shoreface facies may also display high (>6000 mD) permeability, but are characterized by upward-increasing permeability and are significantly more heterogeneous than distributary channel facies. Lagoon-fill successions are the most heterolithic studied, comprising sandstones, siltstones and mudstones with a wide range in permeability.

(3) Outcrop permeability data gathered at 5 cm spacing (or less) displays 'average' permeabilities several orders of magnitude larger than

those within similar facies encountered within subsurface reservoir sections from offshore Bruneian Oilfields. Subsurface strata display increased permeability variability compared to surface outcrop data. However, analysis of semivariograms suggests that the spatial distances over which permeability data are correlateable are broadly comparable for surface and subsurface data. Permeability correlation lengths approximate average bed or bed-set thickness, whereas periodicities within data reflect the scale of stacking of genetic facies units (coarsening or fining upward sediment packages.

(4) Facies packages observed at outcrop are of sufficient thickness (4–10 m) to provide potentially useful scaled elements for flow simulation.

(5) Laboratory analyses indicate the studied outcrop facies are heterogeneous with respect to grain and pore size distribution, porosity and capillary pressure characteristics. In particular, the capillary pressure curves and Hg withdrawal efficiencies determined for outcrop samples suggest that hydrocarbon recovery will vary strongly between the different facies types present. These contrasts will also occur at laminar scale within individual facies.

(6) Simple viscous flood simulations of hypothetical reservoir segments constructed using permeability data input from the outcrop analogue confirm that the studied facies are heterogeneous with respect to waterflood performance. Preferential imbibition within upper shoreface facies and the lower parts of distributary channel fills is predicted.

(7) Assessment of the impact of intra-facies permeability variation upon recovery within trough cross-stratified and laterally accreted distributary channel fills indicates preferential imbibition into the lower parts of the channel fills. Both channel facies are heterogeneous with respect to waterflood performance, but the modelled trough cross-stratified channel fill floods far more rapidly than the laterally accreted channel fill, in which significantly more fingering of the flood front occurs.

(8) Heterolithic lagoon fill deposits display variable waterflood characteristics dependent upon their sand content. If sand content is high and laterally continuous, reasonable horizontal sweep may occur within parts of this facies. Heterolithic lagoon fill facies appear to effectively compartmentalize modelled reservoir successions with respect to vertical cross flow.

(9) The gross results of coarse and fine scale grid block simulation are comparable, e.g. both indicate that channelized sandstone facies will be heterogeneous with respect to waterflood performance, and preferential imbibition of water is likely to occur along the base of the channel fills. However, the coarse grid simulation underestimates the degree to which heterogeneous waterflood is likely to occur within laterally accreted distributary channel facies.

(10) Outcrop analogues are useful for determination of parameters such as facies dimensions, possible magnitude and frequency of permeability variation. They may also provide insight into likely variations in rock capillary pressure characteristics in the subsurface, but should not be used other than in the broadest relative sense.

The authors would like to thank Brunei Shell Petroleum for permission to publish this data. The study described represents first project in a collaborative initiative between the University of Aberdeen, Brunei Shell Petroleum, and the University of Brunei Darussalam. Project funding was supplied by Brunei Shell Petroleum. D.J.P. is grateful to S. Apong for help in gathering field data. Figures were drafted by B. Fulton.

Appendix

The semivariogram function $\gamma^*(h)$ at a sample spacing or *Lag Distance* of h is estimated by

$$\gamma^*(h) = [f(x+h) - f(x)]^2/2N(h)$$

where

$f(x)$ is the value of the sample at a point x in the core
$f(x+h)$ is the value of the sample at a distance h away from x
$N(h)$ is the number of pairs of samples located at distance h from each other.

Experimental semivariograms are constructed by plotting the quantity $\gamma^*(h)$ along the y-axis, versus the spatial separation of sample pairs (h) along the x-axis. The distance at which spatial correlation of the semivariogram value ceases and the experimental semivariogram levels off to a Sill is termed the Correlation Length or Range, which quantifies the distance over which there will be reasonable correlation between sample permeability values. Note, not all semivariograms level off to sill value, some may display continued correlation of data over ever increasing scales.

References

ABRAHMS, A. 1975. The influence of fluid viscosity, interfacial tension, and flow velocity on residual oil saturation left by waterflood. *Journal of the Society of Petroleum Engineers*, Oct. 437–447.

ARCHER, J. S. & WALL, C. G. 1986. *Petroleum Engineering: Principles and practice*. Graham and Trotman, London.

CHEEL, R. J. & LECKIE, D. A. 1993. Hummocky cross-stratification. *In*: WRIGHT, V. P. (ed.) *Sedimentology Review*, **1**. Blackwell Scientific Publications, Oxford, 123–138.

CONOVER, W. J. 1971. *Practical Nonparametric Statistics*. Wiley, New York.

CORBETT, P. W. M. & JENSEN, J. L. 1992*a*. Variation of reservoir statistics according to sample spacing and measurement type for some intervals in the Lower Brent Group. *The Log Analyst*, January–February 1992, 22–39.

—— & ——1992*b*. Estimating the mean permeability: How many measurements do you need. *First Break*, **10**, 89–94.

—— & ——1993*a*. Quantification of heterogeneity, a role for the minipermeameter in reservoir characterisation. *In*: NORTH, C. & PROSSER, D. J. (eds) *Characterisation of Fluvial and Aeolian Reservoirs*. Geological Society, London, Special Publications, **73**, 443–442.

—— & ——1993*b*. Application of probe- permeametry to the prediction of two-phase flow performance in laminated sandstones (Lower Brent Group, North Sea). *Marine and Petroleum Geology*, **10**, 335–346.

——, RINGROSE, P. S., JENSEN, J. L. & SORBIE, K. S. 1992. Laminated lastic reservoirs: The interplay of capillary pressure and sedimentary architecture. *In*: *Proceedings of 6th Annual Technical Conference and Exhibition of the Society of Petroleum Engineers*. Paper No. SPE 24699, Washington, DC, 4–7 October, 365–376.

DALTABAN, T. S., WANG, J. S. & ARCHER, J. S. 1991. Understanding the physics of minipermeameter measurements through the use of the minipermeameter simulation program MIN_PER. *Conference Paper, Minipermeametry in Reservoir Studies*: Petroleum Science and Technology Institute, 27th June 1991, Edinburgh.

DAWE, R. A., WHEAT, M. R. & BINDER, M. S. 1992. Experimental investigation of capillary pressure effects upon immiscible displacement in lensed and layered porous media. *Transport in Porous Media*, **7**, 83–101.

DAWS, J. A. & PROSSER, D. J. 1992. Scales of permeability heterogeneity within the Brent Group. *Journal of Petroleum Geology*, **15**, 397–418.

DRYER, T. 1990. Sandbody dimensions and infill sequences of stable humid climate deltaplain channels. *In*: BULLER, A. T. (ed.) *North Sea Oil and Gas Reservoirs*, **2**. Graham and Trotman, London, 337–351.

——, SCHEIE, A. & WALDERHAUG, O. 1990. Minipermeameter-based study of permeability trends in channel sand bodies. *AAPG Bulletin*, **74**, 359–374.

ECKERT, H. R. 1970. Planktonic foraminifera and time-stratigraphy in well Ampa 2. *Brunei Museum Journal*, **2**, 320–326.

GOGGIN, D. J., CHANDLER, M. A., KOCUREK, G. & LAKE, L. W. 1988. Patterns of permeability in aeolian deposits: Page Sandstone (Jurassic), North-eastern Arizona. *SPE Formation Evaluation*, June, 297–306.

HARMS, J. C., SOUTHARD, J. B., SPEARING, D. R., & WALKER, R. G. 1975. *Depositional environments as interpreted from primary sedimentary structures and stratification sequences*. SEPM, Short Course, **2**.

HARTKAMP-BAKKER, C. A. 1993. *Permeability heterogeneity in cross bedded sandstones: Impact upon water/oil displacement in fluvial reservoirs*. PhD Thesis, Technical University Delft, Netherlands.

HO KIAM FUI 1978. Stratigraphic framework for oil exploration in Sarawak. *Geological Society of Malaysia Bulletin*, **10**, 1–13.

HOHN, M. E. 1988. *Geostatistics and Petroleum Geology*. Van Nostrand Reinhold, New York.

HURST, A. 1993. Sedimentary flow units in hydrocarbon reservoirs: some shortcomings and a case for high resolution permeability data. Special Publications of the International Association of Sedimentologists, **15**, 191–204.

JAMES, D. M. D. (ed.) 1984. *Hydrocarbon Resources of Negara Brunei Darussalam*. Special Publications of the Musium Brunei and Brunei Shell Petroleum. Pub. Kok Wah Press.

JOURNEL, A. G., & HUIJBREGTS, CH. J. 1978. *Mining Geostatistics*. Academic Press Inc., London.

KOCBERBER, S & COLLINS, R. E. 1990. Impact of reservoir heterogeneity on initial distributions of hydrocarbons. *In*: *Proceedings of 65th Society of Petroleum Engineers Annual Technical Conference*, New Orleans. SPE20547, Sept. 23–26.

KORTEKAAS, T. F. M. 1985. Water/oil displacement chartacteristics in cross bedded sandstone reservoir zones. *Journal Society Petroleum Engineers*, **25**, 917–926.

LEICHTI, P., ROE, F. N., HAILE, N. S. & KIRK, H. J. C. 1960. The geology of Sarawak, Brunei and the western part of North Borneo. *Bulletin of the Borneo Geological Survey*, **3**, 360.

MOSS, B. 1992. The petrophysical characteristics of the Brent sandstones. *In*: MORTON, A. C., HASZELDINE, R. S., GILES, M. R. & BROWN, S. (eds) *Geology of the Brent Group*. Geological Society, London, Special Publications, **61**, 471–496.

NYBRATEN, G., SKOLEM, E. & OSTBY, K. 1990. Reservoir simulation of the Snorre Field. *In*: BULLER, A. T., BERG, E., HJELMELAND, O., KLEPPE, J., TORSAETER, O. & AASEN, J. O. (eds) *North sea Oil and Gas Reservoirs – II*. Graham and Trotman, London, 91–114

PEMBERTON, S. G., FREY, R. W., RANGER, M. J. & MACEACHERN, J. A. (1992). The conceptual framework of ichnology. *In*: PEMBERTON, S. G. (ed.) *Applications of Ichnnology to Petroleum Exploration – A Core Workshop*. SEPM Core Workshop, **17**, 1–32.

POTTER, T. L., JOHNS, D. R. & DE NATRIS, T. G. B. 1984. Lithostratigraphy. *In*: JAMES, D. M. D. (ed.) *Hydrocarbon Resources of Negara Brunei Darussalam*. Special Publications, Musium Brunei and Brunei Shell Petroleum. Pub. Kok Wah Press, 43–75

QUI YINAN, CHEN ZIQI & XU SHICE 1982. Waterflooding of channel sandstone reservoirs. *In*: *Proceedings of SPE International Meeting on Petroleum Engineering*, Paper SPE 10559, Beijing, China, March 19th–22nd, 15–42.

RICHARDSON, J. G., SANGREE, J. B. & SNEIDER, R. M. 1987. Permeability distributions in reservoirs. *Journal Petroleum Technology*. October, 1197–1199.

RINGROSE, P. S., SORBIE, K. S., CORBETT, P. W. M. & JENSEN, J. L. 1993. Immiscible flow behaviour in laminated and cross bedded sandstones. *Journal of Petroleum Science and Engineering*, **9**, 103–124.

SHARMA, B., HONARPOUR, M. M., JACKSON, S. R., SCHATZINGER, R. A. & TOMUTSA, L. 1990. Determining the productivity of a barrier island sandstone deposit from integrated facies analysis. *SPE Formation Evaluation*, December, 413–420.

SLATT, R. M. & GALLOWAY, W. E. 1993. Geological Heterogeneities. *In*: MORTON THOMPSON, D. & WOODS, A. M. (eds) *Development Geology Reference Manual*. AAPG Methods in Exploration Series, **10**, 278–281.

SUTHERLAND, W. J., HALVORSEN, C., HURST, A., MCPHEE, C. A., ROBERTSON, G., WHATTLER, P. R. & WORTHINGTON, P. F. 1993. Recommended practice for probe permeametry. *Marine and Petroleum Geology*, **10**, 309–317.

TYLER, N., BARTON, M. D. & FINLAY, R. J. 1991. Outcrop characterization of flow unit and seal properties and geometries. Ferron Sandstone, Utah. *In*: *Proceedings of SPE Annual Technical Conference and Exhibitione*, Paper SPE 22670, Dallas, Texas, October 6th–9th, 127–134.

VAN DE GRAAF, W. J. E. & EALEY, P. J. 1989. Geological modelling for simulation studies. *AAPG Bulletin*, **73**, 1436–1444.

VAN WAGONER, J. C., MITCHUM, R. M., CAMPION, K. M. & RAHMANIAN, V. D. 1990. Siliciclastic sequence stratigraphy in well logs, cores and outcrops. *AAPG Methods in Exploration Series*, **7**.

VAVRA, C. L., ALDI, J. G. & SNEIDER, R. M. 1992. Capillary Pressure. *In*: MORTON THOMPSON, D. & WOODS, A. M. (eds) *Development Geology Reference Manual*. AAPG Methods in Exploration Series, **10**, 221–225.

WARDLAW, N. C. & TAYLOR, R. P. 1976. Mercury capillary pressure curves and the interpretation of pore structure and capillary behaviour in reservoir rocks. *Bulletin of the Can. Pet. Geol.*, **24**, 225–262.

WASHBURN, E. W. 1921. Methods of determining the distribution of pore sizes in porous material. *Proceedings of the National Academy of Sciences, USA*, **7**, 115–116.

WEBER, K. J. 1986. How heterogeneity effects oil recovery. *In*: LAKE, L. W & CARROL, H. B. JR. (eds) *Reservoir Characterization*. Academic Press, 487–544.

WERREN, E. G. SHEW, R. D. ADAMS, E. R. & STANCLIFFE, R. J. 1990. Meander belt reservoir geology, Mid-Dip Tuscaloosa, Little Creek field, Mississippi. *In*: BARWIS, J. H., MCPHERSON, J. G., & STUDLICK, J. R. J. (eds) *Sandstone Petroleum Reservoirs*. Springer Verlag, New York, 000–000.

WILFORD, G. E. 1961. *The geology and mineral resources of Brunei and adjacent parts of Sarawak with descriptions of the Seria and Miri oilfields*. Memoir **10**, Borneo Geological Survey, Publications Brunei Government.

WOLCOTT, D. S. & CHOPRA, A. K. 1993. Incorporating reservoir heterogeneity with geostatistics to investigate waterflood recoveries. *SPE Formation Evaluation*, March, 26–32.

XUE PEIHUA 1986. A point bar facies reservoir model – semi comunicated sandbody. *In*: *Proceedings of SPE International Meeting on Petroleum Engineering, Paper No. 14837*. Beijing, China (17–20 March), 103–115.

The tectonic evolution and associated sedimentation history of Sarawak Basin, eastern Malaysia: a guide for future hydrocarbon exploration

I. C. MAT-ZIN & R. E. SWARBRICK

Department of Geological Sciences, University of Durham, South Road, Durham, DH1 3LE, UK

Abstract: A seismic-stratigraphic study of the reprocessed regional lines for the offshore Sarawak area was undertaken with the aim of reviewing the present understanding of the tectonics and the palaeo-depositional environments of the Sarawak Basin. The study was integrated with biostratigraphy and wireline-log data from the wells drilled throughout the basin.

Seven unconformities were identified within the Tertiary sediments and these were used as the markers for the seismic correlations. Where the unconformities become conformable, well data were used to guide the correlation of the conformities. Palaeo-environment maps were generated which document the interaction of tectonics and sediments throughout the basin history. The development of the Sarawak Basin commenced in late Oligocene times with deposition along a coastline running in a NW–SE direction, which is almost perpendicular to the present-day coastline. The coastline was oriented to the present-day NE–SW during late Miocene times.

Geochemistry of Sarawak Basin oils shows generation of hydrocarbons from land plant dominated source rocks. The distribution of these types of source rocks are favoured in the Coastal Fluvial, Coastal and Lower Coastal Plain environments. The quantity and quality of the source rocks deteriorates with the distance from the coastline. The palaeo-environment maps illustrate the likely distribution of Sarawak Basin source rocks which will help in effectively planning for future exploration in Sarawak.

The study also revealed that the Sarawak Basin was formed as a result of NW–SE-trending right-lateral fault movement during late Oligo–Miocene times. This dextral movement was responsible for creating the NW–SE coastline and divided the offshore Sarawak area into two sub-basins. Deposition and preservation of coastal plain and shallow marine sediments continued in the eastern area while the western area remained as a 'high' until late Miocene times. The dextral strike-slip movement which controlled the evolution of the Sarawak Basin is sub-parallel to a number of lineaments elsewhere in Sarawak. The timing of movement of these suggests progressive younging in an eastward direction. It is also believed the late structuration of the sediments and the formation of structural traps is linked to these tectonic movements.

The Sarawak Basin is located on the northern margin of Borneo extending about 300 km offshore (Fig. 1). It is bounded to the north by the South China Sea, interpreted as mid-Tertiary oceanic crust, and to the south by the deformed early Tertiary and older sediments and basement rocks of Borneo. Late Tertiary volcanic rocks are also found as intrusives near the southern limit of the basin, and in the older deformed basement.

The main aim of this study is to understand the tectonic setting of the Sarawak Basin, and the origin of the structural traps which now form productive oil and gas fields. We also aim to understand the sedimentation history and its relationship to the tectonic development of the basin, which requires a redefinition of its stratigraphic framework. The availability of deep penetration 2D and high resolution 3D seismic data has presented the opportunity to test earlier models for the tectonic and sedimentation histories of the Sarawak Basin.

Data and techniques

A total of 4600 km of regional seismic lines combined with interpreted wireline log data from 45 deep boreholes, located both offshore and onshore throughout the Sarawak Basin, were used in this study. Data quality of the seismic lines is good to very good in most cases. The distribution of regional seismic lines selected gave an approximate line spacing of 75 km, with infill grid of 10 km in selected areas.

The study commenced with conventional seismic structural interpretation and subdivision of the stratigraphy using seismic-stratigraphic techniques. The identification of sequence boundaries is based on recognition of regionally

From Fraser, A. J., Matthews, S. J. & Murphy, R. W. (eds), 1997, *Petroleum Geology of Southeast Asia*, Geological Society Special Publication No. 126, pp. 237–245.

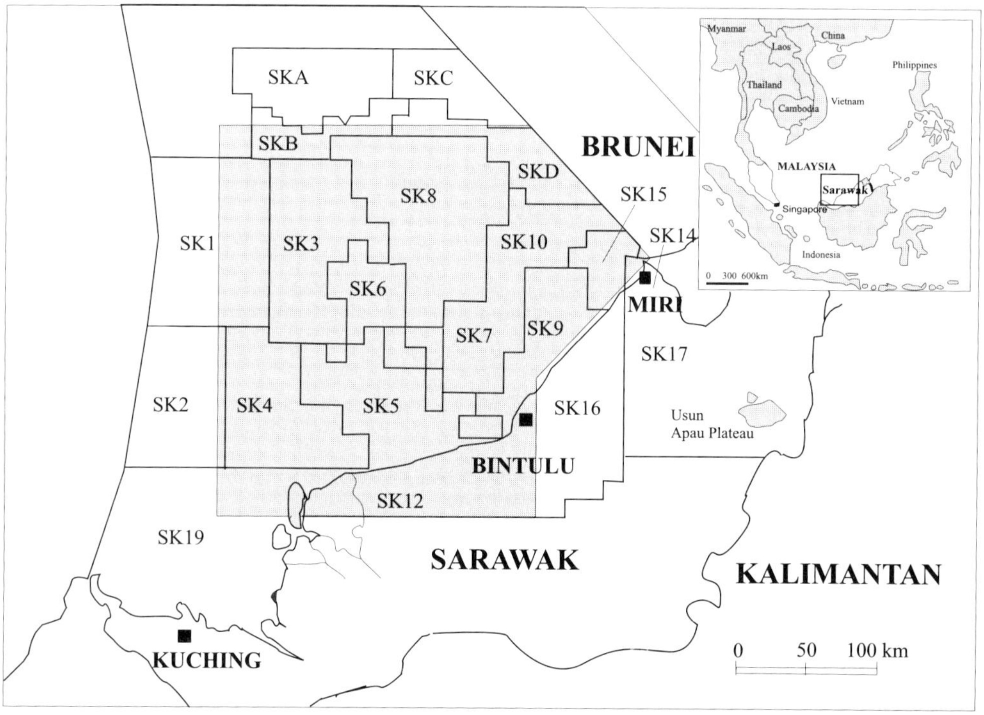

Fig. 1. Map showing the present concession blocks in the Sarawak Basin and the location of the study area.

correlatable seismic reflectors as unconformities, and further confirmed by interpretation of the well data. Within the Sarawak basin, seven sequences are recognized which range from early Oligocene to Holocene in age, and their correlation with the eight cycles of previous workers, based on Ho (1978), is shown in Fig. 2.

A series of structure maps were created at Top Basement (Fig. 3) and at each of the major unconformities recognized regionally across the basin. These were used to create isochore maps of each sequence, which indicated the relative thickness with respect to the fault pattern. Analysis of the available well data has allowed us to develop a series of palaeogeographic maps for each of the sequences recognized. The stability of the sedimentological regime through the sedimentation history allows us to comment briefly on the likely distribution of reservoirs and source rocks.

Tectonic and sedimentary development of the Sarawak Basin

The basement map for the offshore Sarawak Basin (Fig. 3) shows that the basin can be subdivided into a basement high and two basinal areas based on the present day basement topography. The basement high area is located to the SW and the two basinal areas to the NW and E with sediment thicknesses of up to 13.0 km (seismic times exceeding 6 s TWT). These sub-basins are separated by a narrow basement high which coincides with a pronounced NW–SE structural lineament termed the West Balangian Line (WBL), one of several NW–SE lineaments (Fig. 3).

The eastern sub-basin formed during late Oligocene time, based on the age of the oldest Tertiary sediment deposited in the basin. In contrast, the sub-basin to the NE of the basement high is younger in age, having Early Miocene and early Mid-Miocene sediments at its base. Differences in the basement characteristics and the sedimentation patterns and ages are illustrated in two N–S geoseismic lines (Figs 4 & 5).

Eastern sub-basin

Thick T1 and T2 sediments were deposited across the eastern sub-basin (Fig. 4), controlled

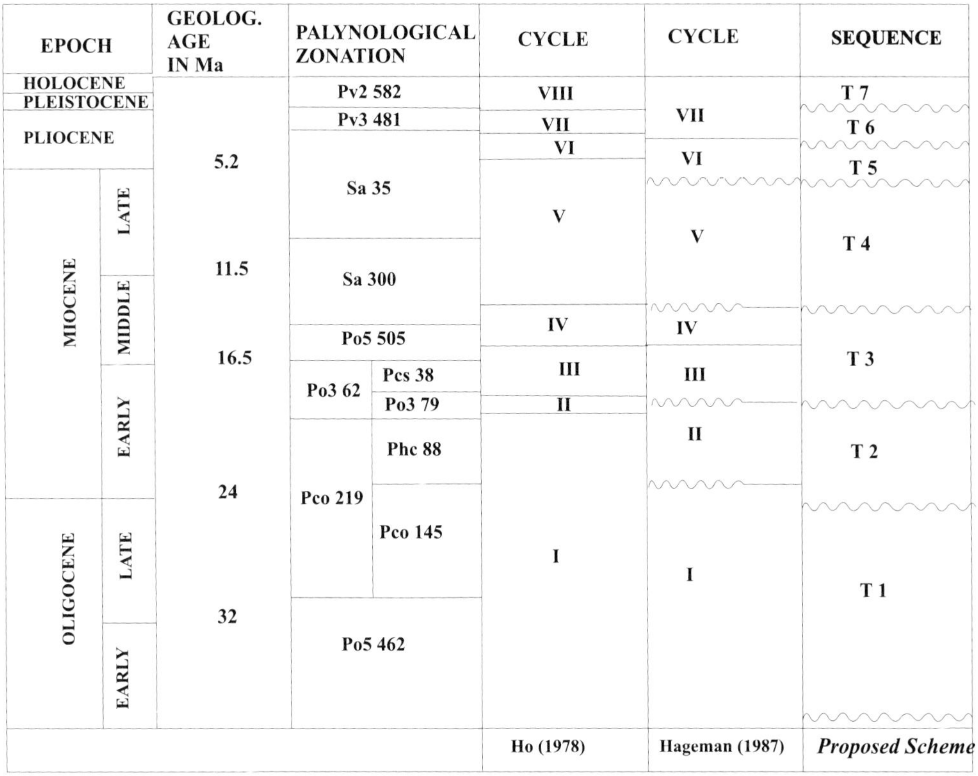

Fig. 2. Stratigraphic scheme used for the Sarawak Basin. The Tertiary succession in the basin was subdivided by Ho (1978) into eight cycles. The base of each cycle is marked by the maximum transgressive unit and the top is represented the maximum regressive unit. The subdivision proposed in this study is based on the recognition of regional unconformities.

initially by extension across the WBL to the west (Fig. 5). Extensive basement-induced faulting exerted a marked control on the distribution and thickness of T3 sediments. Faults are steep to vertical, with both extensional and compressional structure evident (Fig. 4). Continued faulting, possibly coupled with differential subsidence during a hiatus at the end of T3 sequence times, led to the sea-floor topography which controlled localized development of reef carbonates. Faulting continued to influence sedimentation and local erosion in the southern part of the sub-basin. Faults rarely penetrate above the T2–T3 sequence boundary in the north of the sub-basin. Here basement subsidence was minor during T4 and T5 times leading to thicker carbonate development than occurred further south. Rapid subsidence in the north during T6 and later times drowned the carbonates and has led to thicknesses of up to 1.5 km of clastic sediments deposited over 3 Ma.

Northwestern sub-basin

The pattern of sedimentation of the northwestern sub-basin is illustrated in Fig. 6. Sediments deposited in sequences T1 to T2 are absent. T3 sediments are deposited across the basement, with local changes in sediment thickness controlled by faulting. However, major thickness changes occur in T4 times, controlled by fault activity on both the WBL to the east and the Igan–Oya Line (Fig. 3) to the south and west. Basement control is still evident in T5 times with thinning of the sediments onto a local basement high to the NNW. However there was strong basement subsidence occuring from south to north, which led to thick accumulations of sediment of T6 and younger (last 3 Ma).

West Balangian Line

The basement and the eastern basinal area are separated by the WBL fault zone. This major

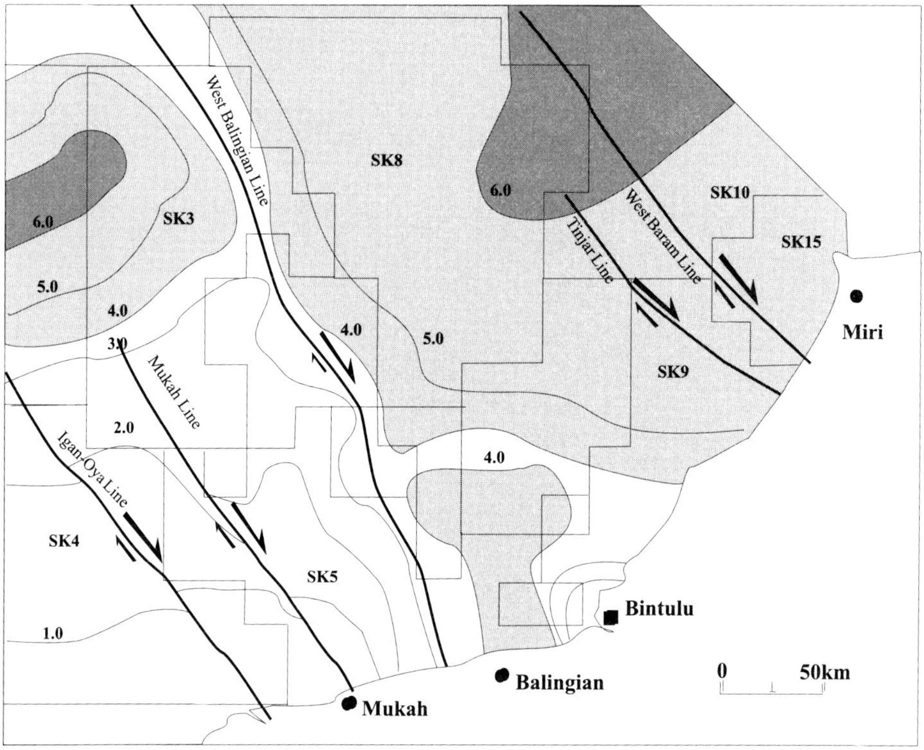

Fig. 3. Basement topography map and the major structural lineaments in the Sarawak Basin, interpreted as dextral strike-slip faults. Contours are for seconds of two-way time. Area in white is basement high and the areas in grey are structural lows.

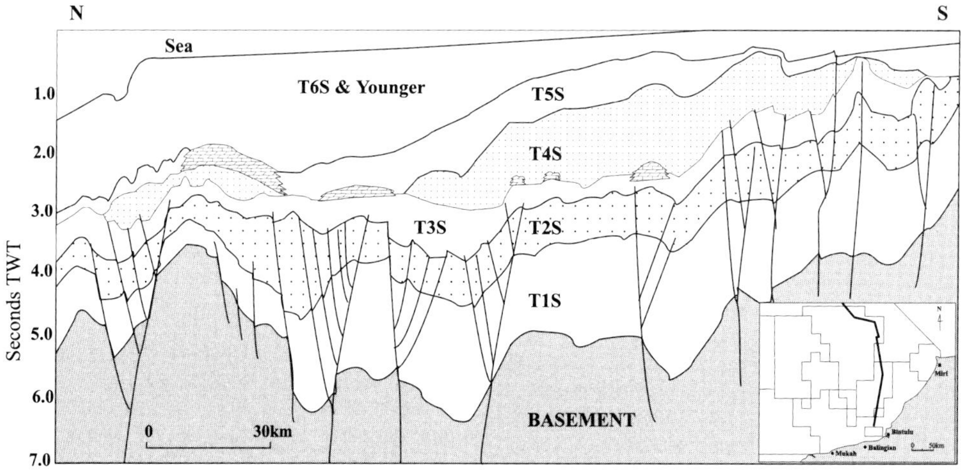

Fig. 4. Geoseismic section from south to north across the eastern sub-basin of the Sarawak Basin, which shows basement topography and the distribution of the Late Oligocene to Holocene sediments. Note the occurrence of late Miocene carbonates that grew on the basement highs.

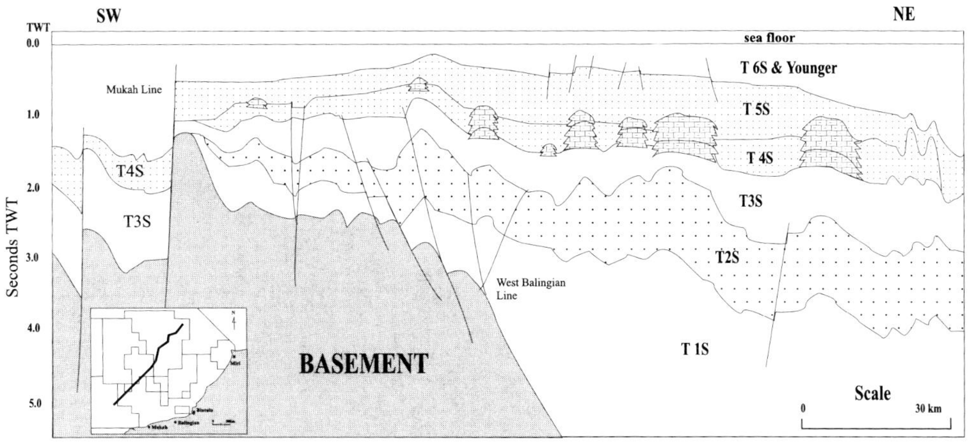

Fig. 5. Geoseismic section oriented SW to NE across the Sarawak Basin. The early sedimentation history is dominated by growth on the West Balingian Line (WBL) which was later inverted. This boundary between the basement high and the eastern sub-basin is characterised by a zone of faults with flower structures suggesting a strike-slip origin. The area to the SW started to open during early Mid-Miocene times (T3S). The carbonates grew on middle Miocene sequences in the inverted area, normally associated with basement highs.

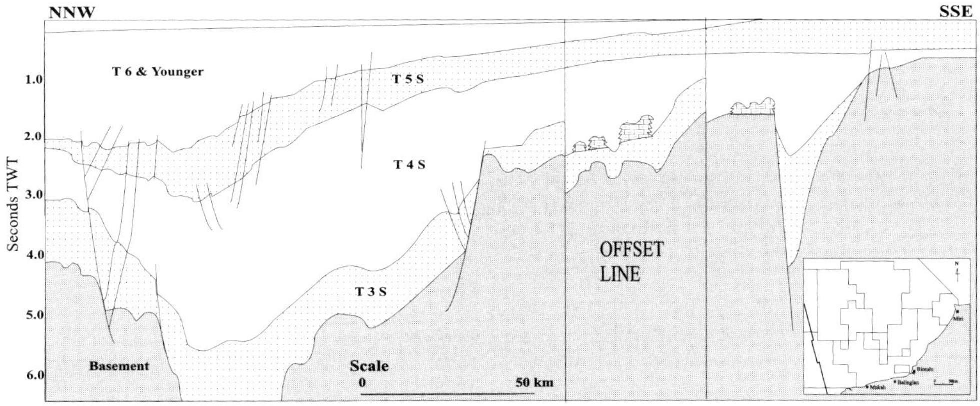

Fig. 6. Geoseismic section from south to north across the western sub-basin of the Sarawak Basin. No TI and T2 sediments are present in this portion of the basin. Pronounced basement subsidence and deposition of T4 sediments is interpreted as pull-apart or over-step tectonics. (Note breaks in the continuity of reflectors are due to incomplete seismic coverage and use of an offset line.)

fault zone trends NW–SE, similar to known major strike-slip faults in the region including the Tinjar Line (Tan & Lamy 1990). The faults within the zone are characterised by variable fault throw and age of fault movement along strike. Features include negative flower structures, local fault inversion, evidence in the stratigraphy for both extension and compression through time, and change in the direction of fault throw along strike. Figure 5 illustrates the change from extension in TI times, to compression during T3 and T4 times. Elsewhere along the fault zone facies boundaries and fault offsets in the onshore area indicate lateral movements of up to 15 km, with dextral strike-slip sense of displacement (Fig. 7). The lineament has been the strongest influence on the thickness and distibution of Tertiary sequences throughout the Sarawak Basin.

The West Balingian Line can be traced on seismic from the offshore area to the near shore and the onshore areas. In the onshore area the fault lineament has been detected by Synthetic Aperture Radar (Kang & Kadir 1990), as well as being evident on both magnetic and gravity data (Levesque & Ooi 1989). In the onshore area the

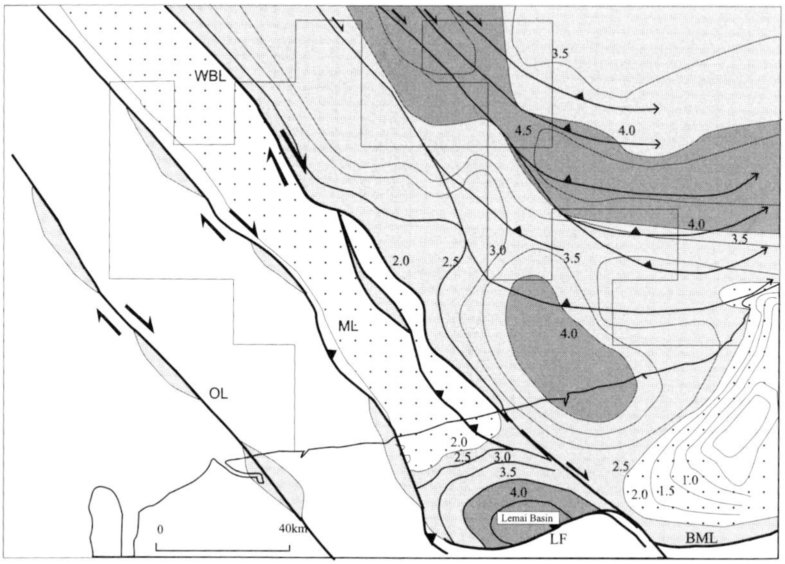

Fig. 7. Map showing the basement topography for the near shore and the onshore areas. The major strike-slip lineaments identified in the area include the Igan–Oya Line (OL), Mukah Line (ML) and West Balingian Line (WBL). The lateral offset of basement topography across the WBL, and the sequential folding and faulting within the eastern sub-basin, and the small grabens along the OL and ML, are used to argue for dextral strike-slip origin for these faults.

WBL offsets the late Oligocene sediments of the TI sediments in the Lemai Basin (Fig. 7) and the outcrops of Tatau Formation in the Bukit Mersing area for about 15 km to the SE (Fig. 7). The orientation of the Lemai Basin relative to the strike of the fault, coupled with the location and shape of the small basins created within and adjacent to the WBL suggest that the line is a dextral strike-slip fault zone. We believe that other sub-parallel lineaments, such as the Mukah Line and Igan Oya Line (Figs 3 & 7), also originated due to dextral strike-slip tectonics.

It is known that the West Baligian Line started to form during Eocene-Oligocene times, based on the oldest sediment, which abutt the fault. The timing of onset of fault movement relates to the opening of the South China Sea, (Briais *et al.* 1993) where the earliest oceanic crust is dated as 32 Ma (Mid-Oligocene). The WBL was initially active with net extension until early Miocene times, and reactivated with dominant compression thereafter (Fig. 5). It is also believed that the Mukah Line was formed in late Oligocene time, and reactivated during the mid-Miocene. The line also appears to be a dominant control on the deposition of the early Tertiary sediments in the Sarawak Basin. The line formed as the western boundary for the deposition of the late Oligocene (TI) sediments.

Other NW–SE fault lineaments

Further to the East, in the vicinity of Miri, two other lineaments have been identified, called the Tinjar Line and West Baram Line (Fig. 3). The Tinjar line is interpreted to be formed in Late Miocene times (T4) and continued to late Pliocene (T6). The line continued to the onshore area, where James (1984), Hutchinson (1988) and Tan & Lamy (1990) interpreted the line as a dextral strike-slip fault. In the subsurface the line is characterized by a negative flower structure, also suggesting strike-slip fault origin. The West Baram Line is almost parallel to the Tinjar line. The deep depression evident at Top Basement (Fig. 3) could have originated as a pull-apart or over-step basin by dextral strike-slip movements.

The West Baram Line extends landward, passing through to the east of the Usun Apau Plateau (Fig. 1). In this area there is a lateral offset to the southeast of the Oligocene–Miocene Setap Shale, based on the surface geological map (Heng 1992). We also note that the line is coincident with major changes in the axes of anticlines and synclines in the onshore Tertiary sedimentary formations.

The West Baram Line marks the shelf-edge margin and eastern boundary of the Sarawak Basin. The water depth changes abruptly across the fault zone, from about 50 m to 1000 m within a distance of only about 50 km from the West Baram Line (see Agostinelli *et al.* 1990). The line also marks the western boundary to the Sabah Trough (Tan & Lamy 1990). We do not interpret the Baram Delta as part of the Sarawak Basin, but include it with the Sabah Basin.

Discussion

Structural traps

We believe that the formation of the Sarawak Basin was strongly influenced by right lateral strike-slip movement that has taken place during Late Oligocene–Miocene times, involving both extension and compression. The West Balangian Line (WBL) is the most prominent fault zone which has subdividing the basin into two main sub-basins. The timing of movement of these faults, based on the stratigraphy, suggests progressive younging in an easterly direction, with the latest movements in the Baram area to the east of the Sarawak Basin as late as Pliocene times. These faults are mainly NW–SE in orientation.

The strike-slip tectonism is responsible for the generation of the structural traps in the Sarawak Basin. Most of the wells drilled to date in the basin have been located on structural highs with TI clastic sediments and T3 carbonates as the principal targets. In the nearshore area the main target is TI clastics, and structures are broadly coincident with the compressional folding and faulting (Fig. 7) that occurred during early to mid-Miocene (T2 and T3) times. We interpret the structural origin of these fold/faults, which produce a distinctive fan shape, as sequential folding (Woodcock & Fisher 1985) related to dextral strike-slip movement along the WBL.

The en-echelon folds formed along the strike-slip faults have produced relatively small structures. In the deeper water area (Luconia Province of James 1984), carbonate reefs are the main target for gas exploration and production. The origin of the structures here are principally related to inversion associated with the strike-slip tectonism. In addition, there may be a component of structural development due to differential subsidence (compactional drape) over both earlier formed inversion features, and basement topography.

Palaeogeography

Depositional environment maps of the Sarawak Basin from Late Oligocene to Late Miocene times are illustrated in Fig. 8. The thickness of the late Oligocene sediment (sequence TI) was strongly controlled by the West Balangian Line, but the limit of deposition was controlled by the Mukah Line that formed as the western boundary for the sequence (Fig. 8a). The coastal plain and coastal belt environment are oriented almost north-south, reflecting the strong underlying tectonic control. The deeper marine environments were located in the NE part of the basin during this time.

The pattern of N–S depositional environments is repeated during the early Miocene times when sequence T2 was deposited (Fig. 8b). However the coastal plain, coastal and inner neritic belts were thinner at this time, with a broader area of outer neritic (deep) marine environment to the NE, inferred from seismic reflection characteristics.

The orientation of the coastline changed during the middle Miocene (sequence T3). The NW–SE and N–S orientation is no longer evident from the pattern of sediment distribution and depositional environments. The dominant orientation became E–W (Fig. 8c), almost parallel to the present-day coastline. This is the result of cessation of basement fault movements and the major basement subsidence in the Sarawak Basin to the north. As a consequence the majority of the basin was in the inner neritic environment, favouring the production of carbonate reefs. Carbonate production continued until the late Miocene time after which the Baram delta started to prograde into the area and shut down carbonate production in the area influenced by clastic progradation. The initial location of the Baram Delta is most likely caused by strike-slip movement along the Tinjar Line.

Geochemical studies conducted on the source rocks and crude oils from Sabah and Sarawak by several workers including Jamil *et al.* (1991) and Muhammad & Hoesni (1992) suggested that most of the hydrocarbons in this region were derived mainly from a land-plant source. No major contribution from a marine source has so far been reported. The study on the distribution of the land-plant source in the present day Baram Delta area shows that the most prolific area for the source rock is restricted to coastal fluvial, coastal and coastal plain areas (Nagtegaal 1989). The source quality and quantity deteriorate with distance from coastline. This model can be applied to all the sediments in the Sarawak

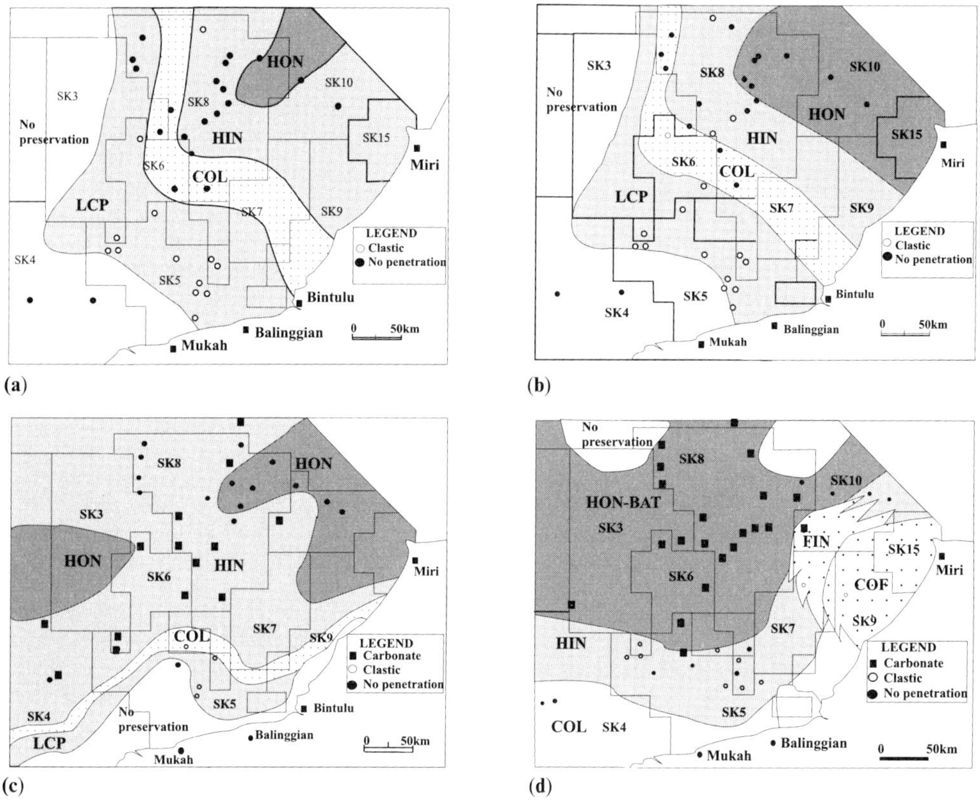

Fig. 8. Depositional environment maps of the Sarawak Basin from Late Oligocene to Late Miocene. (**a**) Late Oligocene (T1S), (**b**) Early Miocene(T2S), (**c**) Middle Miocene (T3S) and (**d**) Late Miocene (T4S). The abbreviations for the environments are: LCP, lower coastal plain; COL, coastal; HIN, holomarine inner neritic; HON, holomarine outer neritc; BAT, bathyal; COF, coastal fluvial; FIN, fluvial inner neritic.

Basin to determine the distribution of the primary sources for the hydrocarbons in the basin.

Conclusions

(1) The formation of the Sarawak Basin is due to strike-slip tectonism. The orientation of the faults and their subsidiary structures, including pull-apart basins, strongly suggests right lateral (dextral) displacement. The age of fault movement, based on analysis of the sequence stratigraphy, is primarily late Oligocene to mid-Miocene along the West Balingian Line, but later to the east.

(2) The sedimentary succession can be divided into eight sequences, based on regionally correlatable unconformities and their relative conformities. The distribution of sediment is strongly controlled by the strike-slip tectonics until mid-Miocene times when basement-controlled subsidence of the entire basin took place, with a hinge-line located close to the present coastline.

(3) The structural traps and the formation of highs where carbonate were preferentially generated, were the mainly the product of the strike-slip deformation.

(4) The creation of palaeogeographic maps can be used in the prediction of source rock and reservoir intervals, and the distribution of depositional environments can be linked to the relative timing and orientation of the principal fault lineaments.

We wish to thank PETRONAS for the permision to present and to publish this paper.

References

Agostinelli, E., Tajuddin, M. R., Antonielli, E. & Aris, M. M. 1990. Miocene-Pliocene paleogeographic evolution of a tract of Sarawak offshore between Bintulu and Miri. *Bulletin of the Geological Society of Malaysia*, **27**, 117–136.

BRIAIS, A., PATRIAT, P. & TAPPONIER, P. 1993. Updated interpretation of magnetic anomalies and sea floor spreading ages in the South China Sea. *Journal of Geophysical Research*, **98**, 6299–6328.

HENG, Y. E. 1992. *Geological map of Sarawak, second edition*. Director General of Geological Survey of Malaysia.

HO, K. F. 1978. Stratigraphic framework for oil exploration in Sarawak. *Bulletin of the Geological Society of Malaysia*, **10**, 1–14.

HUTCHINSON, C. S. 1988. Stratigraphic-tectonic model for eastern Borneo. *Bulletin of the Geological Society of Malaysia*, **22**, 135–151.

JAMES, D. M. D. 1984. *The geology and hydrocarbon resources of Negara Brunei Darulsalam*. Brunei Museum and Brunei Shell Petroleum Company.

JAMIL, A. S. A., ANUAR, M. A. & ERIK SEAH, P. K. 1991. Geochemistry of selected crude oils from Sabah and Sarawak. *Bulletin of the Geological Society of Malaysia*, **28**, 123–150.

KANG, C. S. & KADIR, M. K. A. 1990. The use of SAR imegery for hydrocarbon exploration in Sarawak. *Bulletin of the Geological Society Malaysia*, **27**, 161–182.

LEVESQUE, R. & OOI, C. M., 1989. *Block SK-5: Structural framework based on Regional seismic interpretation*. Sarawak Shell Berhad (SSB) (unpublished).

MUHAMMAD, A. J. & HOESNI, M. J. 1992. Possible source from the Tembungo field oils: evidences from biomarker finger prints. *Bulletin of the Geological Society of Malaysia*, **32**, 213–232.

NAGTEGAAL, P. J. C. 1989. A century of petroleum exploration in Sarawak and Sabah. *Paper SCOPE*, Singapore.

TAN, D. N. K. & LAMY, J. M. 1990. Tectonic evolution of the NW Sabah continental margin since the Late Eocene. *Bulletin of the Geological Society of Malaysia*, **27**, 241–260.

WOODCOCK, N. H. & FISHER, M. 1985. Strike-slip duplexes. *Journal of Structural Geology*, **8**, 725–735.

Platform-top and ramp deposits of the Tonasa Carbonate Platform, Sulawesi, Indonesia

M. E. J. WILSON & D. W. J. BOSENCE

SE Asia Research Group London University, Geology Department, Royal Holloway University of London, Egham, Surrey TW20 0EX, UK

Abstract: This study presents a detailed facies analysis of shallow-water platform and ramp deposits of an extensive Tertiary carbonate platform. Temporal and spatial variations have been used to construct a palaeogeographic reconstruction of the platform and to evaluate controls on carbonate sedimentation The late Eocene to mid-Miocene shallow-water and outer ramp/basinal deposits of the Tonasa Carbonate Platform, from the Pangkajene and Jeneponto areas of South Sulawesi respectively, formed initially as a transgressive sequence in a probable backarc setting. The platform was dominated by foraminifera and had a ramp-type southern margin. Facies belts on the platform trend east–west and their position remained remarkably stable through time indicating aggradation of the platform-top. In comparison outer ramp deposits prograded southwards at intervals into basinal marls.

Tectonics, in the form of subsidence, was the dominant control on accommodation space on the Tonasa Carbonate Platform. The location of 'barriers' and the resultant deflection of cross-platform currents, together with the nature of carbonate producing organisms also affected sedimentation, whilst eustatic or autocyclic effects are difficult to differentiate from the affects of tectonic tilting.

Moderate- to high-energy platform top or redeposited carbonate facies may form effective hydrocarbon reservoirs in otherwise tight foraminifera dominated carbonates, which occur widely in SE Asia, and have not been affected by extensive porosity occlusion.

Tertiary carbonate deposits occur widely throughout SE Asia and comprise a major target for hydrocarbon exploration (Fig. 1). Much of SE Asia has remained in near tropical latitudes throughout the Tertiary, and shallow-water areas away from clastic input often developed thick carbonate successions. Regional and local tectonics often affected the location, initiation, development and diagenesis of carbonate successions in this complex plate tectonic area (Fulthorpe & Schlanger 1989). Hydrocarbon-bearing and potentially hydrocarbon-bearing carbonates occur in many locations in SE Asia (Fig. 1), excellently placed with respect to source and seal lithologies, and very definitely form worthwhile exploration targets. However, carbonate reservoirs are notoriously fickle and unpredictable (Park *et al.* 1995) and relatively few detailed onshore analogue studies of carbonates have been undertaken in the region. It is only through a knowledge of processes of carbonate sedimentation, the development and 'growth' of carbonate platforms, and the subsequent diagenesis of carbonate lithologies that a better understanding of the inheterogeneity of carbonate reservoirs can be achieved.

This paper provides a detailed description and interpretation of shallow-water carbonate platform and ramp deposits of the Eocene to middle Miocene Tonasa Limestone Formation of South Sulawesi. The tectonostratigraphic context of the Tonasa Carbonate Platform and the geology of South Sulawesi are reviewed below. The nature and distribution of platform top, and the more southerly interbedded mid/outer ramp and bathyal lithologies in the Pangkajene and Jeneponto areas respectively (Fig. 2) are described and interpreted. Temporal and spatial variations are used to construct a facies model for the Tonasa Carbonate Platform, to reconstruct facies belts and to evaluate controls affecting carbonate sedimentation. The facies associations and platform model developed may be of use as a template to help better understand subsurface platforms of a similar age in SE Asia, and the implications this study has for hydrocarbon exploration are discussed.

Tectonostratigraphic setting of South Sulawesi

Throughout the Cenozoic, Sulawesi has been located along the eastern margin of Sundaland, the relatively stable cratonic area of SE Asia (Hamilton 1979). The geological history of Sulawesi is inextricably linked to the accretion of microcontinental and oceanic material onto the eastern margin of Sundaland and to the resultant development of volcanic arcs (Hamilton 1979; van Leeuwen 1981). During the late Cretaceous, deep marine clastics and shales of

From Fraser, A. J., Matthews, S. J. & Murphy, R. W. (eds), 1997, *Petroleum Geology of Southeast Asia*, Geological Society Special Publication No. 126, pp. 247–279.

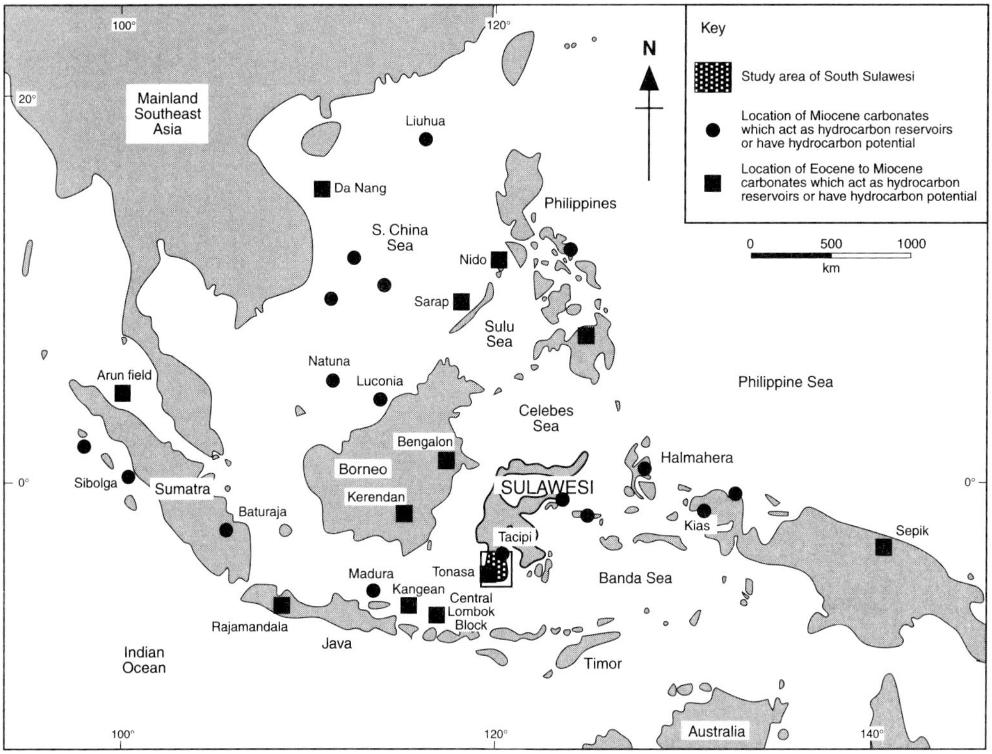

Fig. 1. Map of SE Asia showing the location of carbonate formations with hydrocarbon potential or hydrocarbon reservoirs and the location of South Sulawesi.

the Balangbaru and laterally equivalent (van Leeuwen 1981) Marada Formations were deposited over tectonically intersliced metamorphic, ultrabasic and sedimentary basement lithologies in western South Sulawesi (Figs 2 & 3). These sediments are inferred to have been deposited in a forearc setting to the east of a west-dipping subduction zone (Hasan 1991). By Eocene times, subduction had moved further to the east and marginal marine siliciclastics of the Malawa Formation, which pass transgressively upwards into carbonates of the Tonasa Limestone Formation (Fig. 3), were deposited over western South Sulawesi. During the Eocene/Oligocene, the calc-alkaline, arc-related, volcanic and volcaniclastic Salo Kalupang Formation was deposited in eastern South Sulawesi (Fig. 3; Sukamto 1982; Yuwono *et al.* 1985). Tholeiitic igneous lithologies of the Kalamiseng Formation also occur only in eastern South Sulawesi (Fig. 2) and may be part of an ophiolitic sequence accreted onto western Sulawesi in the Oligo/Miocene (Yuwono *et al.* 1985). Therefore, deposits of the Tonasa Limestone Formation are inferred to have been deposited initially as part of a trangressive sequence in a backarc setting. The Tonasa Limestone Formation is overlain by volcaniclastic deposits of the middle to late Miocene Camba Formation (Fig. 3), which were derived from a north-south trending volcanic arc located in western South Sulawesi (Sukamto 1982; Yuwono *et al.* 1985).

Introduction to the Tonasa Carbonate Platform

The carbonates of the Eocene to middle Miocene Tonasa Limestone Formation outcrop to the south of a NW–SE-trending depression passing through Lake Tempe and mainly to the west of the NNW–SSE-trending Walanae Depression in South Sulawesi (Figs 2 & 3). Deposits of the Tonasa Limestone Formation include up to 600 m of shallow-water deposits in the Pangkajene area (Fig. 2) and over a kilometre of basinal deposits currently located to the north and south of these shallow-water carbonates. This paper documents shallow-water deposits of the Tonasa Carbonate Platform in the Pangkajene area and

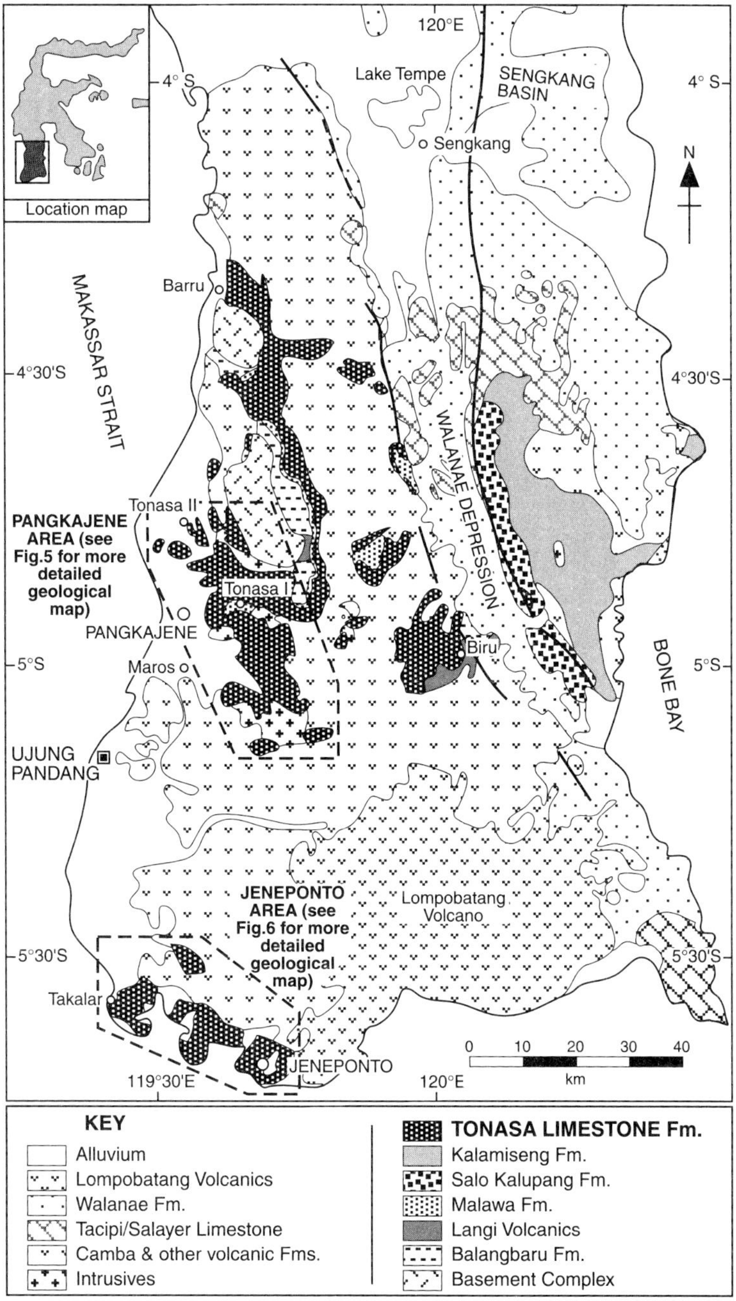

Fig. 2. Geological map of the South Sulawesi, showing the location of the Pangkajene and Jeneponto areas. Modified after Sukamto (1982) and Sukamto & Supriatna (1982).

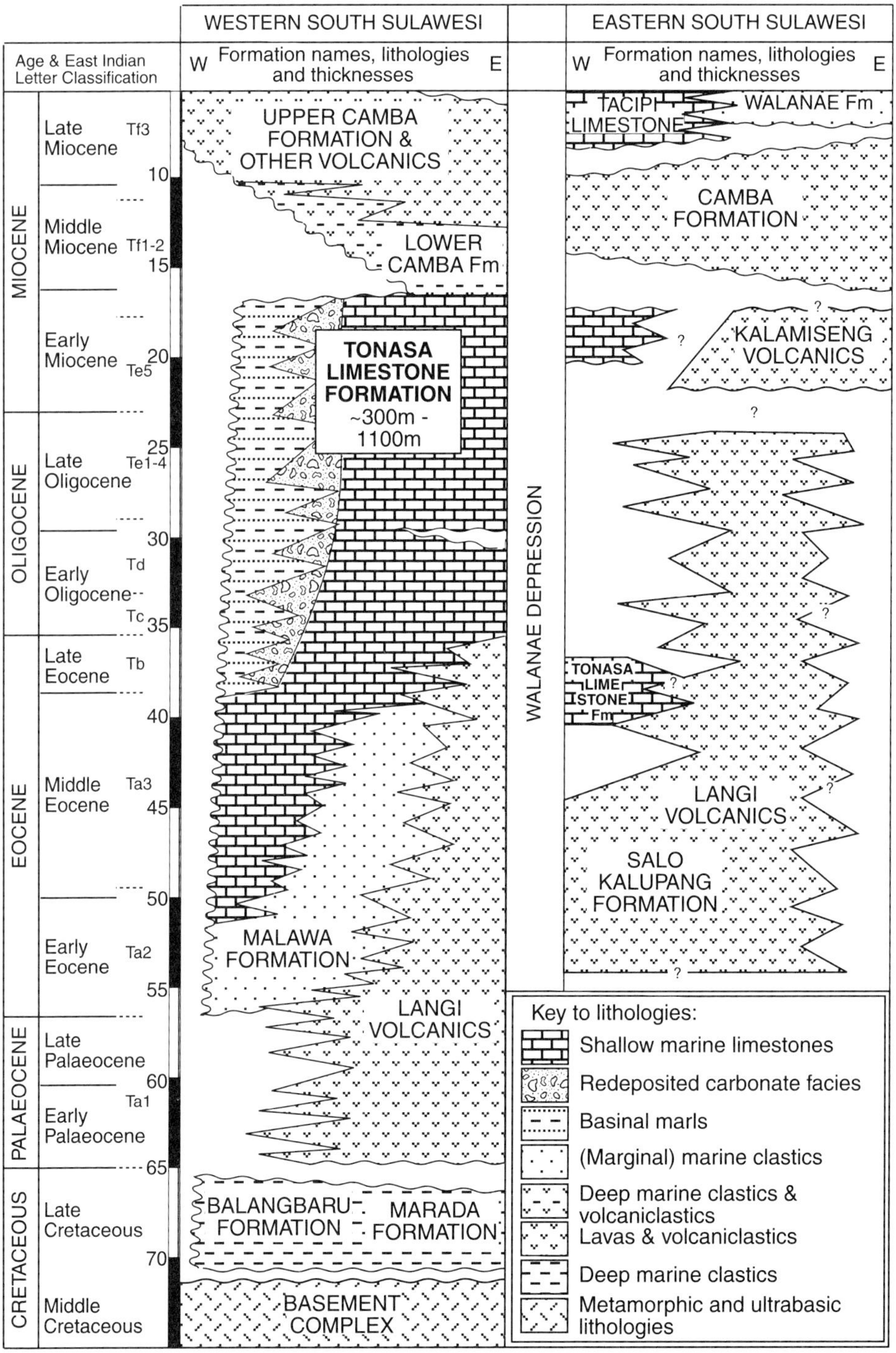

Fig. 3. Stratigraphic correlation chart of the units to the west and east of the Walanae Depression in South Sulawesi. Based on van Leeuwen (1981); Sukamto (1982); Sukamto & Supriatna (1982) and Grainge & Davies (1983).

mid/outer ramp and basinal deposits from the southern margin of the Tonasa Carbonate Platform in the Jeneponto area (Fig. 2). The northern margin of the Tonasa Carbonate Platform is inferred to have been a segmented, faulted escarpment margin which was tectonically active from the late Eocene through to the mid-Miocene (Wilson & Bosence 1996). Areas to the east and west of the Pangkajene area were regions of carbonate sedimentation, affected by more complex syn-depositional block-faulting (Wilson 1995). Since the nature of the margins of the isolated Tonasa Carbonate Platform varied both temporally and spatially from ramps to escarpment margins (Wilson 1995); shallow-water carbonate deposits in the Pangkajene area have not been bracketed into a particular depositional system, and are referred to have as platform-top deposits (cf. Tucker & Wright 1992; James & Kendall 1992).

The lithological and palaeoecological evidence presented below suggests that facies belts trended E–W and that the platform during carbonate production sloped gently southwards in the Pangkajene and Jeneponto areas (Fig. 4). A number of lines of evidence suggest that the shallow and deeper-water carbonates in the Pangkajene and Jeneponto areas respectively were not deposited on a contiguous southward sloping ramp-type platform, but that some form of shallow-water, or in places sub-aerially exposed margin, separated platform-top deposits from the southern ramp-type margin (Fig. 4). This evidence, outlined below, has been gleaned from the deposits in the Pangkajene and Jeneponto areas, since carbonate outcrops in these two areas are separated by volcaniclastic outcrops of the overlying Camba Formation (Fig. 2).

The geology of the Pangkajene and Jeneponto areas are described below. Detailed facies descriptions and interpretations, and their temporal and spatial variations, together with faunal content, are then combined to present a palaeogeographic model of, and to evaluate factors affecting deposition on, the Tonasa Carbonate Platform in the Pangkajene and Jeneponto areas. Facies associations, faunal content and depositional environment of four major E–W-trending facies belts of the Tonasa Carbonate Platform are described below.

Geology of the Pangkajene and Jeneponto areas

Outcrops of the Tonasa Limestone Formation in the Pangkajene area occur as extensive karstic areas, located to the east of the towns of Pangkajene and Maros (Figs 2 & 5). Geomorphologically, outcrops of the carbonate sequence in this area are characterised by the development of tower-karst, which rise almost vertically from flat coastal plains. In comparison, outcrops of the carbonate sequence in the Jeneponto area (Figs 2 & 6), which is located along the south coast of South Sulawesi between the towns of Takalar and Jeneponto, are characterized by gently undulating low-lying topography. The karstic outcrops in the Pangkajene area comprise rather continuous, but often inaccessible exposures. In the Jeneponto area the more competant lithologies are well-exposed as upstanding ridges, whilst more friable lithologies are poorly exposed and only occasionally seen in stream cuttings.

The attitude of bedding of the Tonasa Limestone Formation in the Pangkajene area is virtually horizontal (Fig. 5). In some localities the regional dip of the carbonate sequence is towards the west by a few degrees. A large number of faults segment the carbonate sequence in the Pangkajene area (Fig. 5; Sukamto 1982; Sukamto & Supriatna 1982). From the geological map of South Sulawesi (Sukamto 1982; Sukamto & Supriatna 1982) two sets of apparently near vertical faults, trending NW–SE and between NNW–SSE and NNE–SSW (Figs 2 & 5), can be differentiated.

In the Jeneponto area the Tonasa Limestone Formation and lower unit of the Camba Formation are folded into a series of kilometre scale NW–SE-trending anticline–syncline pairs (Fig. 6). The limbs of these folds have dip angles varying between 5 and 55° (Sukamto & Supriatna 1982). In a number of places NW–SE-trending faults cut the carbonate sequence. Other faults appear to radiate from the extrusive centres of younger volcanoes and some of these cut the carbonate sequence (Fig. 6; Sukamto & Supriatna 1982).

The contact between the Tonasa Limestone Formation and underlying Malawa Formation is exposed only in the central part of the Pangkajene area around the locality of the Tonasa-I Quarry (Fig. 5). In this area the Malawa Formation consists of interbedded quartzose sandstones, thin coal bands, clays and some thin limestone beds. This formation records a transgressive sequence from marginal marine deposits passing conformably upwards into shallow marine carbonates of the Tonasa Limestone Formation. Although the change from clastics to carbonates occurs over a few metres and is therefore inferred to have been relatively rapid, the presence of thin limestone

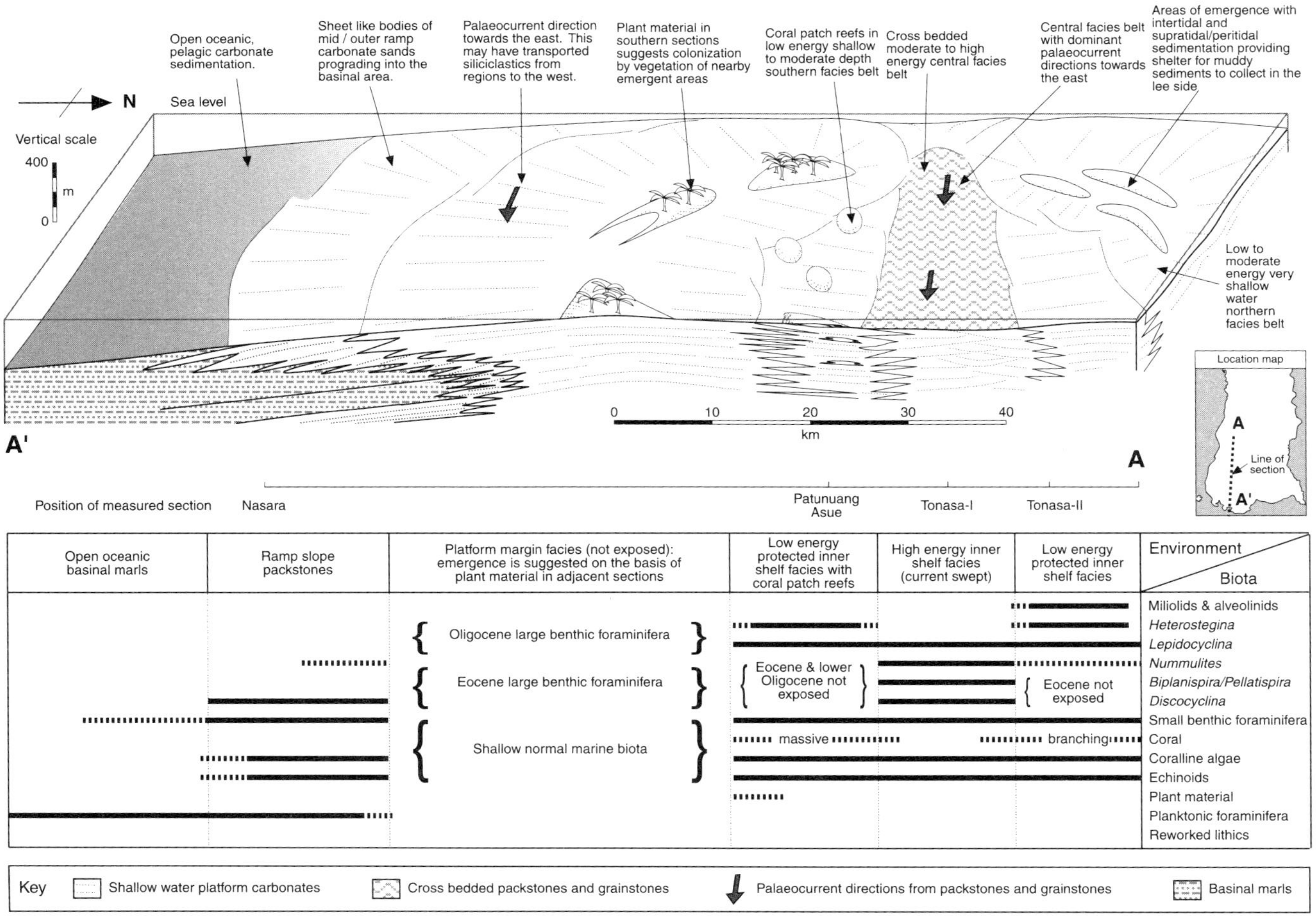

Fig. 4. Oligocene reconstruction of the Tonasa Carbonate Platform and its southern ramp-type margin in the Pangkajene and Jeneponto areas showing the distribution of biota.

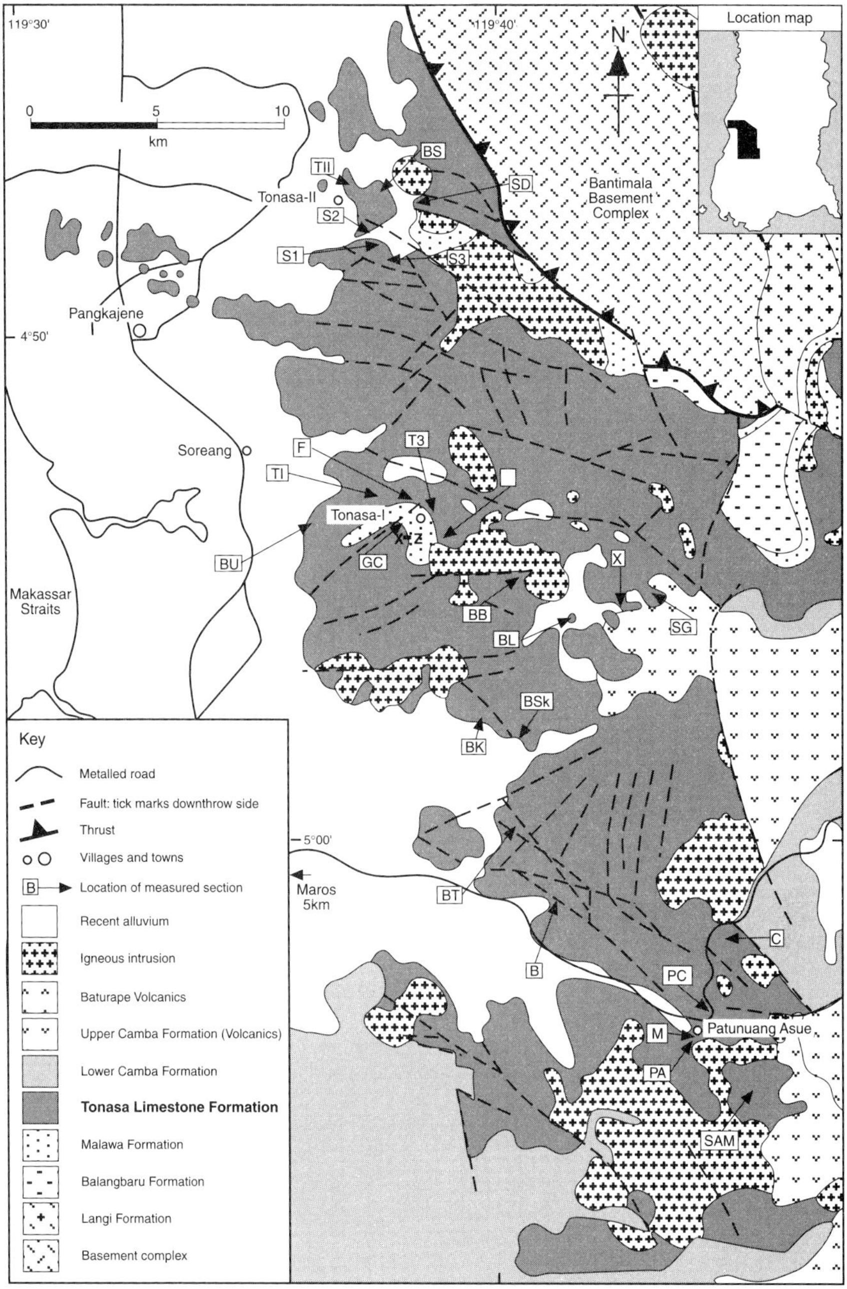

Fig. 5. Geological map of the Pangkajene area (after Sukamto, 1982; Sukamto & Supriatna, 1982) showing the location of measured sections. See Table 2 for the names of measured sections. The location of the photograph and sketch of karstic outcrops with palaeocurrent indicators (Fig. 11) are also shown (X–Z).

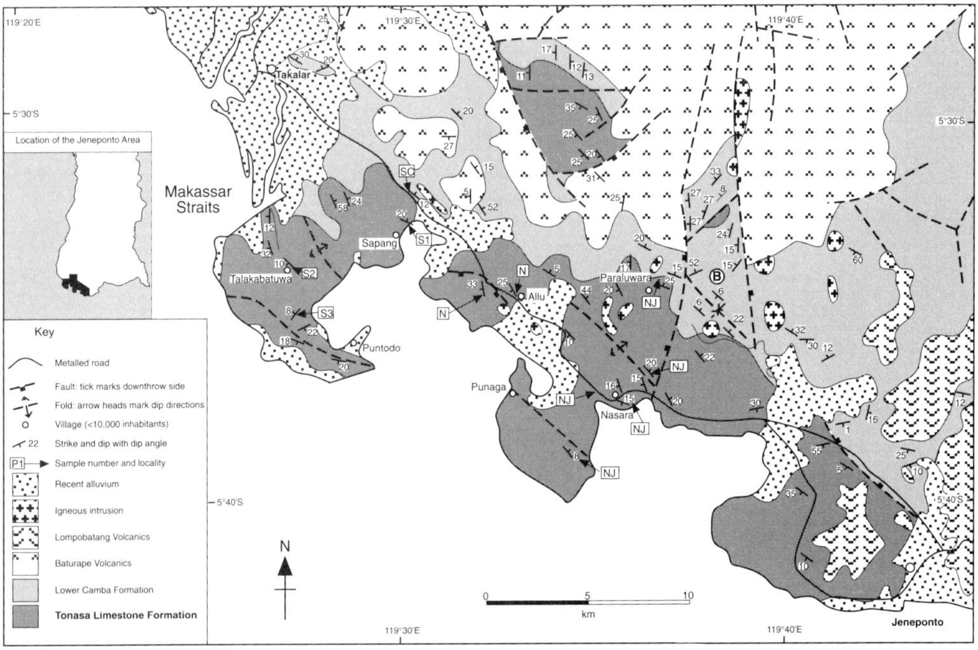

Fig. 6. Geological map of the Jeneponto area (modified after Sukamto & Supriatna, 1982) showing the location of measured sections and samples collected.

beds within the upper part of the Malawa Formation suggests that some interdigitation of the two formations may have occurred. Crotty & Engelhardt (1993) suggested a mid–late Eocene age for the Malawa Formation in the Tonasa-I Quarry area due to the occurrence of well preserved palynomorphs of *Retitribrevicolporites matamanadhensis* and abundant *Verrucatosporites usmensis* respectively. The base of the carbonate sequence is not exposed in the Jeneponto area (Fig. 6).

The contact between the Tonasa Limestone Formation and overlying volcaniclastics of the Camba Formation is exposed along the eastern margin of the Pangkajene area (Fig. 5) and the northern margin of the Jeneponto area (Fig. 6). This contact, although rather poorly exposed, appears to be conformable in the Pangkajene area (Fig. 5) and a possible low-angle unconformable contact occurs in the Jeneponto area (Fig. 6). The lower part of the Camba Formation consists of marginal marine to continental sediments and deep marine shales both interbedded with volcaniclastics in the Pangkajene and Jeneponto areas respectively. Igneous dykes, sills and stocks of basaltic and dioritic composition, with a middle to late Miocene age (Sukamto 1982; Sukamto & Supriatna 1982), have intruded the carbonate sequence in a number of localities in the Pangkajene and Jeneponto areas (Figs 5 & 6).

Methodology: facies descriptions and interpretations

An environmental interpretation of the Tonasa Carbonate Platform linked to facies distribution, faunal content and stratigraphic correlation across the Pangkajene and Jeneponto areas has been undertaken following an extensive programme of logging, sample collection and petrographic work. The methodology for facies descriptions and interpretation of depositional environment, together with dating of lithologies is given below.

Facies classification and descriptions

In this paper a threefold classification scheme has been adopted for the description of carbonate lithologies of the Tonasa Limestone Formation (Table 1). Carbonate lithologies are described as informatively as possible in terms of Dunham's (1962) textural classification, their structures and/or faunal content. Carbonate lithologies are grouped into facies and facies which share important common characteristics have in turn have been collected into facies groups (Table 1). It should be stressed that these facies groups do not necessarily correspond to facies associations, which have some environmental significance (Collinson 1969; Walker 1992). For more discussion on the facies nomenclature used within this paper, and detailed facies descriptions and interpretations, the reader is referred to Wilson (1995). Perhaps due to a complex arrangement of microenvironments, both through time and in space, on the Tonasa Carbonate Platform, facies associations have not often been identified within the carbonate succession. Where these occur they are described and interpreted.

Biostratigraphy

Lithologies of the Tonasa Limestone Formation are often dominated by foraminifera and sometimes by coralline algae. Where planktonic organisms are almost entirely absent, such as in the Pangkajene area, large benthic foraminifera, identified either to genera or species level, have been used to date the carbonate succession. However, dating of foraminifera rich carbonate sequences using the East Indian letter classification scheme (van der Vlerk & Umbgrove 1927; Adams 1970) is not always straightforward, such as where age diagnostic foraminifera may be restricted to certain environments or reworked into younger sediments (van Gorsel 1988). Therefore some of the time tie-lines shown on the stratigraphic correlation diagram (Fig. 7) are necessarily rather tentative. Planktonic foraminifera (cf. Blow 1969) and nannofossils (cf. Martini 1971) were used to date carbonate samples from the Jeneponto area. The geological time scale used and correlation with the various biozonation schemes is from Harland *et al.* (1990) and van Gorsel (1988).

Palaeoecology

In order to interpret the palaeoenvironmental differences between the shallow-water facies of the Tonasa Limestone Formation it was necessary to have an understanding of the palaeoecology of the dominant biota: the benthic foraminifera. Palaeoecology of the larger benthic foraminifera is based on comparisons with extant genera (Reiss & Hottinger 1984), studies of shape and form (Hallock & Glenn 1986), and by integrating this information with sedimentological characters, such as texture or sedimentary structures. From the Eocene to Recent a progressive shallow to deep water shift of foraminifera assemblages has been inferred by Buxton & Pedley (1989). A range of factors, such as energy, light, substrate, temperature and salinity may affect a biotic assemblage and the form of its constituent organisms. It would therefore seem that any palaeoenvironmental scheme for the Tertiary is likely to be a relative one.

Carbonate facies

An environmental interpretation of the Tonasa Carbonate Platform linked to facies distribution and stratigraphic correlation across the Pangkajene and Jeneponto areas has been undertaken following an extensive programme of logging, sample collection and petrographic work. The subdivision of lithologies, found in the Pangkajene and Jeneponto areas, into their respective facies and facies groups is summarized on Table 1. Three main facies groups have been defined in the Pangkajene and Jeneponto areas; these are micrite and bioclast dominated facies groups and a planktonic foraminifera facies group (Table 1). The distribution of the various lithologies, and variations in their constituent components, within measured sections from north to south is shown on Table 2. This spatial distribution of the lithologies reflects fundamental variations in the depositional environment across the Tonasa Carbonate Platform and is discussed below. The occurrence, bed thickness, nature of bed contacts, lateral bed continuity, structures, faunal content, nature of the groundmass and an environmental interpretation for each facies is also shown on Table 1. It is beyond

---

Abbreviations to Table 1

Texture: M, mudstone; W, wackestone; P, packstone; G, grainstone; Fl, floatstone; R, rudstone; Bi, bindstone; brec, brecciated; m, micrite; b, bioclastic.

Bioclasts: mi, miliolid; A, alveolinid; If, imperforate foraminifera; Co, coral L, *Lepidocyclina*; H, *Heterostegina*; S, *Spiroclypeus*; CoA, coralgal; CA, coralline algae; li, lithic; N, *Nummulites*; D, *Discocyclina*; Mo, mollusc; E, echinoid; F, foraminifera; Ha, *Halimeda*; Ga, gastropod; Cyclo, *Cycloclypeus*; Pellat, *Pellatispira*; Oper, *Operculina*; frag, fragmented; PF, Planktonic foraminifera; LBF, Large benthic foraminifera (undifferentiated).

Location and occurrence: N, north; C, central S, south; Pang, Pangkajene; Jene, Jeneponto; c, common; r, rare; v, very.

Bed contacts: P, planar; S, sharp; G, gradational; U, undulose; E, erosive.

Table 1. *Facies description and interpretation of the Tonasa Limestone Formation from the Pangkajene and Jeneponto areas*

Facies	Lithologies	Location & occurrence	Bed thickness	Bed contacts	Lateral continuity
Micrite dominated facies group					
Algal laminite facies		N Pang (r)	40 cm	P & S	10 s m
Mud/wackestone facies	mM bM/W brec bM/W	N & S Pang (r)	0.25–10 m	P & S/G	ms/10 s m
Bioclast dominated facies group					
Bioclastic facies	bW bW/P bP bP/G bG	W(/P) – N & S Pang (c) G(/P) – C Pang (c)	m/dm	P & S/G	ms/10 s m
Miliolid/alveolinid facies	mi bW(/P) mi bP(/G) A bW(/P) A bP(/G) mi & A bW(/P) mi & A bP(/G)	all c N Pang rare m BP in lower TLF C Pang rare A BP in upper TLF S Pang	dm	W(/P) – P & G/S P(/G) – P/U & S	m
Alveolinid & coral/*Lepidocyclina* facies	lf, Co & L bP/Fl lf & Co bP/Fl lf & L bW/P/G Co & L bP/Fl	N Pang (c) rare A & Co BW/P S Pang	dm	P & S/G	m/10 s m
Heterostegina wacke/packstone facies	H b W/P H & L bW/P H, L & S bW/P H & S bW/P	N & S Pang	dm	P & S	m
Coralgal & *Nummulites* facies	CoA bBi/R Co bBi CA bP/Bi CA & N bP	C & S Pang (r)	dm	P & S	m
Coral & *Heterostegina*/lighic facies	Co R Co bP/Fl Co & H bP/R Co, H & li bR Co & li R	S Pang (c) rare lower TLF C Pang	dm	P & S/G	m
Lepidocyclina/*Nummulites* wacke(/pack)stone facies	L bW(/P) N bW(/P) L & N bP/W	Pang (r)	cm/dm	P/U & S/G	m
Lepidocyclina/*Spiroclypeus* pack(/grain)stone facies	L bP(/G) L & S bP(/G) S bP(/G)	C & S Pang (r)	dm	P & S	m/10 s m
Spiroclypeus wacke/packstone facies	S W/P S bW/P	S Pang (r)	dm	P & S/G	m
Discocyclina pack(/grain) stone facies	D bP(/G) D & P bP(/G)	C & N Pang (c)	dm	P/E & S	m/10 s m
Planktonic foraminifera facies group					
Marl facies		Jene (r)		P & S	m/10 s m
Graded bioclastic packstone facies		Jene (r)	5–30 cm	P & S/G	m
Planktonic foraminifera bioclastic packstone facies		SW Pang (v.r) Jene (c)	cm/dm	P & S	m/km
Lithic bioclastic packstone facies		Jene (c)	cm/dm	P & S	m/km
Planktonic foraminifera wacke/packstone facies	Upper Jene (c)	dm	P & S	m	

Table 1 (*continued*)

Biota	Groundmass	Structures	Environmental Interpretation
Cyanobacteria – other fauna rare	Micrite	Algal horizon erosionally truncated and overlain by micrite horizon Bundles of laminae form domal structures	Intertidal algal mats forming ?hypersaline pools or channels
mM – fauna v. rare (mi & Mo) bM/W – rare (CA, E, F, Mo & Co)	mM – micrite bM/W – micrite/bio	mM – gas escape structures brec bM/W erosionally truncated	mM – intertidal/low energy bM/W – low energy brec bM/W – reworking of low energy deposits
W(/P) – whole unabraded fossils G(/P) – fragmented, abraded fossils large & small F, CA & Mo – common Gas, Ha, Co – rare	W – micrite P – frag bioclasts G – Equant & iso cements	G(/P) – low angle cross bedding & parallel orientation of platy fossils	W – W/P – P – P/G – G – progressively higher energy depositional setting
large & small IF dominate (30–75%) – whole mi, A, *Praehapidionina*, *Austrotrillina* & Soritids Other bioclasts – Mo, Ga, E, CA, *Kuphus* & in P(/G) agglutinating foraminifera	W(/P) – micrite & frag bioclasts G(/P) – Equant cement & neo replacements		W(/P) – lagoon or sheltered shoal environment G(/P) – higher energy near crest shoal/bar abundant m < 24 m water depth. Normal marine or restricted
If mostly m & A (15–50%), frag branching Co & massive Co <35% & <20% *Lepido* Other bioclasts – Mo, Ga, E, CA, *Kuphus*, *Amphistegina*	Micrite & frag bioclasts	Fossils random orientation	Shallow-water, moderately turbulent marine conditions
Dominant whole Hetero (30–50%), Lepido (<30%) & Spiro (<20%) Other bioclasts – Cyclo, Amphistegina, E, CA & other small benthic forams	Micrite dominates & frag bioclasts		Normal marine low energy conditions
CoA bBi/R – massive Co may be encrusted by CA CA & N bP – <30% encrusting & frag CA, <18% Numm, E, Mo, Ha, & small benthic forams	Micrite & frag bioclasts		Moderate energy normal marine conditions, CoA Bi may have been small patch reefs in S Pang
S Pang – bored & abraded Co clasts, lithics, *Hetero* (<18%) & less common E, Mo, *Cyclo*, CA C Pang fragmented Acropora, E, CA, Mo, A, *Lepido* & small benthic forams	Micrite & frag bioclasts		Low to moderate energy, shallow to moderate water depths under normal marine conditions. Perhaps storm reworking of some Co & lithic clasts
Numm (<18%), *Lepido* (<25), E, CA & small benthic forams. Sometimes *Disco*, m, A, *Spiro* & Mo	Micrite & frag bioclasts		Moderate energy normal marine conditions
Lepido (<20%), *Spiro* (<20%), E, CA, Mo, small benthic forams, *Hetero*, *Oper*, *Amphistegina*, & rare Co frags.	Frag bioclasts & equant calcite		Moderate to high energy normal marine conditions
Spiro dominate (<20%) & less common Lepido, Cyclo, Oper, Amphistegina, E, CA, small benthic forams & rare Co frags.	Micrite dominates & frag bioclasts		Normal marine, low energy, moderate depth in the photic zone
Abundant *Disco* (15–25%), *Pellat* (5–20%), E, CA. Also Numm, Biplan, Astero, E & small benthic forams.	Frag bioclasts	Low angle trough and planar cross bedding	Moderate to high energy normal marine conditions
PF, small benthic forams & nannofossils	Clays & micrite	mm laminations or bioturbated	Low energy normal marine basinal deposits
PF, small benthic forams, CA, E, frags LBF & sponge spicules	Frag bioclasts & equant calcite	Normal grading and mm parallel lamination in the upper part of the bed	Redeposited distal calciturbidite
PF, small benthic forams, CA, E, frags LBF & sponge spicules	Frag bioclasts	Rare ripples, bioturbation and dissolution seams	Outer shelf to slope normal marine deposits with open oceanic influence
PF, small benthic forams, CA, E, frags LBF, qz, feld & sponge spicules	Frag bioclasts	Low angle trough cross-bedding	Outer shelf to slope normal marine deposits with open oceanic influence and input of lithics
PF, small benthic forams, & rare CA frags	Micrite	Rare normal grading	Possible distal redeposited calciturbidite deposited into low energy basinal area

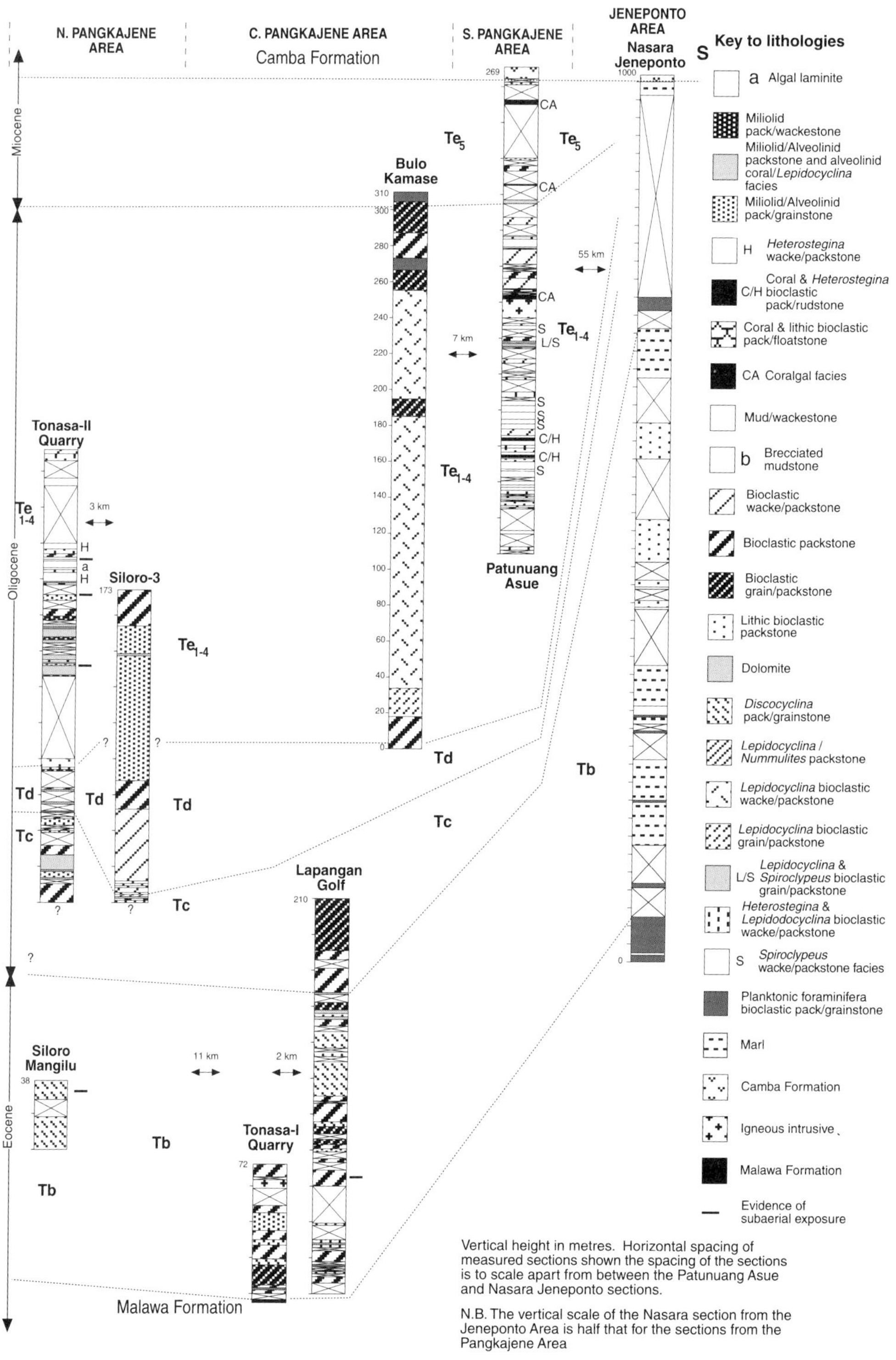

Fig. 7. Stratigraphic correlation and facies distribution between sections in the Pangkajene and Jeneponto areas.

Table 2. *Facies and biota distribution in the Pangkajene and Jeneponto areas of the Tonasa Limestone Formation*

Facies or lithologies

Abbreviations
Alv. - Alveolinid
Mil. - Miliolid
IF - Imperforate foraminifera
Amph. (Am) - *Amphistegina*
Cyclo. (C) - *Cycloclypeus*
CA - Coralline algae
Hetero. (H) - *Heterostegina*
Spiro. (S) - *Spiroclypeus*
Lepido. (L) - *Lepidocyclina*
PF - Planktonic foraminifera
Disco. (D) - *Discocyclina*
Numm. (N) - *Nummulites*

Location (see figures 5 & 6)	Dolomite	Evidence for sub-aerial exposure	Marl	Planktonic foraminifera wacke/packstone	Lithic bioclastic packstone	Graded bioclastic packstone	*Spirocipeus* wacke/packstone	Coralalgal bioclastic bind/rudstone	Spiro. (& Lepido.) bioclastic pack(/grain)stone	Coral & *Heterostegina* / lithic facies	Lepido. bioclastic grain(/pack)stone	PF bioclastic pack(/grain)stone	*Discocyclina* pack(/grain)stone	CA (& Numm.) bioclastic pack(/rud)stone	*Nummulites* bioclastic packstone	*Lepidocyclina* / *Nummulites* packstone	*Discocyclina* (grain) packstone	Bioclastic (pack/)grainstone	Bioclastic packstone	(Bioclastic pack/)wackestone	Miliolid bioclastic pack(/grain)stone	Coral bioclastic pack/floatstone	Coral & IF bioclastic pack/floatstone	Alveo. (& milio.) bioclastic grain(/pack)stone	Lepido. bioclastic pack(/wacke)stone	Mud/wackestone	Miliolid / Alveolinid pack(/wacke)stone	Lepido. & IF bioclastic pack(/wacke)stone	Coral, lepido. & IF bioclastic pack(/wacke/float)stone	H & (Lepido. & Spiro.) bioclastic pack(/wacke)stone	*Heterostegina* bioclastic wacke(/pack)stone	Algal laminite
TII - Tonasa II Quarry		3u																	13	2	2	5b	8	3	4	5u	1	6u	3	7u	9u	1u
BS - Bulo Sela																		1	10		5		2	3			7	3				
SD - Siloro Mangilu		1u															3				2			3								
S2 - Siloro 2																			2	1	1		1	4	1		7	1u				
S1 - Siloro 1																		1	3	5			2			1	1	5				
S3 - Siloro 3	1															2			7	9	7		1	4	4		3					
TI - Tonasa - I																		1	6	1	4		2									
F - Flag Tonasa	1																	12	2	2						2						
T3 - Tonasa																		13	6	2			1									
X, SG - Senggareng																			2													
T - Tonasa-I Quarry															1		2	4	9		4											
BU - Bulo Ujungbulo	1													1				12	1	2												
GC - Lapangan Golf	4	1b											11	1	1		9	6	20		6b	1b										
BB - Bulo Bodong																			2	1												
BL - Balocci																			1													
BSk - Bulo Sinkalong	1																	5	3					1								
BK - Bulo Kamase											1	2t						3	2						8							
BT - Bulo Tedong									7	1	2					1		2				2			6				1			
B - Batimurung	1								3										7	1		2			1	2			2			
C - Karaengta								3											2			1										
PC - Pangea (Karaengta)	1																	1	4			1				2						
M, PA, SAM - Patunuang Asue							6		3	4b				1				1	7	18		6	1t	2t		18			7	8		
NJ - Nasara Jeneponto			20	11	40	2						13																				
# of beds of each facies studied	10	5	20	11	40	2	6	3	13	5	3	15	11	3	2	3	14	62	109	44	31	16	18	20	24	30	19	15	3	17	17	1

N → S: TII to S3 – Northern sections; TI to BL – Central sections; BSk to M, PA, SAM – Southern sections (PANGKAJENE AREA); NJ – JENEPONTO AREA

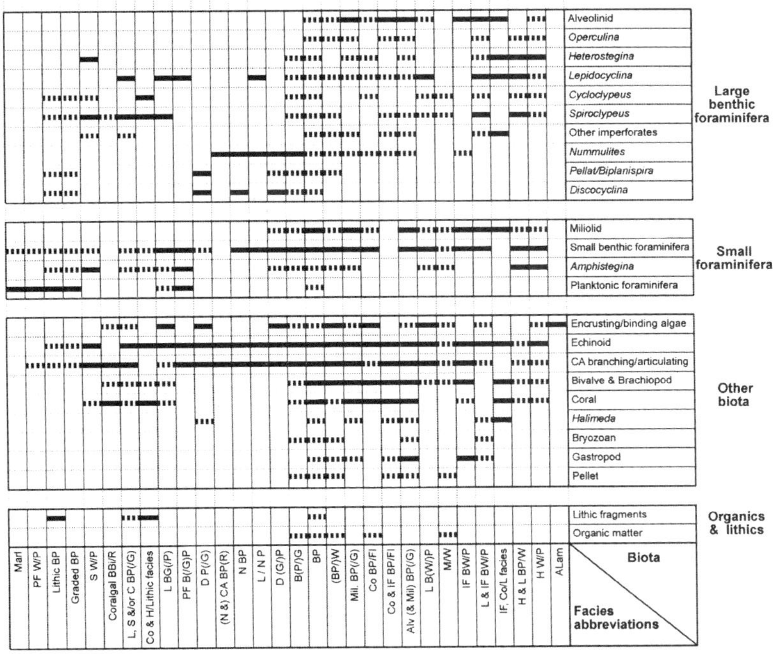

NB Fossils names prefixed to facies names indicates the aforementioned fossil comprises ≥15% by volume of the facies.

▬ Bioclast present in ≥30% of beds of a particular facies; ▪▪▪▪ Bioclast present in <30% of beds of a particular facies } Although this is often correlatable with, it is not a direct measure of how abundant a bioclast is within a particular facies. Abundance of each bioclast within each facies varies from bed to bed.

■ Number of beds studied: b,u & t - base, upper part and throughout section respectively

Table 3. *Inferred relative depositional conditions of carbonate lithologies in the Pangkajene and Jeneponto areas*

Increasing energy →

Restricted	Low energy	Moderate energy	High energy	Open oceanic	Increasing depth in the photic zone ↓
Subaerial karstification Mudstone with laminoid fenestrae					Supratidal
Algal laminite		Brecciated mudstone			Inter/subtidal
Miliolid pack /wackestone	Miliolid/Alveolinid packstone and Alveolinid & coral/*Lepidocyclina* facies.	Miliolid/Alveolinid pack /grainstone			Very shallow marine (probably <24 m)
	Heterostegina wacke /packstone Coral bioclastic pack/floatstone	Coral & *Heterostegina* bioclastic pack/rudstone Coralgal facies	Coral & lithic bioclastic pack/floatstone		Shallow marine
Mud/wacke stone	Bioclastic wacke/packstone	Bioclastic packstone	Bioclastic grain /packstone		Not depth related
Late Eocene lithologies		Coralline algal &*Nummulites* bioclastic packstone	*Discocyclina* pack/grainstone facies	Planktonic foraminifera bioclastic pack/grainstone	Shallow to moderate depths
Early Oligocene lithologies		*Lepidocyclina* / *Nummulites* packstone		Not age dependant	
Late Oligocene lithologies	*Heterostegina* & *Lepidocyclina* bioclastic wacke/packstone	*Lepidocyclina* bioclastic wacke/packstone	*Lepidocyclina* bioclastic grain/packstone	Lithic bioclastic packstone	
	Spiroclypeus wacke/ packstone facies	*Lepidocyclina* & *Spiroclypeus* bioclastic grain/packstone *Spiroclypeus* bioclastic grain/packstone		Graded bioclastic packstone (redeposited)	Moderate to deeper depths
				Planktonic foraminifera wacke/packstone Marl facies	Bathyal depths

the scope of this paper to give a detailed description and environmental interpretation for each facies, and this information can be found in Wilson (1995). Figure 7 shows the stratigraphic correlation and facies distribution between sections in the Pangkajene and Jeneponto areas. The inferred relative depositional environments of the various lithologies of the Tonasa Limestone Formation in the Pangkajene and Jeneponto areas are summarized on Tables 1 and 3.

Micrite-dominated facies group

The micrite-dominated facies group includes mud/wackestone facies and algal laminite facies, which occur most commonly in the northern and southern parts of the Pangkajene area (Tables 1 & 2). These facies contain limited biota and are interpreted to have been deposited in a low energy and in some cases restricted setting (Tables 1 & 3; Fig. 4).

A single bed of the algal laminite facies (Fig. 8a) was seen in the upper part of one of the sections in the northern part of the Pangkajene area (Tables 1 & 2). The algal laminate facies consists of millimetre-scale laminae containing poorly preserved micritic walled cyanobacteria filaments infilled or replaced by microcrystalline calcite, interbedded with and erosionally overlain by unfossiliferous micritic laminae (Fig. 8b). Upwards the micritic horizons become increasingly more filament rich until they are again truncated by an erosive surface. Laminae are grouped into bundles forming low-angle domal structures (wavelength: >60 cm, height: 10–20 cm), which sometimes pinch out laterally over a few metres (Fig. 8a). These low-angle domal, laminated algal mats, which lack desiccation features correspond to the smooth mats of Kinsman & Park (1976). Extensive smooth algal mats formed of cyanobacteria occur in less than 10 m water depth (Park 1977) and are best developed in permanently wet sites, such as shallow pools or channel sites, in the upper intertidal zone of low energy, tropical carbonate dominated areas where hypersaline conditions may develop (Logan *et al.* 1964; Shinn 1968; Kinsman & Park 1976; Park 1977). The truncational surfaces and micritic laminae seen in this algal laminite facies (Fig. 8b) are inferred to result from erosion and subsequent deposition during periods of increased energy, probably due to storm or wave reworking (cf. Gebelein 1974).

Micritic mudstones, which occur in the northern part of the Pangkajene area, are composed almost entirely of micrite, and apart from very rare miliolids or mollusc fragments are devoid of fossils or recognizable fossil fragments. The micritic mudstone bed which overlies the algal laminite facies contains voids, up to 8 cm across, with flat bases parallel to bedding and 'pustulose' upper surfaces (Fig. 8c), interpreted as gas escape structures (cf. Logan *et al.* 1974; Shinn 1983). The fine grained nature of this facies indicates deposition in a low energy environment. The virtual lack of fossils in the micritic mudstones, which occur in otherwise fossiliferous sequences, together with gas escape structures and an association with beds of the algal laminite facies, suggests deposition in a low energy, shallow water setting where both hypersalinity and subaerial exposure may have occurred.

Other mud/wackestones, which occur mainly in the southern part of the Pangkajene area, are composed of very finely fragmented bioclasts and micrite (Tables 1 & 2). The upper surface of some beds of this lithology are erosive, and irregular and angular-pebble sized clasts of the mud/wackestone have been reworked into overlying beds of the bioclast-dominated facies group to form a brecciated bioclastic mud/wackestone (Fig. 8d). Normal marine, low-energy conditions are inferred from the biota and texture of the bioclastic mud/wackestones. The low energy setting may be due to deposition in an area protected from wave or current activity, perhaps by a barrier, or below the zone of wave or current activity. An erosive top to one of the mudstones and reworking of clasts into the overlying packstone bed suggests that the some of the mudstones were at least partially lithified soon after deposition. Beds above and below erosion surfaces contain normal marine fauna and erosive 'events' are therefore inferred to have occurred in a submarine environment, perhaps as the result of reworking by storms or waves.

Bioclast-dominated facies group

Wackestones, packstones, grainstones, floatstones, rudstones and any combination of these textural types comprise the bioclast dominated facies group. Bioclast dominated facies are the most abundant facies types and occur throughout the Pangkajene area (Tables 1 & 2). All the lithologies of the bioclast dominated facies group contain abundant shallow water bioclasts and are inferred to have been deposited within the photic zone.

Bioclastic facies (Figs 8e & f) are common throughout the Pangkajene area and contain abundant and sometimes fragmented shallow

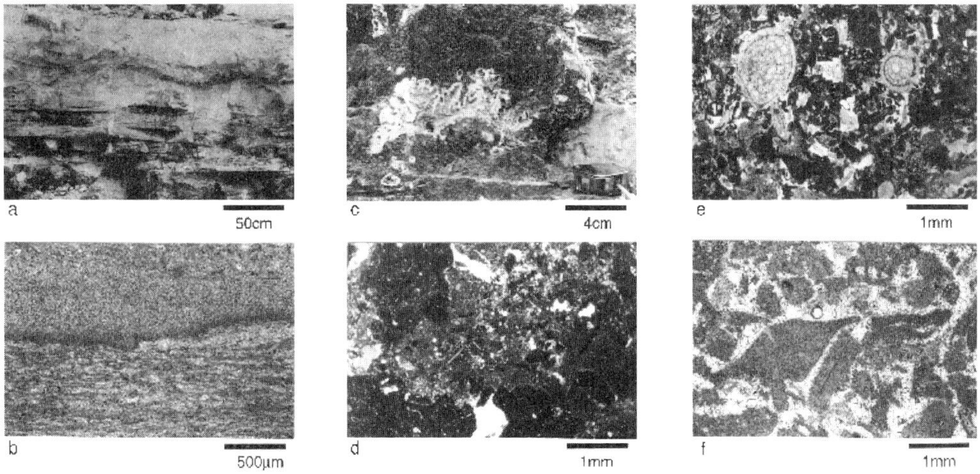

Fig. 8. (**a**) Field photograph of algal laminite from the Tonasa-II Quarry section. (**b**) Thin-section photomicrograph of an algal lamina from the Tonasa-II Quarry section under plane-polarized light. (**c**) Field photograph of probable gas escape structures in a micritic mudstone in the Tonasa-II Quarry section, showing a rather flat base and pustulose upper surface. (**d**) Thin-section photomicrograph of a brecciated bioclastic mudstone from the Patunuang Asue section, under plane-polarized light. Thin-section photomicrographs of (**e**) bioclastic packstones and (**f**) grainstones from the Lapangan Golf section, under plane-polarized light.

marine bioclasts (Table 1). Bioclastic wacke(/pack)stones and grain(/pack)stones (Fig. 8f) tend to be more common in the southern/northern and central parts of the Pangkajene area (Tables 1 & 2), indicating lower and higher energy marine depositional environments for these areas respectively (Fig. 4; Table 3).

Large benthic foraminifera are the dominant constituent particles (greater than 15%) in many of the bioclastic facies (Tables 1 & 2). Studies of the type and form of these organisms (Reiss & Hottinger 1984; Hallock & Glenn 1986), together with considerations of the texture and other biota in the lithologies are important tools for evaluating variations in the depositional environment on the platform top (Tables 3 & 4).

Imperforate foraminifera, including miliolids, alveolinids, *Praerhapidionina* cf. *delicata*, and less commonly *Austrotrillina* and soritiids (Fig. 9a & b) occur abundantly today in shallow parts of the photic zone and can tolerate some salinity fluctuations (Table 4). Miliolid/alveolinid and alveolinid & coral/*Lepidocyclina* facies occur frequently in the northern part of the Pangkajene area with wacke(/pack)stones and pack/grainstones dominating in northern and southern sections in this area respectively. A general increase in the energy of the depositional environment towards the south across the northern Pangkajene area is therefore suggested (Fig. 4). In some parts of the central Pangkajene area miliolid bioclastic packstones occur close to the contact with the Malawa Formation, whilst in the southern Pangkajene area alveolinid bioclastic packstones occur close to the contact with the overlying Camba Formation (Table 2; Fig. 7). A depositional environment in the shallower parts of the photic zone, with lithologies dominated by alveolinids, probably occurring at deeper depths or in more turbulent environments than those dominated by miliolids, is inferred for these lithologies (Tables 3 & 4). Beds containing abundant miliolids and textulariid foraminifera may have been deposited during periods of fluctuations in salinity, although the more diverse stenohaline biota in most of the other lithologies suggests deposition at, or very close to, normal marine salinities (Table 4).

Facies containing abundant, large, benthic perforate foraminifera and other stenohaline bioclasts, such as corals (Fig. 9c), are interpreted to have been deposited in normal marine conditions within the photic zone. The type and forms of the foraminifera varies according to the time period and the depositional environment. Wackestones and packstones, such as the *Heterostegina* wacke/packstone (Fig. 9d), *Lepidocyclina*/*Nummulites* wacke(/pack)stone and *Spiroclypeus* wacke/packstone (Fig. 9e) facies occur in the northern and southern Pangkajene areas (Tables 1 & 2). These contain rather thin, flat, forms of foraminifera (Fig. 9d & e), abundant micrite and are inferred to have been deposited in a low energy environment (Table 3). A shallow water protected setting is postulated for the depositional setting of the *Heterostegina* wacke/packstone facies and the curved forms of many of the foraminifera (Fig. 9d) in some beds of the facies suggests an association with seagrass (cf. Chaproniere 1975). In comparison, the *Spiroclypeus* wacke/packstone (Fig. 9e), which contain *Spiroclypeus*, *Operculina* and *Cycloclypeus* and only occur in the southern Pangkajene area, are inferred to have been deposited in deeper, but still low-energy, parts of the photic zone (cf. Ghose 1977; Hottinger 1983). Fragmentation of bioclasts and the more robust forms of large benthic foraminifera in the *Lepidocyclina*/*Nummulites* wacke(/pack)stone facies compared with other wacke/packstone lithologies, suggests a low- to moderate-energy setting for beds of this facies (Table 3). Lithologies of the *Lepidocyclina*/*Spiroclypeus* and *Discocyclina* grain/packstones (Fig. 9f) occur mainly in the central Pangkajene area and more rarely in the southern Pangkajene area (Tables 1 & 2). These lithologies contain robust and often fragmented large benthic foraminifera (Fig. 9f), and this together with their texture suggests deposition in a moderate to high energy setting.

Many of the large benthic foraminifera are age diagnostic (van der Vlerk & Umbgrove 1927; Adams 1970). Lithologies containing *Discocyclina* and *Pellatispira* are late Eocene (Tb) in age. Beds in the lowermost portions of sections in the northern Pangkajene area containing *Nummulites fichteli* have been dated as Tc (early Oligocene), whilst overlying beds in which *Nummulites fichteli* and *Eulepidina ephippoides* co-occur have been dated as upper lower Oligocene (Td; Crotty & Engelhardt 1993). In this area the last occurrence of *Nummulites fichteli,* below samples containing abundant *Eulepidina ephippoides,* is taken as the boundary of Td/e (early/late Oligocene; cf. Adams 1970).

Corals are rare in deposits of the Tonasa Limestone Formation, but a few lithologies do occur containing massive colonial corals and delicate branching corals in the northern and southern Pangkajene areas respectively (Tables 1 & 2). Lithologies containing in situ or fragmented massive colonial corals (Figs. 9c & 10a) include coralgal bind/rudstones and coral and

Table 4. *Distribution of large benthic foraminifera in the Pangkajene area and the implications this has for platform reconstructions*

Miliolids Abundant in many beds in the northern Pangkajene sections and in some beds close to the contact with the Malawa Formation in central Pangkajene sections. The abundant occurrence of miliolids indicates shallow (<24 m) depositional depths (Murray 1973; Hallock & Glenn 1985). During the deposition of some beds, which contain little other fauna, restricted conditions may have prevailed. However, it is not clear if salinity, nutrient levels or another control was the main restricting factor (cf. Hirshfield *et al.*, 1968; Murray 1973; Lee & Anderson 1991; Hallock & Peebles 1993).

Alveolinids Abundant in many beds in the northern Pangkajene sections and close to the contact with the overlying Camba Formation in the southern Pangkajene sections. Alveolinids can thrive in higher energy conditions or slightly greater depths than miliolids (Ghose 1977; Hottinger 1983; Reiss & Hottinger 1984) and their occurrence may indicate either of these two factors

Heterostegina Only common in wackestones in the northern and southernmost sections in the Pangkajene area. This genera inhabits a range of substrate types (Rttger & Berger 1972) in shallow to moderate water depths in the photic zone (Lutze *et al.* 1971; Hottinger 1977; Röttger *et al.* 1986) and seems to prefer low energy conditions (Newell 1956; Cole 1957; Maxwell *et al.* 1961). In the Tonasa-II Quarry section the curved forms of the *Heterostegina* suggest an association with seagrass communities (cf. Chaproniere 1975).

Thin *Spiroclypeus*, *Operculina* and *Cycloclypeus* Present in many of the lower energy wackestones and packstones in northern and southern Pangkajene sections, although *Operculina* and *Cycloclypeus* are never abundant. These genera are non-specialist and their thin and flat forms suggest moderate to deep depositional depths in the photic zone. The fact that *Operculina* and *Cycloclypeus* are never abundant, suggests that deeper depths within the photic zone never occurred in the Pangkajene area

Robust *Lepidocyclina*, *Spiroclypeus*, *Nummulites*, *Discocyclina* and *Pellatispira* Although the genera change with time, robust perforate large benthic foraminifera dominate the faunal assemblage in the central Pangkajene area. These genera also occur in some beds in northern and southern Pangkajene sections although their form is often not as robust as those in central Pangkajene sections. The robust forms of the foraminifera indicate deposition in moderate to high energy conditions probably in shallow parts of the photic zone.

Heterostegina/lithic facies and are inferred to have been deposited in or close to a patch reef setting in shallow parts of the photic zone (Table 3). However, the limited occurrence of these lithologies indicates that patch reef development was not extensive (Fig. 4). The rubbly, bored and encrusted surface of some the head corals and intraclasts (Fig. 10a), reworked under moderate to high energy conditions is inferred to have provided sufficient protection and a hard substrate on which *Heterostegina* could exist (cf. Hottinger 1977). The coral bioclastic pack/floatstones and alveolinid & coral/*Lepidocyclina* facies in the northern Pangkajene sections contain rather delicate branching corals (Fig. 9c) and are thought to have been deposited in a low- to moderate-energy shallow-marine setting (Table 3).

Planktonic foraminifera facies group

Lithologies of the planktonic foraminifera facies group, which contain more than 2% planktonic foraminifera and include marls and packstone lithologies, occur almost exclusively in the Jeneponto area and are very rare in the Pangkajene area (Tables 1 & 2).

In the Jeneponto area thick sequences of laterally extensive marls (Fig. 10b), up to 90 m thick, occur interbedded (Fig. 7) with thick packages of decimetre-bedded, planktonic foraminifera, bioclastic pack(/grain)stone (Fig. 10c) or lithic bioclastic packstone facies (Fig. 10d & e). The marls contain abundant pelagic marine organisms, may be laminated or bioturbated and were deposited in a low energy bathyal setting where aerobic conditions occurred at least in some areas (Tables 1 & 3).

The planktonic foraminifera bioclastic pack(/grain)stone (Fig. 10c) and lithic bioclastic packstones (Fig. 10d & e) contain abundant well preserved planktonic foraminifera and fragmented shallow marine bioclasts. Glauconite grains are present in some beds and the lithic bioclastic packstones contain up to 5% angular grains of strained quartz and euhedral feldspar (Fig. 10e). The clastic grains suggest derivation from a continental or igneous source and the significance and possible source areas are discussed below. Rare low-angle, trough cross-bedding, indicating transport direction to

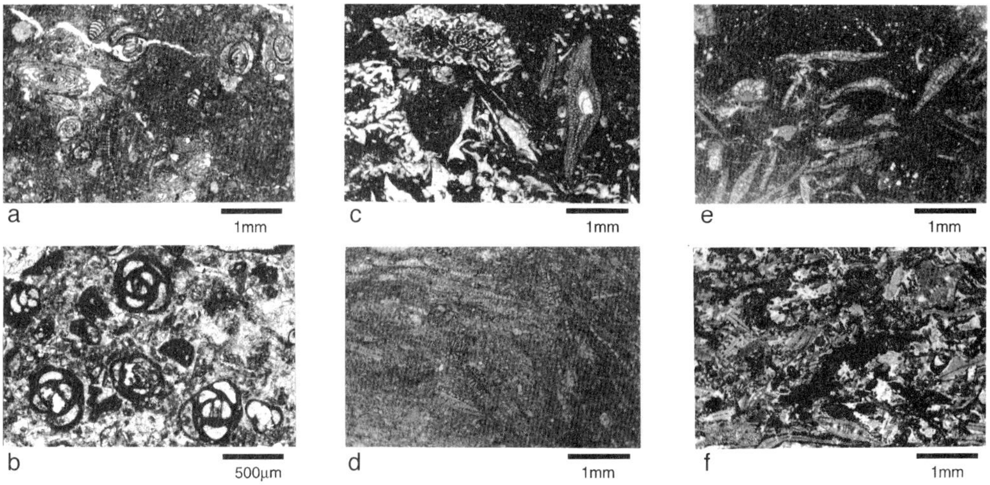

Fig. 9. (**a**) Thin section photomicrographs of a miliolid and *Lepidocyclina* bioclastic packstone, (**b**) a miliolid bioclastic pack/grainstone and (**c**) a coral (*Acropora*) and *Lepidocyclina* bioclastic pack/floatstone from the Tonasa-II Quarry section, under plane-polarized light. (**d**) Acetate-peel photomicrograph of a *Heterostegina* wackestone from the Tonasa-II Quarry section, under plane-polarized light. Thin-section photomicrographs of a (**e**) *Spiroclypeus* wacke/packstone from the Patunuang Asue section and (**f**) a *Discocyclina* bioclastic pack/grainstone from the Lapangan Golf section under plane-polarized light.

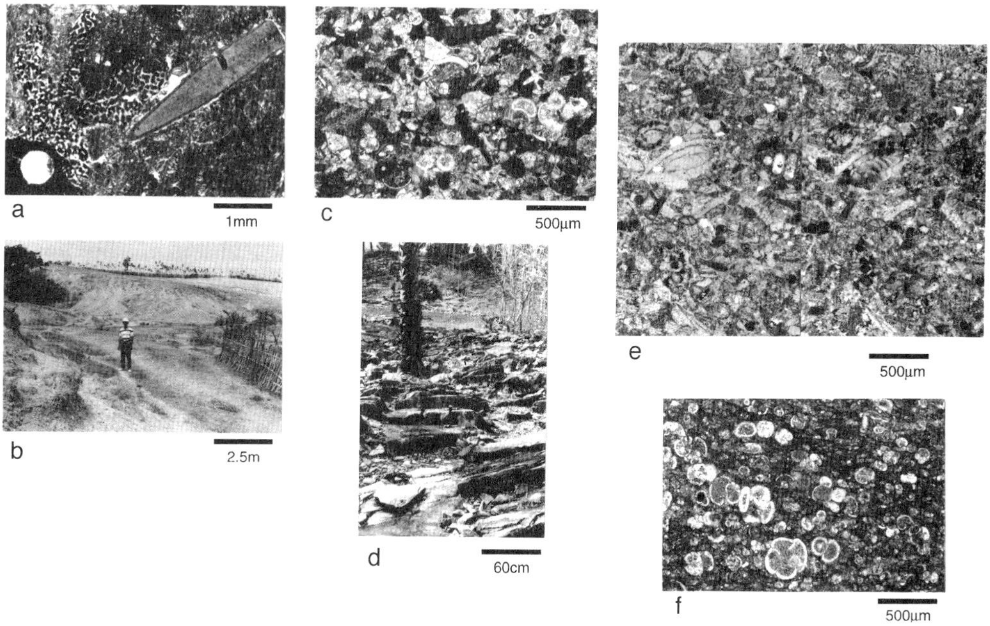

Fig. 10. (**a**) Thin-section photomicrograph of a coral, lithic and *Heterostegina*, bioclastic rudstone from the Patunuang Asue section, under plane polarised light. (**b**) Field photograph of the poorly exposed greeny-grey marls from the Jeneponto area. (**c**) Thin-section photomicrograph of a planktonic foraminifera, bioclastic packstone from the Jeneponto area, under plane polarized light. (**d**) Field photograph of outcrop of well-bedded lithic bioclastic packstones, with possible low-angle cross bedding from the Jeneponto area. (**e**) Thin-section photomicrograph of a lithic bioclastic packstone from the Jeneponto area under plane- (left) and cross-polarized (right) light. (**f**) Thin-section photomicrograph of a planktonic foraminifera packstone cut by dissolution seams from the Nasara Jeneponto section under plane-polarized light.

the east, was observed in the lithic bioclastic packstones (Fig. 10d). A sublittoral marine environment, close to or below the zone of wave or current activity, with an open oceanic influence is the envisaged depositional setting of these facies (Tables 1 & 3). Planktonic foraminifera are extremely rare in the lithologies in the Pangkajene area, with the exception of the upper part of the Bulo Kamase section (Fig. 5). An open oceanic influence is therefore only inferred for the southwestern part of the Pangkajene area for the late Oligocene to early Miocene.

Extremely rare beds of the graded bioclastic packstones occur interbedded with the marls in the Jeneponto area and are interpreted as distal calciturbidites (Tables 1 & 3; cf. McIlreath & James 1984; Mullins *et al.* 1988). In the uppermost part of the section in the Jeneponto area decimetre thickness beds of planktonic foraminifera wacke/packstones (Fig. 10f), composed of abundant planktonic foraminifera (70–90%), occur interbedded with marls. Centimetre-scale, spaced dissolution seams lined with insoluble material are a common feature of this facies. The increased percentage of planktonic foraminifera within this facies compared with the interbedded marls is thought to be due to a combination of reworking or winnowing and/or via dissolution of the matrix.

Palaeocurrent data

Palaeocurrent indicators in the Pangkajene area include cross bedding, truncation surfaces, bed thickening and progradation of beds (Fig. 11). These sedimentary structures vary from metre scale features, to structures tens to hundreds of metres across seen in the near vertical faces of tower-karsts (Fig. 11). Most palaeocurrent data comes from the central sections in the Pangkajene area, where moderate- to high-energy, shallow-marine facies prevail. Dense vegetation and dripstone coatings often render the identification of palaeocurrent indicators in karstic outcrops impossible. For palaeocurrent data obtained from the relatively two-dimensional faces of tower-karsts (Fig. 11), care was taken to view the karsts from a number of different angles in order to ascertain the most accurate transport directions possible. However, despite care taken with measurements errors in this method may be up to 15°.

Figure 11 shows sketches and photographs of tower-karst outcrops in the Tonasa-I Quarry area which include palaeocurrent indicators. Palaeocurrent data was obtained from a number of other karstic locations in a similar way. A compilation rose diagram of all the palaeocurrent data from the Pangkajene area is shown in Fig. 12. This rose diagram reveals that although there was some variability in palaeocurrent directions, the dominant transport direction in the Pangkajene area was clearly towards the east.

Only one small-scale structure; some low angle cross-bedding, also indicating a transport direction to the east was found in the Jeneponto area.

Probable Tertiary subaerial exposure

Evidence for subaerial exposure is generally scarce, localised and mostly occurs in the northern sections in the Pangkajene area (Fig. 7). A number of features, although difficult to date, suggest subaerial or near subaerial exposure. These include lithified carbonate sediment and sparry calcite cement infilling earlier centimetre to decimetre-scale karstic cavities (Fig. 13), and pinky/red irregular dissolution surfaces probably formed in the soil forming horizon. Evidence for near subaerial exposure or peritidal conditions include algal laminites(Fig. 8a & b) and in the overlying bed, voids interpreted as gas escape structures (Fig. 8c). It should be noted that none of these features represent angular disconformities and most of these features cannot be traced laterally for more than a few metres. Possible reasons for subaerial exposure are discussed below.

The Tonasa Carbonate Platform

The presence of well-preserved, shallow-marine bioclasts throughout the Tonasa Limestone Formation in the Pangkajene area suggest deposition occurred entirely within the photic zone. With the exception of the southwestern Bulo Kamase section (Fig. 5), planktonic foraminifera constitute <0.5% of the bulk of the carbonate sequence in the Pangkajene area. Throughout the Pangkajene area lithologies of the Tonasa Limestone Formation are virtually devoid of quartz grains or lithic non-carbonate grains. The exception to this is that some bioclastic facies close to the contact with the underlying Malawa Formation, contain up to 10% by volume quartz. Therefore the Tonasa Limestone Formation in the Pangkajene area represents a widespread area of shallow-water carbonate deposition, named here as the Tonasa Carbonate Platform. This platform area can

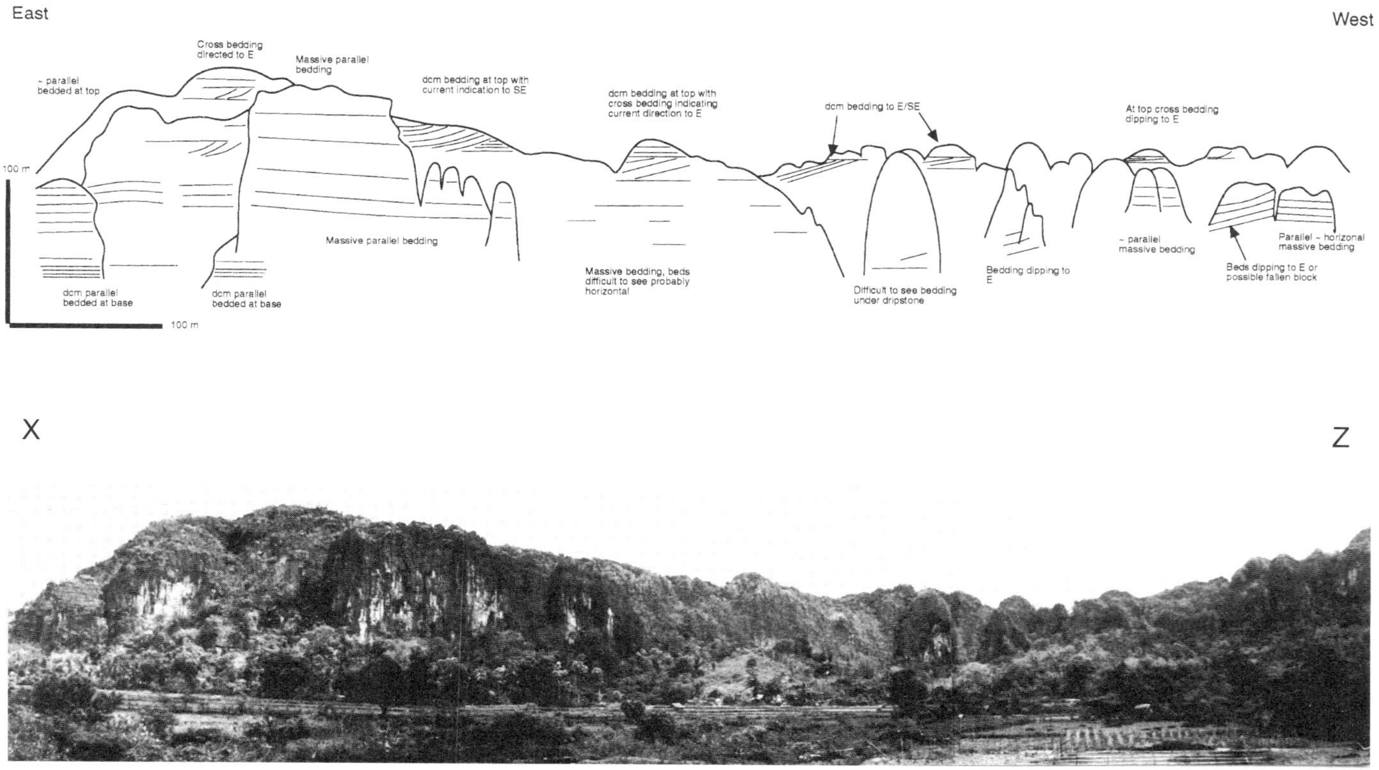

Fig. 11. Sketch and photograph of karstic outcrops showing palaeocurrent data and prograding and aggrading nature of the succession looking southwards in the Tonasa-I Area (see Fig. 5 for location XZ).

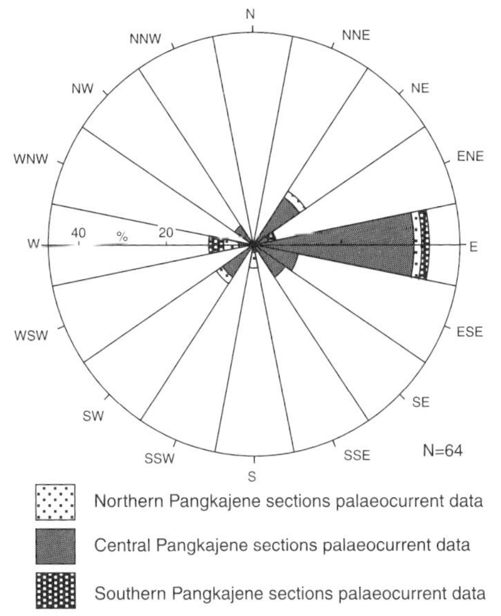

Fig. 12. Rose diagram showing all palaeocurrent data from the Pangkajene area.

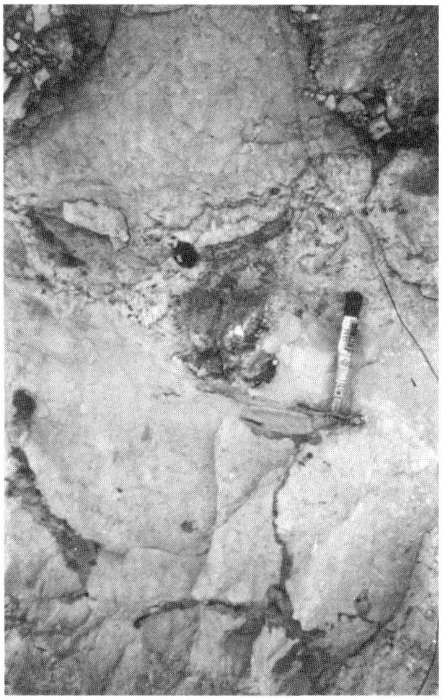

Fig. 13. Probable Tertiary karstic dissolution cavity infilled with fine sediment and a sparry calcite cement.

effectively be regarded as an isolated platform, due to the lack of siliciclastics, which was little affected by open oceanic influences.

In the Pangkajene area, the margins of this widespread region of shallow-water carbonate accumulation are not exposed. However, in the Jeneponto area, the deposition of thick sequences of marls interbedded with packages of carbonate sands, containing no evidence for significant reworking via sediment gravity flow mechanisms, is suggestive of deposition on very low slope angles (Fig. 4; cf. Gawthorpe 1986; Burchette & Wright 1992). The most likely source for shallow-water bioclasts reworked into these carbonate sands is from the Tonasa Carbonate Platform and a southern ramp-type margin is therefore inferred (Fig. 4; see below). From the presence of coarse redeposited facies interbedded with basinal marls in the Barru area it has been inferred that the northern margin of the Tonasa Carbonate Platform was a syndepositional faulted margin with faults periodically active from the late Eocene until the middle Miocene (Wilson & Bosence 1996). Localized outcrops of shallow and deeper-water limestones of the Tonasa Limestone Formation to the east of the Pangkajene area indicate that the Tonasa Carbonate Platform may at times have been laterally much more extensive. The westward lateral extent of the platform is more difficult to constrain since there is little published information on the nature of deposits offshore South Sulawesi in the Makassar Straits.

Spatial variations on the Tonasa Carbonate Platform

Figure 7 shows a biostratigraphic correlation between measured sections in the Pangkajene and Jeneponto areas. Despite the variations in the age ranges of the measured sections in the these areas (see below & Fig. 7), and lack of lithological correlation between sections, certain textures, fauna and lithologies appear to be related to specific regions which trend east/west. These spatial variations are described below and their implications for platform reconstruction are discussed:

Texture

Northern and southern Pangkajene sections tend to be dominated by wackestone, wacke/packstone and packstone textures, whilst pack/grainstone and grainstone textures occur most commonly in central Pangkajene sections (Tables 1 & 2). Mud/wackestone facies were only found in northern and southern Pangkajene sections (Tables 1 & 2). This segregation of the textural types, regardless of age variations, immediately indicates that facies in the northern and southern Pangkajene sections were generally deposited under lower energy conditions than those in central Pangkajene sections (Fig. 4). Thick sequences of marls with a wackestone texture occur interbedded with packages of packstone or very rarely grainstone units in the Jeneponto area (Fig. 7) suggesting fluctuations from a low- to moderate-energy depositional setting.

Biota

Well-preserved large benthic foraminifera dominate the faunal assemblage in the Pangkajene area and their occurrence indicates deposition within the photic zone (Tables 1, 2 & 4). Echinoid plates and fragments of coralline algae are also common in the majority of facies in the Pangkajene area (Tables 1 & 2). Corals, *Halimeda* and bryozoans, although present in a number of different facies, generally occur in very minor amounts (Tables 1 & 2). A wide variety of often fragmented shallow marine organisms also occur in the packstone units in the Jeneponto area. The occurrence of these stenohaline organisms indicates that normal marine conditions prevailed on the Tonasa Carbonate Platform. Well-preserved planktonic foraminifera occur in abundance in the lithologies in the Jeneponto area (Fig. 10b–f), but are extremely rare in lithologies in the Pangkajene area suggesting an open oceanic influence for the more southerly Jeneponto area.

Both specialist and generalist types of large (and small) benthic foraminifera have evolved to fill specific environmental 'niches'. Studies of the distribution and test form of these foraminifera in the Pangkajene area prove to be extremely useful indications of palaeoenvironments on the Tonasa Carbonate Platform. The distribution of the main foraminifera groups (Tables 1 & 2) and their implications for platform reconstructions are outlined in Table 4.

Lithology

Despite the fact that no single section spans the whole stratigraphic sequence of the Tonasa Limestone Formation in the Pangkajene area (Fig. 7), specific lithologic associations do occur in northern, central and southern parts of the Pangkajene area. Tables 1 and 2 show the facies or lithologies present within each section and the variations between sections.

Northern Pangkajene sections. Dominated by shallow-water, low- to moderate-energy mudstones, wackestones and packstones containing imperforate foraminifera, *Heterostegina*, *Lepidocyclina* and branching coral fragments (Tables 1 & 2). Intertidal facies and localised evidence for probable Tertiary subaerial exposure also occur in northern sections (Fig. 7).

Central Pangkajene sections. Dominated by moderate- to high-energy, shallow- to moderate-water depth packstones and grainstones, containing abundant, robust *Discocyclina*, *Pellatispira* and *Lepidocyclina*, depending on their age (Tables 1 & 2). Close to the contact with the Malawa Formation, shallow-water facies containing abundant miliolids are interbedded with pack/grainstones (Fig. 7).

Southern Pangkajene sections. Low-energy mudstones, wackestones and wacke/packstones, including thin forms of *Heterostegina*, *Lepidocyclina* and *Spiroclypeus* and coral fragments (Tables 1 & 2), suggest shallow to moderate depositional depths. Up-section, towards the contact with the overlying Camba Formation, packstones and pack/grainstones containing imperforate foraminifera (Fig. 7) suggest an increase in the energy and perhaps a shallowing of the environment.

Jeneponto area. Lithologies of the planktonic foraminifera facies group dominate in the Jeneponto area and the sequence consists of thick marl units interbedded with packages of planktonic foraminifera or lithic bioclastic packstones and more rarely grainstones (Fig. 7). An oxygenated, but deep, open, oceanic depositional setting is inferred for the marls, whilst the packages of packstones are interpreted as outer shelf or slope deposits.

The frequency of facies changes in the lower energy northern and southern Pangkajene sections is greater than in higher energy central Pangkajene sections. In low-energy, shallow-water settings storms may cause the deposition of high-energy facies, whilst inhibited circulation may result in ponding and the deposition of 'restricted' facies. In comparison, an area washed by constant currents will not be affected so much by storms and ponding. Therefore rapid facies changes are less likely in shallow water areas influenced by high energy conditions.

Temporal variations on the Tonasa Carbonate Platform

A biostratigraphic correlation between measured sections in the Pangkajene area and Jeneponto areas is shown on Fig. 7. The succession in the Pangkajene area is dated mainly on the large benthic foraminifera East Indian Letter Classification (van de Vlerk & Umbgrove 1927; Adams 1970), whilst nannofossils and planktonic foraminifera have been used to date the succession in the Jeneponto area.

Pangkajene area

The Tonasa Limestone Formation in the Pangkajene area spans the late Eocene (Tb) to early Miocene (probably Te_5), although no complete single section occurs (Fig. 7). The juxtaposition of sections of different age ranges is the result of post-depositional faulting. Northern sections in the Pangkajene area generally span the early to late Oligocene (Tc–Te; Fig. 7), although the northeasterly Siloro Mangilu section (Fig. 5) is late Eocene (Tb; Fig. 7) in age. Most of the central sections are late Eocene (Tb) and some pass upwards into the early Oligocene (Tc). Southerly central sections and southern sections range from late early Oligocene (Td) into the early Miocene (Te_5, Fig. 7). It is not usually possible to correlate the same lithology between adjacent sections (Fig. 7). Although spatial variations between sections in the Pangkajene area are important (see above), some up-sequence changes do occur. However, the most obvious pattern is that sections are remarkable for a lack of large-scale up-section changes in facies and environments (Fig. 7). This indicates that the carbonate succession in the Pangkajene area was dominantly aggradational in nature (see below).

In the northern Pangkajene area, up-section changes include variations in the abundance and type of imperforate foraminifera. Textural variations or changes from imperforate foraminifera facies to bioclastic packstones, containing a more diverse faunal assemblage, occur on a scale of metres to tens of metres in the lower part of the Tonasa-II section and other northern sections in the Pangkajene area (Fig. 7). Similar variations also occur between bioclastic pack/grainstones and imperforate foraminifera facies in the lower parts of central sections in the Pangkajene area (Fig. 7). In the Bulo Kamase section there is an alternation on a scale of tens of metres between *Lepidocyclina* bioclastic wacke/packstones and *Lepidocyclina*/*Spiroclypeus* pack/grainstones (Fig. 7). These variations may be related to climatic fluctuations or to variations in the depositional depth, perhaps caused by tectonic tilting, autocyclicity or eustasy. Other variations in the central sections include textural variations and a faunal change from *Discocyclina* to *Lepidocyclina* related to age.

In the upper 50 m of the Tonasa-II Quarry section imperforate foraminifera are scarce and *Heterostegina* and *Lepidocyclina* dominate. This change in fauna may indicate an up-section deepening of the environment. However, the presence of algal laminites and mudstones with probable fenestrae suggest intertidal/supratidal conditions during the deposition of the upper part of the Tonasa-II Quarry section. Alternative reasons for this switch in fauna might include changes in the nutrient level, substrate or depositional energy.

In the southernmost Patunuang Asue section there are metre-scale variations in texture, faunal composition and abundance (Fig. 7). These variations may be related to intrinsic or extrinsic changes in the depositional depth or to energy variations related to climatic fluctuations.

Jeneponto area

The carbonate succession in the Jeneponto area ranges from late Eocene to mid-Miocene, although much of the Oligocene and early Miocene is poorly exposed (Fig. 7). The compilation stratigraphic section from the Jeneponto area for the late Eocene and early Oligocene shows that three packages of planktonic foraminifera or lithic bioclastic packstones are interbedded with two, thick marl successions (Fig. 7). The middle and upper packstone units coarsen-up rather rapidly from fine sand grade to medium(/coarse) sand grade deposits. This increase in grain size suggests the deposits may shallow upwards. The packstone units are interpreted as mid/outer ramp deposits (see below), which are inferred to have prograded out into basinal marls on at least three occasions. The apparent lack of interfingering of marl and packstone units, and the rapid grain size change in the packstones, suggests that when progradation occurred it was a rapid process. The lower and middle packages of packstones contain *Discocyclina* and *Asterocyclina* and are late Eocene in age. The upper bedded packstone unit contains *Lepidocyclina* and is Oligocene in age.

Palaeoreconstruction of the Tonasa Carbonate Platform

Platform top: Pangkajene area

All the facies in the Pangkajene area indicate that sedimentation on the Tonasa Carbonate Platform occurred in the middle to upper parts of the photic zone. Figure 4 shows a reconstruction of the Tonasa Carbonate Platform (and its southern margin) during the Oligocene. The justification for the proposed reconstruction and implications for platform reconstructions during the Eocene or Miocene are outlined below.

The textural, faunal and lithological distribution all indicate that the northern and southern parts of the Pangkajene area were, for much of their depositional history, areas of low to moderate energy sedimentation (Fig. 4). In contrast, moderate to high energy conditions prevailed for much of the time in the central part of the Pangkajene area (Fig. 4). Open-oceanic conditions are inferred to have existed to the southwest of the Pangkajene area due to the abundant occurrence of planktonic foraminifera only in the southwestern Bulo Kamase section (Fig. 5). These facts, and the dominance of east directed palaeocurrents (see above), suggest that northern and southern parts of the Pangkajene area were probably protected from open oceanic influences and sheltered by some form of barrier located to the west (Fig. 4). This barrier is not thought to have been as effective in protecting the central part of the Pangkajene area, since higher energy conditions prevail in this region.

Abundant miliolids in the northern part of the Pangkajene area (Table 4) indicate that very shallow-water depositional depths prevailed in this region. Localised evidence for intertidal sedimentation and subaerial exposure occur in the most northerly sections (Fig. 7). However, some facies in northern sections suggest that shallow to moderate depositional depths also occurred in this region (Fig. 7; Tables 1 & 2). A region of low relief islands or periodically emergent bars with intervening deeper-water, protected areas is therefore inferred for the northern part of the Pangkajene area (Fig. 4). The lack of plant material in northern sections perhaps suggests that subaerial areas did not remain emergent for a sufficient length of time to allow vegetation to develop. Facies in the Siloro Mangilu section suggest that higher energy conditions probably occurred in the northern part of the Pangkajene area during the late Eocene (Fig. 7). Subaerial exposure and karstification of the Siloro Mangilu section also occurred, although it is not clear if this was caused by emergence during the Eocene.

Through much of the late Eocene and Oligocene sedimentation in the central part of the Pangkajene area occurred under moderate to high energy conditions (Table 3; Fig. 4). Some sections close to the contact with the underlying Malawa Formation contain facies in which typical stenohaline fauna is absent and miliolids and textulariids dominate (see earlier). Interbedded with, and lateral equivalents to these facies, are beds deposited under moderate to high energy conditions containing abundant normal marine bioclasts. Initial carbonate sedimentation in the central part of the Pangkajene area is therefore thought to have occurred as a series of moderate/high energy bars (Fig. 4). Intervening shallow-water areas may at times have become ponded, with the possibility of restricted conditions developing (Fig. 4).

Southerly sections in the Pangkajene area are dominated by low-energy facies deposited in shallow to moderate depths within the photic zone. This is the only region in which massive, colonial coral framestones and rudstones occur, and for which patch reef development is inferred (Fig. 4). Plant matter is present in these sections, especially close to the contact with the overlying Camba Formation. Although there is no evidence for subaerial exposure of southern sections, colonization by vegetation of emergent areas close to this region is inferred (Fig. 4). Prior to the development of marginal-marine or terrestrial conditions and the influx of volcaniclastics during the deposition of the Camba Formation, shallow-water, moderate-energy conditions developed during final early Miocene carbonate sedimentation.

Platform-top facies belts

Three major east–west-trending facies belts were present throughout much of the period of carbonate sedimentation the Pangkajene area (Fig. 4). These were:

(i) a northern rather low-energy, shallow-water belt with evidence for intertidal sedimentation and subaerial exposure;
(ii) a central moderate- to high-energy belt deposited in shallow to moderate water depths;
(iii) a southern low-energy belt deposited in shallow to moderate water depths.

Facies interpretations of northern sections reveal a general increase in depositional energy towards the south (Fig. 4). This suggests that the boundaries between the main facies belts were probably gradational rather than sharp changes. Due to outcrop constraints, east–west facies changes are difficult to evaluate. In the central section there appears to have been little change in the depositional setting of facies in an east–west sense. Although it was only possible to record a rather thin stratigraphic sequence at Bantimurung, comparison between this section and the Patunuang Asue section suggests that higher energy conditions may have prevailed towards the west.

Southern ramp-type margin: Jeneponto area

In the Jeneponto area, thick sequences of pelagic marls interbedded with packages of carbonate sands (Fig. 7), containing no evidence of significant reworking via sediment gravity flow mechanisms, are suggestive of deposition on very low slope angles (Fig. 4; cf. Gawthorpe 1986; Calvet & Tucker 1988; Watts & Blome 1990). Typical slope angles of modern carbonate ramps which lack sediment gravity flows are usually less than 1–2°, but may be as steep as 5° (Mullins *et al.* 1988; Ivany *et al.* 1994). Fragmented fauna, within the packstones, derived from areas of shallow-water carbonate production, indicates an absence of organisms which could form organic build-ups. The inferred low slope angle and lack of framework building organisms are characteristic features of homoclinal carbonate ramps (Ahr 1973; Read 1982, 1985; Calvet & Tucker 1988). Burchette & Wright (1992), in their review of carbonate ramps, suggested that the mid-ramp zone extends from fair-weather wave-base to normal storm wave-base. The packages of decimetre bedded packstones, with limited evidence for reworking by waves or currents, are therefore interpreted as mid/outer ramp deposits.

The orientation of the carbonate ramp is difficult to deduce from analysis of the carbonate succession in the Jeneponto area in isolation, since sedimentary structures which might reveal palaeoslope or palaeocurrent directions are very rare or not present. The low-angle cross-bedding in one of the lithic bioclastic packstone beds indicates a transport direction to the east. This is consistent with most of the palaeocurrent data from the Tonasa Carbonate Platform (see above) and may therefore represent reworking by 'along-slope' currents rather than a 'down-slope' indication. Wells drilled offshore South Sulawesi to the south and west of the Jeneponto area pass through the Tonasa Limestone Formation. In these wells there is no evidence for any significant *in-situ* production of shallow-water carbonates, from the late Eocene onwards, and the lithologies are described dominantly as marls or mudstones. In the Jeneponto area there is no noticeable change in the packstone units along strike in an ENE–WSW direction. It would therefore seem that the fragmented, shallow-water biota in the outer ramp facies, deposited in the Jeneponto area, was not derived from the east, west or south, but from the north i.e. from the area of the Tonasa Carbonate Platform in the Pangkajene area.

No outcrops of the Tonasa Limestone Formation occur between the Pangkajene and Jeneponto areas since the carbonate succession lies buried beneath volcaniclastic sediments. However, the nature of the change from platform top to mid/outer ramp and bathyal deposits can be deduced from the sedimentary successions in the southern Pangkajene and Jeneponto areas (Fig. 4; Table 5). In the Patunuang Asue section (Fig. 5), the southernmost section through the platform deposits in the Pangkajene area (located 50 km to the north of the Jeneponto area), the lithologies are low energy mudstones, wackestones and packstones deposited within the photic zone (Table 2). There is an apparent lack of open oceanic material in these inner platform sediments. In comparison, mid/outer ramp deposits in the Jeneponto area contain abundant well-preserved planktonic foraminifera and fragmented shallow-water bioclasts. Some form of 'barrier' may have separated platform deposits of the Pangkajene area from the mid/outer ramp deposits of the Jeneponto area (Fig. 4; Table 5). Alternatively, the wide extremely shallow area of a minimal slope, present on ramps, could protect a 'lagoonal' area from up-ramp migration of waves or currents by dissipating their energy (Pedley *et al.* 1992). The southern margin of the Tonasa Carbonate Platform is inferred to have been an area of large benthic foraminifera and coralline algae production, since these faunal elements dominate the mid/outer ramp deposits (Table 5). The occurrence of plant matter in the late Oligocene deposits in the Patunuang Asue section perhaps suggests that the southern margin of the Tonasa Carbonate Platform may at times have become emergent (Fig. 4).

Little evidence is available for the mode of derivation of the shallow-water material from the areas of carbonate production, whether by sediment gravity flows (development of Bouma sequences) or reworking by waves and currents (hummocky and swaley cross bedding). It may be that gross ramp geometries would only show

Table 5. *Reasons for the inferred nature of the southern margin of the Tonasa Carbonate Platform*

	Nasara section, Jeneponto Area	Patunuang Asue section, Pangkajene Area	Implications for both areas
Textural considerations	Basinal marls interbedded with packstones of inferred mid/outer ramp setting. Packstones contain fragmented shallow normal marine biota suggesting reworking.	Relatively low-energy mudstones, wackestones and some packstones. The shallow, normal marine biota present is mostly unbroken.	Higher-energy deposits with evidence for reworking in Jeneponto Area and lower energy deposits in Pangkajene Area suggests some form of 'barrier' separated the two areas. Alternatively a gently sloping ramp might dissipate energy.
Pelagic biota assemblage	Contains planktonic foraminifera throughout the sequence. Abundant and well-preserved specimens are present in the marls and planktonic foraminifera wacke/packstone facies (70–90%), whilst the other packstone facies contain up to 15% planktonic foraminifera.	Planktonic foraminifera are not present or are very rare.	Open oceanic influence in Jeneponta Area, but some form of 'barrier' prevented an open oceanic influence in the Pangkajene Area.
Large benthic foraminifera assemblage	The normal marine fauna and presence of fragmented *Discocyclina*, *Nummulites* and *Asterocyclina* in the lower part of the section and rotund *Lepidocyclina* in Oligocene deposits suggest a foreslope environment (van Gorsel 1988).	Barren samples or samples with a rather restricted fauna of flat *Lepidocyclina*, *Spiroclypeus* and in some samples monospecific *Heterstegina* suggest a low energy lagoonal or inner platform area (cf. Chaproniere 1975).	The large benthic foraminifera assemblages suggest some form of energy 'barrier' between the higher-energy foreslope' environment in the Jeneponto Area and the lower-energy central inner platform area.
Coral	Framework building corals are not present or very rare. Rare fragments of branching corals are present.	Framework building coral debris is present in very low amounts.	The inferred southern margin of the Tonasa Carbonate Platform is not formed of framework building coral. Patch reefs were probably not present on the southern ramp type margin, although they may have been present in the central 'back-barrier' region.
Lithic grains	Strained quartz & euhedral feldspar grains are present in late Eocene lithic bioclastic facies only.	Although no lithic grains were observed in the section at Patunuang Asue, they might occur in older non-exposed sediments at this locality. However, older platform sediments exposed in other localities contain no lithic grains.	The lack of observed lithic grains in the central platform area suggests they were derived from older formations that were exposed at the platform margin during the late Eocene. Exposure could result from faulting. Another possibility is that the quartz may have been transported from a different area (Fig. 4).
Conclusions	Open oceanic mid to outer ramp setting.	Low-energy back barrier/inner platform sediments deposited in shallow to moderate depths in the photic zone.	Some form of barrier or energy barrier separated the southern ramp deposits from the central inner platform area. Foraminifera shoals are inferred to have formed the margin deposits of the Tonasa Carbonate Platform. During the late Eocene older formations may have been exposed to erosion at the platform margin (Fig. 4).

up on the scale of regional seismic lines rather than on outcrop scale (Burchette & Wright 1992). The absence of reworked deeper water carbonate material suggests that this sequence was deposited on a homoclinal ramp rather than a distally steepened ramp setting. Although impossible to tell from these deposits, it is conceivable that the slope may at one time have steepened to the south of the Jeneponto area.

Controls on sedimentation

Aggradation v. progradation and accommodation space on the platform top

The three major facies belts in the Pangkajene area appear to have remained relatively stationary during the Oligocene and in some areas during the late Eocene. In the central facies belt, bed geometries suggest some progradation towards the east. However, bed geometries in the northern and southern facies belts, and the stationary nature of the facies belts in a north–south sense, indicate that sediment accumulation on the Tonasa Carbonate Platform was mainly aggradational.

Due to the lack of complete sections through the Tonasa Limestone Formation, the total thickness of carbonate sediments at any location in the Pangkajene area is impossible to ascertain (Fig. 7). By combining the stratigraphic thicknesses deposited during different periods in different parts of the Pangkajene area, an estimate for the total thickness of the carbonate sequence of about 600 m is obtained. This figure is unlikely to be accurate, since time correlation between sections is rather tentative (Fig. 7) and thicknesses may vary across the Pangkajene area. However, this figure clearly illustrates the fact that considerable thicknesses of shallow marine carbonate sediments accumulated in the Pangkajene area during the Eocene to Miocene. Apart from lithologies close to the contact with the Malawa Formation there is no evidence to suggest that emergence occurred in either the central or southern facies belts during the deposition of the Tonasa Limestone Formation. Facies in these two regions indicate that deposition occurred in shallow to moderate depths within the photic zone. The great thickness of aggradational sediments of the Tonasa Carbonate Platform, together with a lack of extensive emergent surfaces indicating relative sea level falls, in the Pangkajene area suggests that the dominant control on accommodation space must have been subsidence rather than rises and falls of sea-level.

The nature and position of the three east-west oriented platform top facies belts in the Pangkajene area is probably controlled by east directed palaeocurrents, perhaps driven by prevailing wind directions, and the location of barriers which may have modified water circulation patterns (Fig. 4). Barriers or islands are inferred to have been located in the northern facies belt and to the west of the southern facies belt, thereby providing shelter during deposition in these two regions (Fig. 4). This is similar to the modern Caicos Platform, where its location in the easterly-trade belt and the west-flowing Antilles current cause waves and currents to move across the bank in a westerly direction. Facies belts on the Caicos Platform are approximately east-west trending and islands at the northern margin provide enough shelter for muddy sediments to collect on their lee side (Wanless & Dravis 1989).

It is not possible to distinguish the main factor(s) controlling most up-section changes in texture or fauna, which take place on a scale of metres or tens of metres, in the Pangkajene area. Factors which might influence lithological changes include tectonics, autocyclicity, eustasy or climate.

Location and causes of probable Tertiary subaerial exposure in the Pangkajene area

Evidence for probable Tertiary subaerial exposure is seen only in northerly sections and close to the contact with the underlying Malawa Formation in the central part of the Pangkajene area. In both of these localities shallow sedimentation in water depths of <24 m is inferred due to the abundant occurrence of miliolids. There are three main possible causes for subaerial exposure in the regions:

(i) a global eustatic sea-level fall might result in emergence and karstification of shallow carbonate depositional areas;
(ii) tectonics could result in uplift and subaerial exposure of certain areas;
(iii) autocyclicity may result in emergence if carbonate sediments completely fill the available accommodation space; subsequent sedimentation will occur in the intertidal or supratidal zone.

It is impossible to differentiate between the these three possible causes of subaerial exposure, just by studying carbonate sediments in the Pangkajene area. This is because probable Tertiary subaerial exposure surfaces in the

Pangkajene area are localised and cannot be accurately dated. Studies in other outcrop areas of the Tonasa Limestone Formation suggest that tectonics strongly controlled sedimentation patterns (van Leeuwen 1981; Wilson & Bosence 1996) and may therefore have influenced emergence in the Pangkajene area. The localized occurrence of subaerial exposure and in some cases abrupt facies changes might suggest a primary tectonic or autocyclic control on emergence. However, areas affected by emergence are those in which deposition is inferred to have occurred in the shallowest water conditions in the Pangkajene area.

Reasons for the distribution pattern of very shallow-water facies might result from antecedent topography, sediment accumulation rates, water-circulation patterns or accommodation space. Areas of high antecedent topography in the northern part of the Pangkajene area might have allowed carbonate sedimentation to 'keep-up' with relative sea level. The basal part of the carbonate sequence in the Pangkajene area is inferred to have been deposited under very shallow-water conditions. High-energy conditions inferred for the central facies belt, perhaps caused by the location of barriers may have caused the removal and transport of carbonate material away from this region, with the result that sediments could not 'keep-up' with relative sea level. Alternatively, tectonic tilting could produce greater accommodation space in central and southern parts of the Pangkajene area and carbonate production in these regions was not sufficient to keep pace with relative sea level. Unfortunately outcrop constraints in the Pangkajene area preclude the possibility of differentiating these possible controls on sedimentation and subsequent emergence. A forthcoming paper will discuss these and other factors which may have influenced sedimentation of the Tonasa Limestone Formation.

Controls on mid/outer ramp sedimentation and source of late Eocene lithics

Lithic grains are present in the mid/outer ramp deposits of the Jeneponto area only during the late Eocene. Since the Tonasa Carbonate Platform is almost completely devoid of non-carbonate grains, derivation of the lithics from one of the underlying formations in South Sulawesi is inferred. This may be the result of erosion of the platform margin down to the underlying formations. However, unless a pre-existing topography existed in the underlying formations reworked clasts from the inner platform deposits would be expected, and these do not occur.

The late Eocene was a time of tectonic activity within the region (Bransden & Matthews 1992; Wilson & Bosence 1996). It is possible that the location of the southern margin of the Tonasa Carbonate Platform may have been controlled by pre-existing structures (cf. Ahr 1989; Eichenseer & Luterbacher 1992; Pedley *et al.* 1992), which were reactivated during the late Eocene (Fig. 4). Faulting on the southern margin may have resulted in exposure and erosion of the earlier formations. However, the lack of abundant sediment gravity flows and slide deposits (cf. Pedley *et al.* 1992) argues against this. If faulting did occur on the southern margin it must have been of low relief and in areas of low energy preventing large scale reworking of the coarser debris (Fig. 4). Alternatively, it may be that currents parallel to the slope may have deflected and reworked material from sand grade sediment gravity flows. Another possibility is that the lithic grains may have been transported from a different area. Easterly directed palaeocurrents suggest that the most likely source area for the siliciclastics would have lain to the west (Fig. 4). The Makassar Straits was an area of active faulting during the late Eocene (Bransden & Matthews 1992; Situmorang 1982).

With a thickness of up to 1200 m the mid/outer ramp and bathyal carbonate sequence in the Jeneponto area is considerably thicker than the shallower platform deposits (located 50 km to the north), which have an estimated thickness of 600 m. This variation in thickness is inferred to be related to high differential subsidence in the Jeneponto area. The period of maximum differential subsidence occurred during the late Eocene when sediments in the Jeneponto area reach a thickness of at least 700 m (Fig. 7). In comparison, in the Tonasa-I section in the Pangkajene area (Fig. 5), late Eocene shallow-water platform deposits reach a thickness of only 160 m (Fig. 7). It may be significant that this large variation in thickness, 540 m occurred during the late Eocene, a period of known tectonic activity (cf. Wilson & Bosence 1996).

Progradation of the ramp deposits is thought to result from carbonate production on the southern margin of the Tonasa Carbonate Platform outpacing accommodation space. It is not clear whether this is related to highstand shedding, in which a relative rise in sea level creates an increased area for carbonate production in the photic zone (Schlager 1992). Alternatively, a relative sea-level fall might result in a basinward shift in ramp facies belts across the gently sloping surface (Burchette & Wright 1992;

Aurell *et al.* 1995). It would therefore seem that ramp progradation may result from a complex interplay of a range of factors. These include carbonate production rates, eustatic or base-level changes, shape of slope and oceanic regime (see Burchette & Wright 1992 for review). Differentiating between these various controls is difficult in this area, due to the lack of precise correlation between sections of the Tonasa Limestone Formation.

In the Jeneponto area there is no evidence for the ramp margin developing into a rimmed shelf (cf. Gawthorpe 1986; Fernandez-Mendiola & Garcia-Mondéjar 1989; Burchette & Wright 1992). As in other documented examples, ramp survival depends on relatively high subsidence rates and generation of accommodation space of the carbonate platform without growth into a rimmed platform (Read 1985; Ramsay 1987; Wright & Faulkner 1990; Burchette & Wright 1992; Aurell *et al.* 1995).

Implications for hydrocarbon exploration

Carbonate successions are widespread throughout SE Asia and have considerable potential as hydrocarbon reservoirs (Fig. 1). Miocene carbonate buildups identified from seismic reflection profiles, when drilled are often found to have good porosity and permeability and may yield hydrocarbons (Grainge & Davies 1983; Mayall & Cox 1988). Coral-rich framestones or rudstones, with the exception of very high-energy platform margins deposits containing abundant shallow-marine cements, usually form the best hydrocarbon reservoirs in Miocene deposits (McArthur & Helm 1982; Grainge & Davies 1983; Rose 1983; Longman 1985; Mayall & Cox 1988). In comparison Eocene/Oligocene carbonate successions, similar to the Tonasa Limestone Formation, are commonly found to be tight and are dominantly composed of large benthic foraminifera and other non aragonitic bioclasts (Adams 1965; Kohar 1985; Siemers *et al.* 1992; Cucci & Clark 1993). This study of the Tonasa Limestone Formation has implications for hydrocarbon exploration in other foraminiferal dominated Tertiary carbonate systems.

The Tonasa Limestone Formation is similar to many other Tertiary carbonate successions in SE Asia in that it initially occurs as part of a trangressive sequence and overlies marginal marine deposits containing coals that comprise potential hydrocarbon source lithologies (cf. Phillips *et al.* 1991; Coffield *et al.* 1993). The carbonate succession is itself overlain by marine clays which may form effective seals. In order to evaluate the reservoir potential of the Tonasa Limestone Formation, or to provide useful analogues for hydrocarbon exploration of other subsurface SE Asian carbonates, it is essential to have as full an understanding of carbonate facies distribution, controls on sedimentation and diagenesis as possible.

Low energy platform-top deposits of the northern and southern Pangkajene facies belts and the mid/outer ramp and bathyal deposits of the Jeneponto area include abundant micrite and very little primary porosity occurs. In the high-energy east–west-trending central Pangkajene facies belts, although considerable primary porosity occurred in the grainstone units (up to 25%), this has now been occluded by the pervasive development of blocky to equant non-ferroan calcite cements. The Tonasa Limestone Formation is overlain by three kilometres of volcaniclastic sediments and the cements seen probably formed in a burial environment. However, it may be that in similar deposits, which have not been affected by adverse pore occluding diagenetic processes, primary porosity may be preserved in moderate to high energy units. The lack of abundant aragonitic bioclasts, from corals or green algae, together with only localised subaerial exposure, result in little secondary leaching and associated porosity and permeability development in shallow-water deposits of the Tonasa Limestone Formation. The depositional environment, biota, factors affecting sedimentation and diagenesis of the carbonate succession have all combined to make platform top and ramp deposits of the Tonasa Carbonate Platform poor hydrocarbon reservoirs.

Redeposited carbonate facies of the Tonasa Limestone Formation, derived from block-faulted footwall areas include minor porosity and permeability, and indeed traces of hydrocarbons do occur. This porosity and permeability is due to circum-granular stylolites and some preserved primary intergranular porosity. Although concentrations of argillaceous material around clasts in some beds may lead to reduced permeability, redeposited facies, abutting impermeable basement and platform lithologies, are thought to form the most suitable hydrocarbon reservoir within the Tonasa Limestone Formation (Wilson & Bosence 1996). Fulthorpe & Schlanger (1989) recognized that in seismically active areas, carbonate megaturbidites could provide pathways for the migration of hydrocarbons, or act as reservoirs.

This study suggests that moderate- to high-energy platform-top or redeposited carbonate facies may form effective hydrocarbon reservoirs in otherwise tight foraminiferal dominated carbonates, which have not been affected by extensive porosity occlusion by shallow marine or burial cements. Other factors which may affect secondary porosity and permeability development in tight foraminiferal limestones include tectonic fracturing and the development of stylolites (Siemers *et al.* 1992).

Conclusions

Late Eocene to mid-Miocene carbonate outcrops of the Tonasa Limestone Formation, from the Pangkajene and Jeneponto areas of South Sulawesi, provide a remarkable opportunity to study shallow-water and ramp/basinal deposits of a laterally extensive and long-lived SE Asian carbonate platform. The Tonasa Carbonate Platform was a foraminifera dominated carbonate platform with a ramp-type southern margin. A number of conclusions, which have implications for regional carbonate studies and hydrocarbon exploration in the region, can be drawn from this study.

(1) Facies belts on the Tonasa Carbonate Platform trend east–west and their position remained remarkably stable through time indicating aggradation of the platform top. There was some eastward progradation of the central rather high-energy facies belt.

(2) Although deposits of the southern margin of the Tonasa Carbonate Platform are not exposed, through detailed analysis of sections in the Pangkajene and Jeneponto areas a ramp-type margin has been inferred for the southern margin of the Tonasa Carbonate Platform. Outer ramp deposits prograded southwards at intervals into basinal marls.

(3) Tectonics in the form of subsidence was the dominant control on accommodation space on the Tonasa Carbonate Platform. Other factors which affected sedimentation on the platform include the location of 'barriers' and how these deflected cross-platform currents, together with the nature of carbonate producing organisms. Although the Tonasa Limestone Formation was deposited in a tectonically active setting, the affects of tectonic tilting, versus eustatic or autocyclic effects are impossible to differentiate by studying platform-top and ramp deposits.

(4) Studies of foraminiferal shape and genera, when combined with textural and lithological considerations, provide a powerful tool for the environmental interpretation of shallow-water carbonates.

(5) Although little porosity and permeability occurs within shallow-water deposits of the Tonasa Limestone Formation this study suggests that moderate to high energy platform top or redeposited carbonate facies may form effective hydrocarbon reservoirs in otherwise tight foraminiferal dominated carbonates, which have not been affected by extensive porosity occlusion by shallow marine or burial cements. The facies associations and depositional model developed for the Tonasa Carbonate Platform may be used as a template to understand better subsurface platforms of similar ages in SE Asia.

M. E. J. W. gratefully acknowledges BP Exploration, UK for their generous financial support during the course of her PhD study, of which this work forms a part. The SE Asia Research Group London University, especially A. Barber and D. Cameron, are thanked for their administrative and technical support. In Indonesia, A. Limbong, my counterpart from GRDC, Bandung deserves special thanks. GRDC, Bandung, Kanwil, South Sulawesi, particularly Darwis Falah and family, BP offices in Jakarta and Ujung Pandang and LIPI all provided technical and practical support. T. Finch and F. Banner, both at University College London, and T. Wonders, Robertson Research, are thanked for their excellent biostratigraphic work. K. de Souza is thanked for the production of the photographic plates. The constructive comments made by R. Bond, S. Moss and particularly N. Deeks and two anonymous referees towards improving this paper were much appreciated.

References

ADAMS, C. G. 1965. The foraminifera and stratigraphy of the Melinau Limestone, Sarawak, and its importance in Tertiary correlation. *Quarterly Journal of the Geological Society of London,* **121**, 283–338.

——1970. A reconsideration of the East Indian Letter Classification of the Tertiary, *British Museum of (Natural History), Geology,* **19**(3), 87–137.

AHR, W. M. 1973. The carbonate ramp: an alternative to the shelf model. *Transactions of the Gulf Coast Association of Geological Societes*, **23**, 221–225.

——1989. Sedimentary and tectonic controls on the development of an early Mississippian carbonate ramp, Sacramento mountains area, New Mexico. *In*: CREVELLO, P. D., WILSON, J. L., SARG, F. & READ, J. F. (eds) *Controls on carbonate platform and basin development*. SEPM Special Publications, **44**, 203–212.

AURELL, M., BOSENCE, D. W. J. & WALTHAM, D. 1995. Carbonate ramp depositional systems from a late Jurassic epeiric platform (Iberian Basin, Spain): a combined computer modelling and outcrop analysis. *Sedimentology*, **42**, 75–94.

BLOW, W. H. 1969. Late Eocene to Recent planktonic foraminiferal biostratigraphy. *Proceedings of the 1st Plankton Conference, Geneva 1967*, **1** 199–422.

BRANSDEN, P. J. E. & MATTHEWS, S. J. 1992. Structural and stratigraphic evolution of the east Java Sea, Indonesia. *Proceedings of the Indonesian Petroleum Association, 21st Annual Convention*, October, 417–453.

BURCHETTE, T. P. & WRIGHT, V. P. 1992. Carbonate ramp depositional systems. *Sedimentary Geology*, **79**, 3–57.

BUXTON, M. W. N. & PEDLEY, H. M. 1989. A standardized model for Tethyan Tertiary carbonate ramps. *Journal of the Geological Society, London*, **146**, 746–748.

CALVET, F. & TUCKER, M. E. 1988. Outer ramp cycles in the upper Muschelkalk of the Catalan basin, northeast Spain. *Sedimentary Geology*, **57**, 185–198.

CHAPRONIERE, G. C. H. 1975. Palaeoecology of Oligo-Miocene larger foraminiferida, Australia. *Alcheringa*, **1**, 37–58.

COFFIELD, D. Q., BERGMAN, S. C., GARRARD, R. A., GURITNO, N., ROBINSON, N. M. & TALBOT, J. 1993. Tectonic and stratigraphic evolution of the Kalosi PSC area and associated development of a Tertiary petroleum system, south Sulawesi, Indonesia. *Proceedings of the Indonesian Petroleum Association, 22nd Annual Convention*, October, 679–706.

COLE, W. S. 1957. Larger foraminifera from Eniwetok Atoll drill holes. *US Geological Survey, Professional Paper*, **260-V**, 743–784.

COLLINSON, J. D. 1969. The sedimentology of the Grindslow Shales and the Kinderscout Grit: a deltaic complex in the Namurian of northern England. *Sedimentology*, **39** 194–221.

CROTTY, K. J. & ENGELHARDT, D. W. 1993. Larger foraminifera and palynomorphs of the upper Malawa and lower Tonasa Formations, southwestern Sulawesi Island, Indonesia. *In*: THANASUTHIPITAK, T. (ed.) *Symposium on biostratigraphy of mainland Southeast Asia: Facies & Paleontology*. 31 January–5 February, Chiang Mai, Thailand, 71–82.

CUCCI, M. A. & CLARK, M. H. 1993. Sequence stratigraphy of a Miocene Carbonate Buildup, Java Sea. *In*: LOUCKS, R. G. & SARG, J. F. (eds), *Carbonate sequence stratigraphy, recent developments and applications*. American Association of Petroleum Geologists, Memoirs, **57**, 291–303.

DUNHAM, R. J. 1962. Classification of carbonate rocks according to depositional texture. *In*: HAM, W. E. (ed.) *Classification of carbonate rocks*. American Association of Petroleum Geologists, Memoirs, **1**, 108–121.

EICHENSEER, H. & LUTERBACHER, H. 1992. The marine Palaeogene of the Tremp region (NE Spain) – Depositional sequences, facies history, biostratigraphy and controlling factors. *Facies*, **27**, 119–152.

FERNANDEZ-MENDIOLA, P. A. & GARCIA-MONDÉJAR, J. 1989. Evolution of a Mid-Cretaceous carbonate platform, Gorbea (northern Spain). *Sedimentary Geology*, **64**, 111–126.

FULTHORPE, C. S. & SCHLANGER, S. O. 1989. Paleo-oceanographic and tectonic settings of early Miocene reefs and associated carbonates of offshore southeast Asia. *American Association of Petroleum Geologists Bulletin*, **73**, 729–756.

GAWTHORPE, R. L. 1986. Sedimentation during carbonate ramp-to-slope evolution in a tectonically active area: Bowland Basin (Dinantian), northern England *Sedimentology*, **33**, 185–206.

GEBELEIN, C. D. 1974. *Guidebook for modern Bahamian platform environments*. Geological Society of America Annual Meeting Fieldtrip Guide.

GHOSE, B. K. 1977. Paleoecology of the Cenozoic reefal foraminifers and algae – a brief review. *Palaeogeography, Palaeoclimatology, Palaeoecology*, **22**, 231–256

GRAINGE, A. M. & DAVIES, K. G. 1983. Reef exploration in the east Sengkang basin, Sulawesi, Indonesia. *Proceedings of the Indonesian Petroleum Association. 12th Annual Convention*, June, 207–227.

HALLOCK, P. & GLENN, E. C. 1985. Numerical analysis of foraminiferal assemblages: A tool for recognizing depositional facies in lower Miocene reef complexes. *Journal of Paleontology*, **59**, 1382–1394.

——1986. Larger foraminifera: a tool for palaeoenvironmental analysis of Cenozoic carbonate depositional facies. *Palaios*, **1**, 55–64.

—— & PEEBLES, M.W. 1993. Foraminifera with chlorophyte endosymbionts: habitats of six species in the Florida Keys. *Marine Micropaleontology*, **20**, 277–292.

HARLAND, W. B., ARMSTRONG, R. L., COX, A. V., CRAIG, L. E., SMITH, A. G. & SMITH, D. G. 1990. *A geologic time scale 1989*. Cambridge University Press.

HAMILTON, W. 1979. *Tectonics of the Indonesian region*. United States Geological Survey Professional Paper, **1078**.

HASAN, K. 1991. The Upper Cretaceous flysch succession of the Balangbaru Formation, South-west-Sulawesi. *Proceedings of the Indonesian Petroleum Association 20th Annual Convention*, October, 183–208.

HIRSHFIELD, H. I., CHARMATZ, R. & HELSON, L. 1968. Foraminifera in samples taken mainly on Eniwetok Atoll in 1956. *Journal of Protozoology*, **15**, 497–502.

HOTTINGER, L. 1977. Distribution of larger Peneroplidae, *Borelis* and Nummulitidae in the Gulf of Elat, Red Sea. *Utrecht Micropalaeotological Bulletin*, **15**, 35–109.

——1983. Processes determining the distribution of larger foraminifera in space and time. *Utrecht Micropalaeontological Bulletin*, **30**, 239–253.

IVANY, L. C., NEWTON, C. R. & MULLINS, H. T. 1994. Benthic invertebrates of a modern carbonate ramp: a preliminary survey. *Journal Paleontology*, **68**, 417–433.

JAMES, N. P. & KENDALL, A. C. 1992. Introduction to carbonate and evaporite facies models. *In*: WALKER, R. G. & JAMES, N. P. (eds) *Facies models, response to sea level change*. Geological Association of Canada, 265–274.

KINSMAN, D. J. J. & PARK, R. K. 1976. Algal belt and coastal sabkha evolution, Trucial Coast, Persian Gulf. *In*: WALTER, M. R. (ed.) *Stromatolites*, 421–433.

KOHAR, A. 1985. Seismic expression of late Eocene carbonate build-up features in the JS25 and P. Sepanjang trend, Kangean Block. *Proceedings Indonesian Petroleum Association, 14th Annual Convention*, 437–452.

LEE, J. J & ANDERSON, O. R. 1991. *Biology of Foraminifera*. Academic Press, London.

LOGAN, B. W., HOFFMAN, P. & GEBELEIN, C. D. 1974. Algal mats, crypt-algal fabrics and structures, Hamelin Pool, Western Australia. *In*: LOGAN, B. W. *et al.* (eds) *Evolution and diagenesis of Quaternary Carbonate Sequences, Shark Bay, Western Australia*. American Association of Petroleum Geologists, Memoirs, **22**, 140–193.

——, REZAK, R. & GINSBURG, R. N. 1964. Classification and environmental significance of algal stromatolites. *Journal of Geology*, **72**, 68–83.

LONGMAN, M. W. 1985. Fracture porosity in reef talus of a Miocene pinnacle-reef reservoir, Nido B Field, the Philippines. *In*: ROEHL, P. O. & CHOQUETTE, P. W. (eds) *Carbonate Petroleum Reservoirs*. Springer-Verlag, Berlin, 547–561.

LUTZE, G.F., GRABERT, B. & SEIBOLD, E. 1971. Lebendbeobach – tungen an Gross – Foraminiferen (Heterostegina) dus dem Persischen Golf. *'Meteor' Forsch Ergbn*, **6**, 21–40.

MCARTHUR, A. C. & HELM, R.B. 1982. Miocene carbonate buildups, offshore North Sumatra. *Proceedings of the Indonesian Petroleum Association, 11th Annual Convention*, 127–145.

MCILREATH, I. A. & JAMES, N. P. 1984. Carbonate Slopes. *In*: WALKER, R. G. (ed.) *Facies Models*, **12**, Geoscience, Canada, 245–257.

MARTINI, E. 1971. Standard Tertiary and Quaternary calcareous nannoplankton zonation. *In*: FARINACCI, A. (ed.) *Conference on Planktonic Microfossils, Proceedings*, Rome 1970, **2**, 739–785.

MAXWELL, W. G. H., DAY, R. W. & FLEMING, P. J. G. 1961. Carbonate sedimentation on the Heron Island reef, Great Barrier Reef. *Journal of Sedimentary Petrology*, **31**, 215–230.

MAYALL, M. J. & COX, M. 1988. Deposition and diagenesis of Miocene limestones, Senkang Basin, Sulawesi, Indonesia. *Sedimentary Geology*, **59**, 77–92

MULLINS, H. T., GARDULSKI, A. F., HINE, A. C., MELILLO, A. J., WISE, S. W. & APPLEGATE, J. 1988. Three-dimensional sedimentary framework of the carbonate ramp slope of central west Florida: A sequential seismic stratigraphic perspective. *Geological Society of America Bulletin*, **100**, 514–533.

MURRAY, J. W. 1973. *Distribution and Ecology of Living Benthic Foraminifera*. Heinemann, London.

NEWELL, N. D. 1956. Geological reconnaissance of Raroia (Kon Tiki) Atoll, Tuamotu Archipelago. *Bulletin of the American Museum of Natural History*, **109**, 317–372.

PARK, R. K. 1977. The preservation potential of some recent stromatolites. *Sedimentology*, **24**, 485–506.

——, MATTER, A. & TONKIN, P. C. 1995. Porosity evolution in the Batu Raja carbonates of the Sunda Basin – Windows of Opportunity. *Proceedings of the Indonesian Petroleum Association, 24th Annual Convention*, October 1995, 165–184.

PEDLEY, H. M., CUGNO, G. & GRASSO, M. 1992. Gravity slide and resedimentation processes in a Miocene carbonate ramp, Hyblean Plateau, southeastern Sicily. *Sedimentary Geology*, **79**, 189–202.

PHILLIPS, T. L., NOBLE, R. A. & SINARTIO, F. F. 1991. Origin of hydrocarbons, Kangean Block, Northern Platform, offshore N.E. Java Sea. *Proceedings of the Indonesian Petroleum Association, 20th Annual Convention*, 637–661.

RAMSAY, A. T. S. 1987. Depositional environments in the Dinantian Limestones of Gower, South Wales. *In*: MILLER, J., ADAMS, A. E. & WRIGHT, V. P. (eds) *European Dinantian Environments*. John Wiley, Chichester, 265–308.

READ, J. F. 1982. Carbonate platforms of passive (extensional) continental margins: types, characteristics and evolution. *Tectonophysics*, **81**, 195–212.

READ, J. F. 1985. Carbonate platform facies models. *American Association of Petroleum Geologists Bulletin*, **69**, 1–21.

REISS, Z. & HOTTINGER, L. 1984. *The Gulf of Aqaba, ecological micropaleontology*. Springer-Verlag Ecological Studies, **50**.

ROSE, R. 1983. Miocene carbonate rocks of Sibolga Basin, Northwest Sumatra. *Proceedings of the Indonesian Petroleum Association, 12th Annual Convention*, 108–125.

RÖTTGER, R., & BERGER, W. H. 1972. Benthic Foraminifera: morphology and growth in clone cultures of *Heterostegina depressa*. *Marine Biology*, **15**, 89–94.

——, FLADUNG, M., SCHMALJOHANN, R., SPINDLER, M. & ZACHARIES, H. 1986. A new hypothesis: the so-called megalospheric schizont of the larger foraminifer, *Heterostegina depressa* d'Orbigny, 1826, is a separate species. *Journal of Foraminiferal Research*, **16**, 141–9.

SCHLAGER, W. 1992. *Sedimentology and sequence stratigraphy of reefs and carbonate platforms: A short course*. American Association of Petroleum Geologists, Continuing education course note series, **34**.

SHINN, E. A. 1968. Practical significance of birds eye structures in carbonate rocks. *Journal of Sedimentary Petrology*, **38**, 215–223.

——1983. Tidal flat environment. *In*: SCHOLLE, P. A., BEBOUT, D. G. & MOORE, C. H. (eds), *Carbonate Depositional Environments*. American Association of Petroleum Geologists, Memoirs, **33**, 173–210.

SIEMERS, C. T., DECKELMAN, J. A., BROWN, A. A. & WEST, E. R. 1992. Characteristics of the fractured Ngimbang carbonate (Eocene), West Kangean-2 well, Kangean PSC, East Java Sea, Indonesia. *In*: SIEMERS, C. T., LONGMAN, M. W., PARK, R. K. & KALDI, J. G. (eds) *Carbonate rocks and reservoirs of Indonesia*. Indonesian Petroleum Association Core Workshop Notes, **1**, Ch. 10.

SITUMORANG, B. 1982. The formation of the Makassar basin as determined from subsidence curves. *Proceedings Indonesian Petroleum Association, 11th Ann. Convention*, June, 83–107.

SUKAMTO, R. 1982. *Geologi lembar Pangkajene dan Watampone bagian barat, Sulawesi*. Geological Research and Development Centre.

—— & SUPRIATNA, S. 1982. *Geologi lembar Ujung Pandang, Benteng dan Sinjai quadrangles, Sulawesi*. Geological Research and Development Centre.

TUCKER, M. E. & WRIGHT, V. P. 1990. *Carbonate Sedimentology*. Blackwell Scientific Publications.

YUWONO, Y. S, BELLON, H., SOERIA-ATMADJA, P. & MAURY, R. C. 1985. Neogene and Pleistocene volcanism in South Sulawesi. *Proceedings Ikatan Ahli Geologi Indonesia*, **14**, 169–179.

VAN DE VLERK, I. M. & UMBGROVE, J. H. F. 1927. Tertiaire gidsforaminifern van Nederlandsch Oost-Indië. *Wetenschappelijne Mededeelinsen Dienst vanden Mijnbouw Nederlandsch Oost-Indië*, **6**, 3–31.

VAN GORSEL, J. T. 1988. Biostratigraphy in Indonesia: Methods, pitfalls and new directions. *Proceedings of the Indonesian Petroleum Association, 17th Ann. Convention*, October, 275–300.

VAN LEEUWEN, T. M. 1981. The geology of southwest Sulawesi with special reference to the Biru area. *In*: BARBER, A. J. & WIRYOSUJONO, S. (eds) *The Geology and Tectonics of Eastern Indonesia*. Geological Research and Development Centre, Bandung, Special Publications, **2**, 277–304.

WALKER, R. G. 1992. Facies, facies models and modern stratigraphic concepts. *In*: WALKER, R. G. & JAMES, N. P. (eds) *Facies models: Response to sea level change*. Geological Association of Canada, 1–14.

WANLESS, H. R. & DRAVIS, J. J. 1989. *Carbonate environments and sequences of Caicos Platform*. 28th International Geological Congress, Field Trip Guidebook, **T374**.

WATTS, K. F. & BLOME, C. D. 1990. Evolution of the Arabian carbonate platform margin slope and its response to orogenic closing of a Cretaceous ocean basin, Oman. *In*: TUCKER, M. E., WILSON, J. L., CREVELLO, P. D., SARG, J. R. & READ, J. F. (eds), Carbonate Platforms: Facies, sequences and evolution. Special Publications of the International Association of Sedimentologists, **9**, 291–323.

WILSON, M. E. J. 1995. *The Tonasa Limestone Formation, Sulawesi, Indonesia: Development of a Tertiary Carbonate Platform*. Ph.D. Thesis, University of London.

—— & BOSENCE, D. W. J. 1996. The Tertiary Evolution of South Sulawesi: A Record in Redeposited Carbonates of the Tonasa Limestone Formation. *In*: HALL, R. & BLUNDELL, D. J. (eds) *Tectonic evolution of SE Asia*. Geological Society, London, Special Publications, **106**, 365–389

WRIGHT, V. P. & FAULKNER, T. J. 1990. Sediment dynamics of early Carboniferous ramps: a proposal. *Geological Journal*, **25**, 139–144.

A review of the petroleum potential of Papua New Guinea with a focus on the eastern Papuan Basin and the Pale Sandstone as a potential reservoir fairway

P. J. BOULT

Gartrell School of Mining, Metallurgy and Applied Geology, Levels campus, University of South Australia, South Australia, SA 5095, Australia

Abstract: Substantial oil seeps were first reported from the eastern Papuan Basin in 1911, and oil has since been produced from the sub-surface, but not in commercial quantities. A significant discovery of gas in 1986 sparked an upsurge of activity within the adjacent western Papuan Basin. From 1986 to 1992, 13 new hydrocarbon discoveries were made, which included four major oil fields, and an oil pipeline for export purposes was then opened in 1992. This activity did not spread to the eastern Papuan Basin due to a perceived lack of good reservoir and the last well drilled there was over 30 years ago. In 1986, a potentially excellent reservoir, the Pale (Pah leh) Sandstone was discovered in outcrop in the eastern Papuan Basin. Thus the Papuan Basin as a whole remains an extremely prospective basin. However, due to high infrastructure costs most activity has recently been focussed on the already proven reservoir potential of the Late Jurassic Imburu and Toro Formations in the western section.

The extent of the Imburu–Toro Formation reservoir in the western section of the Papuan Basin is well documented. It was deposited as several regressive episodes within an overall transgressive, passive-margin, clastic sequence. In the eastern section, the Late Cretaceous Pale Sandstone was deposited in a barrier island or strand-plain environment. Lateral equivalents of the Pale Sandstone and the potential for a new reservoir fairway have been documented for a strike length of approximately 400 km. This fairway is coincident with the northeast edge of the western Papuan Basin and on the northern edge of the Aure and Port Moresby Troughs of the eastern Papuan Basin.

All the commercial oil accumulations and seeps found in the Papuan Basin before 1992 have been attributed to the Jurassic, with peak generation occurring in the Late Cretaceous prior to the formation of existing structures. In 1992 an oil seep was discovered which is geochemically very different to all the other oils found in the Papuan Basin. This was found to the north and east of the present discoveries that occur within the western Papuan Basin.The oil seep is indicative of undiscovered Tertiary source rock, which is currently at or near peak generation. However, most of this source rock in the north may be covered by overthrusted allochthonous terrain.

The Aure Trough was a Late Cretaceous to Tertiary rift basin that later developed into an Oligocene to Mid-Miocene foreland basin depocentre occurring on the northern and eastern margin of the Papuan Basin. It is proposed in this paper that mature Tertiary source rocks which occur within this foreland basin and the Pale Sandstone reservoir may form a significant petroleum system on the northeastern rim of the western Papuan Basin and within the eastern Papuan Basin.

PNG exploration history

The search for oil in PNG has been predominantly focused on the Papuan Basin which lies to the south of the present-day New Guinea Orogen. Other potentially prospective basins occur to the north of the New Guinea Orogen and among the offshore islands. For example the Sepik (Doust 1990), Aitape (Kugler 1990) and New Ireland Basins (Exon & Marlow 1990) have periodically been explored, but no significant oil shows have been found. Figure 1 shows the current extent of exploration and production licenses in PNG and that currently no exploration permits exist within these areas (Haumu 1995). At the time of writing all petroleum exploration permits and production licenses are concentrated in the northern section of the Papuan Basin along the fold belt and adjacent parts of the Fly Platform to the south of the fold belt. Most exploration permits in the eastern Papuan Basin were relinquished in 1994.

Exploration for petroleum within PNG was sparked off in June 1911 (Pawih & Halstead 1993) when gold prospectors reported significant oil seeps along the Vailala River (Fig. 2) which flows into the Gulf of Papua. During the subsequent 35 years many exploration wells were drilled in the eastern Papuan Basin in positions related to the occurrence of previously

From Fraser, A. J., Matthews, S. J. & Murphy, R. W. (eds), 1997, *Petroleum Geology of Southeast Asia*, Geological Society Special Publication No. 126, pp. 281–291.

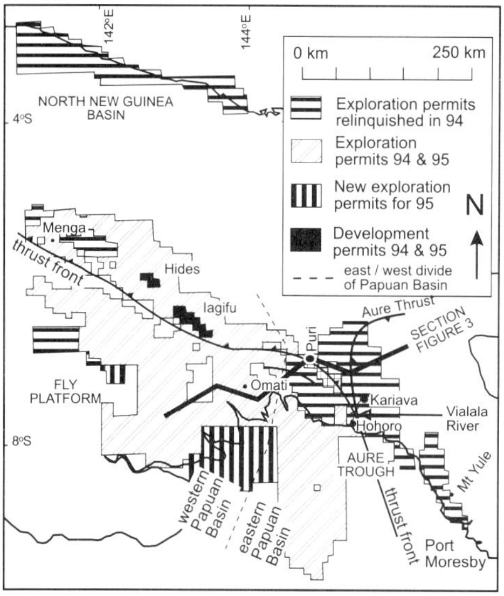

Fig. 1. Map of PNG showing licence areas and locations mentioned in the text. Currently all licence areas are in the Papuan Basin.

known oil seeps and mapped structural closure at the surface. Most of these wells were shallow, being less than 1500 m with two notable exceptions. Kariava-1 (Fig. 1), which reached a total depth of 3847 m in March 1948 after eight years of drilling and Hohoro-2 (Fig. 1), which reached a total depth of 3243 m in March 1952. Hydrocarbon shows were encountered in many of these wells but no commercial flows of hydrocarbons were discovered. This was caused by the lack of a good reservoir within (what is now known as) the Middle Miocene to Pliocene Foreland Basin Megasequence of the Aure Trough (Fig. 3) (Carman 1993).

The Mesozoic, which contains all the commercial and potentially commercial hydrocarbons found to date within the Toro and Imburu Formations, was first confirmed to exist in outcrop in the Whagi valley (Fig. 2) in 1938 (Rickwood 1992). Interest in the Mesozoic was stimulated in 1955 when the Omati well (Fig. 1) produced non-commercial gas and condensate from sandstones in the Jurassic. At the same time as the Omati well was being drilled, many thrust related anticlines farther west were delineated by field mapping. These are where the main commercial accumulations of oil and gas were discovered more than thirty years later. Because of a thick cover of often cavernous limestone and the very rough terrain, seismic methods have met with very little success within the fold belt. Structures are still delineated by field mapping, but these days mapping is often supplemented by the use of airborne radar images, strontium age dating of the limestone to determine structural position and occasionally the use of magnetotelluric modelling (Grainge 1993).

In the late 1950s fractured limestones of the Tertiary carbonate platform in the eastern Papuan Basin temporarily became the focus of attention. This was partly caused by the inaccessibility of the large anticlines of the more prospective thrust belt farther west. In 1958 the Puri-1 well (Fig. 1), which is in the eastern Papuan Basin, flowed 1610 barrels of oil per day before watering out after nine days. This lack of success and the difficult operating conditions did not encourage further exploration in the thrust belt until more powerful helicopters became available in 1970. These helicopters could lift drilling equipment onto the top of the large anticlines which had been mapped twenty years before and the Jurassic, Toro and Imburu Formations became a viable target.

In 1982 the Toro Formation produced 14.9 million cubic feet of gas and 938 barrels of condensate a day from the Juha structure (Rickwood 1990). Further appraisal of the Juha structure in 1985 produced an estimate that it contained an estimated 1.3 trillion cubic feet of gas and 75 to 100 million barrels of condensate. However, this was still not enough to make it commercially viable given the remoteness of the area. Then, in early 1986 almost 8000 barrels of oil per day were produced from a test of the Toro Formation in the Iagifu anticline. An oil pipeline for export purposes was subsequently opened in 1992 (Fig. 4). From1983 to 1993 some 13 discoveries in the Toro Formation were confirmed containing 19 trillion cubic feet of gas and 400 million barrels of oil (Pawih & Halstead 1993). Many of these hydrocarbon accumulations have significantly tilted oil water contacts and an understanding of the hydrodynamic flow in the Jurassic sandstones has enabled more oil to be discovered and the optimisation of production (Eisenberg 1993).

Since 1992 (up to 1 June 1995), only three wildcat wells have been drilled in Papua New Guinea and no significant discoveries have been made. Wildcat well costs have dropped dramatically from the usual $20 million plus in the early eighties, to less than $1 million for the most recent 1200 m slim-hole well at Menga-1 (Victoria Petroleum 1993). Despite this drop in costs a perceived political instability has discouraged investment particularly with more opportunities opening up in other south east

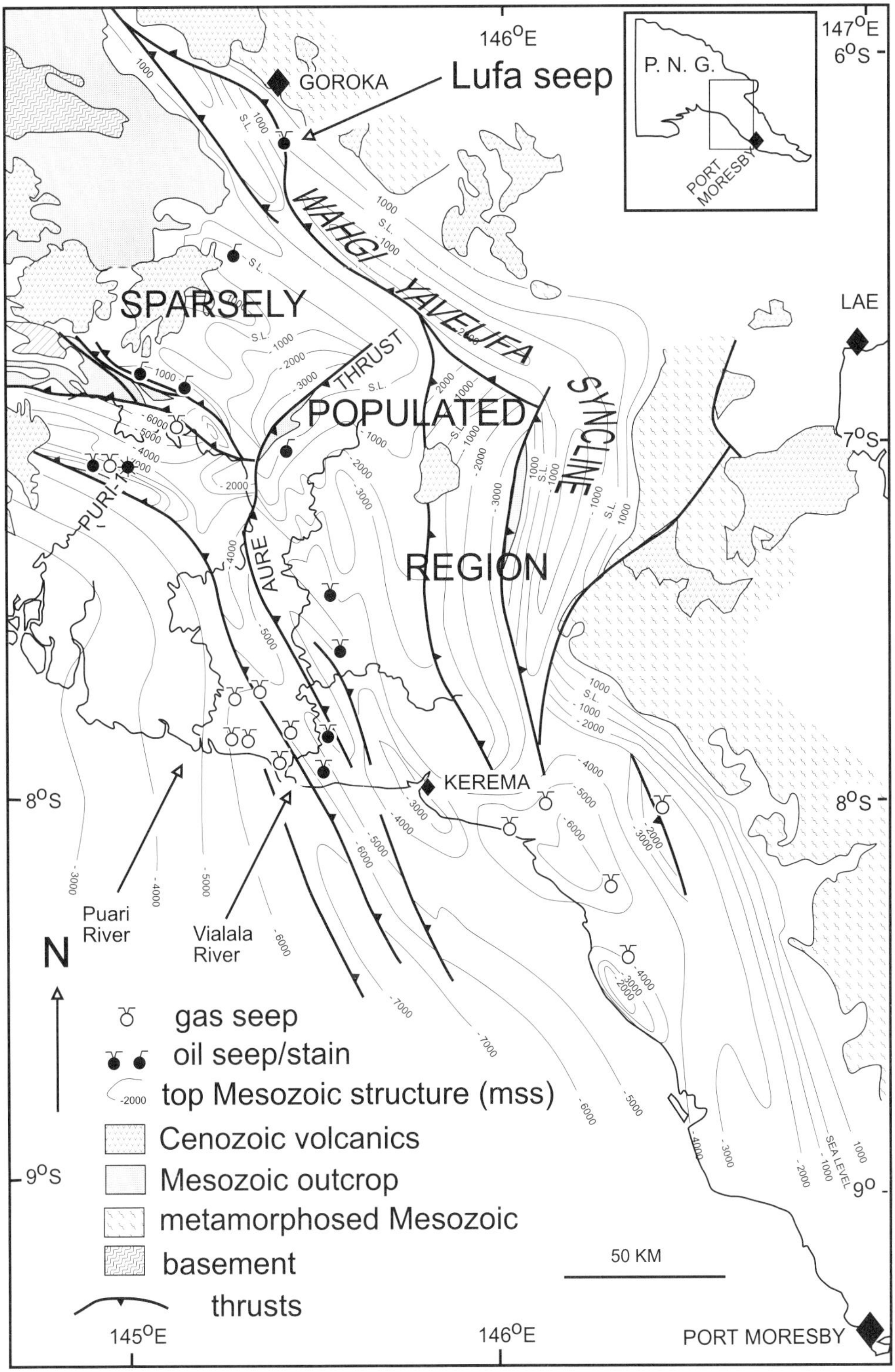

Fig. 2. Structural contours for the top of the Mesozoic, constructed from surface geology and regional geophysical cross sections of St John (1990). Also shows significant hydrocarbon seeps. (modified from Boult & Carman 1993).

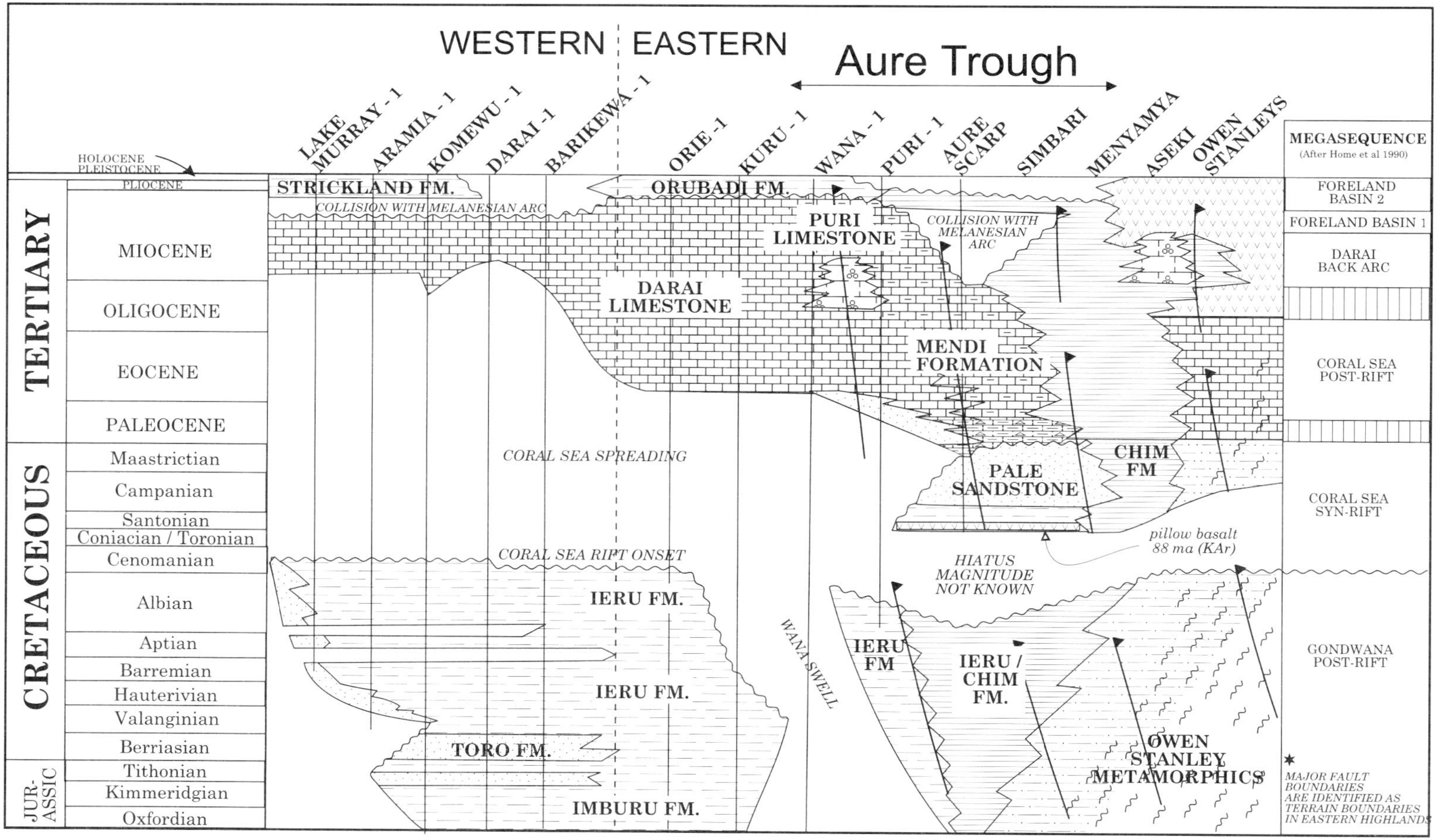

Fig. 3. Chronostratigraphic diagram of the Papuan Basin. For the location of cross section see Fig. 2 (modified from Home 1990 and Carman 1993).

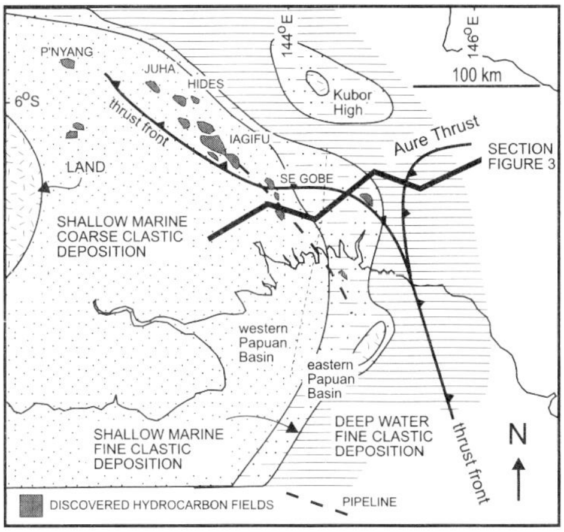

Fig. 4. Palaeogeography at the end of the Kimmeridgian. Equivalent of the deposition of the Imburu Formation within the Gondwana Post-Rift Megasequence (modified from Home 1990).

Asian countries. However, Chevron who operate the Kutubu project, which is exporting up to 150 000 barrels of oil per day, have had very good cooperation from the PNG government and people. BP, Exxon and Chevron are currently discussing the feasibility of exporting liquefied natural gas with the PNG government (Pawih pers. comm. 1995).

Other areas of the onshore Papuan Basin which remain prospective, besides the Jurassic of the Papuan fold belt, are the Toro Formation on the Fly Platform to the south of the fold belt and the Late Cretaceous Pale Sandstone in the eastern Papuan Basin which is discussed in more detail later.

Stratigraphy and palaeogeography of the Papuan Basin

Home (1990) divided the Papuan Basin into eight time-correlative megasequences of which the last six megasequences are shown in Fig. 3. Two major rifting events and one compressional event have controlled the palaeogeography, sedimentation and structure of this basin. The first rift took place in the mid- to late Triassic when the Papuan Basin existed as a passive margin on the southern edge of the Tethyan Ocean. Volcanics and continental to marginal-marine sediments were deposited in the Gondwana Syn-Rift Megasequence. Above the break-up unconformity the passive-margin, Gondwana Post-Rift Megasequence is dominated by clastic sediment deposition as part of an overall transgressive megasequence. Occasional, but economically significant for reservoir development, regressive episodes are represented by the Toro Formation and sandier parts of the Imburu Formation. The palaeo-coastline is postulated (Home 1990) to have followed the present day thrust belt to within 50 km the Aure Thrust (Fig. 4), and then swung to the south. However, definite proof of its absence beneath the present eastern Papuan Basin has not been proved by the drill (Slater & Dekker 1993). Offshore to the north and east of the palaeo-coastline the deeper water organic rich Maril Shale and Chim Formations were deposited.

The second rift that affected the Papuan Basin started towards the end of the Cretaceous and is related to the eventual opening of Coral Sea to the southwest (Fig. 5). Smith (1990) hypothesized that this rift opened in a back-arc setting on the western edge of the ancestral Pacific plate. Uplift occurred in the vicinity of the Fly Platform and southern Papuan Basin. Here much of the Gondwana Post-Rift Megasequence was removed and the Coral Sea Syn-Rift Megasequence was possibly never deposited (Home 1990). Here this period is represented by an unconformity at the base of the Tertiary (Fig. 3). In the eastern Papuan Basin, closer to

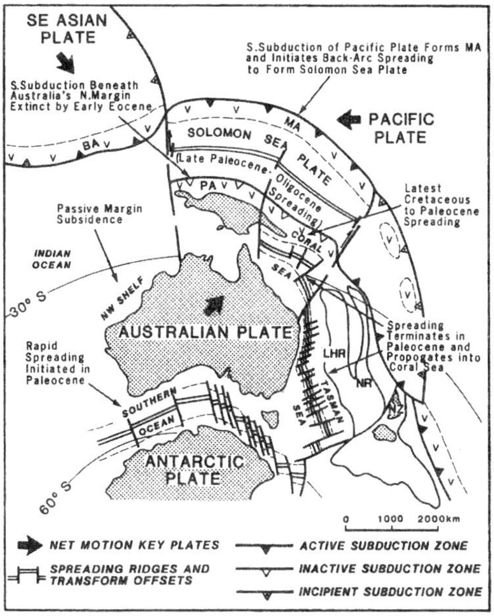

Fig. 5. Schematic Paleocene – Early Eocene plate reconstruction (65–50 Ma). NR, Norfolk Ridge; LHR, Lord Howe Rise; MA, Melanesian Arc; BA, Banda Arc; PA, Palaeocene Arc; NZ, New Zealand (from Smith 1990, Fig. 6).

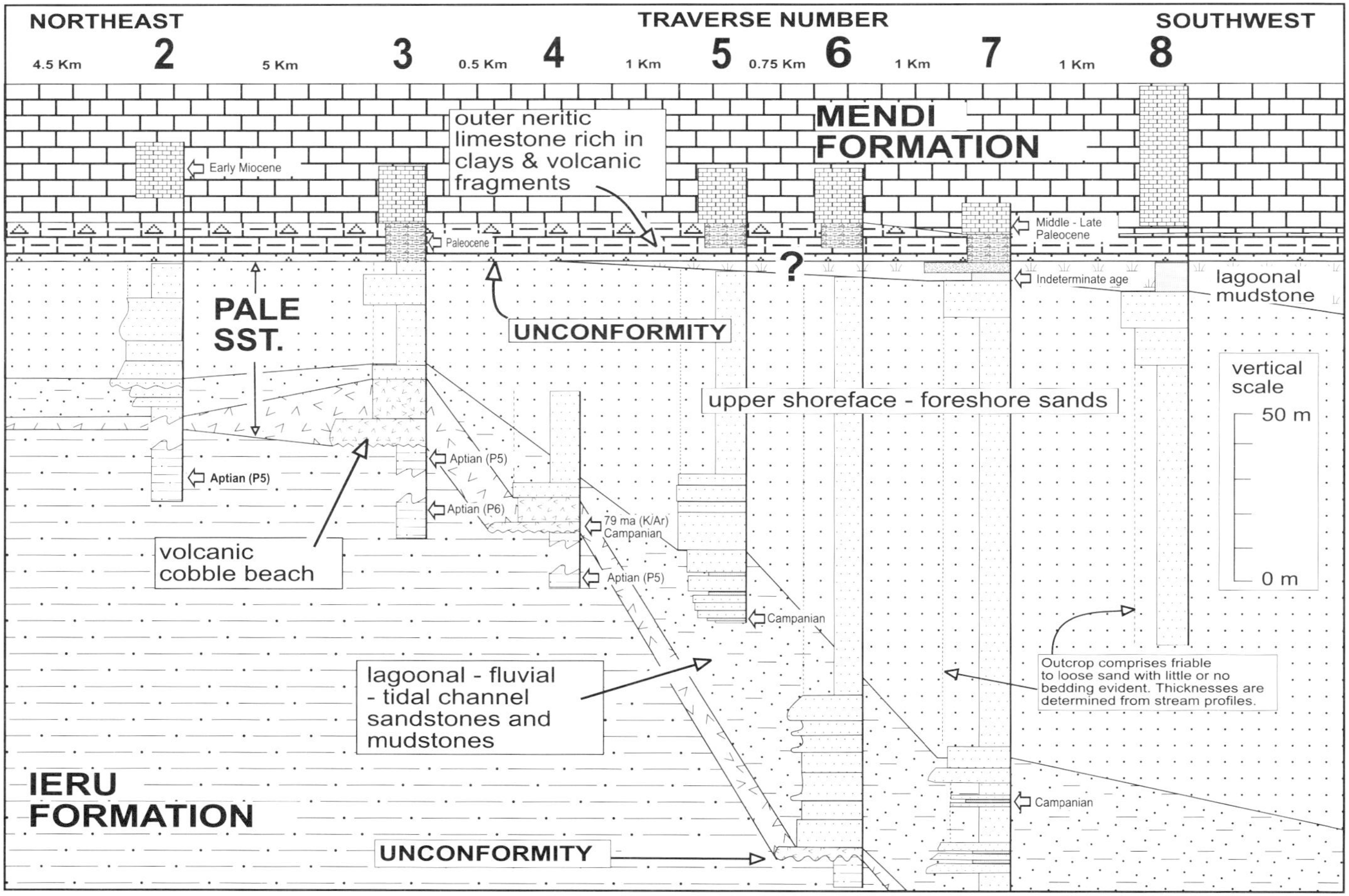

Fig. 6. Stratigraphic correlation of the Pale Sandstone at the Aure Scarp showing the correlation of facies between logged traverses. The Pale Sandstone occurs between the two indicated unconformities. Note the interpreted thickening of the Pale Sandstone and its mudstone seal from north-east to south-west. Dashed lines on traverse logs indicate very friable sand is present but bedding can not be seen. Assigned ages are based on micro fossil analysis except for one radiometric (K/Ar) date (modified from Boult & Carman 1993).

the rift, excellent reservoir quality, shoreline sediments represented by the Pale Sandstone occur beneath this unconformity where a significant amount of the Coral Sea Syn-Rift Megasequence has been preserved (Carman 1987).

The extent and thickness of the Campanian Pale Sandstone are debateable. It has been mapped for a strike distance of 22 km along the Aure Scarp (Boult & Carman 1993) where it is up to 200 m thick. Figure 6 shows a correlation constructed from traverses up the northwest edge of the Aure Scarp. Pale Sandstone boulders have also been found in creeks that run off the Aure Scarp up to 6 km northeast of traverse 2 and 4.5 km southwest of traverse 8. Slater & Dekker (1993) produced a map (Fig. 7) with the Pale Sandstone fairway stretching over 400 km, describing it as 'possibly a fringing deposit around the Fly Platform'. Carman (1993) postulated the existence of the Coral Sea rift north of the Aure Scarp, which is farther north than previously thought. He also linked the distribution of the Pale Sandstone more directly to the Coral Sea rift with the Pale Sandstone only being deposited within the actual rift itself. The provenance of the Pale Sandstone was attributed by Boult & Carman (1993) to longshore drift from nearby Palaeozoic metamorphic rocks exposed in the Kubor Anticline. Offshore in the deeper parts of the rift, undifferentiated pelagic shales still known as the Chim Formation were contemporaneously deposited (Fig. 3).

As PNG drifted northwards during the Tertiary and became the location of a passive margin (Fig. 8), clastic input from the Australian continent waned and carbonate deposition (Mendi Formation) started to dominate on the shallow shelf during the Coral Sea Post-Rift Megasequence. The rate of deposition and aerial extent of the carbonate platform greatly increased (Darai Limestone) during the Oligocene when the passive margin of the Coral Sea Post-Rift Megasequence changed into the convergent margin of the Darai Back Arc Megasequence. This increase was caused by the extra sediment accommodation space on the shelf of the foreland basin that developed as the leading edge of the Solomon Sea Plate (Fig. 9) encroached upon the area (Pigram *et al.* 1990). In the deeper areas to the north and east of the platform, which to some extent were controlled by the previous location of the Coral Sea Rift, only deep water pelagic mudstones were deposited. Eventually the Foreland Basin Megasequence volcaniclastics prograded across the foreland basin from an emergent landmass to the north, which was associated with the collision with the Melanesian Arc in the Late Miocene.

Today the Papuan Orogen is one of the most active in the world. This is perhaps best exemplified by Mount Yule, the top of which

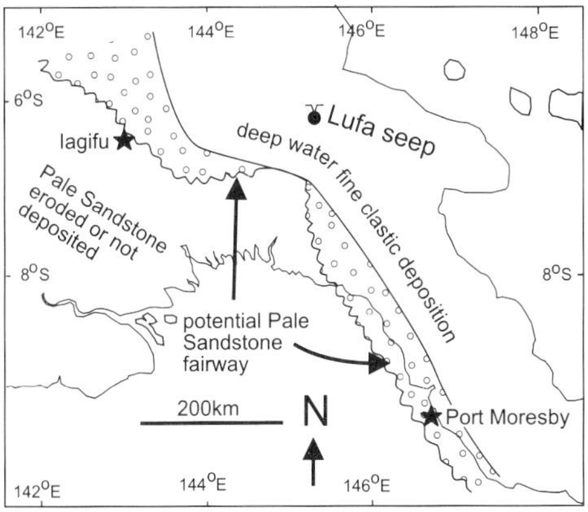

Fig. 7. Potential Campanian Pale Sandstone fairway and location of most hydrocarbon discoveries (after, Slater & Dekker 1993, fig. 8)

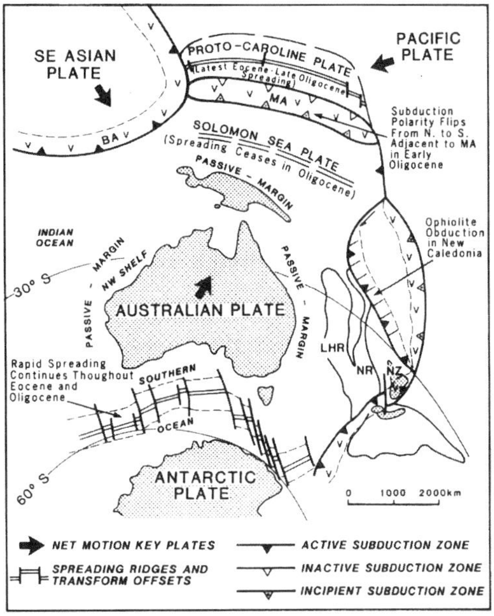

Fig. 8. Schematic Mid-Eocene–Early Oligocene plate reconstruction (50–35 Ma). NR, Norfolk Ridge; LHR, Lord Howe Rise; MA, Melanesian Arc; BA, Banda Arc; NZ, New Zealand; OJP, Ontong Java Plateau (from Smith 1990, fig. 7).

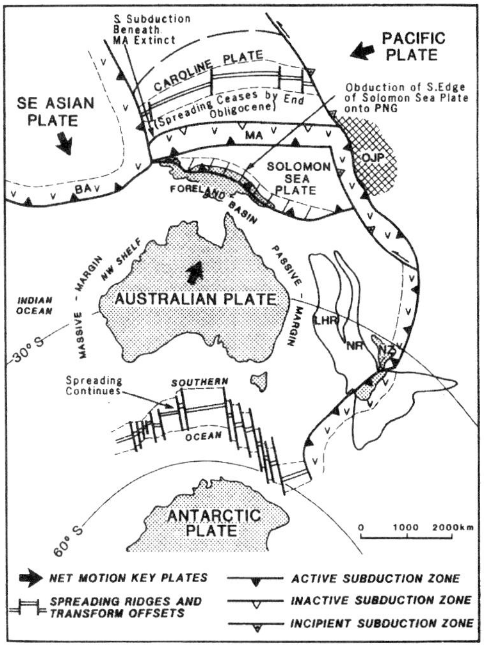

Fig. 9. Schematic late Oligocene–Mid-Miocene plate reconstruction (35–15 Ma). NR, Norfolk Ridge; LHR, Lord Howe Rise; MA, Melanesian Arc; BA, Banda Arc; NZ, New Zealand; OJP, Ontong Java Plateau (from Smith 1990, fig. 8).

is less than two million years old, but it is over 3000 m above sea level and only 60 km from the coast (Rickwood 1992). Denudation and associated sedimentation in the present day Gulf of Papua are perhaps the highest in the world, with rainfall in excess of 10 m per year over much of the Papuan Orogen. A present day depositional depocentre is located within the current Gulf of Papua into which all the major rivers flow which drain the PNG Highlands (Fig. 1).

Source rocks, maturation, migration, reservoirs and entrapment

Western Papuan Basin

Evidence that there are good source rocks in the basin is provided by the abundance of hydrocarbon seeps along the Papuan Fold Belt and the fact that 13 out of the 26 structures drilled by 1993 in the Toro Formation fairway (Grainge 1993) have been discoveries. All the hydrocarbons found to date in the western Papuan Basin have had their origin attributed (Osborne 1990) to the Jurassic Gondwana Post-Rift Megasequence (Fig. 3). Oil generation is estimated (Earnshaw *et al.* 1993) to have started in the Early Cretaceous within the deeper more northern and distal parts of the basin. Source rocks there are estimated to have completely passed through the oil window by the end of the Cretaceous. Uplift occurred from then until the end the Oligocene when the basin started to subside again and the oil generation window spread farther onto the shelf to reach its maximum aerial extent at the beginning of the Pliocene. This second subsidence event is probably not a major contributor to the total oil generation from the Jurassic due to limited source rock potential of the shelf. However, minor amounts of oil may have been generated in pre-existing extensional troughs that exist farther south on the Fly Platform if these did not fall through the oil window during the first subsidence event of the Late Cretaceous. Fluid inclusion data (Earnshaw *et al.* 1993) indicates that petroleum charge of structures took place during the period of active thrusting from Late Pliocene to Recent. Thus much of the oil that is trapped in thrust anticlines within the Mesozoic of the thrust belt was probably previously trapped within extensional structures. It has since migrated into the present structures during deformation. Cracking of previously trapped oil into gas and gas generation during the late Tertiary subsidence event probably accounts for the large gas accumulations along the fold belt. Moreover, given the abundance of seeps that are present today, it would appear that hydrocarbon re-migration is still occurring.

Sands of the Imburu and Toro Formations, which form the main reservoirs of the western Papuan Basin, were deposited as several regressive episodes within an overall transgressive, passive-margin clastic sequence. The regional seal to these reservoir sands is of excellent quality, being the conformable, transgressive mudstone of the Ieru Formation.

Eastern Papuan Basin

The hydrocarbon seeps that are abundant in the Vailala River (Fig. 2) area of the eastern Papuan Basin, which originally initiated the search for oil in PNG, have recently been analysed and have also been attributed to the Mesozoic by BP (Slater & Dekker 1990). However Slater & Dekker believe this attribution is speculative and that oil seeps in the area may be derived from multiple sources in the Mesozoic and the Tertiary. Their belief is not unreasonable because the Coral Sea Rift Basin/Aure Trough was active as a depositional trough throughout the late Cretaceous to Recent and pelagic, organic-rich shales are probably very common within it.

In 1992 there was a landslip which revealed a new oil seep (Fig. 2) within the New Guinea Mobile Belt, known as the Lufa seep (Murray *et al.* 1993). This is a region where collision between the Australian and Pacific plates took place and sediments are highly sheared and steeply dipping. The Lufa seep has a very different geochemistry to all the other seeps that occur in the fold belt to the south. It is an oil of Cainozoic source rock origin. All previous oil analyses have attributed produced oils and seeps to Mesozoic source rocks. The Lufa seep is also very fresh, whereas most other seeps in PNG are degraded. Thus Murray *et al.* (1993) reasoned that this oil has migrated directly to the surface, perhaps via the faults that triggered the landslip, rather than having been previously trapped within a reservoir where degradation could take place. They proposed that the oil does not come from the Tertiary sediments that outcrop near to the seep but from sediments which are now concealed beneath overthrusted allochthonous terrain. A tectonic history for the area in agreement with Murray *et al.* is discussed in the next section.

The Pale Sandstone of the Coral Sea Megasequence is considered to be the most important potential reservoir in the eastern Papuan Basin but other potential reservoirs, particularly within the Darai and Foreland Megasequence do exist (Carman 1993). The Pale Sandstone at the Aure Scarp is up to 200 m thick with an estimated average net to gross ratio in excess of 90% (Boult & Carman 1993). Analyses of cores taken from outcrop indicate porosity values are commonly greater than 20% and that permeability values of two to three Darcies are common. It comprises a texturally mature quartz arenite overlying a very pure volcaniclastic arenite/conglomerate (Fig. 6).

The Pale Sandstone is interpreted as being deposited in a barrier-island to strand-plain environment. Correlations of facies within the Pale Sandstone along the Aure Scarp, where it has been mapped in detail (Fig. 6), are very tentative because of the lack of outcrop in the heavily boulder strewn creeks. However, there is some evidence for the following:

(a) volcanic beach facies at the base in traverses 3, 4 and 6;
(b) more fluvial and lagoonal dominated, better cemented facies occur towards the base as in traverses 4 5 and 7;
(c) more upper shoreface to foreshore, less well cemented sands occur towards the top.

The latter sandstone (c) is responsible for the very poor outcrop in the major creeks but it is well exposed in smaller creeks where narrow canyons often form due to its high susceptibility to erosion (e.g. in traverse 5). Despite sporadic outcrops of the shoreface sands in the major creeks, their regressive erosional characteristic allows a stratigraphic thickness to be estimated by surveying the obvious creek profile gradient change.

The stratigraphic interpretation in Fig. 6 of the mapped area indicates some thickening of the Pale Sandstone from the north towards the south. This may be due to either syndepositional tectonics with greater subsidence to the south providing more sediment accommodation space or post depositional tectonics and erosion in the north. A conglomerate made of reworked oyster shell fragments with rounded quartz grains occurs at the base of the Tertiary and is well exposed in traverse 7. This conglomerate, together with the missing Maastrictian and Early Palaeocene section suggests that post depositional tectonics may be the primary reason for the northward thinning of the Pale Sandstone.

The major regional seal to the Pale Sandstone, which is apparent at the outcrop site at the Aure Scarp, is a non-conformable, transgressive, volcanic fragment and clay rich (up to 30%) micritic limestone (Boult & Carman 1993). Due to this high clay content, there is some chance that this seal may have remained plastic during deformation and is still intact today. Trapping beneath the unconformity may also have resulted from the preservation of lagoonal mudstone seals that have been recorded within the Campanian section itself. If the major north to south thickness change of the Pale Sandstone, which is apparent at the Aure Scarp, is a result of tilting prior to erosion on the base Tertiary unconformity, then the chances are fair to good for a thick intraformational seal to exist further south which is similar to the lagoonal mudstone found at the southern end of the Aure Scarp in traverses 7 and 8. Also deeper into the Aure Trough it is possible that pelagic mudstones of the Chim Formation will onlap across the top of the Pale Sandstone forming an excellent seal.

Tectonic palaeogeography of the eastern Papuan Basin

During the Mesozoic the eastern Papuan Basin represented the offshore section of the Papuan

Basin as a whole with the western Papuan Basin representing the more onshore section (Fig. 4). Then during the late Cretaceous to Tertiary rapid deepening of the eastern Papuan Basin occurred with the formation of the Aure Trough. Carman (1993) suggests that the Aure Trough started out as a Late Cretaceous to Tertiary rift basin representing the most northerly expression of the Coral Sea spreading. Pigram & Symonds (1993) suggest that the Aure Trough is a pull-apart basin related to a complex wrench fault system that extended northwards from the Coral Sea Rift. The Aure Trough then further developed into a Late Oligocene to Mid-Miocene foreland basin depocentre on the northern margin of the Papuan Basin (Rogerson & Hilyard 1990). This development is today represented in part by the Whagi–Yaveufa syncline (Fig. 2) which was delineated by the regional bouguer anomalies map of St John (1990). The Whagi–Yaveufa syncline is located between the Kubor uplift and the metamorphic terrain to the NE running in a NW–SE direction. Much of this Tertiary foreland basin may be concealed beneath overthrust terrain from the north. Currently the Aure and Moresby Troughs are the morphological expression of the present day foreland basin resulting from the New Guinea Orogen.

Discussion

The perceived lack of good reservoir within the eastern Papuan Basin as an impediment to active exploration has been dispelled in recent years by the discovery of the Pale Sandstone at the Aure Scarp. All wells that have been drilled in the eastern Papuan Basin are now known to have failed to reach either the top of the oil window (Slater & Dekker 1990) or the level of the Pale Sandstone, which explains the lack of success there. With current oil generation being proven to be active within the Tertiary of the eastern Papuan Basin then it is very possible that up dip migration into the stratigraphically lower Pale Sandstone could have occurred or is still occurring.

There is a gap in the recording of oil seeps between the Vailala river area and the Lufa seep to the north (Fig. 2). However 'the recording of oil seepages suggests that they are most abundant in areas which are well populated and are known to local people. In unpopulated areas, their scarcity may be due to non-discovery rather than non-existence' (Rickwood 1992). Thus the gap in seep recordings may be explained because the area surrounding the Aure Scarp and the southern extension of the Whagi–Yaveufa syncline is extremely sparsely populated and very rough terrain even by PNG standards (Carman 1987).

A comparison of the possible petroleum system of the eastern Papuan Basin with the proven productivity of the Imburu–Toro Formation of the western Papuan Basin reveals the following very important facts. Firstly that peak oil generation and migration took place in the western Papuan Basin before major structuring occurred, but in the eastern Papuan Basin there are indications that peak generation is post structuring. The seal (the Ieru Formation) in the western Papuan Basin must therefore be, and had to be, very good. The seal to the reservoir of the western Papuan Basin is conformable whereas the mapped regional seal to the potential new reservoir of the eastern Papuan Basin is unconformable. This means that considerable onlap could be occurring and that in the more distal parts of the basin, deeper water mudstones such as the Chim Formation may form the seal as well as the source rock. It also possible that there may be potential seals within the pre-unconformity rocks such as lagoonal or even offshore mudstones. If the generation and migration of oil is still active within the eastern Papuan Basin today, then there is a fair to good chance that this oil will be trapped within the Pale Sandstone or its equivalent.

References

BOULT, P. J. & CARMAN, G. 1993. The sedimentology, reservoir potential and seal integrity of the Pale Sandstone at the Aure Scarp, Papua New Guinea. *In*: CARMAN, G. J. & CARMAN, Z. (eds) *Petroleum Exploration in Papua New Guinea*. Proceedings of the Second PNG Petroleum Convention, Port Moresby, 31 May–2 June 1993, 125–137.

CARMAN, G. J. 1987. The stratigraphy of the Aure Scarp, Papua New Guinea. *Petroleum Exploration Society of Australia Journal*, September, 26–35.

——1993. Palaeogeography of the Coral Sea, Darai and Foreland Megasequences in the eastern Papuan Basin, *In*: CARMAN, G. J. & CARMAN, Z. (eds) *Petroleum Exploration in Papua New Guinea*. Proceedings of the Second PNG Petroleum Convention, Port Moresby, 31 May–2 June 1993, 291–309.

DOUST, H. 1990. Geology of the Sepik Basin, Papua New Guinea. *In*: CARMAN, G. J. & CARMAN, Z. (eds) *Petroleum Exploration in Papua New Guinea*. Proceedings of the First PNG Petroleum Convention, Port Moresby, 12–14 February 1990, 461–478.

EARNSHAW, P. J., HOGG, A. J. C., OXTOBY, N. H. & CAWLEY, S. J. 1993. Petrographic and fluid inclusion evidence for the timing of diagenesis and petroleum entrapment in the Papuan Basin, *In*: CARMAN, G. J. & CARMAN, Z. (eds) *Petroleum Exploration in Papua New Guinea.* Proceedings of the Second PNG Petroleum Convention, Port Moresby, 31 May–2 June 1993, 459–476.

EISENBERG, L. I. 1993. Hydrodynamic character of the Toro Sandstone, Iagifu/Hedinia area, Southern Highlands Province, Papua New Guinea. *In*: CARMAN, G. J. & CARMAN, Z. (eds) *Petroleum Exploration in Papua New Guinea.* Proceedings of the Second PNG Petroleum Convention, Port Moresby, 31 May–2 June 1993, 447–458.

EXON, N. F. & MARLOW, M. S. 1990. The New Ireland Basin: a frontier basin in Papua New Guinea. *In*: CARMAN, G. J. & CARMAN, Z. (eds) *Petroleum Exploration in Papua New Guinea.* Proceedings of the First PNG Petroleum Convention, Port Moresby, 12–14 February 1990, 513–534.

GRAINGE, A. 1993. Recent developments in prospect mapping in the Hides/Karius area of the Papuan Fold Belt. *In*: CARMAN, G. J. & CARMAN, Z. (eds) *Petroleum Exploration in Papua New Guinea.* Proceedings of the Second PNG Petroleum Convention, Port Moresby, 31 May–2 June 1993, 527–537.

HAUMU, J. 1995. *1994 annual report on petroleum activity in Papua New Guinea.* Papua New Guinea Department of Mining and Petroleum, Petroleum Division. Petroleum Licences, Location Map revised 1st May 1995.

HOME, P. C. 1990. Geological evolution of the Western Papuan Basin. *In*: CARMAN, G. J. & CARMAN, Z. (eds) *Petroleum Exploration in Papua New Guinea.* Proceedings of the First PNG Petroleum Convention, Port Moresby, 12–14 February 1990, 107–118.

KUGLER JR, A. 1990. Geology and petroleum plays of the Aitape Basin, New Guinea. *In*: CARMAN, G. J. & CARMAN, Z. (eds) *Petroleum Exploration in Papua New Guinea,* Proceedings of the First PNG Petroleum Convention, Port Moresby, 12–14 February 1990, 479–490.

MURRAY, A. P., SUMMONS, R. E., BRADSHAW, J. & PAWIH, B. 1993. Cainozoic oil in Papua New Guinea – evidence from geochemical analysis of two newly discovered seeps. *In*: CARMAN, G. J. & CARMAN, Z. (eds) *Petroleum Exploration in Papua New Guinea.* Proceedings of the Second PNG Petroleum Convention, Port Moresby, 31 May–2 June 1993, 489–498.

OSBORNE, D. G. 1990. The hydrocarbon potential of the western Papuan Basin foreland – with reference to worldwide analogues. *In*: CARMAN, G. J. & CARMAN, Z. (eds) *Petroleum Exploration in Papua New Guinea.* Proceedings of the First PNG Petroleum Convention, Port Moresby, 12–14 February 1990, 197–213.

PAWIH, B. & HALSTEAD, P. H. 1993. Investments and reserves in the Papuan Basin, Papua New Guinea, *In*: CARMAN, G. J. & CARMAN, Z. (eds) *Petroleum Exploration in Papua New Guinea.* Proceedings of the Second PNG Petroleum Convention, Port Moresby, 31 May–2 June 1993, 27–44.

PIGRAM, C. J. & SYMONDS, P. A. 1993. Eastern Papuan Basin: a new model for tectonic development and implications for petroleum prospectivity, *In*: CARMAN, G. J. & CARMAN, Z. (eds) *Petroleum Exploration in Papua New Guinea.* Proceedings of the Second PNG Petroleum Convention, Port Moresby, 31 May–2 June 1993, 213–232.

——, DAVIES, P. J., FEARY, D. A. SYMONDS, P. A. & CHAPRONIERE, G. C. H. 1990. Controls on Tertiary carbonate platform evolution in the Papuan Basin: new play concepts. *In*: CARMAN, G. J. & CARMAN, Z. (eds) *Petroleum Exploration in Papua New Guinea.* Proceedings of the First PNG Petroleum Convention, Port Moresby, 12–14 February 1990, 185–195.

PHILLIPS, S. 1990. Petrology Report, Pale Sandstone, PNG. *In*: BOULT, P. J. (ed.) *The Pale Sandstone, Papua New Guinea, Sedimentology, Reservoir Potential and Seal Integrity.* unpublished report for Austin Oil NL.

RICKWOOD, F. K. 1990. Towards Development – The Long History of Petroleum Exploration in Papua New Guinea. *In*: CARMAN, G. J. & CARMAN, Z. (eds) *Petroleum Exploration in Papua New Guinea.* Proceedings of the First PNG Petroleum Convention, Port Moresby, 12–14 February 1990, 1–13.

——1992. *The Kutubu Discovery: Papua New Guinea, its people, the country and the exploration and discovery of oil.* Brown Prior Andersen Pty Ltd, Australia.

ROGERSON, R. S. & HILYARD, D. B. 1990. Scrapland: a suspect composite terrane in Papua New Guinea. *In*: CARMAN, G. J. & CARMAN, Z. (eds) *Petroleum Exploration in Papua New Guinea.* Proceedings of the First PNG Petroleum Convention, Port Moresby, 12–14 February 1990, 271–282.

SLATER, A. & DEKKER, F. 1993. An overview of the petroleum geology of the Eastern Papuan Fold Belt, based on recent exploration. *In*: CARMAN, G. J. & CARMAN, Z. (eds) *Petroleum Exploration in Papua New Guinea.* Proceedings of the Second PNG Petroleum Convention, Port Moresby, 31 May–
2 June 1993, 499–516.

SMITH, R. I. 1990. Tertiary plate tectonic setting and evolution of Papua New Guinea, *In*: CARMAN, G. J. & CARMAN, Z. (eds) *Petroleum Exploration in Papua New Guinea.* Proceedings of the First PNG Petroleum Convention, Port Moresby, 12–14 February 1990, 229–244.

ST JOHN, V. P. 1990. Regional gravity and structure of the eastern Papuan Basin. *In*: CARMAN, G. J. & CARMAN, Z. (eds) *Petroleum Exploration in Papua New Guinea.* Proceedings of the First PNG Petroleum Convention, Port Moresby, 12–14 February 1990, 311–318.

VICTORIA PETROLEUM, N. L., 1993. *Drilling and Logistics Cost Estimates for Menge No.1 and Tumuli No.1.* PPL 106 Onshore Papua New Guinea, unpublished.

An overview of the hydrocarbon potential of the Spratly Islands archipelago and its implications for regional development

J. B. BLANCHE & J. D. BLANCHE

Blanche International Oil and Gas Consultants, Scotland, 'Tara', Laighill Place, Dunblane, Perthshire FK15 0BJ, UK

Abstract: The hydrocarbon potential of the Spratly Islands archipelago is relatively unknown and can thus only be evaluated via analogues. This area of the South China Sea has been, and remains, a region of many disputes and conflicts which has further confined the availability of hard technical information, in particular seismic and well data, to the confidential vaults of the claimant states. In addition, the authors have been constrained by the area's political sensitivity and the commercial confidentiality of technical data. The Spratly Islands are claimed, either in whole or in part, by the People's Republic of China (PRC), the Republic of China (ROC/Taiwan), Vietnam, the Philippines, Brunei and Malaysia.

The archipelago is surrounded by oil-producing areas, i.e. the Nam Con Son (Wan'an Bei) basin of Vietnam, the East and West Natuna basins of Indonesia, the Northwest Palawan (Nansha Trough) basin of the Philippines, the productive Luconia Shelf (Zengmu Shelf) offshore Sarawak and the Brunei/North West Sabah basins which provide regional analogues. This frontier region may yield considerable reserves, the range of which varies from 1–2 billion barrels to 225 billion barrels of oil equivalent. It is anticipated that the Spratly Island archipelago and Dangerous Grounds area of the South China Sea will become the focus of exploration for hydrocarbons over the next decade, once the multi-national boundary disputes are resolved, hopefully by negotiation and peaceful means by the claimant states.

This paper presents an overview of the hydrocarbon potential of this area within the context of regional political relationships.

Strategic position

The Spratly Islands archipelago, named after a nineteenth century British whaling Captain, is a group of reefs in the South China Sea located midway between Vietnam and the Philippines and claimed by several countries. The archipelago covers an area 1 000 km wide across the southern reaches of the South China Sea and lies between latitudes 4° and 11°30′N and longitudes 109° and 117°50′E. The archipelago is composed of 150 reefs together with 13 shoals and atolls, 4 banks and 16 islands which are permanently above sea level in water depths of 1000–5000 m (Fig. 1). They are referred to in China as the Nansha Islands, in Vietnam as the Truong Sa, and in the Philippines as the Kalayaan (Freedom) Group.

The Islands are claimed, either in whole or in part, by the People's Republic of China (PRC), the Republic of China (ROC/Taiwan), Vietnam, the Philippines, Brunei and Malaysia (Fig. 2; Table 1). The South China Sea has been, and still is, a region of many disputes and conflicts, where piracy is endemic. The Islands are located strategically close to the sea lanes of communication in the South China Sea between Singapore, Hong Kong and the Chinese mainland. A quarter of the world's maritime trade, including almost 70% of Japan's oil supplies, passes through the Spratly Islands area. The area is rich in fish stocks and potential hydrocarbon deposits. There are many disputes over fishing rights, continental-shelf boundaries, reefs and islands.

Claims to the Spratly Islands are historic and complicated. China claims sovereignty on historic grounds and in 1974 China and Vietnam had conflict over the islands. Since 1982, naval clashes have continued periodically between China and Vietnam over the Paracel Islands 700 km to the north and, more recently and triggered by oil and gas developments, between China and the Philippines over the Spratly Islands. The People's Republic of China Liberation Army Navy (PLAN) frequently exercises in the vicinity of both the Spratly and Paracel Islands.

Regional geology and basinal development

The megatectonic history and evolution of the South China Sea has been well documented by Emery & Ben-Avraham (1972), Katili (1981), Holloway (1982), Taylor & Hayes (1983), Kudrass *et al.* (1986), Ru & Pigott (1986), Li (1988), Liu & Yang (1988), Roberts (1988), Letouzey *et al.* (1988), Liang and Liu (1990),

From Fraser, A. J., Matthews, S. J. & Murphy, R. W. (eds), 1997, *Petroleum Geology of Southeast Asia*, Geological Society Special Publication No. 126, pp. 293–310.

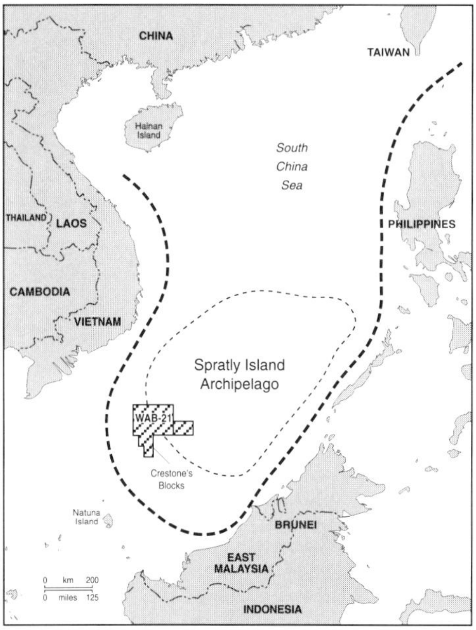

Fig. 1. Spratly Islands archipelago, South China Sea. The Spratly Islands are referred to in China as the Nansha islands, in Vietnam as the Trumg Sa, and in the Philippines as the Kalayaan (Freedom) Group. The islands are claimed either in whole or in part, by the People's Republic of China (PRC), the Republic of China (ROC/Taiwan), Vietnam, the Philippines, Brunei and Malaysia. **Hainan Special Administration District.

Daly *et al.* (1991) and Lee & Lawver (1992) and will not be discussed further in this paper (Fig. 3a and b).

The Spratly Island archipelago is located in the eastern sub-basin of the South China Sea which formed as the result of seafloor spreading active from the Late Oligocene to the Late Mid-Miocene and is associated with roughly east–west-trending magnetic lineations (Taylor & Hayes 1983) implying a north–south spreading system. In the SW arm of the South China Sea, NE–SW lineations are also present, indicating a second phase of NW–SE extension (Daly *et al.* 1991).

Table 1. *Spratly Islands archipelago: claimant states occupation*

Claimant state	No. occupied islands	No. islands, cays etc. claimed
Philippines	8	8
Vietnam	23 (+2?)	25 (+2?)
Taiwan	1	All
Malaysia	3	3
Brunei	0	1
China	7	All

The southern margin of the South China Sea includes the Nansha (Spratly) area, which includes the Reed Bank (Liyue Bank of China) and the reef-and-shoal area to the west of the Dangerous Grounds, the Palawan Trough, the Palawan Block and the Sarawak Shelf (Zengmu) (Figs 3 and 4).

Chinese geoscientists from the South China Sea Institute of Oceanology (Academia Sineca) identify six basins in the Nansha (Spratly) Islands archipelago. These are the West Wan'an (Nam Con Son), Zengmu (Sarawak Shelf), Liyue (Reed Bank) Nansha Trough (NW Sabah/SW Palawan) and two uniquely Chinese basins; the Nanhua and Nanwei, which are frontier and located within the Dangerous Grounds/Spratly Islands area (Jiang & Zhou 1994).

The recognition of a Late Eocene compressional event in the Reed Bank area has significant geological implications on the evolution of the South China Sea. Recent Chinese research differs from previous models in that the Nansha Block collided with palaeo-Borneo and was uplifted in Late Eocene times prior to the opening of the South China Sea (Xia & Zhou 1992).

The Chinese consider the Nansha Block prior to the collision to have been a shallow sea bordering the palaeo-South China continent, where marginal rifting had been occurring since the late Cretaceous. During the Late Oligocene extensional stress extended southwards toward the Nansha Block and initiated the opening of the South China Sea (Xia & Zhou 1992).

The Spratly Islands archipelago, the Dangerous Ground and the Reed Bank are all located on the Reed Bank terrane (Fan 1992). The average water depth of the region is 1400–2000 m. Sea-bed samples obtained via dredging and coring, together with seismic data, indicate that the Reed Bank terrane consists of continental crust with 5–6 km of marine Mesozoic basin fill and 8–9 km of marine Cenozoic basin fill. The northwestern part of the Reed Bank terrane is a passive continental margin, whilst the southwestern part is a colliding convergent margin. The Reed Bank terrane comprises 14 Cenozoic sub-basins with over 2 km of sediments (Fan 1992).

The Reed Bank terrane was a part of a proto-China/Indonesia which rifted during Palaeocene time and continued to expand during the Early Oligocene, ceasing by the Mid-Oligocene and coinciding with the collision of the Reed Bank terrane with the North Palawan terrane at the

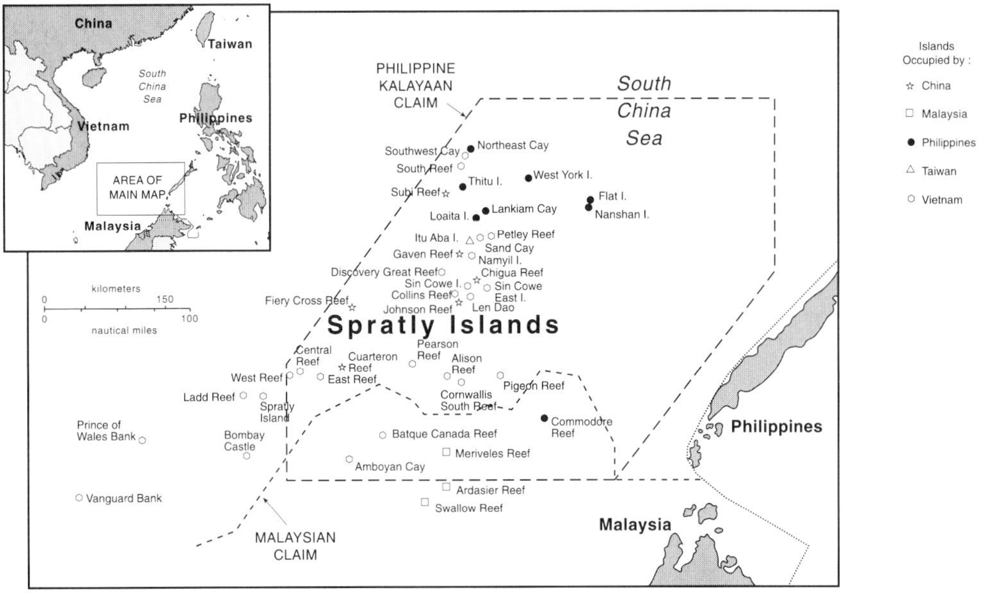

Fig. 2. Occupation of the Spratly Islands by claimant states. Spratly Islands dispute (after Thomas 1989).

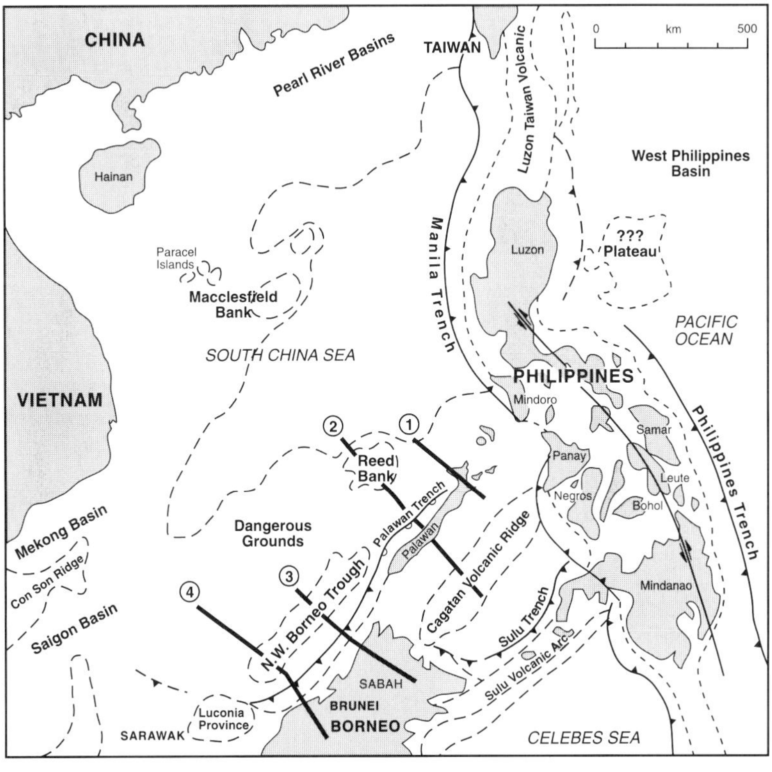

Fig. 3. (**a**) South China Sea: location of geological cross sections (after Letouzey *et al.* 1988). (**b**) Reed Bank/ Dangerous Grounds: geological and structural cross section.

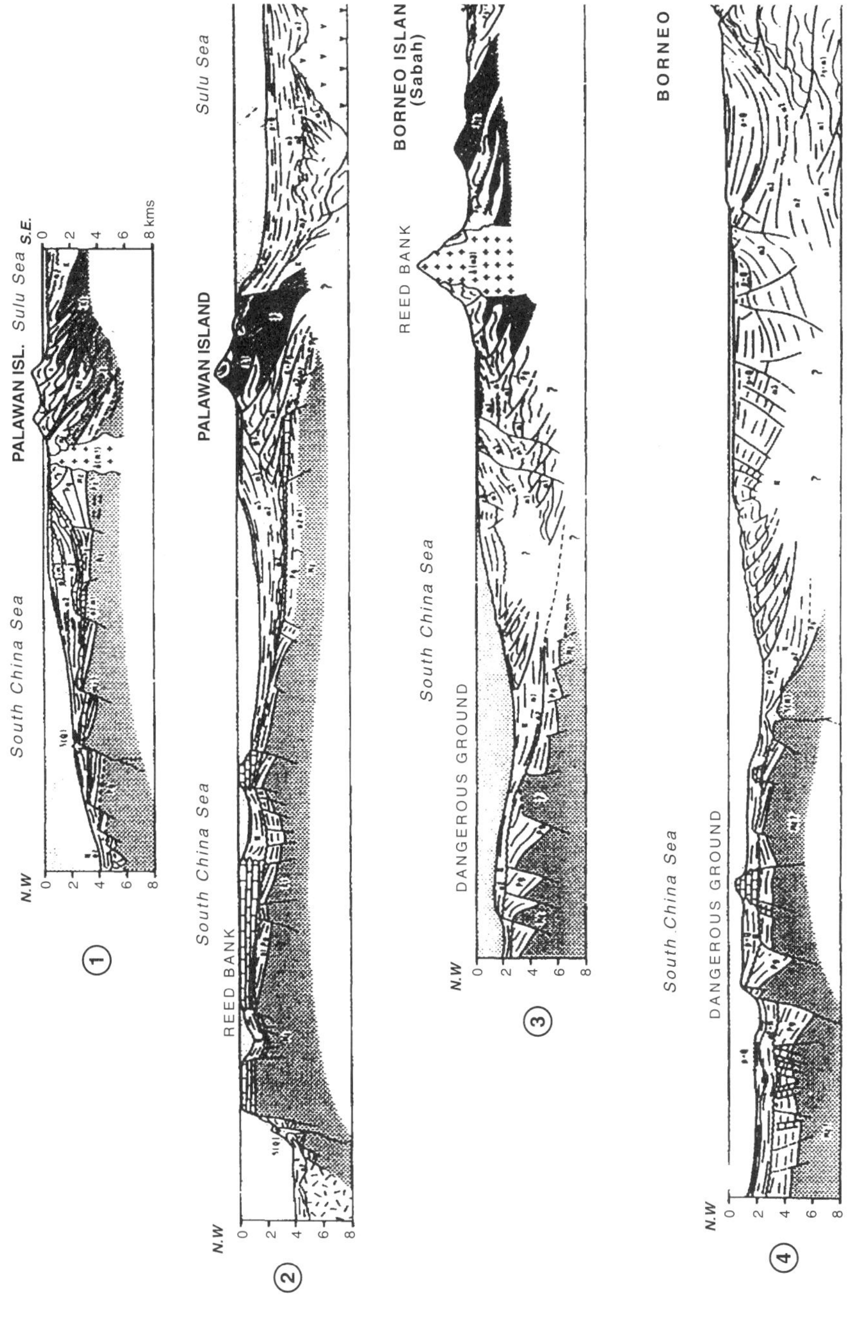

Figure 3. Continued.

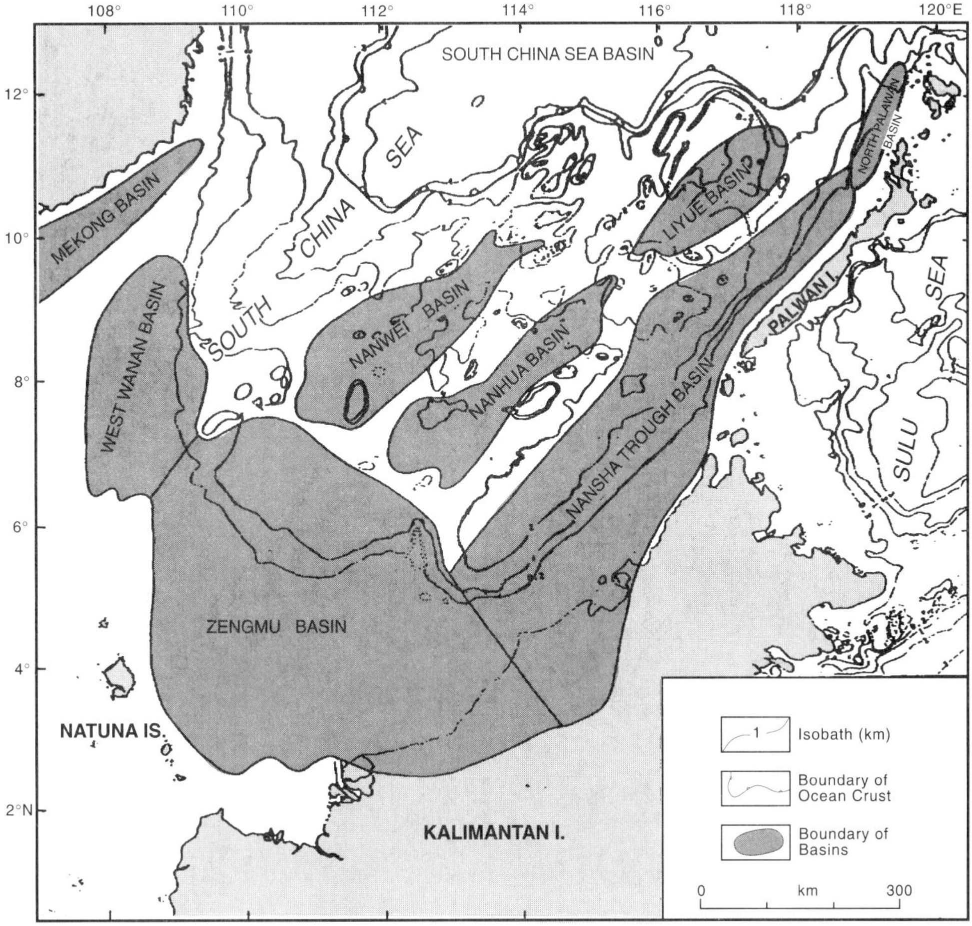

Fig. 4. South China Sea: basins of the Nansha archipelago identified by the Chinese (after Jiang & Zhou 1994).

subduction boundary of Borneo. Shallow marine carbonates of Early to Late Miocene age are present from the Luconia Platform to the south, across the Reed Bank terrane to the oil-producing Nido Formation of South Palawan terrane to the north (Fan 1992).

Stratigraphy

The stratigraphy of the Spratly Islands archipelago is unknown as no exploration drilling in the archipelago proper has been undertaken. However, the stratigraphy of the analogous basins bordering the Spratly Islands archipelago is documented from exploration drilling by Du Bois (1985), Hirayama (1991) and Xia & Zhou (1992). These analogue basins are the Reed Bank, NW Palawan, SW Palawan, NW Borneo and Sarawak Shelves and the Nam Con Son basin (Figs 4 and 5).

Geophysical surveys undertaken by the Chinese Ministry of Geology and Mineral Resources and by other foreign institutions during the period 1958–1993 have given indications of stratigraphic events. This is particularly true of surveys undertaken post-1980.

Chinese multi-channel seismic profiles illustrate a thick Palaeogene sequence overlying a package of thick Mesozoic sediments. The Palaeogene sequence comprises neritic clastics with clear stratification and widespread distribution. This sequence was later strongly deformed into asymmetric folds and eroded. The unconformity above this sequence has been recognised in several seismic profiles by the Chinese in the

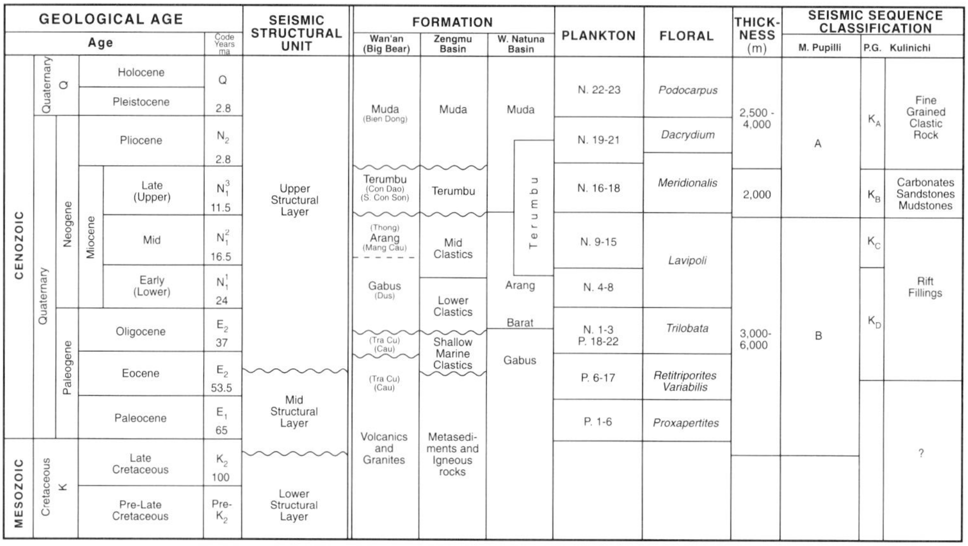

Fig. 5. Generalized stratigraphy: Wan'an Bei (after Pupilli 1973).

Dangerous Grounds area. The Chinese tentatively date this unconformity as Late Eocene, based on correlation with distant wells (Xia & Zhou 1992).

The Dangerous Grounds is the name for numerous shallow and exposed reefs located at the southern end of the South China Sea between Reed Bank and Natuna Islands, north of Sarawak, Borneo. Above the unconformity, the Neogene sequence is thin and mainly occurs as infills of discrete half-grabens in the Dangerous Grounds area, but thickens and becomes continuous over the Reed Bank, Palawan Shelf and the Great Sarawak (Zengmu) basin. This indicates that the Dangerous Grounds area has stayed as a 'high' since the Late Eocene, whilst the shelf area to the south and east have become depocentres.

The oldest rocks sampled by dredging during the Sonne cruises of 1982/83 were of Upper Triassic/Jurassic slightly metamorphosed deltaic sandstones/siltstones containing plant fragments, and dark green mudstones containing shells of Late Triassic/Early Jurassic age. The results from RV Sonne cruises SO-23 in 1982 and SO-27 in 1983 (Hinz & Schluter 1985) reveal that there are five regional unconformities in the Dangerous Grounds that can be recognized on the basis of reflection seismic and geological sampling: a Cretaceous/Palaeocene unconformity, a Middle to Upper Oligocene unconformity, a Lower Miocene unconformity, a Middle Miocene unconformity which correlates with the opening if the South China Sea and a Miocene/Pliocene unconformity. These findings were corroborated by Qiu Yan (1995) in a paper presented in Manila at the Oil and Gas Asia Conference. The Mesozoic pre-rift sediments in this region are highly faulted by a complex system of NE–SW-striking horsts and half-grabens. The faulting presumably developed in pre-Mid-Oligocene times and indicates extensional tectonics. The trend of the depth-contour lines for the Middle to Upper Oligocene unconformity surface off South Palawan and the Balabac Strait, showing pronounced changes that are interpreted by Hinz & Schluter (1985) as having been caused by NW–SE-trending left-lateral transform faults (Lee & Lawver 1992).

An unpublished paper by Jiang & Zhou (1994) of the South China Sea Institute of Oceanology, Academia Sineca, Guangzhou, reports that over the period 1987–1991 the Nansha Investigation Team completed four geophysical cruises covering 19 seismic lines with a total length of 6883 line km. The seismic data indicated a number of basins of Mesozoic and Cenozoic age with multiple depositional cycles. Based on this data, and applying seismo-stratigraphic and tectono-stratigraphic techniques, the Chinese have identified three seismic sequences, separated by two regional unconformities within the Upper Cretaceous and Upper Eocene.

Exploration history

The South China Sea, and the Spratly Islands archipelago in particular, has been the focus of geophysical surveys by both the People's Republic of China, who have reportedly also drilled a well, Nan Yang-1, on Fiery Cross Reef in the western Spratlys (Crestone Energy Corporation pers. comm.), and other claimant states as well as non-exclusive seismic surveys by western geophysical companies and academic/oceanographic institutions over the period 1958–95 (Blanche & Blanche 1995*a*, *b*; Fig. 6).

In the 1970s the prospect of large offshore oil deposits brought renewed interest in the Spratly Islands archipelago. South Vietnam announced the incorporation of the group into Phuoc Tuy province in 1973 and let contracts to foreign companies to explore for oil. When China reacted by asserting its claim to the Spratlys, as well as claiming and occupying the Paracels Islands to the north, South Vietnam occupied three of the Spratly Islands (including Spratly Island itself) to forestall a Chinese occupation. Taiwanese troops remained on Itu Aba.

Since 1980, larger and more frequent geophysical surveys have been conducted as interest in the hydrocarbon potential of the area increased.

During 1982–83, the Federal Institute for Geosciences and Natural Resources in Hanover, Germany, conducted a multi-channel reflection-seismic, magnetic and gravity data survey along

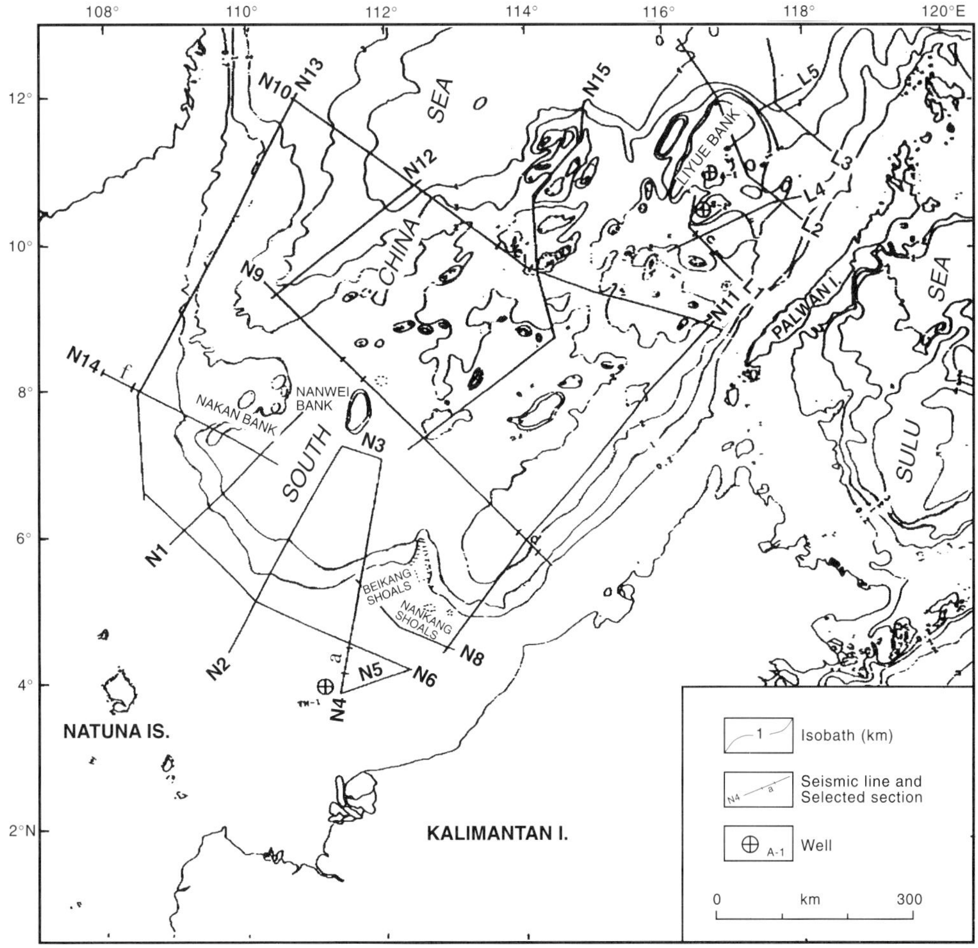

Fig. 6. Institute of Geophysical Cruises 1987–91 (after Jiang & Zhou 1994).

51 lines, with a total length in excess of 10 000 km, in the southeastern part of the South China Sea, predominantly in the Dangerous Grounds and Palawan Trough areas during the Sonne cruises SO-23 (1982) and SO-27 (1983). The results produced good quality data which enhanced greatly the knowledge of the structural development of the southeastern sector of the South China Sea (Hinz & Schluter 1985; Fig. 7).

Seismic has been acquired by the China National Offshore Oil Corporation (CNOOC) and geological surveys have been undertaken by the Chinese Ministry of Geology and Mineral Resources and the Chinese Academy of Sciences over the period 1987–91. During this period they completed four geophysical surveys acquiring 6883 km (19 lines) of seismic data. These data were processed by the Geophysical Exploration Institute of the National Petroleum and Gas

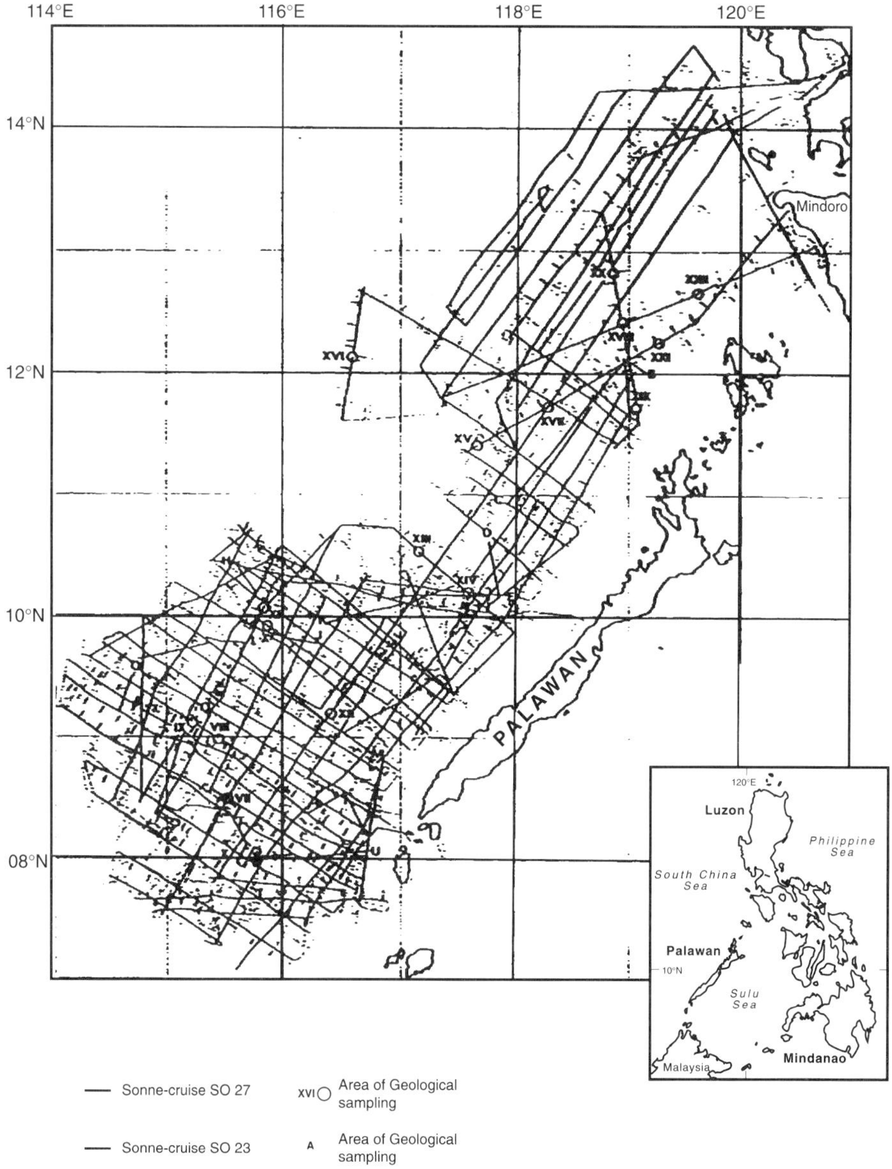

Fig. 7. BGR seismic lines.

General Corporation. These data are of good quality and illustrate potential oil- and gas-trapping mechanisms (Jiang & Zhou 1994).

In 1987, a Chinese survey vessel (Lance #11) acquired 103 400 km of gravity data and also undertook sea-bottom sampling.

International attention on the hydrocarbon potential of the area was focused by the award of the 25 155 km^2 WAB-21 Block in the Wan'an basin located 100 km to the SW of the Spratly Islands which are claimed by China and Vietnam, with an additional adjacent Contingent Contract Area covering 5076 km^2 in water depths ranging from 0 to 700 m (Fig. 8). The award was to the Crestone Energy Corporation from the CNOOC on 8 May 1992 and covers one of the largest concession areas in the South China Sea, adjacent to a large province of significant oil and gas discoveries offshore Vietnam. During 1992, Crestone Energy Corporation reprocessed seismic data acquired in 1978 by Petty-Ray Geophysical and Western Geophysical. The award in April 1994 to the Mobil group of the Blue Dragon (Thanh Long) structure in the Nam Con Son basin by Vietnam, in waters disputed between China and Vietnam, where substantial oil and gas reserves have been discovered, further focused attention on the area (Fig. 9).

During August 1995, the Crestone Energy Corporation announced the signing of a joint venture with the Exploration Development Research Centre (EDRC); the scientific and research subsidiary of the CNOOC. The project will help to determine the hydrocarbon potential of the Wan'an Bei WAB-21 Block. Ultimately, the results obtained will better define the large prospective geophysical structures already identified and mapped from seismic data. Existing seismic data acquired by Crestone connects the Wan'an Bei prospects to recently drilled discoveries in Vietnamese waters in the Nam Con Son basin by BP and Mobil which have confirmed several trillion cubic feet of natural gas reserves.

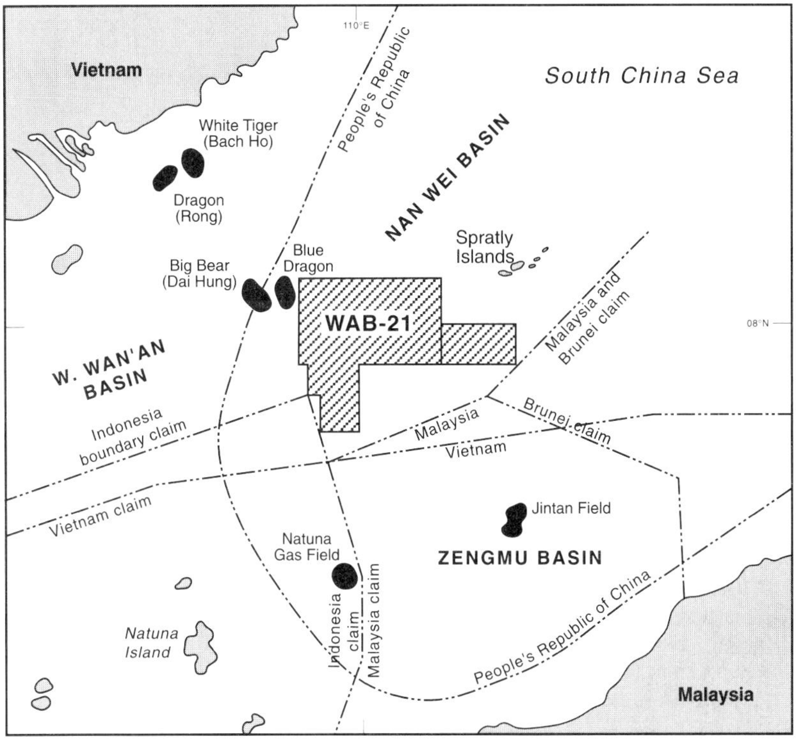

Fig. 8. Location of Crestone's WAB-21 block and boundary claims. Sovereignty over the Spratly Islands is in dispute, claims are historic and complicated. The Islands are claimed, either in whole or in part, by the People's Republic of China (PRC), the Republic of China (ROC/Talwan), Vietnam, the Philippines, Brunei and Malaysia.

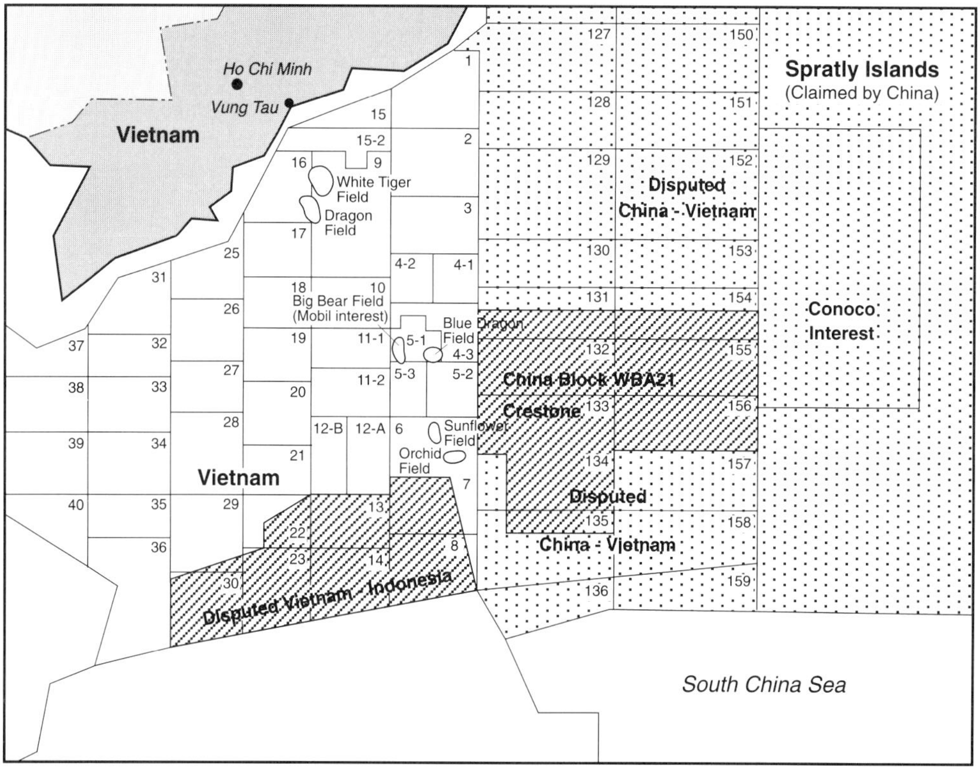

Fig. 9 Location of Mobil and BP interests to west of WAB-21 block.

However, in April 1996, PetroVietnam awarded Conoco Blocks 133 and 134 covering 14 285 km^2, under terms in which Conoco holds 70% interest and the operatorship. The Blocks are located within the current Crestone acreage area awarded by the People's Republic of China in the Wan'an Bei Basin, thus illustrating further the nature of the dispute between the two countries. There is little seismic coverage over these deep water blocks, with only one well being drilled by PetroVietnam in 1994 within the area of Block 134 on the Vanguard Bank which lies within the Crestone acreage and was completed as a tight hole.

The oil and gas potential of the Philippine Kalayaan Group was further highlighted when the Philippines Department of Energy approved a joint application with VAALCO Energy, a Houston-based oil and gas company, on 13 May 1994 for a non-exclusive geophysical (NEGP) permit covering a portion of the Kalayaan area. VAALCO's subsidiary, Alcorn Petroleum and Minerals Corporation, is currently exploring for and producing hydrocarbons in the Philippines. The NEGP includes seven of the eight islands occupied by the Philippines in the Kalayaan Group, including Pagasa, Lawak (Nanshan) and Patag (Flat) Islands.

On 2 November 1993, the Taiwan government stated that a technical group would be sent to the disputed Spratly Islands to prepare for oil and gas exploration. The group would also study the feasibility of building an airstrip and port on Taiping Island (Itu Aba). The team would include geoscientists from the Chinese Petroleum Corporation (CPC) and experts from the Ministry of Transport and Communications.

BP has drilled two wells on its Vietnamese Block 06 located immediately to the west of Crestone's WAB-21 block, and claimed by China (PRC). The wells, named Lan Do-1 (Red Orchid) and Lan Tay-1 (West Orchid), are also located in disputed waters (Fig. 9) The Lan Tay-1 well is reported to have penetrated a thick gas column in a Miocene carbonate and to have had a gas blow-out from this objective.

In the Reed Bank area, 5985 km of seismic have been acquired over the period 1970–1980. Of these, 2329 km had been reprocessed by

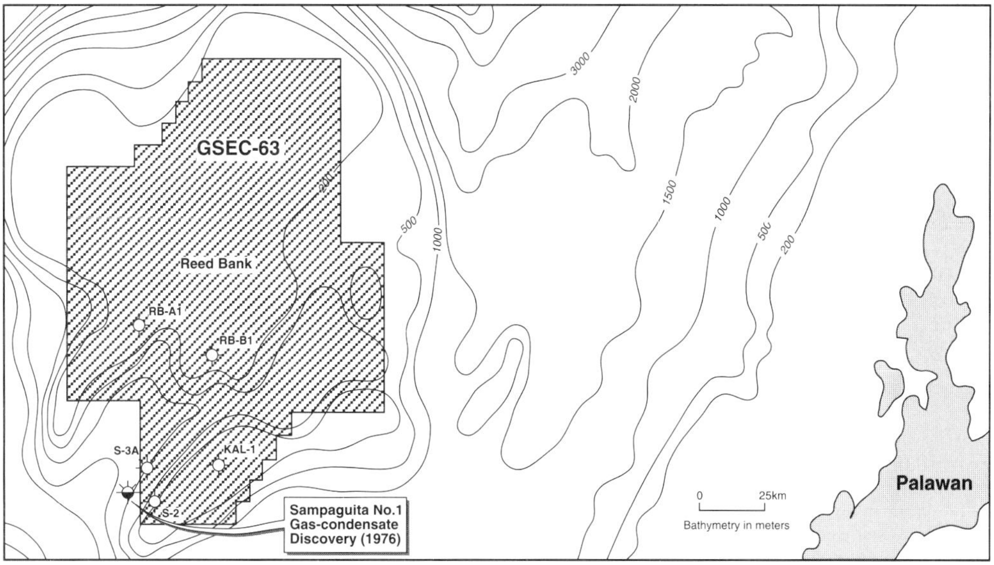

Fig. 10. GSEC-63: location of Sampaguita wells (after Dragon Oil plc).

PCIAC and Kirkland AS/Dragon Oil plc (Quinton 1993). Seven new field wildcats were drilled over the period 1976–1984, two of which encountered hydrocarbons. The West Sampaguita-1 (1976) encountered the upper part of graben-filling sands over the interval 3078–3161 m before crossing the easterly boundary fault of the graben into the next fault block and reaching total depth at 4069 m. The lower 24 m of these sands tested 3.6×10^6 scfg/d and 24 bcpd of 45.8° API gravity. The well is interpreted to have tested in an area of drape over the most southeasterly of the three Cretaceous highs.

The Sampaguita-3A well (1984) reached a total depth of 3898 m in Lower Cretaceous and tested 3.707×10^6 scfg/d and 80 bcpd of 44.6° API gravity from an interval 3162–3141 m.

Kirkland AS/Dragon Oil plc are the operators for GSEC 63, covering 15 000 km^2, in the Reed Bank (Liyue Bank) area which was awarded on 25 May 1992.(Quinton 1993). Gas condensate potential over the area is considered to range 5–15 tcf (Fig. 10).

Hydrocarbon potential

The littoral states of the South China Sea enjoy a regional economy second to none for predicted economic growth, and there is every indication that oil and gas reserves are there to be exploited. Thus there can be no doubt that the South China Sea's high profile as a focus for exploration and production of hydrocarbons will continue well into the next century with exploration drilling in the Spratly Islands archipelago. Analysts predict 40 new offshore developments coming onstream by the year 2005 in South East Asia, and the consequent recovery of more than 10 billion barrels of oil equivalent (75% gas) representing US$18 billion investment. Ever greater volumes of oil and gas are required to keep pace with burgeoning domestic demand among the littoral states of the South China Sea, including all of the Spratly Islands archipelago claimants. This is particularly true in the case of Indonesia, Malaysia and Brunei who will strive to maintain their current leading positions in liquified natural gas (LNG) production, primarily exported to Japan.

China (PRC)'s demand for LNG is expected to rise 72% over the period 1995–2000 to 18.13 million metric tons per year. The demand within Asia grew 4.2% during 1993, with the Chinese increase of 4.6%. China (PRC)'s oil demand jumped 11.1% during 1993 and China became an importer of oil for the first time since the early 1970s. In 1993, China (PRC) produced approximately 3 billion barrels of crude per day, but had to provide some 2 billion barrels per day for home consumption. The demand for home consumption is increasing daily. In 1993, China was importing considerable quantities of crude oil to satisfy refining needs in coastal provinces. By the year 2010, China will be at least 20% short of demand.

Whereas China could export 640 000 bopd in 1985, it has steadily had to cut this to 420 000 bopd in 1992, severely affecting its foreign currency earnings from oil exports. These factors all affect China's attitudes to oil reserves that may be available to it via its claim to the South China Seas, and may explain the militant attitude taken by Beijing in 1992 when the National People's Congress declared the whole of the South China Sea to be Chinese territorial waters (Law of the People's Republic of China on the Territorial Sea and the Contiguous Zones, adopted 25 February 1992) and stated that China would defend its claim to the disputed islands by force (Fig. 1).

Petroleum geology

The Spratly Islands archipelago is surrounded by prolific oil-producing areas, all within a 500 km radius of Spratly Island, i.e. the Nam Con Son (Wan'an) basin of Vietnam, the East and West Natuna basins of Indonesia, the Northwest Palawan (Nansha Trough) basin of the Philippines, the productive Luconia Shelf (Zengmu Shelf) offshore Sarawak and the Brunei/North West Sabah basins (Table 2; Fig. 4). These basins have been reviewed and discussed by Du Bois (1985), Liang & Liu (1990), Giao *et al.* (1991), Areshev *et al.* (1992) and Hirayama (1991).

By analogue with these areas, the petroleum geology of this frontier region is assessed as follows.

Trapping mechanisms

Traps include both stratigraphic and structural plays. The stratigraphic traps, particularly reefs similar to those encountered on the Luconia Shelf (Sarawak) of Malaysia (Zengmu basin), together with mudstone diapirs and reefal structures encountered in the East Natuna (Indonesia) basin, are likely to be developed. Stratigraphic drape and pinchout against palaeo-highs may also provide suitable trapping mechanisms, as evidenced by the Bach Ho and Dai Hung fields, offshore South Vietnam.

Structural traps are likely to be characterized by anticlines, horst blocks, tilted fault blocks, rollover anticlines and growth faults as are proven in the Reed Bank area (Liyue basin), Northwest Sabah Shelf (Malaysia) and Nam Con Son basin (Vietnam).

In addition, weathered and fractured pre-Tertiary granitic basement highs may provide suitable trapping mechanisms, as in the case of Dai Hung and Bach Ho fields in the Nam Con Son and Mekong basins of south Vietnam respectively (Areshev *et al.* 1992).

All of these features can be recognised on the seismic acquired to date by the CNOOC over the Spratlys archipelago and by western contractors in the Philippines, Vietnam and Malaysia.

In the Cuu Long, Nam Con Son and Phu Khanh basins, multiple structural and stratigraphic plays have been identified in Blocks 122–130, offshore Vietnam, which are currently being offered for award. These plays include numerous

Table 2. *Reserves and production details summary: major oil and gas fields adjacent to the Spratly Islands*

Country	Basin	Field	Reserves and production
Vietnam	Mekong	Bach Ho/ Rong	300 MMbbls, 140 000 bopd, 100 MMbbls, 30 000 bopd
	Nam Con Son	Dai Hung	80–100 MMbbls, 11 000 bopd
	Nam Con Son	Blue Dragon	5–500 MMbbls, 1–3 tcf
Indonesia	West Natuna	7 fields	15–150 MMbbls, 141 300 bopd 0.1–45 tc
	East Natuna	Natuna	45–60 tcf
Malaysia	Sarawak	'M' fields	6–9 tcf
	Sarawak	Jintan *et al.*	6 tcf 75 MMbbls condensate
	Sabah	Kinabalu	200 MMbbls 1.5 tcf
Philippines	Northwest Palawan	West Linapacan	71–106 MMbbls
	Northwest Palawan	Malampaya	70–300 MMbbls 1.6 – 2 tcf
	Northwest Palawan	Camago	60 MMbbls 3–6 tcf

horst blocks, rotated fault blocks, elongated and faulted anticlines formed by normal and reversed faults, drape features and flower structures demonstrated by seismic to occur in the pre-Middle Miocene section. Stratigraphic plays include carbonate reefal complexes of Oligocene to Pliocene age within which barrier reefs, build-ups and bioclastic aprons occur. Purely stratigraphic plays include pinchouts, submarine fan complexes and traps related to slope turbidites. These occur throughout the section, ranging in age from Palaeocene to Upper Miocene/Lower Pliocene.

Reservoirs

Potential reservoirs include well-developed prograding deltaic clastics and marine shelf carbonates of Oligocene to Pliocene age. These reservoirs, Oligocene–Upper Miocene clastics and Upper Miocene carbonates, are exemplified in the productive Dai Hung field in the Nam Con Son basin (Fig. 11).

Eocene to Oligocene prograding fluvio/deltaic encountered in the Reed Bank are of regional extent, massive in character and are excellent reservoirs with individual sand units having a net pay thickness ranging 15–30 m and a gross thickness of 229–682 m. Porosities range from 10–24% and permeabilities are assessed to be high. In the Wa'an Basin of the South China Sea, potential reservoir sandstones of Oligocene to Lower Miocene age have been reported to have porosities in the range of 17–23% (Qiu Yan 1995).

The Sampaguita sands, drilled in the Reed Bank, are also well developed and have tested gas and condensate from the Sampaguita-1 and -3A wells. The Sampaguita-1 well encountered a gross sandstone interval of 110 m and a net sand of 58 m. The Sampaguita-3A well encountered a gross sandstone interval of 69 m and a net of 20 m. Reservoir quality was measured at an average of 23% porosity and 20 mD permeability respectively.

Cretaceous sections encountered by all wells in the Reed Bank displayed thick sandstones with fair reservoir potential and porosities ranging from 19 to 26%. Permeabilities were low. Gross sand thicknesses were in the region of 183 m and net sand thicknesses 76 m. Facies types are deltaic, neritic-littoral and abyssal turbiditic.

Weathered and fractured basement of pre-Tertiary age may also provide suitable reservoirs, as exemplified by production from the Bach Ho and shows from Dai Hung fields offshore south Vietnam (Areshev *et al.* 1992). The effective porosity of the weathered granitic basement in the latter field is created by tectonic fractures, weathering and hydrothermal caverns and pores. Based on well core data 60% of the total porosity consists of pores with an average diameter of 0.1 mm (Giao *et al.* 1991).

	Source Rocks			Reservoir Rocks			Seal Rocks
					Properties		
Basin	**Formation & Age**	**Kerogen Types**	**Total Organic Carbon (%)**	**Formation & Age**	**Porosity (%)**	**Permeability (mD)**	**Regional Distribution**
Zengmu	Setap (E_3-N_1^2)	II - III		South Luconia reefal limestone (N_1^{2-3})	15-32	0-4000	Transgressive shales of cycle V (N_1^3)
Wan'an	Arang (N_1^{1-2})	II - III	0.69-0.93	Terumbu reefal limestone (N_1^{2-3})	15-24		Muda mudstones (N_2-Q)
	Gabus (E_3^2)			Medium-grained clastic rocks (N_1^{1-3})			Shale (N_1^3)
Brunei-Sabah	Setap (E_3-N_1^2)	II - III		Seria and Miri deltaic sandstone (N_1^{2-3})	12-35	100-1900	Pliocene mudstones
Reed Bank	Marine Shale (E_3-E_2)		0.12-1.9	Paleocene to Eocene deltaic sandstone (E_{1-2})			Eocene-Oligocene mudstones
Palawan	Pagasa Shale, Miocene limestones (N_1^{1-2})	II - III	0.50-2.48	Pagasa turbidites (N_1^{1-2}) Nido reefal limestone (N_1^1)	8-30	120-2000	Pagasa mudstones, shales and claystones

Fig. 11. Characteristics of source, reservoir and seal rocks of basins in southern South China Sea.

Seal

Sealing mechanisms are provided by faults and regionally developed intraformational transgressive mudstones and shales (Fig. 11). In the Reed Bank area these consist of shales and claystones of Lower Cretaceous to Upper Oligocene age, whilst in NW Palawan these consist of Lower Oligocene to Middle Miocene shales of the Pagasa Formation. Offshore Sarawak, in the Luconia Shelf, intraformational shales (Cycle V), faults and tight carbonates provide the sealing mechanism.

In addition, the marine Upper Miocene to Holocene mudstones and shales form a regional seal of uniform thickness.

Source rocks

Source rocks are principally marine and syn-rift lacustrine shales (Crestone Energy Corporation, pers. comm.), mudstones and reefal limestones in the analogue basins (Fig. 11). Palaeocene/Miocene stratified carbonates may also provide adequate source.

The results of geochemical evaluation are only available from two wells, Reed Bank A-1 and B-1, and may not be representative of the entire area. Nevertheless, these results give indications about the source potential of the area. The Lower Cretaceous section exhibits TOC values which suggest fair potential with kerogen composition being primarily gas prone (humic) with minor oil prone organic matter (sapropelic). The Palaeocene–Oligocene section constitutes the best oil and gas source potential for the area. The kerogen composition consists of mixed oil-prone and gas-prone organic matter.

In the NW Palawan shelf, source rocks are clays and shales of the Pagasa Formation and calcilutites of the Nido Formation, whilst, in SW Palawan, gas-prone Eocene transitional to shallow marine lignites and carbonaceous shales are developed, together with oil-prone Lower Miocene marine claystones. The Luconia Shelf, offshore Sarawak, the organic-rich shales of Cycles I-V provide gas-prone source rocks.

In the South Con Son basin of Vietnam, Lower Oligocene to Lower Miocene intraformational shales provide both gaseous and oil-prone source rocks. Geochemical studies by the Vietnamese (Giao *et al.* 1991) indicate that total organic carbon content increases from 0.5% to 2.5% from peripheral highs towards the depocentres in the Nam Con Son basin. The basin is characterised by Type III and Type II kerogens, the former on the highs; the latter may have existed in the troughs, as exemplified by the Dai Hung field.

In the Cuu Long and Phu Khanh basins, and the eastern part of the Nam Con Son basin covered by open Vietnamese Blocks 122–130, the main source is believed to be Oligocene shales containing organic matter of lacustrine and lagoonal origin. Possible marine ?Lower Miocene source rocks may also occur. Modelling studies over Blocks 122–130 (Nam Con Son and Cuu Long basins) suggest that the source rocks tend to be oil prone, whilst the northern area (Phu Khanh basin) may be gas prone. Results also suggest that oil may exist down to 5000 m.

In the Sarawak Shelf, the carboniferous mudstones and claystones of coastal plain facies which constitute upper Cycle I and Cycle II (Lower Oligocene–Upper Miocene) provide good source rocks.

Hydrocarbon type

The API gravity range is anticipated to be 28–52°, covering the full spectrum from light oils to condensates. However, Chinese and Russian research indicates that the Spratly Island region is likely to be gas prone (Vysotsky *et al.* 1995; Valencia 1995).

Maturation, generation and migration

The Lower Cretaceous section encountered in the Reed Bank A-1 and B-1 wells is thermally mature for hydrocarbon generation. In the Reed Bank A-1 well, sediments are thermally mature for oil generation below 2133 m. The level of thermal maturity is sufficient for gas generation in the same well and is attained below 2743 m. In the Reed Bank B-1 well significant low gravity oil generation occurs below 3353 m.

The Palaeocene–Oligocene sediments constitute the best oil and gas source rocks for the Reed Bank area. However, the section is thermally immature in both the Reed Bank A-1 and B-1 wells. This does not necessarily downgrade the source rock potential of the area as both wells were drilled on Cretaceous highs. Based on the two wells, sediments attain a level of thermal maturity for significant generation of low to medium gravity oil below 2743 m. This depth of burial can easily be attained in the troughs and half-grabens where the thickness of the section is at its maximum.

A study of vitrinite reflectivity provides maximum palaeo-temperatures of approximately 195°C in the Early Oligocene (Areshev *et al.* 1992) in the Wan'an Bei basin area. Studies by the Chinese Academia Sineca indicate that the top limit of the oil generation window in the

Wan'an Bei area lies at 3000 m. The horizon above the sedimentary Cycle V (Upper Miocene) has not yet entered the oil generation window. However, the organic-rich sequences in Cycles I–V (Oligocene–Lower Miocene) usually form good source rocks for oil and gas generation in the Sarawak Shelf. Carboniferous mudstones and claystones of a coastal plain facies from Upper Cycle I and Cycle II (Lower Oligocene – Upper Miocene) may constitute good source rocks in the Wan'an Bei area. This section entered the oil generation window during the period of rapid basin subsidence in the Early to Mid-Miocene.

In addition, Vietnamese geothermal analysis of the Nam Con Son basin, which is adjacent to the Crestone acreage, indicates that the depth, sufficient for maturation of organic matter, is 2590 m (Giao *et al.* 1991).

The main conduit for hydrocarbon migration, particularly from Oligocene source rocks, could be vertically along fractures and laterally along unconformities between the Palaeogene basement and the Oligocene/Miocene.

Geothermal gradient

The abyssal floor lies at water depths of 3.7–4.4 km. Heat flow values average 2.1 HFU in the northern part of the South China Sea and average 2.8 HFU in the southern part (Hutchison 1986).

The geothermal gradient from the Sarawak Shelf, averages 4.35°C 100 m^{-1} (Rutherford & Qureshi 1981). A study of vitrinite reflectivity by the Russians (Areshev *et al.* 1992) over the sedimentary basins offshore southern Vietnam provided maximum palaeo-temperature values of 195°C in the Early Oligocene.

The average geothermal gradient for NW Palawan is 2.95°C 100 m^{-1}.

Studies by Ru & Pigott (1986) indicate high heat flow over the southwestern quadrant of the South China Sea, which could assure a good oil generation process from the two recognized oil-generating kitchen areas, located to the north and the south of the Wan'an Bei basin. The Crestone acreage, designated WAB-21 in the Wan'an Bei basin, straddles these two areas of high heat flow, which could be favourable for the generation and migration of hydrocarbons into potential reservoirs.

Hydrocarbon occurrence

The NE and SW quadrants of the Spratly Islands archipelago are adjacent to recently established areas of oil production, in particular, the oil and gas fields of the Nam Con Son basin, offshore southern Vietnam, the East and West Natuna basins of Indonesia, the Luconia Shelf, offshore Sarawak, East Malaysia and Northwest Palawan, offshore Philippines (Table 2).

Potential reserves

By analogue with the areas discussed, this frontier region may yield considerable reserves. The range of potential reserves varies from 1–2 billion barrels to 225 billion barrels of oil equivalent.

The Chinese Ministry of Geology and Mineral Resources, following the results of numerous surveys over the Nansha Islands (Spratly Islands) archipelago during the late 1970s to early 1980s, estimated that 25 billion cubic metres of gas and 105 billion barrels of oil may be present. This estimate was revised upwards to 225 billion barrels of oil equivalent in 1994 (Hutchings 1994*a*). The Chinese surveys also stated that 91 billion barrels may be present in the vicinity of James' Shoal, off the coast of East Malaysia (Sarawak). In 1988, US geologists estimated reserves of 2.1–15.8 billion barrels (Ritcheson 1994).

Russian estimates of reserves in the Spratly Island archipelago are 7.5 billion barrels, 70% of which (5.25 billion barrels of oil equivalent) are probably gas resources leaving approximately 2.25 billion barrels oil (Vysotsky *et al.* 1995).

By contrast, the international oil industry has a less optimistic perception of potential reserves. The authors consider that published reserve estimates by both the People's Republic of China and Russia are highly optimistic and that reserves in the order of 0.5–5 billion barrels of oil equivalent may be present, and that any hydrocarbons discovered have a high probability of being gaseous.

The East West Centre Programme on Resources based in Honolulu in Hawaii assess that there is potential, but that the number and size of known structures, and the depth of water over these structures, lead them to believe that the area is not likely to become a major and gas producing area given current technology and price levels (Valencia 1995). They do concede that a couple of commercial accumulations of modest size may be discovered, but the chance of big-league reserves is quite low. Although the seismic data over the Crestone blocks is of good quality and illustrates good structuring, the East West Centre considers that the area is an extension of the Vietnamese shelf/slope.

The East West Centre states that the cost of drilling exploration wells can exceed US$25 million in some areas of the South China Sea due to severe weather conditions, complex geology and difficult drilling conditions. They also assess that any potential development could exceed US$1 billion. This, they argue, raises the minimum commercial field size and reduces the number of exploration targets. Their analysts believe that more than 90% of the commercial oil and gas reserves in the South China Sea will be discovered on the shelves and slopes of the surrounding states, i.e. Malaysia, Vietnam, Philippines, Brunei.

Influences on regional stability

The primary influence on regional stability in the South China Sea which may impact on its littoral states is the foreign policy and power projection of the People's Republic of China (PRC). In February 1992, China enacted a law to enable it to exercise sovereignty over its Territorial Sea and Contiguous Zone which were defined in relation to the country's land territory. This law, in essence, laid claim to the whole of the South China Sea (Fig. 1). This sweeping annexation encompasses some of the most prolific hydrocarbon provinces along the South China Sea littoral. This, coupled with the many competing claims to sovereignty, particularly in the region of the Spratly Islands, may threaten the exploration and exploitation and subsequent development of the hydrocarbon resources of the region.

This has prompted a reaction from the claimant and other littoral states (Hutchings 1994*b*), not only to enforce their own claims, but also to embark on an arms build-up to counter any perceived threat (Schofield 1994).

Implications for regional development

The successful exploitation and development of the hydrocarbon resources of the Spratly Islands archipelago to the mutual benefit of the claimant states can only be undertaken in an atmosphere of regional security and mutual respect via the ASEAN Regional Forum (ARF) to promote:

(i) demilitarization of the area;
(ii) settlement of boundary disputes;
(iii) creation of joint development areas.

Demilitarization of the area

Bilateral agreements may give hope for ultimate peaceful resolution of the Spratly Islands dispute, but the complexity of the issue as it now stands will not allow much progress through such negotiations.

Peaceful resolution would require the demilitarization of the Spratlys. It would also require the freezing of all claims, and demand of the claimants what, under the present circumstances, would be an extraordinary political will to settle the issue.

A military resolution is unfortunately a distinct possibility. China is the only claimant with enough military assets to attempt such an action.

As demonstrated by incidents in the Paracels (1974) and Mischief Reef (1995), China has shown little aversion to demonstrating its claim militarily. Such a move, regardless of peaceful overtures to other claimants would, however, certainly foul China's effort to curry favour with ASEAN. More critically, it could also imperil the strategic interests of extra regional powers, i.e. USA, Japan, and run the risk of inviting their intervention.

Settlement of boundary disputes

After withdrawing all military forces from the islands, the claimants could agree to negotiate a maritime subdivision of the South China Sea (Figs 1, 8 and 9) (Valencia 1995). Such a subdivision could be made by China, Vietnam and the Philippines, while agreeing to respect the EEZ boundary already declared by Malaysia. If based on median lines only, this scheme would exclude China, and Taiwan, from the Spratly Islands, and thus from the more likely area of resource occurrence (Dzurek 1995). A delimitation based on equitable principles, taking into account historic claims, would probably be the only one with a chance of success.

Although bilateral treaties would serve to fix segments of the sub-division, the process of necessity would have to be a multilateral one. However, China (PRC) has stated that it will only negotiate bilaterally. Taiwan announced on 11 May 1995 that it was willing to take part in international talks on the Spratly Islands. It further stated that it would co-operate fully with claimant states to avoid triggering potential conflicts in the Spratly Islands, but it added that its territorial claims would still stand.

Not only would agreement with third countries be needed on the tripoints, but all parties to the South China Sea dispute would have to agree to accept this scheme and the bases for delimitation to be used (Thomas 1989).

Thus the claimant states would undertake to resolve boundary dispute problems via negotiation and peaceful means, as per the Manila Declaration of July 1992.

Creation of joint development areas (JDAs)

The claimants could agree to share the resources of the sea in a co-operative regime (Valencia 1988). All claims and military positions could be frozen, and a multi-lateral Spratly Management Authority established to administer exploration and development of resources (Valencia 1995). Such a treaty would also guarantee passage of vessels and would promote international co-operation in scientific research, probably through the United Nations. Strategic powers, such as the US and Japan, might be admitted as associate members, as well as marginally located Indonesia, which clearly has interests in the area. Allowance would also have to be made for the membership of Brunei (which, up to now, has not occupied any of the Spratly Islands) (Thomas 1989).

The precedent has been set for such undertakings via the recently ratified Thai/Malaysian Joint Development Area, the Malaysia/Vietnam agreement for exploration and exploitation of the South China Sea, the Thai/Vietnamese demarcation and joint development of resources in the Gulf of Thailand. Recent (Hanoi) press reports would indicate that China and Vietnam have already begun the dialogue to settle their boundary disputes regarding the Spratly archipelago, with particular reference to joint development of disputed areas.

A model for negotiation could be the accord between Indonesia and Australia on a Joint Development Area for oil and gas exploration in the Timor Sea. In this agreement, there was no definition of international boundaries, no mention of sovereignty; it was pragmatic and sensible apportionment of commercial resources (Cloughley 1995).

Joint development is clearly not the optimal or permanent solution to the problem of unresolved boundaries. However, in some situations it may be the only alternative to no action and, thus, no hydrocarbon development – or worse, to confrontation and conflict (Morgan & Valencia 1983; Valencia 1985, 1988, 1995).

Conclusions

The Spratly Islands archipelago and Dangerous Grounds area of the South China Sea will become the focus of exploration for hydrocarbons over the next two decades, offering a diversity of hydrocarbon plays. However, first the multi-national boundary disputes need to be resolved by the claimant states, hopefully by negotiation and peaceful means.

References

Areshev, E. G., Tran Le Dong, Ngo Thuong San & Shnip, O. A. 1992. Reservoirs in fractured basement on the continental shelf of Southern Vietnam. *Journal of Petroleum Geology*, **15**, 451–464

Blanche, J. B. & Blanche, J. D. 1995*a*. *The Spratly Islands archipelago: influence on regional stability in the South China Sea*. Non-Exclusive Report.

—— & ——1995*b*. Chinese bureaucrats draw the line in South China Sea. *Petroleum Economics*, **62**, 16–18

Cloughley, B. 1995. No need for war in South China Sea. *International Defense Review*, **28**, 22–25.

Daly, M. C., Cooper, M. A., Wilson, I., Smith, D. G. & Hooper, B. G. D. 1991. Cenozoic plate tectonics and basin evolution in Indonesia. *Marine and Petroleum Geology*, **8**, 2–21.

Du Bois, E. P. 1985. Review of principal hydrocarbon-bearing basins around the South China Sea. *Bulletin of the Geological Society of Malaysia*, **18**, 167–209.

Dzurek, D. J. 1995. China occupies Mischief Reef in latest Spratly gambit. *Boundary and Security Bulletin*, **3**, Apr, Univ. of Durham, England, 65–71.

Emery, K. O. & Ben-Avraham, Z. 1972. Structure and stratigraphy of China Basin. *American Association of Petroleum Geologists Bulletin*, **56**, 839–859.

Fan Pow-Foong 1992. Geology of Spratly Islands and vicinity. (abs). *Symposium on Tectonic Framework and Energy Resources of the Western Margin of the Pacific Basin*, Circum-Pacific Conf, Kuala Lumpur, Nov–Dec.

Giao, N., Tin, N. T. & Dang, H. N. 1991. *Petroleum Geology of the South Con Son Basin*. Vietnam Petroleum Institute, Ho Chi Minh City.

Hinz, K. & Schluter, H. U. 1985. Geology of the Dangerous Grounds, South China Sea, and the continental margin off South West Palawan: results of Sonne Cruises SO-23 and SO-27. *Energy*, **10**, 297–315.

Hirayama, J. (ed.) 1991. Total sedimentary isopach maps offshore East Asia, with basin descriptions. *CCOP Technical Bulletin*, **23**, 1–114.

Holloway, N. H. 1982. North Palawan Block, Philippines – its relation to Asian mainland and role in evolution of South China Sea. *American Association on Petroleum Geologists Bulletin*, **66**, 1355–1383.

Hutchings, G. 1994*a*. China's hopes for islands oil bonanza. *Daily Telegraph*, 7 September.

——1994*b*. China claim to the sea alarms its neighbours. *Daily Telegraph*, 29 July.

HUTCHINSON, C. S. 1986. Formation of marginal seas in Southeast Asia by rifting of the Chinese and Australian margins and implications for the Borneo region. *In*: *Geosea V Proceedings* **II**. *Geological Society of Malaysia Bulletin*, **20**, 201–220.

JIANG, S. & ZHOU, X. 1994. *Seismic Sequence, Geotectonics and Petroleum Geology in the Nansha Region*. South China Sea Institute of Oceanography, Academia Sineca

KATILI, J. A. 1981. Geology of Southeast Asia with particular reference to the South China Sea. *Energy*, **6**, 1077–1091.

KUDRASS, H. R., WIEDICKE, M., CEPEK, P., KRUZER, H. & MULLER, P. 1986. Mesozoic and Cenozoic rocks dredged from the South China Sea (Reed Bank Area) and Sulu Sea and their significance for plate tectonic reconstructions. *Marine and Petroleum Geology*, **3**, 19–30.

LEE, T. Y. & LAWVER, L. A. 1992. Tectonic evolution of the South China Sea region. *Journal of the Geological Society of China*, **35**, 353–388.

LETOUZEY, J., SAGE, L. & MULLER, C. 1988. *Geological and Structural Map of Eastern Asia*. Scale 1:2,500,000. *American Association of Petroleum Geology*, Tulsa.

LI, Y. 1988. On the synchronism of extensional subsidence of the China South Sea basin and compressional uplift of Qinghai-Xizang plateau. *SEAPEX Proceedings*, **8**, 92–99.

LIANG DEHUA & LIU ZONGHUI. 1990. The genesis of the South China Sea and its hydrocarbon-bearing basins. *Journal of Petroleum Geology*, **13**, 59–70.

LIU, Z. & YANG, S. 1988. [The tectonics of the South China Sea and spreading of the continental margins]. South China Sea Institute of Oceanology, Academia Sineca [in Chinese].

MORGAN, J. R. & VALENCIA, M. J. 1983. *Atlas for maritime policy in Southeast Asian seas*. Univ. of California Press.

PUPILLI, M. 1973. Geological evolution of South China Sea area: tentative reconstruction from borderland geology and well data. *Proceedings of the Indonesian Petroleum Association.*

QIU YAN. 1995. Depositional architecture of Wa'an Basin, South China Sea (abs). *Oil and Gas Asia Conference*, Manila, Jan.

QUINTON, N. 1993. Hydrocarbon potential of the Reed Bank Area. *In*: *Proceedings of Oil and Gas Asia '93*, Manila, Nov.

RITCHESON, P. L. 1994. China's impact on South East Asian security. *Military Review*, **74**, May.

ROBERTS, D. G. 1988. Basin evolution and hydrocarbon exploration in the South China Sea. *In*: WAGNER, H. C., WAGNER, L. C., WANG, F. F. H. & WONG, F. L. (eds) *Petroleum Resources of China and Related Subjects*. Circum-Pacific Council for Energy and Min. Res., Earth Science Series, **X**, 157–177.

RU KE & PIGOTT, J. D. 1986. Episodic rifting and subsidence in the South China Sea. *American Association of Petroleum Geologists, Bulletin*, **70**, 1136–1155.

RUTHERFORD, K. J. & QURESHI, M. K. 1981. *Geothermal Gradient Map of Southeast Asia*. SEAPEX/IPA.

SCHOFIELD, C. 1994. An arms race in the South China Sea? *Boundary and Security Bull*, **2**, Univ. of Durham, England, 39–48.

TAYLOR, B. & HAYES, D. E. 1983. Origin and history of the South China Sea Basin. *In*: HAYES, D. E. (ed.) *The Tectonic and Geological Evolution of South East Asian Seas and Islands*, Pt. II. AGU Geophysical Monograph Series, **27**, 23–56.

THOMAS, B. L. 1989. The Spratly Islands imbroglio: a tangled web of conflict. *In*: *International Boundaries and Boundary Conflict Resolution*. Proceedings of IBRU Conference, Univ. of Durham, England, 413–428.

VALENCIA, M. J. 1985. *South East Asian seas:* Oil Under Troubled Waters. Oxford University Press.

——1988. The Spratly Islands: dangerous ground in South China Sea. *Pacific Review*, **1**, 382–395.

——1995. China and the South China Sea disputes. *International Institute of Strategic Studies Adelphi Paper*, **298**. Oxford University Press, 3–75.

VYSOTSKY, V. I., RODNIKOVA, R. D., SHLEFER, V. M. & LARKOVA, T. N. 1995. The Spratly Islands region: petroleum potential (abs). *Oil and Gas Asia Conference*, Manila, Jan.

XIA KANG-YUAN & ZHOU DI 1992. The geophysical characteristics and evolution of the northern and southern margins of South China Sea (abs). *Symposium on Tectonic Framework and Energy Resources of the Western Margin of the Pacific Basin*. Circum-Pacific Conf, Kuala Lumpur, Nov–Dec.

The tectonostratigraphic evolution of SE Asia

IAN M. LONGLEY

Woodside Offshore Petroleum Pty Ltd, 1 Adelaide Terrace, Perth, Western Australia 6000, Australia (e-mail: ian.longley@woodside.com.au)

Abstract: A new model for the Tertiary tectonic and stratigraphic evolution of SE Asia is presented integrating stratigraphic data from many of the major sedimentary basins in the area. The model can be divided into four phases.

(1) Stage I (50–43.5 Ma) during which the India–Eurasia collision was initiated and proceeded contemporaneously with oceanic subduction beneath southern Eurasia. The continental collision caused a slow-down in the oceanic spreading rates in the Indian Ocean reducing the convergence velocity along the Sunda Arc subduction system and resulting in a phase of extension in the adjacent fore-arc and back arc areas. The isolated rift basins in the fore-arc and the East Java area were filled with transgressive then open marine sediments since these basins were on the low-lying edge of the Sunda craton whereas the mid-Eocene sea was not able to penetrate in to the more cratonic backarc rift basins of Sumatra or NW Java which were filled with fluvio-lacustrine sequences.

(2) Stage II (43.5–32 Ma) was triggered by the termination of oceanic subduction beneath the India-Eurasia collision zone. This locked up the spreading system in the Indian Ocean and caused a major plate reorganization effective in the Indian, Southern and Pacific oceans. The plate reorganization in the Indian Ocean slowed convergence rates yet again along the Sunda Arc producing a second phase of rifting and rift basin deposits. The plate reorganization in the Pacific Ocean resulted in the onset of extension in the South China Sea and the deposition of fluvio-lacustrine sequences in the isolated rift basins whereas in the East Kalimantan area a failed rift system in the Makassar Straits resulted in isolated rift basins filled with deltaic and marine sequences which were overlain by more extensive post-rift marine shales. The first major collision of the Luconia Shoals block with a subduction system along the NW Borneo margin resulted in the deposition Balingian delta sequence.

(3) Stage III (32–21 Ma) is contemporaneous with the first phase of seafloor spreading in the South China Sea during which the entire continental area south of a line bisecting and joining the South China Sea and the Gulf of Thailand rotated clockwise about a pole near the head of the Gulf of Thailand. This block rotation formed the Malay Basin within which a large fluvio-lacustine sequence was deposited. The rotation also produced a phase of increased plate convergence rates and inversion along the Sunda Arc ending the rifting in these basins. The subsequent sedimentary fill in these basins records a marine transgression which is interpreted to be partially eustatic in origin and partly a result of the subsidence associated with the onset of post-rift thermal sag phases. Elsewhere in East Kalimantan extensive marine and carbonate sequences were deposited whilst in the South China sea the incursion of marine environments led to the deposition of inboard transgressive deltaic and marine deposits and marine and carbonate deposits offshore in more distal settings.

(4) Stage IV (21–0 Ma) was initiated by the cessation of the first phase of seafloor spreading in the South China Sea caused by the collsion of the Baram block with the NW Borneo subduction system. This event is interpreted to be contemporaneous with the onset of continental shortening in Tibet via block rotation and lateral extrusion along strike-slip faults. Major collisions, an inevitable result of the particular configuration established by the plate-reorganization at 43.5 Ma, occurred throughout this stage in the NW Borneo, Sulawesi and Timor areas and together with the rotation of Sumatra, and extrusion-related wrench faulting in the Malay–West Natuna and Bac Bo/Yinggehai Basin areas resulted in a extensive structural invertsion and highly variable basin fill sequences. A eustatic high in the middle Miocene marks a period of maximum marine transgression onto the Sunda craton and the deposition of extensive marine shale deposits. This was followed by dominantly regressive sequences puctuated by major eustatic sea-level falls at about 5 and 10 Ma.

Particular structural styles and eustatic models are discussed in relation to the proposed model and the distribution of hydrocarbons in the Tertiary basins of southeast Asia is presented. This distribution is explained in terms of the different types of source rocks present in the different basins.

This paper describes a model for the Tertiary tectonic and stratigraphic evolution of Southeast Asia and explains the distribution of the discovered hydrocarbons in the region based on the presence of different source rock types within the different Tertiary sedimentary basins

From Fraser, A. J., Matthews, S. J. & Murphy, R. W. (eds), 1997, *Petroleum Geology of Southeast Asia*, Geological Society Special Publication No. 126, pp. 311–339.

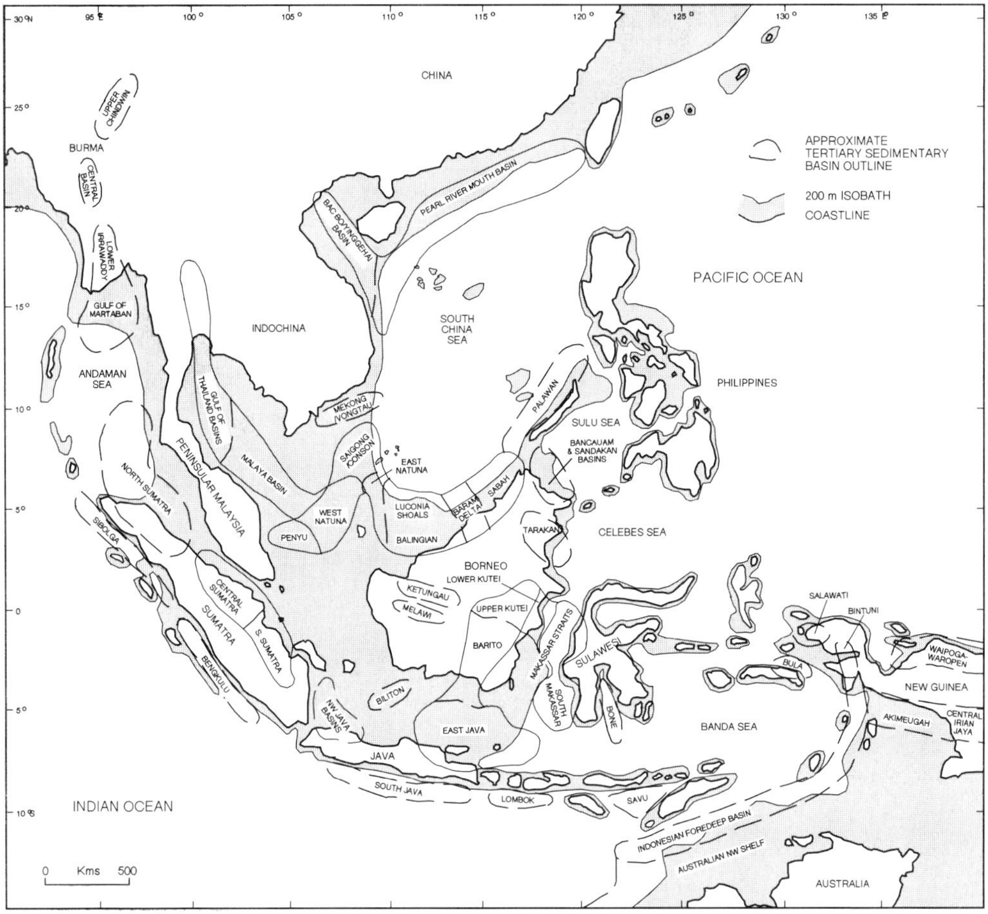

Fig. 1. A simplified map of the main Tertiary basins in SE Asia (modified after Daly *et al.*,1991 and Hutchison, 1989).

contained within the study area (Fig. 1). These basins contain estimated total oil reserves of between 31 and (Shaw 1990) and 35 billion bbls (Nation 1994) and estimated total gas reserves of 281 TCF (Nation 1994).

Previous studies

Previous regional tectonic syntheses of the Tertiary evolution of SE Asia can be split into two types. The first are those such as Briais *et al.* (1993) which are based on the extrusion (or escape) tectonics model of Tapponier *et al.* (1982, 1986) which suggest that the Eocene Indian continental collision and penetration into Asia has rotated and extruded large crustal blocks to the east along major strike-slip faults as summarized on Fig. 2. The second type includes syntheses such as those by Rangin *et al.* (1990) and Daly *et al.* (1991) which place greater emphasis on kinematic modelling of numerous microplates constrained by the present plate motions, marine magnetic anomalies and selective geological data.

None of the above studies integrates the volumous geological data from the major sedimentary basins in the area principally documented in the oil and gas industry literature. This paper attempts to fill this void and the resulting model is an attempt to explain why the disparate sedimentary basins in the region often contain similar sequences bounded by similar-aged unconformities.

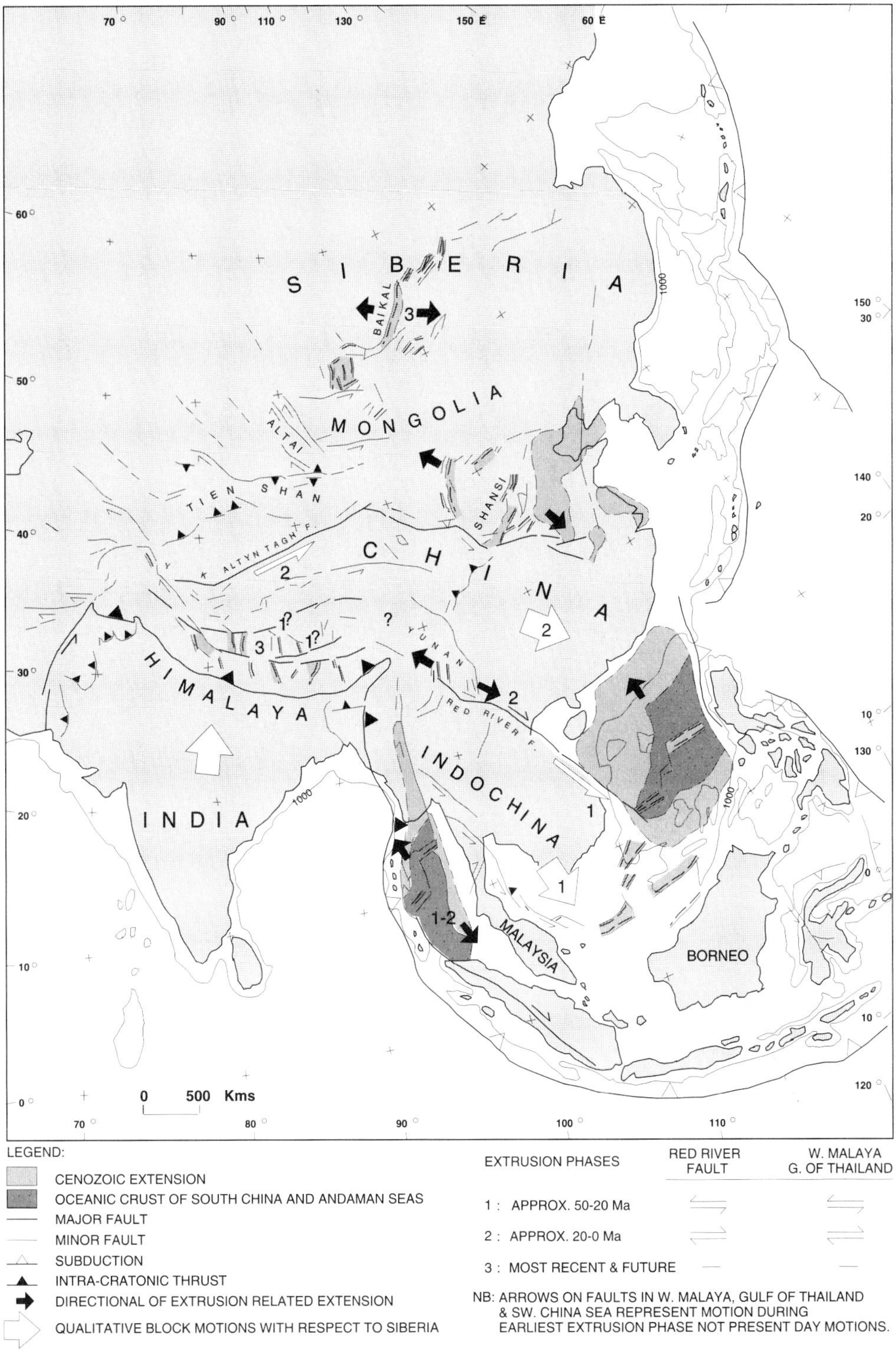

Fig. 2. Summary of the Tapponnier *et al.* (1982) extrusion tectonics model.

Proposed tectonic model

The major tectonic elements of the study area, as best described by Hutchison (1989) and Hamilton (1979, 1988), are shown on Fig. 3 and the interpreted constituent continental microplates (herein called 'blocks') and plates together with the nomenclature system used in this paper for describing combinations of these plates are shown on Fig. 4. As is evident from these figures the SE Asian region contains a complex of minor oceanic and continental plates sandwiched between the relatively stable Eurasian continental plate to the northwest, the rapidly northward moving Australian (and Indian Ocean) plate to the south and the convergent Philippine and Pacific Ocean plates to the east. These plate motions have produced active subduction and arc/trench systems along the southwest and northeastern flanks of SE Asia which are known as the Sunda and Philippine arc systems respectively. This simple structural picture is complicated in the southeastern portion of the study area where the Australian continent has collided with the northwesterly moving Philippine plate, producing a left-lateral strike-slip margin from which micro-continental blocks have been shaved off the top of the Australian plate and carried westwards.

The oceanic palaeomagnetic record in the region is shown on Fig. 3 and is summarized chronostratigraphically on Fig. 5. The marginal seas of SE Asia (Fig. 4) include, the Andaman, Banda, Celebes, Sulu and South China Seas (Fig. 1) plus the undated possible oceanic crust beneath 5 km of sediments in the Makassar Straits; the presence of which is based on a strong positive gravity anomaly as described by Hutchison (1989). The Miocene to present-day oceanic spreading history of the Andaman Sea is described by Curray *et al.* (1979) and the interpreted Cretaceous spreading history of the Banda Sea follows the interpretations of Lee & McCabe (1986) (Fig. 5). In the Celebes and Sulu Sea areas the more recent DSDP drilling data Silver *et al.* (1991) supports a mid-Eocene interpretation of the magnetic anomalies in the Celebes Sea by Weissel (1980) and is equivocal about the age of the Sulu Sea. As described by Rangin & Silver (1991) the available data constrain the age of the Sulu Sea to be older than 19 Ma (Early Miocene). The tentative interpretation of the Sulu Sea magnetic anomalies as 30–10 Ma by Roeser (1991) is not utilized here. Instead the interpretation is that the seafloor is of Late Eocene to Earliest Oligocene age (38–35 Ma).

The interpretation of marine magnetic anomalies in the main Eastern Sub-basin and smaller Northwestern Sub-basin areas of the South China Sea follows the Oligocene to Early Miocene interpretations of Taylor & Hayes (1983) and Briais *et al.* (1993) (Figs 4 & 5). Based on the work of the Pautot *et al.* (1986) the central portion of the main Eastern sub-basin is interpreted to represent a more restricted second phase of early to mid-Miocene spreading consistent with the dating of dredged volcanics (10–15 Ma) described by Hekinian *et al.* (1989). The Southwestern Sub-basin of the South China Sea contains non-diagnostic magnetic anomalies (Briais *et al.* 1993) as is evident by the variation in ascribed ages by Ru & Pigott (1986), Hayes & Spangler (1987), Liang & Liu (1990), Daly *et al.* (1991) and Briais *et al.* (1993), which range in age from early Cretaceous to mid-Miocene. It is interpreted here (Figs 4 & 5) to be a continuation of the earlier spreading system in the Eastern and Northwestern Sub-basins.

The spreading history in the adjacent Indian and Pacific Ocean areas is also summarized on Figs 4 & 5. More detail concerning the history of spreading in the Indian Ocean is shown on Figs 6 based on Veevers *et al.* (1991) and Veevers & Li (1991). The pre-Tertiary spreading history of this area relates directly to the fragmentation of the Gondwana supercontinent and documents the rifting of continental fragments from the Australian plate in the Jurassic, the separation of India and Australia in the Early Cretaceous and the separation of Australia and Antarctica in the Late Cretaceous. In the Pacific region magnetic anomalies in the Philippine Plate utilized here are after Hilde & Lee (1984) and the history of the Pacific Plate and the region in general follows that described by Hilde *et al.* (1977).

The synthesized stratigraphic data for nineteen sedimentary basins around SE Asia are shown as chronostratigraphic charts on Figs 8, 9 and 10 and a legend for these figures and the subsequent plate reconstruction maps is shown on Fig. 7. The 19 basins selected encompass most of the main hydrocarbon-bearing basins in the region and were selected from the 72 basins identified by Hutchison (1989) and the 97 basins identified by Shaw (1990) within the study area. The stratigaphic nomenclature and original source data for these summaries are detailed in the Appendix. One element highlighted on these figures is the interpreted level of the oldest reliable biostratigraphic age which typically corresponds either to the base of the deepest drilled section or to the base of the section with a marine influence since non-marine sequences are

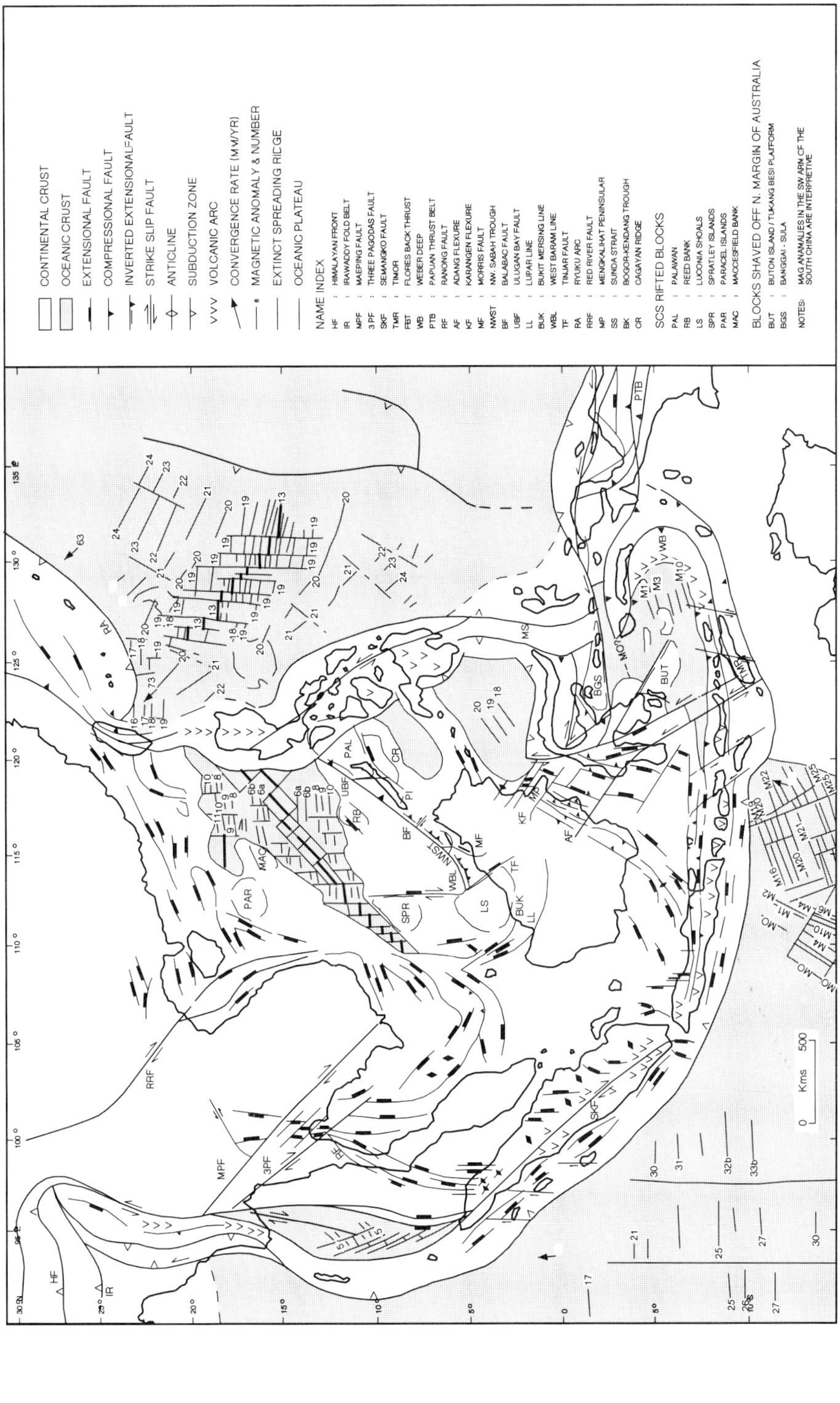

Fig. 3. SE Asia tectonics element map.

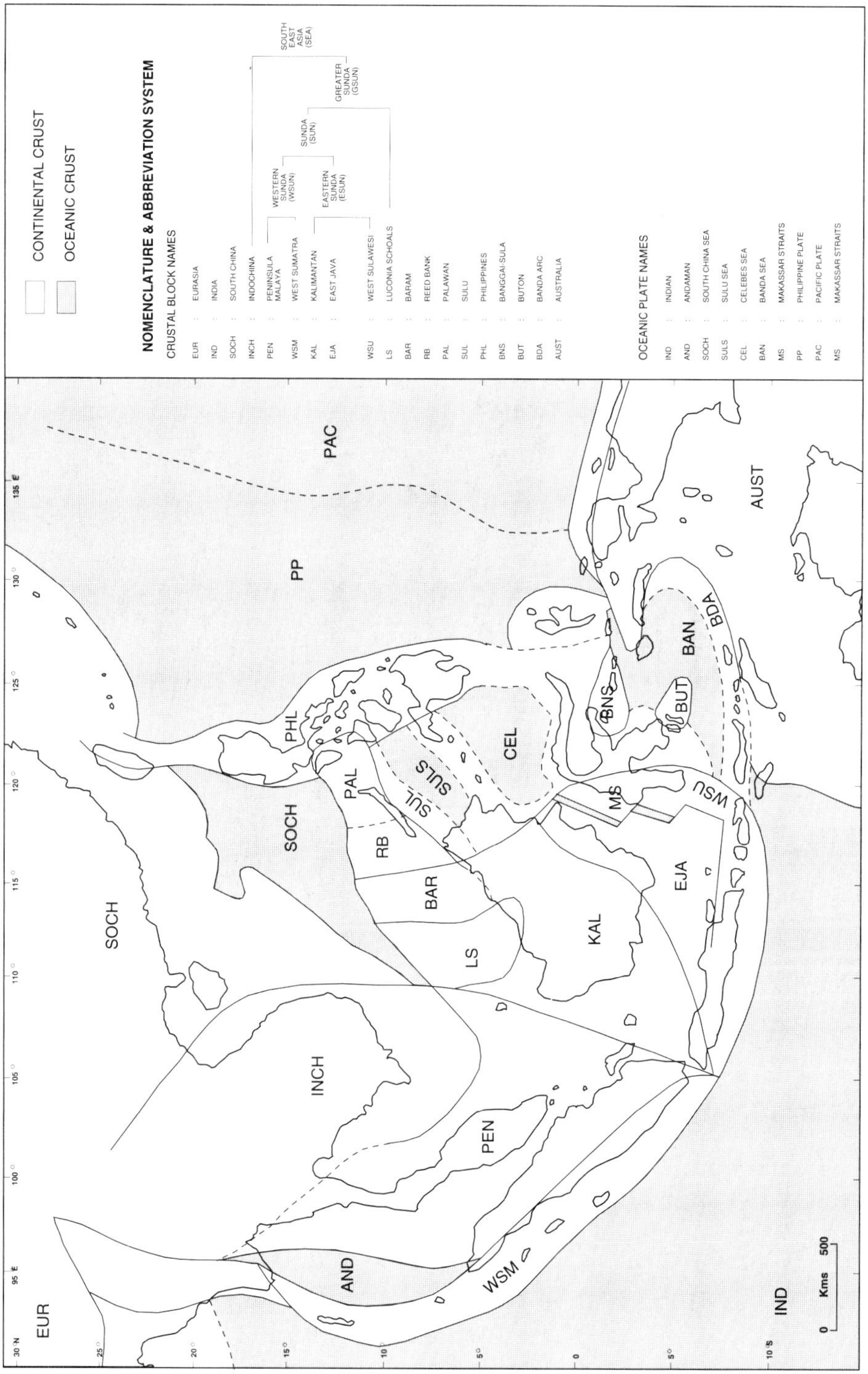

Fig. 4. Simplified block areas and interpreted block boundaries used in the tectonic model.

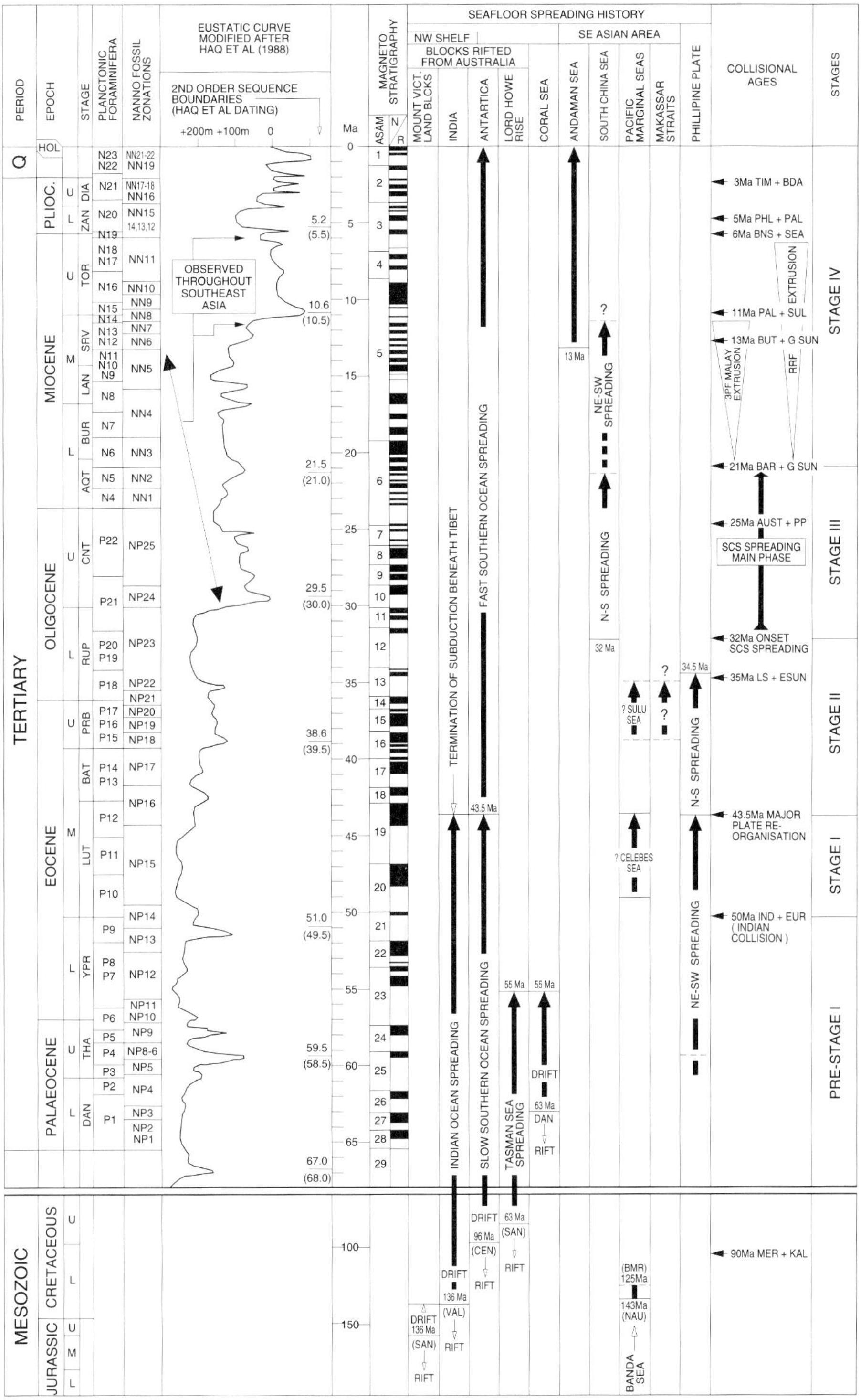

Fig. 5. Chronostratigraphic summary of key SE Asian geological events (timescale after Harland *et al.* 1990).

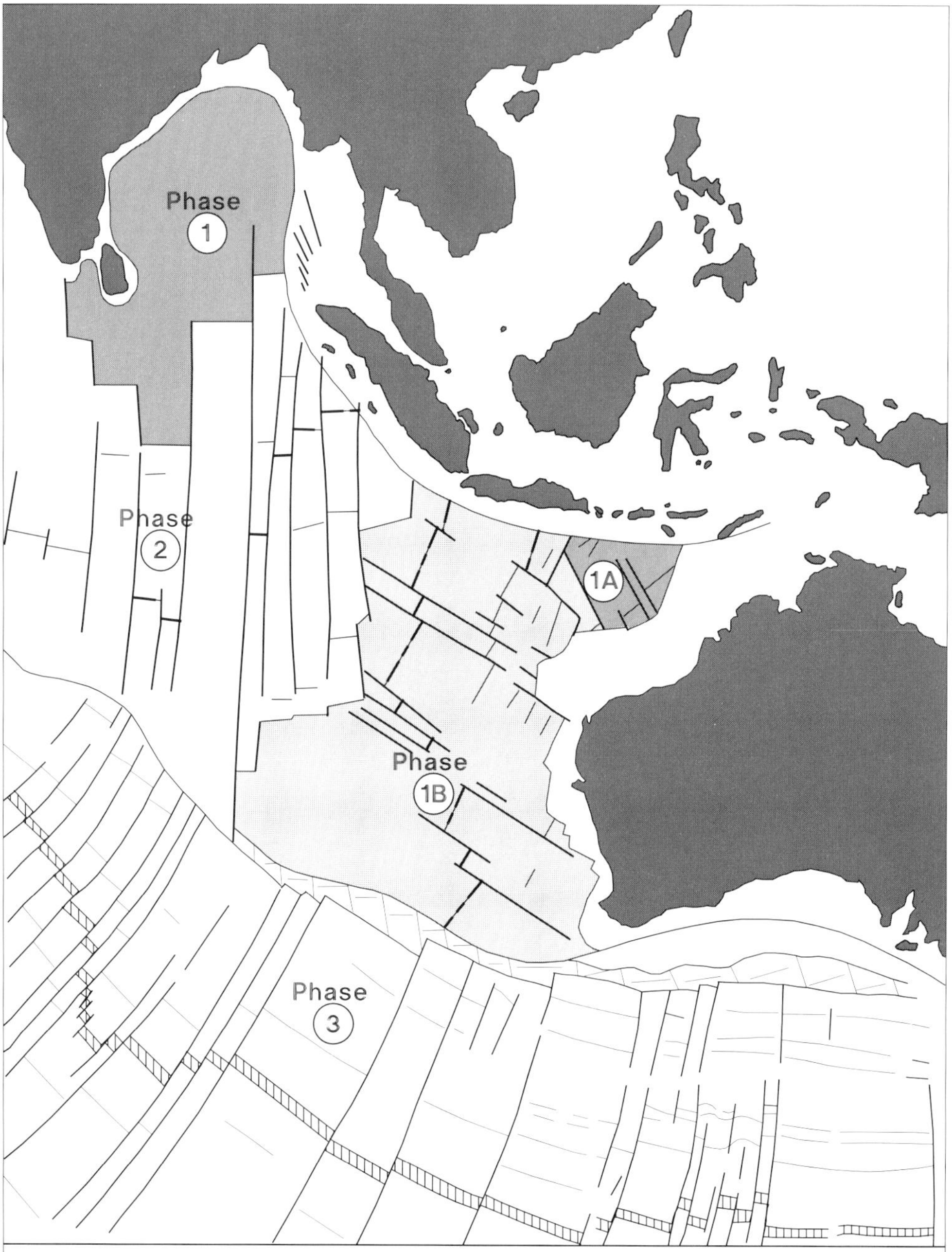

PHASE (Rifted Blocks)	MAG ANOMALY RANGE	AGE (Ma)		
1A (MVL BLOCKS)	M 26 - M 11	155 - 136	LATE JURASSIC (OXF)	- EARLY CRET (VAL)
1B (INDIA)	M 10 - M 34	136 - 96	EARLY CRET (VAL)	- EARLY CRET (CEN)
2 (ANTARCTICA)	A 34 - A 20	96 - 43.5	EARLY LATE CRET (CEN)	- MIDDLE EOCENE
3 -	A 20 - PRESENT	43.5 - 0	MIDDLE EOCENE	- PRESENT

Abandoned Spreading Ridge; Active Spreading Ridge; Fracture zone; Magnetic Anomaly

Fig. 6. Summary of Indian Ocean seafloor-spreading history (modified after Packham 1990).

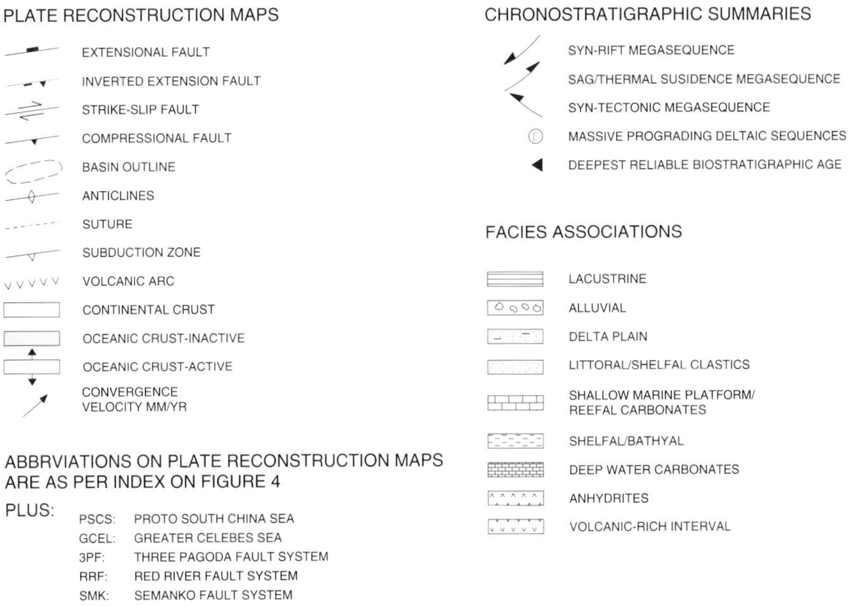

Fig. 7. Key to plate reconstruction maps and chronostratigraphic summaries.

difficult to date in SE Asia due to long ranging spore-pollen biozones. Below these marked ages the stratigraphic sequence and their ages are purely interpretive or at best weakly constrained by radiometric or other dating techniques.

The plate tectonic model presented below assumes that within-plate basin histories relate directly to plate margin activity and that the effects of plate margin processes can be transmitted great distances as described by Ziegler (1992). In addition the extensional or compressional nature of arc systems can be related to the relative velocities of the trench 'rollback velocity' and the overriding plate as described by Dewey (1981) and Daly *et al.* (1991). As summarized on Fig. 11, it follows that an extensional arc or a phase of fore-arc and back arc extension can be caused either by a relative decrease in the rate of convergence or a relative decrease in the convergence velocity of the overriding plate. Similarly a compressional arc or a phase of fore-arc and back arc compressional tectonism can be caused by a relative increase in the convergence velocity of the subducting or overriding plates. This model assumes there is little variability in the age and consequently the density of the subducted oceanic crust, as is, and probably was, the case along the Sunda Arc system throughout the Tertiary.

The plate-tectonic model

The model for the Tertiary plate tectonic evolution of SE Asia presented here is related to the direct and indirect effects resulting from the Indian collision with Eurasia. The Model is grouped into five different time intervals (or 'stages') comprising a description of the situation prior to the India–Eurasia collision (Pre-I Stage) and four stages which are contemporaneous with the collision (Stages I–IV) separated by periods of fundamental changes to the plate tectonic regime across the region. Each of these stages is described in detail below utilising a total of eleven plate reconstruction maps which are modifications of those presented by Daly *et al.* (1991). These maps are not palinspastically restored or geometrically rigorous and should be considered as schematic sketch maps.

Pre-Stage I (pre-50 Ma)

The plate-tectonic situation prior to the India–Eurasia collision is poorly known and very speculative. Immediately prior to 90 Ma the interpreted plate tectonic setting (Fig. 12) comprised the Indochina, South China and the

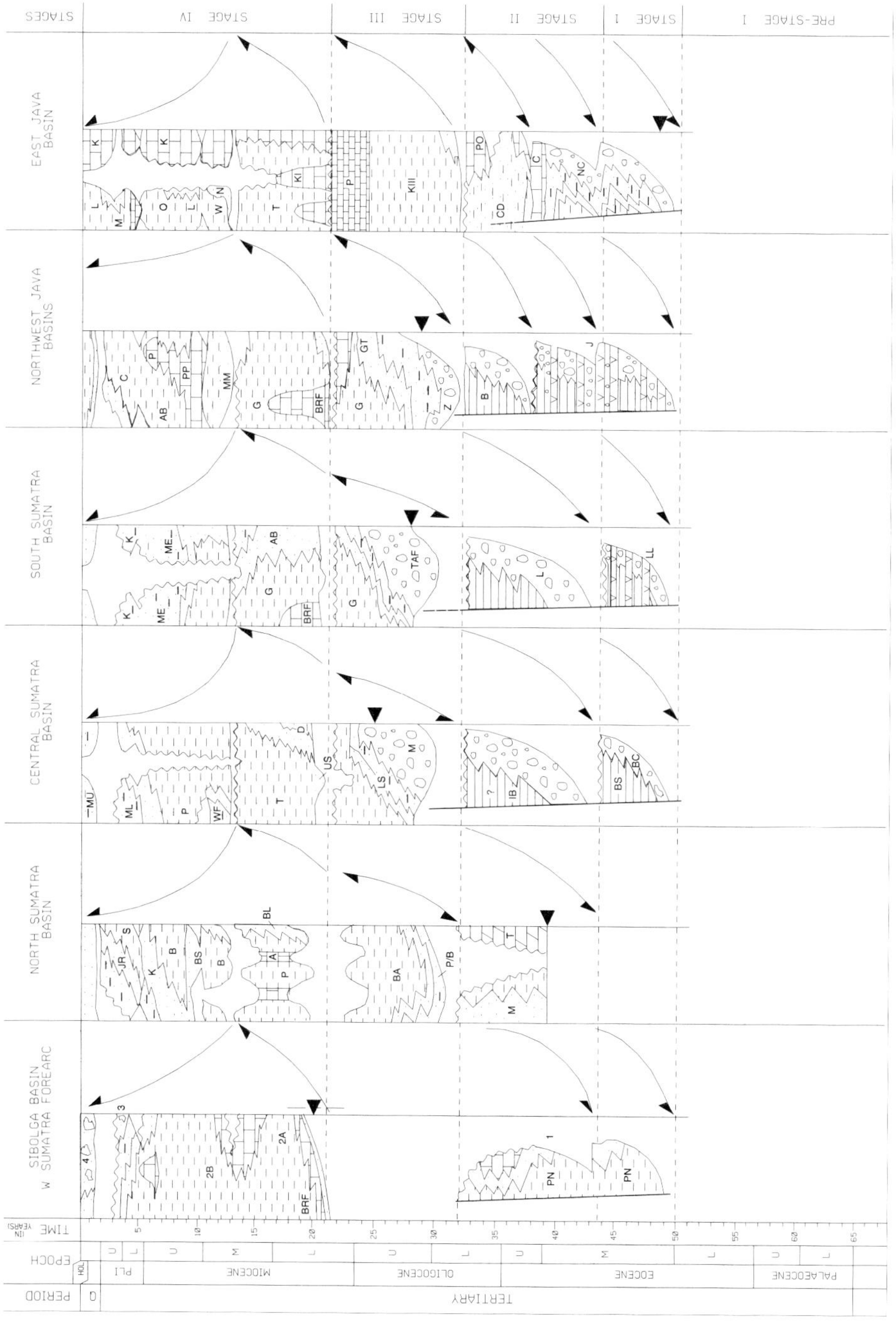

Fig. 8. Sunda Arc basins stratigraphic summary (see Fig. 7 for key).

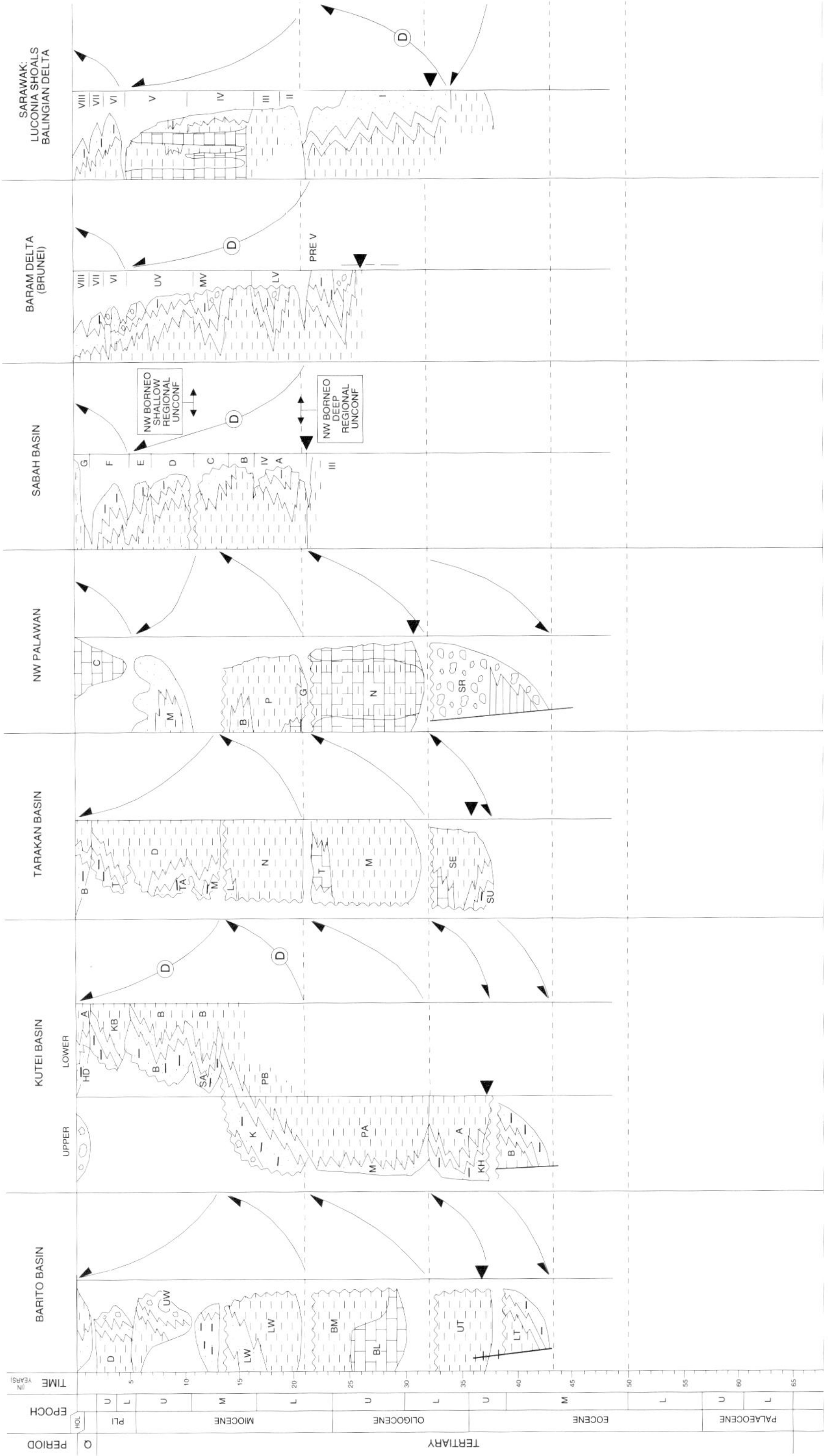

Fig. 9. East Kalimantan and NW Borneo areas stratigraphic summary (see Fig. 7 for key).

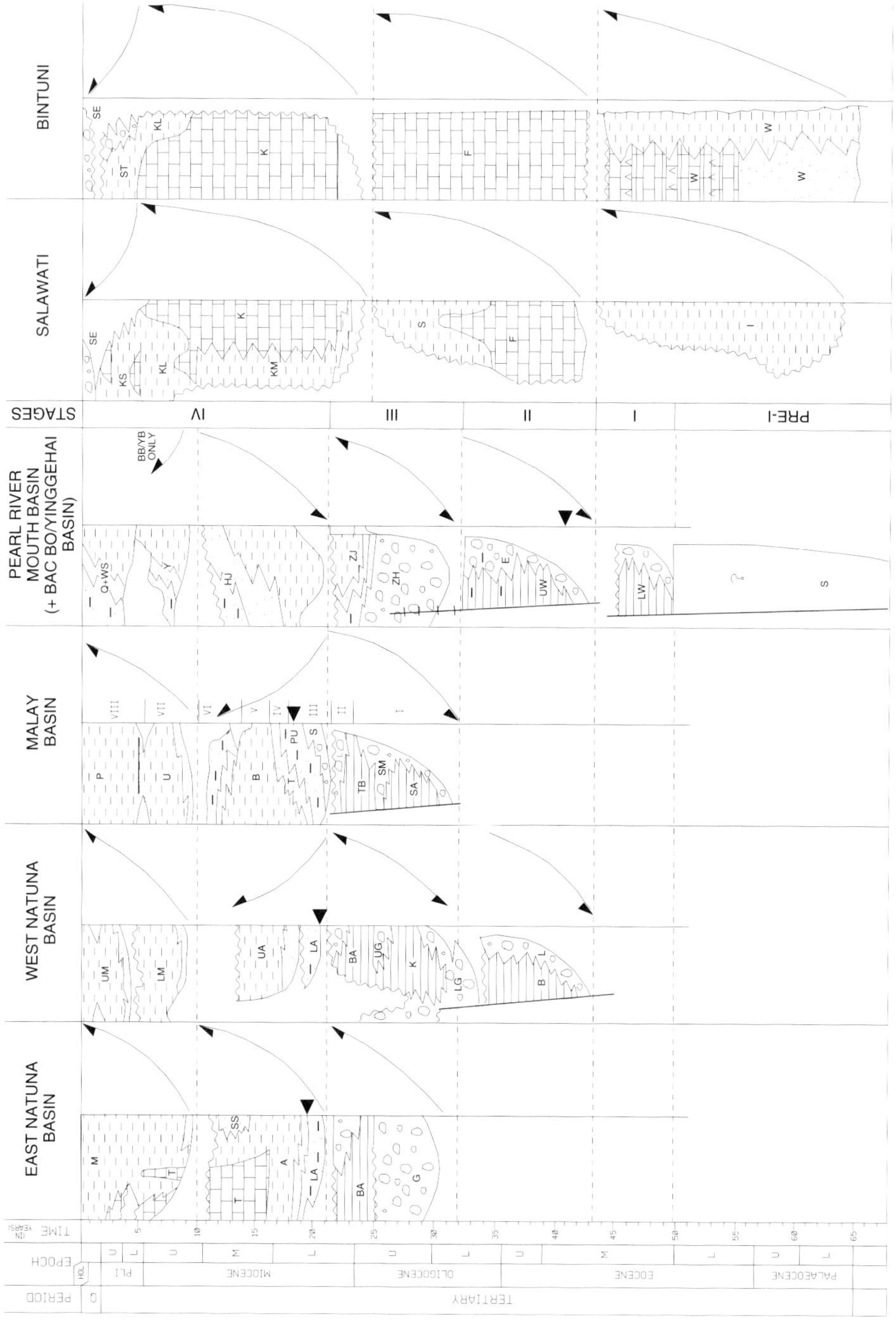

Fig. 10. Natuna–Malay, South China Sea and Birdshead area stratigraphic summary (see Fig. 7 for key).

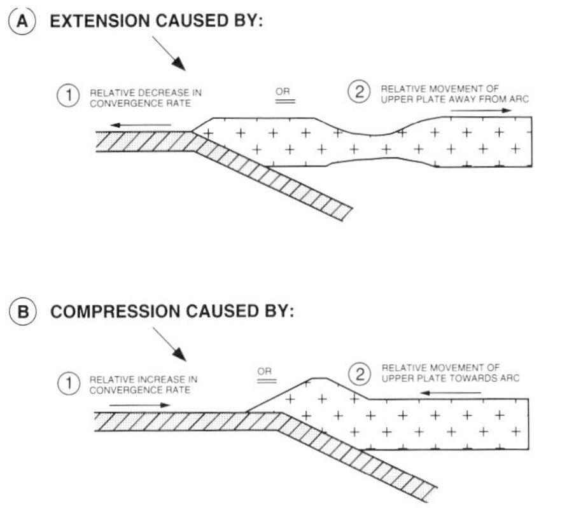

Fig. 11. (top) Schematic summary of the kinematics of subduction rollback extension and compressional events.

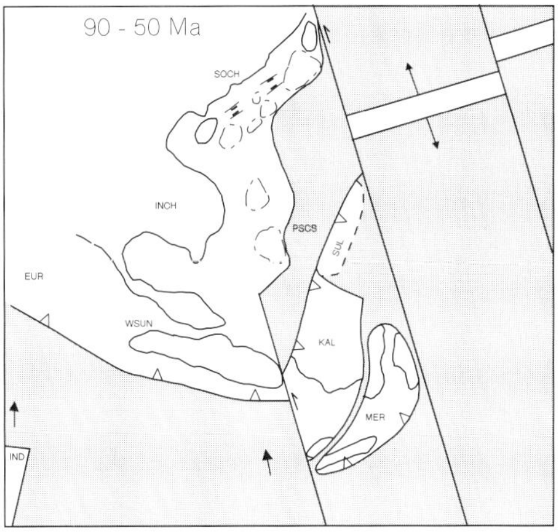

Fig. 13. (top) 90–50 Ma plate reconstruction map (see Fig. 7 for key).

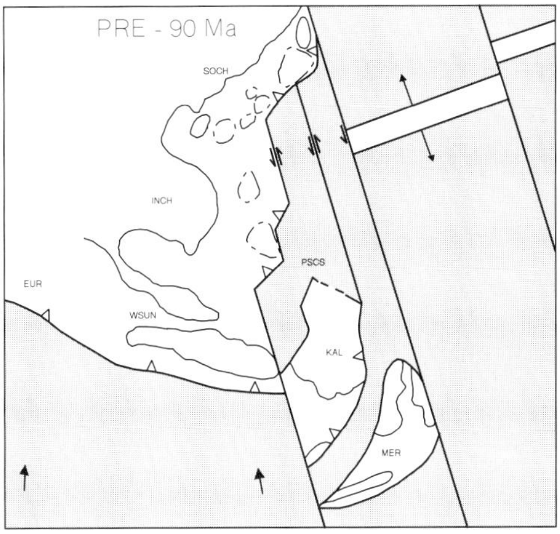

Fig. 12. (bottom) Pre-90 Ma plate reconstruction map (see Fig. 7 for key).

Western Sunda Blocks (Fig. 4) as portions of a stable Eurasian continental mass with active oceanic subduction along the southern Eurasian margin. Offshore two rifted fragments are interpreted namely the Kalimantan Block (Fig. 4) previously rifted from the Eurasian margin during the early Cretaceous rifting in the Proto-South China Sea (Hutchison 1989) and the Meratus Block rifted from the NW margin of Australia as part of the fragmentation of Gondwana. The latter block is interpreted to have been moving northwards as a consequence of the east–west-aligned ridge system in the Indian Ocean (Fig. 6, Phase 2 rifting) much of which has subsequently been subducted beneath the Sunda Arc.

The collision between the Meratus and Kalimantan blocks is shown on Fig. 13 occurred in the Late Cretaceous at about 90 Ma as described by Bransden & Mathews (1992) and summarized by Kusuma & Darin (1989). This collision resulted in the amalgamation of the Kalimantan and Meratus blocks into the Eastern Sunda block with the possible establishment of a new subduction system along it's the southern margin. The collision is also interpreted to be associated with the onset of oceanic subduction of the proto-South China Sea beneath NW Borneo consistent with the description of the geology of northwest Borneo by Williams *et al.* (1988), the cessation of active oceanic subduction beneath the Indochina and South China Blocks of Eurasia as evidenced by the end of the second thermal episode of the Yenshanian orogeny at about 85 Ma based on the dating of volcanics (Yu 1990) and the onset of minor extension and basin formation in the South China NE–SW-trending rift basins (Figs 10 and 13) as described by Ru & Pigott (1986) and Chen *et al.* (1993).

Except for minor deposition in localized rift basins in the Pearl River Mouth Basin area and passive margin marine deposition around the northwestern margin of the Australian continent (including the Irian Jayan basins of Salawati and Bintuni; Fig. 10) the Late Cretaceous to mid-Eocene represents a period of non-deposition over the study area (Figs 8, 9 & 10).

A key element of the suggested plate reconstructions during this phase is the postulated presence of a major strike-slip faulting along the western flank of the Kalimantan Block. which is based on figs 4.14 and 4.15 of Hutchison (1989). This fault system corresponds to the active strike-slip fault marking the western limit of subduction beneath NW Borneo as described by Williams *et al.* (1988).

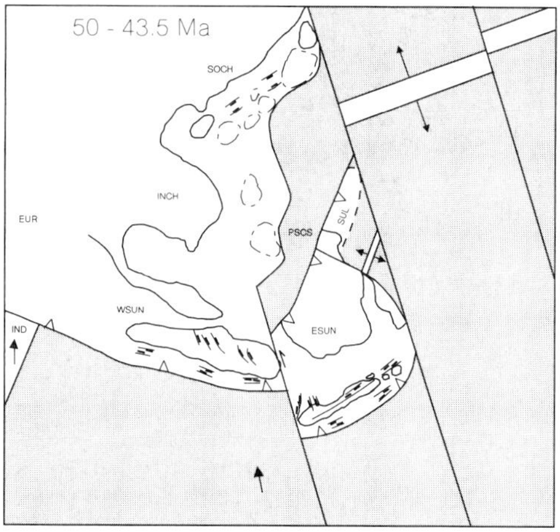

Fig. 14. 50–43 Ma plate reconstruction map (see Fig. 7 for key).

Stage I (50–43.5 Ma)

The northern India outcrop data summarized by Gaetani & Garzanti (1991) suggests that India collided with the southern Eurasian margin close to the Palaeocene–Eocene boundary (56 Ma). The collisional process has subsequently continued through to the present day as the Indian block has driven northwards deeper into the Eurasian continent. The first measurable effect on plate convergence rates in the Indian Ocean was not until the time of magnetic Chron A22 at about 50 Ma (Patriat & Achache 1984) where a decrease in convergence rates is noticed (see Fig. 15). During Stage I the India–Eurasia collision proceeded contemporaneously with crustal shortening and the subduction of oceanic crust beneath the southern margin of Eurasia.

The effect of the decrease in spreading rates due to Indian collision was to decrease the convergence velocity along the Sunda arc system causing a phase of extension in the fore-arc and back-arc regions (Fig. 11: Case A1). The key basin which dates this phase of extension is the East Java Basin (Figs 1 and 8) which contains numerous grabens filled with dominantly marine sediments containing definite Middle Eocene P9/P10 or older biozone ages (Bransden & Matthews 1992) which equate to 47–52 Ma or older (Fig. 6). The remainder of the back arc basins along the Sunda Arc contain poorly dated non-marine lacustrine and tuffaceous sediments and the Palaeogene sequences in the fore-arc basins are very lightly drilled and remain poorly defined. Elsewhere non-marine sediments continued to be deposited in the Pearl River Mouth Basin area and marine deposition continued on the margins of the Australian continent (Fig. 10). The remainder of the study area was characterized by non-deposition.

The Celebes Sea was spreading during this stage and is interpreted to be either a continuation of the pre-Stage I seafloor spreading pattern or a result of the northerly migration of the Eastern Sunda Block towards the Eurasian margin as shown on Fig. 14.

Stage II (43.5–32 Ma)

Stage II was triggered by the termination of oceanic subduction beneath the India–Eurasia collision zone. This 'locked up' the spreading system in the Indian Ocean (Fig. 6) and caused a major plate reorganization effective not only in the Indian Ocean but also in the Southern and Pacific Oceans (Patriat & Achache 1984). The continuing northward movement of India was subsequently driven by a new spreading system in the Southern Ocean (Fig. 6) and was accommodated in the collision zone by continental underthrusting and internal deformation. The plate reorganisation in the Indian Ocean slowed convergence rates yet again and produced a second phase of extension along the Sunda Arc (Fig. 15) whereas in the Pacific Ocean it resulted in the first phase of regional extension in the South China Sea and East Kalimantan areas (Fig. 16).

The Pacific and Philippine plates are interpreted to have comprised a single contiguous plate prior to 43.5 Ma moving in a NNW direction (Hilde *et al.* 1977). Following the major plate reorganisation at 43.5 Ma the plate motion changed to WNW and a major transform fault is thought to have developed into a subduction zone dividing the plate into the westerly Philippine plate and the easterly Pacific plate. This plate reorganization is evident in the Philippine plate by a change in spreading direction recorded by oceanic magnetics data (Fig. 3) and is also evident in the Pacific plate by the

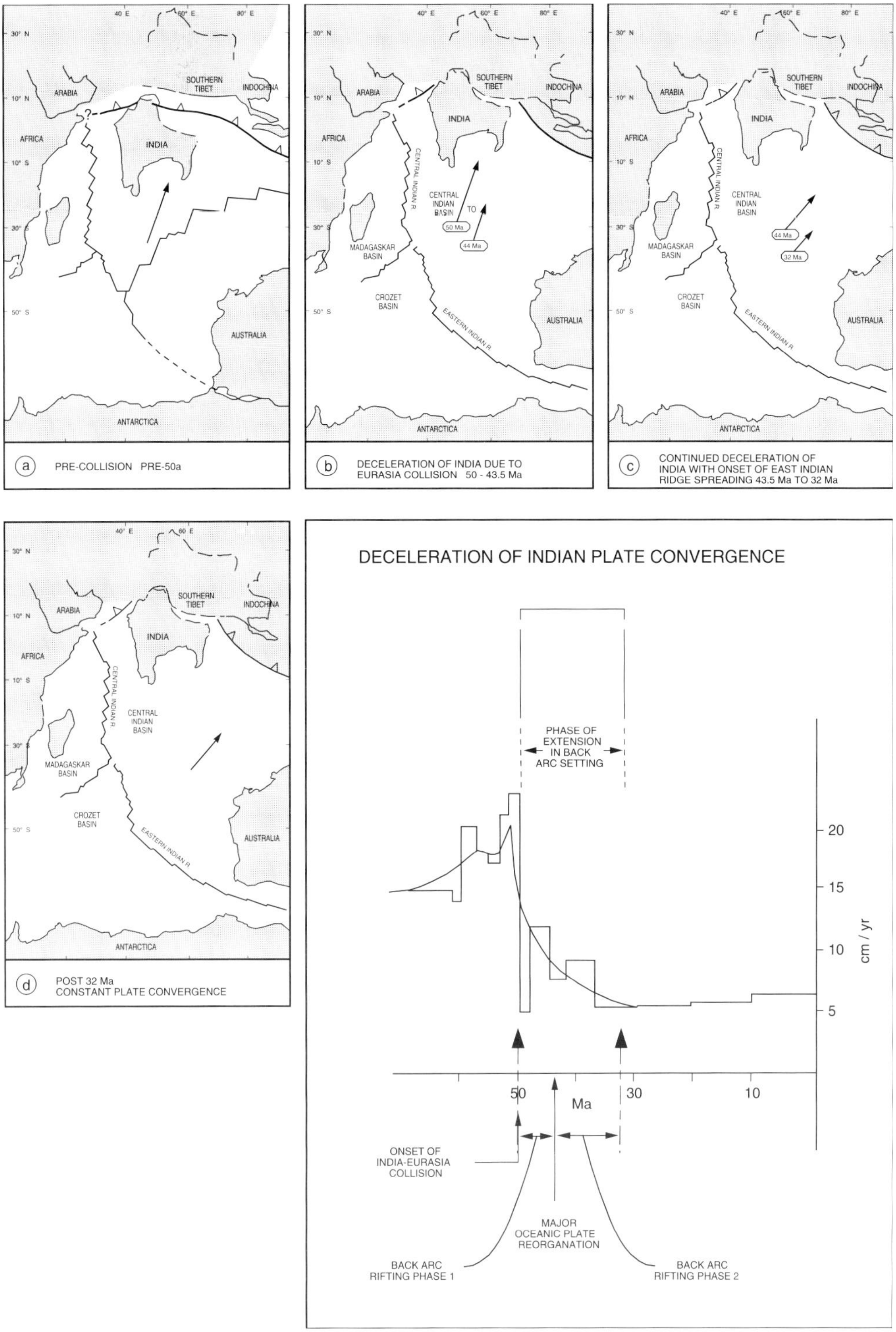

Fig. 15. A summary of Indian and Southern Ocean spreading history (modified after Daly *et al.* 1987 and Patriat & Achache 1984).

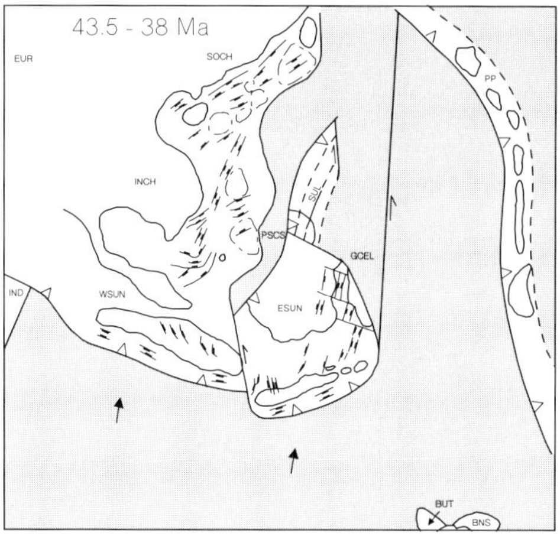

Fig. 16. (top) 43.5–38 Ma plate reconstruction map (see Fig. 7 for key).

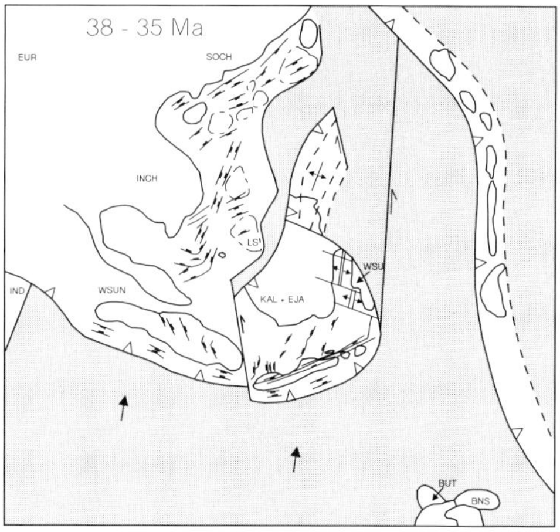

Fig. 17. (bottom) 38–35 Ma plate reconstruction map (see Fig. 7 for key).

change in orientation of the Emperor–Seamount chain, as described by Patriat & Achache (1984). The plate motions since this reorganization have been constant to the present day.

The Stage II rifting along the Sunda Arc is reflected in additional rift sequences (Fig. 8) although there is no direct biostratigraphic support for this interpretation due to the predominantly non-marine facies of the sedimentary sequences deposited. The top Brown Shale unconformity described by Williams *et al.* (1985) from the Central Sumatra Basin (Fig. 1) is believed to mark the transition from Stage I to Stage II deposition.

The onset of sedimentation in East Kalimantan Basins is evident from rift sequences in the Barito, Kutei and Tarakan basins (Fig. 9). These are interpreted to be associated with a rift system established in the Makassar Straits (Fig. 16). This rift system, although initiated by the regional plate re-organization at 43.5 Ma, is interpreted to be a product of the northward drift of the Eastern Sunda and Sulu Blocks associated with the subduction of the proto-South China Sea beneath NW Borneo (Fig. 16). This drift initially resulted in back arc spreading in the Sulu Sea (Fig. 16) and eventually seafloor spreading in the Makassar Straits (Fig. 17). The timing of this latter event (at about 38 Ma) is inferred from the intra-Late Eocene rapid increase of waterdepths recorded in the Eastern Kalimantan Basins (Fig. 9) and the change from rift to sag structural styles in the Barito Basin during the Upper Eocene (Kusuma & Darin 1989). These changes are considered to be characteristic of a 'break-up' unconformity caused by the onset of oceanic crust formation in the Makassar Straits.

The rapid drift of the West Sulawesi Block away from the combined Kalimantan and East Java Blocks (Fig. 4) following the onset of seafloor spreading in the Makassar Straits described above is interpreted to have formed the Bogor-Kendang rift system which runs the length of Java Island (Figs 3 and 17). This event is dated within the P15 biozone (Fig. 8) by Bransden & Matthews (1992) in the East Java Basin and corresponds to a phase of major rifting and associated with a rapid increase in water depth.

The onset of Stage II is marked by more extensive rifting in the South China Sea area and rifting is noted for the first time in the West Natuna Basin area (Fig. 16) although again because of the nature of the non-marine rift fill (Fig. 10) and the fact that the rift sequence is rarely drilled the sequences are difficult to date biostratigraphically.

The key event which occurred at the end of Stage II was the collision of the Luconia Shoals block with NW Borneo (East Sunda block *sensu* Fig. 4). This collision is recorded in outcrop in Sarawak within the Upper Eocene (Haile 1992) and is shown as latest Eocene on the chronostratigraphic summary of Benard *et al.* (1990) and is interpreted as Late Eocene in age by Murphy (1994). The space formed by the downwards buckling of the southern margin of the Luconia block into the NW Borneo trench was filled by sediments of the Balingian Delta

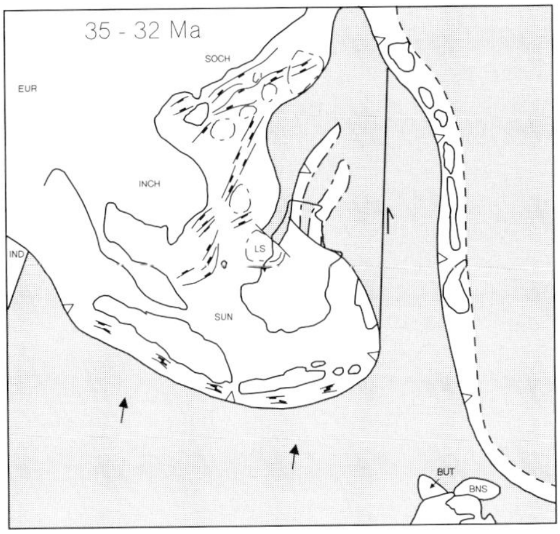

Fig. 18. (top) 35–32 Ma plate reconstruction map (see Fig. 7 for key).

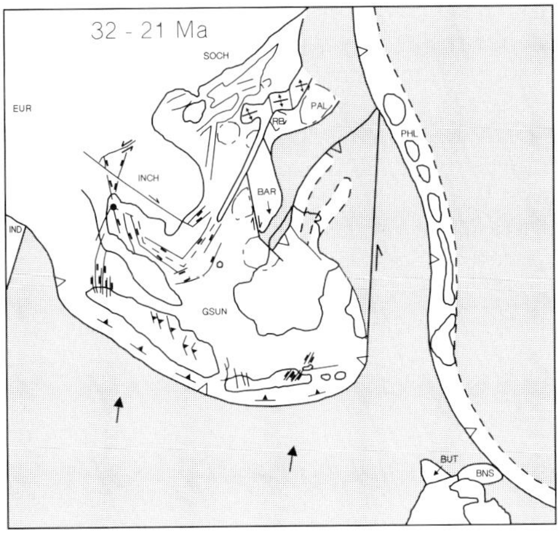

Fig. 19. (bottom) 32–21 Ma plate reconstruction map (see Fig. 7 for key).

sourced from the uplifted fore-arc sediments to the southeast. The first influx of sand into this delta is shown as earliest Oligocene (*c.* 35 Ma) by Doust (1981) (Fig. 9). The interpreted effect of this collision (Fig. 18) was to stop the northward movement of the Eastern Sunda block and end the spreading in the Makassar Straits and Sulu Sea areas. The 'intra P18' (35 Ma) tectonic event in the East Java Basin noted by (Bransden & Matthews 1991) is thought to be a reflection of these changes to the regional stress field.

The key difference in the above Luconia collision model to previous models such as Daly *et al.* (1991) is the northwards drift of the Eastern Sunda block rather than the southwards drift of the Luconia block since the latter is difficult to ascribe to spreading in the South China Sea as the initiation of spreading (32 Ma) post-dates the collision event.

Stage III (32–21 Ma)

The onset of Stage III is synchronous with the first phase of seafloor spreading in the South China Sea (32–21 Ma). This seafloor spreading event was the result of continuous extension and rifting in the area initiated by the plate reorganisation at 43.5 Ma. The onset of seafloor spreading is interpreted here to have resulted in the clockwise rotation of the *entire* Greater Sunda Block (Fig. 19) about a pole of rotation located in the present day Gulf of Thailand. The particular configuration of this block rotation is believed to have been controlled by (but not driven by) the continuing collision of India with Eurasia since the block rotation to the east of the pole of rotation was accommodated by continued continental underthrusting of the Indian Plate beneath the southern Eurasian margin and internal deformation of both blocks in the west.

The onset of rotation of the Greater Sunda Block caused a relative increase in the convergence rate along the Sunda Arc (Fig. 11: Case B2) producing a minor but significant phase of compressional tectonism and ending the main rifting phase in both the back-arc and fore-arc basins (Fig. 8). This was followed by a post-rift thermal sag phase which in combination with a general eustatic rise in sea level resulted in the deposition of alluvial, deltaic and finally marine sequences in most basins. The exception is the North Sumatra Basin where a series of north-south oriented major horst and graben structures were developed (Courtney *et al.* 1990) related to the basin's unique position south of the pole of rotation. A similar set of faults to the north of the pole of rotation resulted in the establishment of a series of lacustrine graben onshore Thailand described by Polacahan *et al.* (1991).

The rotation of the Greater Sunda block also opened the Malay rift basin and produced another phase of rifting in the West Natuna Basin which were both filled with broad lacustrine sequences (Fig. 10) whereas in the South China Sea area the onset of oceanic spreading led to a passive margin subsidence phase of basin development and the eventual encroachment of marine environments (Fig. 10). In the

East Kalimantan area the basins were largely unaffected by the rotation and marine shale deposition dominated (Fig. 9).

The first collisional contact of the northward moving Australian craton (as part of the combined Indian Ocean and Australian plate) with the Philippine plate occurred at about 25 Ma based on the work of Ali & Hall (1993). This event marks the onset of the 'Salami Slicer' whereby the transcurrent motion of the Philippine plate resulted in the shaving of blocks from the northerly protrusion of the Australian continent (Figs 10 & 19).

Stage IV (21–0 Ma)

Stage IV was initiated by the cessation of the first phase of seafloor spreading in the South China Sea and the contemporaneous onset of continental shortening in Tibet via block rotation and lateral extrusion along strike-slip faults. The termination of the first phase of seafloor spreading in the South China Sea is considered to have been caused by the collision of the Baram block with NW Borneo, The Baram block comprises an area of attenuated crust to the east of the West Baram Line (Fig. 3) and to the west of the Reed Bank Block (Fig. 3). This collision event marks the end of subduction of the proto-South China Sea oceanic crust beneath NW Borneo (Fig. 20). The event can be dated from both the oceanic magnetic anomalies and the onset of major deltaic deposits in the Sabah and Baram Delta described by James (1984) and dated by Benard *et al.* (1990) deposited into the space formed by the downwarping of the Baram

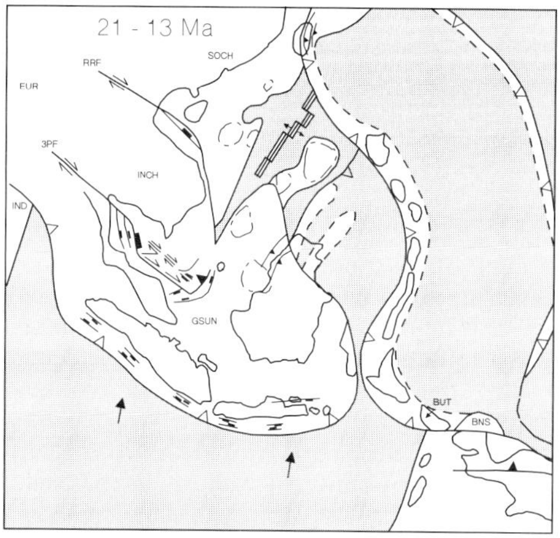

Fig. 20. (top) 21–13 Ma plate reconstruction map (see Fig. 7 for key).

block into the NW Borneo Trench and sourced by sediments shed off uplifted fore-arc sediments to the southeast. It can also be dated by the age of deltaic sediments shed southeastwards from the uplifted area into the Kutei Basin area as described by van de Weerd & Armin (1993) (Fig. 9).

Following the cessation of the main phase of spreading in the South China Sea, the continuing Indian collision was no longer able to exploit the rotation of the Greater Sunda Block as a mechanism whereby it could continue its northward motion. The Indian–Eurasian collision had presumably reached a point where crustal shortening via crustal underthrusting and internal deformation in the Himalayas had reached its limit. Consequently the motion of India was preserved by developing a new structural solution: extrusion tectonics comprising rotation of crustal blocks along strike-slip faults ahead of the Indian 'indentor' as described by Tapponier (1982). The principal difference between the model presented here and that presented by Tapponier (1982) is in both (i) the timing of the *major* displacements along these faults since Tapponier considered the strike-slip faults to be intimately associated with the genesis of the South China Sea (*c.* 37–32 Ma) whereas in the model presented here the *major* strike-slip movements do not occur until after 21 Ma and thus post-date the formation of the South China Sea and (ii) the scale of the displacements along these extrusion related faults since the Tapponier (1982) model predicts major fault displacements (*c.* 100–1000 km) whereas the model presented here suggests much smaller displacements (*c.* 10–100 km) more consistant with the models of Dewey *et al.* (1989) who argue against major Tertairy lateral extrusion fault movements.

The cessation of the rotation of the greater Sunda block produced a relative decrease in the convergence velocity along the Sunda Arc and a phase of related extension (Fig. 11: Case A2). This extension, evident as a phase of block faulting in the Sunda back arc basins, produced the subtle structures which formed the nuclei of numerous carbonate build-ups (Fig. 8). In the clastics dominated Central Sumatra Basin this (21 Ma) event led to the development of subtle low-relief structures described by Praptono *et al.* (1992) which are related by Heidrick & Aulia (1993) to a minor phase of tectonism. Marine inundation due to continuing subsidence plus a eustatic high in the mid-Miocene (Fig. 5) resulted in extensive marine shale sedimentation throughout the Sunda basins during the early stages of Stage IV (Fig. 8).

The initial extrusion tectonics phase of Stage IV (Fig. 20) is characterized by onset of major right-lateral fault systems along the Red River and Three Pagodas Fault systems. These fault movements formed the Bac Bo/Yinggehai Basin at a releasing bend along the Red River Fault system and numerous small basins offshore in the Gulf of Thailand related to the releasing bend of the Three Pagodas Fault system. The lateral displacement along the Three Pagodas and Malay Fault System in the Malay Basin area resulted in major transtensional strike-slip fault movement along the southeastern flank of the basin which was ultimately accommodated via the generation of spectacular inversion structures in the West Natuna Basin area described by Ginger *et al.* (1993). The depositional history of all the basins in the Natuna, Malay and Gulf of Thailand areas records a marine transgression throughout this early phase (Fig. 10) with the deposition of alluvial, deltaic deposits and eventually marine shales.

At about 13 Ma the SE Sulawesi Buton block, originally shaved off the northern protrusion of the Australian continent, collided with the Greater Sunda Block as described by Smith & Silver (1991) and Davidson (1991) (Fig. 21). Davidson interprets the onset of the collision as Latest Oligocene biostratigraphic N3 age (approx 25 Ma) with the main collision recorded by a N11 to N13 hiatus (13.5–11 Ma). The collision age here is taken as the base of this hiatus within the N11 biozone at approximately 13 Ma (Fig. 5). This event brought the fast-moving Philippine Plate into contact with the Sunda Craton for the first time and added momentum to the clockwise rotation of the Great Sunda Block which eventually led to the point where the oblique angle of subduction beneath Sumatra was of sufficient magnitude to produce the Semanko Fault system and the formation of the West Sumatra Block (Fig. 21). As described by Mount & Suppe (1992) once this fault was emplaced it accommodated all the strike-slip component of the arc-induced compression such that the back arc basins of Sumatra only experienced compressional tectonism with the principal stress direction orientated orthogonal to the Semangko fault system. This stress field resulted in the formation of large anticlinal structures sub-parallel to the Semangko fault system throughout the entire back arc area of Sumatra. These anticlines, best described by Eubank & Makki (1981), have grown progressively as Sumatra rotated further clockwise contemporaneous with deposition of sediments which record a marine regression interpreted to be partly eustatic and partly driven by sedimentary input from the uplifted hinterland.

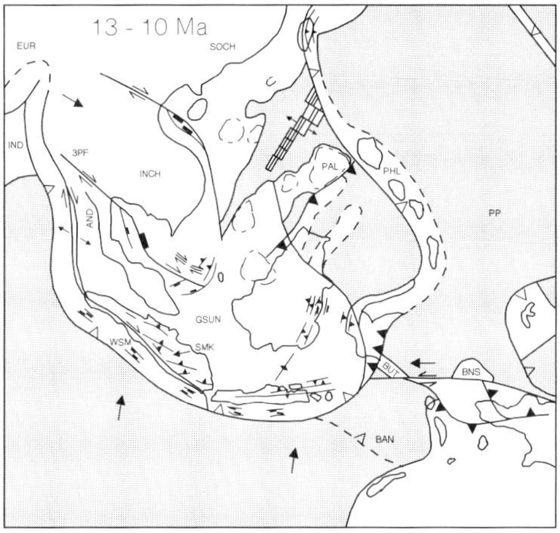

Fig. 21. (bottom) 13–10 Ma plate reconstruction map (see Fig. 7 for key).

The Buton collision in interpreted to have resulted in the onset of major compressional tectonism throughout the eastern portion of the Greater Sunda Block as described for the East Java area by Letouzey (1990), producing a foldbelt along the line of the Meratus Suture and subsequent deposition of the Mahakam Delta sequence to the east of this zone of uplift (Figs 9 & 21). It is also interpreted to have resulted in the formation of the Banda arc. The reason for the location and formation of this oceanic subduction system is poorly understood but it is probably related to the conservation of the northward movement of the combined Australia and Indian Ocean plate following the Buton collision with the Greater Sunda Block. It was possibly initiated along an old pre-Tertiary transform fault of spreading phase 1A on Fig. 6.

The Palawan–Sulu Block collision is interpreted by Williams (1992) to be within the N14 Biozone and is interpreted by Gallagher (1987) to be of magnetic Anomaly A5 age. These consistent interpretations indicate a collision age of approximately 11 Ma (Fig. 5). This collision marks the cessation of the second and final phase of seafloor spreading in the South China Sea. In NW Borneo the unconformity associated with this collision is termed the 'shallow regional' unconformity to differentiate it from the 'deep regional' unconformity interpreted here to be associated with the Baram block collision (Fig. 9).

At about 12 Ma the accelerating strike-slip movements along the Semanko fault system,

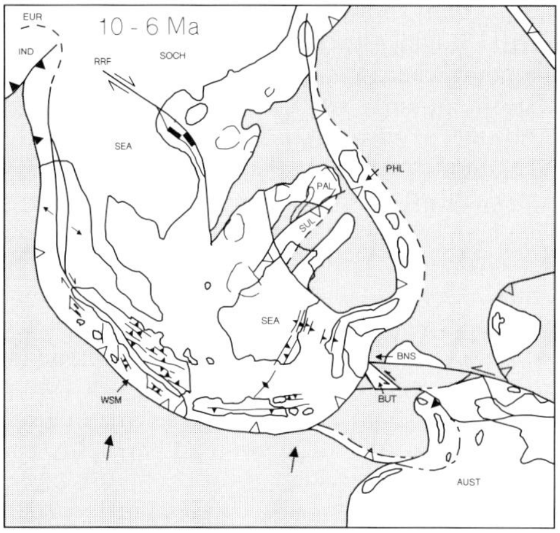

Fig. 22. (top) 10–6 Ma plate reconstruction map (see Fig. 7 for key).

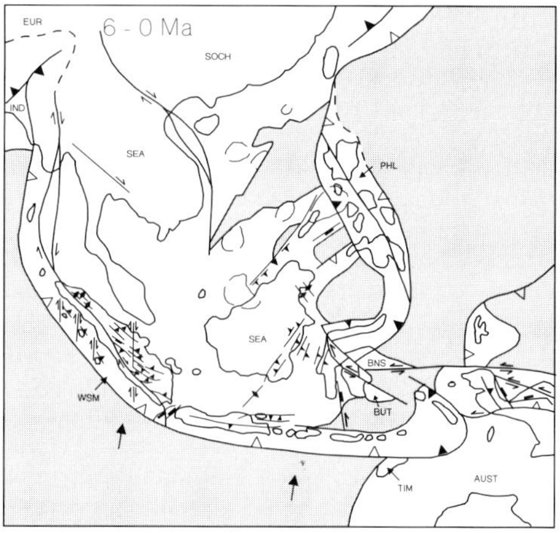

Fig. 23. (bottom) 6–0 Ma plate reconstruction map (see Fig. 7 for key).

related to the continuing clockwise rotation of the Greater Sunda Block, had propagated the Andaman Sea seafloor spreading zone sufficiently north to neck-off the extrusive movements along the Three Pagodas and Malay Fault System (Figs 21 & 22). This formed the SE Asian Block (Fig. 4) and it marks the end of active fault movements within the Malay–Natuna basins (Fig. 10). It also resulted in the extrusion movements resulting from the Indian collision to be principally concentrated further northwards along the Red River Fault system forming the deep Bac Bo/Yinggehai Basin which was filled with thick marine sediments. The fault movements along the Red River Fault are believed to have been terminated in the late Miocene (*c.* 6 Ma) presumably extinguished by a similar mechanism as India indented futher northwards and the Red River Fault was necked-off.

The Banggai-Sula Microcontinental Block collided with the Greater Sunda Block at about 6 Ma (Fig. 22) during the latest Miocene N17 biozone (*c.* 6 Ma) based on the detailed dating of pre- and post-collisional sediments described by Handiwiria (1990) and Abimanyu (1990) respectively. This dating is also consistent with the late Miocene age inferred for this collision by Smith & Silver (1991). The geology of the microcontinental block is best described by Garrard *et al.* (1988). The effect of this event was to intensify both anticlinal structures in the Kutei basin and inversion structures in the East Java Basin.

The Philippines Arc collided with the Palawan Block at about 5 Ma causing a terminal phase of tectonism along the adjacent NW Borneo area (Fig. 23). The collision is dated by Gallagher (1987) to be of Chron A3 age (6 Ma) which is consistent with the migration of the active volcanic arc from the western side of Luzon Island to the eastern flank at about 5 Ma as described by Hutchison (1989). This migration of the volcanic arc system is interpreted to be a direct result of the collision. The effect of the collision is evident as minor structuring in the Sabah, Baram Delta and NW Palawan Basins. Subsequent marine and deltaic deposits in these areas are unstructured.

The Timor area of the Australian continent collision with the Sunda Arc is dated at about 3 Ma although the collision history dates back to the late Miocene. As best described by Daly *et al.* (1991) the collision can be divided into three phases. The early phase (11–7 Ma) comprised the collision of the northwest portion of the Australian craton with the south facing Banda Arc resulting in the development of an area of thickened crust, and a fold–thrust belt dated by radiometric dates at 11–6 Ma (Hutchison 1989). A similar age obtained from dating of outcropping units on Timor Island is described by Sawyer *et al.* (1993). The main collision phase (7–3 Ma) is associated with continuing northwards subduction of oceanic crust beneath the Banda Arc and continental shortening contemporaneous with a phase of uplift (and extension) caused by isostatic rebound of thickened crust. The cessation of northwards subduction of oceanic crust occurred at about 3 Ma as evidenced by the termination of arc volcanism in the area (Abbott & Chamalaun 1981; Hutchison 1989). The main collision phase is recorded on Timor Island by the deposition of the Viqueque Sequence as described by Sawyer *et al.* (1993). The third (post-collision) phase (3–0 Ma) relates to the development of the Flores Backthrust Fault system located to the north of the Banda Arc (Fig. 3) which developed to accommodate the continuing northward motion of the Australian continent of 50 km Ma^{-1}. Based on fig. 2A of Charlton (1992) it is interpreted here that this fault system is in the process of developing into a north-facing subduction system beneath which the Banda Sea oceanic crust will be consumed. Isostatic uplift of Timor Island continued throughout this third phase.

It is interpreted here that the stresses associated with the main Timor collision (*c.* 3 Ma) propagated deep into the Greater Sunda block along the northwesterly oriented faults near Timor on Fig. 3. Movement along these and similar faults are interpreted to have caused the upper-Pliocene (intra N21 biozone, 2–3 Ma) tectonic event in the Banggai area of Sulawesi as described by Abimanyu (1990) and the 'bevelling T65' Pliocene unconformity in the East Java Basin (Bransden & Matthews 1992), and uplift within the Kutei basin area of Kalimantan dated at 3 Ma by Burrus *et al.* (1994).

It is clear from the above that with five major collision events within Stage IV and the effect of major structuring associated with major strike-slip movements and the rotation of Sumatra that the sedimentary fill throughout this phase is highly variable. Common elements to the stratigraphic record throughout SE Asia are summarised on Fig. 5 namely a eustatic high during the middle Miocene and major sea-level changes at about 5 and 10 Ma.

Summary and discussion

The model presented here is an iterative balance between the vagaries of palaeomagnetic, stratigraphic and structural data and is summarized on Fig. 24 which shows the age of the principal block rotations and the collisional events. The impliactions of the model in relation to structural styles and eustatic curves is described in more detail below.

Southeast Asia is well known for, and characterized by, a particular style of structural inversion known as the Sunda Fold. As originally defined by Eubank & Makkie (1981) and described from the Eastern Sunda Shelf by Letouzey (1990) and from the Natuna Sea by Ginger *et al.* (1993) the Sunda Fold comprises the structural inversion of an extensional fault whereby the depositional thick portion of the rift is structurally elevated to form an anticlinal structure. This style of structural inversion is interpreted here to be common in SE Asia due to the combination of two favourable circumstances namely the prevalence of listric extensional faults with a shallow depth of detachment (which are relatively easier to invert than steep planar faults) and a situation where the orientation of the principal extensional and compressional stress directions are alligned as is predicted by the proposed tectonic model for the areas mentioned above.

The utilization of a eustatic sea level curve to explain synchronous events in widely separated basins has been frequently used in SE Asia, e.g. Bauman (1982), Courtney *et al.* (1990) and Chen *et al.* (1993). This intepretation is interpreted here to be flawed since based on the model presented here many tectonic events in SE Asia are regional events which transcend basin boundaries.

The specific 'Exxon' eustatic sea-level curve of Haq *et al.* (1988) (Fig. 5) as described by Vail *et al.* (1977) is based on inclusion of data from the Malay, South Sumatra and NW Borneo Basins plus other basins not within SE Asia. The application of this reference curve as a benchmark has recently come under severe criticism by Miall (1992) and Hubbard (1988) amongst others whereas the application of the Exxon eustatic curve to the basins of SE Asia received some qualified support from Hutchison (1986). The data presented in this paper summarized on Fig. 5 shows a suspiciously high level of correlation between the timing of major collision events and the age of the major second order sequence boundaries of the Haq *et al.* (1988) curve. This tends to support Hutchison's (1986) qualifying statement that 'the excellent correlation of eustatic sea levels with transgressions and regressions in SE Asia maybe an artifact of overdependence upon SE Asian basins during the compilation of eustatic curves'.

It is the stated *opinion* here that given the tectonostratigraphic model described in this paper the stratigraphic record in SE Asia supports a gradual rise in relative sea level between about 30 Ma and 13 Ma as recorded in almost all Southeast Asian sedimentary basins (Figs 8–10) and major eustatic drops and subsequent transgressions at about 10 Ma and 5 Ma. The Chattian lowstand event (*c.* 29.5 Ma, Fig. 5) maybe recorded in SE Asian basins but its effect is interpreted to have been confused with the onset of Stage III of the proposed tectonic model (at 32 Ma). All other more subtle transgressive and regressive events will certainly

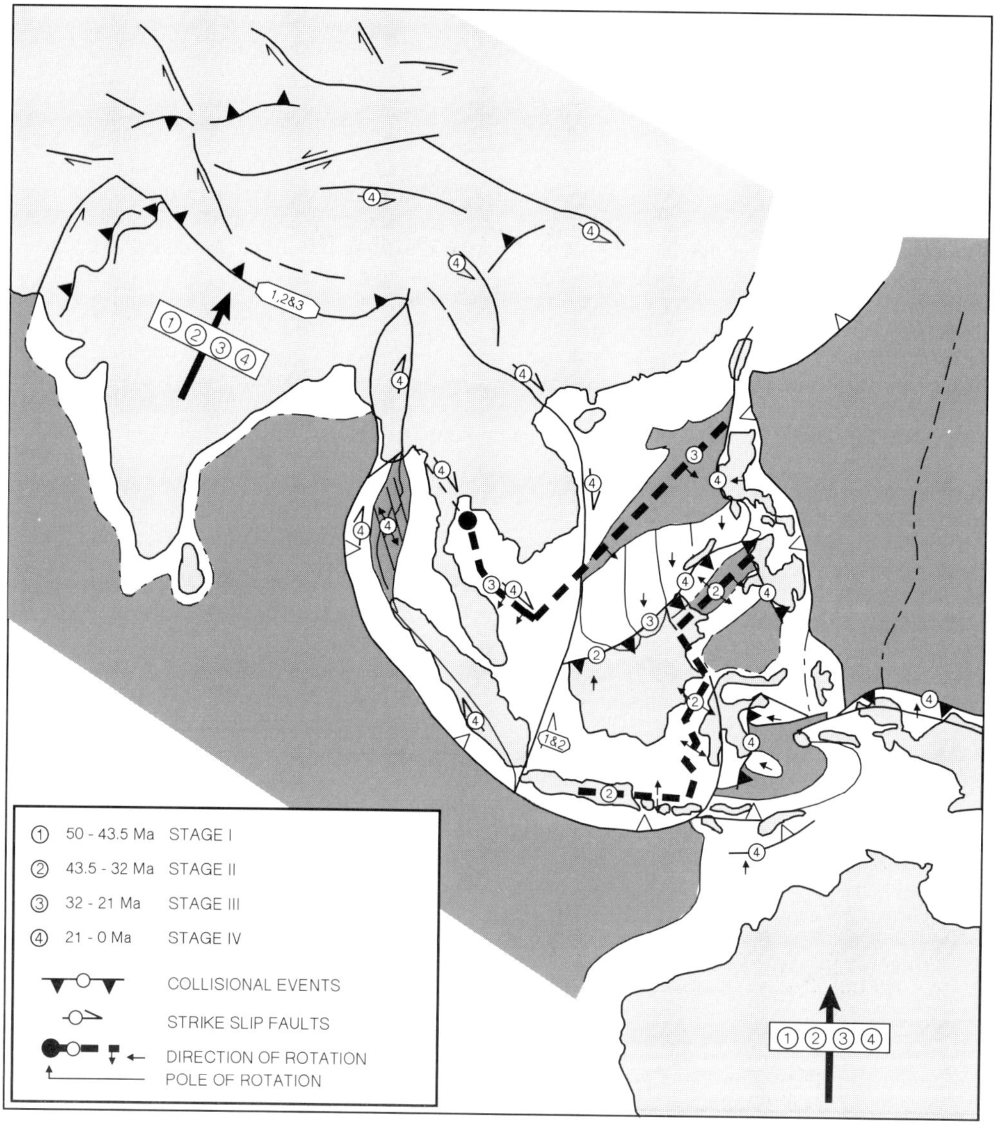

Fig. 24. Schematic summary of plate tectonic model.

be able to be correlated with events on the Exxon eustatic curve since, as argued by Miall (1992), the number of options combined with the relative large error bar on the dating always provides a correlation. Given the frequency of significant tectonic Tertiary events in SE Asia (Fig. 5) it is clear a similar charge could well be levelled at any model attributing these changes to purely tectonic causes.

In summary, the geology of SE Asian supports the premise that eustatic variations have had a major influence on the stratigraphic record but the data does not either support or disprove the specific detailed curve postulated by Haq *et al.* (1988). Detailed dating of tectonic events across the region is still required to differentiate between the tectonic vs eustatic origins of the observed sedimentary cycles in the region.

SE Asian petroleum distribution

The spatial distribution of discovered hydrocarbons within southeast Asia is shown schematically on Fig. 25 based on the data shown on

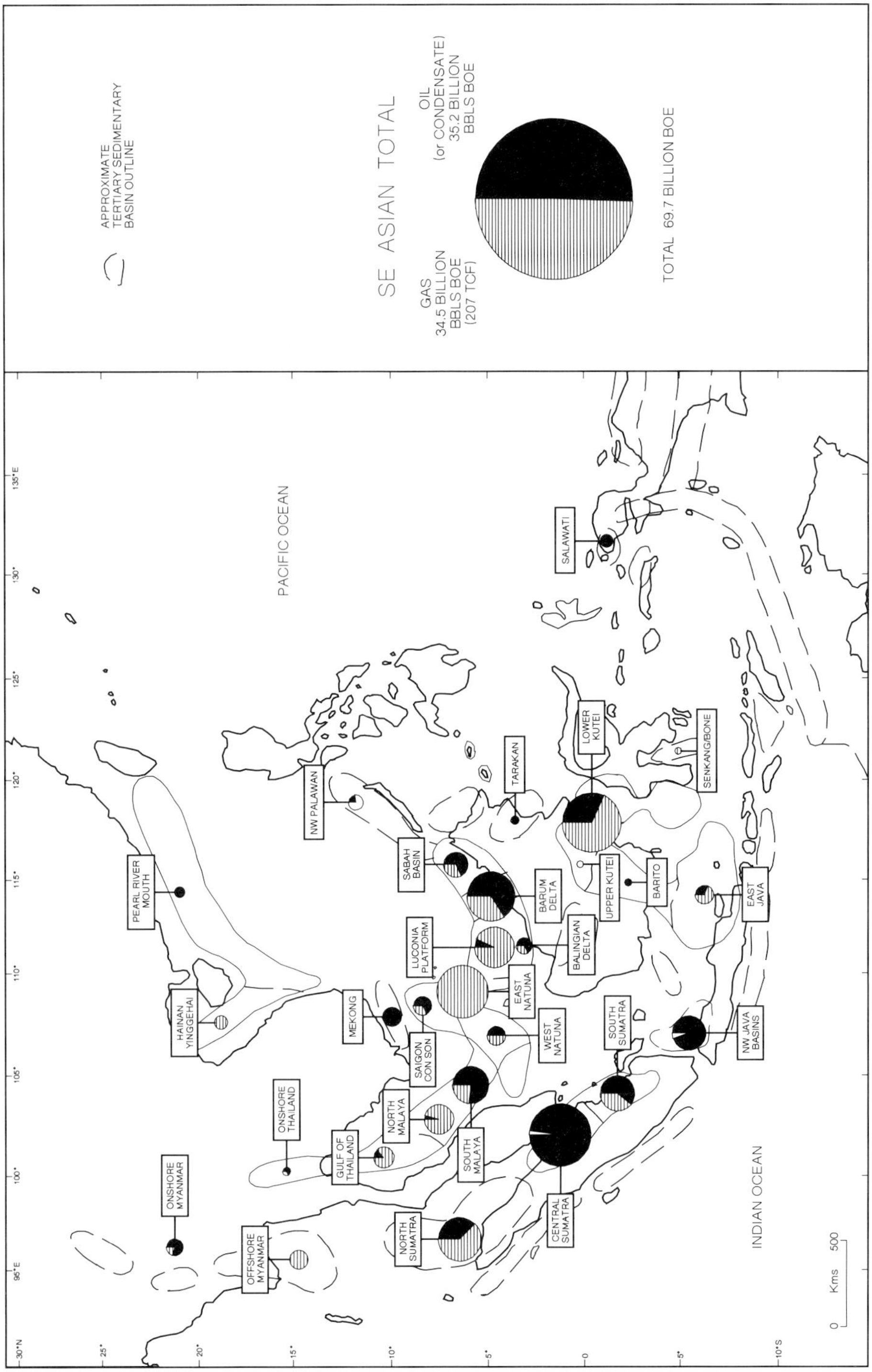

Fig. 25. Distribution of hydrocarbon reserves in SE Asia.

Table 1. The area of the circles on Fig. 25 is proportional to the oil equivalent recoverable reserves of the basin using a conversion ratio of 6 bcf of gas equivalent to 1×10^6 bbls of oil. The condensate reserves in the region are included with the oil. The data source is largely confidential and the numeric values are in many cases only guestimates with error bars of up to ±50% in some cases. The intention is merely to give a broad overview of the hydrocarbon

Table 1. *Speculative SE Asian reserves by interpreted source rock sequence*

Basin or province	Oil reserves ($\times 10^9$)	Gas reserves TCF	Total reserves ($\times 10^9$)	Oil by BOE (%)	Interpreted source rock type	Source rock name	Stage of deposition
Central Sumatra	10.5	1.3	10.72	98	C	Brown shale	I
Lower Kutei	3.0	39	9.54	32	D/E	Miocene deltaics	IV
East Natuna	0	45	7.50	0	D/E	Balingian deltaics	II
Barum Delta	4.31	13.5	6.56	66	D/E	Miocene deltaics	IV
North Sumatra	2	20	5.33	38	B	Bampo	III
Luconia Platform	0.3	24.6	4.40	7	D/E	Balingian deltaics	II
South Malaya	3	5.5	3.91	77			
	3	0	3.00		C	Banua	II
	0	5.48	0.91		D/E	Arang	IV
South Sumatra	2.1	8.1	3.45	61			
	2.1	0	2.10		C	Benakat-Lahat	I+II
	0	8.1	1.35		D/E	Talang akar	III
Northwest Java	3.1	1	3.27	95			
NW Java	1.5	1	1.67				
	1.5	0	1.50		C	Banuwati	I+II
	0	1	0.17		D/E	Talang akar	III
Sunda	1.4	0	1.40		C	Banuwati	I+II
Asri	0.2	0	0.20		C	Banuwati	I+II
North Malaya	0.1	14	2.43	4	D/E	Pre-Bekok sequence	IV
Sabah	1.2	3.4	1.77	68	D/E	Miocene deltaics	IV
Gulf of Thailand/ Pettani Trough	0.16	6.4	1.23	13	D/E	Pre-Bekok seq. equiv.	IV
Mekong	1.1	0	1.10	100	C	Unknown	II
Saigon/Con Son	0.7	2	1.03	68	C	Unknown	II
West Natuna	0.5	3	1.00	50			
	0.5	0	0.50		C	Benua	II
	0	3	0.50		D/E–F	Arang	IV
East Java	0.31	4.1	0.99	31			
	0.08	2.6	0.51		D/E	Ngimbang	I+II
	0.23	1	0.40		B	Kujung I shales	IV
	0	0.5	0.08		G	Late Mio-Plio seq.	IV
Gulf of Martaban/ Irrawaddy Delta	0	5.8	0.97	0	D/E	Unknown	IV
Myanmar-Onshore	0.68	1	0.85	80	D/E–F	Miocene deltaics	III
Balingian	0.5	1.5	0.75	67	D/E	Balingian deltaics	II
NW Palawan	0.15	3	0.65	23	C	Unknown	II
Hainan/Yinggehai	0	3	0.50	0	?G	Post-Meuhan seq.	IV
Salawati	0.5	0	0.50	100	A	Klasafet shales	IV
Pearl River Mouth/ South China Shelf	0.35	0	0.35	100	C	Yacheng seq.	IV
Tarakan	0.23	0.08	0.24	95	D/E	Mio-Plio deltaics	IV
Onshore Thailand	0.14	0.4	0.21	68	C	Various	III
Barito	0.2	0	0.20	100	D/E	Lower Tanjung	II
Upper Kutei	0	0.8	0.13	0	D/E	Runtu deltaics	IV
Senkang/Bone	0	0.5	0.08	0	G	Late Mio-Plio seq.	IV
Totals	35.2	207.0	69.7	50			

Stage gives the stage in which the source rocks were deposited.

distribution and for this purpose the data is considered adequate. The recent significant discoveries of gas in Vietnam and in the pre-Tertiary in the Bintuni Basin are not included in the data.

The southeast Asian total of 35 billion barrels of oil and 207 Tcf of gas (Table 1) compares favourably with previous published estimated total oil reserves of 31 billion bbls by Shaw (1990) and 35 billion bbls by Nation (1994) and estimated total gas reserves of 281 Tcf (Nation 1994).

The distribution of hydrocarbons shows there is a strong concentration of hydrocarbons surrounding the Sunda Craton and within this ring of productive basins there is a high degree of variability both in the productivity (total reserves) of the basins and in the relative proportion of oil and gas. The former variability is not believed to be due to lack of exploration since most of the major basin's are 'mature' by modern exploration standards.

The explanation for this distribution is interpreted to be due to the distribution of source rock types within each basin with the volume and quality of the source rock determining the volume of hydrocarbons generated (and trapped) and the type of source rock (oil v. gas prone) determining the amount of oil v. gas. To demonstrate this link the source rock scheme of Pepper & Corvi (1994) is used. This scheme identifies five different source rock types to which a sixth has been added here (Type G) to incorporate possible biogenic gas source rocks:

- Type A source rocks deposited in aquatic marine clastic-poor and carbonate/evaporite-rich environments which produces about 75% of it's hydrocarbon potential as oil;
- Type B source rocks deposited in aquatic marine clastic-rich environments which produces about 70% of it's hydrocarbon potential as oil;
- Type C source rocks deposited in aquatic non-marine lacustrine environments which produces about 75% of it's hydrocarbon potential as oil;
- Type D/E source rocks deposited in non-marine terrigenous environments with high wax contents typically in Mesozoic and younger 'ever-wet' coastal plain/deltaic settings which produces about 35% of it's hydrocarbon potential as oil;
- Type F source rocks deposited in non-marine terrigenous environments with low wax content typically in Late Palaeozoic and younger coastal plain/deltaic setting which have no liquids potential;
- Type G source rocks comprising marine biogenic dry gas source rocks which have no liquids potential.

The interpreted source rock type and the source rock interval is shown for all the major hydrocarbon basins in SE Asia on Table 1. These interpretations are based on data presented by Robinson (1987), Durkee (1992), Duval *et al.* (1992), Livingstone (1992), Scherer & Hitam (1992), Williams (1992) and Woodroof (1992), and on additional proprietary data. The interpretations presented here are tentative at best and in many cases are outright guesses. One clear fact from these data is the interpreted sparcity of marine source rock units in the region and the dominance of non-marine lacustrine and deltaic source rock types.

These interpretations together explain the distribution of hydrocarbons in SE Asia since for example the rich back arc basins of Central and South Sumatra and SW Java are attributed to numerous rift basins filled with oil-prone lacustrine (Type C) shales (with or without a gaseous contribution from overlying Type D/E or Type F deltaics) and the major Kutei and Baram mixed oil and gas basins are attributed to large volumes of stacked delta top (Type D/E) oil and gas prone shales and coals. Other basins such as the East Java Basin has a number of interpreted source units none of which are interpreted to be very productive.

It is interesting to note the increase in oil content in a northeasterly direction for the NW Borneo Basins (from the East Natuna through the Luconia and Baram basins into the Sabah basin) on Fig. 25. This may be an artefact of poor data or it may be related to the progressive younging of the collision ages along the same trend (Fig. 24). It is also clear from the above interpretation that the deltaic sediments deposited in 'point-sourced' depositional systems adjacent to continental collision zones (e.g. Kutei and Baram Deltas) are more oil prone than the deltaics deposited in the broader regressive backstepping systems typically deposited in rift settings (e.g. Gulf of Thailand, Northern Malay basins). This is presumably related to more rapid burial in the former systems.

There is clearly a strong link between the distribution of source rocks in SE Asia and the region's tectonic evolution. The challenge now is to use the tectonic models to better predict the presence, volume and nature (oil v. gas prone) of source rocks in undrilled (or under-explored) basins or develop new plays based on new source rock models in the existing hydrocarbon basins.

The author wishes to thank Woodside Offshore Petroleum and Lasmo plc for permission to publish this paper. Thanks are due to M. Welland for the vision to support this work, T. Thackeray and A. Pepper for their technical input and to P. Richardson and L. Small-Baczkiewicz for the drafting.

Appendix 1

West Sumatra Forearc: stratigraphic sequences as numbered. PN, pre-Neogene. Modified after Pertamina/Beicip (1985), Beadry & Moore (1985) & Rose (1983).
North Sumatra: JR, Julurayeu; S, Seurula; K, Ketapang; B, Baong; BS, Baong Sands; P, Pelutu; BL, Belumai; A, Arun; BA, Bampo; P/B, Parapat/Brukash; M, Meucampli; T: Tampur. Modified after Pertamina/Beicip (1985), Courtney *et al.* (1990), Sosromihardjo (1988).
Central Sumatra: MU, Upper Minas; ML, Lower Minas/Nilo; P, Petani; WF, Wingfoot Mbr; D, Duri; T, Telisa; US, Upper Sihapas; LS, Lower Sihapas/Bekasap/Bangko; M, Menggala/Lakat/Upper Red Beds; IB, Interbedded/Lake Fill Fm; BS, Brown Shale; BC, Basal Clastic/Lower Red Beds. Modified after Pertamina/Beicip (1985), Courtney *et al.* (1990).
South Sumatra: K, Kasai; ME, Muara Enim/Middle Palembang; AB, Air Benakat/Lower Benakat; G, Gumai/Telisa; BRF, Baturaja; TAF, Talang Akar; L, Lemat/Upper Lahat; LL, Lahat/Lower Lahat. Modified after Pertamina/Beicip (1985), Courtney *et al.* (1990).
NW Java: C, Cisbubuh; P, Parigi; PP, Pre-Parigi; AB, Air Benakat; MM, Mid Main; G, Gumai; BRF, Baturaja; TAF, Talang Akar comprising; GT, Gita; Z, Zelda members; B, Banuwati; J, Jatibarang. Modified after Pertamina/Beicip (1985), Courtney *et al.* (1990), Atkinson & Sinclair (1992), Butterworth & Atkinson (1993).
East Java: K, Karren; L, Lidah; M, Mundu; O, Ok; L, Ledok; W, Wonocolo; N, Ngroyong; T, Tuban; KI, Kujung I Limestone; P, Prupuh/Kujung II; KIII, Kujung III; PO, Poleng Limestone; CD/Ngimbang; C, Ngimbang Carbonate; NC, Ngimbang Clastics. Modified after Pertamina/Beicip (1985), Courtney *et al.* (1990), Bransden & Mathews (1992).
Barito: D, Dahor; UW, Upper Warukin; LW, Lower Warukin; BM, Berai Marl; BL, Berai Limestone; UT, Upper Tanjung; LT, Lower Tanjung. Modified after Pertamina/Beicip (1985), Courtney *et al.* (1990).
Kutei: HD, Handil Dua; A, Attaka; KB, Kampung Baru; B, Balikpapan Gp; SA, Sangatta Fm; PB, Palau Balang; K, Klinjau; PA, Pamaluan; M, Marah; A, Atan Beds; KH, Kiham Haloq; B, Boh Beds. Modified after Pertamina/Beicip (1985), Courtney *et al.* (1990).
Tarakan: B, Bunyu; T, Tarakan; D, Domarang; TA, Tabul; M, Meliat; L, Latih; N, Naintupo; T, Tabalar; M, Mesaloi; SE, Seilor; SU, Sujau. Modified after Pertamina/Beicip (1985), Courtney *et al.* (1990).
NW Palawan: SR, unnamed syn-rift; N, Nido; P, Pagasa; G, Galoc; B, Batas; M, Matinloc; C, Carcar. Modified after Gallagher (1987), Holoway (1982).
Sabah: letter scheme a marked. Modified after Tan & Lamy (1990).
Baram Delta: letter scheme as marked. Modified after James (1984), Rijks (1981).
Sarawak: cycle Numbers as marked. Modified after Doust (1981).
East Natuna: M, Muda; T, Trembu; A, Arang Shales; SS, Sokang Sandstone; LA, Lower Arang; BA, Barat; G, Gabus. Modified after Pertamina/Beicip (1985), Courtney *et al.* (1990).
West Natuna: UM, Upper Muda; LM, Lower Muda; UA, Upper Arang; LA, Lower Arang; BA, Barat; UG, Upper Gabus; K, Keras; LG, Lower Gabus; B, Benua; L, Lama/Lumpur. Modified after Pertamina/Beicip (1985), Courtney *et al.* (1990), Ginger *et al.* (1993), Jonklaas (1991).
Malay: cycles as marked plus: P, Pilong (Groups A&B); U, unnamed sequence (Groups D/E&F); B, Bekok (Groups H&I); T, Tapis (Group J); PU, Pulai (Group K); S, Seligi (Group L); L, Ledang (Group M); TB, Teluk Butun; SM, Samala; SA; Sambas. Modified after Murphy (1992), Ramli (1988).
Pearl River Mouth (+Bac Bo/Yinggehai Basin), Q, Quaternary; WS, Wanshan; Y, Yehail; HJ, Hanjiang; ZJ, Zhujiang; ZH, Zhuhai; E, Enping; UW, Upper Wenchang; LW, Lower Wenchang; S, Shenhu. Modified after Chen *et al.* (1993), Su *et al.* (1989) and Wang & Sun (1994).
Salawati: SE, Sele; KS, Klasaman; K, Kais; KL, Klasafet; KM, Klamogun; S, Surga; F, Faumai; I, Imskin. Modified after Pertamina/Beicip (1982), Courtney *et al.* (1990).
Bintuni, SE, Sele; ST, Steengkool; KL, Klasafet; K, Kais; F, Fumai; W, Waripi. Modified after Pertamina/Beicip (1982), Courtney *et al.* (1990).

References

Abbott, M. J. & Chamalaun, F. H. 1981. Geochronology of some Banda arc volcanics. *In*: Barber, A. J. & Wiryosujono, S. (eds) *The Geology and Tectonics of Eastern Indonesia*. Geological Research & Development Centre, Bandung, **2**, 253–268.

Abimanyu 1990. The Stratigraphy of the Sulawesi Group in Tomori PSC, East arm of Sulawesi. *Proceedings PIT XIX Ikatan Ahli Geologi Indonesia, Bandung*, 11–13 December 1990.

Ali, J. R. & Hall, R. 1995. Evolution of the boundary between the Philippine Sea Plate and Australia; palaeomagnetic evidence from eastern Indonesia. *Tectonophysics*, **251**, 251–275

Atkinson, C. D. & Sinclair, S. W. 1992. Early Tertiary rift evolution and its relationship to hydrocarbon source, reservoirs and seals in the Offshore Northwest Java Basins, Indonesia (abs). *Bulletin American Association of Petroleum Geologists*, **76**, 1088.

Bauman, P. 1982. Depositional Cycles on Magnetic and back arcs: An example from Western Indonesia. *Revue de l'Institut Français du Petrole*, **37**, 3–17.

BEADRY, D. & MOORE, G. F. 1985. Seismic Stratigraphy and Cenozoic Evolution of West Sumatra Forearc Basin. *AAPG Bulletin*, **69**, 742–759.

BENARD, F., MULLER, C., LETOUZEY, J., RANGIN, C. & TAHIR, S. 1990. Evidence of Multi phase deformation in the Ronjong–Crocker Range (Northern Borneo) from Landsat imagery interpretation: Geodynamic implications. *Tectonophysics*, **183**, 321–339.

BRANSDEN, P. J. E. & MATTHEWS, S. J. 1992. Structural and Stratigraphic Evolution of the East Java Sea, Indonesia. Proceedings of the 21st Annual Convention of the Indonesian Petroleum Association, **I**, 417–454.

BRIAIS, A., PATRIAT, P. & TAPPONNIER, P. 1983. Updated interpretation of magnetic anomalies and seafloor spreading stages in the South China Sea: implications for the Tertiary of Southeast Asia. *Journal of Geophysical Research*, **98**, B4, 6299–6328.

BURRUS, J. & BROSSE, E., DE CHOPPIN, J. & GROSJEAN, Y. 1994. Interactions between Tectonism, Thermal History and Palaeohydrology in the Mahakam Delta, Indonesia: Model Results, Petroleum Consequences (abs). Proceedings of the American AAPG. *Southeast Asian Basins Conference, Kuala Lumpur*, 21–24 Aug., 1994.

BUTTERWORTH, P. J. & ATKINSON, C. D. 1993. Syn-rift deposits of the Northwest Java Basin. Fluvial Sandstone Reservoirs and lacustrine Source Rocks. *Proceedings of the Indonesian Petroleum Association Clastic Core Workshop*, October 1993, 211–229.

CHEN, P. R. H., CHEN, Z. H. & ZHANG, Q. M. 1993. Sequence stratigraphy and continental margin development of the Northwestern Shelf of the South China Sea. *AAPG Buletin*, **77**, 842–862.

COURTNEY, S., COCKCROFT, P., MILLER, R., PHOA, R. S. K. & WIGHT, A. W. R. 1990. Indonesia oil and gas field atlas. **I–VI**, Indonesian Petroleum Association.

CURRAY, J. R., MOORE, D. G., LAWVER, L. A., EMMEL, F., RAITT, J., HENRY, R. W. & KEIKHEFER, R. 1979. Tectonics of the Andaman Sea and Burma. *In*: WATKINS, J. S., MONTADERT, L. & DICKERSON, P. W. (eds) *Geological and Geophysical Investigations of Continental Slopes and Rises*. AAPG Memoirs, **24**, 189–98.

DALY, M. C., COOPER, M. A., WILSON, I., SMITH, D. G. & HOOPER, B. G. D. 1991. Cenozoic plate tectonics and basin evolution in Indonesia. *Marine and Petroleum Geology*, **8**, 2–21.

DAVIDSON, J. W. 1991. The geology and prospectivity of Buton Island, S.E. Sulawesi, Indonesia. *Proceedings of the 20th Annual Convention of the Indonesian Petroleum Association*, **I**, 209–233.

DEWEY, J. F. 1981. Episodicity, sequence, and style at convergent plate boundaries. *In*: STRANGWAY, D. W. (ed.) *The Continental Crust and its Mineral Deposits*, Special Papers of the Geological Association of Canada, **20**, 553–573.

——, CANDE, S. & PITMAN, W. C., III 1989. Tectonic evolution of the India/Eurasia collision zone. *Ecologae Geologicae Helvetiae*, **82**, 717–734.

DOUST, H. 1981. Geology and exploration history of offshore central Sarawak. *In*: HALBOUTY, M. T. (ed.) *Energy Resources of the Pacific Region*. AAPG Studies in Geology, **12**, 117–132.

DURKEE, E. F. 1992. Oil geology and changing concepts in the southwest Philippines (Palawan and the Sulu Sea). *Proceedings of the 9th Offshore South East Asia Conference*, 469–490.

DUVAL, B. C., CHOPPIN DE JANVRY, G. & LORIET, B. 1992. The Mahakam Delta Province: An Ever-Changing Picture and a Bright Future. Proceedings of the 24th Offshore Technology Conference, 393–404.

EUBANK, R. T. & MAKKI, A. C. 1981. Structural Geology of the Central Sumatra Back-arc Basin. *Proceedings of the 10th Annual Conference of the Indonesian Petroleum Association*, **I**, 153–196.

GAETANI, M. & GARZANTI, E. 1991. Multicyclic History of the Northern India Continental Margin (Northwestern Himalaya). *AAPG Bulletin*, **75**, 1427–1446.

GALLAGHER, J. J. 1987. Philippine Microplate Tectonics and Hydrocarbon Exploration. *In*: HORN, M. K. (ed.) *Transactions of the 4th Circum Pacific Energy and Mineral Resources Conference*, Singapore 1986, 103–119.

GARRARD, R. A., SUPANDJONO, J. B. & SURONO 1988. The geology of the Banggai–Sula/Microcontinent/Eastern/Indonesia. *Proceedings of the 17th Annual Convention of the Indonesian Petroleum Association*, **I**, 23–52.

GINGER, D. C., ARDJAKUSUMAH, W. O., HEDLEY, R. J. & POTHECARY, J. 1993. Inversion history of the West Natuna Basin: Examples from the Cumi-Cumi PSC. *Proceedings of the 22nd Annual Convention of the Indonesian Petroleum Association*, **I**, 635–658.

HAILE, N. S. 1992. Evidence of Multiphase deformation in the Rajang – Crocker Range (northern Borneo) from Landsat imagery interpretation: geodynamic implication – comment(2). *Tectonophysics*, **204**, 178–180.

HAMILTON, W. 1979. Tectonics of the Indonesian region. US Geological Survey Professional Paper, **1078**.

——1988. Plate tectonics and island arcs. *Geological Society of American Bulletin*, **100**, 1503–1527.

HANDIWIRIA, Y. E. 1990. The Stratigraphy and hydrocarbon occurrences of the Salodik Group, Tomori PSC area East Arm of Sulawesi. *Proceedings of PIT XIX Ikatan Ahli Geologi Indonesia*, Bandung 11–13 December, 1990.

HARLAND, W. B., ARMSTONG, R. L., COX, A. V., CRAIG, L. E., SMITH, A. G. & SMITH, D. G. 1990. *A geologic time scale 1989*. Cambridge University Press.

HAQ, B. U., HARDENBOL, J. & VAIL, P. R. 1988. Mesozoic and Cenozoic chronostratigraphy and cycles of sealevel change. *In*: WILGUS, C. K. *et al.* (eds) *Sea Level Changes: an Integrated Approach*. SEPM Special Publications, **42**, 71–108.

HAYES, D. E. & SPANGLER, S. E. 1987. Age and Evolution of the South China Sea Southwest Subbasin. *Eos*, **68**, 1496.

HEIDRICK, T. L. & AULIA, K. 1993. A structural and tectonic model of the Coastal Plains Block, Central Sumatra Basin, Indonesia. *Proceedings of the 22nd Annual Convention of the Indonesian Petroleum Association*, **I**, 285–318.

HEKINIAN, R., BONTE, P., PAUTOT, G., JACQUES, L., LABEYRIE, L., MIRRELSEN, N. & REYSS, J. 1989. Volcanics from the South China Sea Ridge System. *Oceanologica Acta*, **12**, 101–105.

HILDE, T. W. C. LEE, C. 1984. Origin and evolution of the West Philippine Basin: A new interpretation. *Tectonophysics*, **102**, 85–104.

——, UYEDA, S. & KROENKE, L. 1977. Evolution of the Western Pacific and its margin. *Tectonophysics*, **38**, 145–165.

HOLLOWAY, N. H. 1982. North Palawan Block, Philippines – Its relation to Asian mainland.

HUBARD, R. J. 1988. Age and significance of sequence boundaries on Jurassic and Early Cretaceous rifted continental margins. *AAPG Bulletin*, **72**, 49–72.

HUTCHISON, C. S. 1986. Tertiary basins of S.E. Asia – their diseparate tectonic origins and eustatic stratigraphic similarities. *Geological Society of Malaysia Bulletin*, **19**, 109–122.

——1989. *Geological Evolution of Southeast Asia.* Oxford Monographs on Geology & Geophysics, **13**. Clarendon Press.

JAMES, D. M. D. 1984. *The geology and hydrocarbon resources of Negara Brunei Darusslam.* Special Publications of Brunei Museum.

JONKLAAS, P. 1991. Integration of depth conversion, seismic inversion and modelling over the Belida Field, South Natuna Sea Block 'B', Indonesia. *Proceedings of the 20th Annual Convention of the Indonesian Petroleum Association*, **I**, 557–585.

KUSUMA, I. & DARIN, T. 1989. The Hydrocarbon Potential of the Lower Tanjung Formation, Barito BASIN, S. E. Kalimantan. *Proceedings of the 18th Annual Convention of the Indonesian Petroleum Association*, **I**, 107–138.

LEE, C. S. & MCCABE, R. 1986. The Banda–Celebes–Sulu basin: a trapped piece of Cretaceous–Eocene oceanic crust? *Nature*, **322**, 51–53.

LETOUZEY, J., WEARNER, P. & MARTY, A. 1990. Fault reactivation and structural inversion. Backarc and intraplate compressive deformations. Example of the eastern Sunda shelf (Indonesia). *Tectonophysics*, **183**, 341–362.

LIANG, D. LIU, Z. 1990. The genesis of the SCS and it's hydrocarbon-bearing basins. *Journal of Petroleum Geology*, **13**, 59–70.

LIVINGSTONE, H. J. 1992. Hydrocarbon Source and Migration Salawati Basin, Irian Jaya. *Proceedings of the Eastern Indonesia Exploration Symposium*, 29 April, 1992.

MIALL, A. D. 1992. Exxon global cycle chart: An event for every occasion? *Geology*, **20**, 787–790.

MOUNT, V. S. & SUPPE, J. 1992. Present-day Stress Orientations adjacent to Active Strike-Slip Faults: California and Sumatra. *Journal of Geophysical Research*, **97**, 11 995–12 013.

MURPHY, R. W. 1992. *The Petroleum Geology of Southeast Asia.* Text Figures to accompany courses notes. Unpublished.

——1994. Tertiary Tectonic events, Southwestern South China Sea (abs). *Proceedings of the AAPG Southeast Asian Basins Conference*, Kuala Lumpur, 21–24 August 1994.

NATION, L. 1994. Ideas shared at K.L. Meeting. *AAPG Explorer Magazine*, October 1994, 28–29.

PACKHAM, G. H. 1990. Plate Motions and Southeast Asia: Some Tectonic sequences for Basin development. *Proceedings of the 8th offshore Southeast Asia conference*, 4–7 Dec., Singapore.

PEPPER, A. S. & CORVI, P. J. 1995. Simple kinetic models of petroleum formation. Part I: oil and gas generation from kerogen. *Marine and Petroleum Geology*, **12**, 291–319.

PATRIAT, P. & ACHACHE, J. 1984. India–Eurasia collision chronology has implications for crustal shortening and driving mechanism of plates. *Nature*, **311**, 615–621.

PAUTOT, G., RANGIN, C., BRIAIS, A., TAPPONNIER, P., BAUZART, P., LERICOLAIS, G., MATHIEU, X., WU, J., HAN, S., LI, H., LU, Y. & ZHAO, J. 1986. Spreading direction in the central South China Sea. *Nature*, **321**, 150–154.

PERTAMINA/BEICIP 1982. *Petroleum Potential of Eastern Indonesia.* Unpublished.

——1985. *Hydrocarbon Potential of Western Indonesia.* Unpublished.

POLACHAN, S., PRADIDTAN, S., TONGTAOW, C., JANMAHA, S., INTARAWIJITR, K. & SANGSUWAN, C. 1991. Development of Cenozoic basins in Thailand. *Marine and Petroleum Geology*, **8**, 84–96.

PRAPTONO, S. H., DWIPUTRO, R., LONGLEY, I. M. & WARD, R. W. 1991. Kurau: an example of the low-relief structural play in the Malacca Straits PSC, Sumatra, Indonesia. *Proceedings of the 20th Annual Convention of the Indonesian Petroleum Association*, **I**, 299–318.

RAMLI, M. N. 1988. Stratigraphy and Palaeofacies Development of Carigali's Operating Areas in the Malay Basin, South China Sea. *Geological Society of Malaysia Bulletin*, **22**, 153–187.

RANGIN, C. & SILVER, E. A. 1991. Neogene Tectonic evolution of the Celebes-Sulu Basins: new insights from Leg 124 drilling. *In*: SILVER, E. A., RANGIN, C. & VON BREYMANN, M. T. (eds) *Proceedings of the Ocean Drilling Program, Scientific Results*, **12**, 51–64.

——, JOLIVET, L., & PUBELLIER, M. 1990. A simple model for the tectonic evolution of southeast Asia and Indonesia region for the past 43 m.y. Bullletin de Société Geologique de France, **(8)**, **VI**, **6**, 889–905.

RIJKS, E. J. H. 1981. Baram Delta geology and hydrocarbon occurrence (Sarawak). *Geological Society of Malaysia Bulletin*, **14**, 1–18.

ROBINSON, K. 1987. An overview of Source Rocks and oils in Indonesia. Proceedings of the 16th Annual Convention of the Indonesian Petroleum Association, **I**, 97–122.

ROESER, H. A. 1991. Age of the southeast Sulu Sea basin based on magnetic anomalies and age determined at site 768. *In*: SILVER, *et al.* (eds) *Proceeding Ocean Drilling Program, Scientific Results*, **124**, 339–343.

ROSE, R. 1983. Miocene carbonate rocks of Sibolga Basin, Northwest Sumatra. *Proceedings of the 12th Annual Convention of the Indonesian Petroleum Association*, **I**, 107–114.

RU, K. & PIGOTT, J. D. 1986. Episodic rifting and subsidence in the South China Sea. *AAPG Bulletin*, **70**, 1136–1155.

SAWYER, R. K., SANI, K. & BROWN, S. 1993. The stratigraphy and sedimentology of West Timor, Indonesia. *Proceedings of the 22nd Annual Convention of the Indonesia Petroleum Association*, **I**, 533–574.

SCHERER, M. & HITAM, R. 1992. Distribution and maturation of source rocks in Brunei Darussalam. *Proceedings of the AAPG International Conference & Exhibition, Sydney*, 2–5 Aug., 71.

SHAW, R. D. 1990. Frontier Basins of Southeast Asia: A review of their hydrocarbon potential. *Proceeding Southeast Asia Petroleum Exploration Society (Seapex)*, **IX**, 69–80.

SILVER, E. A., RANGIN, C. & VON BREYMANN, M. T. (eds) 1991. *Proceedings of the Ocean Drilling Program, Scientific Results*, **124**.

SMITH, R. B. & SILVER, E. A. 1991. Geology of a Miocene Collision complex, Buton, eastern Indonesia. *Geological Society of America Bulletin*, **103**, 660–678.

SU, D., WHITE, N. & MCKENZIE, D. 1989. Extension and subsidence of the Pearl River Mouth Basin, northern South China Sea. *Basin Research*, **2**, 205–222.

SOSROMIHARDJO, S. P. C. 1988. Structural Analysis of the North Sumatra Basin with emphasis on SAR data. *Proceedings of the 17th Annual Convention of the Indonesian Petroleum Association*, **I**, 321–340.

TAN, N. K. LAMY, J. M. 1990. Tectonic Evolution of the NW Sabah Continental margin since the Late Eocene. *Geological Society of Malaysia Bulletin*, **27**, 241–260.

TAPPONNIER, P., PELTZER, G., LE DAIN, A. Y. & ARMIJO, R. 1982. Propagating extension tectonics in Asia: New insights from simple experiments with plasticine. *Geology*, **10**, 611–616.

—— & ARMIJO, R. 1986. On the Mechanism of collision between India and Asia. *In*: COWARD, M. P. & RIES, A. C. (eds) *Collision Tectonics*. Geological Society, London, Special Publications, **19**, 115–157.

TAYLOR, B. & HAYES, D. E. 1983. Origin and history of the South China Sea Basin. *In*: The tectonic and geologic evolution of southeast Asian seas and islands, part 2. *American Geophysical Union Geophysical Monograph*, **27**, 23–56.

VAIL, P. R., MITCHUM, R. M. & THOMPSON, S. 1977. Seismic stratigraphy and global changes in sea level, part 4: Global cycles of relative changes of sea level. *In*: PAYTON, C. E. (eds) *Seismic Stratigraphy-Application to hydrocarbon exploration. AAPG Memoirs*, **26**, 83–89.

VAN DE WEERD, A. & ARMIN, R. A. 1992. Origin and Evolution of the Tertiary hydrocarbon bearing basins in Kalimantan (Borneo), Indonesia. *AAPG Bulletin*, **76**, 1778–1803.

VEEVERS, J. J. & LI, Z. X. 1991. Review of Seafloor spreading around Australia. II. Marine magnetic anomaly modelling. *Australian Journal of Earth Science*, **38**, 391–408.

——, POWELL, C.MCA. & ROOTS, S. R. 1991. Review of seafloor spreading around Australia. I. Synthesis of the patterns of spreading. *Australian Journal of Earth Science*, **38**, 373–389.

WANG, C. & SUN, Y. 1994. Development of Palaeogene Depressions and Deposition of Lacustrine Source Rocks in the Pearl River Mouth Basin, Northern Margin of the South China Sea. *American Association of Petroleum Geologists Bulletin*, **78**, 1711–1728.

WEISSEL, J. K. 1980. Evidence for Eocene oceanic crust in the Celebes Basin. *In*: HAYES, D. E. (ed) *The Tectonic and Geologic Evolution of Southeast Asian Seas and Islands*. American Geophysical Union, Geophysical Monograph Series, **23**, 37–47.

WILLIAMS, H. H. 1992. Geochemistry of Palawan Oils, Philippines: Source implications. *Proceedings of the 9th Offshore Southeast Asia Conference*, Singapore 1–4 Dec. 1992.

——, JOHNSON, C. R., ALMOND, R. A. & SIMAMORA, W. H. 1988. Late Cretaceous to Early Tertiary structural elements of West Kalimantan. *Tectonophysics*, **148**, 279–297.

——, KELLY, P. A., JANKS, J. S. & CHRISTENSEN, R. M. 1985. The Palaeogene Rift Basin Source Rocks of Central Sumatra. *Proceedings of the 17th Annual Convention of the Indonesian Petroleum Association*, **I**, 57–90.

WOODROOF, P. B. 1992. Source rock and hydrocarbon geochemistry, Offshore NW Sabah, Malaysia. *Proceedings of the AAPG International Conference & Exhibition, Sydney*, 2–5 Aug., 78.

YU, H. S. 1990. The Pearl River Mouth Basin: a rift basin and its geodynamic relationship with the Southeastern Eurasian Margin. *Tectonophysics*, **183**, 177–186.

ZIEGLER, P. A. 1992. Plate Tectonics, plate moving mechanisms and rifting. *Tectonophysics*, **215**, 9–34.

Constraints on strike-slip motion from seismic and gravity data along the Vietnam margin offshore Da Nang: implications for hydrocarbon prospectivity and opening of the East Vietnam Sea

D. ROQUES, S. J. MATTHEWS[1] & C. RANGIN[2]

Département de Géotectonique (CNRS URA 1759), Université Pierre et Marie Curie, T.26 E1, 4 place Jussieu, 75252 Paris cedex 05, France

[1] *BP Exploration, BP Plaza, 200 WestLake Park Boulevard, Houston, Texas, 77079, USA*

[2] *Laboratoire de Geologie de l'Ecole Superieure (CNRS URA 1316), 24 rue Lhomond, 75231 Paris cedex 05, France*

Abstract: The Vietnamese continental margin records the tectonic processes which have controlled opening of the East Vietnam Sea basin. The geometry and chronology of Cenozoic structures linked to the opening of this marginal basin have been studied offshore Da Nang (central Vietnam), based on unpublished data acquired by BP Exploration from 1989 to 1991. These data demonstrate that a N160° E fault trend is the main structural orientation recognised on both gravity and seismic data. Important crustal extension on shallow-dipping fault planes striking N060° E and connected to the N160° E faults is observed also. This horse-tail map-pattern is compatible with transtensional dextral motion along the N160° E faults. The discrepancy between the amount of motion observed onshore and offshore suggests the presence of a major shear zone which trends subparallel to the east Vietnam margin. This shear zone is located in the region of the offshore Qui Nhon ridge, which is blanketed by carbonate sediments. These carbonates have been drilled and dated as Middle Miocene. Dextral motion along this shear zone diminished during the late Early Miocene (around 20 Ma.), as evidenced by the onset of carbonate deposition on top of the ridge. This is broadly synchronous with propagation of the East Vietnam Sea basin towards the southwest. This major shear zone has been a fundamental part of the fault network which developed during opening of the East Vietnam Sea.

The structural development of the offshore Da Nang region has resulted in the development of contrasting sub-basin geometries; those within the major shear zone appear to be relatively restricted laterally whereas those to the east display geometries more typical of extensional sub-basins. Both sub-basins show a variety of potential hydrocarbon trapping styles and by analogy with comparable areas in SE Asia have the potential to contain important oil reserves, in addition to the considerable volume of gas already proven by recent drilling.

The offshore Da Nang area (Fig. 1) is in a key location for understanding development of the Vietnam margin during the Cenozoic. It is located at the junction of the offshore continuation of the Red River fault system with a N–S-trending fault system offshore central Vietnam. Several sedimentary basins have developed within these fault systems, between and 14 and 21° N, and they are typically blanketed by relatively young Neogene sediments interpreted to have been sourced from the Vietnamese hinterland.

The Tonkin Gulf basin (Fig. 1) is located offshore to the SE of the Red River delta; the eastern part of this basin system extends north of Hainan island where it forms the structurally distinct Beibu Gulf basin. To the SE of the Tonkin Gulf basin is the NW–SE-trending Song Hong basin which tapers southward and is separated from the southerly Quang Ngai graben by a structural high, the Da Nang high. The Quang Ngai graben is located adjacent to the N160° E orientated Qui Nhon ridge (Fig. 1). Several major basins of the East Vietnam Sea have formed to the east and SE of the Qui Nhon ridge; two of these, the Hoang Sa and the deep Nha Trang basin, are shown in Fig. 1.

Between 11°30′ N and 14° N, the Vietnam margin is defined by a relatively continuous N–S-trending fault system with approximately 2000 m cumulative vertical offset from west to east, which has been blanketed by Neogene sediments. This fault system has been interpreted previously as a dextral transform fault along the western margin of the East Vietnam Sea (Taylor & Hayes 1983). Alternative regional interpretations (e.g. Tapponnier *et al.* 1986) have concentrated on assessment of the motion to the north of this, along the southern extension of the Red River fault system. The Tonkin Gulf basin is located within this southern extension

From Fraser, A. J., Matthews, S. J. & Murphy, R. W. (eds), 1997, *Petroleum Geology of Southeast Asia*, Geological Society Special Publication No. 126, pp. 341–353.

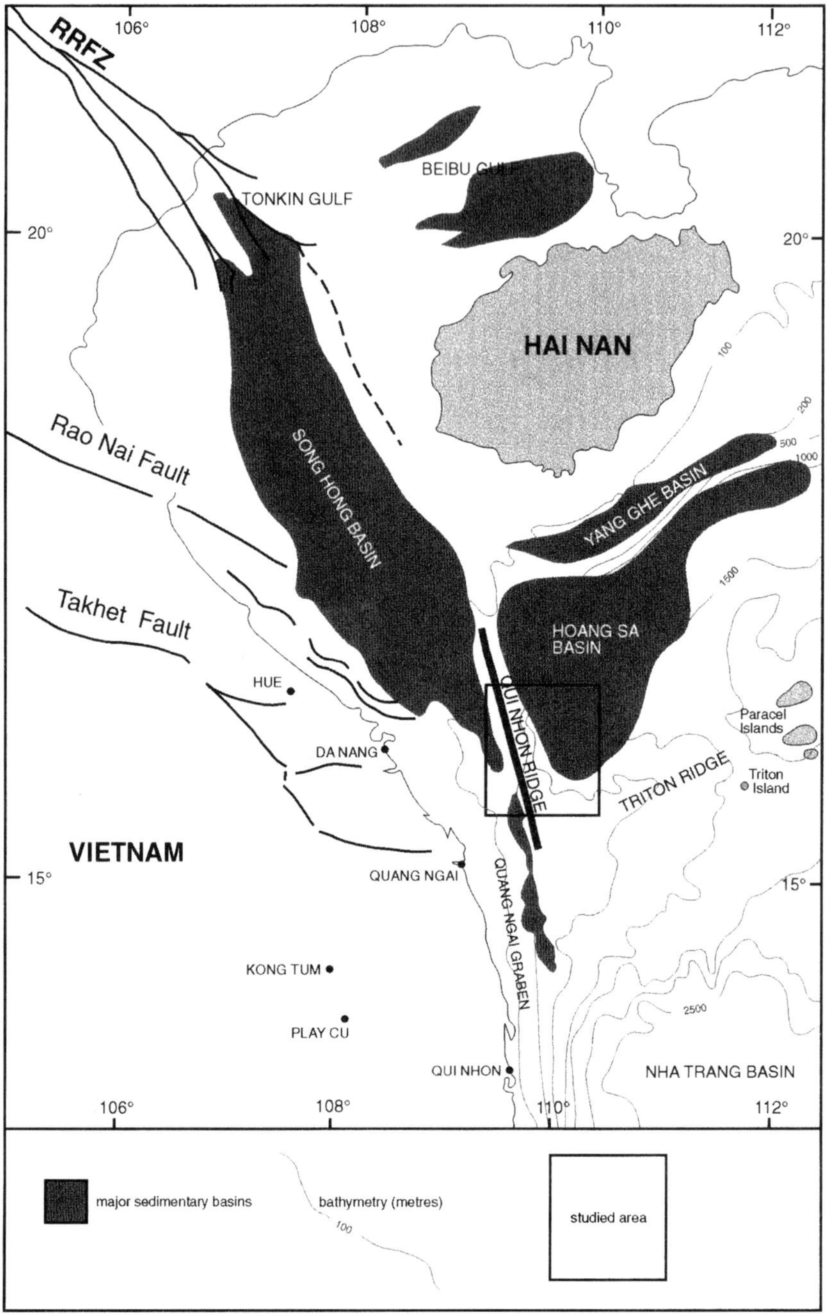

Fig. 1. Major structures and basins of NE Vietnam and the East Vietnam Sea. The Qui Nhon ridge is a sub-continuous N160° E-trending ridge. RRFZ, Red River Fault Zone.

and appears to have formed in response to left lateral motion within the Red River fault system. This interpretation of movement sense contrasts with our preferred model for the Nha Trang basin to the south, which we interpret to be an extensional basin linked directly to dextral motion along the N–S-trending fault system offshore central Vietnam (Marquis *et al.* 1995;

Roques *et al.* 1995). Between these two basin systems lies the Hoang Sa basin which is located close to the junction of the two major strike-slip fault zones; the Red River fault system in the north and the N–S-trending Vietnam fault system in the south. Fault geometry and timing offshore Da Nang (Fig. 1), and also within the Hoang Sa basin, is compatible with right lateral motion along the east Vietnam margin during the latest Palaeogene to Early Miocene. The structural geometrical basis for this kinematic interpretation is derived from seismic and gravity data acquired since 1989, and is presented later in this paper.

Regional gravity field and structural setting offshore Da Nang

Two major structural zones are developed offshore Da Nang; a region of NE–SW-trending tilted fault blocks which formed during thinning of East Vietnam Sea continental crust, and a sub-meridian (longitudinal) (N160° E)-trending ridge and basin complex located to the west of the tilted fault blocks. This map-view geometry is recognizable on the Sandwell gravity map (Fig. 2).

The Hoang Sa basin (Fig. 2) is characterized by a negative gravity anomaly (60 mgal) with an

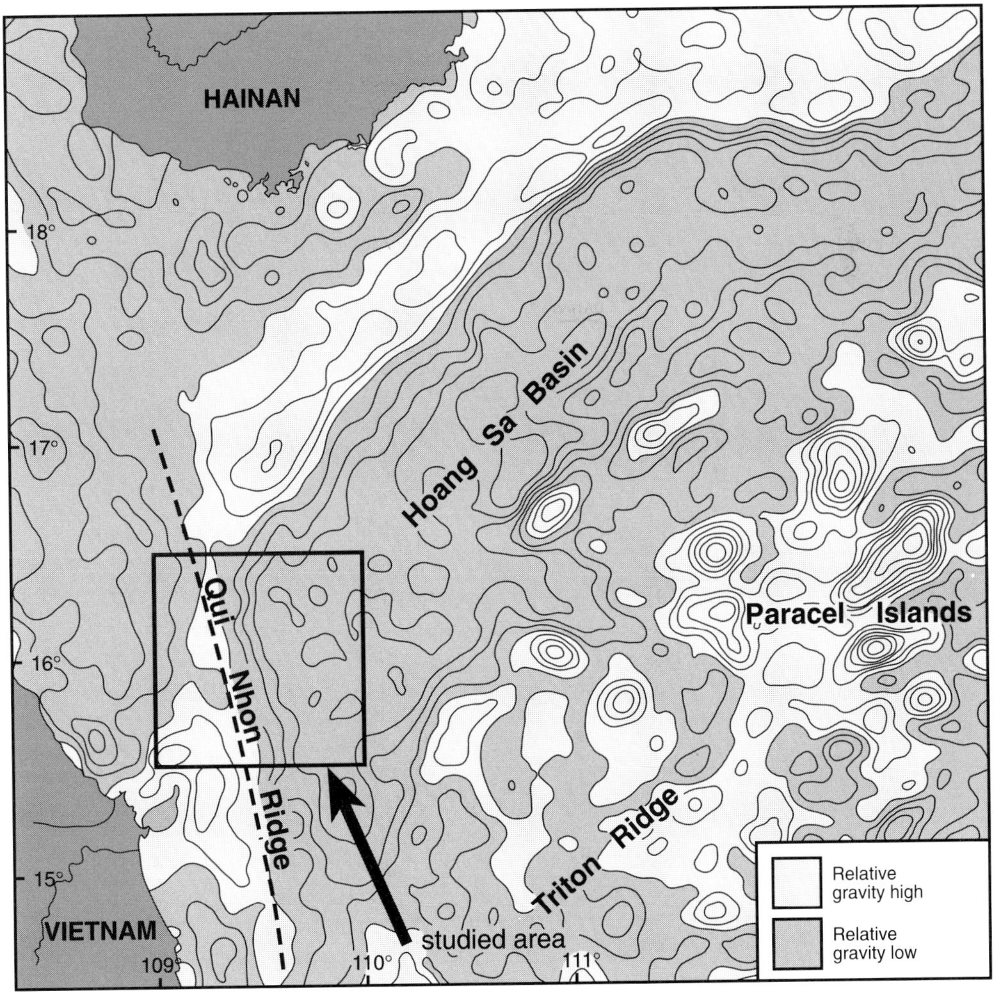

Fig. 2. Regional gravity map in the area of the Hoang Sa basin (based on Sandwell gravity data). The Hoang Sa basin is represented by an arcuate gravity low (reaching −60 mgals) and the Qui Nhon ridge by a linear gravity high (5–10 mgals).

arcuate southern margin which can be traced from 15° N to 18°30′ N, north of the Paracel Islands. This negative anomaly connects westward with the Qui Nhon ridge, which is marked by a sub-meridian (longitudinal) positive anomaly. The negative anomaly of the Hoang Sa basin is over 150 km in extent westward but narrows rapidly to the east. This gravity field is interpreted to be a response to the structural geometry which evolved during East Vietnam Sea crustal thinning. South of the Hoang Sa basin, crustal thinning increases southward and is interpreted to attain highest extension values in the Nha Trang basin where a beta value of 2.3 has been calculated, which is compatible with a 13 km thick crust (Marquis *et al.* 1995).

The Qui Nhon ridge is shown on the Sandwell map (Fig. 2) to connect northward with a N140° E trending positive anomaly in the Song Hong basin (Fig. 1). This connection is not imaged clearly on seismic data. The ridge is blanketed by an eastward prograding and

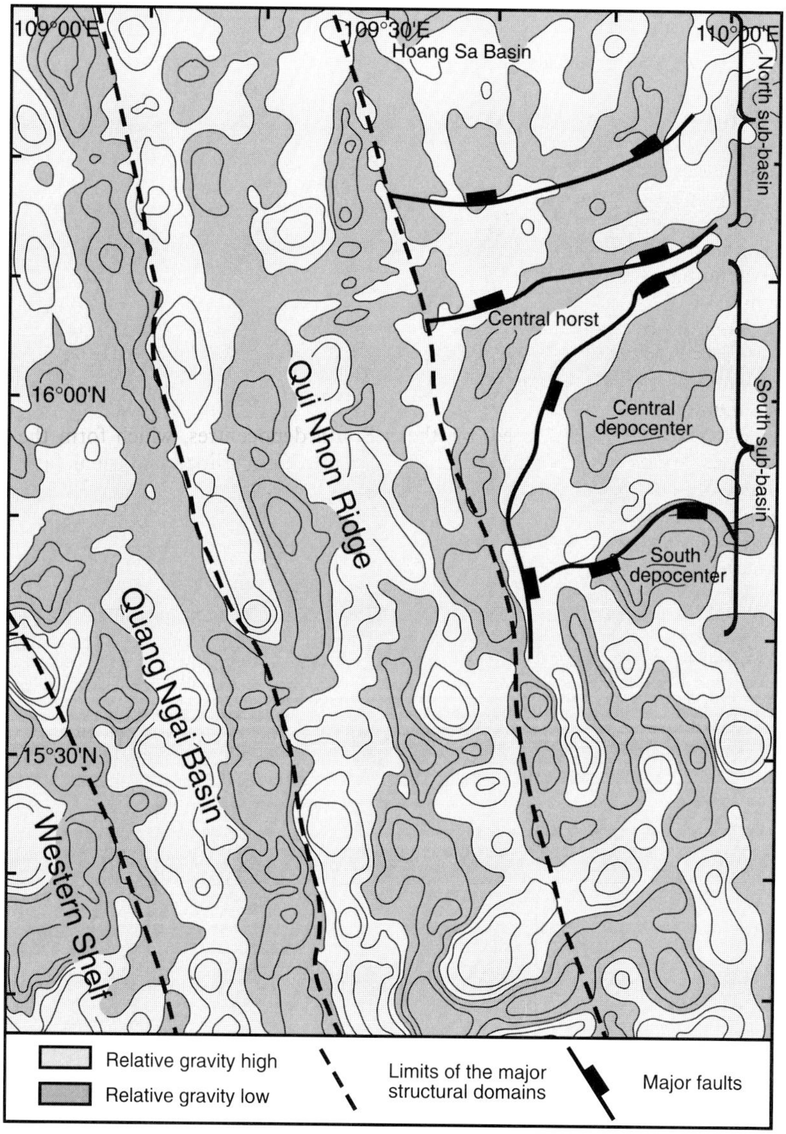

Fig. 3. Simplified gravity map offshore Da Nang based on data acquired by BP Exploration. Note two distinct trends (N160° E and N050° E) that are clearly represented on the map.

undeformed clastic depositional system, which is 1–2.5 s (two-way time) thick at the latitude of Qui Nhon and Hue (Fig. 1) respectively. The ridge reduces progressively as a topographic feature northward; seismic data demonstrates that the Qui Nhon ridge is only 20 km wide near to 17° N, but much wider southward. The junction of the Qui Nhon ridge with the Triton ridge (Fig. 1) is close to the southern termination of the Hoang Sa basin.

Structural geometry

Basin-scale structural geometry based on gravity and seismic data

Simplified gravity data offshore Da Nang (Fig. 3) demonstrates the presence of a N160° E fabric over a large part of the area. In addition, this map shows N050° E to N060° E trending gravity highs and lows in the Hoang Sa basin. Minor N–S-trending gravity highs are also locally present.

The western margin of the Qui Nhon ridge is defined by a N160° E trending anomaly, but its eastern flank is not as clear on this gravity map (Fig. 3). The Quang Ngai basin, west of the Qui Nhon ridge, is also well defined by the N160° E fabric as far west as the shelf break (Western Shelf on Fig. 3). Subparallel fault systems have also been identified onshore (Rangin *et al.* 1995). East of the Qui Nhon ridge, the southern termination of the Hoang Sa basin is outlined by a series of N050° E-trending anomalies. These anomalies reduce in width and curve westward at the junction with the Qui Nhon ridge.

The southeastern portion of the map (Fig. 3) does not illustrate clearly any dominant trend, although there are some minor gravity highs trending N160° E. This trend is also detectable in the southern part of the Hoang Sa basin (Fig. 3). It appears that the N160° E trend is a regional anomaly trend, which becomes less dominant in the eastern part of this map area (Fig. 3) where a N050° E anomaly trend is more strongly developed.

The regional structure offshore Da Nang is illustrated by two seismic lines (Fig. 5a & b) which are located in Fig. 4. Line A (Fig. 5a) (oriented NNW–SSE) displays tilted fault block geometries to the east of the Qui Nhon ridge in the Hoang Sa basin and to the south. These tilted blocks correspond to N050° E-trending gravity anomalies observed east of the Qui Nhon ridge on Fig. 3.

Rift-fill basin geometries are clearly shown on the profile and they are bounded by two major south-dipping normal fault systems. These faults are coincident with the limits of the two major gravity lows observed on Fig. 3 (Central and South depocentres, which form the South sub-basin). North-dipping normal faults visible on the northern part of Fig. 5a cannot be correlated as clearly with the gravity data (Fig. 3). Line B (Fig. 5b) shows the Qui Nhon ridge near latitude 16° 15′ N. The ridge is located beneath the major

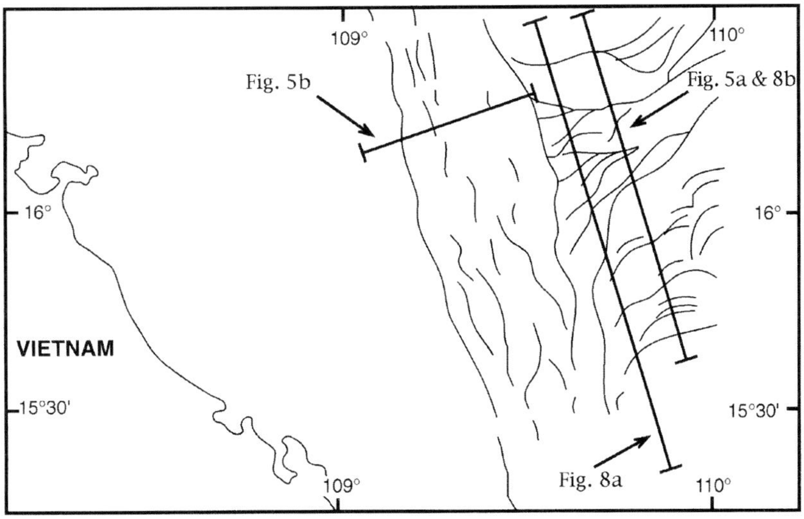

Fig. 4. Location of seismic lines illustrated in Fig. 5 and structural cross-sections illustrated in Fig. 8.

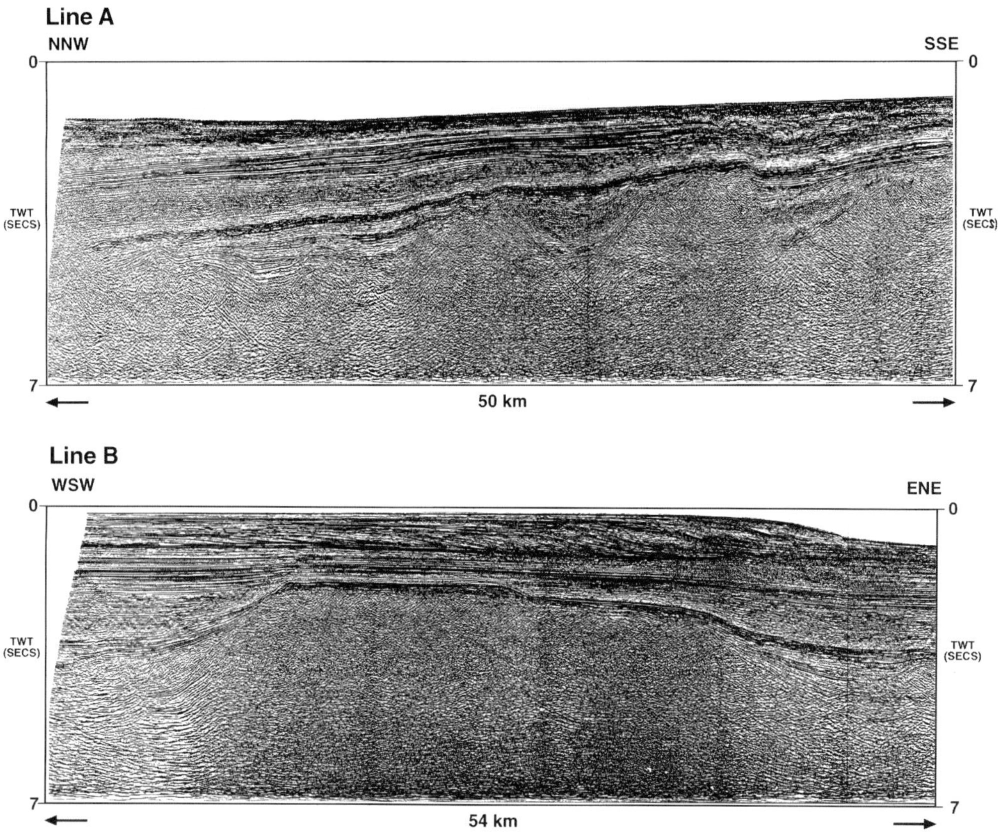

Fig. 5. (**a**) Line A: a NNW–SSE seismic line across the tilted fault blocks of the Hoang Sa basin and South basin, located in Fig. 4. (**b**) Line B: a WSW–ENE seismic line across the Qui Nhon ridge, located in Fig. 4. Note the carbonate build-up which overlies the ridge and the easterly prograding clastic system which forms the uppermost part of the section.

carbonate build-up. Small tilted fault-block geometries are imaged locally below the carbonates on some profiles.

The structural and stratigraphic geometries observed on seismic data are compatible with early (possibly Palaeogene) rifting of 'pre-rift' basement, with a transition to the 'post-rift' phase of basin development during the latest Palaeogene to Early Miocene. This interpretation is based on a combination of well data (which constrains the post-rift history) and regional integration which supports Palaeogene rifting around the East Vietnam Sea margin.

Fault geometry in map-view and cross-section

A structural map (Fig. 6) shows the main faults interpreted at top 'pre-rift' basement, and a sequence of line drawings and seismic examples shows the cross-sectional geometry of some of the major faults (Figs. 7 & 8).

The Qui Nhon ridge is imaged clearly on both gravity and seismic data (Figs 3 and 5b). It is bounded to the west by a major fault zone with consistent net-normal movement sense. This fault zone is continuous for over 160 km from north to south (Fig. 6). A discontinuous east-dipping fault zone controls the eastern flank of the ridge. This fault zone is apparently hard-linked eastward to the Hoang Sa basin fault system. Many of the N050° E-trending faults join abruptly with this east-dipping N160° E fault zone. However there are several examples of major SE-dipping N050° E normal faults which curve progressively into alignment with the Qui Nhon ridge (for example the fault which controls the Central depocentre of the South sub-basin Fig. 7), in contrast to displaying an apparently truncational relationship. These SE-dipping faults commonly correspond with

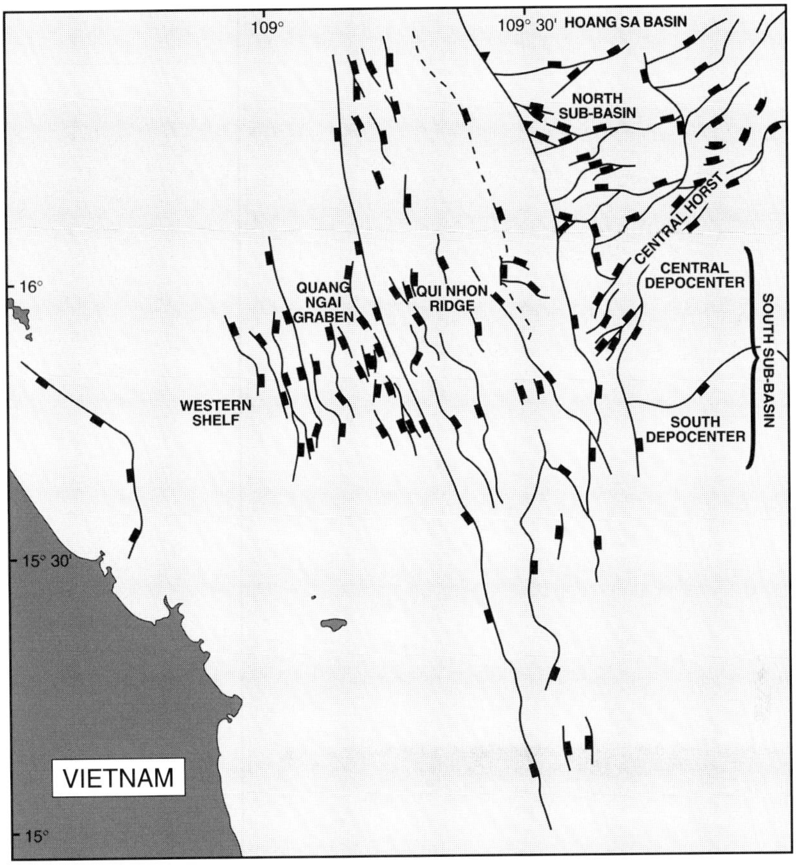

Fig. 6. Structural map of the offshore Da Nang area (south western tip of the Hoang Sa basin).

positive gravity anomalies (Fig. 3). The gravity high which corresponds to the Central depocentre fault shows the same map-view shape as the fault, and is the boundary to the central depocentre which is observed clearly on the gravity map (Fig. 3).

The detailed geometry of this major fault is illustrated in Fig. 7; over 3 s TWT of rift-fill sediment is present in the hanging wall of the fault on sections A and B. This major fault has a strongly curved shape, with a listric geometry (approximately normal movement sense) along the N050° E trend and a steeper geometry (possible transtensional movement sense) southward. This geometry is compatible with dextral strike-slip motion along the eastern flank of the Qui Nhon ridge which appears to be linked to dip-slip movement to the east.

The structural cross-sections of Fig. 8 illustrate the relationship between the SE-dipping listric faults described above and a NW-dipping normal fault system. The footwall to these two fault systems is a curved horst, the central horst, which separates the North sub-basin from the South sub-basin. In map view (Fig. 6), the NW-dipping faults terminate against the eastern edge of the Qui Nhon ridge. These faults are interpreted to be broadly contemporaneous with the SE-dipping normal faults. A comparable geometrical relationship, but at a larger scale, is observed to the south where the Triton ridge forms a horst between the Hoang Sa basin and the Nha Trang basin (Roques *et al.* 1995).

NW-dipping faults have been identified to the north, offshore Hue (Fig. 1), in Central Vietnam. Here these normal faults connect with the Rao Nai sinistral fault zone (Fig. 1), a fault subparallel to the Red River fault system, and they can be interpreted as part of the southern termination of this strike-slip fault system. If this interpretation is correct then the Da Nang area may be interpreted as the connection between two major and distinct strike-slip fault systems; a sinistral strike-slip fault zone

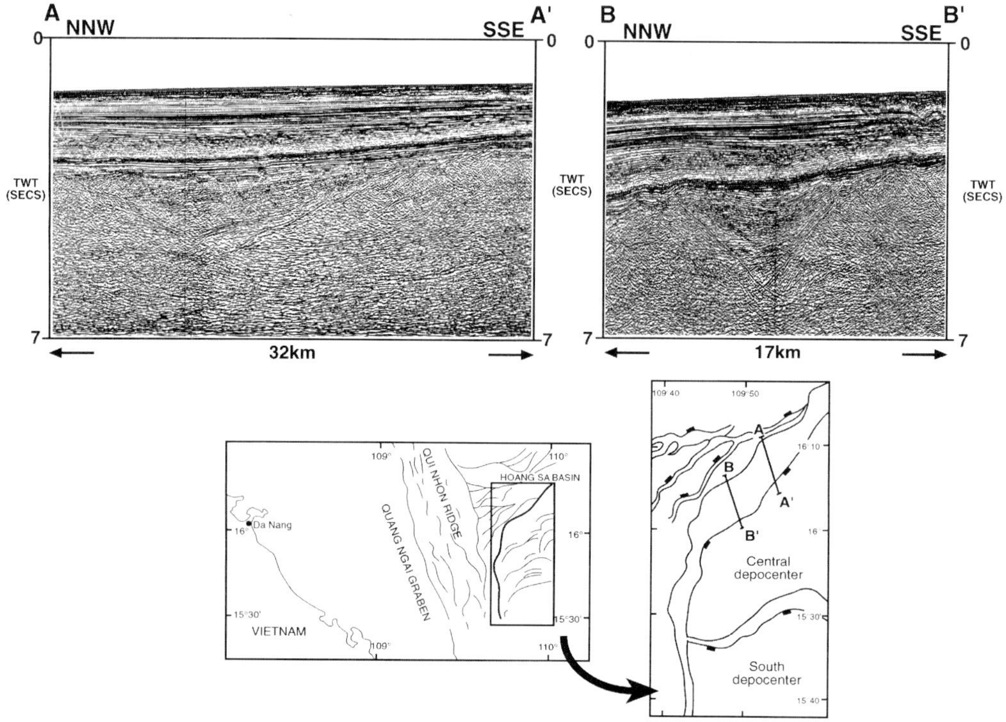

Fig. 7. Geometrical variations along-strike of fault geometry in the southern Hoang Sa basin. (The location map of seismic line segments illustrates only that part of the line which crosses the main central depocentre controlling fault).

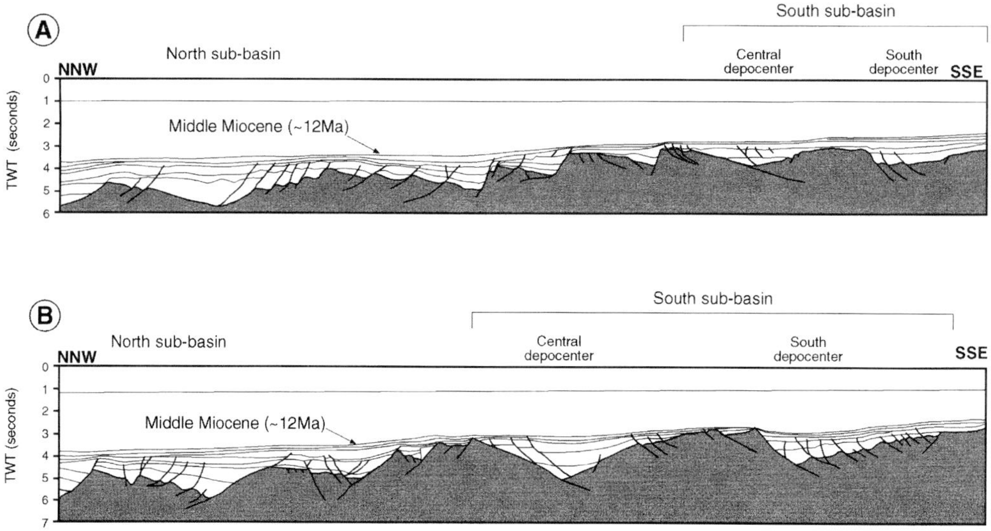

Fig. 8. NNW–SSE structural cross-sections of the southwestern part of the Hoang Sa basin (North and South sub-basins), located in Fig. 4.

connected with the Red River fault system and a system of N–S-trending dextral strike-slip faults to the south.

Timing of structural development

The Qui Nhon ridge

Miocene carbonates drilled by BP Exploration on the Qui Nhon ridge, between 15° N and 16° N, range in age from Burdigalian to Langhian. Carbonate facies suggest a variation from relatively shallow to deeper water depositional bathymetry up-section, and the overlying clastic sediments are deep marine (Holland *et al.* 1992). This variation in facies is a response to the rapid subsidence of the east Vietnam margin from the Burdigalian onwards.

Above the ridge, three phases of progradation are recognised which represent pulsed clastic sedimentation on top of the carbonates. These progradational episodes commenced near to 12 Ma, and may be linked to uplift of the Vietnamese peninsula during the Mid- to Late Miocene. This is coincident with the age of some of the basalts that have been sampled onshore (Rangin *et al.* 1995). These basalts are bracketed between 11 Ma and the present day (Bellon *et al.* 1995).

The Hoang Sa basin

Existing exploration drilling has been concentrated on the Qui Nhon ridge, and therefore the age of the seismic reflectors in the Hoang Sa basin is not constrained directly. However, a profile across the Qui Nhon ridge and the adjacent tilted blocks (Fig. 9 B–B′) shows a sedimentary infill apparently synchronous with block tilting. This rift-fill occurs below an horizon which can be traced from the Qui Nhon ridge at the base of the carbonates, into the basin. This horizon is dated at approximately 21 Ma, and overlies the majority of the normal fault displacement. The age of the

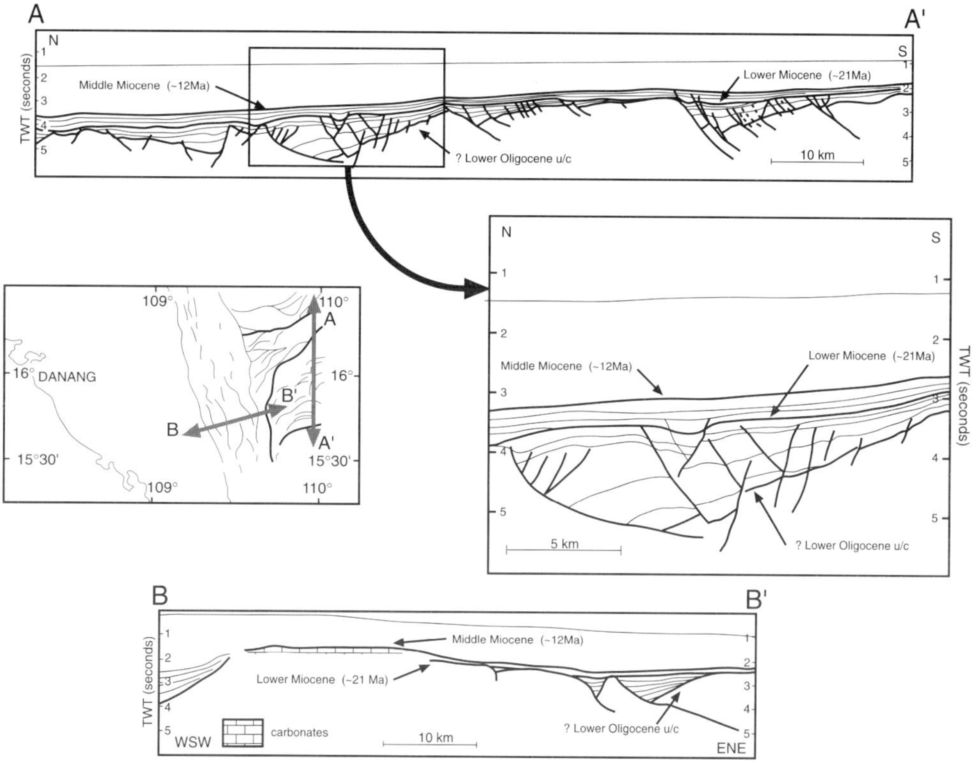

Fig. 9. Line drawings (based on seismic data) across the Hoang Sa basin and the Qui Nhon ridge showing the main structural–stratigraphic relationships.

underlying horizons is interpreted therefore to be earliest Miocene and older.

The line drawing of Fig. 9 (A–A′) illustrates the eastern terminations of the SE-dipping faults in the basin. In the hanging wall to these faults, the rift-fill is pre-Burdigalian. Note that the dip-slip displacement on the northernmost fault is actually sealed by an intra-Burdigalian horizon. However, the more southerly faults show minor dip-slip displacement affecting younger horizons. These dip-slip displacements are interpreted to young slightly and progressively to the south. Note that the southernmost fault affects a seismic horizon dated at near to 12 Ma, synchronous with the cessation of carbonate deposition.

The cross-sectional fault geometry of the North sub-basin is shown in Fig. 8. The geometries on these cross-section suggest a migration in the timing of cessation of deformation from north to the south, based on the relative age of reflectors offset by the major faults. It appears therefore that extension was concentrated prior to the Burdigalian but did continue in a series of very minor and pulsed movements through to the Langhian.

Geodynamic interpretations and implications for hydrocarbon potential

The southern termination of the Hoang Sa basin (Fig. 10) shows complex structural geometries which are interpreted to have evolved during basin opening, mainly during the Palaeogene to earliest Miocene. Major structures within this area such as the N160° E trending Qui Nhon ridge and the adjacent Quang Ngai graben (Fig. 10) define the eastern boundary of a wide pervasively deformed area, recognisable both on- and offshore, and which is characterised by a dominant N160° E fault system. Onshore central Vietnam, interpretation of satellite images (in conjunction with new fieldwork) demonstrate that Mesozoic rocks are deformed by a pervasive N160° E fault system exhibiting numerous dextral striations, but with no demonstrable large-scale displacement (Rangin *et al.* 1995; Roques *et al.* 1995). It is suggested that these structures formed during the same Cenozoic tectonic phase responsible for formation of the structures observed offshore.

The Qui Nhon ridge is also the western boundary of the N050° E-trending Hoang Sa basin. In this basin, N050° E trending NW-dipping normal faults cannot be correlated convincingly with gravity anomalies. These faults appear to be cross-cut by the N160° E-trending faults which subparallel the eastern border of the Qui Nhon ridge, but this may be high-angle hard-linkage rather than the result of superimposition of fault motion. N050° E SE-dipping normal faults in the Hoang Sa basin are precisely correlated with gravity anomalies. These faults define small tilted basins, which also correlated with clear negative gravity anomalies. The major SE-dipping faults and their associated gravity anomalies, curve southward and coalesce with the eastern flank of the Qui Nhon ridge. This horse-tail geometry of the fault system (Fig. 10), is interpreted as evidence for dextral motion along this part of the N160° E-trending fault system.

To the south and at a larger scale, from the Triton ridge to the deep Nha Trang basin, we observe a comparable horse-tail structural geometry (Roques *et al.* 1995). The amount of extension, calculated by using local isostasy constrained by 2D gravity modelling, gives an extension factor (beta) of 2.3 for the deep basin (Marquis *et al.* 1995). This magnitude of extension, observed in the Hoang Sa basin and also in the Nha Trang basin, far exceeds the amount of Tertiary extension onshore, and emphasizes the importance of a major N160° E dextral shear zone along this part of the Vietnam margin.

The majority of the normal fault activity associated with the N160° E dextral shear zone is sealed by post-rift sediments dated as approximately 21 Ma and younger. This is compatible with interpretations from the East Vietnam Sea ridge magnetic anomalies (Taylor & Hayes 1983; Briais *et al.* 1993). This extension would have been initiated by 28 Ma but could very possibly be older (Eocene?). However, very minor extension is still recorded in the basin between 21 and 12 Ma.

The North Borneo subduction zone is one possible drive for the most recent pulses of extension in the East Vietnam Sea marginal basin. Young extensional pulses may have resulted from minor differences in relative location, timing and magnitude of shortening due to subduction and extension due to spreading; it can be proposed that spreading did not precisely balance the total amount of shortening along the trench during the Mid-Miocene and one consequence of this may have been minor pervasive rifting along the Vietnam margin up to approximately 12 Ma.

The N160° E dextral shear zone offshore central Vietnam has partly controlled the extensional and spreading history in the East Vietnam Sea, at least between 28 and 21 Ma. The apparent coincidence of the southern termination of this N160° E fault system with the western termination of the Borneo Trench suggests a kinematic

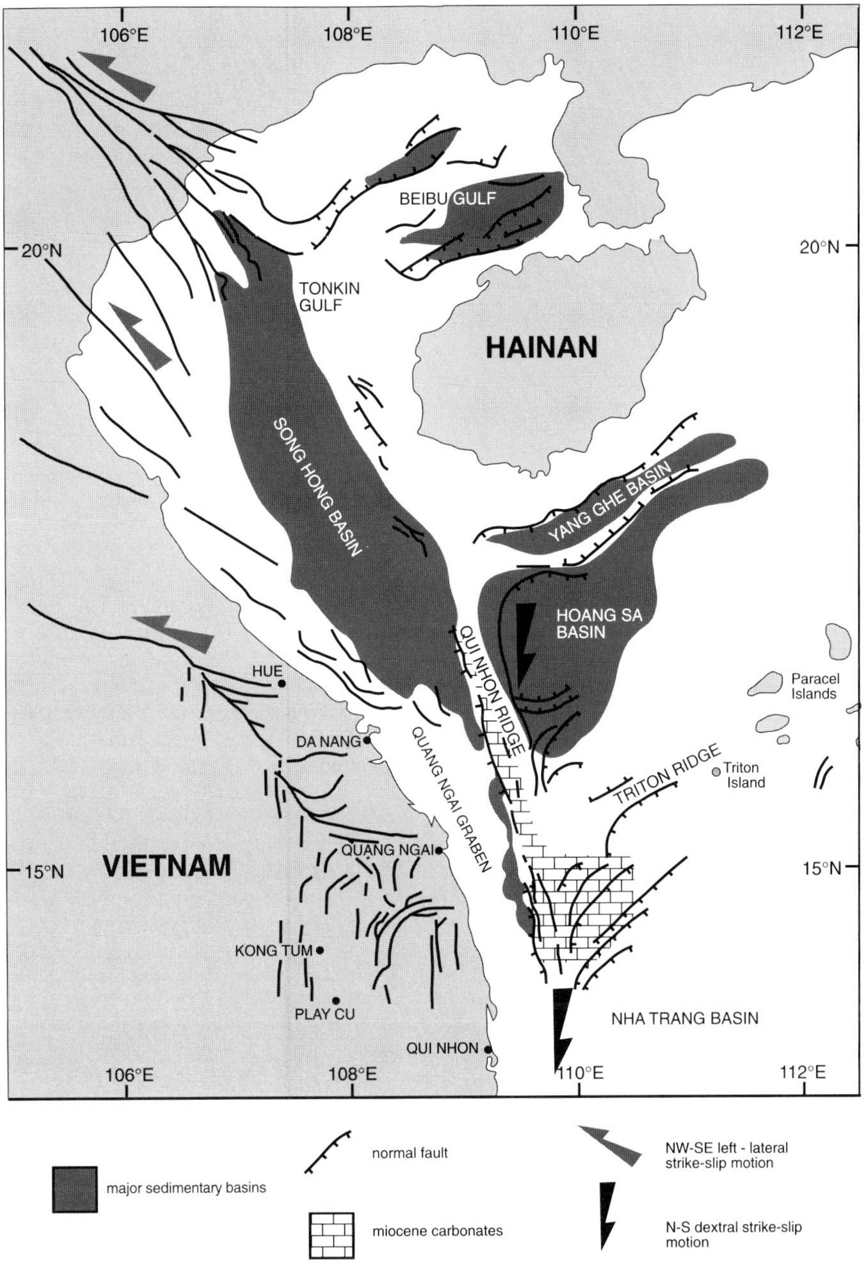

Fig. 10. The map-pattern and interaction of strike-slip and extensional fault systems onshore and offshore central and north Vietnam.

link between the two crustal-scale features and supports the hypothesis that subduction along the Borneo Trench may be the dominant mechanism which has driven extension and spreading in the East Vietnam Sea.

The complex interaction of strike-slip and extension offshore Da Nang has resulted in the development of a range of potential hydrocarbon trapping geometries. Older plays are predicted to be Palaeogene with the possibility of a lacustrine basin-fill within either the transtensional or extensional sub-basins of the Da Nang area. A regional analogue is located offshore southern Vietnam in the Cuu Long basin, where

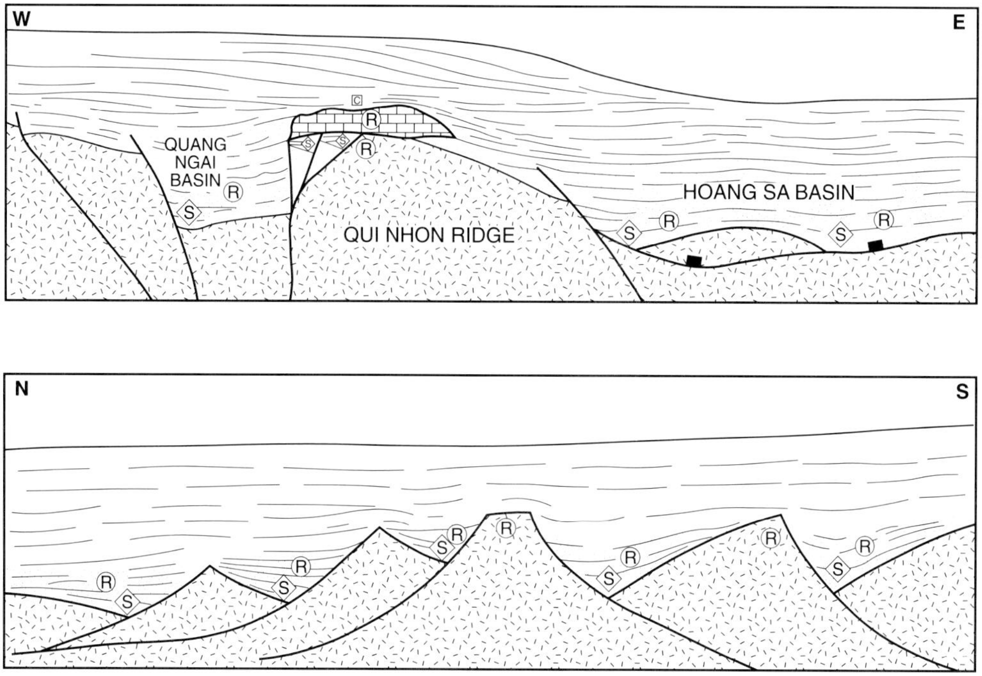

Fig. 11. Petroleum plays offshore Da Nang. Potential source, reservoir and cap rocks are illustrated by S, R and C respectively. Rift-fill is indicated by stipple.

the source system is interpreted to be lacustrine and importantly the main producing reservoir is 'pre-rift' basement. The same basement play may exist offshore Da Nang (Fig. 11), and in addition there is clastic reservoir potential in the pre-Burdigalian rift-fill of both the North and South sub-basins, and the Quang Ngai Basin. The Miocene carbonates of the Qui Nhon ridge are a proven gas play; the key to future exploration success in this fairway is advanced understanding of the distribution of carbon dioxide. Progress in understanding prospectivity offshore Da Nang therefore requires the effective integration of detailed structural geometrical and kinematic interpretation with both the regional magmatic and depositional history during the Oligo-Miocene.

The authors are very grateful to PetroVietnam for permission to publish this paper. BP Exploration provided seismic and gravity data for discussion during completion of a post-graduate research project by D. Roques under the supervision of C. Rangin; they are thanked for provision of the seismic and gravity examples illustrated in the paper. Steve Matthews worked within the BP Exploration Da Nang team during 1991–92 and acknowledges the considerable assistance of J. Dickens, D. Holland, G. Flanagan and H. Upshall in assessment of models for structural evolution. C. Sladen is thanked for helpful comments on an earlier version of this paper.

References

Bellon, H., Huchon, P. & Rangin, C. 1995. Late Miocene to Quaternary volcanism of Vietnam: time and space of geochemical variations. *In*: *Cenozoic evolution of the Indochina peninsula. Tectonics and sedimentary basins*. Hanoi-Do Son, Viet Nam, 47.

Briais, A., Patriat, P. & Tapponnier, P. 1993. Updated interpretation of magnetic anomalies and seafloor spreading stages in the South China Sea: Implications for the tertiary tectonics of Southeast Asia. *Journal of Geophysical Research*, **98**, 6299–6328.

Holland, D. C., Dickens, J. S. & Horbury, A. D. 1992. A case of Drowning – The death of a Carbonate Platform in the South China sea. *In*: *AAPG International Conference and Exhibition*, Sydney, Australia.

Marquis, G., Roques, D., Chamotrooke, N., Huchon, P., Le Pichon, X., Phan Truong, T., Rangin, C. & The Ponaga Team 1995. Structure and evolution of the Triton ridge-Nha Trang basin extensional system. *In*: *Cenozoic evolution of the Indochina peninsula. Tectonics and sedimentary basins*. Hanoi-Do Son, Viet Nam, 66–67.

RANGIN, C., HUCHON, P., LE PICHON, X., BELLON, H., LEPVRIER, C., ROQUES, D., NGUYEN, DINH, H. & PHAN VAN, Q. 1995. Cenozoic deformation of central and south Vietnam. *Tectonophysics*, **251**, 179–196.

ROQUES, D., RANGIN, C., LEPVRIER, C. & HUCHON, P. 1995. Da Nang area structural analysis. An onshore/offshore integrated study of Central Vietnam. *In*: *Petroleum Geology of South East Asia Conference*. Geological Society, London, 10.

TAPPONNIER, P., PELTZER, G. & ARMIJO, R. 1986. On the mechanics of the collision between India and Asia. *In*: COWARD, M. P. & RIES, A. C. (eds) *Collision Tectonics*. Geological Society, London, Special Publications, **19**, 115–157.

TAYLOR, B. & HAYES, D. E. 1983. Origin and history of the South China Sea basin. *In*: HAYES, D. E. (eds) *The tectonic and geologic evolution of Southeast Asian Seas and Islands, Part 2*. AGU Geophysical Monograph Scrics, **27**, 23–56.

Cenozoic deformation in Sumatra: oblique subduction and the development of the Sumatran Fault System

A. J. McCARTHY[1] & C. F. ELDERS

Southeast Asia Research Group, Geology Department, Royal Holloway, University of London, Egham, Surrey TW20 0EX, UK

[1] *Present address: Enterprise Oil plc, Grand Buildings, Trafalgar Square, London WC2N 5EJ, UK*

Abstract: Sumatra is situated at a long-lived active continental margin. At the present day, subduction beneath the margin is strongly oblique, leading to subduction of Indian Oceanic crust at the Sunda Trench and right-lateral strike-slip along the onshore Sumatran Fault System (SFS). A pre-Tertiary history of accretion against the margin was followed by Palaeogene basin formation throughout Sumatra. A strong mid-Miocene inversion event is recorded in the onshore part of the forearc basin in southern Sumatra. The mid-Miocene also marked the inception of seafloor spreading in the Andaman Sea and the probable inception of major strike-slip movement along the SFS, possibly following clockwise rotation of Sumatra towards it present NW–SE trend. The SFS exhibits a wide variety of geometries and deformational styles and is characterised by an extremely complex deformation history including polyphase reactivation of fault surfaces and contemporaneous strike-slip and orthogonal compression or extension. A new estimate of approximately 150 km offset of Mesozoic units across the SFS in central Sumatra is proposed. Several basins have formed along the SFS during the Quaternary. Some of these are not readily explained in terms of classic models of strike-slip basin formation and it is proposed that thermally induced uplift along the axis of the volcanic arc of the Barisan mountains was of primary importance in their formation.

Sumatra is situated on the SW edge of the Sundaland craton and has formed part of the active continental margin of Southeast Asia since at least the early Mesozoic (McCourt & Cobbing 1993). At the present day, subduction of Indian Ocean lithosphere beneath SE Asia is strongly oblique in the region of Sumatra (Fig. 1). The Indo-Australia/Asia plate convergence vector has an azimuth of 003–020° (DeMets *et al.* 1990; McCaffrey 1991) and a magnitude of approximately 70 mm a^{-1} (Curray 1989), whilst the Sunda Trench is oriented 120–145°, resulting in an obliquity of subduction of 20–45°. Consequently, the plate convergence vector is resolved into subduction at the Sunda Trench and right-lateral transcurrent movement along the onshore Sumatra Fault System (SFS), an arrangement which is believed to date from the mid-Miocene.

The SFS runs the 1650 km length of the Barisan mountains which lie along the western side of Sumatra. They consist of uplifted basement blocks intruded by Mesozoic and Tertiary granitoids and are capped by the active volcanic arc, with elevations exceeding 3000 m.

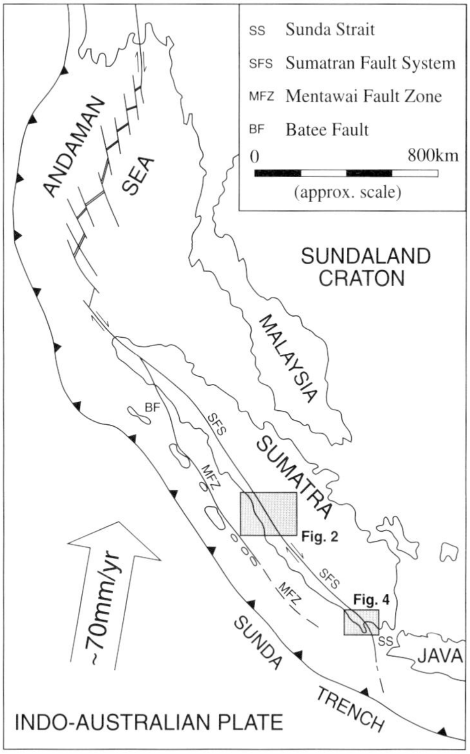

Fig. 1. Tectonic setting of Sumatra, showing locations of Figs 2 & 4.

History of deformation

Through both map and field studies it is clear that Cenozoic deformation is superimposed

From Fraser, A. J., Matthews, S. J. & Murphy, R. W. (eds), 1997, *Petroleum Geology of Southeast Asia*, Geological Society Special Publication No. 126, pp. 355–363.

upon a strong pre-Tertiary structural grain. Most basement units are elongate NW–SE, contain predominantly NW–SE-trending structures (at least some of which are of pre-Tertiary age) and there is, in general, a progression from the oldest (Carboniferous) units in the NE to the youngest (Cretaceous) units in the SW (see Fig. 2). This pattern is consistent with a pre-Tertiary history of accretion against the continental margin.

The early part of the Tertiary was characterised by basin formation throughout Sumatra (Fig. 3). Transgressive sedimentary sequences were deposited in basins in what now correspond to forearc, intermontane, and backarc positions (Fig. 2). These basins were later inverted and dissected. In the forearc basin, compression resulted in predominantly NW–SE-trending faults and folding of varying intensity (Fig. 2). Unconformities are common, where renewed (transgressive) sedimentation followed a compressional event. A strong angular unconformity in the onshore forearc basin of southern Sumatra (Bengkulu Basin) dates inversion there as Mid-Miocene (Fig. 3). Post-Mid-Miocene deformation in the fore arc is largely limited to gentle tilting (Fig. 3). In the back-arc basins, a strong mid-Miocene tectonic and igneous event (Eubank & Makki 1981) was followed by numerous gentler phases of inversion throughout the late Tertiary and Quaternary. Eubank & Makki (1981) and Williams & Eubank (1995) suggest that almost all folding in the Central Sumatra Basin is related to wrenching, but it seems likely that at least the largest, basin-scale, NW–SE-trending folds are a product of simple compression (de Coster 1974) or forced draping over uplifted basement blocks (Moulds 1989).

There is no clear evidence of major strike-slip along the SFS prior to the mid-Miocene (Fig. 3), although the SFS may locally have been active in an extensional mode prior to this (Rosidi *et al.* 1976). The SFS was almost certainly active in a strike-slip mode by the end of the mid-Miocene, following the initiation of seafloor spreading in the Andaman Sea (Curray *et al.* 1979). Since that time, right-lateral strike-slip has been the dominant mode of movement along the SFS.

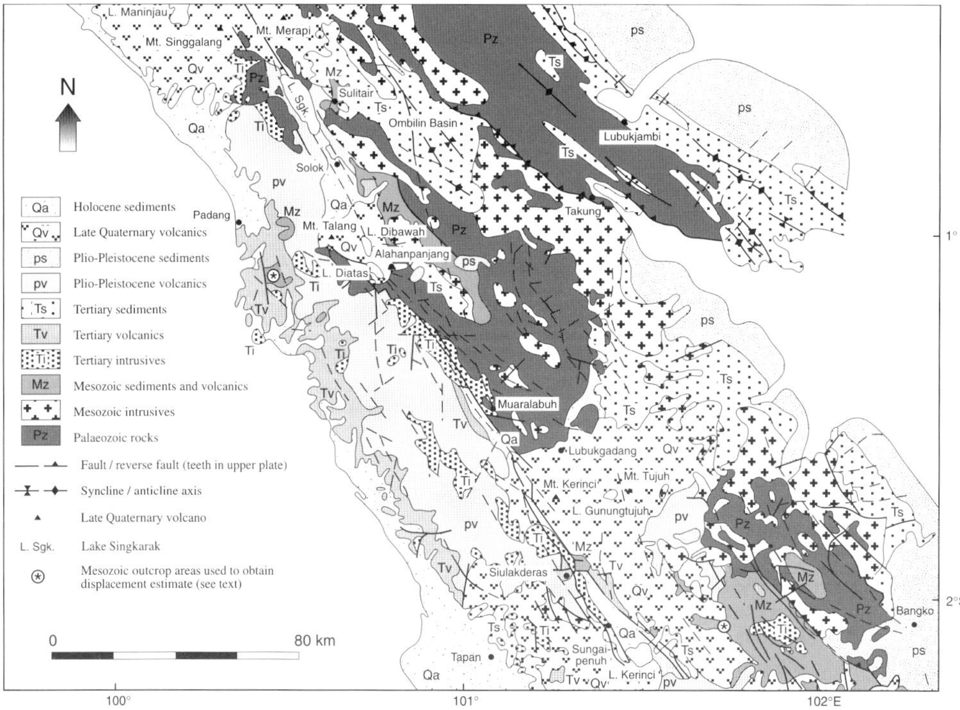

Fig. 2. Geological sketchmap of west central Sumatra, showing distribution of basement units; Tertiary sediments in forearc, intermontane and back-arc positions; active volcanoes; and the trace of the Sumatran Fault System (SFS). Adapted from Gafoer *et al.* (1992).

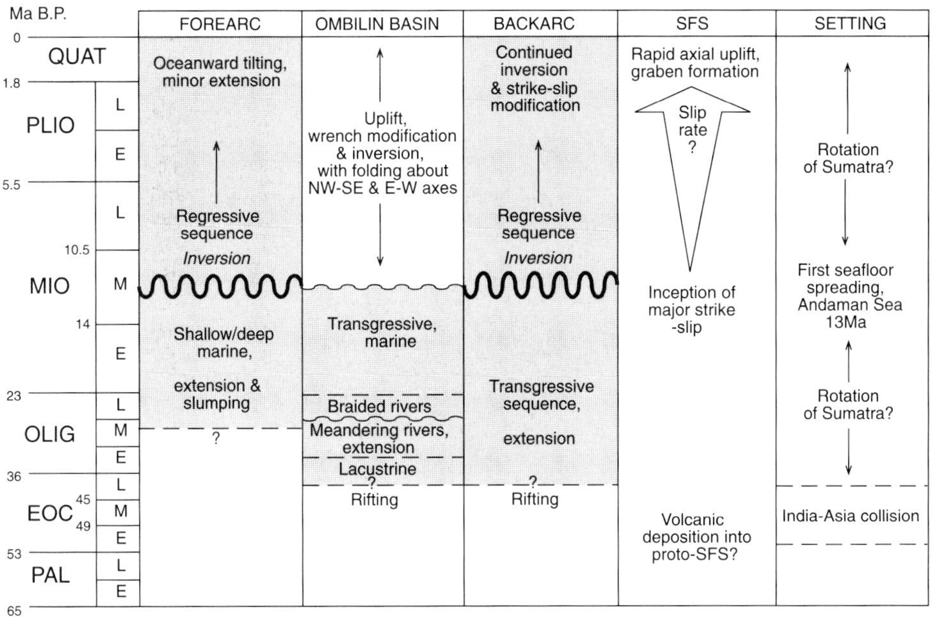

Fig. 3. Simplified correlation of Tertiary sedimentary and tectonic events in onshore forearc, intramontane Ombilin Basin, and back-arc areas, and relation to development of the SFS and regional setting of Sumatra. Grey shading represents generalized sedimentary record. Note that ages of onset of sedimentation are poorly constrained. See text for details.

Geometry and styles of deformation along the SFS

Individual segments of the SFS, cutting across rocks of all ages, have clearly experienced complex movement histories. Single fault surfaces commonly display multiple lineations indicating activation under pure strike-slip and pure dip-slip (both normal and reverse) as well as oblique-slip movement.

The SFS displays a variety of widths and geometries along its length. Along some segments, the active SFS is restricted to a single narrow trace (e.g. between Alahanpanjang and Siulakderas, Fig. 2). Elsewhere, as described later, the system widens to enclose Quaternary basins (e.g. the Singkarak and Kerinci Rifts, Fig. 2; and the Semangka Rift and Suwoh basins, Fig. 4). In northern Sumatra the width of the SFS increases to 40 km, enclosing a lens of complexly faulted Palaeozoic limestone intruded by Mesozoic granitoids (Rock *et al.* 1983). There appears to be no correlation between age of surface geology and width of the fault system.

Many lineaments observable on SLAR images of the SFS correspond to major mapped faults. Others, even relatively faint, correspond directly in the field to individual faults, fault systems, or areas of structural disturbance. This association was particularly striking in Quaternary volcanic rocks which are not pervasively deformed. The mapping of these lineaments (Fig. 4) produces two key observations. Firstly, there is an abundance of lineaments lying at significant distances (tens of kilometre) from the active principal trace of the SFS. Secondly, those lineaments lying outside the trace of the SFS fall into a number of differently oriented sets, the two most prominent being those lying subparallel with the SFS, i.e. trending NW–SE, and those trending at very high angles to the SFS. This latter set lie closest to the antithetic or 'anti-Riedel' shear orientation of classical models of structural elements associated with strike-slip systems (e.g. Wilcox *et al.* 1973). As such they are unusually well developed. At outcrop scale this set of faults is very poorly represented. Furthermore, evidence of antithetic i.e. left-lateral movement on this set of faults is rare on all scales from map scale to outcrop scale.

Synthetic or 'Riedel' lineaments would be predicted to lie in orientations around NNW–

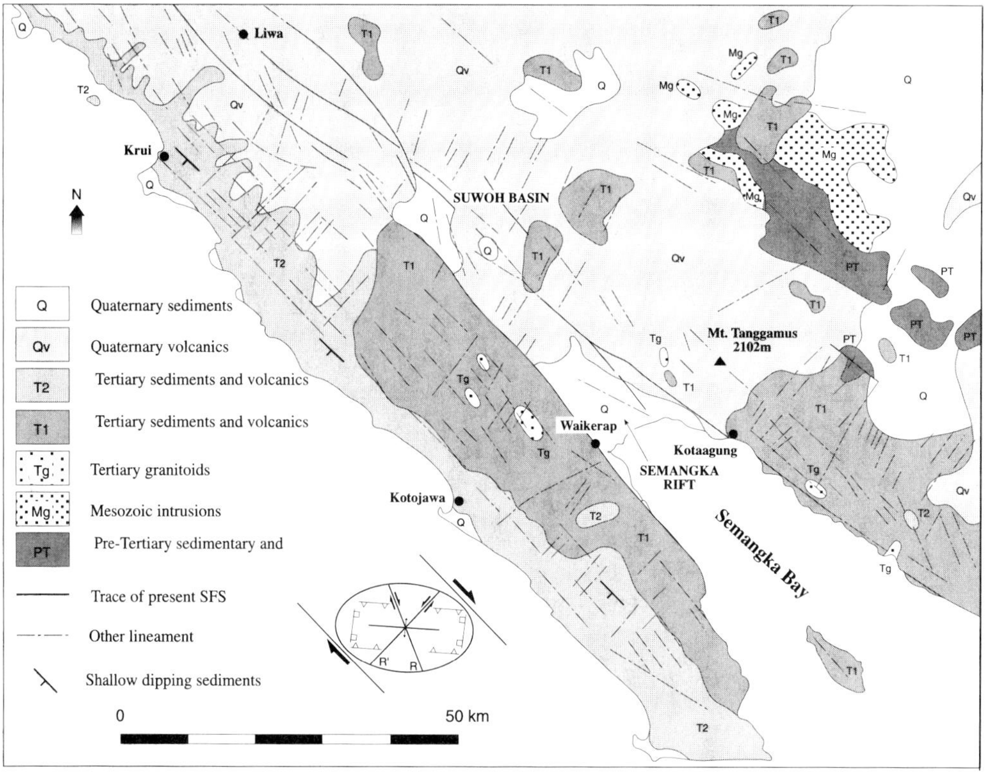

Fig. 4. Geology and SLAR image interpretation of southernmost part of the Sumatra Fault System. Adapted from Amin *et al.* (1993), with additional lineaments from SLAR.

SSE to N–S. In marked contrast to the antithetic lineaments, this set is very poorly represented at the scale of published geological and topographic maps and SLAR images. However, at outcrop scale this set is abundant and is invariably observed to record right-lateral movement. Once more, lineament sets are apparently equally well developed in rocks of all ages from Carboniferous through to late Tertiary and even Quaternary.

There is a great variety of deformational style along the SFS. Many outcrops, particularly of younger units, show pervasive brittle faulting, indicative of deformation at high crustal levels. Others outcrops, particularly of granitoid rocks, record ductile deformation. While some fabrics recorded in granitoid lithologies represent pre-full-crystallization (PFC) deformation prior to or during the final cooling and solidification of the body, in many instances early fabrics of this nature are cut by more localized zones of ductile simple shear. Fabrics recorded in granitoids range from very coarsely developed foliations reminiscent of pure flattening or pure shear to localized mylonite zones of intense grain size reduction and evident simple shear. In all fabric types, NW–SE strikes and steep dips dominate. Few sense of shear indicators were found.

Besides recording complex deformation histories, the granitoid bodies lying along or close to the SFS display evidence of long and complex intrusive histories. The bodies range in age from Permian through to late Tertiary, with main phases of activity in the mid–late Permian (264–256 Ma), Jurassic to early Cretaceous (203–130 Ma), mid–late Cretaceous (117–80 Ma), early Eocene (57–53 Ma) and Miocene (20–5 Ma) (McCourt *et al.* 1996). Compositions range from quartz diorite to granite. Several bodies, such as the Sulitair suite (northern part of Fig. 2) demonstrably contain intrusions of significantly varying ages (McCourt & Cobbing 1993). Outcrops of the Lassi pluton, east of Solok (Fig. 2) contain evidence of at least five phases of intrusion and veining interrupted by at least one phase of foliation-forming ductile deformation and at least one later phase of faulting and folding.

Quantification of displacement on the SFS

Several attempts have been made to quantify horizontal displacements along the SFS, by different methods. Curray's (1979) estimate of 460 km based on the extent of opening of the Andaman Sea since the mid-Miocene can be regarded as the absolute maximum displacement which might be expected to be accounted for in the system. River offsets of 20–35 km have been documented by Katili & Hehuwat (1967), while Posavec *et al.* (1973) deduced estimates of 90–135 km from offset of volcanic centres and aeromagnetic anomalies in West Sumatra. In northernmost Sumatra, Page *et al.* (1979) observed that lithium geochemical provinces had apparently been offset by more than 200 km.

Hahn & Weber (1981) propose 42 km offset of granite bodies in northern West Sumatra province. The granites are considered to be of Palaeogene age by Hahn & Weber (1981) but 'probably Permian to Mesozoic' by Rock *et al.* (1983). The estimate is based on correlation of a simple petrological zoning pattern between the two bodies. Fieldwork conducted during the present study revealed that the considerable petrological complexity and variety of even small outcrop areas of the two granitic bodies invalidates this simple correlation.

Other estimates have been based upon correlation of Mesozoic units in west central Sumatra, particularly that of Yancey & Alif (1977). The present study proposes that there may be a valid correlation between the Siguntur Formation, 20 km south of Padang, known only to be 'probably Jurassic in age' (Rosidi *et al.* 1976); and the Asai Formation of Jambi Province, 50 km west of Bangko, known to be of mid-Jurassic age (Kusnama *et al.* 1993) (Fig. 2). Both units are dominated by fine-grained siliciclastic lithologies, with weak metamorphism adjacent to intrusions, and very minor carbonate contents, in strong contrast to the majority of Mesozoic units in the region, which are carbonate-dominated. Deformation of both units is characterised by steeply dipping strata, with NW–SE strikes dominant. This correlation, which assumes limited lateral continuity of the units, results in an offset estimate of approximately 150 km.

Recent slip rates along the SFS have been deduced by Bellier & Sebrier (1995) from stream offsets observed on SPOT images. They show a clear increase from less than 10 mm a^{-1} in the Sunda Strait at the southern end of the system, through 11–19 mm a^{-1} in central Sumatra, to 23 ± 2 mm a^{-1} adjacent to Lake Toba in northern Sumatra. Furthermore, they deduce a slip rate of 12 ± 5 mm a^{-1} on the Batee Fault, a major subsidiary structure which runs onshore from the forearc basin and joins the SFS 300 km NW of Lake Toba in northern Sumatra (Fig. 1). It has been suggested that this fault links with the Mentawai Fault Zone (MFZ) in the forearc which runs parallel with the SFS for 'at least 600 km' (Diament *et al.* 1992). Significant strike-slip movement along the MFZ has been proposed but remains unproven. The sum of the slip rates on the SFS and the Batee Fault in northern Sumatra (37 ± 7 mm a^{-1}) is a significant proportion of the approximately 40–50 mm a^{-1} which might be predicted from a simple consideration of the partitioning of the plate convergence vector between subduction and transcurrent displacement.

Origin of Quaternary basins

Besides continued major strike-slip, the Quaternary history of the SFS has been characterised by the formation of a number of small basins along its length. They show a great variety of geometries but few, if any can be regarded as classic rhombohedral or subrhombohedral pull-aparts (e.g. Crowell 1974).

The majority of these basins are long and narrow rifts enclosed by strands of the SFS. Some of these narrow rifts are parallel-sided (Singkarak Rift, northern part of Fig. 2), while others have a gently tapering geometry (Kerinci Rift, southern part of Fig. 2). The shapes of the lakes which occupy parts of these basins may be taken as superficial evidence of rhombohedral pull-aparts but are not an accurate guide to the shapes of the basins, since the lakes have decreased in size, and particularly in length, due to sedimentation in these basins during and since their formation. Furthermore, these basins are now partly volcanically blocked. More 'classical' strike-slip basins occur at releasing bends, fault wedges and complex fault junctions. Several lake basins are volcanotectonic in origin, occupying calderas or vents which have either formed in tectonic accommodation zones or have been later modified by faulting (e.g. Lakes Dibawah and Diatas, (central part of Fig. 2); Lake Ranau in southern Sumatra).

Figure 4 shows two Quaternary basins at the southern end of the SFS. The Suwoh Basin is perhaps the closest example to a classic pull-apart, although it is partly volcanically blocked.

The principal active trace of the SFS entering the basin from the north runs along the northeast side of the basin, while the active trace entering from the south runs along the southwest side of the basin. The basin has opened approximately 10 km and is the site of mild volcanic activity.

The Semangka Rift at the southern end of the basin diplays a tapered geometry, the SFS widening out to the south into Semangka Bay and the extensional Sunda Strait. The basin is bounded to the southwest by a very sharp recent 500 m fault scarp, and to the northeast by a similar, slightly older feature. Figure 4 also shows that the strata and topography of the entire block to the southwest of the rift are gently tilted to the southwest. Observations in the field and from SLAR images show that rivers flowing southwest from the axis of the SFS towards the Indian Ocean incised into the incipient scarp and continued to do so during its development, so that rivers still flow southwestwards across a northeast facing escarpment in the region of Liwa (northern part of Fig. 4). Nearby, very fresh fault scarps indicate that uplift of the block to the southwest of the SFS is continuing. These observations suggest that rapid Quaternary uplift of the fault scarp and southwestward tilting of the crustal block to the southwest of the SFS were important in the formation of both the Semangka Rift and the Suwoh Basin.

While the rapid vertical movements inferred at the southern end of the SFS may be partly attributable to the proximity of the extensional Sunda Strait system, it is suggested here that such movements are of primary importance for the formation of other basins along the length of the SFS, particularly the long, narrow basins described earlier. Further evidence for rapid recent uplift comes from geomorphological observations on streams along the SFS, including widespread incision, abandoned meanders and straightened channels. These basins appear not to have formed as classic pull-apart or transtensional basins, but rather as blocks which have subsided between two major subparallel traces of the SFS. The setting of the SFS, running along the axis of the volcanic arc of the Barisan Mountains, leads the authors to suggest that Quaternary uplift and arching along this axis led to tilting of crustal blocks and the subsidence of narrow crustal blocks or slivers along the crest of the Barisan Mountains, facilitated by (possibly pre-existing) faults of the SFS.

These basins contain little evidence of pervasive extension or the occurrence of cross-faults which might be expected to develop in order to reduce the high aspect ratios of these basins. The Kerinci Rift (Fig. 2) contains two further structures of interest. A low hill consisting of horizontally bedded Plio-Pleistocene pumiceous sediments, which appears to represent a horst produced by extensional block faulting during subsidence, lies in the centre of the rift. However, nearby a small hill (approximately 20 m wide × 8 m high) consists of very steeply bedded Quaternary rift fill sands and gravels. This is interpreted as a minor transpressive åpop-upí structure related to recent movements on faults transecting the valley.

In the centre of the Singkarak Rift, 6.5 km south of Lake Singkarak (Fig. 2), an anticlinal structure is developed in Quaternary rift-fill fluvio-lacustrine sediments and displays very clear growth strata which thicken away from the crest of the structure. It has an amplitude of 2 m, a half-wavelength of 15 m, and it remains a positive topographic feature. Its axis strikes perfectly parallel with the basin-bounding faults, i.e. NW–SE. Such a structure indicates periodic compression during sedimentation and is incompatible with a transtensional origin for the basin. Its parallelism with the SFS suggests that its development is not directly related to transcurrent movements along the SFS, rather its growth is a product of arc-perpendicular (i.e. NE–SW) compression.

Discussion

It has been shown that Sumatra has suffered a long and complex deformation history. Pre-Tertiary folding, faulting and localized metamorphism was followed by Palaeogene basin formation and Neogene–Quaternary inversion and strike-slip faulting. This last stage at least has occurred in an oblique subduction setting.

Currently popular tectonic models (e.g. Tapponnier *et al.* 1986) propose that Palaeogene collision of India with the southern margin of Asia resulted in the eastward and southeastward extrusion of large crustal blocks and the clockwise rotation of large parts of Southeast Asia including Sumatra. Sparse palaeomagnetic data (Haile 1979; Siagian *et al.* 1995) suggest that there may have been clockwise rotations of 20–40° since the Mesozoic, or possibly 34° since the Oligocene (Haile 1979). Davies (1984) suggests that Sumatra has rotated anticlockwise during the Tertiary, while in the reconstructions of Hall (1996) Sumatra is tied to the South Malaya Block with its larger palaeomagnetic dataset indicating anticlockwise rotation of 15° between 20 Ma and 10 Ma.

Clockwise rotation of Sumatra from a more E–W trend to its present NW–SE trend, beginning in the Eocene or Oligocene, would have resulted in increasing obliquity of the normal to the active margin with respect to the plate convergence vector. Consequently the tendency towards transcurrent movements parallel to the margin would have increased with time. This can be regarded as consistent with mid-Miocene inception of the SFS as a right-lateral transcurrent system.

Such a model of clockwise rotation also implies that, assuming continued rotation, slip rates along the SFS may increase with time. Measured Recent slip rates of up to 23 mm a^{-1} along the SFS (Bellier & Sebrier 1995) ,when extrapolated back to the mid-Miocene, imply displacements of the same order of magnitude of those observed (100–200 km, e.g. Posavec *et al.* 1973; Page *et al.* 1979, this study) but are always 20–50% too high. This may suggest that slip rates have increased since the mid-Miocene, although it should be noted that given the observed present day width of distributed deformation across Sumatra, estimates of displacement based on offsets across single or even several strands of the SFS should be treated with caution.

Evidence of earlier strike-slip deformation in Sumatra is more tentative. The origin and tectonic setting of the N–S-striking Palaeogene graben of the 'backarc' basins in particular remain enigmatic (see Williams & Eubank 1995, and references therein). Several authors suggest a wrench component to pre-mid-Miocene deformation in the Sumatran backarc (e.g. Eubank & Makki 1981; Davies 1984), and the similarly oriented Mergui Basin north of Sumatra (Polachan 1988) and it is possible that the Palaeogene graben originated in a regional dextral wrench setting (Packham 1993). Wrenching may later have been isolated along the SFS following establishment of the modern arc system. Similarly, the Palaeogene origin of the Ombilin Basin (Fig. 2) is also poorly understood (see Howells this volume), although strike-slip movements may have contributed to the dissection of the basin during the latter part of its sedimentary history.

The pre-Tertiary structural fabric of Sumatra has undoubtedly been an important control on Tertiary deformation. Major fold and fault trends have been at least partly inherited, while some major faults (possibly sutures) have been reactivated under different modes of faulting or have served as magmatic conduits. Basement lineaments may be responsible for 'unusual' geometries, such as the lineaments striking at high angles to the SFS. This set is unusually well developed, particularly in the absence of the commonly more abundant synthetic shear set. There is no field evidence to suggest that, for example, these lineaments, are related to arc-parallel extension of the forearc to accommodate the northwestward increase in slip rate along the SFS.

Another important control on the positioning and development of the SFS and related structures has been the magmatic arc. Only in northernmost Sumatra does the principal active trace of the SFS stray more than 20 km from the active volcanic arc. At the largest scale, the position of the magmatic arc is governed by the dip angle of the subducted slab of Indian Ocean lithosphere. Magmatic softening (i.e. weakening of the lithosphere as a result of the large scale intrusion of hot magmatic bodies) along the line of the arc serves to concentrate strain so that the SFS tends to run along it. On a more local scale however, the positions of volcanic centres or individual vents may be influenced by the presence of accommodation zones along the SFS. The SFS also acts as a conduit for the ascent of granitic melts before deforming and dissecting the resulting, often polyphase, plutons.

Observations of multiple phases of deformation and intrusion imply that the area within and near to the active SFS has a long history of both, suggesting that the SFS is a fundamental, long-lived structure on the lithospheric scale.

Stress regimes and strain patterns, both regional and local, may be more complex and variable in regions of oblique subduction than they are in regions of orthogonal subduction or regions of non-subduction-related continental transcurrent deformation. The majority of fault surfaces studied in the field bore multiple lineations, and there is abundant evidence of strike-slip occurring broadly contemporaneously with arc-perpendicular extension or shortening.

A variety of stress systems act in Sumatra, and along the axis of the Barisan Mountains in particular. Subduction-related stresses interact with stresses related to magmatic emplacement, uplift and doming, while localized stress and strain patterns may be set up around individual structures of the SFS.

This complexity is particularly well illustrated by the variety of Quaternary basin types along the SFS. Many of these basins do not resemble classic strike-slip basins and it is suggested here that their origin may be due in part to thermal doming along the axis of the Barisan Mountains – a process unique to strike-slip systems situated within volcanic arcs. The presence of 'anomalous' young compressional or transpressional structures within these basins, as described above, further illustrates how both extensional and contractional structures can develop broadly contemporaneously.

Mount & Suppe (1992) present evidence, mostly from borehole breakouts, that the maximum horizontal stress direction in Sumatra is oriented NE–SW, i.e. perpendicular to the SFS. Such an orientation is consistent with the dominant NW–SE orientation of fold axes throughout Sumatra and implies that crustal scale transcurrent faults may be inherently weak surfaces which may be activated by low shear stresses. This orientation is also consistent with the presence of compressional structures within basins along the SFS, but not with the presence of the basins themselves. It is proposed here that it is necessary to invoke another driving force, namely the uplift of the axial zone of the Barisan Mountains.

This work forms part of A.J.M.'s PhD study of the Sumatran Fault System, funded by the Consortium for Geological Research in Southeast Asia and instigated by A. Barber. Helpful discussions were held with C. Howells, S. Moss and R. Hall among others.

References

AMIN, T. C., SIDARTO, SANTOSA, S. & GUNAWAN, W. 1993. *Geological map of the Kotaagung Quadrangle, Sumatra (Quadrangle 1010) Scale 1 : 250,000*. Geological Survey of Indonesia, Directorate of Mineral Resources, Geological Research and Development Centre, Bandung.

BELLIER, O. & SEBRIER, M. 1995. Is the slip rate variation on the Great Sumatran Fault accommodated by fore-arc stretching? *Geophysical Research Letters*, **22**, 1969–1972.

CROWELL, J. C. 1974. Origin of Late Cenozoic basins in southern California. *In*: DICKINSON, W. R. (ed.) *Tectonics and sedimentation*. Society of Economic Paleontologists and Mineralogists Special Publications, **22**, 190–204.

CURRAY, J. R. 1989. The Sunda Arc: a model for oblique plate convergence. *Netherlands Journal of Sea Research*, **24**, 131–140.

——, MOORE, D. G., LAWVER, L. A., EMMEL, F. J., RAITT, R. W., HENRY, M. & KIECKHEFER, R. M. 1979. Tectonics of the Andaman Sea and Burma. *In*: WATKINS, J., MONTADERT, L. & DICKENSON, P. W. (eds.) *Geological and Geophysical Investigations of Continental Margins, CCOP–SEATAR meeting, Bandung 1978*. American Association of Petroleum Geologists Memoirs, **29**, 189–198.

DAVIES, P. R. 1984. Tertiary structural evolution and related hydrocarbon occurrences, North Sumatra Basin. *Indonesian Petroleum Association, Proceedings 13th annual convention, Jakarta 1984*, **I**, 19–50.

DE COSTER, G. L. 1974. The geology of the Central and South Sumatra Basins. *Indonesian Petroleum Association, Proceedings 3rd Annual Convention, Jakarta 1974*, 77–110.

DEMETS, C., GORDON, R. G., ARGUS, D. F. & STEIN, S. 1990. Current plate motions. *Geophysical Journal International*, **101**, 425–478.

DIAMENT, M., HARJONO, H., KARTA, K., DEPLUS, C., DAHRIN, D., ZEN, M. T., GERARD, M., LASSAL, O., MATIN, A. & MALOD, J. 1992. Mentawai fault zone off Sumatra: a new key to the geodynamics of western Indonesia. *Geology*, **20**, 259–262.

EUBANK, R. T. & MAKKI, A. C. 1981. Structural geology of the Central Sumatra back-arc basin. *Indonesian Petroleum Association, Proceedings 10th Annual Convention, Jakarta 1981*, 153–196.

GAFOER, S., HERMANTO & AMIN, T. C. 1992. *Geological Map of Indonesia, Padang sheet, Scale 1 : 1,000,000*. Geological Survey of Indonesia, Geological Research and Development Centre, Bandung.

HAHN, L. & WEBER, H.-S. 1981. The structure system of west central Sumatra. *Geologisches Jahrbuch*, **B47**, 21–39.

HAILE, N. S. 1979. Palaeomagnetic evidence for rotation and northward drift of Sumatra. *Journal of the Geological Society of London*, **136**, 541–545.

HALL, R. 1996. Reconstructing Cenozoic SE Asia. *In*: HALL, R. & BLUNDELL, D. J. (eds) *Tectonic evolution of Southeast Asia*. Geological Society, London, Special Publications, **106**, 153–184.

HOWELLS, C. G. 1997. Tertiary response to oblique subduction and indentation in Sumatra, Indonesia: new ideas for hydrocarbon exploration. *This volume*.

KATILI, J. A. & HEHUWAT, F. 1967. On the occurrence of large transcurrent faults in Sumatra, Indonesia. *Journal of Geosciences, Osaka City University*, **10**, 5–16.

KUSNAMA, R., PARDEDE, S., ANDI MANGGA, S. & SIDARTO 1993. *Geological map of the Sungaipenuh & Ketaun Quadrangles, Sumatra (Quadrangles 0812 & 0813) Scale 1 : 250,000*. Geological Survey of Indonesia, Directorate of Mineral Resources, Geological Research and Development Centre, Bandung.

MCCAFFREY, R. 1991. Slip vectors and stretching of the Sumatran forearc. *Geology*, **19**, 881–884.

MCCOURT, W. J. & COBBING, E. J. 1993. *The geochemistry, geochronology and tectonic setting of granitoid rocks from southern Sumatra, western Indonesia*. Southern Sumatra Geological and Mineral Exploration Project Report Series, **9**.

——, CROW, M. J., COBBING, E. J. & AMIN, T. C. 1996. Mesozoic and Cenozoic plutonic evolution of SE Asia: evidence from Sumatra, Indonesia. *In*: HALL, R. & BLUNDELL, D. J. (eds) *Tectonic evolution of Southeast Asia*. Geological Society, London, Special Publications, **106**, 321–335.

MOULDS, P. J. 1989. Development of the Bengkalis depression, central Sumatra and its subsequent deformation – a model for other Sumatran grabens? *Indonesian Petroleum Association, Proceedings 18th Annual Convention, Jakarta 1989*, **I**, 217–246.

MOUNT, V. S. & SUPPE, J. 1992. Present-day stress orientations adjacent to active strike-slip faults: California and Sumatra. *Journal of Geophysical Research*, **97**(B8), 11 995–12 013.

PACKHAM, G. H. 1993. Plate tectonics and the development of sedimentary basins of the dextral regime in western Southeast Asia. *Journal of Southeast Asian Earth Sciences. Special Volume: Geosea VII Proceedings*, **8**(1–4), 497–511.

PAGE, B. G. N., BENNETT, J. D., CAMERON, N. R., BRIDGE, D. M., JEFFREY, D. H., KEATS, W. & THAIB, J. 1979. A review of the main structural and magmatic features of northern Sumatra. *Journal of the Geological Society, London*, **136**, 569–579.

POLACHAN, S. 1988. *The geological evolution of the Mergui Basin, SE Andaman Sea, Thailand.* PhD thesis, University of London.

POSAVEC, M., TAYLOR, D., VAN LEEUWEN, T. & SPECTOR, A. 1973. Tectonic controls of volcanism and complex movements along the Sumatran Fault System. *Bulletin of the Geological Society of Malaysia*, **6**, 43–60.

ROCK, N. M. S., ALDISS, D. T., ASPDEN, J. A., CLARKE, M. C. G., DJUNUDDIN, A., KARTAWA, W., MISWAR, THOMPSON, S. J. & WHANDOYO, R. 1983. *The geology of the Lubuksikaping Quadrangle, Sumatra (Quadrangle 0716) Scale 1:250,000.* Geological Survey of Indonesia, Directorate of Mineral Resources, Geological Research and Development Centre, Bandung.

ROSIDI, H. M. D., TJOKROSAPOETRO, S. & PENDOWO, B. 1976. *The Geology of the Painan and Northeastern part of the Muarasiberut Quadrangles, Sumatra (Quadrangle 5/VIII) Scale 1:250,000.* Geological Survey of Indonesia, Directorate of Mineral Resources, Geological Research and Development Centre, Bandung.

SIAGIAN, H. P., MUBROTO, B. & WAHYONO, H. 1995. Penelitian magnet purba di daerah Baturaja dan Sekitarnya, Sumatera Selatan (Palaeomagnetic investigations in the area of Baturaja, South Sumatra). *'Proceedings' Seminar hasil penelitian/pemetaan geologi dan geofisika, Pusat Penelitian dan Pengembangan Geologi, Bandung, 22–23 Mei 1995 [Proceedings of the seminar on results of geological and geophysical research, Geological Research and Development Centre, Bandung, 22–23 May 1995]*, 339–359.

TAPPONNIER, P., PELTZER, G. & ARMIJO, R. 1986. On the mechanism of collision between India and Asia. *In*: COWARD, M. P. & RIES, A. C. (eds) *Collision Tectonics.* Geological Society, London, Special Publications, **19**, 115–157.

WILLIAMS, H. H. & EUBANK, R. T. 1995. Hydrocarbon habitat in the rift graben of the Central Sumatra Basin, Indonesia. *In*: LAMBIASE, J. J. (ed.) *Hydrocarbon Habitat in Rift Basins.* Geological Society, London, Special Publications, 80, 331–371.

WILCOX, R. E., HARDING, T. P. & SEELY, D. R. 1973. Basic wrench tectonics. *American Association of Petroleum Geologists Bulletin*, **57**, 74–96.

YANCEY, T. E. & ALIF, S. A. 1977. Upper Mesozoic strata near Padang, West Sumatra. *Bulletin of the Geological Society of Malaysia*, **8**, 61–74.

Tertiary response to oblique subduction and indentation in Sumatra, Indonesia: new ideas for hydrocarbon exploration

CHRIS HOWELLS

ARCO British Limited, London Square, Cross Lanes, Guildford, Surrey GU1 1UE, UK

Abstract: The Tertiary basins of Sumatra, which lie along the leading edge of the Eurasian Plate, comprise a mature hydrocarbon province whose origin and subsequent evolution are closely related to the processes of oblique subduction and indentation that have resulted from the collision of the Indo-Australian and Eurasian plates. The Sumatran basins exhibit a general rift-sag geometry with broad plate-margin parallel sag basins overlying a series of N–S-oriented grabens. It is these N–S-oriented grabens, whose origin and subsequent development are still debated despite extensive oil industry data, that control the distribution of the lacustrine sediments which are the primary source-rock for hydrocarbons in Sumatra. The Ombilin Basin provides high quality outcrops of sediments of equivalent age and depositional setting that may be studied, providing new insights into petroleum exploration of the basin itself that may be applied to the more mature areas of Sumatra.

The Ombilin Basin shows a three-fold evolution. Initial sedimentation in the Eocene appears to have been controlled by normal fault displacements and not strike-slip motions. This suggests a close genetic relationship with the N–S-oriented early Tertiary grabens currently located within the back-arc region (North, Central and South Sumatra Basins), and that the Ombilin Basin did not originate as a local pull-apart related to the Sumatra Fault Zone. Subsequent sedimentation in the basin (Oligocene) was dominated by fluvial deposition which is shown to have had a tectonic control at a time of active volcanism and strike-slip faulting. This is taken to indicate modification of the initial basin style by strike-slip activity along the Sumatra Fault Zone. The final phase of sedimentation in the basin (Early Miocene) is dominated by marine deposits which form the lower part of a transgressive-regressive wedge of regional extent. This wedge of sediments buries the earlier graben systems and is related to thermal subsidence. Uplift of the basin to its present intramontane setting and differentiation from the Central and South Sumatra Basins took place in the Mid-Miocene or later.

The model for the Ombilin Basin proposed here is that of a wrench-modified rift basin and is quite different from previous models giving the possibility for development of new play concepts. A similar genetic origin to the Central and South Sumatra Basins is clearly suggested, highlighting the applicability of the Ombilin Basin as an exploration model for the Sumatra graben and other analogous basins in SE Asia.

This paper presents some of the results of PhD research undertaken by the author whilst with the University of London SE Asia Research Group at Royal Holloway College. The work presented here is not intended as an exhaustive treatment of this research, but more as an introduction to the main aspects of it. It is hoped that it will stimulate discussion about the region and help direct the reader to some of the key related works.

SE Asia is geologically a very complex area that includes the boundaries of several tectonic plates including the Indo-Australian, Pacific, Eurasian and Philippine Sea plates. A complex history of convergence between the Indo-Australian and Eurasian plates has dominated the Tertiary to Recent geological evolution of Sumatra (Fitch 1972; Hamilton 1979; Peltzer & Tapponnier 1988; Daly *et al.* 1987, 1991; Packham 1993), which lies at the western end of the Indonesian archipelago. The present structural and topographic grain of Sumatra is oriented NW–SE, parallel to the adjacent plate margin (Fig. 1). The NW–SE orientation of this plate margin, combined with the northwards movement of the Indo-Australian Plate at between $50\,mm\,a^{-1}$ (Patriat & Achache 1984) and $80\,mm\,a^{-1}$ (Stock & Molnar 1982) has created a regime of oblique subduction beneath Sumatra. This has in turn given rise to a component of dextral strike-slip motion parallel to the plate margin which is expressed as the active Sumatra Fault System (SFS) (Fitch 1972). This fault system runs the entire length of

From Fraser, A. J., Matthews, S. J. & Murphy, R. W. (eds), 1997, *Petroleum Geology of Southeast Asia*, Geological Society Special Publication No. 126, pp. 365–374.

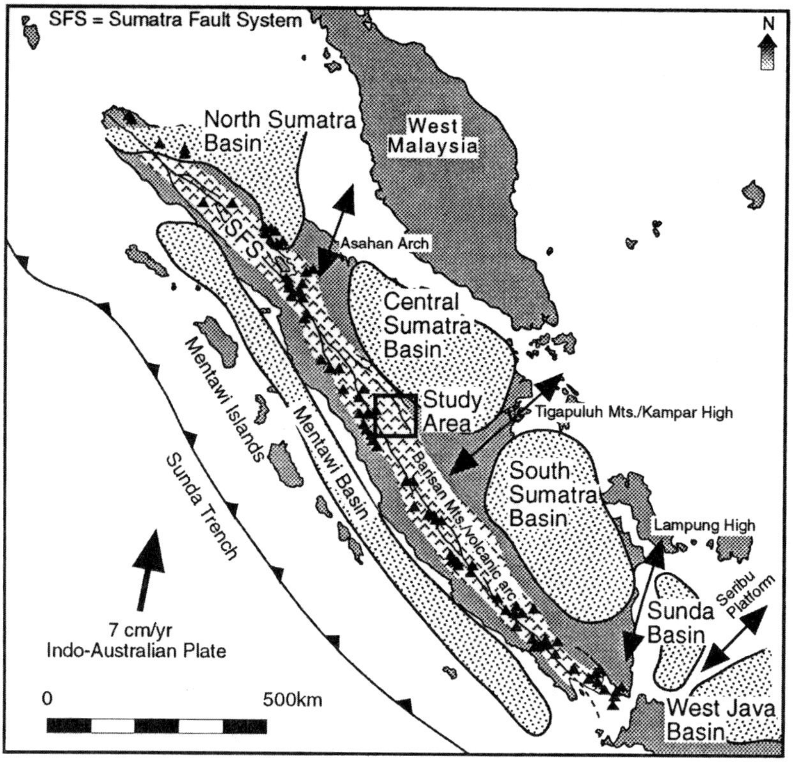

Fig. 1. Location and present tectonic setting of the Ombilin Basin within the Barisan Mountains of Sumatra. Note the strong NW–SE structural trend controlled by the adjacent subduction zone, and the coincidence of the Barisan Mountains with the trace of the dextral strike-slip Sumatra Fault System and active volcanic arc. The present configuration of basins in Sumatra is controlled by the NW–SE orientation of the subduction zone.

Sumatra, near the west coast, and is coincident with the Barisan Mountains and currently active volcanic arc.

The present, broadly NW–SE, configuration of the major basins of Sumatra (Fig. 1) is quite different from their configuration in the early Tertiary. Below the broad, plate margin-parallel outlines of the North, Central and South Sumatra Basins (NSB, CSB and SSB respectively), lie a system of N–S oriented Palaeogene grabens (Fig. 2) in which important lacustrine source-rocks were deposited. Whilst the geometry and origin of the broad sag basins can be related to the present subduction system, the origin of these earlier grabens remains enigmatic despite abundant data obtained from hydrocarbon exploration. Previous models explaining the origin and evolution of these graben have included various forms of back-arc extension (White & Wing 1978; Daly *et al.* 1987, 1991), crustal rotation (Ninkovich 1976; Davies 1984; Hutchison 1992), north–south-oriented compression (Moulds 1989) and plate inertia coupled with rotation resulting from the indentation of India into Eurasia (Packham 1993).

One reason for the continued debate about the origin and evolution of the grabens is the largely sub-surface nature of the Lower Tertiary sediments comprising them. Exposures of Lower Tertiary sediments are largely controlled by the presence of depressions along the Barisan Mountains (e.g. Ombilin Basin, Payakumbuh Basin and the Kampar Kanan area to the north of the Ombilin Basin, Fig. 2) which are separated from the present back-arc area by uplifted basement forming the Eastern or Scheifer Barisan (van Bemmelen 1949). The study of these outcrops may provide part of the answer. To do this, however, the origin of these depressions must first be resolved. The Ombilin Basin has been reported variously to be a pull-apart basin related to the SFS (Koning & Aulia 1985; Situmorang *et al.* 1991), a southerly continuation of the N–S grabens under the CSB since uplifted by the Barisan Mountains (Moulds 1989; de Smet 1991) or an

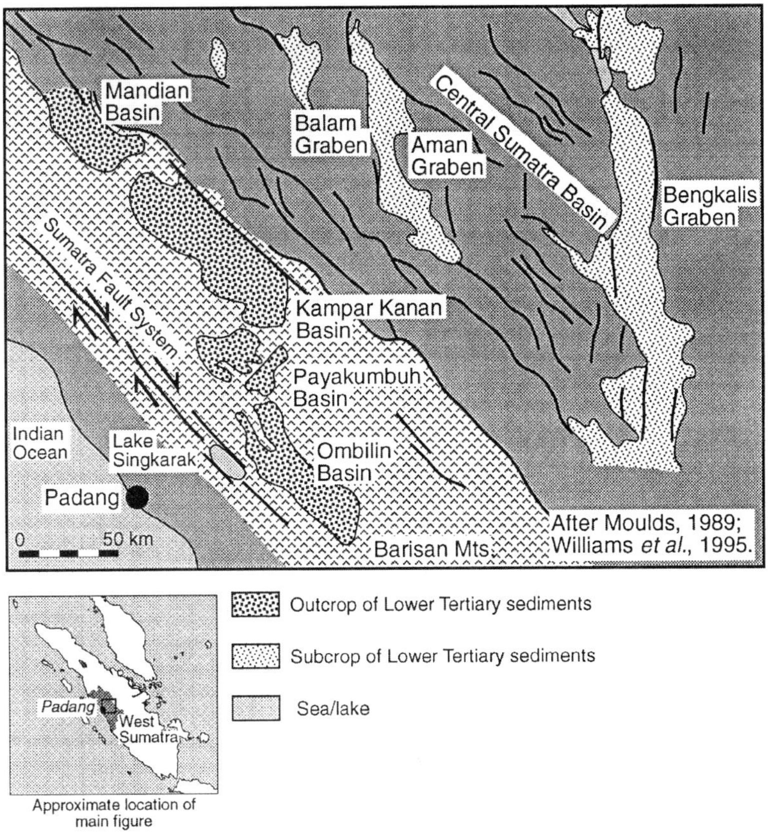

Fig. 2. Configuration of Lower Tertiary sediments in the Central Sumatra area showing a strong N–S trend which is not easily explained by the present orientation of the subducting plate margin.

intermontane basin related purely to Tertiary uplift of the Barisan Mountains (Koesoemadinata & Matasak 1981). It is proposed here that the Ombilin Basin resulted from a combination of rifting and strike-slip processes and is a wrench-modified rift basin.

Ombilin Basin Setting and Tertiary Stratigraphy

The Ombilin Basin is a small ($20 \times 60\,km^2$) basin, located in an intramontane setting. The currently active trace of the Sumatra Fault System lies 30 km to the west of the Ombilin Basin, near the town of Solok, where it forms a rift valley about 10 km wide. The basin is filled with greater than 3020 m of Tertiary (Eocene–Middle Miocene) sediments deposited in terrestrial and marine environments. The surrounding lithologies consist of pre-Tertiary metasedimentary units of Permian to Triassic age and granitic intrusions of Permian and younger ages.

Formal definitions and type sections for the Tertiary sediments were given by Koesoemadinata & Matasak (1981). However, it has been recognized by several workers (de Smet 1991; Humphreys *et al.* 1991; Cameron & Howells 1993) that there are problems with the stratigraphic nomenclature as it stands. This is seen particularly in relation to the diachronous basin margin fan deposits and the transition from meandering to braided fluvial sediments. For this reason, a refined stratigraphy based on that of Koesoemadinata & Matasak (1981) is used which resolves the primary problems and confusion of the formally defined stratigraphy. The two schemes are presented in Fig. 3 along with the lithologies and interpreted depositional environments. A correlation of the Tertiary stratigraphy of the Ombilin Basin with that of the NSB, CSB and SSB is presented as Fig. 4.

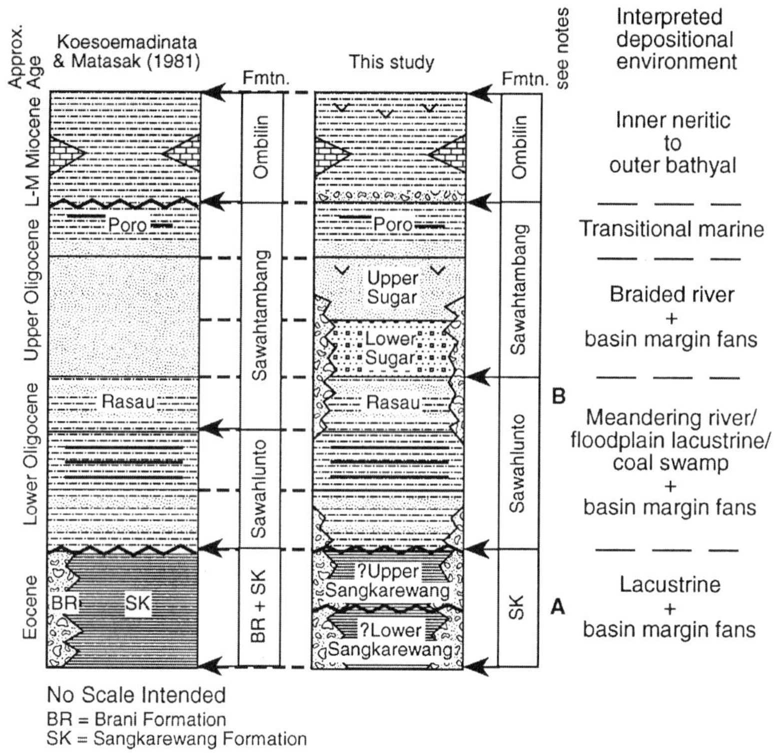

Fig. 3. Correlation of the Tertiary stratigraphy used in the present study with the formally defined stratigraphy of Koesoemadinata & Matasak (1981). The primary differences are as follows: (A) Discontinuation of the term Brani Formation. This formation was defined as the basin margin fan deposits at the stratigraphic level of the Sangkarewang Formation. Subsequent workers recognised the diachronous nature of these sediments and their occurrence above and below the unconformity separating the Sangkarewang and Sawahlunto Formations. In addition the term was originally applied (de Haan 1942) to similar sediments from the Harau Valley to the north of the Ombilin Basin where the sediments are believed to be of a different age. (B) Inclusion of the Rasau Member within the Sawahlunto Formation on the basis of the interpreted depositional environment. As noted in the text, a fundamental change in fluvial style is recognised within the basin and the lithologies of the Rasau Member have closer affinities with those of the Sawahlunto Formation. In this interpretation, the Sawahlunto and Sawahtambang Formations are not considered to interfinger as was indicated by Koesoemadinata & Matasak (1981).

Initial Tertiary sedimentation

A study of clast types in the basin margin fan deposits was made in an attempt to distinguish any syn-sedimentary strike-slip activity along the basin margins and resolve the origin of the basin as either a rift or pull-apart. The results of pebble counts made from the conglomerates are shown in Fig. 5 which documents the dominant clast type and the distribution of pre-Tertiary lithologies. Where available, palaeocurrent data have been utilized to suggest possible sediment dispersal paths.

Figure 5 clearly shows that most of the clasts within the Tertiary conglomerates and breccias of the Talawi sub-basin can be explained by derivation from adjacent or nearby pre-Tertiary basement lithologies. Most of these conglomerates are from the Sangkarewang Formation and represent the initial input into the newly-formed Tertiary basin. No lateral offset of sediment and source needs to be invoked to explain clast types. Such an offset or displacement of sediment and source lithology has been documented from many basins that were actively undergoing strike-slip faulting during deposition (e.g. Ridge Basin, Crowell & Link 1982; Horneln Basin, Steel 1988; Steel & Gloppen 1980; Lake Hazar, Hempton & Dunne 1984) and it is a recognized characteristic of such basins (Reading 1980;

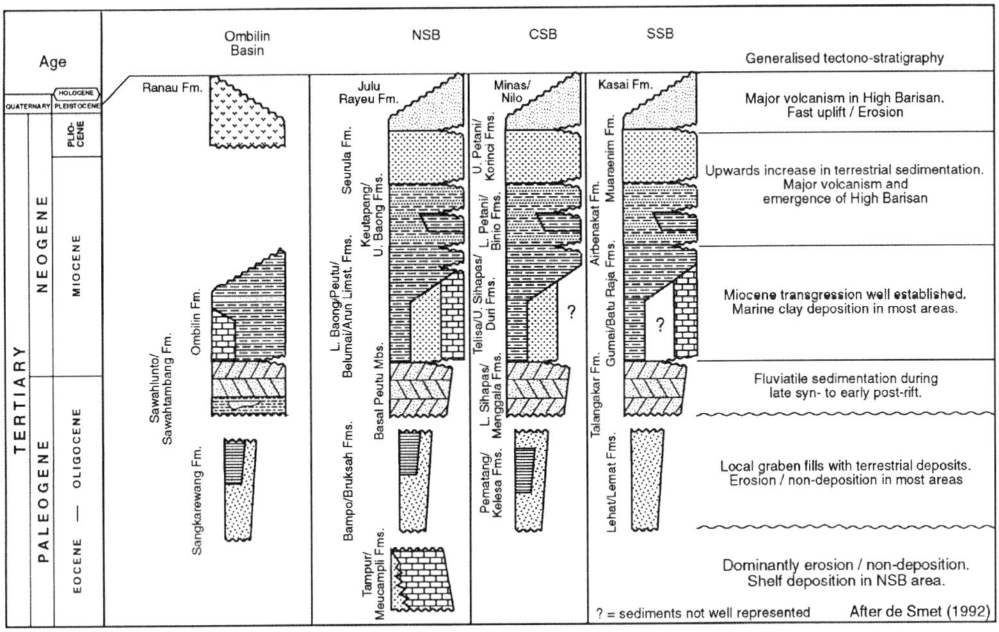

Fig. 4. Correlation of the Tertiary stratigraphies of the Ombilin, North Sumatra, Central Sumatra and South Sumatra basins highlighting similarities in lithology and depositional environment.

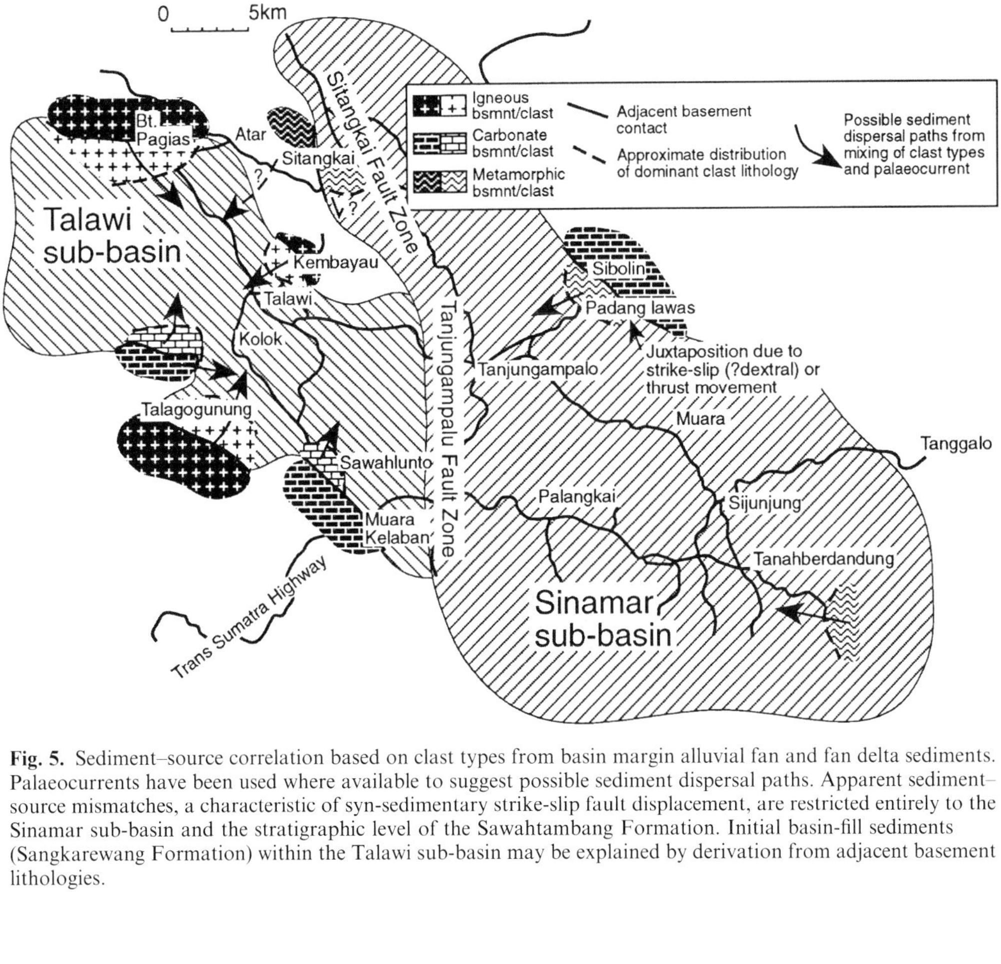

Fig. 5. Sediment–source correlation based on clast types from basin margin alluvial fan and fan delta sediments. Palaeocurrents have been used where available to suggest possible sediment dispersal paths. Apparent sediment–source mismatches, a characteristic of syn-sedimentary strike-slip fault displacement, are restricted entirely to the Sinamar sub-basin and the stratigraphic level of the Sawahtambang Formation. Initial basin-fill sediments (Sangkarewang Formation) within the Talawi sub-basin may be explained by derivation from adjacent basement lithologies.

Christie-Blick & Biddle 1985). However, it must be noted that this observation does not rule-out strike-slip faulting during deposition, but rather provides a lack of positive evidence for active strike-slip fault movements during deposition.

Outcrops which do show some apparent mismatch between sediment and source are restricted in distribution to the margins of the Sinamar sub-basin and the stratigraphic level of the Sawahtambang Formation (Upper Oligocene). At Sitangkai the apparent mismatch may be post-depositional or related to a simple scenario involving unroofing of the basement. The section at Sibolin however, strongly suggests syn-sedimentary fault movement and sediment-source mismatch. The vertical sequence of sediments indicates a change through distal–proximal–distal alluvial fan environments coincident with a clear change in provenance from a metasedimentary source to a granitic source.

Significant post-Sawahtambang Formation fault movements along the basin bounding fault are indicated by the proximity at present (less than 500 m) to extensive outcrops of pre-Tertiary carbonate lithologies and the complete absence of carbonate clasts in the Tertiary section. It is unclear if these displacements are strike-slip in nature or are related to high angle reverse faulting as presently found along the basin margin.

Meandering–braided fluvial transition

A sedimentary log drawn through the Sawahlunto/Sawahtambang Formation boundary near Kayu Gadang in the Talawi sub-basin shows a systematic change in clast type (Fig. 6). Clasts within the Sawahlunto Formation are dominated by metamorphic types (mostly quartzite with some phylite), with minor reworked sedimentary material. Plutonic and volcanic components are very minor or absent. The abundance and composition of clast types noted from the Sawahtambang Formation are quite different. Well-rounded pebbles of milky quartz, absent in the underlying sediments are abundant, the percentage of metamorphic clasts decreases and the percentage of reworked sedimentary material increases. Volcanic material also becomes a significant component of the sediments.

This indicates a significant change in the provenance of the sediment in the Ombilin Basin, matched by a change in fluvial style and palaeocurrent direction. The braided fluvial sediments of the Sawahtambang Formation were deposited by a transverse fluvial system flowing across the basin towards the NE and E and not an axial system as during deposition of the Sawahlunto Formation.

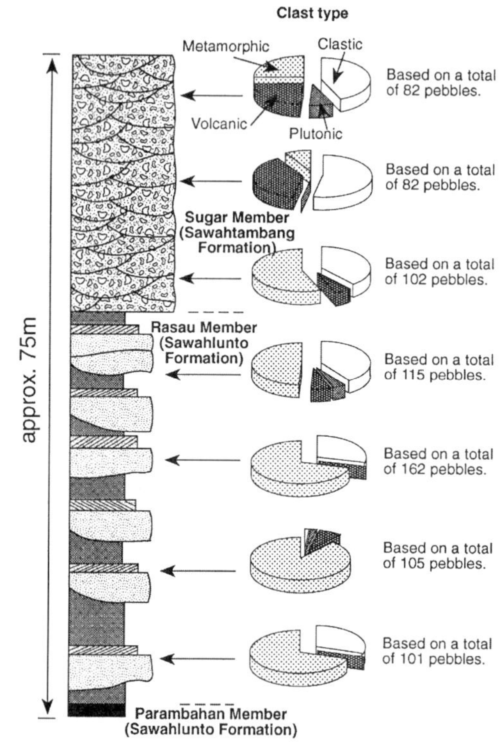

Fig. 6. Sedimentary log through the meandering (Sawahlunto Formation) to braided (Sawahtambang Formation) fluvial transition in the Ombilin Basin indicating a distinct change in provenance. In addition to the upwards increase in volcanic and re-worked sediment indicated on the log, well-rounded quartz pebbles, absent in the underlying Sawahlunto Formation, become abundant within the lower portions of the Sugar Member, Sawahtambang Formation.

This change in palaeoslope indicates a tectonic control on the meandering to braided transition and associated changes in character of the sediment (composition and architecture). The new palaeoslope direction is consistent with the interpretation of uplift along a NW–SE trend to the west of the basin, parallel to the present structural and topographic grain of Sumatra. In connection with the observation of the first volcanic detritus within the Tertiary sediments at this stratigraphic level, it is interesting to note that volcanic centres in Sumatra often lie at the intersection of E–W basement lineaments with NW–SE-oriented strike-slip faults (Katili 1969).

Significant uplift related to major strike-slip faults is observed on South Island, New Zealand which is also a region of oblique convergence between tectonic plates (Kamp *et al.* 1989, 1992). Here rapid uplift associated with the Alpine Fault has created a palaeoslope perpendicular to the trend of the fault system. This rapid uplift is also responsible for the creation of an extensive braid plain, the Canterbury Plains, running directly into the sea. This is considered analogous to the Sawahtambang–Ombilin Formation transition.

Seismic data

Figure 7 shows a stratigraphic breakdown of fault patterns interpreted from seismic data over the Sinamar sub-basin. The faulting at a stratigraphic level within the Sawahtambang Formation shows a pattern that may be closely related to a theoretical strain ellipse for a NW–SE-oriented dextral strike-slip shear couple. The orientation of faults cutting the base of the Ombilin Formation however, is quite different and is not clearly explained by

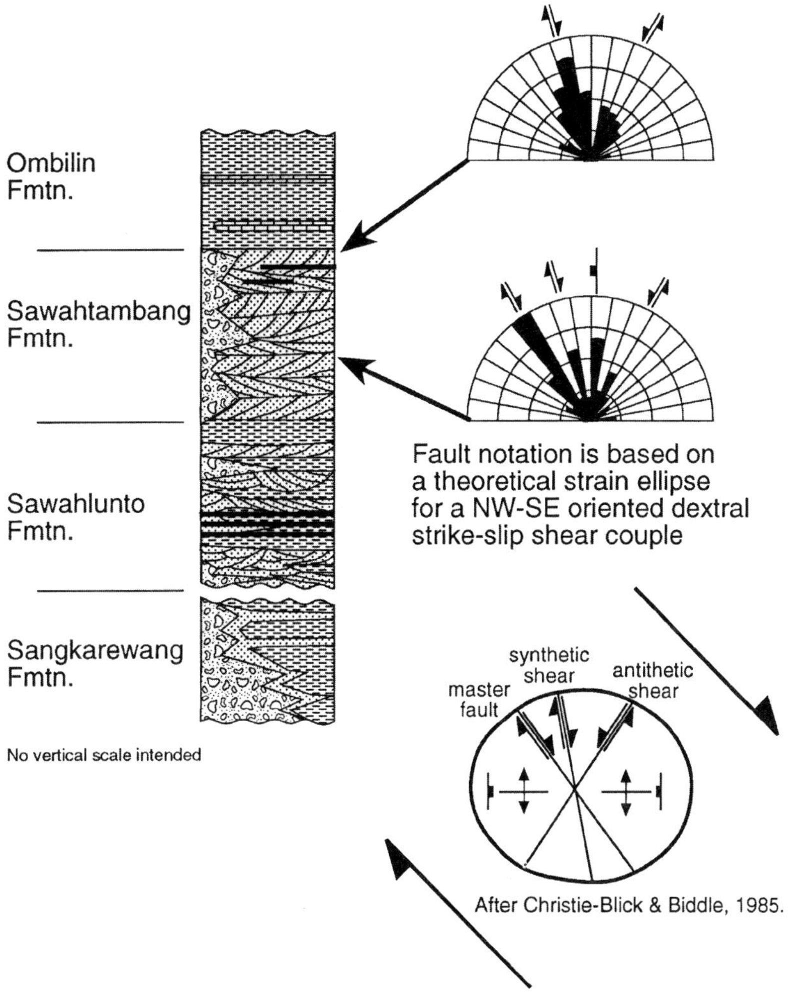

Fig. 7. Stratigraphic breakdown of changing fault patterns interpreted from seismic data over the Sinamar sub-basin. The rose diagrams are plotted at the same scale with radius measuring cummulative length of fault traces having a given orientation. The faults are near vertical and so this is believed to be a good representation of actual fault trend. A distinct change in both the magnitude and orientation of faulting is seen that is related here to a change from a dextral strike-slip regime within the sub-basin to one controlled by thermal subsidence. The structural interpretation on which this analysis and conclusion are drawn is that of Steve Deetz, Apache International Oil Company.

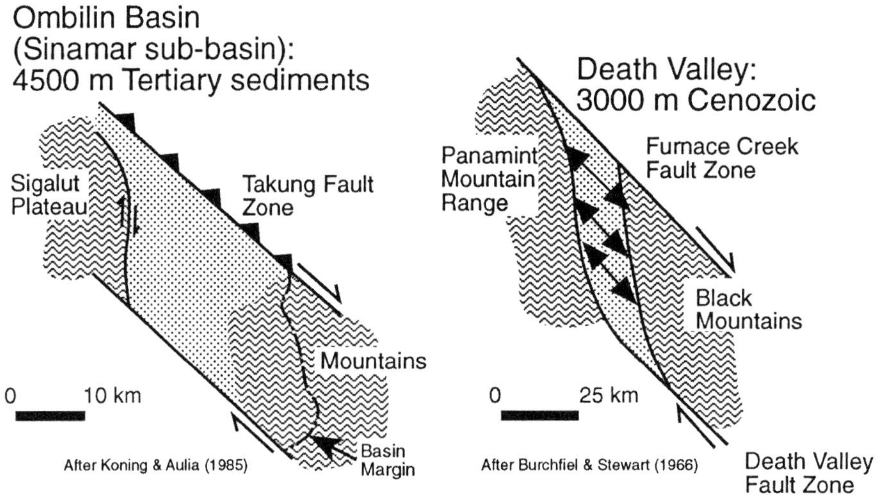

Fig. 8. Comparison of the structure of the Sinamar sub-basin with the strike-slip basin of Death Valley, California from Koning & Aulia (1985). Observations outlined in the accompanying text suggest that this analogy is valid for the Sinamar sub-basin at the stratigraphic level of the Sawahtambang Formation, but the not the Ombilin Basin as a whole as proposed by some workers (e.g Koning & Aulia 1985; Situmorang *et al.* 1991).

applying strain ellipse theory. The magnitude of the faulting penetrating the base of the Ombilin Formation is also different, being significantly less than that within the Sawahtambang Formation. The rose diagrams in Fig. 7 are plotted at the same scale and the radius measures the total length of the fault traces penetrating the horizon. This is consistent with sedimentological observations and regional considerations indicating the Ombilin Formation was deposited during the initial phases of thermal subsidence.

Discussion of Ombilin Basin model

Previous models of the Ombilin Basin thought of the basin either as a rift/intermontane basin (Koesoemadinata & Matasak 1981) related to the CSB and SSB rift system (Moulds 1989; de Smet 1991), or as a pull-apart related to the SFS (Koning & Aulia 1985; Situmorang *et al.* 1991). The pull-apart models consider the evolution of the Ombilin Basin to be unrelated to the other early Tertiary basins of Sumatra, with the Ombilin Basin developing under purely wrench stresses. These models imply that the oblique motion of the Indo-Australian plate being subducted under Sumatra was de-coupled into a wrench component and extensional component early in the Eocene in the area of the Ombilin Basin. The rift hypotheses consider the Ombilin Basin to have evolved as part of the Sumatra rift system up until the Plio-Pleistocene when the Barisan Mountains were uplifted and the Ombilin Basin with them. This implies that de-coupling of the stress system resulting from oblique subduction and, therefore, the origin of the SFS (Fitch 1972), took place relatively recently, in post-Late Miocene times.

Koning & Aulia (1985) drew a comparison between the geometry of the Ombilin Basin and Death Valley, California (Fig. 8) relating them to theoretical geometries for pull-apart basins. Similarities were also pointed out between fault orientations and fault type expected from application of strain ellipse theory, a valid comparison for the Sawahtambang Formation of the Sinamar sub-basin as shown in Fig. 7. In this comparison, however, Koning & Aulia (1985) only considered the Sinamar sub-basin (the target for hydrocarbon exploration), ignoring the Talawi sub-basin (lacking any seal or trap and with the source rocks exposed at outcrop).

This is not in conflict with the present interpretation of a wrench-modified rift which also draws an analogy between the structure and geometries of the Sinamar sub-basin and theoretical strike-slip models. Consideration of basin margin clast type suggests strike-slip movements later in the basin history (Late Oligocene) at a time when the Sinamar sub-basin was the primary depocentre. The sedimentological considerations of the meandering-braided transition (Sawahlunto–Sawahtambang Formation) coupled with structural observations provide strong evidence of wrenching at this time (Mid–Late Oligocene). However, the lack of sediment-source mismatch in the initial Tertiary sediments (Sangkarewang Formation) of the Talawi sub-basin, suggests a dominance of normal faulting in the initial stages of basin-filling. This implies a common genetic origin with the N–S-oriented grabens of the present back-arc area and that de-coupling of stress associated with oblique subduction had occurred by mid-Oligocene at the latest.

The model proposed here for the Ombilin Basin clearly discounts the proposal that the Ombilin Basin is a pull-apart basin (Koning & Aulia 1985) related to the Sumatra Fault System or its Lower Tertiary equivalent and that it originated separately from the Lower Tertiary grabens. Although the evidence presented here suggests a common genetic origin with the Lower Tertiary graben of the present back-arc area, the origin of these grabens, particularly the N–S orientation, remains problematical. It is beyond the scope of this paper to discuss in detail the various proposals for the origin of these basins, however, the model does not discount the possibility that the graben system originated within a zone of distributed dextral shear several hundred kilometres in width.

References

BURCHFIEL, B. C. & STEWART, J. H. 1966. "Pull-apart" origin of the central segment of Death Valley, California. *Geological Society of America Bulletin*, **77**, 439–442.

CAMERON, N. & HOWELLS, C. G. 1993. *Ombilin Basin – new play concepts following from the integration of the outcrop geology with the Sinamar-1 well.* Ombilin Associates report for Apache International Oil Company.

CHRISTIE-BLICK, N. & BIDDLE, K. T. 1985. Deformation and basin formation along strike-slip faults. *In*: BIDDLE, K. T. & CHRISTIE-BLICK, N. (eds) *Strike-slip deformation, basin formation and sedimentation.* SEPM Special Publications, **37**, 1–34.

CROWELL, J. C. & LINK, M. H. (eds) 1982. *Geologic history of the Ridge Basin, southern California.* SEPM (Pacific Section) Special Publications.

DALY, M. C., HOOPER, B. G. D. & SMITH, D. G. 1987. Tertiary plate tectonics and basin evolution in Indonesia. *Proceedings of the Indonesian Petroleum Association 16th Annual Convention*, 399–428.

——, COOPER, M. A., WILSON, I., SMITH, D. G. & HOOPER, B. G. D. 1991. Cenozoic plate tectonics and basin evolution in Indonesia. *Marine and Petroleum Geology*, **8**, 2–21.

DAVIES, P. R. 1984. Tertiary structural evolution and related hydrocarbon occurrences, North Sumatra Basin. *Proceedings of the Indonesian Petroleum Association 13th Annual Convention*, 19–49.

DE HAAN, W. 1942. Over de stratigraphie en tectoniek van het Mangani gebied, Sumatra's west kust. *Geologie en Mijnbouw*, **4**, 65–77.

DE SMET, M. E. M. 1991. *The geology of the Ombilin Basin, preliminary report.* University of London Group for Geological Research in Southeast Asia, Reports, **99**.

—— 1992. *A guide to the stratigraphy of Sumatra, part 2: Tertiary.* University of London Group for Geological Research in Southeast Asia, Reports, **108**.

FITCH, T. J. 1972. Plate convergence, transcurrent faults and internal deformation adjacent to Southeast Asia and the western Pacific. *Journal of Geophysical Research*, **77**, 4432–4460.

HAMILTON, W. 1979. *Tectonics of the Indonesia Region.* USGS Professional Papers, **1078**.

HEMPTON, M. R. & DUNNE, L. A. 1984. Sedimentation in pull-apart basins: active examples in eastern Turkey. *Journal of Geology*, **92**, 513–530.

HUMPHREYS, B., MATCHETTE-DOWNES, C. J. & MORLEY, R. J. 1991. *Geological reconnaissance of the Ombilin Basin, central Sumatra, Indonesia.* LEMIGAS/University of London/BGS Indonesian Hydrocarbon Basin Assessment Project, Reports, **91/1**.

HUTCHISON, C. S. 1992. The Eocene unconformity on Southeast Asia and East Sundaland. *Geological Society Malaysia Bulletin*, **32**, 69–88

KAMP, P. J. J., GREEN, P. F. & WHITE, S. H. 1989. Fission track analysis reveals character of collisional tectonics in New Zealand. *Tectonics*, **8**, 169–195.

——, GREEN, P. F. & TIPPET, J. M. 1992. Tectonic architecture of the mountain front-foreland basin transition, South Island, New Zealand, assessed by fission track analysis. *Tectonics*, **11**, 98–113.

KATILI, J. A. 1969. Large transcurrent faults in Southeast Asia with special reference to Indonesia. *Bulletin National Institute of Geology and Mining, Bandung*, **2**, 1–20.

KOESOEMADINATA, R. P. & MATASAK, TH. 1981. Stratigraphy and sedimentation, Ombilin Basin, central Sumatra (West Sumatra Province). *Proceedings Indonesian Petroleum Association 10th Annual Convention*, 217–249.

KONING, T. & AULIA, K. 1985. Petroleum geology of the Ombilin intermontane basin, West Sumatra. *Proceedings Indonesian Petroleum Association 14th Annual Convention*, 117–137.

MOULDS, P. J. 1989. Development of the Bengkalis Depression, Central Sumatra and its subsequent deformation – a model for other Sumatran grabens? *Proceedings Indonesian Petroleum Association 18th Annual Convention*, 217–245

NINKOVICH, D. 1976. Late Cenozoic clockwise rotation of Sumatra. *Earth and Planetary Science Letters*, **29**, 269–275.

PACKHAM, G. H. 1993. Plate tectonics and the development of sedimentary basins of the dextral regime in western Southeast Asia. *Journal of Southeast Asian Earth Sciences*, **8**, 497–511.

PATRIAT, P. & ACHACHE, J. 1984. India-Eurasia collision chronology has implications for crustal shortening and driving mechanisms of plates. *Nature*, **311**, 615–621.

PELTZER, G. & TAPPONNIER, P. 1988. Formation and evolution of strike-slip faults, rifts, and basins during the India-Asia collision: an experimental approach. *Journal of Geophysical Research*, **93**, **B12**, 15085–15117.

READING, H. G. 1980. Characteristics and recognition of strike-slip fault systems. *In*: BALLANCE, P. F. & READING, H. G. (eds) *Sedimentation in oblique-slip mobile zones*, Special Publications of the International Association of Sedimentologists, **4**, 7–26.

SITUMORANG, B, YULIHANTO, B., GUNTAR, A., HIMAWAN, R. & GAMAL JACOB, T. 1991. Structural development of the Ombilin Basin West Sumatra. *Proceedings Indonesian Petroleum Association 20th Annual Convention*, 1–16

STEEL, R. J. 1988. Coarsening-upward and skewed fan bodies: symptoms of strike-slip and transfer fault movement in sedimentary basins. *In*: NEMEC, W. & STEEL, R. J. (eds) *Fan deltas: sedimentology and tectonic settings*. Blackie & Son, 75–83

—— & GLOPPEN, T. G. 1980. Late Caledonian (Devonian) basin formation, western Norway: signs of strike-slip tectonics during infilling. *In*: BALLANCE, P. F. & READING, H. G. (eds) *Sedimentation in oblique-slip mobile zones*. Special Publications of the International Association of Sedimentologists, **4**, 79–103.

STOCK, J. & MOLNAR, P. 1982. Uncertainties in the relative position of the Australia, Antarctica, Lord Howe, and Pacific plates since the late Cretaceous. *Journal of Geophysical Research*, **87**, 4697–4714.

VAN BEMMELEN, R. W. 1949. *The Geology of Indonesia, Volume 1a: General geology of Indonesia and adjacent archipelagos*. Government Printing Office, Nijhoff, The Hague

WHITE, J. M. JR & WING, R. S. 1978. Structural development of the South China Sea with particular reference to Indonesia. *Proceedings Indonesian Petroleum Association 7th Annual Convention*, 159–177

WILLIAMS, H. H., KELLY, P. A., JANKS, J. S. & CHRISTENSEN, R. M. 1985. The Paleogene rift basin source rocks of Central Sumatra. *Proceedings Indonesian Petroleum Association 14th Annual Convention*, 57–90

WILLIAMS, H. H., FOWLER, M. & EUBANK, R. T. 1995. Characteristics of selected Palaeogene and Cretaceous lacustrine source basins of Southeast Asia. *In*: LAMBIASE, J. J. (ed.) *Hydrocarbon habitat in rift basins*. Geological Society, London, Special Publications, **80**, 241–282

A tectonic model for the onshore Kutai Basin, East Kalimantan

JOHN L. C. CHAMBERS[1] & TIMOTHY E. DALEY[2]

[1] *LASMO Runtu Ltd, Ratu Plaza Office Tower, Jalan Jenderal Sudirman No. 9, Jakarta 10270, Indonesia*

[2] *LASMO plc, 101 Bishopsgate, London EC2M 3XH, UK*

Abstract: Past structural geological models for the Samarinda Anticlinorium have included gravity-driven slumping, shale diapirism and thrusting related to regional tectonic events. New geological cross-sections utilizing LASMO-acquired data over structures in the Runtu Block have been constructed from both surface geological measurements and 1993–1994 vintage seismic data. Relatively rigid deltaic and shelf sediments have been deformed into box-folds, above disharmonically folded shale-rich pro-delta and bathyal sediments. Detachment of these structures appears to be at the top of, or within, over-pressured shales at the base of Lower Miocene deltaic packages.

Gravity data, acquired along with seismic data, suggests semi-regional uplifts of the underlying over-pressured strata which would have been initially deposited as a flat or gently inclined unit. Basement configuration is not visible on seismic but gravity and aeromagnetic data show that it is deep throughout the study area, ranging between 7 and 14 km. The implication of the structural models is that relatively small amounts of shortening are interpreted to have occurred across these near-surface structures and that the amount of relative uplift is large. This model suggests that basin inversion may be the cause of the structural styles illustrated in this paper.

Data presented supports a tectonic model for the central Kutai Basin involving inversion of a deep Palaeogene rift basin which gives rise to apparent broad regional folding of the shale rich over-pressured section. Closer spaced folding in the near surface, normally pressured and thus less ductile deltaic and shelf section of the Samarinda Anticlinorium, is a result of the same inversion.

The study area for this paper is located within the Samarinda Anticlinorium of the onshore Kutai Basin of East Kalimantan (Fig. 1). The southern part of the Runtu Block covers most of the study area and data collected within this area by LASMO Runtu Limited makes up the bulk of the input to this paper. This study investigates the nature of structural styles in this part of the Kutai Basin using the combination of gravity, magnetic, surface geological and seismic data.

The Kutai Basin represents one of the most economically important sedimentary basins in Indonesia (Fig. 1). It covers an area of approximately 60 000 km^2 and contains a Tertiary sedimentary section of up to 14 km at its thickest part. The basin is the largest and the deepest Tertiary basin in western Indonesia.

East of the study area are the prolific Kutai Basin gas and oil fields which produce from Middle Miocene to Pliocene age reservoirs. The study area contains outcrop and well data defining Oligocene to Lower Miocene stratigraphy and can be regarded as a separate megasequence from the younger, currently productive intervals. In fact, the Middle Miocene sediments appear to be largely derived from erosion of the Lower Miocene succession.

Regional geology

Kutai Basin sedimentation was fairly continuous throughout the Tertiary beginning with rifting and the formation of a series of half grabens that filled with syn-rift sediments during, or prior to the Upper Eocene. Post-rifting Upper Eocene to lower Lower Miocene sediments are of outer shelf to deep marine facies and are shale dominated. On the basin margin, or above local highs, thick carbonate platforms or build-ups developed during the Oligocene to Early Miocene. Basin margin stratigraphy has been discussed in detail by Wain & Berod (1989).

The area considered in this paper is close to the centre of the basin, the Palaeogene section does not crop out but is thought to be shale dominant throughout (see Fig. 2). The oldest sediments for which there are data are Upper Oligocene bathyal shales drilled by the Segihan-1, Murung-1, and Kahala-1 wells.

During the Early Miocene, major eastward deltaic progradation began, and in excess of 4000 m of shelf and delta plain section was deposited at any one place. Rapid burial combined with the shale-rich nature of the underlying pro-delta and bathyal sediments resulted in the

From Fraser, A. J., Matthews, S. J. & Murphy, R. W. (eds), 1997, *Petroleum Geology of Southeast Asia*, Geological Society Special Publication No. 126, pp. 375–393.

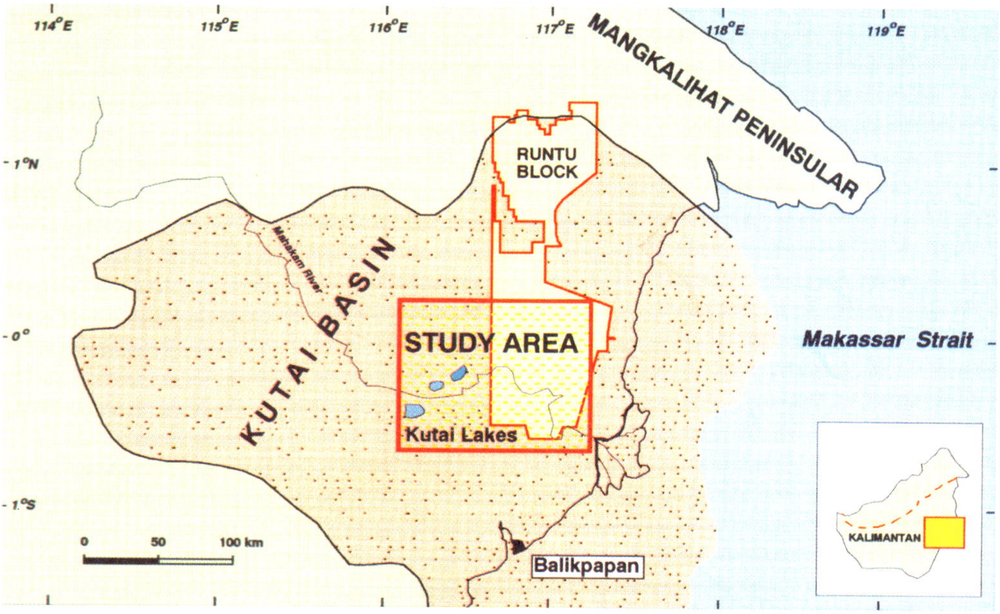

Fig. 1. Location map showing Kutai Basin, East Kalimantan.

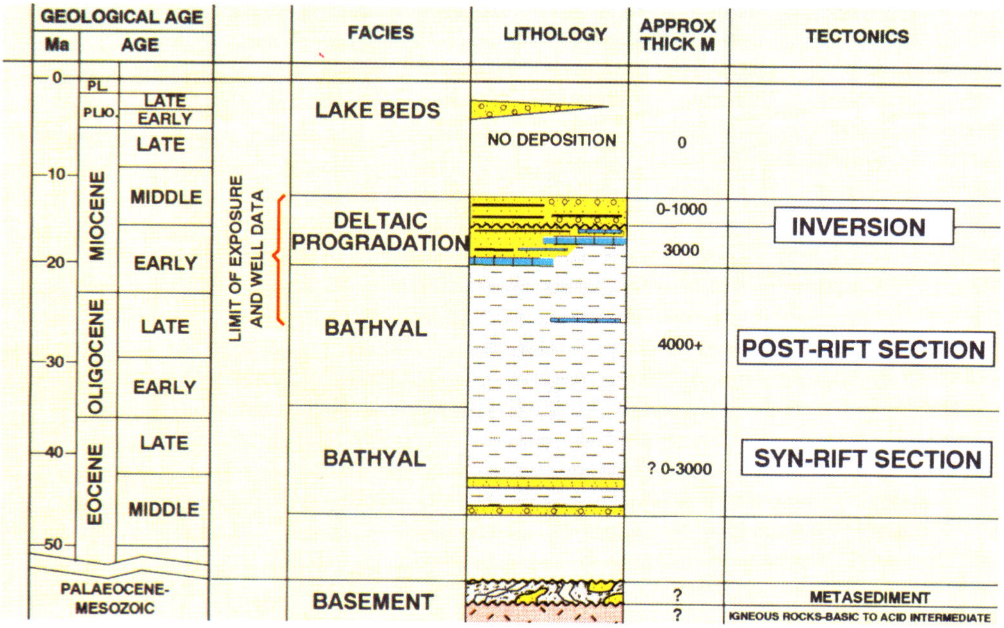

Fig. 2. Summary of southern Runtu PSC stratigraphy.

early onset of over-pressuring due to under-compaction. The boundary between normal and over-pressured sediments is strongly controlled by sedimentary facies and is likely therefore to be diachronous throughout the Lower Miocene. However it is important to realise that the transition to over-pressured strata is inherited from the depositional and burial history of the sediments and would originally have been a flat or gently inclined boundary. This boundary has

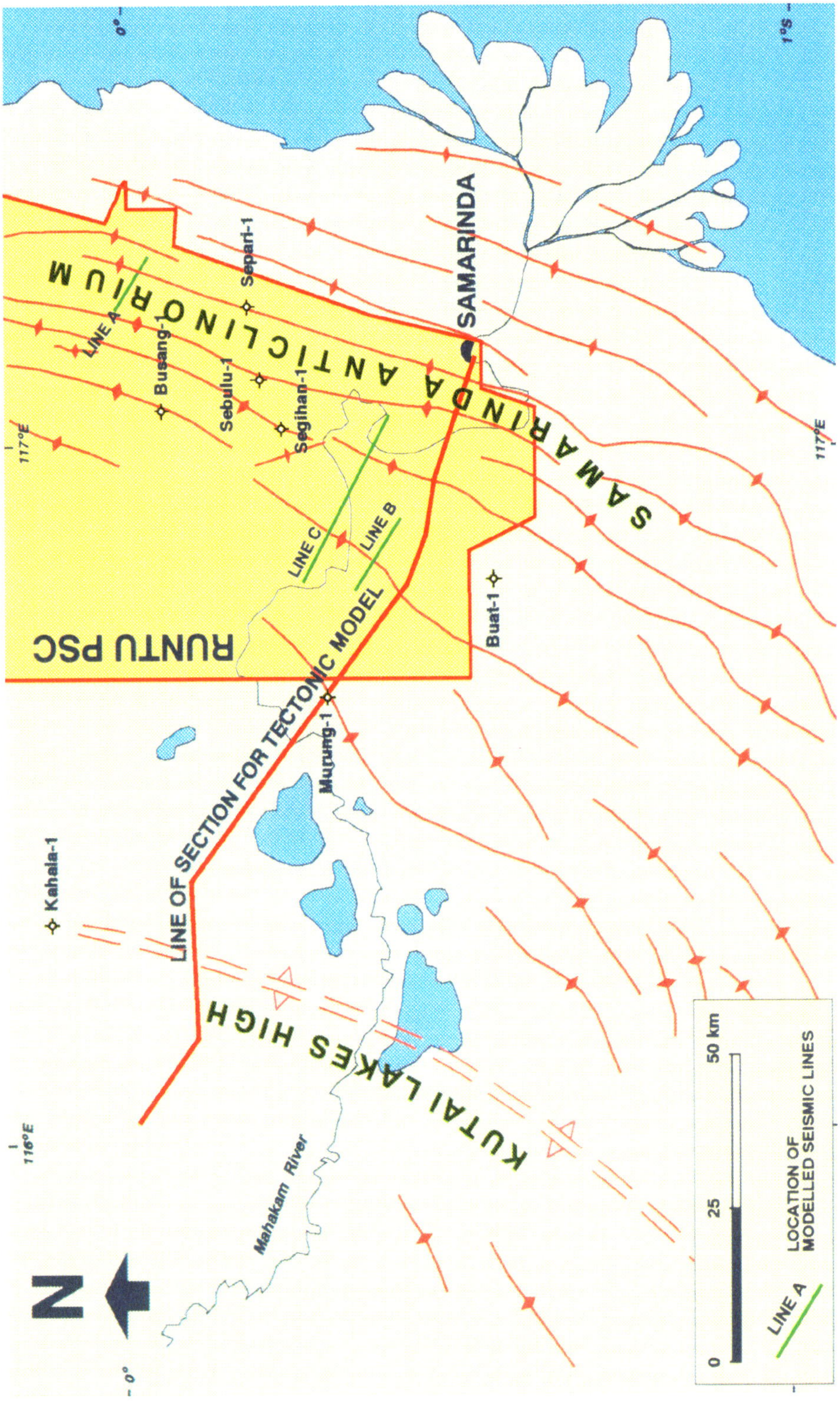

Fig. 3. Tectonic elements of southern Runtu Block area.

important consequences for generation and detachment of structures and may provide clues to overall basin deformation.

Deltaic sedimentation continued into the Mid- and Late Miocene and Pliocene east of the Runtu Block, punctuated by periods of compressional deformation, uplift and erosion. It is these sediments outside of the study area that make up the present day hydrocarbon productive intervals. Some thin intervals of Middle Miocene are exposed in the centers of synclines within the study area as a syn-inversion section. To the west, up to 400 m of Pliocene gravels and lake deposits of the Anap Formation directly overlie Oligocene bathyal shales as observed in the Kahala-1 well. Sedimentation in the Kutai Basin continues to the present day at the currently active Mahakam Delta depocentre to the east of the study area.

The complete Lower Miocene section has been folded and faulted and is referred to as the Samarinda Anticlinorium. It is dominated by a series of NNE–SSW-trending tight linear anticlines and broad open synclines (see tectonic elements on Fig. 3 and the regional seismic display on Fig. 4). Anticlinal ridges are often laterally continuous for in excess of 100 km and are characterised by abrupt changes in axial dip from steep east to steep west. Flat dips at the crest of these anticlines are generally absent leading previous interpreters to frequently describe these structures as being faulted. Facies in the cores of anticlines are generally shelf to bathyal and shale dominated whereas the broad synclines have sand-rich deltaic sediments exposed at the surface.

Previous explanations of these structures have invoked concepts as diverse as vertical diapirism, growth faulting and inversion through regional wrenching (Biantoro *et al.* 1992) as well as regional gravity sliding (Ott 1987). Compressional tectonic models producing duplex-cored anticlines have also at times been considered but never formally published. In this paper, a close examination of the structural style as visible on modern seismic and gravity data is presented and an inversion tectonic model is proposed.

Method

During 1993–1994 LASMO Runtu Limited acquired over 1500 km of regional and prospect specific seismic data across the southern Runtu Block. Detailed surface geological mapping was undertaken to collect surface structural measurements and to identify the distribution of prospective facies. Palynological analysis of seismic shot-hole samples combined with facies boundaries identified in the field has enabled a consistent geological interpretation to be made across the study area. The methodology and results of this stratigraphic investigation were presented by Carter & Morley (1996).

Gravity data were routinely collected along LASMO seismic lines and has been merged with digitized regional data available to the west by ARK Geophysics/Geoservices. Two dimensional modeling by ARK/LASMO has combined seismic interpretation, well data and estimated depths to basement from aeromagnetic surveys.

Migrated seismic lines have been interpreted incorporating well and surface geological data, the interpretations were then transferred to the GEOSEC software (a Cogniseis Proprietary package) for structural geological modelling.

Limitations of time and space allow us to present only a few examples in this paper, but those chosen are representative of many of the structures within the study area.

Structural geological modeling

Tight shale-cored anticlines and broad sand-rich synclines typical of the Samarinda Anticlinorium characterize the regional seismic line across the southern part of the Runtu Block PSC shown in Fig. 4. Basement is too deep to be resolved on seismic, but gravity and magnetic modelling described below suggests that basement is between 7 and 14 km depth.

A consistent structural geological model is difficult to construct from seismic data alone as seismic resolution at depth and in the cores of tight anticlines is generally poor. Surface geological measurements and sedimentary facies data has been important in constraining structural interpretations. Sufficient examples exist to indicate that surface structures are detached at varying levels, most likely controlled by the local depth of over-pressured sediments.

Rock competence during deformation is directly related to the sand/shale ratio, which is in turn a product of the sedimentary process within the basin. Thick shale-rich pro-delta sections in the Kutai Basin are generally over-pressured when drilled at depth. Over-pressured shale lacks shear strength and will deform disharmonically, whereas the overlying competent sand-rich deltaic section is assumed to deform according to the laws of flexural slip, i.e. constant bed length. This scenario is likely to give rise to detachment folds with the thick, over-pressured section in the core of structures

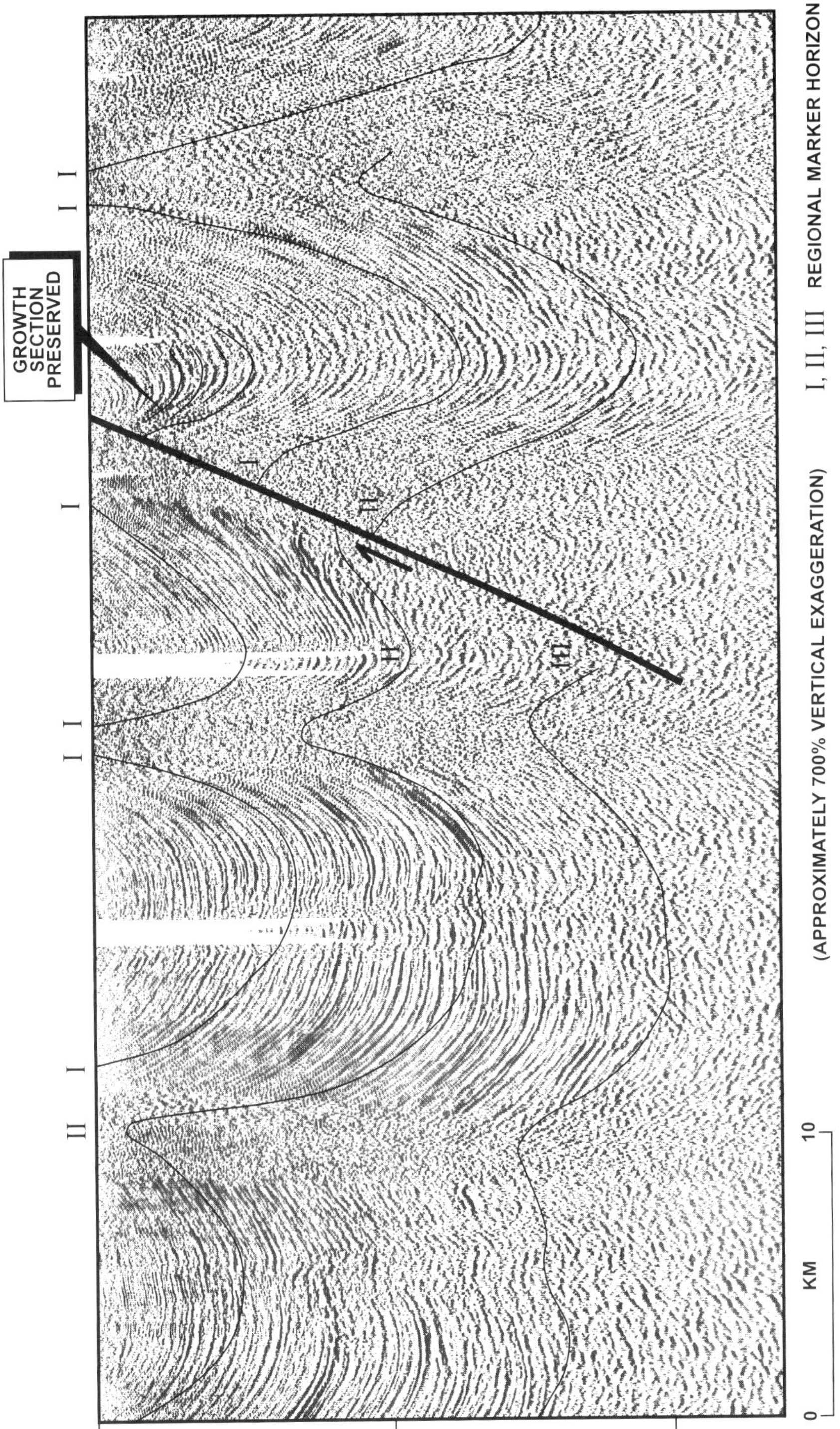

Fig. 4. Regional seismic line across the Samarinda Anticlinorium.

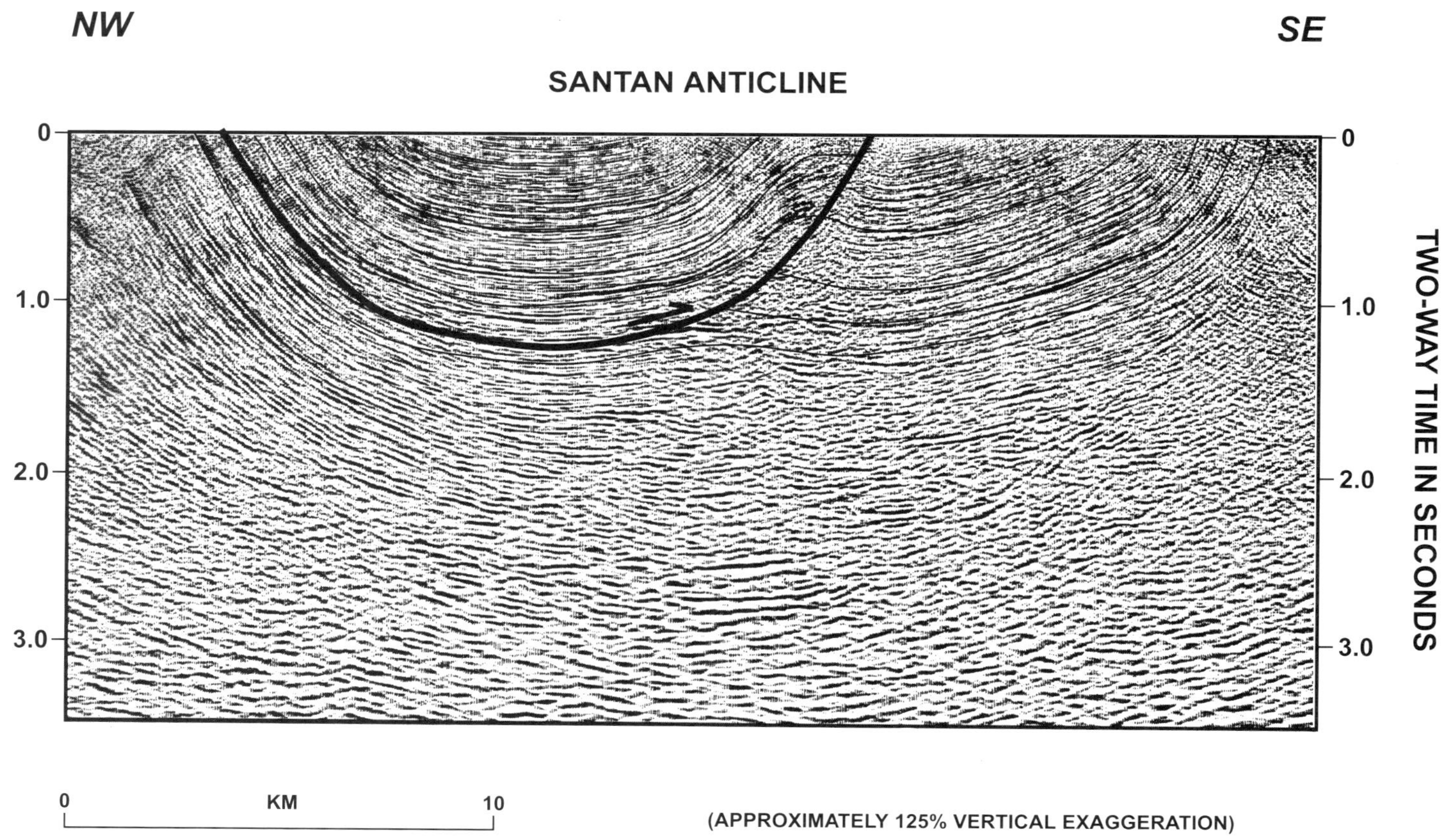

Fig. 5. Seismic Line A across Santan Anticline.

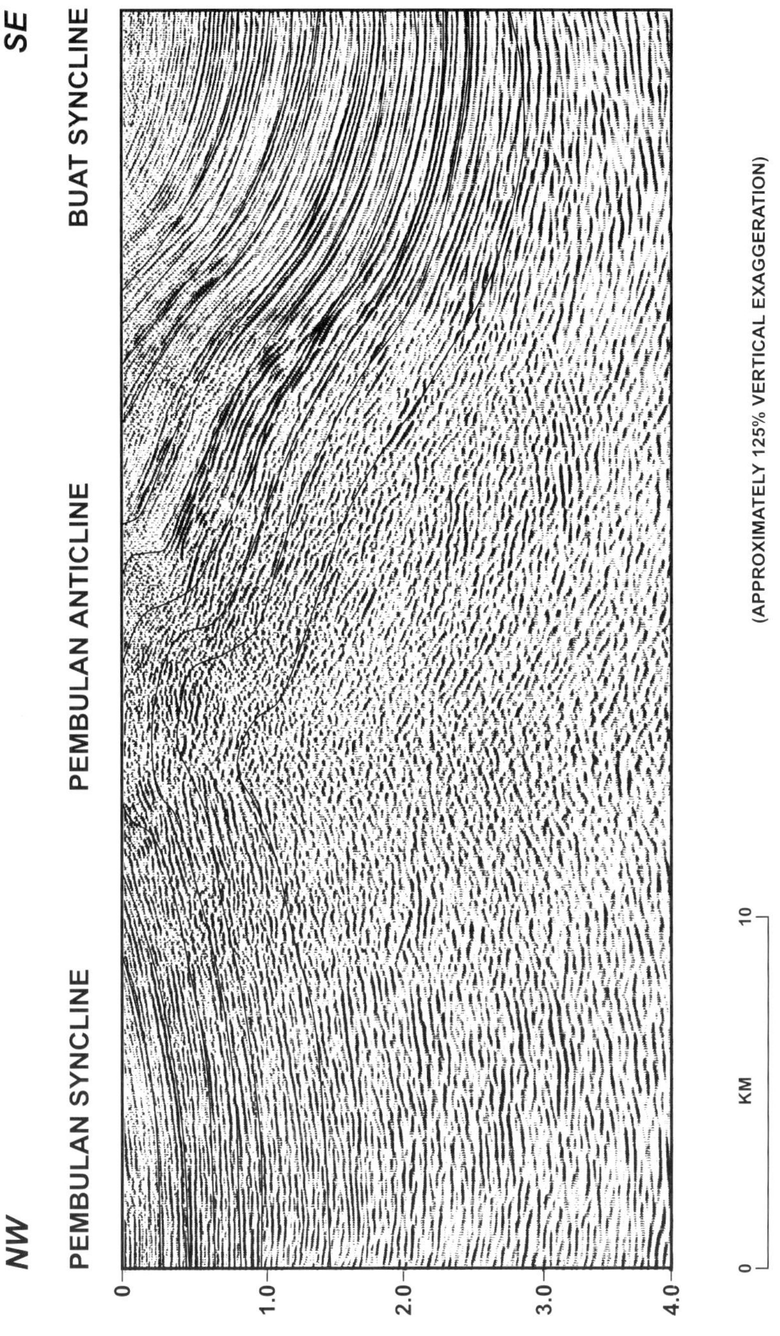

Fig. 7. Seismic Line B across the Pembulan Anticline.

unlikely to deform coherently. It is thus not believed to be possible to balance structurally the cores of structures using standard techniques. As a result, we have attempted structurally consistent models for the higher stratigraphic levels only.

Seismic Line A

In the study area there are several examples of small anticlines and faults within synclinal areas between major anticlines. The Santan Thrust Fault and associated anticline is one such example, lying between the Maritan and Sebulu Anticlines (see Seismic Line A, Fig. 5). The Santan Thrust Fault soles out at approximately 1.2 s two-way time (TWT). Below this level continuous reflectors consistent with a broad syncline between the Maritan and Sebulu Anticlines are apparent.

A depth section taken from an interpretation of this seismic line is shown in Fig. 6. This section has been restored using a fault-slip fold mechanism to reveal a symmetric detachment fold, which shows the characteristic thickening into the core of the anticline. Stratigraphic units

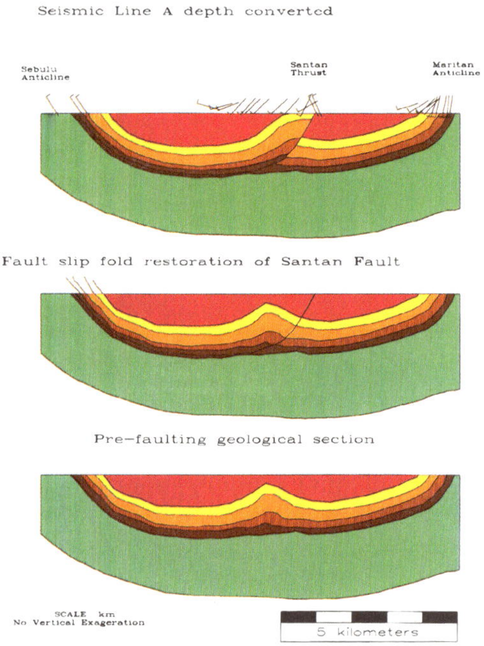

Fig. 6. Seismic Line A depth converted and structurally restored.

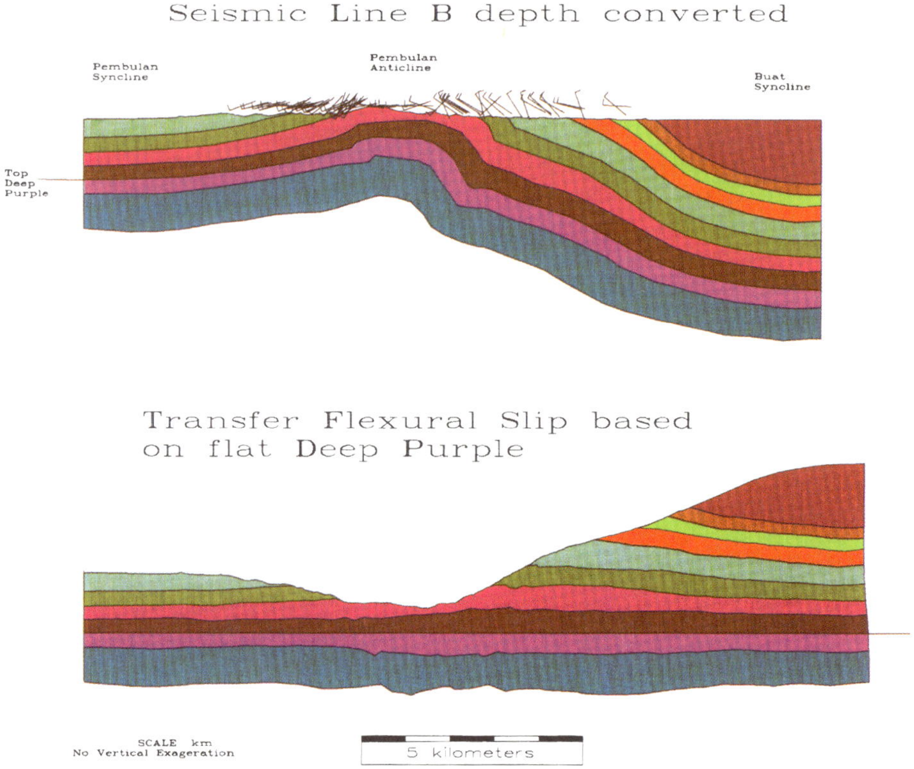

Fig. 8. Seismic Line B depth converted and structurally restored.

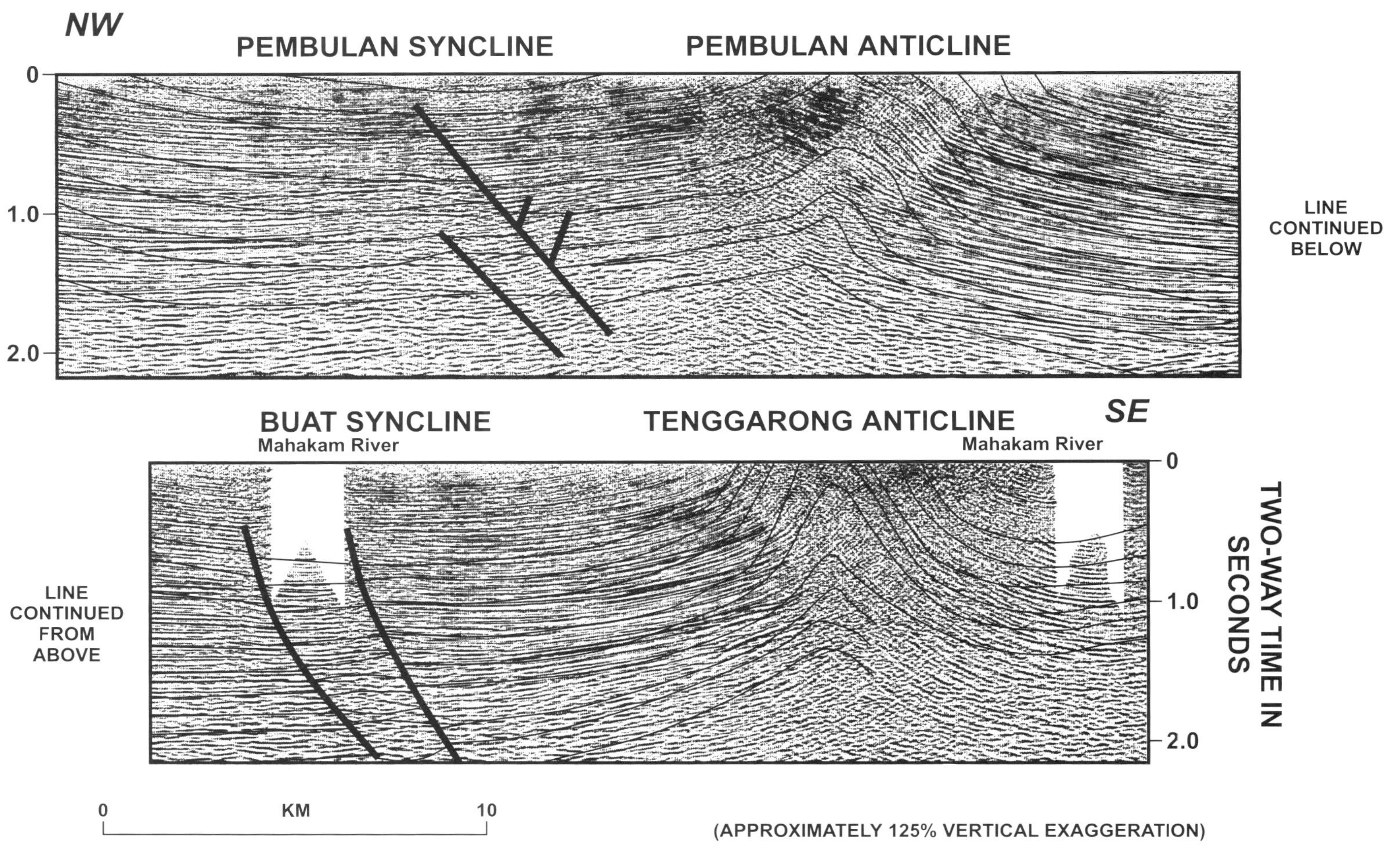

Fig. 9. Seismic Line C across the Pembulan and Tenggarong Anticlines.

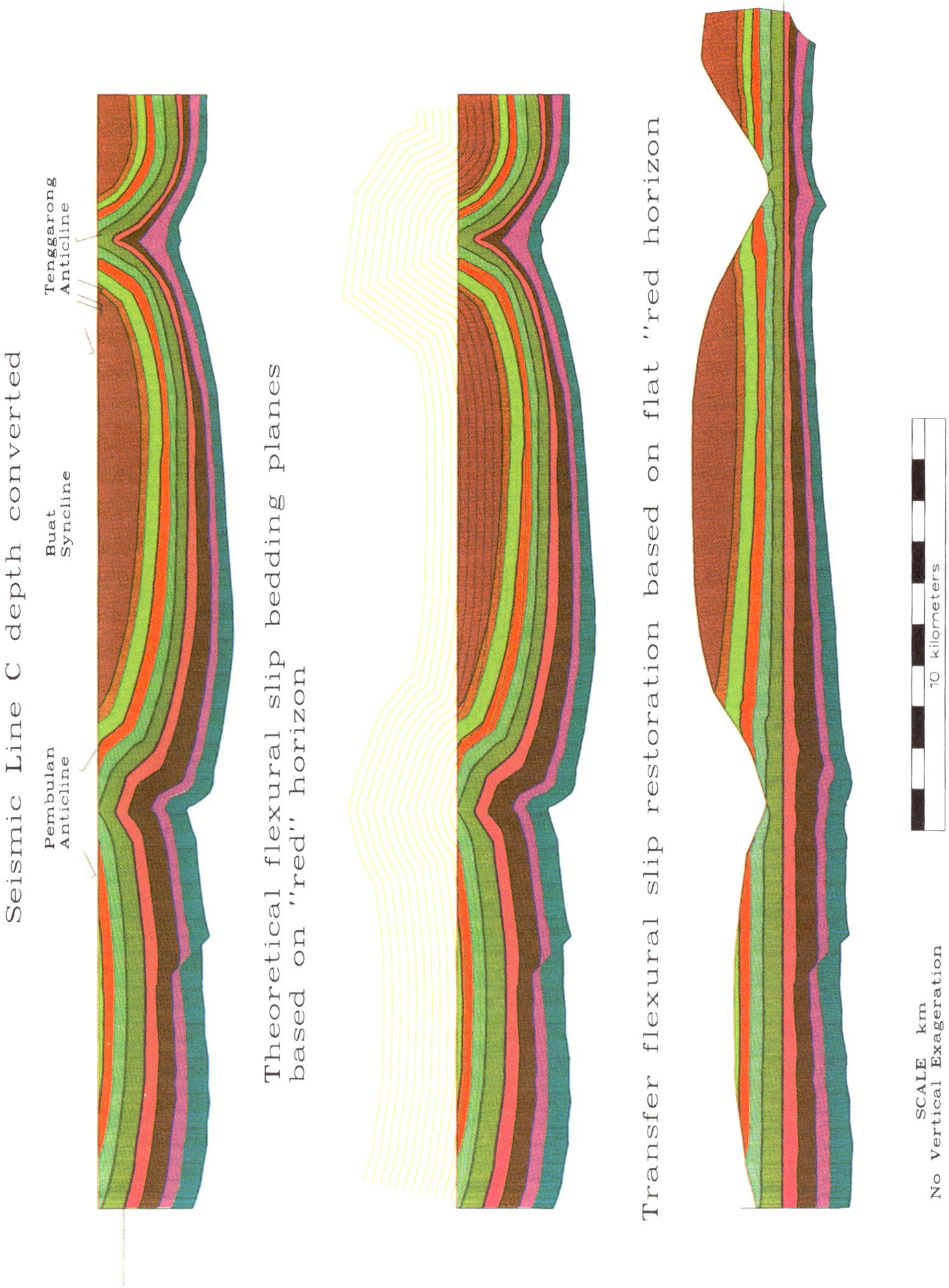

Fig. 10. Seismic Line C depth converted and structurally restored.

involved in the thickened section can be traced to outcrop and are found to be made up of shale-rich pro-delta to bathyal facies. The higher stratigraphic levels contain deltaic facies, which it is assumed were deposited on a near flat deposurface suggesting that a layer parallel stratigraphy was likely prior to deformation.

Compressive deformation appears to have formed a detachment fold at the over-pressure to normal-pressure boundary. Continued, or later compression, caused the rocks to fail along the Santan Fault. Although the Santan Fault is a relatively minor structure compared to surrounding features, the structural style is interpreted to reflect processes that commonly influence structures throughout the area.

Seismic Line B

Seismic Line B (Fig. 7) crosses the Pembulan Anticline in the southwestern part of the study area. Seismic data within the anticline is of moderate quality but when combined with surface structural geological measurements, this feature can be interpreted as an asymmetric box fold. Figure 8 is a depth section constructed from Seismic Line B. A flexural slip restoration of this section shows a very consistent bed length in all horizons and indicates a shortening of less than 7%. Note that in general, stratigraphic intervals maintain their thickness or thin gradually to the east.

Seismic Line C

Seismic Line C (Fig. 9) is a longer section to the north of Line B linking both the Pembulan and Tenggarong Anticlines. The Pembulan Anticline is clearly visible as a box fold in this example but the Tenggarong Anticline is more typical of other major anticlines in the Samarinda Anticlinorium with outcrop of steep east dips on the east flank abruptly changing to steep west dips

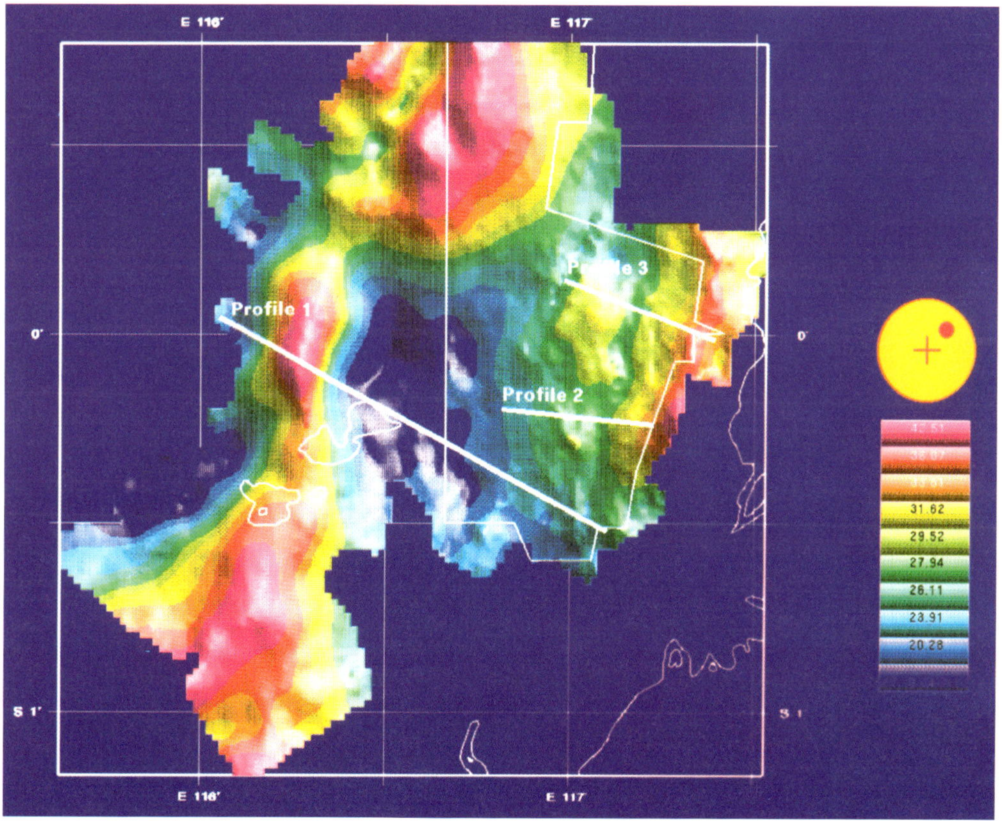

Fig. 11. Bouger gravity map.

on the west flank. This fold may well be a larger detachment fold similar in style to that seen in Seismic Line A but seismic resolution at depth in the core of the anticline is poor.

A depth section constructed from this line is shown in Fig. 10. Theoretical flexural slip bedding planes have been phantomed above the 'red' horizon to demonstrate that at higher stratigraphic levels (now eroded) the Tenggarong Structure may have been a box fold. A flexural slip restoration based on a flat 'red' horizon has been performed and shows consistent bed lengths in the stratigraphy above the levels involved in detachment folding. Interestingly, this restoration suggests that only 7.4% shortening is required over the length of the section to produce the present day folds.

As sediment supply to the basin was from the west, the stratigraphy is more proximal and sand-rich in the Pembulan area than to the east in the Tenggarong area. This may explain why a box fold has formed at Pembulan but to the east folds are considerably tighter because it is the shale-dominated sediments that are exposed.

Gravity data and modelling

As depth to basement is large and not apparent on seismic sections, the mechanisms for fold formation have previously been obscure. Gravity and magnetic data have provided information on the deeper basement structure and hence have had an important role in constraining structural models.

The composite Bouguer gravity map is included as Fig. 11. This map is dominated by a large dissected north–south gravity high in the west, known as the Kutai Lakes Gravity High. Much of the southwestern part of the Runtu Block is occupied by a deep gravity low. East of this gravity minimum there is a regional increase in gravity response, but there is no strong correlation of this response with the individual structures of the Lower Miocene fold-belt.

This Bouguer gravity data was residualized, that is the regional component to the data was removed, to enable higher frequency anomalies to be studied more easily (Fig. 12). The resultant map still shows the Kutai Lakes High Trend as

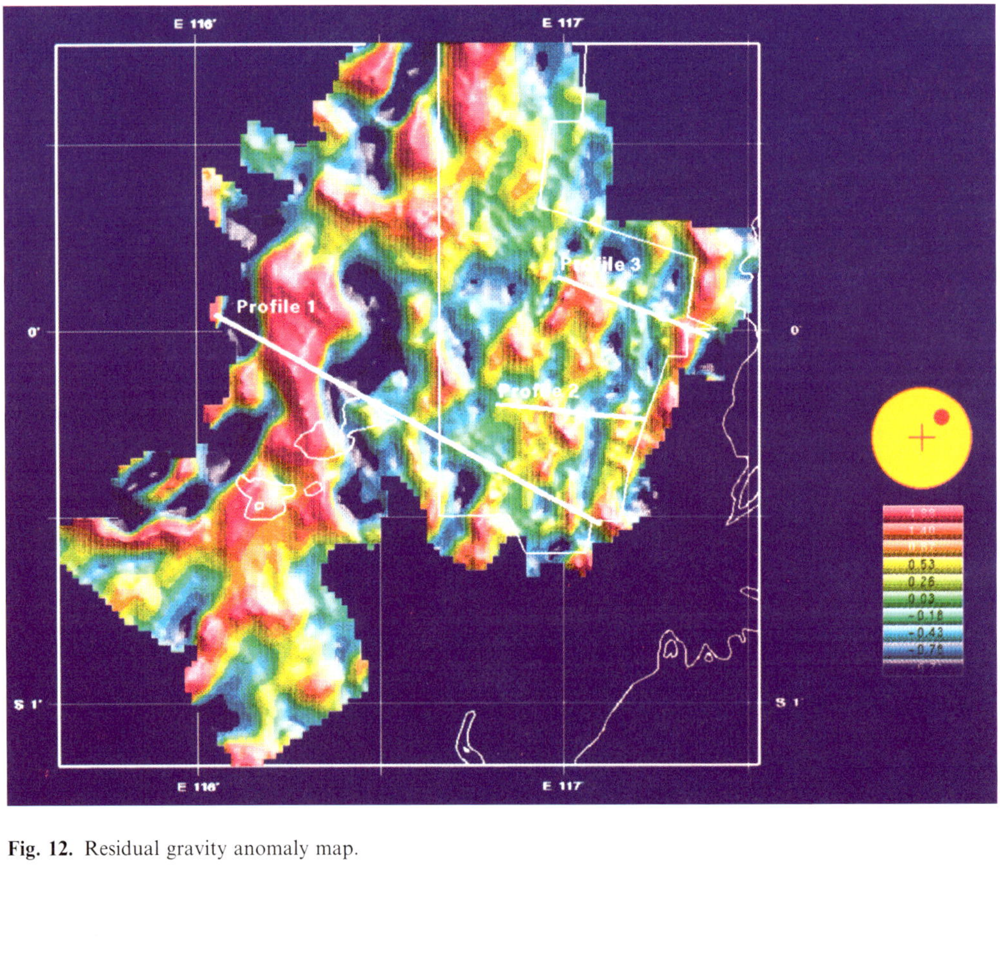

Fig. 12. Residual gravity anomaly map.

the most obvious feature, but in addition there are now several more visible anomalies and trends across the Samarinda Anticlinorium. A series of NNE–SSW anomalies coincident with the dominant folding orientation of surface structures is apparent. These trends are offset along NW–SE lineations. Although the residual gravity anomaly map has similar trends to the surface structures, anomalies are not all correlatable to individual anticline–syncline pairs.

To study the relationship between the gravity response and structure, the data was plotted along a profile 130 km in length which followed seismic lines across the region as shown in the upper part of Fig. 13 and plotted on the map, Fig. 11 (profile 1). The Bouguer gravity profile shows three main elements; the 25 mGal amplitude of the Kutai Lakes Gravity High, a regional gradient increasing from a low of 10 mGal immediately east of the Kutai Lakes to a value of 30 mGal in the east of the study area and finally, shorter wavelength anomalies up to 1–2 mGal in amplitude. The aeromagnetic profile is similar to the regional element of the Bouguer gravity trend, although the response to the Kutai Lakes area is more subdued and no residual anomalies are resolvable.

A depth model was constructed corresponding to the profile in Fig. 13. Seismic data has adequate resolution to define a series of anticline-syncline pairs and density data from the upper 4–5 km are derived from offset well data (see Table 1). As no well in the central part of the Kutai Basin penetrates basement, densities of between 2.62 and 2.72 $g\,cm^{-3}$ were modelled to be consistent with outcrops of metasediments and acid to intermediate igneous rock near the basin margin. Depths to basement estimated from aeromagnetic data indicate generalized structure at deepest levels. The gravity response calculated from this model is plotted against the observed Bouger gravity profile in the upper part of Fig. 13 and a good match is apparent.

The regional decrease in gravity and magnetic values from the east towards the Kutai Lakes minimum corresponds to increasing depth to basement, from 7–8 km deep in the east to 14 km or more (20%) adjacent to the Kutai Lakes area. The Kutai Lakes Gravity High itself is modelled as being due to a combination of a step-up of basement and major uplift bringing older and more compacted sediments near surface.

The small residual anomalies appear to be related to the lateral density contrasts in the upper 4 km caused by variations between the anticlines and synclines. Anticlines tend to be the locus of residual gravity minima, while synclines cause local maxima. Although anticlines are comprised of older, more uplifted strata, they in fact contain relatively less dense over-pressured and under-compacted delta-front and bathyal shales. Depth to over-pressure and

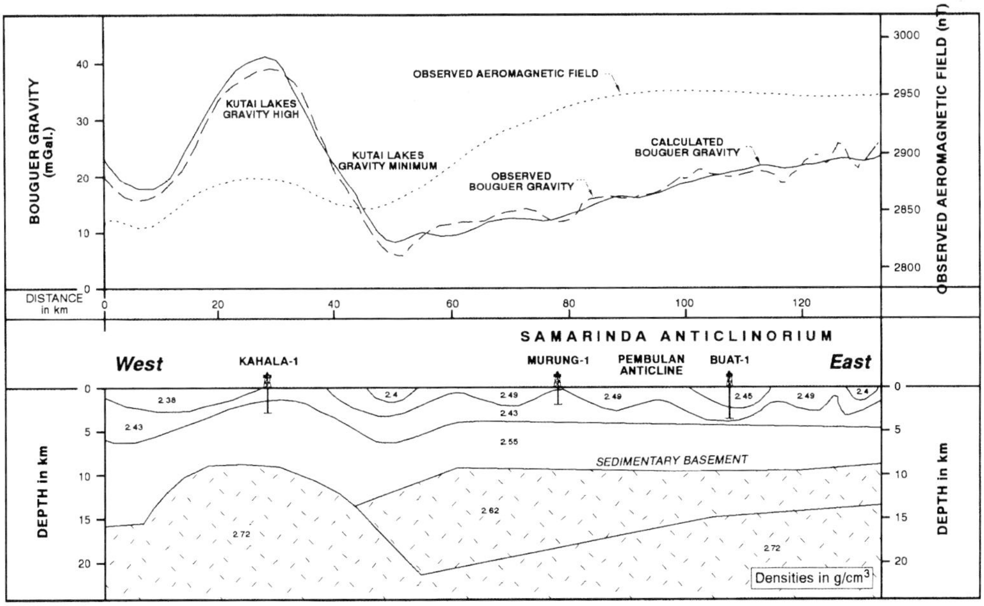

Fig. 13. Regional gravity model across Kutai Basin.

Table 1. *Average velocities and densities for Runtu area wells*

Well	Age	Structural setting	Approx. top over-pressure (m)	TD subsea (m)	Two-way time to TD (s)	Average velocity $m\,s^{-1}$	Pseudo density $(g\,cc^{-1})$*
Buat-1	Lower Miocene	Syncline	2900	3619	2.02	3583	2.55
Busang-1	Lower Miocene	Faulted anticline	750	2091	1.33	3144	2.47
Segihan-1	Lower Miocene/Upper Oligocene	Anticline	600	2407	1.91	2520	2.34
Sebulu-1	Lower Miocene	Faulted anticline	750	1497	1.16	2580	2.35
Separi-1	Lower Miocene	Anticline	900	1374	0.98	2804	2.40
Murung-1	Lower Miocene	Anticline	1500	1500	1.08	2777	2.39
Kahala-1	Lower Miocene/ Oligocene	Highly uplifted	Normal pressure	2978	2.02	2949	2.43

* Average density estimated from Gardners relationship:
Density $= CV^{1/4}$
V = velocity in $m\,s^{-1}$
C was found to be approximately satisfied by the value 0.33 when applied to wells with both sonic and density log data.

average well densities are shown in Table 1. Under-compacted shale within anticlines has been elevated with respect to the normally-compacted and therefore relatively denser shelf and deltaic sediments which are preserved in the synclines on either side.

Two more profiles, numbered 2 and 3 are displayed as Figs 14 and 15, to illustrate further the behaviour of residual gravity anomalies with respect to near surface structuring. In both these profiles the wavelength of the residual anomalies is longer then the wavelength of the surface folding. It can be observed that the deeper, wider synclines are associated with positive residual gravity anomalies, while broader areas of uplift are marked by relative negatives. It is interpreted that the larger wavelength (20–30 km) of residual gravity anomalies, when compared to the wavelength of surface folding (5–10 km), is a response to broad folding of the normal to over-pressure boundary. The near surface folding and faulting has been superimposed onto this apparently broader, semi-regional folding event. This may offer an explanation for the wide variations in the level of detachment of surface structures, since the detachment unit itself has been folded.

Summarizing the potential field interpretation leads to the first observation that basement is

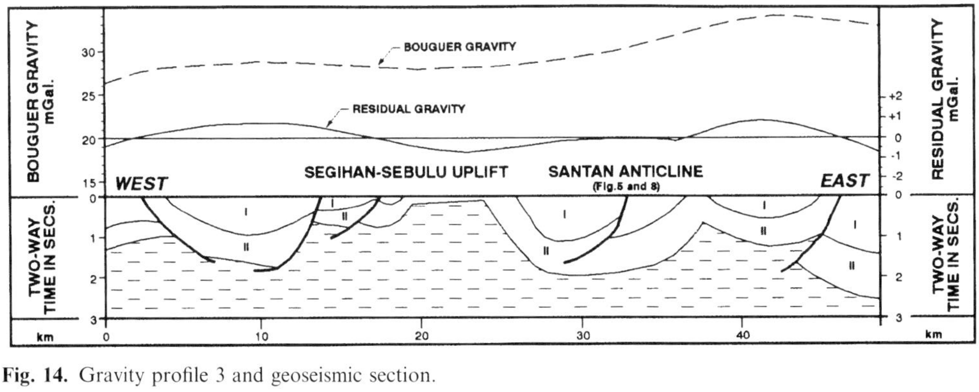

Fig. 14. Gravity profile 3 and geoseismic section.

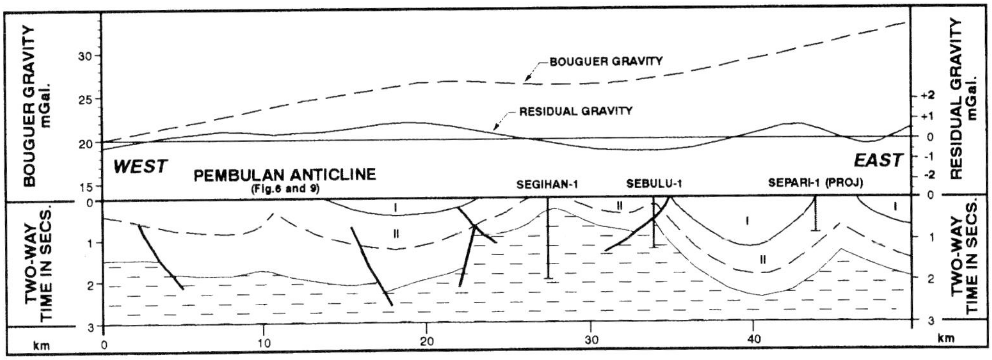

Fig. 15. Gravity profile 2 and geoseismic section.

deep throughout the study area, varying from 7–8 km in the east to 14 km (±20%) adjacent to the Kutai Lakes. It is difficult to resolve a more detailed basement architecture from this data alone. Secondly, the residual component of the Bouger gravity data is due to density contrasts in the upper 4 km of section. This variation is interpreted to be related to uplift of a lower density over-pressured layer at a broader semi-regional scale when compared to the shorter wavelength of near-surface high-amplitude folding.

Discussion

Geological correlations across anticlines of the Samarinda Anticlinorium which have been restored indicate relatively small amounts of shortening, approximately 7%. The structural style appears to be one where a competent overburden of shelf and deltaic sediments is folded in a rigid box-fold style over a weak shale rich over-pressured unit which deforms with a more disharmonic style in the core of structures. The overall effect is to produce detachment folds, with the level of detachment varying and being dependant on the local depth to regional over-pressure. In several cases, these detachment folds are cut by relatively high-angle thrust faults similar in style to those described by Mitra & Namson (1992). In contrast to the modest shortening observed across anticlines, uplift is significant and often in excess of 3 km of erosion is apparent from missing strata.

The combination of significant uplift and relatively little shortening is more indicative of a response to inversion tectonics rather than a response to a horizontally compressive thin-skinned mechanism. Gravity data add further evidence by suggesting broader wavelength semi-regional folding of the over-pressured pro-delta strata. In combination with depth estimates to magnetic basement, a schematic cross section of the Kutai basin has been constructed (Fig. 16). The line of this section is the same as the one used by the gravity model shown in Fig. 13. The near surface structuring is summarised as a single unit with its detachment at the base of the prograding Lower Miocene shelf and deltaic sediments. Broad uplift of over-pressured section above a regional level defined by the deepest preserved synclines, is illustrated. Estimates of depths to magnetic basement have been projected onto the line of section and notional faulted basement depicted to indicate a possible present day configuration of the Palaeogene.

The schematic section in Fig. 16 is designed to illustrate the hypothesis that the Samarinda Anticlinorium is a series of tight, subsidiary structures, located above broad inversion folds over reactivated rifted basement. Typically in western Indonesia the maximum uplift areas are associated with the deepest palaeo-depocentres (Letouzey *et al.* 1990). To try to make this clearer, a pre-inversion restored version of Fig. 16 has been derived (Fig. 17). The Lower Miocene is assumed to have been a 4 to 5 km thick unit across the region comprised of a series of off-lapping sedimentary wedges. The basement configuration of Fig. 16 has been restored (with a vertical adjustment only) to its pre-uplift position using estimates of uplift from the wells, observed erosion and regional uplift of over-pressure. The results show a deep basin under the Samarinda Anticlinorium, interpreted to be rift in style, and further suggests that the Kutai Lakes High is also the result of a large positive inversion.

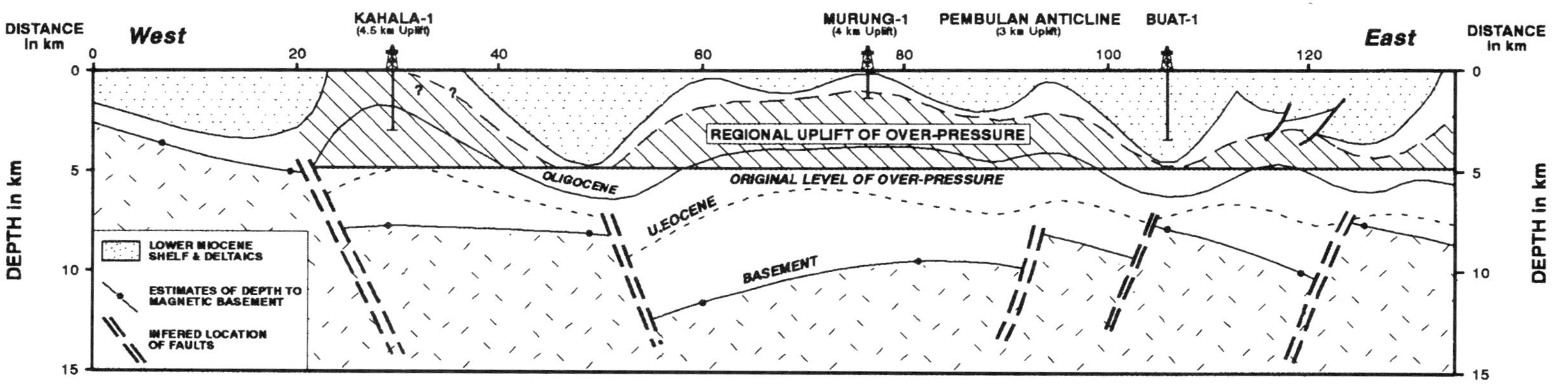

Fig. 16. Schematic regional cross section of Kutai Basin.

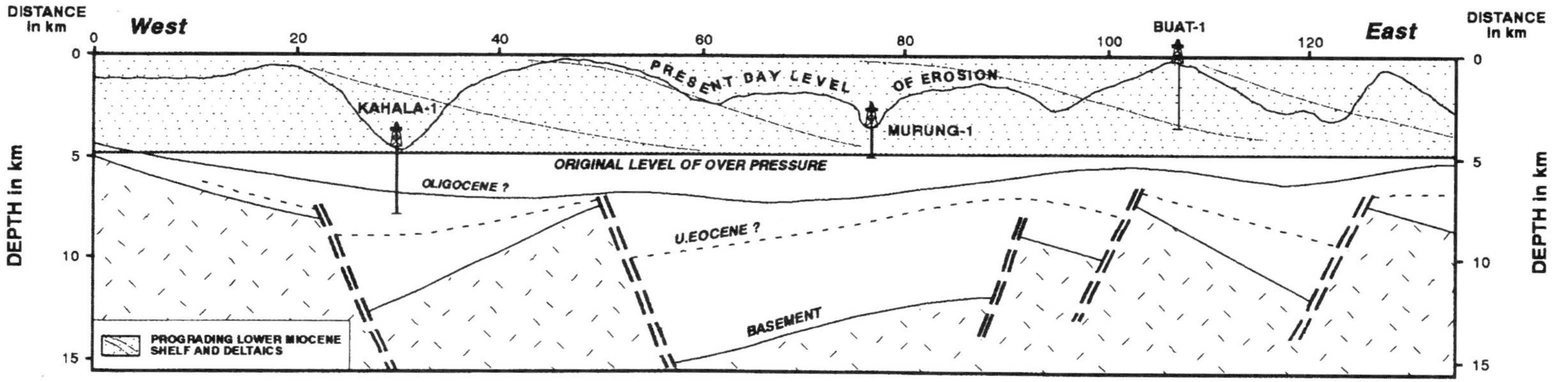

Fig. 17. Schematic (vertical) restoration of Kutai Basin cross section (end Early Miocene).

Figure 18 is a foward model showing the development of a small half graben system that is subsequently overlain by Oligocene bathyal shale and Lower Miocene deltaics. The rapid deposition of the Lower Miocene deltaics is inferred to have caused disequilibrium compaction of underlying Oligocene shale and consequently overpressuring below the facies boundary (here assumed to be horizontal). A fairly typical Sunda Fold (Letouzey *et al.* 1990) is produced at the Eocene to Oligocene levels as a result of basement involved inversion in the bottom part of Fig. 18. The inversion fault will tend to flatten out within the ductile overpressured section at the base of the Lower Miocene deltaics rather than cutting through the normally pressured and less ductile deltaic package. This results in a situation where the line length of the Lower Miocene deltaic package is longer than that of the underlying Oligocene shale. In reality this excess line length does not propagate into space as suggested on the model in Fig. 18 but is

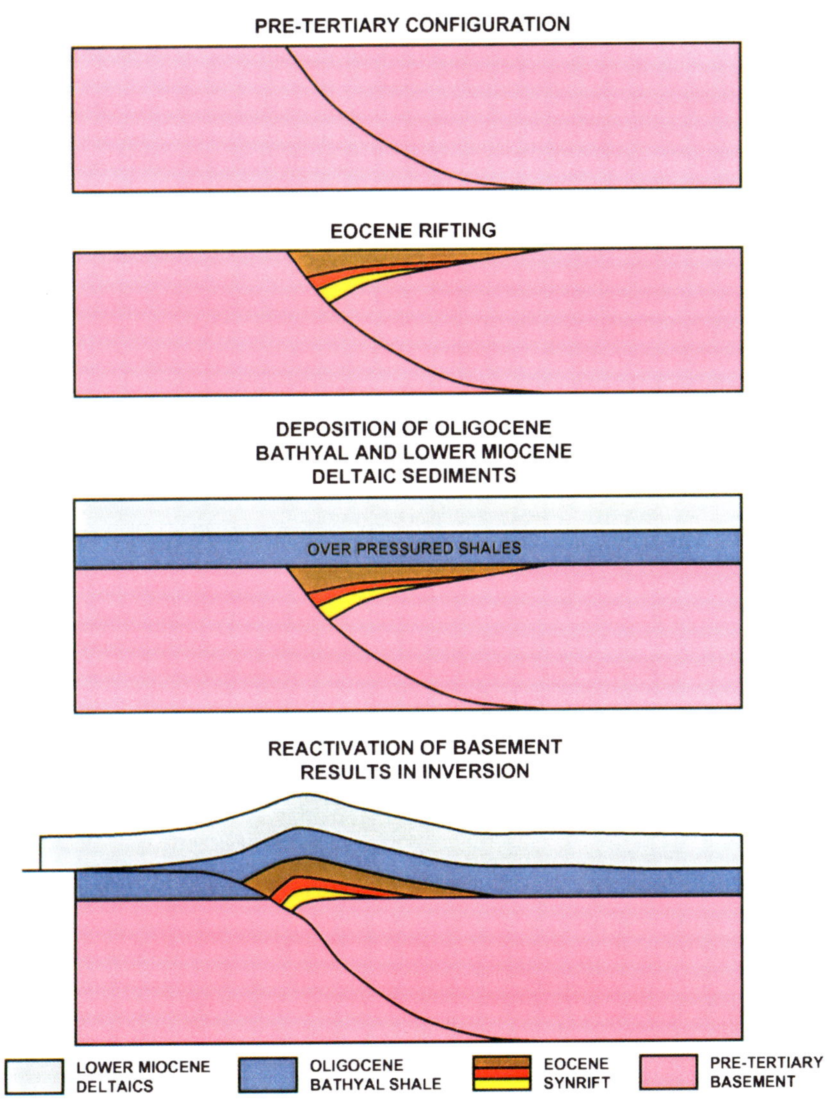

Fig. 18. Development of horizontal shortening in the Miocene as a result of inversion of an Eocene half graben.

translated in the case of the Samarinda Anticlinorium into a series of detachment folds and faults superimposed on the broad uplift that results from the inversion and formation of the Sunda Fold. This model illustrates how a sub-vertical fault movement at depth can be translated to a horizontal compressive force at surface.

In the Natuna Sea it has been found that inversion begins with the largest graben fault with smaller faults being reactivated at progressively younger ages (Ginger *et al.* 1993). Given such a system, inversion would have been progressive across the study area initiating in the Kutai Lakes Area and thereafter shifting eastward. Although there is little control on the number, location and throw of the original rift defining faults, Figs 16 and 17 illustrate a view of the Kutai Basin which is similar to other basins across western Indonesia. There is no requirement to invoke horizontally compressive thin-skinned tectonics to explain the Samarinda Anticlinorium.

There is little direct evidence for the timing of inversion events since much of the initial syn-inversion sediment has been eroded in the study area. However, growth sections are locally visible (see regional seismic line, Fig. 4). Palaeontological and palynological data suggest vertical movements commenced as early as the end of the Early Miocene at this location. Late Early Miocene inversion is likely to have significantly uplifted the Lower Miocene section and subsequently provided the sediment source area for the thick Middle Miocene deltaic wedge found east of the study area.

Conclusion

The Samarinda Anticlinorium is interpreted to be a series of detachment folds over a deep but regionally inverted Eocene rift basin. There is an extreme heterogeneity in the stratigraphy with a competent section of shelf and deltaic sediments deposited on top of thick, regionally overpressured strata with much lower shear strength. The observed surface structural style is different to the classic 'Sunda' fold style of much of western Indonesia due to the great thickness of sediment and the presence of extensive overpressure. At depth it is likely that tectonic processes similar to the inversion documented elsewhere in western Indonesia have occurred, but seismic data is unable to confirm details. However, gravity data suggests broad semi-regional uplifts with a wavelength of 20–30 km similar to that described elsewhere in inverted rift basins as opposed to the observed surface fold wavelength of 5–10 km. Surface folds may thus be seen as tight, detached structures secondary to this broader inversion folding.

Timing of inversion commenced at least as early as the end of the Early Miocene and the most highly inverted part of the study area, the Kutai Lakes High, probably represents the first and largest inverted depocenter. Locus of inversion probably migrated eastward with time as less major depocentres were uplifted. This inversion model has particular significance since it provides a nearby and abundant sediment source during the Mid-Miocene which explains the tremendous thickness of rapidly deposited Mid-Miocene sediments found to the east of the study area.

The authors wish to thank the Management of Pertamina, LASMO and partners in the Runtu Block, UGO, Ross Petroleum, JAPEX and OPIC for the permission to publish this paper. Special acknowledgement is made for help received from M. Parker of Geoservices/ARK Geophysical and T. Marsh of Cogniseis for technical guidance on gravity and structural geological modeling respectively. The results of this paper are the product of extensive discussions amongst the LASMO Runtu Group and the work of all is acknowledged. The authors wish to particularly thank I. Carter, D. Paterson, B. Dingwall and J. Towart for reviewing drafts of this paper. The authors also acknowledge the help of LASMO drafting and secretarial support.

References

BIANTORO, E., MURITNO, B. P. & MAMUAYA, J. M. B. 1992. Inversion Faults as the Major Structural Control in the Northern Part of the Kutai Basin, East Kalimantan. *Proceedings of the Indonesian Petroleum Association*, **21**, 417–453.

CARTER, I. S. & MORLEY, R. J. 1996. Utilising Outcrop and Palaeontological Data to Determine a Detailed Sequence Stratigraphy for the Early Miocene Deltaic Sediments of the Kutai Basin, East Kalimantan. *Proceedings of the Indonesia Petroleum Association Symposium on Sequence Stratigraphy, 1995*.

GINGER, D. C., ARDJAKUSUMAH, W. O., HEDLEY, R. J. & POTHECARY, J. 1993. Inversion History of the West Natuna Basin: Examples from the Cumi-Cumi PSC. *Proceedings of the Indonesian Petroleum Association*, **22**, 635–658.

LETOUZEY, J. J., WERNER, P. & MARTY, A. 1990. Fault Reactivation and Structural Inversion. Back Arc and Intraplate Compressive Deformations. Examples of the Eastern Sunda Shelf (Indonesia). *Tectonophysics*, 183, 341–362.

MITRA, S. & NAMSON, J. 1992. *Balanced Cross Sections in Hydrocarbon Exploration and Production*. AAPG Short Course Notes.

OTT, H. L. 1987. The Kutai Basin – A Unique Structural History. *Proceedings of the Indonesian Petroleum Association*, **16**, 307–317.

WAIN, T. & BEROD, B. 1989. The Tectonic Framework and Paleogeographic Evolution of the Upper Kutai Basin, *Proceedings of the Indonesian Petroleum Association*, **18**, 55–78.

New observations on the sedimentary and tectonic evolution of the Tertiary Kutai Basin, East Kalimantan

STEVE J. MOSS[1,2], JOHN CHAMBERS[3], IAN CLOKE[4], DHARMA SATRIA[4], JASON R. ALI[5], SIMON BAKER[6], JOHN MILSOM[6] & ANDY CARTER[7]

[1] *Present address: School of Applied Geology, Curtin University of Technology, Perth, 6845, WA, Australia (e-mail: rmoss@cc.curtin.edu.au)*

[2] *University of London Southeast Asia Research Group, Department of Geology, Royal Holloway, UK*

[3] *LASMO Runtu Ltd, Jakarta, Indonesia*

[4] *Geological Research & Development Centre, Bandung, Indonesia*

[5] *Department of Earth Sciences, University of Hong Kong*

[6] *University of London Southeast Asia Research Group, Department of Geological Sciences, University College London, UK*

[7] *University of London Fission Track Research Laboratory, Department of Geological Sciences, University College London, UK*

Abstract: The Kutai Basin, Kalimantan, is one of several large Tertiary basins which opened during the mid- to late Eocene in Borneo. Extensional tectonics are responsible for basin formation, and not crustal flexure as has been suggested previously. Extensional faulting did, however, occur within a general foreland setting located to the south of the late Cretaceous/Palaeogene Central Kalimantan fold belt. The basin post-dates late Cretaceous/early Tertiary orogenic activity which affected the Central Kalimantan foldbelt and formed several upper Cretaceous granites. The Palaeogene stratigraphy of the basin includes basal conglomerates, shallow marine clastics and minor clastics and thick sequences of bathyal marine shales. Neogene stratigraphy is dominated by deltaic clastics and carbonate platforms. Several phases of Tertiary igneous activity are seen intruded into and interbedded with Tertiary sediments of the Kutai Basin. Three suites are recognized and have been variously interpreted as the products of melting of an orogenic root, extensional driven melting and/or subduction related melting. We review Tertiary sedimentation in the basin and propose a new model which relates the formation of the Kutai Basin to the opening of Celebes Sea and collapse of an uplifted Late Cretaceous/Palaeogene orogenic belt.

The 60 000 km^2 Kutai Basin of east Kalimantan (Fig. 1) is the largest and deepest (15 000 m) basin in Indonesia (Rose & Hartono 1978; Hutchison 1989; CCOP 1991). The Neogene section in the Samarinda region reaches thicknesses of up to 9000 m. Tertiary sedimentation in the basin has been fairly continuous since its inception in the mid- to late Eocene, although the centre of deposition has gradually moved toward the east (the present-day Mahakam Delta). The eastern part of the Kutai Basin (sometimes called the lower or East Kutai Basin) is a major oil and gas province with reserves of oil and oil-equivalent gas of up to 5 billion bbls, and several giant oil fields. The Sanga-Sanga Field was discovered in 1897 (Verdier *et al.* 1980). This and subsequent successes resulted in the concentrated exploration of the coastal and eventually offshore areas.

The Kutai Basin is bounded to the southwest by the Schwaner Mountains (Figs 1 & 2) and to the northwest by late Cretaceous/early Tertiary turbidites (Rajang and Embaluh Groups), and older basic igneous rocks and chert of the variously named Kalimantan Central Ranges, Kapuas Ranges or Kuching uplift (Fig. 1). On the eastern side of the Kutai Basin is the deep North Makassar Basin (Fig. 1) bounded by two NW–SE-trending faults zones, the Adang and the Sangkulirang to the south and north respectively. To the south are the Meratus Mountains and Tertiary Barito and Pasir Basins (Fig. 2), which are separated from the Kutai Basin by the Adang fault zone (or Cross Barito High). The western (or upper) part of the Kutai Basin is connected to the Barito Basin via a narrow strip of Tertiary sediments in the area southeast of Muara Teweh (Figs 1 & 2). The Muller Mountains form the western basin margin and comprise Cretaceous shelf sandstones and limestones of the Selangkai Formation and granites and metamorphics of the Busang complex. The

From Fraser, A. J., Matthews, S. J. & Murphy, R. W. (eds), 1997, *Petroleum Geology of Southeast Asia*, Geological Society Special Publication No. 126, pp. 395–416.

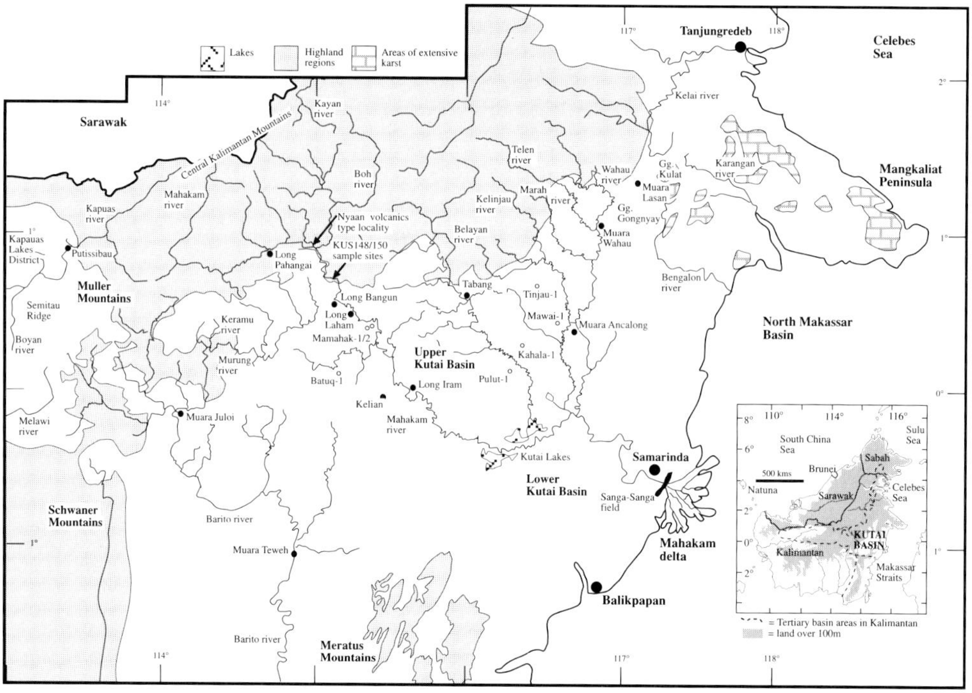

Fig. 1. Location map for the Kutai Basin. Inset map of Borneo. Localities mentioned in the text are shown on this figure.

Muller Mountains also terminate eastern end of the Semitau Ridge (Fig. 2). The Tertiary of the Kutai Basin is continuous with sediments of the Mandai and Melawi Basins via a narrow route through the Muller Mountains (Pieters *et al.* 1987; Pieters & Supriatna 1990; Fig. 2).

The Tertiary succession within the Kutai Basin is divided into a syn-rift phase in the mid-Eocene, a sag phase in the late Eocene to Oligocene and a renewed phase of tectonic activity and subsidence in the late Oligocene to Miocene. Inversion beginning at least as early as the end of the early Miocene and probably as early as the late Oligocene and has been responsible for the reworking of earlier sediments and continuing deposition of 'syn-inversion' deltaic packages. Continuing erosion from the hinterland in response to Miocene and Pliocene tectonic activity results in continuing deltaic deposition.

Geological evolution of the Kutai Basin

Pre-Tertiary events

Within the Kalimantan Central Ranges the Mesozoic sequence comprises of spillites, ultrabasics and serpentinites ('ophiolites') and radiolarian cherts of Jurassic–early Cretaceous age, and turbidites of the Rajang/Embaluh Groups and Mentareng Formation. Rajang/Embaluh Groups and Mentareng Formation are the names given to similar sediments in Sarawak and parts of Kalimantan, we use Embaluh Group here. The Central Kalimantan Ranges (Fig. 1) have been interpreted as an accretionary prism resulting from SW-directed subduction of oceanic crust beneath the continental core of Borneo (Hamilton 1979; Hutchison 1989; Bénard *et al.* 1990; Tongkul 1991). The geology of this area and an alternative interpretation is presented in Moss (in press).

The Schwaner Mountains (Figs 1 & 2) contain numerous granitic and tonalitic batholiths of Cretaceous age (Williams *et al.* 1988; Amirruddin 1989). These are intruded into slates, amphibolites, schist and gneiss of ?Silurian to Permian age. Hutchison (1989) proposed that the Schwaner Mountains formed a continental core to the island and that throughout the (?)Jurassic/Cretaceous this part of Sundaland built outwards by the accretion of marginal basin fill, island arc and ophiolitic material. Tate (1992) has questioned the continental origin of the Schwaner Mountains and alternatively suggested an magmatic arc origin for the

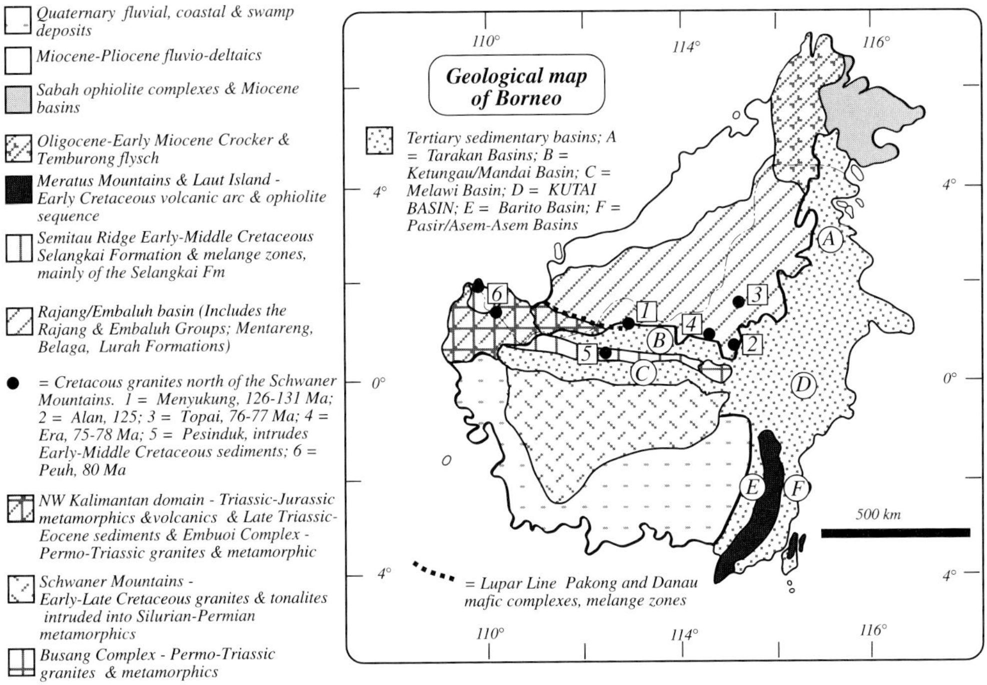

Fig. 2. Tectonic 'provinces' map of Borneo, based on references cited and our fieldwork. Radiometric ages of Cretaceous granites north of the Schwaner Mountains are from Pieters *et al.* (1993*a*, *b*, *c*, *e*).

Schwaner Mountains but the nature of the country rock into which the Schwaner Mountains granites have been intruded and the location of lower to upper Cretaceous granites (Alan, Era, Topai, Pesinduk and Peuh, Fig. 2) and Permian–Triassic granites and metamorphics (Busang and Embuoi complexes, Fig. 2) to the north of the Schwaner Mountains, implying the presence of continental crust north of the Schwaner Mountains, argues against this hypothesis.

In the SE corner of Kalimantan ophiolitic and volcanic island arc rocks exposed in the Meratus Mountains (Fig. 2) were finally emplaced by the late Cretaceous (Sikumbang 1990). Several phases of NNE–SSW-oriented strike-slip deformation are suggested to have affected these rocks during both the Cretaceous and Tertiary (Sikumbang 1990; Kusuma & Darin 1989). Granite generation is also linked with deformation in this region. The Kintap Granite pluton (K–Ar dated at 95 Ma, Sikumbang 1990) and the Rimuh Granodiorite pluton intrude the volcanic arc sequence of the lower Cretaceous Alino Group. The Alino Group is in the footwall to thrust sheets carrying peridotites of the ophiolite (Yuwono *et al.* 1988). The reported Kalimantan ophiolites, apart from the Meratus ophiolite (Sikumbang 1990), have yet to be studied in detail and in outcrop typically contain only part of the classical Penrose sub-divisions of an ophiolite. In particular, pillow lavas and sheeted-dyke complexes, the upper parts of an ophiolite sequence, are noticeably lacking from outcrops and outcrop descriptions of pre-Tertiary rocks in both Kalimantan and Sarawak. The Meratus Complex (SE Kalimantan) and Danau Mafic Complex (West Kalimantan; Fig. 2) may represent true ophiolitic sequences in Kalimantan. Cretaceous granite intrusions and Tertiary rhyo-dacitic volcanics and quartz-rich sediments overlying these mafic rocks all suggest the involvement of continental crust. In the upper Belayan river (Fig. 1), tectonically intercalated basalts, gabbros and microgabbros show several phases of deformation and are intruded by granitic dykes. These dykes only show a single phase of tensional deformation and hence postdate earlier deformation.

The Boyan melange of the Semitau Ridge (Fig. 2) comprises predominantly blocks of shallow marine carbonates and sands from the lower–middle Cretaceous Selangkai Formation with more exotic fragments of metamorphic rocks and granite derived from the Busang and Embuoi complexes enclosed in scaly clay. The scaly clay contains Cenomanian to Santonian nannofossils, although the blocks within the clay yield Aptian–Albian foraminifera. More coherent parts of the Selangkai Formation yield lower to middle Cretaceous diagnostic faunas (Heryanto *et al.* 1993). Both the Semitau Ridge and the Lupar Line (Fig. 2) record several phases of contractional/transpressional deformation. The Semitau Ridge has been thrust southwards onto the Melawi Basin, as shown by north-south seismic sections across the Melawi Basin (Pertamina-Beicip-Franlab 1992). The youngest footwall sediments to these thrusts are upper Eocene/lower Oligocene in age (K. Glazebrook, Can-Oxy, pers. comm. 1994). Fission-track analysis of apatites from a Cretaceous sandstone from the Boyan melange (Boyan river, see Fig. 1) of the Semitau Ridge yielded an lower Oligocene (35 ± 4 Ma) age of cooling and (by inference) uplift.

Two wells in the Kerendan gas field area near Muara Teweh (Fig. 1) penetrated basement granodiorites which yielded K–Ar ages of 68 ± 3 and 63 ± 2 Ma (Van de Weerd *et al.* 1987). Other wells in the same area have penetrated upper Jurassic to lower Cretaceous shales and intrusive rocks and upper Palaeozoic metasediments (Van de Weerd *et al.* 1987). These may be the lateral equivalents of the Selangkai Formation and Busang Complex to the northwest.

Geological data for the pre-Tertiary of Kalimantan is far from complete and many problems remain, not least the geometry of Cretaceous subduction complexes (Moss in press). These problems are the subject of on-going research.

Timing of basin initiation

The timing of initiation of the Kutai Basin is poorly constrained reflecting the absence of reliable biostratigraphic indicators in the terrestrial dominated initial basin fill and the lack of well penetration into the deepest parts of the syn-rift sequence. Onlap of syn-rift and younger sediments onto the basin flanks, and also the possibility that parts of the basin fill have been thrust over the basin margins during inversion, implies that outcrops along the basin margins may not be representative of the oldest sediments in the basin. Most authors suggest that the base of the Tertiary basin fill is middle to upper Eocene in age. Uppermost middle and late Eocene nannofossils (NP16-19) and foraminifera (Tb or P14-16) are present in the basin. This agrees with the ages suggested by Van de Weerd & Armin (1992) and Pieters *et al.* (1993*a*) on the basis of radiometric and biostratigraphic determinations. Sediments of Palaeocene–lower Eocene age have yet to be found within the Kutai Basin. Although a Palaeocene age for the Kiham Haloq Sandstone Formation of the upper Kutai Basin has been suggested (Wain & Berod 1989), the oldest fossils are middle to upper Eocene foraminifera from beds of the Ritan Limestone Member intercalated in the upper part of the formation.

Rhyolite lava flows and associated sub-aerial volcaniclastic rocks occur both below and pass laterally into initial basin sediments in basin margin and intra-basin locations (discussed later). In the upper Mahakam River (Fig. 1) these are called the Nyaan volcanics and have K–Ar radiometric ages of 48–50 ± 1 Ma (early mid-Eocene, Pieters *et al.* 1987, 1993*a*) and at Kelian (Fig. 1) felsic volcanics and lavas are conformable within a sedimentary succession containing limestone lenses with late Eocene (Tb) benthic foraminifera (Van Leeuwen *et al.* 1990). A late mid- to late Eocene timing for the formation of the Kutai Basin is most likely (Fig. 3a & b).

Middle Eocene

The upper middle Eocene represents a syn-rift section and is remarkably similar throughout Kalimantan. A particularly common feature is the presence of a basal conglomerate variably referred to as the Tanjung Formation (Barito basin and southern Kutai Basin, Kusuma & Darin 1989), Kuaro Formation (southern part of the Kutai Basin) and Kiham Haloq Sandstone Formation (upper parts of the Mahakam River) western basin margin, and Melawi Basin, (Van de Weerd & Armin 1992). Similar deposits have also been located along the northern basin margin north of the Wahau River (Figs 1, 3a & b). Deposition of these units may have been rapid and controlled by palaeotopography (Wain & Berod 1989). In outcrops along the upper reaches of the Wahau River thick beds (80 cm to 2 m) of breccio-conglomerate, some of which are channelled, form packages 6–7 m thick. Clasts vary from sub-angular to rounded and lie within the pebble size range. Abundant chert with phyllite, quartzite, milky quartz and other metamorphic rocks are seen within these conglomerates. Intercalated over a distance of several metres

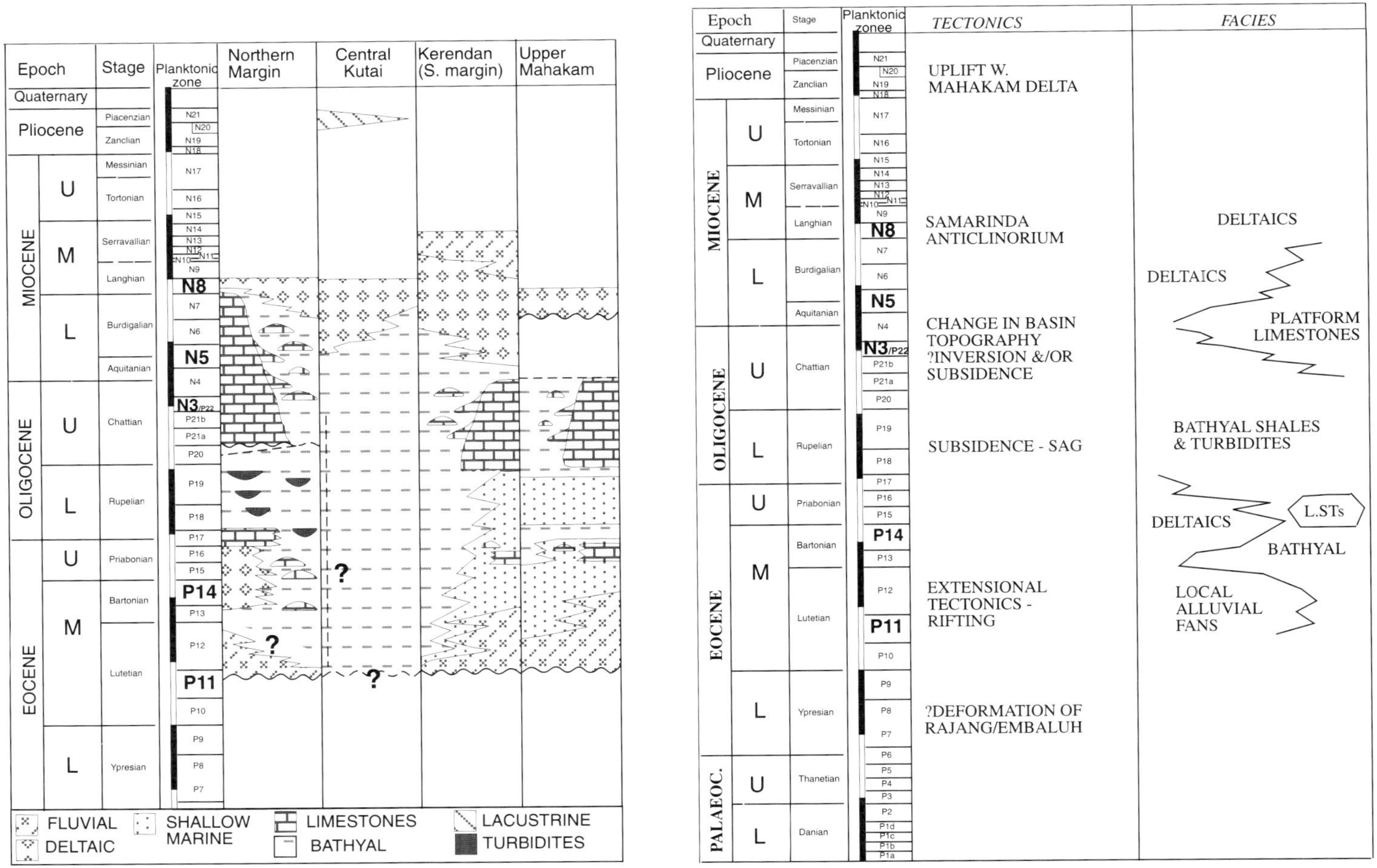

Fig. 3. (**a**) Comparative stratigraphic columns for the northern margin, central region, Kerendan (southern margin) and upper Mahakam River (northwestern margin) of the Kutai Basin. The Kerendan column is based on data from Van de Weerd *et al.* (1987, 1992). It should be noted that for the central region the oldest rocks studied (either from outcrop or well penetration) are Late Oligocene in age. The dotted line indicates where the section has been hypothesised. The black and white bars represent 5 Ma. (**b**) Major tectonic and facies changes for the Kutai Basin. The events are discussed in greater detail in the text.

with the breccio-conglomerates are laterally discontinuous lenses of coarse lithic sands which infill shallow channel features. The breccio-conglomerates are framework supported and downcut into coarse sandstone and/or conglomerate beds. Reddish purple marls with a green mottling and fine red, lithic sands are also interbedded with breccio-conglomerates. These are continental in origin and the product of alluvial fan deposition through sediment gravity flow and fluvial channel processes. Overlying and laterally equivalent in deeper parts of the basin are bathyal shales and turbiditic sands containing upper middle–upper Eocene foraminifera and nannofossils (P15-16 or Tb; NP16-19).

Isolated outcrops of middle to upper Eocene foraminiferal limestones (Tb benthonic forms, F. Banner pers. comm. 1995) occur along the Mahakam River and Wahau River (Figs 1 & 3a). Although direct field observations of their relative positions with respect to conglomerates have not been possible, microfacies analysis of the carbonates indicates a shallow marine environment of deposition. The occasional presence of rounded to sub-rounded basement clasts within these limestones suggests the influence of an uplifted basement high shedding clastic material. To the northeast in the upper reaches of the Karangan River (Fig. 1) fine sands and shales of upper middle Eocene (P14) age are bathyal to abyssal, confirming that the palaeo coast in the Eocene was east of central Kalimantan.

Upper Eocene to Lower Oligocene

Marine influence is increasingly apparent in the sediments towards the end of the middle Eocene and into the upper Eocene (Fig. 3a & b). Locally alluvial sands and conglomerates give way to an interval of deltaic sediments with coal seams, cross bedded channel sands, carbonaceous shales and the occasional high to moderate energy sandy foraminiferal wackestones of upper Eocene age. Along the upper reaches of the Mahakam River, sediments of the Kiham Haloq Sandstone Formation pass up from a coarse fluvial conglomerate-sandstone basal section, with metre scale channel structures, into high energy shallow marine and tidal sands with intercalations of upper Eocene (Tb) *Nummulites* foraminiferal packstones (Ritan Limestone Member). Above and laterally equivalent to these lithologies are lithic and quartz arenites, shales, siltstones and conglomerates. Syn-sedimentary slumping is common within these sediments, as is the amalgamation of sandstone beds and sole structures in the form of groove, prod and load casts are seen on the base of beds. Partial Bouma sequences, often with only Tabc or Ta-c parts present, are common within the sandstone beds. These are interpreted to be turbidites, an interpretation supported by the generally bathyal foraminifera. Turbidite sediments of this age are particularly well exposed along the Telen, Marah and Wahau River sections in the northern part of the central Kutai Basin (Fig. 1) and along the upper Mahakam River in the northwestern part of the upper Kutai Basin. Lithostratigraphic names of these sediments include the names Batu Kelau Formation, Atan Formation, the Bongan Formation (Van de Weerd *et al.* 1987) and Marah Formation (Supriatna 1990). In the upper Mahakam River area the distal turbidites of the Batu Kalau Formation are overlain conformably by upper Eocene to upper Oligocene shelfal quartz sands and mudstones of the Batu Ayau and Ujoh Bilang Formations. The presence of shelfal sediments in this part of the basin and the predominantly terrestrial nature of the Eocene–Oligocene sediments in the Melawi, Ketungau and Mandai Basins to the west is probably linked, with the western part of the basin shallower than eastern areas at this point in time. Marine deposits in the Melawi and Mandai Basins are limited to the extreme eastern edges of the basins and to features such the tidal sands within the basal section of the Mandai Group.

Upper Oligocene

The late Oligocene was a significant time in the evolution of the Kutai Basin. After a period of deep marine sedimentation throughout much of the basin in the late Eocene and Oligocene, a major change in basin topography occurred (Figs 3a & b). In the Kelai River (Fig. 1), on the northern margin of the basin, a basal conglomerate rests unconformably on Cretaceous metasediment. Shales associated with the conglomerate contain N3 planktonic foraminifera and indicate lower sublittoral to bathyal environments of deposition. A disconformable relationship between the lower Oligocene shales and upper Oligocene limestones is suggested from seismic and field evidence in the Kerendan gas field (Fig. 1) (Van de Weerd *et al.* 1987). This relationship may change to a more conformable one within the centre of the basin. Upwards within the Kelai River sequence these conglomerates and shales pass into several hundred metres of platform limestones seen on the northern basin margin in large karstified areas (Fig. 1). The foraminiferal fauna of these carbonates suggest an upper Oligocene to lower Miocene age range (upper and lower Te), although most indicate Aquitanian age (top N4-N5). Within the basal section and fringing these thick carbonates along the Kelai river and other locations, metre scale beds of limestone breccias consist of pebble to cobble sized, rounded to sub-rounded clasts of grainstones, packstones, boundstones, foraminiferal packstones, fragments of coralline organisms and wackestones and clasts of lithic material (phyllites, metasediments, metabasics, chert, milky quartz). Silicic clasts are sub-angular to sub-rounded and are also seen in decimetre beds of calcareous cemented chert-rich conglomerates. These breccias and conglomerates are interpreted as debris flows and outcrops of similar sediments occur in the vicinity of other upper Oligocene to lower Miocene platform limestones. The rounding of the lithic clasts suggests they were transported further than the limestone clasts. No indication of sub-aerial exposure of adjacent platform areas, such as vadose cement fabrics and/or micro-karstic dissolutional features, have been found within any of the limestone clasts. The limestone clasts were at least partially cemented within the marine diagenetic realm prior to erosion and show no sign of soft sediment deformation. These breccias may be the result of reactivation of pre-existing faults bounding highs upon which carbonates were accumulating, allowing erosion of carbonate sediment and an underlying basal conglomerate. This could explain the similarity in the clast composition of the basal conglomerates and the basement rocks seen in the limestone breccias. Alternatively the breccias may be the result of renewed uplift in the hinterland and shedding of basement clasts into marine environment (possibly axially along grabens or at by-pass channels) where it is mixed with proximally derived carbonate material shed off highs. The breccias may also be due to progradation of a highstand systems tract and production of oversteepened slopes. The lack of meteroic features argues against a lowstand formation for these breccias. The breccias differ from those seen in the southern Kutai Basin, which contain only clasts of carbonate material ('lithoclast conglomerates') and are interpreted to be the result of deposition during a highstand of sea-level and backstepping of the platform and production of oversteepened cliffs (Saller *et al.* 1994). The upper section of the Ujoh Bilang Group in the upper Mahakam River area contains olistolithic blocks and debris flows of limestone and volcanic material within bathyal shales and the volcaniclastic turbidites of the Len Muring Sandstone (Wain & Berod 1989).

On the upper reaches of the Mahakam River (Fig. 1), interbedded decimetre thickness beds of mudstones and wackestones with fragments of corals, fish teeth, oyster shells, rotalinid(?) foraminifera and upper Oligocene nannofossils (NP24-25) pass up into thicker (metre-scale) bedded foraminifera–coralline algae packstones exposed in spectacular 100–150 m high cliffs. The southeastern edge of these cliffs show a probable NE–SW-trending syn-sedimentary fault, which cuts the older beds within the cliff face but is itself truncated by younger beds higher up within the cliff section. The nature of the outcrop does not allow close examination of this relationship. The presence of a NE–SW-trending gravity high (Pieters *et al.* 1993*a*) associated with the area of the limestone outcrop indicating limestone developed upon a NE–SW-trending horst block partially reactivated during deposition of the limestones.

At the same time as these limestones were being deposited, major uplift may have occurred in the upper Kutai Basin. Van Leeuwen *et al.* (1990) and Wain & Berod (1989) suggest a period of folding and faulting in the late Oligocene and early Miocene. Apatite fission-track ages of 22 ± 2 and 25 ± 2 Ma and the narrow track length distributions obtained for Santonian–Campanian aged turbiditic sands (KUS148/150) record rapid cooling and therefore by implication rapid uplift and/or erosional denudation (in the sample area at least; Fig. 1) around the Oligocene–Miocene boundary (Moss *et al.* 1997). These sands, which lie unconformably below the middle–upper Eocene Kiham Haloq Sandstone Formation, are less than 5 km north of the present-day northern margin of the upper Kutai Basin.

Lower Miocene

Dating of foraminifera within shallow marine carbonates suggest deposition of platform limestones and related limestone breccias continued from the late Oligocene into early Miocene on both the northern and southern margins of the basin (Fig. 3a & b). Several outcrops along the Wahau River (Fig. 1) expose rubbly breccias containing a variety of limestone and lithic clasts and are of similar origin to those described above. A sequence of deltaic clastics was deposited with the eastward progradation of the delta system starting in the west of the Kutai Basin in the early Miocene (lower Miocene, N5/6, LASMO pers. comm. 1995) and continuing to the present-day Mahakam Delta.

Middle to upper Miocene

Thick middle Miocene sediments east of the Samarinda anticlinorium (Fig. 4) were rapidly deposited following an inversion event at the end of the early Miocene (Chambers & Daley 1995). The deltaic sediments aggraded to form what is now the coastal plain and present-day delta. Over 3000 m of delta plain sediments of middle to late Miocene age are present indicating significant amounts of subsidence occurred in parallel with deposition and sea level change. It is likely that these sediments were derived from erosion of the lower Miocene deltaic sediments (Chambers & Daley 1995) as well as continuing erosion of the central hinterland.

The Neogene stratigraphy and lithostratigraphy for the Lower Kutai Basin has been generally established (Marks *et al.* 1982, Van de Weerd & Armin 1992). However the lithostratigraphic terminology proposed is cumbersome and fails to recognize that the lower Miocene and middle Miocene deltaics are different megasequences (Siemers *et al.* 1994; Carter & Morley 1995). Siemers *et al.* (1994) present a more comprehensive stratigraphy and lithostratigraphy for the area. Several megasequences are recognized which overall represent a shallowing upward sequence from bathyal shales to pro-delta shales with localized shallow marine carbonate buildups, delta sands and delta plain/fluvial sands. This sequence was produced with the eastward progradation of a delta system starting in the west of the Kutai Basin in the lower Miocene (N5/6, LASMO pers. comm. 1995) and continuing to the present day and the Mahakam Delta.

Pliocene to Recent

The Kutai Lakes are located west of Samarinda (Fig. 1), lacustrine sedimentation within these lakes began in the Pliocene (Anap Formation) and continues to the present day. The formation of the lakes area is the result of ponding of the Pliocene Mahakam River drainage basin with the uplift of the Samarinda anticlinorium, a domal area west of Samarinda. Uplift of the

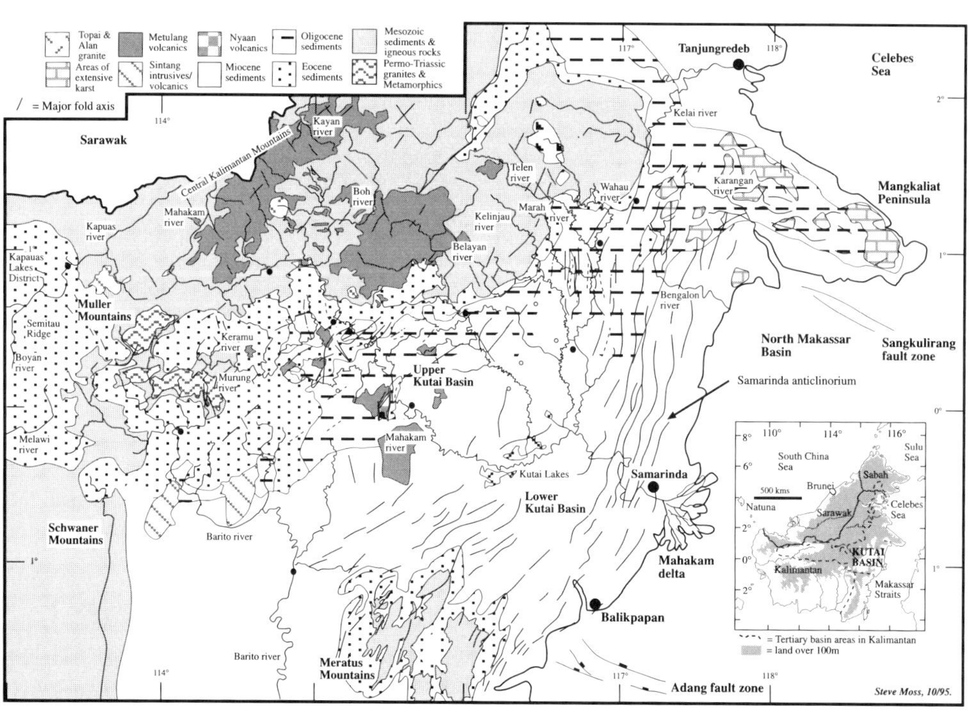

Fig. 4. Geological map of the Kutai Basin and East Kalimantan. The approximate basement–Tertiary contact, Eocene, Oligocene and Miocene positions are shown.

eastern edge of the Samarinda anticlinorium in the Pliocene is also suggested to be responsible for fresh water flushing of deltaic sediments on the edge of the Mahakam delta (Burrus *et al.* 1994) (Fig. 3a & b).

Structure of the basin

Basement

The most prominent feature of the Kutai Basin Bouguer gravity map (Fig. 5) is a north–south-trending high with an amplitude of about 40 mGal. The Kutai Lakes region (Fig. 1), an area of subsidence and Pliocene to Recent lake deposition, roughly coincides with the crest of the high and hence the feature has been referred to as the Kutai Lakes Gravity High (KLGH, Fig. 5). A similar correlation between gravity high and subsidence is observed in the Kapuas Lakes district of West Kalimantan (Fig. 1), where the lakes are associated with an east–west-elongated positive anomaly, the Lubok Antu–Kapuas gravity ridge, which has amplitudes of up to 70 mGal (Williams *et al.* 1988; Foss 1993). The anomaly is associated with the Pakong and Danau mafic complexes which outcrop near the Lupar Line (Fig. 2) in Sarawak and Kalimantan respectively. Foss (1993) suggested that the Quaternary subsidence of the Kapuas Lakes district was due to isostatic readjustment following relaxation of the compressive forces which led to the emplacement of the high density rocks. It seems possible that a similar explanation could be applied in the Kutai Lakes region, with the gravity high marking a subsurface extension of the ophiolitic rocks of the Meratus Mountains; serpentinites and layered gabbros crop out at the northern margin of the basin due north of the Kutai Lakes. However, Wain & Berod (1989) and Chambers & Daley (1995) attribute the high to basin uplift, relating the gravity response to inversion of once deeply buried and compacted Eocene and Oligocene sediments over an uplift in the underlying basement.

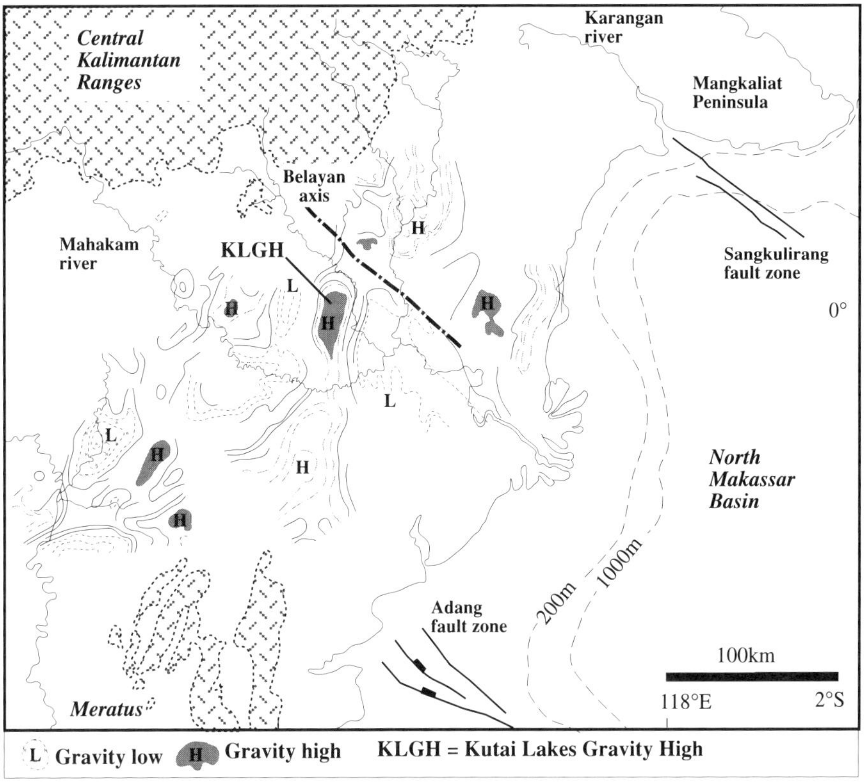

Fig. 5. Simplified Bouguer gravity anomaly map for part of the Kutai Basin. Re-drawn from Chambers & Daley (1995).

Seismic reflection surveys in this area have not achieved penetration to levels deep enough to distinguish between these scenarios.

East of the Kutai Lakes high Bouguer gravity values drop to a low of a little above 0 mGal and then rise gradually towards the coast, where values are of the order of +30 mGal. Bouguer gravity levels are thus slightly to moderately positive throughout the area and not dramatically different for the levels recorded over the areas of basement outcrop to the east (Pieters & Supriatna 1990), despite the presence of a thick Neogene sediment section, which should represent a considerable mass deficient. A balancing mass excess seems to be regional at depth, which could be accounted for by the crust being either thinner or denser than in areas beyond the basin margins. It is possible that the basin developed on top of oceanic crust continuous with that beneath the North Makassar Basin, or on greatly extended and attenuated continental crust. The observations, including the steady rise in gravity levels towards the Makassar Straits, are compatible with either possibility, but not with origin as a foreland (or flexural) basin. Depression of foreland basins occurs because of loading of elastic lithosphere, and the load is, by definition, displaced from the depocentre. Local isostatic compensations do not apply and low gravity is observed over the sediment accumulations. It is probable that the present-day Mahakam delta is prograding out over oceanic crust. Modelling of the gravity anomalies will help to determine whether this is likely or not.

Magnetic surveys along the western edge of the upper Kutai Basin suggest that the pre-Tertiary basement has a general east-west structural trend (Wain & Berod 1989). However, north–south-trending features in these low magnetic latitudes would tend to produce weaker anomalies than features with significant east-west components. In particular the NNE–SSW- to NE–SW-oriented structural highs and lows of the Meratus trend, which formed a significant palaeotopography and exerted a strong control on initial Palaeogene sedimentation in the upper Kutai Basin (Wain & Berod 1989), would be expected to be poorly represented on magnetic maps. The NW–SE to WNW–ESE trends of the Adang fault/Cross Barito High and similar lineaments which separate the Kutai and the Barito depositional areas, may be more clearly defined. Indeed an important feature of this basement configuration along the southern edge of the basin are NW–SE aligned lineaments. However NNE–SSW to NE–SW (Meratus trend) oriented structural highs and lows are important.

The KLGH ridge is noticeably off-set by a narrow, NW–SE-oriented, right lateral shear, (Belayan axis, Wain & Berod 1989) (Fig. 5). NW–SE lineaments within the basin centre off-setting NNE–SSW aligned residual gravity anomalies coincident with surface folds affecting Lower Miocene sediments (Chambers & Daley 1995). A NW–SE fabric seems to be an inherited basement feature possibly related to the Cretaceous and early Tertiary deformation. It affects the late Cretaceous to Palaeocene Embaluh Group and is one of the dominant fracture sets seen within outcrops of Jurassic–Cretaceous gabbro, layered gabbro, micro-gabbro and basalt outcrops along the upper Belayan River and in tributaries of the Wahau River on the northern margin of the Kutai Basin. Left-lateral S–C fabrics within serpentinites south–west of Muara Lasan (Fig. 1), are also oriented NW–SE.

Basin-bounding faults

There are no obvious basin-bounding faults to the Kutai Basin. Along the upper reaches of the Wahau, Belayan, and Mahakam Rivers (Fig. 1) the contact between the Tertiary and pre-Tertiary is an angular unconformity (Fig. 4). The northern and southern margins of the basin are marked by a thinning of sediments. In the north a NW–SE-trending lineament that partially follows the Bengalon River separates the Eocene outcrops of the Gunung Gongnyay (Fig. 1) area to the north from a deep early Miocene basin to the south. This appears to be a normal down to the south fault formed or re-activated during the late Oligocene. It can be considered as a basin bounding fault for the Neogene or lower Kutai Basin, although a thin Neogene interval is present north of this. The Bengalon River fault is oriented similarly to the NW–SE offshore Sangkulirang fault zone (which forms the northern margin to the North Makassar Basin) (Figs 1 & 4). The southern margin of the Kutai Basin is delineated by the onshore continuation of the NW–SE-trending Adang fault (Figs 1 & 4). This fault is a down to the north fault, bordering the northern margin of the Paternoster Platform on the western side of the Makassar Strait. The fault comes onshore just to the north of Adang Bay but its surface expression is limited to a series of short lineaments parallel to the trend of the offshore Adang Fault, recognised from SAR images (Wain & Berod 1989). Seismic data (Unocal pers. comm. 1995) shows that the Adang fault zone splays into a synthetic transfer zone immediately offshore Kalimantan. The down to the north Balikpapan fault zone, located to the north of the Adang fault is one such splay. Splaying of the fault may explain its lack of marked surface expression onshore Kalimantan.

Despite its lack of a marked surface expression the Adang fault clearly has been important in the evolution of the Kutai Basin. The Barito Basin and the Kutai Basin are separated along the onshore continuation of this fault line (hence 'Cross Barito High'). The line of the Adang fault coincides with the boundary between the upper Oligocene Berai Limestone Formation platform margin and deeper water clastics of the Kutai Basin to the north (Van de Weerd & Armin 1992; Saller *et al.* 1992). The similarity to the situation at the northern basin margin described above (Bengalon/Sangkulirang faults) should be noted. The extensive development of carbonates on both the northern and southern basin margins and the rapid deepening of the central part of the basin in the late Oligocene (Figs 3a & b) suggests that this is a period of extension and basin formation and not inversion as is often quoted in the literature (Wain & Berod 1989; Van de Weerd & Armin 1992). Inversion would tend to radically uplift the hinterland and cause a flood of sediments rather than creating relatively stable carbonate shelf areas as observed.

The Adang fault zone is also proposed to be a major strike-slip fault (Woods 1985; Hutchison 1989) with the offset of the North Makassar Basin from the South Makassar Basin indicating sinistral displacement. Both Hutchison (1989) and Biantoro *et al.* (1992) favour dextral displacement on this fault line. Rangin *et al.* (1990) show the Adang and Sangkulirang fault zones as two major normal faults bordering the mid-Eocene to early Miocene depocentre of the North Makassar Basin. Sinistral strike-slip reactivation along the Adang fault zone is suggested to have occurred in the mid-Miocene (Rangin *et al.* 1990). Greater importance of the Adang fault is implied by authors who have linked it across Borneo with the Lupar Line (of west Sarawak) to form a 'trans-Borneo shear' (Woods 1985; Bransden & Matthews 1992; Wain & Berod 1989). In turn a number of the major SE Asian fault lines such as the Red River fault (e.g. Williams *et al.* 1988) and the Three Pagodas fault are advocated to link up with the Lupar Line in west Borneo (Biantoro *et al.* 1992; T. Fraser, pers. comm. 1995). Some authors further suggest that the Sangkulirang fault zone links up with similarly oriented faultlines recognised in Sarawak (Woods 1985; Tate 1992). Linkage of these fault lines does not appear to be supported by any of the geophysical or geological data over Borneo. A major bend or transfer fault and restraining or releasing bend structures may reasonably be expected to have developed if these two faults were joined and horizontal movements took place. Linkage is therefore deemed unlikely, although the data currently available can neither prove or disprove this.

The Tinjau well (Fig. 1) drilled by Mobil was located upon a NE–SW-trending horst block. Upper Oligocene carbonates near Long Laham (Fig. 1), based on their outcrop pattern and shape of associated gravity high (Foss 1993) are developed upon a NE–SW-trending horst. The orientation of the original rift-bounding faults are proposed to be similarly oriented. NE–SW-trending, upright, tight, chevron folds within basement cherts and meta-siltstones and other NE–SW-trending structures within the basement on the northern edge of the basin indicate that basement heterogenieties may have influenced the orientation of basin-forming rifts. The Adang and Sangkulirang fault zones are major NW–SE lineaments that effectively delineate the northern and southern boundaries of the basin; however it is suggested here that the trend of major faults upon which basin opening occurred were oriented NE–SW to NNE–SSW (Cloke *et al.* 1997).

Orientation and nature of the folds

Figure 4 shows the orientation of the folds in the Kutai Basin. The Tertiary sediments of the Kutai Basin contain open folds with a general NE–SW trend, except near to the northern margin the basin where the trend is N–NNE (Fig. 4). Immediately to the north of the Meratus Mountains, folds within the Tertiary are oriented ENE–WSW. Within the central region of the Lower Kutai Basin, referred to as the Samarinda anticlinorium (Fig. 4), folds trend NNE–SSW and form continuous anticlinal ridges for over 100 km. In the vicinity of the Mangkalihat Peninsula an E–W trend dominates (Ott 1987; Sunaryo *et al.* 1988; Biantoro *et al.* 1992; Van de Weerd & Armin 1992) and further inland to the west of the Mangkalihat Peninsula the NNE–SSW trend reappears (e.g. Beriun fault trend, LASMO terminology). The curvature of fold trends is similar in its overall pattern to that seen in the folds affecting the Rajang–Emabaluh Group and Crocker Formation sediments of Sarawak and Sabah. Within the upper Kutai Basin fold trends have a more NE–SW trend than the Samarinda anticlinorium, and several open anticline–syncline pairs exposed along the upper Mahakam River upstream from Long Iram (Fig. 1).

Field and published seismic data show folds of the Samarinda anticlinorium and the east Kutai Basin are tight, generally shale cored, asymmetric to symmetric, anticlines separated by broad, open, sand-rich synclines (Biantoro *et al.* 1992; Bendang 1993; Chambers & Daley 1995). Axial traces of folds are long, between 20 and 100 km along strike (Fig. 4). Anticlines may be cut by relatively high angle thrust faults with similar trends, predominantly dipping toward the west (Ott 1987; Biantoro *et al.* 1992; Bendang 1993). Detachment folds are typical with both box and chevron styles. Folds detach over extensive variably inverted over-pressured ?Oligocene shale with varied depth of detachment suggested (Chambers & Daley 1995).

Fold origin

Many attempts have been made to explain the origin and nature of the prominent fold curved pattern of the Samarinda anticlinorium. Models invoking diapirism, growth faulting, gravity gliding, compression upon a rift shoulder, strike-slip movements, thrusting and related folding and inversion have been proposed, although not all of these have been published. All these models are possible mechanisms of fold formation. However, they do not explain all the features of folding in the Kutai Basin

Diapirism of thick pro-delta fine-grained clastics is described by Ott (1987) and by Biantoro *et al.* (1992). However, it is difficult to explain the linear and continuous nature of the fold trends by diapirism alone. Recent high quality seismic data acquired by onshore operators also casts doubt on the earlier interpretation of diapirs in the anticlinorium. Alternatively, gravity sliding of material with uplift of the Central Kalimantan Ranges to the west was proposed (Ott 1987). This may explain fold formation in some settings but does not explain the variable depth to detachment across the anticlinorium and the decrease in intensity of deformation to both the east and west of the anticlinorium. Bendang (1993) proposed compression on a rift shoulder with the opening and extension of the Makassar Straits, but this timing does not agree with Miocene deformation of the Samarinda anticlinorium. Strike-slip movements along both the Adang and Sangkulirang faults have been invoked as a mechanism of variable inversion (Biantoro *et al.* 1992) but it is difficult to explain the extensive linear nature of the fold axes using such a model. None of the models so far described adequately explain the formation of the folds.

The structural style of the folds, best seen on seismic profiles (Biantoro *et al.* 1992; Chambers & Daley 1995) combined with their geometry in map view and their nature at outcrop is more suggestive of formation through inversion of the fill of an earlier rift system. Inversion would have preferentially affected the deeper parts of the fill of the pre-existing Eocene rift depocentres. Chambers & Daley (1995) advocate a punctuated inversion model for the formation of the anticlinorium, which they describe as a series of detachment folds and breakthrough faults above a variably uplifted regional over-pressured shale. Pre-folding contraction is indicated by the presence of now downward facing cm-scale duplex structures exposed on the limbs of anticlines of upper Eocene sediments on the northern margin of the basin. Steeply dipping reverse faults associated with the folds are later features developed as accommodation structures allowing tightening of folds and eventual breakthrough of folds. If this punctuated inversion model is correct and can be extended into the west Kutai Basin significant uplift, inversion and erosion of the Palaeogene section during the early Miocene is implied.

Inversion history

Assuming inversion is the cause of folding, what is known of the folding (inversion) history of the Kutai Basin is outlined below as an attempt to clarify the situation. The cause of inversion is still unknown and all models proposed still remain in the realm of speculation. The inversion appears to have reactivated and nucleated upon the original NE–SW/NNE–SSW rift-bounding faults. The off-setting of fold axes and thrusts along NW–SE trending lineaments suggests that basement fabric was also reactivated (Cloke *et al.* 1997).

Several periods of inversion are suggested here to have affected the Kutai Basin. A major tectono-volcanic event occurred in the late Oligocene, but its origins are uncertain (Moss *et al.* 1997). Many authors regard this as an inversion event and as marking a switch from predominantly deep marine shale and sand deposition to delta deposition and progradation occurred. This event is recorded by local disconformities and deposition of debris flow sediments followed by carbonate platform deposition on both the north and south basin margins. The extreme northwestern margin of the basin was being uplifted in the late Oligocene

(Moss *et al.* 1997). Deltaic sedimentation did not appear in the basin until N5 time (Van de Weerd & Armin 1992; LASMO pers comm. 1995) but the inversion is dated as N3 or earlier. Inversion events are easier to date within the post N5 deltaic section by observation of thinning of packages of seismic reflectors onto highs. The earliest clearly documented event and the event that was responsible for the bulk of the deformation in the lower Miocene sediments of the Samarinda anticlinorium occurred at the end of the early Miocene (end N8) (Chambers & Daley 1995), but earlier events are possible.

At Kelian the upper Eocene to upper Oligocene sequence has been folded into a broad anticline and faulted, with major structures trending to the north/northeast. Most of this deformation is suggested to have occurred by the time the Kelian andesites had been intruded in the late Oligocene/early Miocene (Van Leeuwen *et al.* 1990). Later inversion events in the mid- to late Miocene and in the Pliocene are documented by sedimentary thinning over anticlines (Allen & Gastsaldo 1987) and growth sections seen on regional seismic lines (Chambers & Daley 1995). Uplift in the Meratus Mountains area is also suggested to have occurred in the late Miocene–Pliocene from the presence of the thick, non-marine upper Miocene Dahor Formation (Hutchison 1992).

Significant amounts of rock have been removed since the end of the early Miocene through denudation. Burial history modelling of the Batuq well (Fig. 1) suggests an estimated 3500 m of sediment has been removed by erosion initiated by uplift (Wain & Berod 1989). The Mawai well (Fig. 1), drilled by Mobil, penetrated Oligocene at surface with an R_0(average) of 1.2, implying burial for several million years at depths of around 3–4 km, assuming average geothermal gradients. Vitrinite reflectance data for the lower Miocene of the upper Kutai Basin also imply uplift and erosion of 'several thousands of metres' of section (Van de Weerd & Armin 1992) suggesting that some of these rocks were buried to 2–3 km depth prior to uplift, whilst others have experienced relatively shallow burial.

A popular explanation for the inversion(s) is the collision of micro-continents and continental fragments, such as Bangai–Sula, Tukang Besi fragments, Australia (Daly *et al.* 1991) and south China derived fragment(s), with Borneo–Sulawesi. Daly *et al.* (1991) invoked NW-directed inversion following the collision of Australia with the Sunda trench. Australian collision with the Banda Arc occurred around 5 Ma (Daly *et al.* 1991) so is too late to have caused inversion at the end of the early Miocene. Collision of Bangai–Sula is suggested to be responsible for the thinned-skinned deformation in the Kalosi PSC area and western Sulawesi (Coffield *et al.* 1993; Bergman *et al.* 1996). However a regional seismic line (CGG-104) across the Makassar Straits shows that the deformation front clearly stops offshore west Sulawesi and that Oligocene and Miocene sections in the northern Makassar Straits are undeformed. A low-angle (flat) detachment would be required to transmit stresses from the Bangai–Sula collision to cause deformation in East Kalimantan. Collision of south-China-derived fragment(s), such as the Reed Bank/Dangerous Grounds, with northern margin of Borneo is another possible source of compressional stresses to induce inversion. Collision of the Reed Bank/Dangerous Grounds micro-continental block with the northern margin of Borneo may have caused contraction from the northwest. This direction is approximately 45° or more to the general direction of original basin-bounding faults in the Runtu Block. Even if the orientation of the principal stress direction was oblique to the original faults, new faults are likely to nucleate upon the earlier structures since a weakness already exists. However, the origin and timing any proposed continental collision remains unclear. The late Oligocene–early Miocene uplift in Borneo may be the result of South China Sea spreading and extension causing contraction in the adjacent regions. Thermal doming, possibly related to the onset of the Sintang Intrusive suite (see below), is another possible cause of this uplift. However, such doming usually induces extensional block faulting rather than inversion, as has been variously described for the North Sea. Doming related uplift rarely exceeds 500 m and is typically localized, but may initiate a phase of erosional rebound that could remove several kilometres of cover. However, if thermal doming is the cause of the inversion, a more extensive area of intrusive and extrusive rocks associated with the main inversion locus might be expected, but is not seen. Isostatic adjustments of the crust following thickening of the crust or removal of a load can also cause uplift but again are more likely to produce extensional features rather than folds. The cause of inversion therefore remains unclear.

Tertiary igneous activity

Igneous activity is an important feature of the Tertiary Kutai Basin, and has previously been used to constrain the Tertiary stratigraphy (cf. Pieters *et al.* 1987). Three suites of volcanic and intrusive rocks are recognized within the Tertiary of Kalimantan (Fig. 6), the Nyaan volcanics, Sintang Intrusives suite and the Metulang Volcanics (or Plateau volcanics).

The felsic Nyaan volcanics in the upper Mahakam River (Fig. 1) have K–Ar ages of 48–50 ± 1 Ma have been obtained (Pieters *et al.* 1987, 1993*a*). On the basis of composition and age the Nyaan volcanics correlate with the

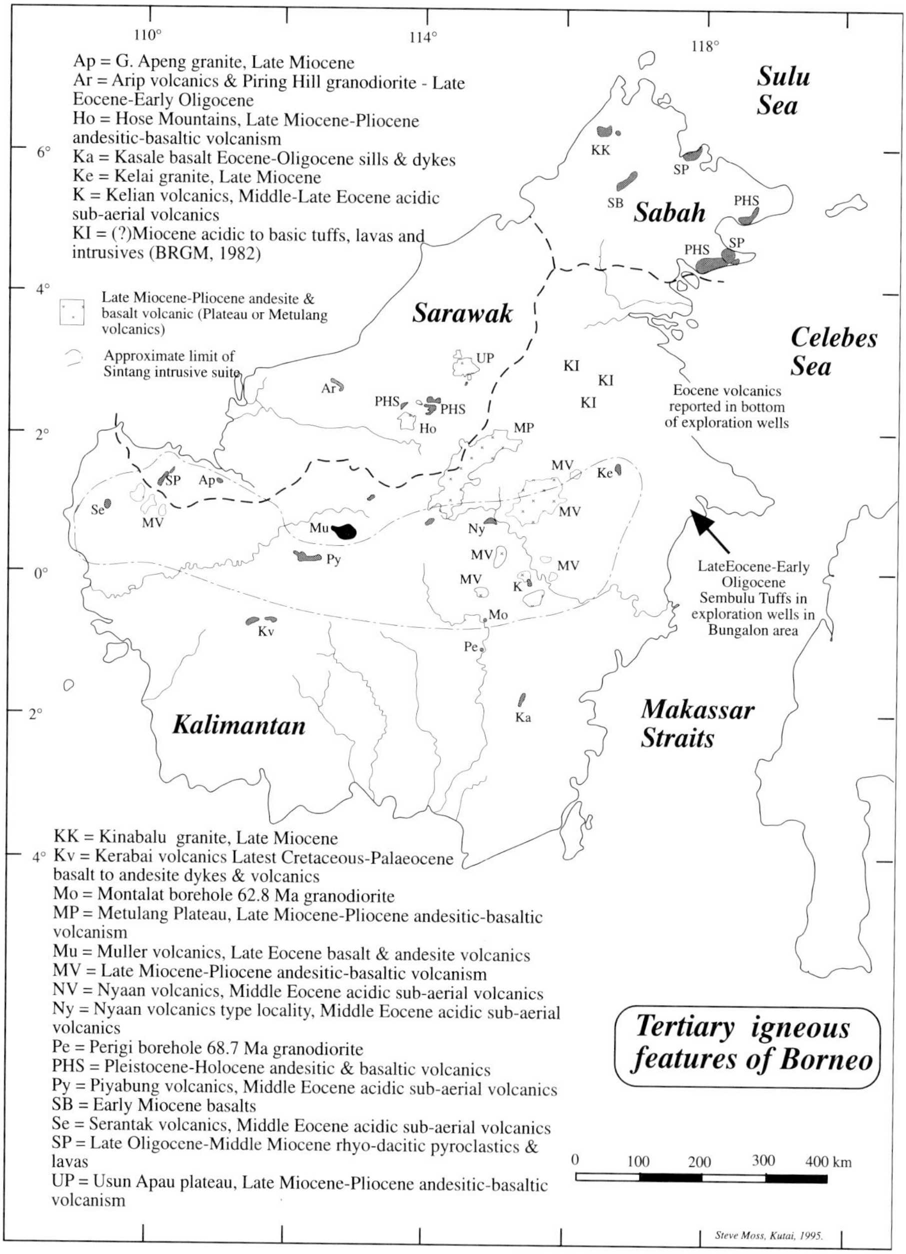

Fig. 6. Distribution of Tertiary igneous activity in Borneo/Kalimantan. The nature, relationships and genesis of the features shown in the diagram are discussed in the text.

Piyabung Volcanics of the Semitau Ridge in western Kalimantan, the Serantek Volcanics on western edge of Kalimantan and the Kelian Volcanics in the centre of the Kutai Basin (Figs 1 & 6). The thick (over 1200 m) Sembulu Tuffs of ?upper Eocene to lower Oligocene age are reported from exploration wells in the Bungalon area in the northeast corner of the Kutai Basin adjacent to the Mangkalihat Peninsula (Sunaryo *et al.* 1988). Acidic volcanics also occur within Sarawak (Arip volcanics and Piring granodiorite intrusion), although these may be slightly younger. Pieters *et al.* (1987) suggest the Nyaan volcanics suite are stratigraphically below the Tertiary succession, although Wain & Berod (1989) place the volcanics within the basal part of their Tanjung Formation. At the Kelian gold mine site (Fig. 1), felsic bedded tuffs, agglomerates, rhyolite lava and reworked pyroclastics are part of an upper Eocene sedimentary succession (Van Leeuwen *et al.* 1990). At the Nyaan volcanics type locality (Fig. 1), flow banded, vesicular rhyoltic lavas and sub-aerial volcanics crop out, but as they are not in contact with any Tertiary sediments, stratigraphic relationships cannot be ascertained. The basaltic to andesitic Muller volcanics, northwest of the Muller Mountains (Fig. 1), have a K–Ar radiometric age of 41 ± 0.4 Ma (Pieters *et al.* 1993*b*) and may represent a late stage event of the Nyaan suite. One explanation for the origin of these volcanics is that they are related to extensional tectonics (Hutchison 1996).

The Sintang Intrusive suite is defined as spanning the upper Oligocene to early Miocene interval, although, a wider range of K–Ar radiometric ages (41–8 Ma) have been obtained from rocks assigned to this suite (Heryanto *et al.* 1993; Pieters *et al.* 1993*a*; Moss *et al.* 1997). The Sintang Intrusives have a mafic to felsic composition and consist of diorites, microdiorites, dacite, micrograntes and andesites. Intrusions of the Sintang Intrusive suite are widely distributed as plugs and other hypabyssal intrusions throughout parts of West, Central and East, Kalimantan, often forming low conical hills (Fig. 6). The finely crystalline nature and limited contact aureole of these intrusions implies high level emplacement. Andesitic breccias and agglomerates are typically associated with the plugs. Pieters *et al.* (1993*a*) correlate the Sintang Intrusives with the compositionally similar coeval Oligo-Miocene intrusives of Kirk (1968) in Sarawak.

Volcanics interbedded with lower to middle Miocene (and younger) sediments, geographically associated with the area of Sintang suite, are likely to the sub-aerial products of Sintang intrusive activity. Volcanogenic tonsteins and waterlain crystal and lithic tuffs are interbedded with lithic arenites, carbonaceous shales and rare foraminiferal limestones of lower Miocene age along part of the Telen River (Fig. 1). Several microgranitic, sometimes glassy, plugs intrude lower Miocene coal measures on the Telen River. The similarities in mineralogy and alteration state of biotite crystal tuffs and glassy microgranitic plugs imply a close relation. An unconformable contact is inferred to exist between welded biotite tuffs interbedded with dipping lower Miocene sediments and sub-horizontal water-lain pumice rich tuffs exposed a short distance downstream. This suggests that volcanism occurred both prior to and subsequent to early Miocene inversion. Addison *et al.* (1983) detail eight laterally continuous tonstein bands within lower Miocene deltaics in the Samarinda area.

Pliocene–Pleistocene volcanics of the Metulang suite are common throughout the centre of Kalimantan and parts of Sarawak and have been termed the Metulang or Plateau Volcanics. The Metulang Volcanics have K–Ar radiometric ages of between 2.4 and 1.7 Ma (Pieters *et al.* 1993*a*). The basalts of the Nuit volcanics in West Kalimantan (Fig. 1), with K–Ar radiometric age of 4.9 ± 0.1 Ma (Supriatna *et al.* 1993), belong to this suite (Supriatna *et al.* 1993). This suite comprises medium to high-K calc-alkaline basalts and andesites forming high-level intrusions (often sub-volcanic), and lava flows from several extinct volcanic centres. The Plateau/Metulang Volcanics may record an evolutionary trend toward more mafic compositions with time from the rhyo-dacitic Sintang Intrusives suite. One proposal is that they are part of an evolutionary cycle with the Sintang Intrusive Suite involving melting of a deep lithospheric root following thickening of the crust (Pieters *et al.* 1993*a*).

The distribution of Tertiary igneous activity in Kalimantan and Borneo is shown in Fig. 6. There would appear to be insufficient material for volcanic activity to be related to subduction, although this impression could be due to the effects of erosion. It has been proposed from the age data, major and trace element signatures of upper Miocene (7.8 ± 0.3, 7.7 ± 0.3 Ma K–Ar ages) minette dykes in the Keramu river (SW Kutai basin; Fig. 1; Bergman *et al.* 1988) that magmatism was not directly associated with subduction. Dykes, which probably correlate with the Sintang Intrusive suite, are viewed by Bergman *et al.* (1988) to have formed due to intraplate alkaline magmatism associated with the melting of subcontinental mantle lithosphere and tentatively related to crustal thickening with late Cretaceous–mid-Tertiary contraction. A similar model for the evolution of the Sintang suite was proposed by Pieters *et al.* (1993*a*). The Pliocene–Pleistocene Metulang suite we regard as the final stages of the Sintang suite.

The Kelian gold mine in the centre of the Kutai Basin (Figs 1 & 6) records all three suites. A wide variety of lithologies, including sediments, sometimes thermally altered, tuffs, rhyolites, andesites, basaltic flows, dykes, sills and stocks are exposed in the mine. There were at least three phases of volcanic and intrusive activity. Flow banded rhyolites and pyroclastic material (welded tuffs, agglomerates) of the initial phase grade into thick sandstones, shales and rare limestones of upper Eocene age (Van Leeuwen *et al.* 1990). Limestone intercalations in outcrops of the upper part of the Kiham Haloq Sandstone Formation along strike to the northeast of the Kelian site yielded upper Eocene foraminifera (Tb or P15-16). The first phase of volcanism may well be correlatable with the Nyaan volcanics suite. The Eocene volcanics and sediments are intruded by small stocks and plugs of andesite which have yielded K–Ar radiometric ages of 22.9 ± 0.5 Ma (Van Leeuwen *et al.* 1990). This phase of volcanic activity belongs to the Sintang Intrusive suite. The whole sequence has been folded into a broad anticline and faulted with major structures trending to the north/northeast, probably in the late Oligocene/early Miocene (Van Leeuwen *et al.* 1990). Several stages of hydrothermal and epithermal gold mineralization occurred in the early Miocene (Van Leeuwen *et al.* 1990). Upper Miocene/Pliocene basalt dykes and basalt flows of the Metulang Volcanics intrude and unconformably

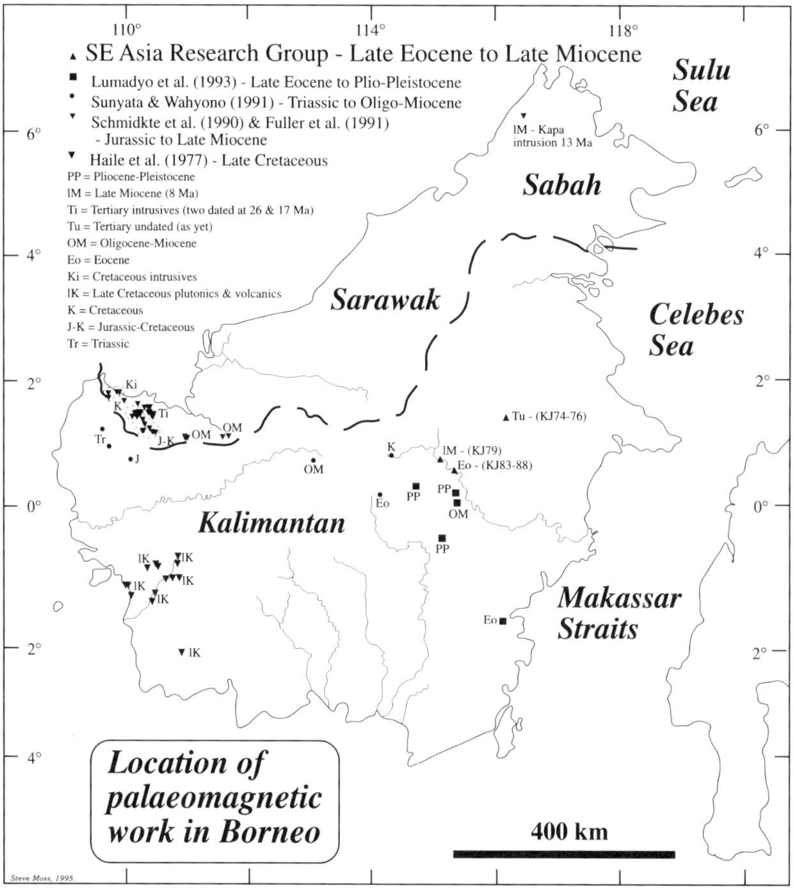

Fig. 7. Location of palaeomagnetic sampling sites in Borneo and the location of University of London SE Asia Group sites in the Kutai Basin.

overlie the entire sequence. It is interesting to note that Van Leeuwen *et al.* (1990) show Kelian and other epithermal gold deposits occur in a remarkably linear NE–SW-trending lineaments, although the significance (if any) of this with respect to other structural features of East Kalimantan is not clear.

Rotation history

The rotational history of Borneo is controversial and a good quality, comprehensive data set is still lacking. Previous palaeomagnetic studies have concentrated upon West Sarawak/West Kalimantan (Fig. 7). A large variation in

Table 1. *Summary of previous and University of London palaeomagnetic work in Borneo*

Age &/or formation	No. of sites*	Direction of rotation	Gross rotation	Timing†
Haile et al. *(1977)*				
Late Cretaceous plutonics & volcanics	48	CCW	50°	Since Mid-Cretaceous
Schmidtke et al. *(1990)*				
Kedadom Fm, Late Jurassic	3	CCW	50°	Since late Jurassic
Pedawan Fm, Late Jurassic–Cretaceous,	4	CCW	90°	Since late Jurassic–Cretaceous
Bau Formation, Late Jurassic–Cretaceous	5	CCW	50°	Remagnetized in Tertiary, timing not constrained
Silantak Fm, Eocene to Miocene	4	CCW (1 site no rotation)	41°	Silantak Fm very poorly dated – no time constraints
'Tertiary igneous intrusions'	10	CCW (4 sites negligible rotation)	−2° to 52°	Only two of intrusions dated (26 & 17 Ma)
Sunyata & Wahyono (1991)				
Kalung Formation Triassic	(?)2	CCW	81–73°	Since the Triassic
Jurassic mudstones	11	CCW	93°	Since the Jurassic
Selangkai Fm, Early/Mid-Cretaceous	39	CCW	76°	Since the Cretaceous
Kiham Haloq &/or Batu Kelau Fm, Late Eocene	38	CCW	37°	Since the Eocene
Undated ?Oligo-Miocene basalt sills	2	Negligible rotation	0°	Since Oligo–Miocene
Lumadyo et al *(1993)*				
Undated Eocene basalt flow	1	Negligible rotation	4°	Since the Eocene
Oligo–Miocene basalt & andesite flows	1	Negligible rotation (slight CW)	1°	Since the Oligo–Miocene
Pleistocene–Pliocene basalt flows	3	Negligible rotation	0°	Since the Pleistocene–Pliocene
University of London SE Asia Research Group				
Late Eocene (NP17-18) sediments	4	CCW	15°	Late Eocene
Early Miocene (18 Ma) diorite plug	4	CCW	16°	Since early Miocene
Undated andesitic lava flow ?Late Miocene	3	CCW	23°	Late Miocene

* 1 Site = 1 horizon.
† Timing = timing of magnetization.

measured rotation for both the Mesozoic and Tertiary is evident (Table 1). Lumadyo *et al.* (1993) question whether Borneo has rotated at all in the Tertiary, but we contest from the limited data available that Borneo underwent counter clockwise (CCW) rotation of no more than 25° in the Neogene. This is in direct contrast to previous results, notably Schmidkte *et al.* (1990); Fuller *et al.* (1991) and Sunyata & Wahyono (1991), who suggest much larger rotations in the Tertiary, particularly in the early Tertiary. Discrepancies between the different studies may be an indication of separate block rotations within the island of Borneo and the influence of major faultlines. Borneo is a very large island which could easily be divided into several blocks. From the dataset available it is not possible to say whether Borneo has rotated as a single block or several semi-independent blocks with similar rotation histories. The view that Borneo behaved as a single crustal block should be questioned certainly for the Mesozoic and probably for the Tertiary. The Lupar Line, Adang fault zone, Semitau Ridge and Sangkulirang fault zone all are probably major structural features of Borneo and even if not linked may have been utilized in the rotation of Borneo. CCW rotation of Borneo conflicts with the extrusion model (Tapponnier *et al.* 1986) in which no rotation of Borneo is implicit. The driving mechanisms and timing of the rotation(s) remain unclear, but it is plausible that periods of inversion in the Kutai Basin and rotations of Borneo may be related. The presence and activity of major faults such as Lupar Line and the West Baram faultline in Sarawak along which deformation associated with the counter-clockwise rotation of Borneo may have occurred. The major oroclinal bending of turbiditic sediments of the Rajang–Embaluh Group and Crocker–Temburong Formations in Sabah and Sarawak (Bénard *et al.* 1990) could also reasonably be related to this counter-clockwise rotation. Further we suggest that the available data imply very little latitudinal change in the position of Borneo throughout the Mesozoic and Cainozoic.

Model of formation of the Kutai Basin

The formation of Kalimantan basins followed phases of late Cretaceous magmatic arc and

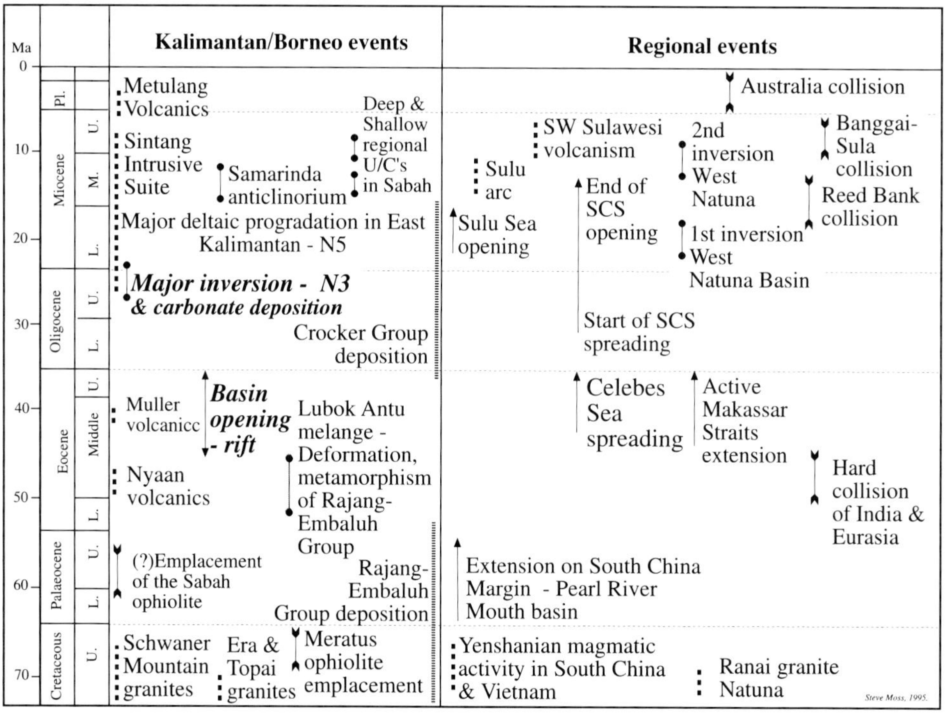

Fig. 8. Correlation of events within Kalimantan and surrounding areas through the Tertiary.

early Palaeogene deformation (Fig. 8). Opening of basins in Kalimantan overlaps with the sea floor spreading in the Celebes Sea, whilst the phases of late Oligocene–Miocene inversion and folding in Borneo overlap with inversions in the West Natuna Basin, continued opening of the South China Sea, opening of the Sulu Sea and several microcontinent and continent collisions.

The Kutai Basin is an extensional basin that formed in a foreland setting relative to the deformed Central Kalimantan ranges (Fig. 9). In common with the Ketungau/Mandai Basins in west Kalimantan, the fill of the Kutai Basin unconformably overlies the deformed Cretaceous–lower Eocene turbidites of the Rajang–Embaluh and older ophiolitic and chert units (Lubok Antu and Keriau melanges) with marked angularity. We suggest basin extension was initially driven by continuation of coeval extension in the Celebes Sea and Makassar Straits into Kalimantan possibly aided by gravitational collapse of the Embaluh/Rajang orogen. The presence of several thermal anomalies within the upper crust from upper Cretaceous granite intrusions and early Tertiary deformation may have aided this process. The Permo-Triassic granites and metamorphics of the Busang complex on the basin edge may be significant, and the complex is interpreted here to be basement core

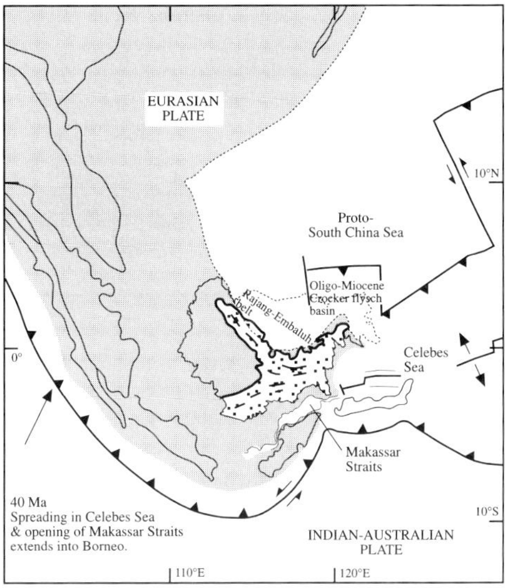

Fig. 9. Opening and extension within the Kutai Basin. Spreading within the Celebes Sea extends into the Sunda Craton and is responsible for generating the Eocene opening of the Makassar Straits and extensional faulting in the Kutai Basin area. Plate-tectonic setting adapted from Hall (1996).

complex unroofed via this extension. The depth of the Kutai Basin (up to 15 km) precludes formation of the basin through flexural mechanisms as advocated by Pieters *et al.* (1993*a*) for Tertiary basins of Kalimantan. The orientation of extensional faults and their evolution in the Paleogene were strongly controlled by the basement structures. Extension had all but ended by the mid-Oligocene, by which time the basin had moved into the sag phase of development. Thermal doming and associated crustal thickening with the onset of the Sintang Intrusive suite may have caused a shorter, less dramatic period of subsidence and is related to late Oligocene tectonism in the area. From the end of the Oligocene to Pliocene the basin was affected by a series of inversion events, most important of which was the early Miocene event, which overlaps with the first phase of inversion in the West Natuna Basin (N8, Ginger *et al.* 1994).

The Kutai Basin, as described, evolved in a way quite different from that reported for other Southeast Asia basins such as those in Sumatra and Vietnam. Basins in Sumatra, for example, show a remarkably similar sequence of (?)upper Eocene syn-rift alluvial and lacustrine deposits (e.g. the Lahat Formation) passing up into fluvial and fluvio deltaic sediments (e.g. the Sihapas and Talang Akar Formations) and a Miocene transgressive–regressive mega-sequence (e.g. Gumai and Air Benakat Formations). The major difference may be the location of the Kutai Basin on the eastern edge of the Sunda craton during the Tertiary.

Conclusions

The Kutai Basin formed in the eastern part of the Sunda craton. During the early Palaeogene the Embaluh and Rajang Groups were deformed and locally intruded by granites. This over thickened crust was rifted with extension driven by coeval extension in the Celebes Sea and Makassar Straits, between 44 and 34 Ma. A deep basin with marked depositional topography formed with initial alluvial fan conglomerates passing up into deep marine bathyal shales and fringing shallower marine clastics and carbonates. The orientation of basin-forming rift faults was strongly controlled by lineaments in the basement. At the end of the Oligocene a major event is recorded by the presence of a basin wide unconformity reflecting a phase of uplift deduced from fission track data (Moss *et al.* 1997). The cause of this uplift/inversion remains unclear. Following this uplift was a period of deep marine shale and platform limestone deposition. A regressive sequence of upper lower Miocene deltaics was then deposited and diachronously migrated from west to east. Deltaic deposition continues today in the form of the Mahakam Delta. Several phases of inversion during the Miocene reactivated the original NE–SW-trending faults. The Samarinda anticlinorium was formed during this stage. Volcanics are intercalated within several parts of the succession, particularly the upper part of the sequence.

I. Bahar, the Director of the Geological Research and Development Centre, Bandung, Indonesia, is gratefully acknowledged for his help and assistance in the organisation of this project. F. Banner and T. Finch are thanked for their excellent biostratigraphic work. Discussions concerning the geology of Borneo and SE Asia with T. Barber, I. Carter, M. Clutson, T. Daley, T. Dooley, T. Fraser, N. Haile, R. Hall, C. Hutchison, K. Glazebrook, M. Keep, K. McClay, S. Mills, D. Murphy and M. Wilson are all gratefully acknowledged. Funding for this project came from the Southeast Asia Consortium (Arco, Canadian Oxy, Exxon, LASMO, Mobil and Union Texas) who are all gratefully acknowledged. CRA Limited are thanked allowing access to the work pits at the Kelian gold mine and providing excellent field accommodation.

References

ADDISON, R., HARRISON, R. K., LAND, D. H. & YOUNG, B. R. 1993. Volcanogenic Tonsteins from Tertiary coal measures, East Kalimantan, Indonesia. *International Journal of Coal Geology*, **3**, 1, 1–30.

AMIRRUDDIN 1989. *The Preliminary Study of the Granitic rocks of West Kalimantan, Indonesia.* MSc thesis Wollonggong University, Australia, (Project B-geol. 951).

ALLEN, G. P. & GASTALDO, R. 1994. The modern Mahakam Delta (Indonesia); sedimentary facies and reservoir geometry in a low wave energy, mixed tide and fluvial dominated delta. *American Association of Petroleum Geologists and Society of Economic Paleontologists and Mineralogists annual convention. Abstracts volume*, 92–93.

BÉNARD, F., MULLER, C., LETOUZEY, J., RANGIN, C. & TAHIR, S. 1990. Evidence of multistage deformation in the Rajang-Crocker Range (northern Borneo) from Landsat imagery interpretation: geodynamic implications. *Tectonophysics*, **183**, 321–339.

BENDANG, F. 1993. *Structure and stratigraphy of the Lower Kutai Basins, Kalimantan, Indonesia.* Independent Project Report, Unpublished MSc dissertation RHUL, Dept. of Geology.

BERGMAN, S. C., DUNN, D. P. & KROL, L. G. 1988. Rock and mineral chemistry of the Linhaisai minette, central Kalimantan, Indonesia, and the origin of Borneo diamonds. *Canadian Mineralogist*, **26**, 23–43.

BERGMAN, S. C., COFFIELD, D. Q., TALBOT, J. P. & GARRARD, R. A. 1996. Late Tertiary tectonic and magmatic evolution of SW Sulawesi and the Makassar Strait: Evidence for a Miocene continental collision. *In*: HALL, R. & BLUNDELL, D. J. (eds) *Tectonic Evolution of SE Asia.* Geological Society, London, Special Publications, **106**, 391–429.

BIANTORO, E, MURITNO, B. P. & MAMUAYA, J. M. B. 1992. Inversion faults as the major structural control in the northern part of the Kutei Basin, East Kalimantan. *Indonesian Petroleum Association, Proceedings 21st annual convention, Jakarta 1992*, **I**, 45–68.

BRGM 1982. *Geological Mapping and Mineral Exploration in Northeast Kalimantan 1979–1982. Final Report.* Report Bureau de Recherches Géologiques et Minières, Orléans, France, 82 RDM 007 AO, October 1982.

BRANSDEN, P. J. E. & MATTHEWS, S. J. 1992. Structural and stratigraphic evolution of the East Java Sea. *Indonesian Petroleum Association, Proceedings 21st annual convention, Jakarta 1992*, **I**, 417–453.

BURRUS, J., BROSSE, E., CHOPPIN, J. D. & GORSJEAN, Y. 1994. Interactions between tectonism, thermal history and paleohydrology in the Mahakam delta, Indonesia. Model results and petroleum consequences. *American Association of Petroleum Geologists Bulletin*, **78**, 7, 1136.

CARTER, I. S. & MORLEY, R. J. 1995. Utilising outcrop and palaeontological data to determine a detailed sequence stratigraphy ofthe Early Miocene deltaic sediments of the Kutai Basin, East Kalimantan. *Indonesian Petroleum Association International Symposium on Sequence Stratigraphy in Southeast Asia*, Jakarta 1995, Abstracts volume,

CCOP 1991. *Total Sedimentary Isopach Maps Offshore East Asia.* CCOP Technical Bulletin, **23**, Sheets 1–6 & Explanatory Notes, CCOP, Bangkok, Thailand.

CHAMBERS, J. L. C. & DALEY, T. 1995. A tectonic model for the onshore Kutai Basin, East Kalimantan, based on an integrated geological and geophysical interpretation. *Indonesian Petroleum Association, Proceedings 24th annual convention, Jakarta 1995*, **I**, 111–130

CLOKE, I. R, MOSS, S. J. & CRAIG, J. C. 1997. The influence of basement reactivation on the extensional and inversional history of the Kutai Basin, East Kalimantan. *Journal of the Geological Society, London*, **154**, 157–161.

COFFIELD, D. Q., BERGMAN, S. C., GARRARD, R. A., GURITNO, N., ROBINSON, N. M. & TALBOT, J. 1993. Tectonic and stratigraphic evolution of the Kalosi PSC area and associated development of a Tertiary petroleum system, South Sulawesi, Indonesia. *Indonesian Petroleum Association, Proceedings 22nd annual convention, Jakarta 1993*, **I**, 679–704.

DALY, M. C., COOPER, M. A., WILSON, I., SMITH, D. G. & HOOPER, B. G. D. 1991. Cainozoic plate tectonics and basin evolution in Indonesia. *Marine and Petroleum Geology (Special issue: South-east Asia)*, **8**, 2–21.

FOSS 1993. Gravity. *In*: PIETERS, P. E., SURONO & NOYA, Y. (eds) *Geology of the Kapuas sheet area, Kalimantan, 1:250 000*. GRDC, Bandung, Indonesia, 000–000

FULLER, M., HASTON, R., LIN, J.-L., RICHTER, B., SCHMIDTKE, E. & ALMASCO, J. 1991. Tertiary palaeomagnetism of regions around the South China Sea. *Journal of Southeast Asian Earth Sciences*, **6**, 161–184.

GINGER, D., ARDJAKUSUMAH, W. O., HEDLEY, R. J. & POTHECARY, J. 1994. Inversion of the West Natuna Basin: Examples from the Cumi-Cumi PSC. *Indonesian Petroleum Association, Proceedings 22nd annual convention, Jakarta 1993*, **I**, 635–658.

HALL, R. 1996. Reconstructing Cainozoic SE Asia. *In*: HALL, R. & BLUNDELL, D. J. (eds) *Tectonic Evolution of SE Asia*. Geological Society of London Special Publications, **106**, 153–184

HAMILTON, W. 1979. *Tectonics of the Indonesian region* USGS Professional Papers, **1078**.

HAILE, N. S., MCELHINNEY, M. W. & MCDOUGALL, I. 1977. Palaeomagnetic data and radiometric ages from the Cretaceous of West Kalimantan (Borneo) and their significance in interpreting regional structure. *Jour-nal of the Geological Society, London*, **133**, 33–44.

HERYANTO, R., WILLIAMS, P. R., HARAHAP, B. H. & PIETERS, P. E. 1993. *Geology of the Sintang sheet area, Kalimantan, 1:250 000*. GRDC, Bandung, Indonesia.

HUTCHISON, C. S. 1989. *Geological evolution of South-East Asia*. Oxford Monographs on Geology and Geophysics, **13**.

——1992. The Eocene unconformity on Southeast and East Sundaland. *Geological Society of Malaysia, Bulletin*, **32**, 69–88.

——1996. The 'Rajang Accretionary Prism' and 'Lupar Line' problem of Borneo. *In*: HALL, R. & BLUNDELL, D. J. (eds) *Tectonic Evolution of SE Asia*. Geological Society of London Special Publications, **106**, 247–261.

KIRK, H. J. C. 1968. *The Igneous Rocks of Sarawak and Sabah*. Geological Survey Borneo Region, Malaysia, Kuching, Malaysia, Bulletin **5**.

KUSUMA, I. & DARIN, T. 1989. The hydrocarbon potential of the Lower Tanjung Formation, Barito BASIN, S. E. Kalimantan. *Indonesian Petroleum Association, Proceedings 18th annual convention, Jakarta 1989*, **I**, 107–138.

LUMADYO, E., MCCABE, R., HARDER, S. & LEE., T. 1993. Borneo: a stable portion of the Eurasian margin since the Eocene. *Journal of Southeast Asian Earth Sciences*, **8**, 1–4, 225–231.

MARKS, E., SUJATMIKO, SAMUEL, L., DHANUTIRTO, H., ISMOYOWATI, T., & SIDIK, B. B. 1982. Cainozoic stratigraphic nomenclature in East Kutai Basin, Kalimantan *Indonesian Petroleum Association, Proceedings 11th annual convention, Jakarta 1982*, 147–180.

MOSS, S. J. In press. Embaluh Group turbidites in Kalimantan: evolution of a remnant oceanic basin in Borneo during the Late Cretaceous to Palaeogene. *Journal of the Geological Society, London*.

——, CARTER, A., BAKER, S. & HURFORD, A. J. In press. A Late Oligocene tectono-volcanic event in East Kalimantan and implications for tectonics and sedimentation in Borneo. *Journal of the Geological Society, London*.

OTT, H. L. 1987. The Kutei Basin – a unique structural history *Indonesian Petroleum Association, Proceedings 16th annual convention, Jakarta 1987*, **I**, 307–317.

PERTAMINA-BEICIP-FRANLAB 1992. *Seismic Atlas of Indonesian Basins*. Unpublished report.

PIETERS, P. E., TRAIL, D. S. & SUPRIATNA, S. 1987. Correlation of Early Tertiary rocks across Kalimantan. *Indonesian Petroleum Association, Proceedings 16th annual convention, Jakarta 1987*, **I**, 291–306.

PIETERS, P. E. & SUPRIATNA, S. 1990. *Geological Map of the West, Central and East Kalimantan Area, 1:1 000 000*. GRDC, Bandung, Indonesia.

——, ABIDIN, H. Z. & SUDANA, D. 1993*a*. *Geology of the Long Pahangai sheet area, Kalimantan, 1:250 000*. GRDC, Bandung, Indonesia.

——, SURONO & NOYA, Y. 1993*b*. *Geology of the Puttissibau sheet area, Kalimantan, 1:250 000*. GRDC, Bandung, Indonesia.

——, BAHARUDDIN & MANGGA, S. A. 1993*c*. *Geology of the Long Nawan sheet area, Kalimantan, 1:250 000*. GRDC, Bandung, Indonesia.

——, SURONO & NOYA, Y. 1993*d*. *Geology of the Kapuas sheet area, Kalimantan, 1:250 000*. GRDC, Bandung, Indonesia.

——, —— & ——1993*e*. *Geology of the Nangaobat sheet area, Kalimantan, 1:250 000*. GRDC, Bandung, Indonesia.

RANGIN, C., JOLIVET, L., PUBELLIER, M. & others 1990. A simple model for the tectonic evolution of Southeast Asia and Indonesia region for the past 43 m.y. *Bulletin de la Société géologique de France*, **8**, VI, 889–905.

ROSE, R. & HARTONO, P. 1978. Geological evolution of the Tertiary Kutei – Melawi Basin, Kalimantan, Indonesia. *Indonesian Petroleum Association, Proceedings 7th annual convention, Jakarta 1978*, 225–252.

SALLER, A., ARMIN, R., LA ODE ICHRAM & GLENN-SULLIVAN, C. 1992. Sequence stratigraphy of upper Eocene and Oligocene limestones, Teweh area, central Kalimantan *Indonesian Petroleum Association, Proceedings 21st annual convention, Jakarta 1992*, **I**, 69–92.

——1994. Sequence Stratigraphy of Aggrading and Backstepping carbonate shelves, Oligocene, Central Kalimantan, Indonesia. *In*: LOUCKS, R. G. & SARG, J. F. (eds) *Carbonate Sequence Stratigraphy: Recent Developments and Applications*. American Association of Petroleum Geologists Memoirs, **57**, 000–000.

SCHMIDTKE, E. A., FULLER, M. D. & HASTON, R. B. 1990. Palaeomagnetic data from Sarawak, Malaysian Borneo, and the Late Mesozoic and Cainozoic tectonics of Sundaland. *Tectonics*, **9**, 123–140.

SIEMERS, C. T., CHAMBERS, J. L. C. & ALLEN, G. P. 1994. *Field Seminar on the Sedimentology and Sequence Stratigraphy of Modern and Ancient Fluvial, Deltaic and Shelf Deposits. Mahakam delta/Kutei Basin fluvial/deltaic depositional systems, East Kalimantan, Indonesia.* American Association of Petroleum Geologists/P. T. Geoservices, 13–20th November.

SIKUMBANG, N. 1990. The geology and tectonics of the Meratus mountains, south Kalimantan, Indonesia. *Geologi Indonesia.* Journal of the Indonesian Association of Geologists, **13**, 1–31.

SUPRIATNA, S. 1990. *Geology of the Muara Wahau sheet area, Kalimantan, 1:250 000.* GRDC, Bandung, Indonesia.

——, MARGONO, U., SUTRISNO, DE KEYSER, F., LANGFORD, R. P. & TRAIL, D. S. 1993. *Geology of the Sanggau sheet area, Kalimantan, 1:250 000.* GRDC, Bandung, Indonesia.

SUNARYO, R., MARTODJOJO, S. & WAHAB, A. 1988. Detailed geological evaluation of the hydrocarbon prospects in the Bungalun area, East Kalimantan. *Indonesian Petroleum Association, Proceedings 17th annual convention, Jakarta 1988*, **I**, 423–446.

SUNYATA & WAHYONO 1991. VI. Palaeomagnetism. *In*: HUTCHISON, C. S. (ed.) *Studies in East Asian Tectonics and Resources (SEATAR): Crustal Transect VII Java-Kalimantan-Sarawak-South China Sea.* CCOP, **26**, 000–000.

TAPPONIER, P., PELTZER, G. & ARMIJO, R. 1986. On mechanics of the collision between India and Asia. *In*: COWARD, M. P. & RIES, A. C. (eds) *Collision Tectonics.* Geological Society, London, Special Publications, **19**, 115–157.

TATE, R. 1991. Correlation of Sarawak and Kalimantan. *Bulletin of the Geological Society of Malaysia*, **28**, 63–95

——1992. The Mulu shear zone – a major structural feature of NW Borneo. *Bulletin of the Geological Society of Malaysia*, **31**, 51–65.

TONGKUL, F. 1991. Tectonic evolution of Sabah, Malaysia. *Journal of Southeast Asian Earth Sciences*, **6**, 395–405.

VAN DE WEERD, A. A. & ARMIN, R. A. 1992. Origin and evolution of the Tertiary hydrocarbon bearing basins in Kalimantan (Borneo), Indonesia. *American Association of Petroleum Geologists Bulletin*, **76**, 1778–1803.

——, ——, MAHADI, S. & WARE, P. L. B. 1987. Geologic setting of the Kerendan gas and condensate discovery, Tertiary sedimentary geology and paleogeography of the northwestern part of the Kutei Basin, Kalimantan, Indonesia. *Indonesian Petroleum Association, Proceedings 16th annual convention Jakarta 1987*, **I**, 317–338.

VAN LEEUWEN, T., LEACH, T., HAWKE, A. A. & HAWKE, M. M. 1990. The Kelian disseminated gold deposit, East Kalimantan, Indonesia. *In*: HEDENQUIST, J. W., WHITE, N. C. & SIDDELEY, G. (eds) *Epithermal gold mineralization of the Circum Pacific: Geology, Geochemistry, Origin and Exploration, I, Journal of Geochemical Exploration*, **35**, 1–62.

VERDIER, A. C., OKI, T. & SUARDY, A. 1980. Geology of the Handil Field (East Kalimantan, Indonesia). *In*: HALBOUTY, M. T. (ed.) *Giant Oil and Gas Fields of the Decade 1968–1978.* American Association of Petroleum Geologists Memoirs, **30**, 399–421.

WAIN, T. & BEROD, B. 1989. The tectonic framework and paleogeographic evolution of the upper Kutei Basin. *Indonesian Petroleum Association, Proceedings 18th annual convention, Jakarta 1989*, **I**, 55–78.

WILLIAMS, P. R, JOHNSTON, C. R., ALMOND, R. A. & SIMAMORA, W. H. 1988. Late Cretaceous to Early Tertiary structural elements of West Kalimantan. *Tectonophysics*, **148**, 279–298.

WOODS, B. G. M. 1985. The mechanics of progressive deformation in crustal plates – a working model for Southeast Asia. *Geological Society of Malaysia Bulletin*, **18**, 55–59.

YUWONO, Y. S., PRIYOMARSONO, S., MAURY, R. C., RAMPNOUX, J. P., SOERIA-ATMADJA, R., BELLON, H. & CHOTIN, P. 1988. Petrology of the Cretaceous magmatic rocks from Meratus Range, southeast Kalimantan. *Journal of Southeast Asian Earth Sciences*, **2**, 15–22.

Gravity anomalies and deep structural controls at the Sabah-Palawan margin, South China Sea

JOHN MILSOM[1], ROBERT HOLT[1], DZAZALI BIN AYUB[2] & ROSS SMAIL[3,4]

[1] *Birkbeck/UCL Research School of Geological Sciences, London WC1E 6BT, UK*
[2] *Geological Survey of Malaysia, Beg Berkunci 2042, Kota Kinabalu, Sabah, East Malaysia*
[3] *Nigel Press Associates, Edenbridge, Kent TN8 6HS, UK*
[4] *Present Address: BHP Minerals International Exploration Inc., London W6 7AN, UK*

Abstract: The hydrocarbon-rich southeastern margin of the South China Sea is divided by NW–SE lineaments into a series of sharply contrasting segments distinguished by, among other things, abrupt changes in gravity patterns. The Sabah segment is bounded to the SW by the Tinjar or West Baram line and to the NE by the Balabac line at the southwest margin of the Sulu Sea. The most prominent gravitational feature of this segment is the strong free-air gravity low associated with the Sabah Trough which lies about 150 km offshore. Seismic reflection data suggest that loading by prograding sediments and gravity driven thrust sheets has depressed the extended continental crust of the South China Sea below its level of local isostatic compensation, producing the trough as a foreland basin in which sedimentation has failed to keep pace with subsidence. The load masses themselves, supported in part by the rigidity of the underlying crust and lithosphere, are above their levels of local compensation and deep Neogene sedimentary basins lie on the flanks of Bouguer and free-air gravity highs. Gravity values decrease across the Sabah coast so that the Crocker Ranges, including Mt Kinabalu, rest in rough isostatic equilibrium on presumably weaker lithosphere. The free-air gravity anomaly associated with the Sabah Trough is smaller than would have been predicted from the thickness of the water column, suggesting crustal thinning beneath the trough axis. This is not a characteristic of normal foreland basins and can therefore be assumed to predate basin formation. It can be concluded that the NE–SW trending belts of parallel gravity anomaly and geomorphology, of which the Sabah Trough is the most obvious, have been controlled by the pre-existing fabric of the crust and lithosphere of the South China Sea since they are discordant to the Palaeogene geological trends in Sabah. Reconstructions of the Tertiary history of the Sabah segment can be based on this assumption, which also suggests that sediments deposited in rift basins formed during the Palaeogene break-up of the South China margin were the source for much, if not all, of the hydrocarbon reserves of the area.

The southeastern margin of the South China Sea (Fig. 1) consists of a complex suture zone which can be divided into three sharply defined segments distinguished by differences in bathymetry and in the associated gravity fields. These differences reflect differences in the thicknesses and compositions of the crustal units on either side of the suture. In the northern (Palawan) section, oceanic and island arc elements from the Sulu Sea margin have been thrust over continental crust derived from southern China (Durkee 1993). This South China crust has been variably but generally very largely extended and attenuated (Holloway 1982; Kudrass *et al.* 1986; Briais *et al.* 1993); positive elements, such as Reed Bank and the islands, shoals and banks of the Spratly Islands region (Hinz & Schlüter 1985) mark the presence of blocks of locally thicker crust. In the southern (Sarawak) segment, the thick crust of Borneo abuts against the almost equally thick crust of the Luconia Platform (Hutchison 1989). The Luconia block may also have been derived from the South China margin but little is known about its basement geology. The possible suture with Borneo is concealed beneath the Balingian Fold and Thrust Belt.

The central, Sabah, segment, which contains most of the known hydrocarbon resources and which is the focus of this study, is bounded in the NE by the Balabac Line or Sabah Shear (which continues the trend of the Sulu Sea coast of Sabah into the South China Sea), and in the SW by the Tinjar or West Baram line. Its extent is defined most obviously by the presence offshore of the NE–SW-trending Sabah Trough, which is associated with a free-air gravity low. Outboard of the trough, the South China Sea is underlain by extended continental crust, as in the Palawan segment, but onshore the similarities of the

From Fraser, A. J., Matthews, S. J. & Murphy, R. W. (eds), 1997, *Petroleum Geology of Southeast Asia*, Geological Society Special Publication No. 126, pp. 417–427.

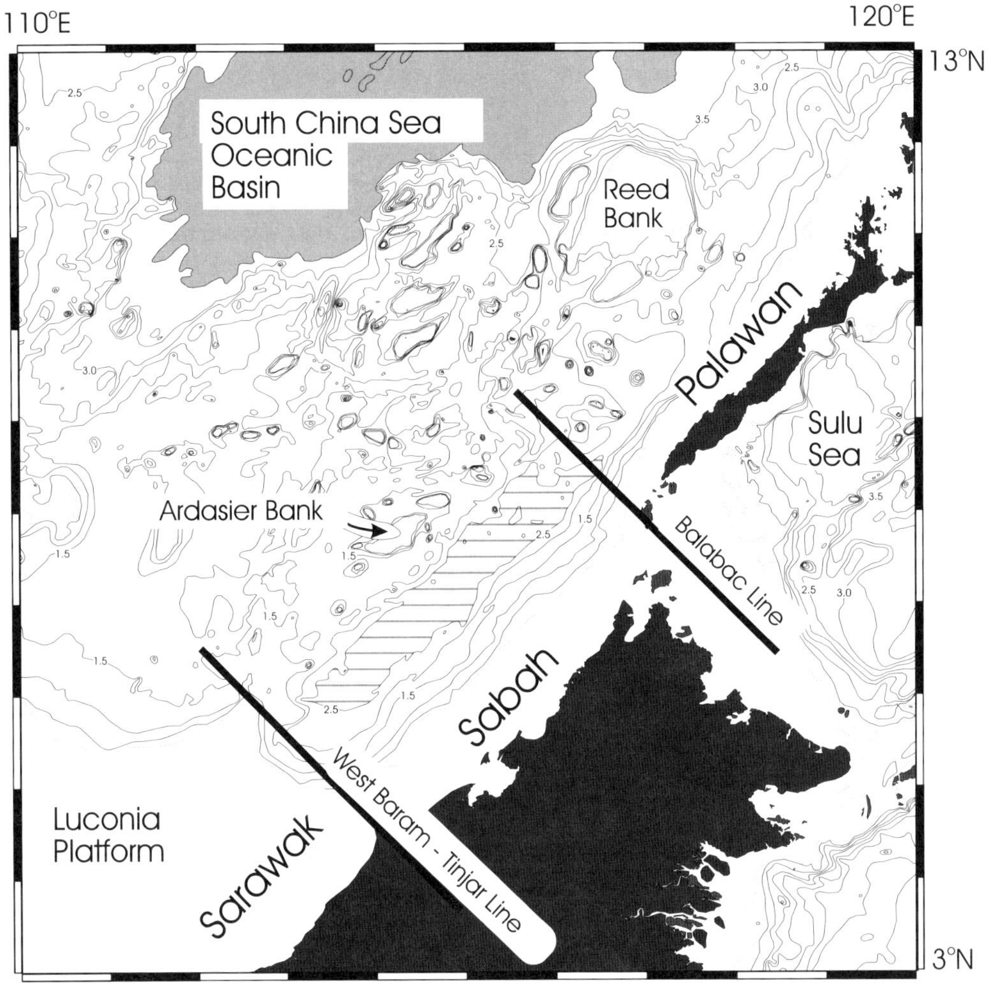

Fig. 1. Segmentation of the southeastern margin of the South China Sea, showing the locations of the Sarawak, Sabah and Palawan segments. Bathymetry taken from the digital version of the General Bathymetric Chart of the Oceans (GEBCO). Light stipple shows area in the South China Sea which is underlain by Late Oligocene – Early Miocene oceanic crust. Horizontal lines show the region of the Sabah Trough with water depths in excess of 2.5 km.

Sabah segment are with Sarawak; Sabah is underlain by thick crust of unknown composition and the geology of its western part is dominated by a thick wedge of Palaeogene sediments similar to those which outcrop over much of Sarawak. Within the segment, the South China Sea and Borneo domains interact over a wide zone of thrust propagation and sediment progradation of which the Sabah Trough marks the northwestern limit (Hinz *et al.* 1989; Hazebroek & Tan 1993).

Geology

Geological summary

The published literature on the Palaeogene geology of NW Borneo is a testimony to the complexity of the deformation, the paucity of biostratigraphic control, the poor quality of the available outcrops and the logistic difficulties of mapping. New road building is reducing the importance of these last two factors but the scarcity of dateable fossils still poses enormous interpretational problems (cf. Tongkul 1991). Some of the correlations which have been made between outcrops may be incorrect while some of the multitude of named formations may be direct equivalents or facies variants of formations assigned different names in other areas. Particular problems are caused by differences in the terminologies applied in Sabah, Sarawak and Indonesian Kalimantan. There seems to be general agreement on a few points only, one of which is that there was a major event, marked by unconformity, deformation and the formation of melange, in the Mid- or Late Eocene. Beneath the unconformity, the Rajang Group sediments (Embaluh Group in Indonesia) have been isoclinally folded with dips generally of between 80 and 90° and have been locally metamorphosed to lower greenschist facies (Hutchison 1996). The break between these and the younger, less deformed and unmetamorphosed Crocker Group sediments has been taken as identifying a 'Sarawak orogeny' but in Sabah the situation is complicated by the existence of an East Crocker Formation which may be a Rajang Group equivalent (Hutchison 1996).

In western Sabah and on the adjacent shelf, the geology can be described in terms of a succession of belts of sediments which become progressively younger and less deformed towards the northwest (Fig. 2). The oldest, Palaeogene and Lower Miocene, sediments have proved difficult to map in detail but can be loosely assigned to a composite Rajang–Crocker group which is generally quartz-arenaceous and commonly turbiditic. Their distribution and history have been summarised by Bénard *et al.* (1990) on the basis of photogeological interpretation with some ground control. Complex deformation and lack of clear marker horizons have contributed to the difficulty of identifying the unconformity associated with the Sarawak orogeny, although. Hutchison (1996) has re-emphasised the distinction which should be drawn between the older Rajang and younger Crocker groups. Although the sediments of the two groups were deposited respectively before and after the period of strong tectonic uplift associated with the orogeny, both

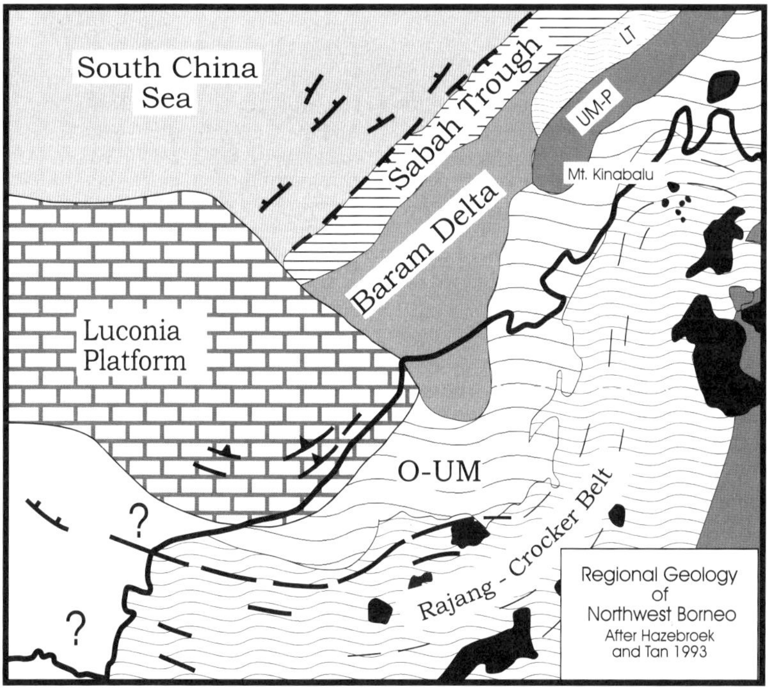

Fig. 2. Geology of NW Borneo, simplified and modified after Hazebroek & Tan (1993). O-UM, Oligocene-Upper Miocene 'Inboard Belt'; UM-P, Upper Miocene–Pliocene 'Outboard Belt'; LT, Lower Tertiary thrust sheet. Igneous rocks are indicated by black shading.

have been widely regarded as constituting trench fills and accretionary complexes associated with subduction beneath Borneo of a 'proto South China Sea' which has now been completely consumed (cf. Hamilton 1979).

In Sabah the top of the Crocker group is marked by a major disconformity at the Early Miocene–Mid-Miocene boundary. Younger sediments have generally not been extensively deformed, although there is a further important break at the close of the Mid-Miocene. The uplift which gave rise to the present-day Crocker Range is even younger and may be correlated with cooling ages of between 6 and 7 Ma obtained from granitic rocks of the Kinabalu massif (Rangin *et al.* 1990). The episodic nature of the Late Neogene uplift is documented offshore by the presence of several unconformity surfaces lying between major sequence boundaries known as the Deep and Shallow Regional Unconformities, dated to 15 Ma (Mid-Miocene) and 9 Ma (Late Miocene) respectively (cf. Hazebroek & Tan 1993). The name 'Sabah Orogeny', which has been suggested for this phase of tectonism (Hutchison 1996), implies a rather local regime, but inversion seems to have been widespread during this period throughout the South China Sea and its margins.

Petroleum geology of Sabah

Published geological maps of western Sabah and its shelf region show the Rajang–Crocker belt trending N to NNE and cutting across the coastline (Fig. 2; cf. Hazebroek & Tan 1993). As a consequence, the environment of the offshore oilfields changes from SW to NE. The characteristics of the various hydrocarbon-bearing zones have been plentifully illustrated by seismic sections published by Shell and by the German BGR in conjunction with Petronas (Bol & Van Hoorn 1980; Hinz *et al.* 1989; Hazebroek & Tan 1993) and also by marine institutes of the Chinese People's Republic (cf. Xia & Zhou 1993). Off southern Sabah the fields are associated with the prograding Baram Delta, which has built out from the flank of the Palaeogene belt in the angle formed between the belt and the Luconia Platform. Seismic images of the Baram region show a classic delta system with thrusting confined to the toe of the sediment wedge where gravity-driven listric extensional faults break through to the sea floor.

Immediately east of the deltaic wedge, there are oil fields in the Oligocene to Upper Miocene sediments of the 'Inboard Belt', a structurally complex zone characterised by tight, broken anticlines separated by wide synclines (O-UM in Fig. 2). Farther to the north, hydrocarbons are reservoired in deep synclinal Late Miocene to Pliocene basins (the Outboard Belt, UM-P in Fig. 2) developed upon thrust sheets of a fold and thrust belt (here termed the Kinabalu Fold and Thrust Belt) which involves Palaeogene and Early Neogene sediments. These older rocks (LT in Fig. 2) form a ridge which separates the Outboard Belt from the Sabah Trough but are generally concealed beneath a thin cover of younger, gently deformed sediments. As in the Baram Delta, the ultimate driving force for the thrusting is probably gravitational, the impetus in this case being provided by the uplift of the Crocker Range. Rapid deposition of sediments within the overlying basins evidently led to overpressuring since clay diapirism is common along the western margin of the Outboard Belt (Hazebroek & Tan 1993).

Regional tectonics

Hamilton (1979) not only suggested a subduction zone setting for Rajang–Crocker deposition but also identified the Sabah Trough as the site of the subduction trench. However, seismic reflection sections clearly show that the attenuated continental crust which underlies the adjacent parts of the South China Sea extends beneath the trough and its landward margin (Hinz *et al.* 1989; Hazebroek & Tan 1993) and the difference in trend between the supposed Palaeogene accretionary prism (Rajang–Crocker group) and the supposed Palaeogene trench (Sabah Trough) can also be cited as an argument against the subduction hypothesis. It has also been claimed that, although vigorous volcanism would be expected at an active margin, there was no volcanic activity at the time of Rajang–Crocker sedimentation (Hazebroek & Tan 1993).

The objections certainly have force as far as the Sabah Trough is concerned but subduction provides a very plausible setting for the deposition and subsequent deformation of the very thick Rajang–Crocker sediments. Moreover, Rangin *et al.* (1990) have reported dates on volcanic rocks in eastern Sabah which appear to define two distinct volcanic episodes separated by a hiatus in the Early and Middle Miocene; the Oligocene volcanics might mark the northeastern end of the belt of Oligocene 'Sintang' igneous activity that has been identified in Kalimantan (Moss *et al.* this volume) and which might have been associated with subduction of a 'proto-South China Sea'. The distance between the volcanic outcrops and the Rajang–Crocker belt is well within the range for observed gaps between modern arcs and their associated trenches. It is only necessary to suppose that the original trench axis lay a significant distance landward of the present trough to preserve both the essential elements of Hamilton's (1979) hypothesis and compatability with more recent geological and geophysical observations.

Gravity

Satellite gravity data

Free-air gravity values calculated from satellite measurents of sea surface elevations have been available for more than a decade. However, the maps produced of equatorial regions have generally been decorative rather than practical, because the separations between adjacent satellite tracks have been of the order of 100 km and features of geological interest have been very poorly resolved. Along the eastern margin of the South China Sea the satellite-derived data were until recently sufficient only to demonstrate the very obvious division into Sarawak, Sabah and Palawan segments and quite inadequate for use in any form of quantitative analysis.

Early in 1995 the quality of satellite derived gravity data improved dramatically with the release of the results of the first 168-day mission of the ERS-1 satellite, on which adjacent tracks were separated, even at the equator, by no more than 10 km. The free-air data from this mission have been combined in Fig. 3 with high-quality shipborne gravity data from the BGR 1986 EXPLORA cruise, with almost no need for adjustment of either data set.

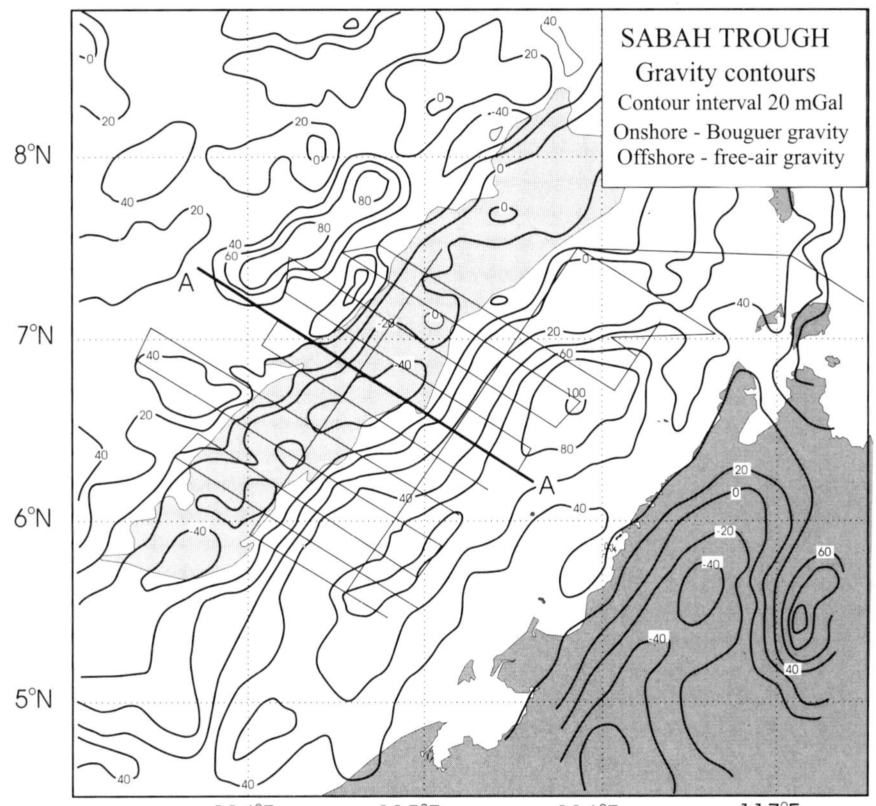

Fig. 3. Gravity field of the Sabah segment, NW Borneo. Free-air gravity contours offshore are based on gravity measurements made along track lines (continuous straight lines) on the 1986 EXPLORA cruisee (Hinz pers. comm. 1995) and altimeter measurements obtained during the first ERS-1 168-day mission. Bouguer gravity contours onshore are based on preliminary data from a 1995 survey by the University of London and the Geological Survey of Malaysia. Line AA shows the location of the gravity profile modelled in Fig. 5. Heavy stipple shows land areas and light stipple the region in the Sabah Trough with water depths of more than 2500 m.

Sabah segment gravity and geomorphology

It is also, of course, possible to link satellite and shipborne data to data obtained onshore by conventional means. The gravity map of the Sabah segment produced by combining data recently acquired in the Crocker Ranges and farther east with the marine data (Fig. 3) is dominated by a series of parallel linear features with NE–SW trends, of which the most extensive is the free-air gravity low associated with the Sabah Trough. To the east of the trough there is a gravity ridge which overlaps in part the area where Miocene basins are associated with major hydrocarbon production. Short-wavelength highs and lows between this feature and the coast are not well resolved by the ERS-1 data but inland there is a third major NE–SW-trending feature in the form of a deep Bouguer gravity low coinciding almost precisely with the steep and rugged Crocker Range which culminates in the northeast in Mt Kinabalu, at 4100 m the highest mountain in SE Asia.

The Crocker Range is one of three significant geomorphological features which share the gravitational NE–SW trend, the other two being the Sabah Trough and the South China Sea coastline of Sabah. Coincidence in trend between gravitational and topographic features is hardly surprising, especially where, as in the case of the Crocker Range, there is a significant isostatic control on the gravity field. There is, however, a small but significant discordance between the gravitational/topographic trend in the Crocker Range and the geological trends, and Mt Kinabalu is a large granitic pluton which is geologically unrelated to the rocks to the southwest.

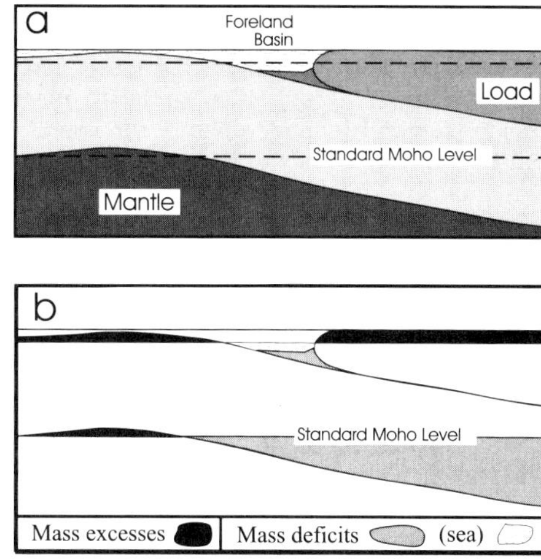

Fig. 4. Simple model of the structure of a foreland basin. (**a**) distribution of main geological units. (**b**) distribution of main mass deficits and excesses. The upper dashed line in (**a**) represents the equilibrium level for the sea floor corresponding to the standard Moho level indicated by the lower dashed line. The positions of these lines correspond in this example to attenuated continental crust, as found beneath much of the South China Sea.

Mass balance in foreland basins

Northwest directed mass transport in both the Baram delta region and the Kinabalu Fold and Thrust Belt implies loading of South China Sea lithosphere and hence the possibility that the Sabah Trough developed as a foreland basin. Such basins can occur wherever the lithosphere supports near-surface loads by elastic bending instead of by local isostatic depression of the Moho and are inevitably associated with gravity anomalies. Simple models (Fig. 4) involve only an impressed load (e.g. a fold and thrust belt or a prograding delta), an elastically bent slab and a resulting basin which may be partly or completely infilled with sediment. Mass excesses giving rise to positive gravity effects are represented by the load and also by an upward bulge of the surface and Moho of the flexed lithosphere on the far side of the basin. Mass deficits are concentrated in the vicinity of the foreland basin, being represented by the water column, the sediment infill and also by the depression of the Moho beneath the basin. The Moho is depressed still more deeply in the region beneath the load but the resulting mass deficit in that region cannot completely balance the impressed load; if it did, i.e. if local isostasy applied, there would be no reason for a flexural basin to form.

The Sabah Trough

Qualitatively, the free-air gravity pattern across the Sabah Trough is as predicted by the foreland basin model. There is a high over the loads represented by the Baram Delta and the thrust sheets, a low along the Sabah Trough and a second high in the Spratly Islands region beyond, with a peak over Ardasier Bank. However, quantitative modelling based on the 2D models appropriate for an area which displays such strong linearity has shown that there are differences between the expected and observed fields. These can be illustrated by comparing the profile recorded along Line 14 of the BGR EXPLORA cruise (Fig. 5a, Curve A) with the

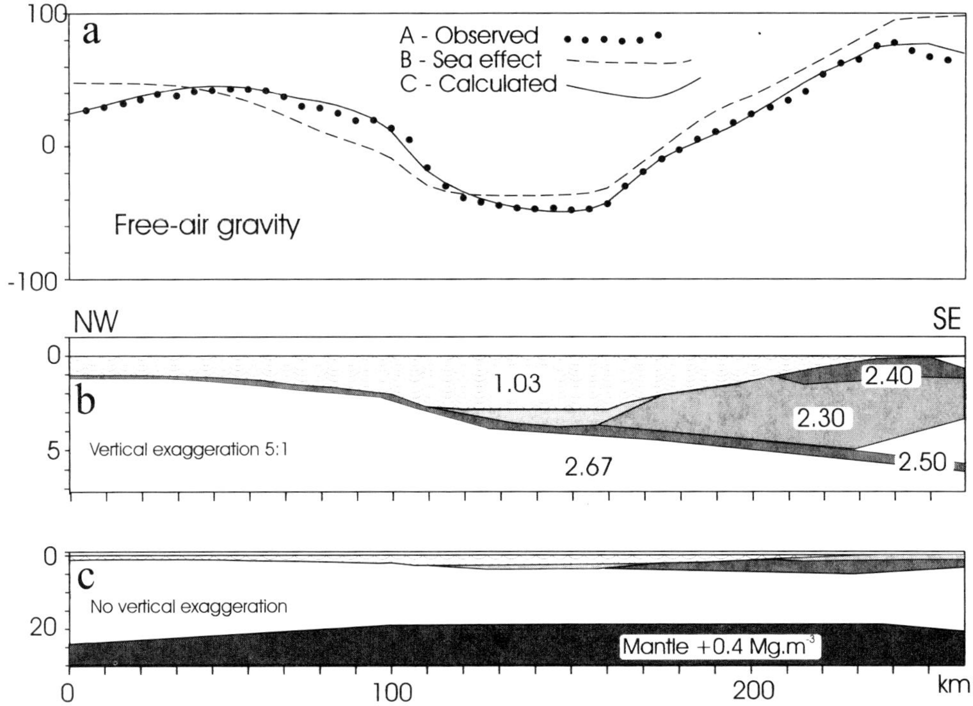

Fig. 5. Gravity models of the Sabah Trough. (**a**) Free-air gravity profile (Curve A), together with the curves computed for a flat Moho and a sea floor topography of density 2.2 Mg m^{-3} (Curve B) and for the model shown in the two views below (Curve C). (**b**) Close up of the near surface part of the two-dimensional density model producing curve C vertical exaggeration 5 : 1. (**c**) Two-dimensional density model for the Sabah Trough which produces Curve C showing elevation of the Moho. No vertical exaggeration.

free-air profile which would have been produced had gravity been controlled simply by variations in the depth of the sea floor. A model in which the water layer was assigned a density of -1.17 Mg m^{-3} (equivalent to using the very low density of 2.2 Mg m^{-3} for the material making up the trough walls) produced the good match in amplitude shown by Curve B in Fig. 5a. To produce the rough coincidence with observation, the Moho has been set everywhere at slightly less than 24 km depth, i.e. a little more than 6 km above the 30 km level usually considered appropriate for continental crust. The same effect could have been obtained by adding 100 mGal to all points on the curve.

The similarity in the shapes of curves A and B on the SE side of the trough indicates the suitability of the low rock density in this region. In contrast, the detailed match to the NW is poor. The need to use higher average densities in this region is unsurprising since the sediments across the Spratly Rise are older, thinner and are commonly carbonates, but obtaining a local match in this way destroys the rough agreement between the total amplitudes of the observed and calculated trough anomalies. Agreement can be restored only by having a more elevated Moho, and hence thinner crust, beneath the trough than beneath the Spratly Rise, as shown in Fig. 5c.

The model illustrated in Fig. 5b and c is still a very simple one but the computed curve (Curve C in Fig. 5a) agrees well with the observed free-air values (Curve A). Basement throughout the profile was assigned zero density contrast with standard crust, defined as having a 2.67 Mg m^{-3} density in its upper parts and a thickness of 30 km. On the Spratly Rise the basement is overlain by only 100–200 m of sediments but thick sediments of varying but generally low densities make up the southeastern trough wall. In the main a density of 2.3 Mg m^{-3} has been used but a small block of slightly higher density (2.4 Mg m^{-3}) near the upper break of slope represents, in schematic form, a slightly denser thrust sheet. Details of the density variation in this region are in any case very poorly controlled, but no plausible changes would affect the main conclusion that can be drawn from the model study, which is that there must be crustal thinning from beneath the Spratly Rise to beneath the Sabah Trough.

The density contrast across the Moho has been assigned a value of 0.4 $Mg\,m^{-3}$. A very large rise in Moho level, to almost 20 km in places, was required to produce the level of agreement shown by Curve C in Fig. 5. Such thinning is normal for extensional basins but is not a usual feature of foreland basins, which are produced by offset loads. Since such loads would not significantly extend the crust, the implication is that the crust beneath the trough is likely to have been already thin before it was loaded. Since there is relatively little variation in gravity pattern parallel to the strike of the trough, this argument can be applied anywhere along its length and not merely in the immediate vicinity of this particular profile. It can be concluded that the Sabah trough overlies an old zone of crustal attenuation with a NE–SW trend and deduced that this widespread trend direction is a characteristic of the South China Sea and has been imprinted on Borneo only relatively recently.

Tertiary history of the Sabah segment

Any attempt to reconstruct the Tertiary history of Sabah must provide some explanation for the discordance between the geological trends in the Crocker group and present-day gravitational and geomorphological trends. Moreover, reconstructions which involve the supposed presence of an active margin in NW Borneo during the Palaeogene should be compatible with current understanding of subduction processes derived from studies of modern examples. The basic geometric relationships in these zones have been described by Isacks & Barzangi (1977) and are illustrated in Fig. 6. Volcanic activity is concentrated above the region where the upper surface of the subducted lithosphere is at depths of between 100 and 150 km. The dip of the slab is shallow beneath the trench and forearc but increases rapidly in a region (the subduction 'hinge') which lies somewhere between the volcanic line and the trench axis. The high density of the slab causes it to sink through the asthenosphere so that there will be a component of motion parallel to its length. This component can lead to migration of the trench away from the volcanic line (subduction roll-back) and is most likely to do so if subduction slows or ceases. However, the presence of continental crust on the subducting plate will halt roll-back because such crust is too buoyant to be taken

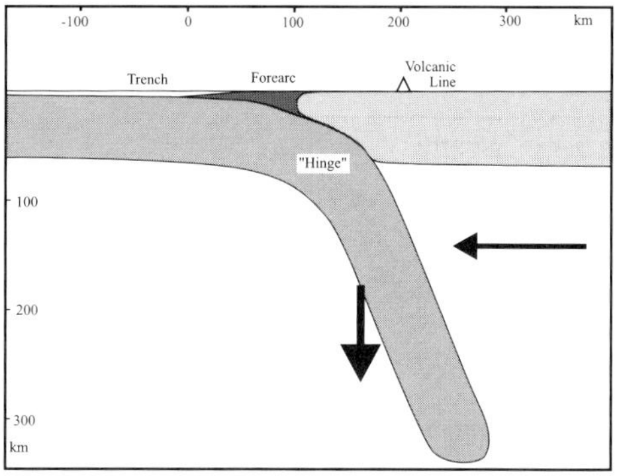

Fig. 6. Simplified subduction zone geometry, showing location of the 'hinge' region where the dip of the subducted slab steepens rapidly.

down very far. Stages in a hypothesised evolution of the Sabah region controlled by these considerations are shown in Fig. 7 and are discussed below. A Borneo-based frame of reference is used; rotation of Borneo would require reorientation of each map as a whole but would not greatly affect postulated relative motions.

Figure 7a shows the hypothesised situation in the earliest Tertiary. An oceanic 'proto-South China Sea' existed off Eurasia and was being subducted at a trench in the Borneo–Palawan region. Eventually (Fig. 7b), the Luconia platform arrived at the subduction zone off Sarawak, initiating the Sarawak Orogeny. The block may have been a part of the Eurasian margin or a separate microcontinent. In either case, the shape of the margin of the block was such that the trench farther to the northeast still faced oceanic crust and there was no orogeny at this time in Sabah.

During the Oligocene, continued slab pull led to continued subduction off NW Sabah, placing the Eurasian crust which lay on the far side of the remaining oceanic fragment under tension and leading to the development of the Tinjar Line as a transform separating extending and normal crust (Fig 7c). Eventually, in the Late Oligocene, the extension of the Eurasian margin culminated in the generation of the oceanic crust of the present-day South China Sea oceanic basin (Briais *et al.* 1993). It was probably at about the same time that thinned continental crust began to arrive at the Sabah subduction zone. By the end of the Early Miocene, the NE–SW-trending edge of the extended, but continental, crust of the Spratly Islands platform

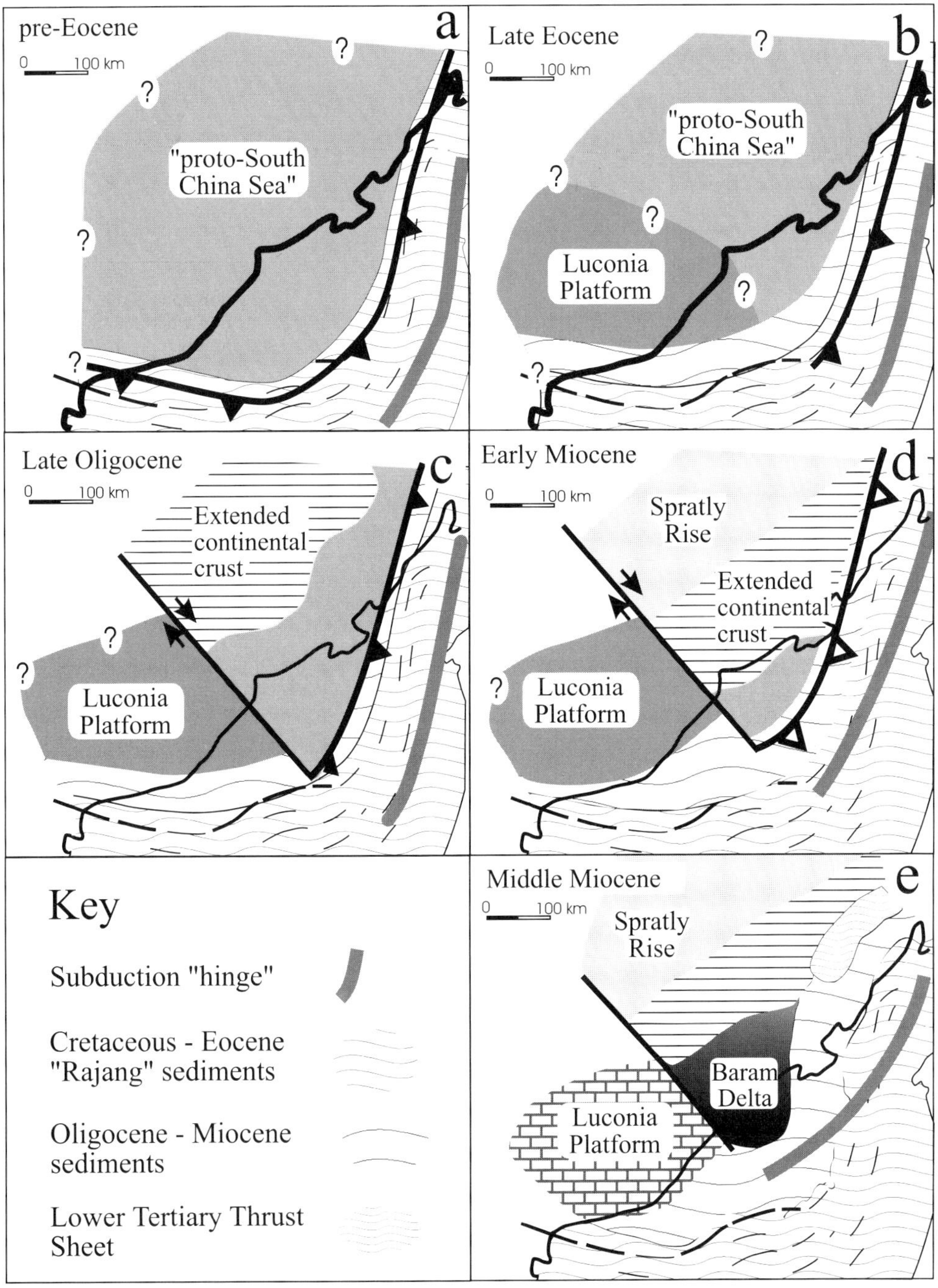

Fig. 7. Stages in the Tertiary evolution of the Sabah and Sarawak segments.

had entered and choked the subduction zone off Palawan and NW Sabah (Fig. 5d). Off Brunei and SW Sabah this thicker crust was still some distance from the subduction zone

Following collision, the subducted slab continued to sink but was pinned in the collision area, although the continental crust was to some degree dragged down to deeper levels and higher temperatures. Farther to the southwest, rollback was still possible, altering the orientation of the subduction hinge from NNE–SSW to NE–SW (Fig. 5e). Subsidence was accompanied by the movement of thrust sheets outwards from the Borneo margin under gravity and the development of a prograding delta in the embayment between the Luconia Block, where platform carbonates were accumulating, and the Borneo margin. Subsequently, melting took place within the underthrusting continental crust and was greatest in the region where collision had first occurred. Ascending granitic magmas formed the Kinabalu pluton. The reoriented subducted slab detached and sank into the mantle, allowing rebound and uplift of the Crocker Range, roughly along the line of the subduction hinge. Uplift of the Crocker Range increased the sediment supply to the Baram Delta and the gravitational drive on the Kinabalu Fold and Thrust Belt. Loading further depressed the thinned crust of the Eurasian margin, accentuating the Sabah Trough and producing the general geological setting shown in Fig. 2.

Conclusions

Gravity modelling suggests that the present-day trend of the Sabah Trough and parallel features as far east as the Crocker Range is inherited from the fabric of the Early Miocene extended margin of Eurasia. However, this fabric must in turn have been determined by, or at least modified by, the pattern of sea-floor spreading in the South China Sea. The primary control may well have been the position of the NE margin of the Luconia Platform, along which a transform fault permitting NW–SE relative motion developed during the Late Palaeogene to accomodate extension in the region farther north. This former transform is now expressed as the Tinjar or West Baram Line, which runs at right angles to, and terminates, the Sabah Trough.

Palaeogene extension of the Eurasian margin produced a horst and graben terrane in which thick sediment piles were deposited in confined basins. In the southern part of the South China Sea these basins were generally lacustrine. Marine conditions probably prevailed at the latitude of Sabah but circulation would have been restricted by the proximity of the Borneo landmass and anoxic conditions were probably common in bottom waters, favouring preservation of organic material. The sediments deposited in these grabens now lie at depth beneath the Outboard Belt, Inboard Belt and Baram Delta basins which rest on the Kinabalu Fold and Thrust Belt. Each of these very different shallow geological domains hosts significant reserves of hydrocarbons, which suggests that the source rocks lie within the deeper older sediments which are common to all three. Migration paths into the Borneo-derived thrusts and sediment wedges will have been complex and hydrocarbons could be reservoired at almost any level, in traps of almost any conceivable type. The deepest fields are likely to be found in regions where diapirism is absent and potential seals have not been breached.

The bathymetry shown in Fig. 1 is reproduced from GEBCO Sheet 5.06, compiled by Y. Iwabuchi and made available through the GEBCO Digital Atlas published by the British Oceanographic Data Centre on behalf of the IOC and IHO, 1994. We thank K. Hinz of the Bundesanstalt fur Geowissenschaften und Rohstoffe for allowing us to make use of gravity data collected on the 1986 EXPLORA cruise, and the Director-General, Geological Survey of Malaysia, for permission to present some of the onshore gravity data acquired in the Crocker Ranges region in 1995.

References

BÉNARD, F., MULLER, C., LETOUZEY, J., RANGIN, C. & TAHIR, S. 1990. Evidence of multiphase deformation in the Rajang–Crocker Range (northern Borneo) from Landsat imagery interpretation: geodynamic implications. *Tectonophysics*, **183**, 321–339.

BOL, A. J. & VAN HOORN, B. 1980. Structural styles in the western Sabah offshore. *Geological Society of Malaysia Bulletin*, **31**, 67–83.

BRIAIS, A., PATRIAT, P. & TAPPONNIER, P. 1993. Updated interpretation of magnetic anomalies and seafloor spreading stages in the South China Sea: implications for the Tertiary of Southeast Asia. *Journal of Geophysical Research*, **98**, 6299–6328.

DURKEE, E. F. 1993. Oil, geology and changing concepts in the Southwest Philippines (Palawan and the Sulu Sea). *Geological Society of Malaysia Bulletin*, **33**, 241–262

HAMILTON, W. 1979. *Tectonics of the Indonesian region.* US Geological Survey, Professional Papers, **1078**.

HAZEBROEK, P. & TAN, D. N. K. 1993. Tertiary tectonic evolution of the NW Sabah continental margin. *Geological Society of Malaysia Bulletin*, **33**, 195–210

HINZ, K. & SCHLÜTER, H. U. 1985. Geology of the Dangerous Grounds, South China Sea, and the continental margin off southwest Palawan: results of Sonne cruises SO-23 and SO-27. *Energy*, **10**, 297–315

——, FRITSCH, J., KEMPTER, E. H. K., MOHAMMAD, A. M., MOHAMED, D., VOSBERG, H., WEBER, J. & BENAVIDEZ, J. 1989. Thrust tectonics along the north-western continental margin of Sabah/ Borneo. *Geologishe Rundschau*, **78**, 705–730

HOLLOWAY, N. H. 1982. North Palawan Block, Philippines – its relation to Asian mainland and role in the evolution of South China Sea. *AAPG Bulletin*, **66**, 1355–1383.

HUTCHISON, C. S. 1989. *Geological evolution of South-East Asia*. Clarendon Press, Oxford.

——1996. The 'Rajang accretionary prism' and 'Lupar Line' problem of Borneo. *In:* HALL, R. & BLUNDELL, D. J. (eds) *Tectonic Evolution of Southeast Asia*. Geological Society, London, Special Publications, **106**, 247–261

ISACKS, B. L. & BARZANGI, M. 1977. Geometry of Benioff Zones: Lateral segmentation and downwards bending of the subducted lithosphere. *In*: TALWANI, M. & PITMAN III, W. C. (eds) *Island Arcs, Deep Sea Trenches and Back-arc Basins*. American Geophysical Union Maurice Ewing Series **1**, 99–114

KUDRASS, H. R., HEIDICKE, M., CEPEK, P., KREUZER, H. & MULLER, P. 1986. Mesozoic and Cenozoic rocks dredged from the South China Sea (Reed Bank area) and Sulu Sea, and their significance for plate tectonic reconstructions. *Marine and Petroleum Geology*, **3** 19–30

MOSS, S. J., CHAMBERS, J., CLOKE, I., DHARMA SATRIA, ALI, J., MILSOM, J., CARTER, A. & BAKER, S. (1997). New observations on the sedimentary and tectonic evolution of the Tertiary Kutai Basin, East Kalimantan. *This volume*.

RANGIN, C., BELLON, H., BÉNARD, F., LETOUZEY, J., MULLER, C. & SANUDIN, T. 1990. Neogene arc-continent collision in Sabah, Northern Borneo (Malaysia). *Tectonophysics*, **183**, 305–319.

TONGKUL, F. 1991. Tectonic evolution of Sabah, Malaysia. *Journal of Southeast Asian Earth Sciences*, **6**, 395–405.

XIA, K.-Y. & ZHOU, D. 1993. Geophysical characteristics and evolution of northern and southern margins of the South China Sea. *Geological Society of Malaysia Bulletin*, **33**, 223–240

Index